VECTORS

		Pages
$\mathbf{e}^i$	a vector in $\mathbf{R}^n$ having a 1 in the ith row and 0's elsewhere	43
$\mathbf{e}_i$	a vector in $\mathbf{R}_n$ having a 1 in the ith column and 0's elsewhere	43
$\mathbf{x}(k)$	vector in $\mathbf{F}^n(\mathbf{N})$	137
$\odot$	scalar multiplication of vectors	122
$\oplus$	addition of vectors	123
span S	set of all linear combinations of vectors in set S	143
dim $\mathbf{V}$	the dimension of vector space $\mathbf{V}$	170
$\#$	coordinate system (ordered basis)	244
$[\mathbf{x}]_\#$	the coordinate vector of vector $\mathbf{x}$ in coordinate system $\#$	245
$\mathbf{V}_\#$	coordinate space for vector space $\mathbf{V}$	245
$(\mathbf{x}, \mathbf{y})$	the inner product of vectors $\mathbf{x}$ and $\mathbf{y}$	112, 119, 202–203
$\|\mathbf{x}\|$	the norm of vector $\mathbf{x}$	110–111, 119, 194
$d(\mathbf{x}, \mathbf{y})$	the distance between vectors $\mathbf{x}$ and $\mathbf{y}$	105, 111, 197

LINEAR TRANSFORMATIONS

L	a linear transformation	229
$E(\mathbf{x}(k)) = \mathbf{x}(k + 1)$	shift transformation	231
kernel $L = \mathbf{K}$	the kernel of linear transformation L	237
range L	the range of linear transformation L	242
$L_\#([\mathbf{x}]_\#)$	a linear transformation on the coordinate space, $L_\#([\mathbf{x}]_\#) = [L(\mathbf{x})]_\#$	255–256

COMPLEX NUMBERS

$z = a + bi$	a complex number	401
$\overline{z} = \overline{a + bi}$	the conjugate $a - bi$ of the complex number $a + bi$	402
$\mathbf{C}_n$	vector space of row vectors of complex numbers	130
$\mathbf{C}_{m,n}$	vector space of matrices of complex numbers	130
$(\mathbf{z}, \mathbf{w})$	inner product for complex numbers	315
A^*	if $A = [a_{ij}]$, $A^* = [\overline{a_{ij}}]^t$	50

MISCELLANEOUS

R_i	row i of a matrix or equation i of a system of equations	12
$\mathrm{R}_i \leftrightarrow \mathrm{R}_j$	interchange operation for rows	12
$c\mathrm{R}_i \to \mathrm{R}_i$	scaling operation for rows	13
$\mathrm{R}_j + c\mathrm{R}_i \to \mathrm{R}_j$	substitution operation for rows	14
C_i	column i of a matrix	80
$\mathrm{C}_i \leftrightarrow \mathrm{C}_j$	interchange operation for columns	80
$c\mathrm{C}_i \to \mathrm{C}_i$	scaling operation for columns	80
$\mathrm{C}_j + c\mathrm{C}_i \to \mathrm{C}_j$	substitution operation for columns	80
$(k) = (k_1, \ldots, k_n)$	a permutation of $(1, 2, \ldots, n)$	73
perm (N)	the set of all permutations of $N = \{1, 2, \ldots, n\}$	73
$s(k)$	sign function for permutation (k)	73
sign (r)	the sign of the number r	72
$\mathrm{E}_i \leftrightarrow \mathrm{E}_j$	interchange of entries i and j of a permutation	76
δ_{ij}	1 if $i = j$ and 0 if $i \neq j$ (Kronecker delta)	43
$A \otimes B$	Kronecker product of matrices	266
$f'(t), p'(t)$	the derivative with respect to t of the named function of t	
□	end of an example	
■	end of a proof	

ELEMENTARY LINEAR ALGEBRA

ELEMENTARY LINEAR ALGEBRA

D.J. HARTFIEL/ARTHUR M. HOBBS
TEXAS A&M UNIVERSITY

PRINDLE, WEBER & SCHMIDT, BOSTON

PWS PUBLISHERS

Prindle, Weber & Schmidt • Duxbury Press • PWS Engineering • Breton Publishers •
20 Park Plaza • Boston, Massachusetts 02116

PWS Publishers is a division of Wadsworth, Inc.

Library of Congress Cataloging in Publication Data
Hartfiel, D.J.
Elementary Linear Algebra
Bibliography: p.
Includes index
1. Algebra. I. Hobbs, Arthur M., joint author.
II. Title.
QA33.2.P716 1986 613 .14 87-23789

ISBN 0-87150-038-8

Printed in the United States of America

87 88 89 90 91—10 9 8 7 6 5 4 3 2 1

Sponsoring Editor: Dave Pallai
Production Coordinator: S. London
Production: Lifland et al., Bookmakers
Interior and Cover Design: S. London
Cover Photo: Greg Bowl
Typesetting: Polyglot Compositors
Cover Printing: New England Book Components
Printing and Binding: Halliday Lithograph, Arcata Graphics

To our wives, Faye Hartfiel and Barbara J. Hobbs, and daughters, Andra and Simone Hartfiel and Melissa and Patricia Hobbs.

PREFACE

The two questions we are most often asked about our linear algebra course are "What is this course trying to do mathematically?" and "Where is this material ever used?" To a great extent this text was written to answer these questions.

To give context to the mathematics, we present linear algebra in a natural and reasonable way. Concepts and results are not simply stated; they are accompanied by explanations and numerous illustrative examples. To show the importance of the material, we point out applications throughout the text. Almost all of the mentions of applications are backed up by exercises that demonstrate how the work is applied. Even if these exercises are not assigned, students can profit from reading through them.

Because of the additional material, this text is a bit longer than standard texts in the area. However, the extra examples are designed to help students master the subject and may be skipped in class presentations, and the numerous exercises are for instructors and students to choose among.

This text is intended for a first course in linear algebra for students with varying majors, including computer science, economics, engineering, mathematics, and the social and physical sciences. The students should be familiar with basic geometry and algebra—a prerequisite for any linear algebra course.

TEXT

A breakdown of the material follows.

CHAPTERS 1 AND 2

These chapters cover systems of linear algebraic equations, the arithmetic of matrices, and the determinant. Here as in the rest of the text, when possible we place the material in a natural setting to show its importance. For example, the technically intricate definitions of matrix multiplication and the determinant are given natural developments rather than just stated. Such developments can, of course be skipped in class presentations.

CHAPTERS 3 AND 4

These chapters develop the study of vector spaces, the centerpiece of any linear algebra book. As is traditional, we consider all vector spaces simultaneously by studying general or abstract vector spaces. We believe, however, that use of the central theme of developing the geometrical concepts of dimension, distance, and angle for all vector spaces has made our presentation more cohesive than many.

Although the work is general, we support it by viewing all concepts and results geometrically in the particular vector space of fixed arrows. Thus, drawings can be made that allow the work to be envisioned.

CHAPTER 5

Using vector spaces for their domains, we define the special functions called "linear transformations." Because in practice linear transformations occur most often in linear equations, we also study linear equations in this chapter. Sufficient material is provided so that a student can decide whether a particular equation is linear and understand how such equations are solved. Coordinate systems for vector spaces are defined, and we show how these can be used to find matrix representations of linear transformations. Representation of linear equations in various coordinate systems is also introduced.

CHAPTERS 6 AND 7

Continuing the approach developed in Chapter 5, we show how eigenvalues and eigenvectors are used to find diagonal matrix representations for linear transformations. Then we show how these representations can be used to study various linear transformations and to solve various linear equations.

CHAPTERS 8 AND 9

These chapters provide a brief introduction to numerical methods and linear programming.

EXERCISES

Exercises which are far more numerous than in most linear algebra texts, are written so that some benefit can be derived simply from reading them. Thus, many problems can be assigned for reading only. In addition to computational exercises, we have the following.

COMPLEX NUMBER EXERCISES

Each chapter contains a set of involving complex numbers. Although the text is written using real numbers, results, with occasional adjustments, are also valid for complex numbers. Any adjustments necessary for the results of the text to hold for complex numbers are described in these exercises.

THEORETICAL EXERCISES

These exercises are designed to help develop reasoning skills. Traditionally, one of the goals of a first course in linear algebra is to develop reasoning skills. There are two excellent reasons for concentrating on reasoning skills:

1. In any text, the results given are mostly basic. Others too numerous to be given are consequences of these basic results. As the need for other results arises in practice, it is most useful to be able to reason them out from the basic results given in the text.

2. **Being** able to reason to conclusions assures correctness, even in calculations. In practice where there are no answers in the back of the book, you know your work is correct when you have thought it out logically.

APPLICATIONS EXERCISES

These exercises are intended to show how linear algebra is used in practice. In order to ensure that these exercises require little to no technical background, we occasionally replaced technical parts of a problem with more familiar parts, thus changing the look of the problems. For example, in Chapter 3 we wrote up control problems in terms of pipes and water towers. The basic principles of the original problem, however, were always preserved.

COMPUTER EXERCISES

These exercises require no previous knowledge of computers. They are designed to show how computers would make the various calculations described throughout the text. Basically, they concern numerical considerations, that one should address when using a computer for calculations—round-off and speed, for example.

FLEXIBILITY

Linear algebra classes vary, even within departments. Differences in presentation are one source of variation. For example, the first author likes to develop the material by appealing to geometry, making numerous sketches and diagrams of the work, whereas the second author prefers to show how the work is done in a step-by-step algorithmic form. Further variation results from differences in concentration. Some instructors emphasize theory, whereas others may have a particular interest in applications or computers.

This book contains sufficient material to permit various forms of development. Basic material is presented within the text, and exercises can be chosen to complement individual approaches.

Pace in courses also varies. We believe that this text can be taught at a pace of two to three sections per week. Several plans for presenting the basic material follow.

A. (22 sections) Chapters 1, 2, 3, 6 and section 5.1 (leaving out references to coordinate spaces).

B. (25 sections) Chapters 1, 2, 3, 5, 6.

C. (28 sections) Chapters 1, 2, 3, 5, 6, 4 (definition of length and orthogonality), 7 (distinct eigenvalue case).

D. (30 sections) Chapters 1, 2, 3, 5, 6, 4, 7 (where the ordering assures that 6 is covered).

E. (30 sections) Chapters 1, 2, 3, 4, 5, 6, 7.

Note that no decision on these plans need be made until after Chapter 3. At that time, some idea of class pace is usually known. Also, the numerical methods and linear programming material can be incorporated into any of the plans by adding Chapters 8 and 9.

ACKNOWLEDGMENTS

The first author would like to thank the students and faculty members, especially those outside the mathematics department, for participating over the past many years in his seminars on matrix theory, matrix computation, economics, control theory, transmission of data, optimization, biological systems, and numerical analysis. Discussions in these seminars provided the background for numerous problems in the text.

The second author would like to thank Jack Thurman and Ray Anthony of Texas A&M University for advice about oil refining, chemistry, and catalysts, which was used in preparing several allocation problems in Chapter 9. Also Jim Baker, John Pittman, and Robert A. Wilkie of Texas A&M University deserve thanks for their help in preparing the computer drawings that appear in the text. And finally, the following students are thanked for working many of the exercises in the first few chapters of the book: Kenneth Coe, Fadi R. Laham, Robert Richter, and Kathleen Young.

We would also like to thank the following reviewers for their contributions: Marshall Anderson, Millersville University; Wayne Barrett, Brigham Young University; Brian Bourgeois, University of Houston, Downtown; Charles Holmes, Miami University; David Hostetler, Mesa Community College; John Karlof, University of North Carolina at Wilmington; Donald F. Linton, University of Louisville; H. W. Pu, Texas A&M University; Erik Scheiner, Western Michigan University; Kirby Smith, Texas A&M University; Marcos Wright, Rutgers University; and David Zitarelli, Temple University.

Contents

5 LINEAR TRANSFORMATIONS AND LINEAR EQUATIONS 227

6 SIMILARITY AND PROBLEM SOLVING 267

7 ORTHOGONAL SIMILARITY AND PROBLEM SOLVING 305

8 NUMERICAL METHODS 342

9 LINEAR PROGRAMMING 361

APPENDIX A: COMPLEX NUMBERS 401

APPENDIX B: REFERENCES 403

ANSWERS TO SELECTED EXERCISES 405

INDEX 431

Introduction

In elementary algebra, you studied the set **R** of real numbers. Functions of a variable x, perhaps called expressions in x, were defined, with the domain of x in **R**. Equations involving these functions were defined, and you learned how to solve linear and quadratic equations.* Later you studied $\mathbf{R}_2$, the set of ordered pairs of real numbers. Functions and equations of two variables x and y were defined, with the domain of (x, y) in $\mathbf{R}_2$, and you learned how to solve two linear equations involving these two variables.

Various applied and geometric problems, called stated problems or word problems, were introduced to show the usefulness of this work. These problems were usually formulated mathematically as equations. The equations were then solved, and the solution was translated into the solution of the original problem. This procedure is depicted in Figure 1.

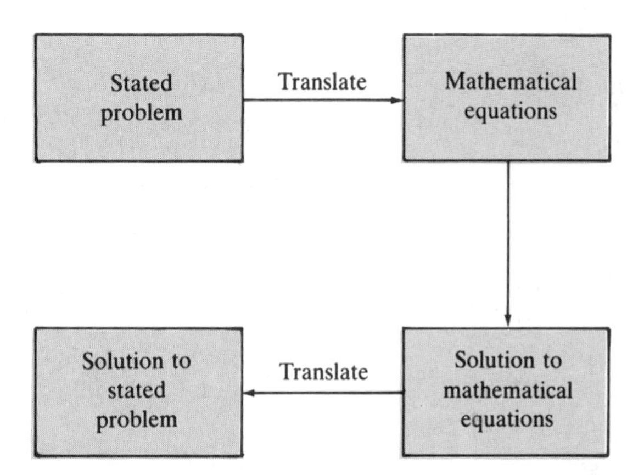

FIGURE 1

Basically, mathematically formulating any problem requires a space such as **R** or $\mathbf{R}_2$ in which the problem can be mathematically stated, as well as functions and equations that can be used in the actual problem description. Many problems, however, cannot be formulated in **R** or $\mathbf{R}_2$. In linear algebra, we develop the more general spaces, functions, and equations required for such problems.

* Other equations, such as polynomial equations of higher degree, are not dealt with in elementary algebra because formulas for their solution are either rather complicated or nonexistent.

In Chapter 1 we define systems of linear algebraic equations of n variables and show some organized methods for manipulating such systems to find a solution. Chapter 2 develops matrix arithmetic, which allows us to write systems of linear algebraic equations, as well as systems of other equations, more compactly as matrix equations. We show that these matrix equations can be mathematically manipulated in much the same way as equations involving one variable are manipulated.

In Chapters 3 and 4, we define general vector spaces. These are spaces on which various applied and geometrical problems can be mathematically formulated. The arithmetic and geometric properties of these spaces are developed. The arithmetic properties describe how we can add, subtract, multiply, and arithmetically manipulate expressions in these spaces. The geometric properties relate to the concepts of dimension, distance, and angle within these spaces.

In Chapter 5 we give a general definition of linear functions and linear equations defined on vector spaces. Further, we describe how, in general, these linear equations can be solved. Then in Chapters 6 and 7 we show how various particular such equations can be manipulated to find a solution.

Finally, Chapters 8 and 9 provide a brief introduction to two supplementary topics, numerical methods and linear programming.

CHAPTER ONE

SYSTEMS OF LINEAR ALGEBRAIC EQUATIONS

When you studied algebra, you learned how to simultaneously solve two linear equations involving two unknowns. You also learned how this work is used to solve various applied problems. The following example will serve as a review of that work.

EXAMPLE 1

A stretch of river 2 miles in length is used by local motorboat enthusiasts for determining boat speeds. If a small motorboat races down this stretch in 3 minutes and back in 3.1 minutes, how fast is the boat in still water?

Letting x = the speed of the boat in still water and y = the speed of the river, we have

$$x + y = \text{the speed of the boat as it goes downstream}$$

and

$$x - y = \text{the speed of the boat as it goes upstream}$$

Thus, using that distance/speed = time,

$$\frac{2}{x+y} = \text{the time required for the trip downstream}$$

and

$$\frac{2}{x-y} = \text{the time required for the trip upstream}$$

Hence the equations relating x and y are

$$\frac{2}{x+y} = 3 \qquad \text{and} \qquad \frac{2}{x-y} = 3.1$$

or, clearing fractions,

$$\begin{aligned} 3x + 3y &= 2 \\ 3.1x - 3.1y &= 2 \end{aligned}$$

Solving these equations yields

$$\begin{aligned} x &= \tfrac{122}{186} \text{ miles per minute} \approx 39.35 \text{ miles per hour} \\ y &= \tfrac{2}{186} \text{ miles per minute} \approx 0.65 \text{ miles per hour} \end{aligned}$$ □

In more complex applied problems, there can be many more unknowns, which in turn are related by larger sets of equations. Some examples of such problems follow.

EXAMPLE 2

A dietician wants to prepare a dinner containing 463 calories, 3530 milligrams of protein, 284 milligrams of calcium, and 5 milligrams of iron. The dietician can use beef, potatoes, green beans, and milk. Recording the desired information from a nutritional value table, the dietician writes

	CALORIES	PROTEIN (mg)	CALCIUM (mg)	IRON (mg)
1 Serving of Beef	214	2470	10	3.1
1 Potato	98	240	13	0.8
1 Cup of Green Beans	27	180	45	0.9
1 Glass of Milk	124	140	216	0.2

To find the equations that show the dietician how to fix the dinner, set

x_1 = the number of servings of beef used in the dinner
x_2 = the number of potatoes used in the dinner
x_3 = the number of cups of green beans used in the dinner
x_4 = the number of glasses of milk used in the dinner

Calculating the calories and amounts of protein, calcium, and iron provided by the dinner, we get a set of four equations with four unknowns:

$$\begin{aligned}
214x_1 + 98x_2 + 27x_3 + 124x_4 &= 463 \\
2470x_1 + 240x_2 + 180x_3 + 140x_4 &= 3530 \\
10x_1 + 13x_2 + 45x_3 + 216x_4 &= 284 \\
3.1x_1 + 0.8x_2 + 0.9x_3 + 0.2x_4 &= 5
\end{aligned}$$

□

EXAMPLE 3

A planet in an elliptical orbit passes through the points (0, 1), (1, 0), (0, −2), (−2, 0), $(-1, \sqrt{2})$, and $(-1, -\sqrt{2})$. Find the equations that determine the constants in the equation of this ellipse.

The equation for an ellipse is $ax^2 + bxy + cy^2 + dx + ey + f = 0$, where $a, b, \ldots, f$ are constants. We will determine these constants by substituting into the equation the given values for x and y. We obtain the following six equations with six unknowns:

$$\begin{aligned}
c + e + f &= 0 \\
a + d + f &= 0 \\
4c - 2e + f &= 0 \\
4a - 2d + f &= 0 \\
a - \sqrt{2}b + 2c - d + \sqrt{2}e + f &= 0 \\
a + \sqrt{2}b + 2c - d - \sqrt{2}e + f &= 0
\end{aligned}$$

□

These examples show how rather large sets of equations with many unknowns can arise. In fact, some sets of equations found in practice contain hundreds of equations and hundreds of unknowns. In this chapter we are concerned with providing formal methods for solving such sets of equations.

1.1 DEFINING SYSTEMS OF LINEAR ALGEBRAIC EQUATIONS

In this section we introduce and define some of the language associated with the type of equations that arose in Examples 2 and 3 above. For this purpose, consider the function $f(x) = ax$, where a is a constant and x is a variable. The graph of this function in the coordinate plane is a line; hence, the function is called a linear function. More generally, any function that can be written as

$$f(x_1, \ldots, x_n) = a_1x_1 + a_2x_2 + \cdots + a_nx_n$$

where $a_1, a_2, \ldots, a_n$ are constants and $x_1, x_2, \ldots, x_n$ are variables, is called a *linear function.* Note that the *terms* of the function are expressions of the form a_ix_i for $i = 1, 2, \ldots, n$, where a_i is the *coefficient* of the single variable x_i.

A *linear algebraic* equation* is an equation that can be obtained by setting a linear function equal to a constant. Thus, a linear algebraic equation is one that can be written as

$$a_1x_1 + a_2x_2 + \cdots + a_nx_n = b$$

where $a_1, \ldots, a_n$ and b are constants and $x_1, \ldots, x_n$ are the unknowns.

A set of such linear algebraic equations is called a *system of linear algebraic equations*:

$$\begin{aligned} a_{11}x_1 + a_{12}x_2 + \cdots + a_{1n}x_n &= b_1 \\ a_{21}x_1 + a_{22}x_2 + \cdots + a_{2n}x_n &= b_2 \\ \vdots \qquad \vdots \qquad \vdots \qquad \vdots \quad &\quad \vdots \\ a_{m1}x_1 + a_{m2}x_2 + \cdots + a_{mn}x_n &= b_m \end{aligned}$$

EXAMPLE 4

The following are all systems of linear algebraic equations:

(*i*)
$$\begin{aligned} 2x_1 - 3x_2 &= 5 \\ -7x_1 + 4x_2 &= 6 \end{aligned}$$

(*ii*)
$$\begin{aligned} 2x_1 + 7x_2 + x_3 &= 6 \\ 5x_1 - 3x_2 + 4x_3 &= 0 \end{aligned}$$

(*iii*)
$$\begin{aligned} -3x_1 + x_2 + 4x_3 \qquad &= 0 \\ 2x_1 - 3x_2 \qquad + 2x_4 &= -1 \\ -4x_1 \qquad + 2x_3 + 6x_4 &= 5 \end{aligned}$$

The examples given in the introduction to this chapter were also systems of linear algebraic equations. □

* We use the word "algebraic" because we will later define other kinds of linear equations.

EXAMPLE 5

The following are not systems of linear algebraic equations:

(*i*) $x_1^2 + x_2 = 6$

(*ii*)
$$\begin{aligned} 2x_1 + x_1x_2 + x_2 &= 4 \\ -3x_1 \qquad + 2x_2 &= -3 \end{aligned}$$

(*iii*)
$$\begin{aligned} x_1 + \quad x_2 &= 2 \\ x_1 + \ln(x_2) &= 3 \end{aligned}$$

□

A *solution* to a system of linear algebraic equations is a sequence of numbers, written $(c_1, c_2, \ldots, c_n)$, such that, when we substitute $x_1 = c_1$, $x_2 = c_2, \ldots, x_n = c_n$, each of the equations in the system is satisfied. The *solution set* of a system of linear algebraic equations is the set of all solutions to the system.

Systems of linear algebraic equations can be solved by the substitution method, which you studied for at least two equations in algebra. This method is as follows: Solve one of the equations for one of the variables. Substitute this expression into the other equation, thus replacing that equation with one involving only the remaining variables. This system and the original one have the same solution set.

The following examples will demonstrate this method and show various kinds of solution sets that can occur when systems of linear algebraic equations are solved.

EXAMPLE 6

SOLUTION SET TO A LINEAR ALGEBRAIC EQUATION DEPICTED GEOMETRICALLY AS A LINE

Consider the linear algebraic equation $x_1 - 3x_2 = 0$. Solving this equation for x_1 yields $x_1 = 3x_2$. Since there is a solution (x_1, x_2) for each choice of x_2, the solution set for this equation is infinite. This infinite set can be described algebraically as

$$\{(3x_2, x_2) : x_2 \text{ is any number}\}$$

For instance, if $x_2 = 1$, (3, 1) is a solution; if $x_2 = 3$, (9, 3) is a solution; and if $x_2 = -2$, $(-6, -2)$ is a solution.

Geometrically the graph of this solution set in the coordinate plane is the line shown in Figure 1.1. The solutions (3, 1), (9, 3), and $(-6, -2)$ are indicated on this line.

□

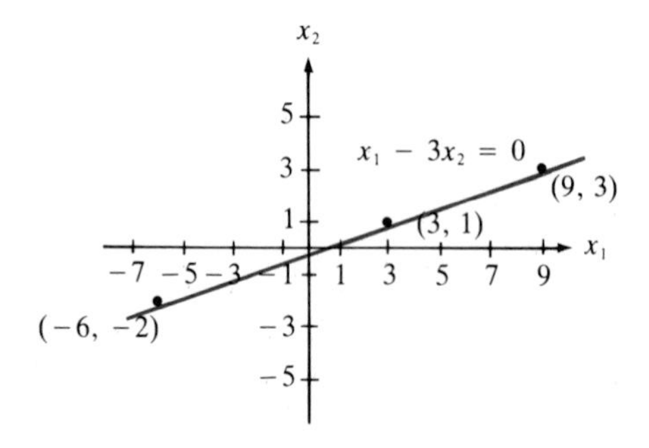

FIGURE 1.1

EXAMPLE 7

SYSTEM WITH EXACTLY ONE SOLUTION

Consider

$$x_1 + x_2 = 2$$
$$x_1 - x_2 = 0$$

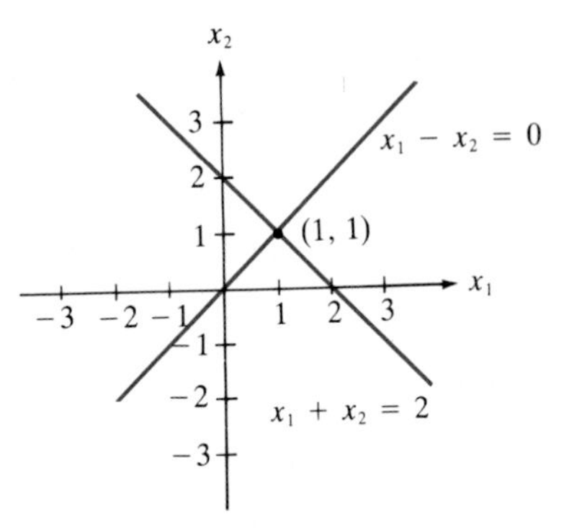

FIGURE 1.2

Geometrically the graph of the solution set to each equation is a line, as shown in Figure 1.2. A solution to the system is a solution to each equation and thus lies at the intersection of these lines.

To compute the solution by the substitution method, solve the first equation for x_1, which yields $x_1 = 2 - x_2$. To eliminate x_1 from the second equation, substitute this expression into the second equation to get the system

$$x_1 + x_2 = 2$$
$$(2 - x_2) - x_2 = 0$$

or

$$x_1 + x_2 = 2$$
$$2x_2 = 2$$

Hence, $x_2 = 1$. Substituting this value into the first equation, we get $x_1 = 1$. Thus the solution set is $\{(1, 1)\}$, and it contains exactly one solution. □

EXAMPLE 8

SYSTEM WITH NO SOLUTIONS

Consider

$$x_1 + x_2 = 2$$
$$2x_1 + 2x_2 = 0$$

Geometrically the graph of the solution set to each of these equations is a line, as shown in Figure 1.3. It is clear that there is no intersection of the lines and thus no solution to this system.

Algebraically, using the substitution method, we can solve the first equation for x_1, which yields $x_1 = 2 - x_2$. We eliminate x_1 from the second equation by substituting this expression into the second equation to obtain the system

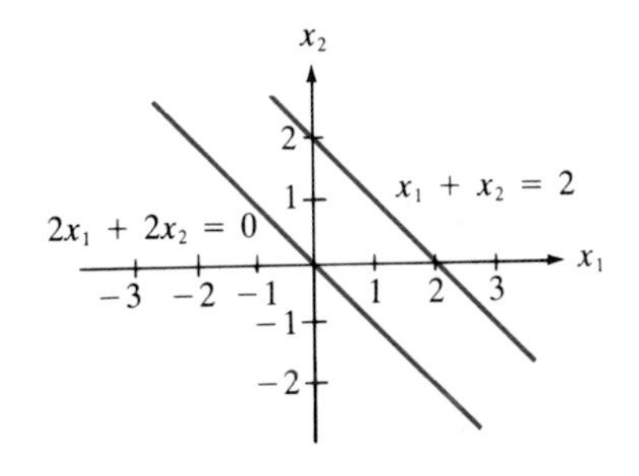

FIGURE 1.3

$$x_1 + x_2 = 2$$
$$2(2 - x_2) + 2x_2 = 0$$

or

$$x_1 + x_2 = 2$$
$$4 = 0$$

Since none of the choices of x_1 and x_2 satisfy the second equation, there is no solution to this system, and consequently none to the original system. We write that the solution set is $\varnothing$, the empty set. □

EXAMPLE 9

SYSTEM WITH INFINITELY MANY SOLUTIONS

Consider

$$\begin{aligned} x_1 + x_2 &= 2 \\ 2x_1 + 2x_2 &= 4 \end{aligned}$$

The graph of each of the solution sets of these equations is the line given in Figure 1.4. We can see from the geometric representation that the solution set to the system also is depicted by this line.

Algebraically, if we solve the first equation for x_1 to get $x_1 = 2 - x_2$ and substitute into the second equation, we have the system

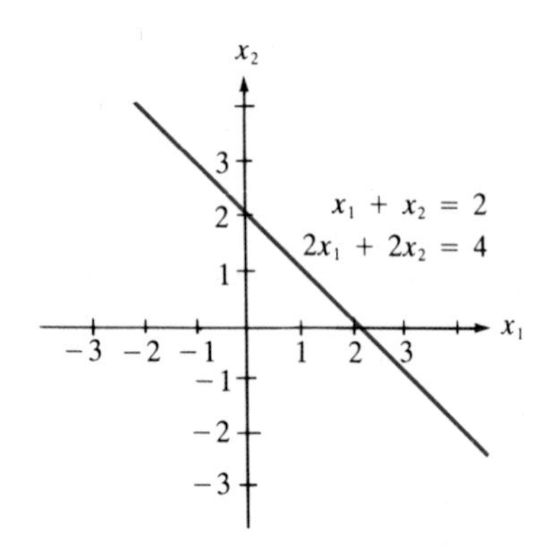

FIGURE 1.4

$$\begin{aligned} x_1 + x_2 &= 2 \\ 2(2 - x_2) + 2x_2 &= 4 \end{aligned}$$

or

$$\begin{aligned} x_1 + x_2 &= 2 \\ 4 &= 4 \end{aligned}$$

Thus, any solution to the first equation is a solution to the system. We can describe this set, which is infinite, as

$$\{(2 - x_2, x_2) : x_2 \text{ is any number}\}$$ □

EXAMPLE 10

A LARGER SYSTEM

Consider the system of linear algebraic equations

$$\begin{aligned} x_1 + x_2 + x_3 + x_4 &= 4 \\ x_1 - x_2 + x_3 - x_4 &= 0 \end{aligned}$$

We cannot depict the solution set to this system geometrically; however, we can compute it algebraically as in the previous examples. Solving the first equation for x_1 yields $x_1 = 4 - x_2 - x_3 - x_4$. Substituting into the second equation gives the system

$$\begin{aligned} x_1 + x_2 + x_3 + x_4 &= 4 \\ (4 - x_2 - x_3 - x_4) - x_2 + x_3 - x_4 &= 0 \end{aligned}$$

or

$$\begin{aligned} x_1 + x_2 + x_3 + x_4 &= 4 \\ x_2 + x_4 &= 2 \end{aligned}$$

Regardless of the choices of x_3 and x_4, the second equation can be solved for $x_2 = 2 - x_4$ and the first equation can be solved for $x_1 = 4 - x_2 - x_3 - x_4 = 4 - (2 - x_4) - x_3 - x_4 = 2 - x_3$. Thus the solution set to the system can be described algebraically in terms of x_3 and x_4 as

$$\{(2 - x_3, 2 - x_4, x_3, x_4): x_3, x_4 \text{ are any numbers}\}$$

Note that if $x_3 = 1$ and $x_4 = 2$, then $(1, 0, 1, 2)$ is a solution, whereas if $x_3 = -1$ and $x_4 = 3$, the solution is $(3, -1, -1, 3)$. Making infinitely many different choices of x_3 and x_4 would show that the solution set is infinite. □

In the remainder of this chapter we will develop systematic methods for solving systems of linear algebraic equations of arbitrary size. These methods are organized extensions of the substitution method shown in Examples 7 through 10.

EXERCISES FOR SECTION 1.1

COMPUTATIONAL EXERCISES

1. Which of the following is a solution to

$$\begin{aligned} x - y - z &= 0 \\ x + y + z &= 0 \end{aligned}$$

(*a*) $(1, 2, -1)$ (*b*) $(0, 1, -1)$

2. Which of the following is a solution to

$$\begin{aligned} 3x - 5y + z &= -16 \\ -x + 3y - 2z &= 6 \end{aligned}$$

(*a*) $(-1, 3, 2)$ (*b*) $(2, 1, -17)$

3. Is $(1 + c, 2 - c, c)$ a solution to

$$\begin{aligned} x + y &= 3 \\ y + z &= 2 \end{aligned}$$

for every number c?

4. Is $(1 + c, c)$ a solution to

$$\begin{aligned} 2x + y &= 2 \\ x - 2y &= 1 \end{aligned}$$

for every number c?

In exercises 5 and 6, show by substitution that each set is a set of solutions to the given system of linear algebraic equations.

5.
$$\begin{aligned} x - y + 2z &= -1 \\ 2x + y - z &= 3 \end{aligned}$$

$$\left\{\left(\frac{2 - z}{3}, \frac{5 + 5z}{3}, z\right): z \text{ is any number}\right\}$$

6.
$$\begin{aligned} 2x + 2y - 4z &= 2 \\ -3x - 3y + 6z &= -3 \end{aligned}$$

$$\{(1 - y + 2z, y, z): y, z \text{ any numbers}\}$$

In exercises 7 and 8, decide which of the sets of equations are systems of linear algebraic equations.

7. (*a*)
$$\begin{aligned} x + xy + y &= 2 \\ 2x - xy + y &= 1 \end{aligned}$$

(*b*)
$$\begin{aligned} x_1 - x_2 &= 4 \\ 2x_1 - 4x_3 &= 8 \\ x_2 - x_3 &= 5 \\ x_1 + x_2 + x_3 &= -1 \end{aligned}$$

(*c*) $-x + y + 2z = 0$

8. (*a*)
$$\begin{aligned} x - y &= z \\ -y &= z^2 \end{aligned}$$

(*b*)
$$\begin{aligned} 3x - y &= 7 \\ 2 - x &= 7y \end{aligned}$$

(*c*)
$$\begin{aligned} x - y^2 &= 3 \\ 2x + y &= 1 \end{aligned}$$

In exercises 9 through 16, solve the systems of linear algebraic equations. Graph the equations as in Examples 6 through 9 and indicate the solution set on your figure.

9. $x + 3y = 2$

10. $x - 2y = 5$

11.
$$\begin{aligned} x + 2y &= 1 \\ 2x + y &= 2 \end{aligned}$$

12.
$$\begin{aligned} 2x + 3y &= -1 \\ -4x - 6y &= 2 \end{aligned}$$

13.
$$\begin{aligned} 2x + y &= 1 \\ 4x + 2y &= 6 \end{aligned}$$

14.
$$\begin{aligned} 2x + 3y &= -1 \\ -x + y &= 2 \end{aligned}$$

15.
$$\begin{aligned} x - y &= 1 \\ -2x + 2y &= -2 \end{aligned}$$

16.
$$\begin{aligned} 2x - 3y &= 1 \\ -6x + 9y &= 0 \end{aligned}$$

The equation for any conic section in the plane is $ax^2 + bxy + cy^2 + dx + ey + f = 0$. For exercises 17 and 18, write the system of linear algebraic equations that can be solved to determine the equation of the given conic section through the given points.

17. Ellipse through the points $(0, 2)$, $(0, -2)$, $(2, 0)$, $(-2, 0)$, $(\frac{2}{3}\sqrt{3}, \frac{2}{3}\sqrt{3})$, and $(-\frac{2}{3}\sqrt{3}, \frac{2}{3}\sqrt{3})$.

18. Hyperbola through the points $(\sqrt{3}, 0)$, $(-\sqrt{3}, 0)$, $(0, \sqrt{3})$, $(0, -\sqrt{3})$, $(1, 2)$, $(1, -1)$.

19. Write out the linear algebraic equations in the unknowns a_0, a_1, a_2, a_3, and a_4 whose solution will give the polynomial

$$p(x) = a_4x^4 + a_3x^3 + a_2x^2 + a_1x + a_0$$

such that $p(1) = 0, p(0) = -3, p(2) = 33, p'(0) = 0, p'(1) = 10$, and $p'(2) = 68$.

In exercises 20 and 21, the variables y_1 and y_2 are given in terms of variables x_1 and x_2. Express the variables x_1 and x_2 in terms of y_1 and y_2. (You will encounter similar equations when solving problems by change-of-variable techniques.)

20. $y_1 = x_1 + x_2$
$y_2 = x_1 - x_2$

21. $y_1 = 2x_1 - 3x_2$
$y_2 = x_1 - 2x_2$

COMPLEX NUMBERS

In exercises 22 and 23, solve each system of linear algebraic equations.

22. (*a*) $(5 - 2i)z = 13 - 11i$
$(1 + 3i)w = -2 + 14i$

(*b*) $iz + (1 - i)w = 4 + 3i$
$(1 + i)z + (3 + 2i)w = -1 + 5i$

23. (*a*) $(3 + 5i)z = 8 + 2i$
$(4 - 3i)w = -5 + 10i$

(*b*) $(1 - i)z + (2 - 3i)w = -5$
$(2 + i)z + (3 - 2i)w = 4 + 12i$

THEORETICAL EXERCISES

24. Consider the system

$$a_{11}x_1 + a_{12}x_2 = b_1$$
$$a_{21}x_1 + a_{22}x_2 = b_2$$

Assume that the constants $a_{11}, a_{12}, a_{21}, a_{22}, b_1$, and b_2 are such that the graph of each of the two equations is a line. Let S be the solution set of this system. How many elements can S have? Explain your answer geometrically.

25. Consider the system

$$x_1 + x_2 = 1$$
$$ax_2 = b$$

of linear algebraic equations. Explain in terms of a and b when the system would have zero solutions, one solution, and infinitely many solutions.

APPLICATIONS EXERCISES

26. (*a*) Suppose that x is the number of units of a commodity that consumers will buy at a price p where $p = mx + b$ for some numbers m and b. In economics this equation is called a *demand equation*. Determine the demand equation—that is, find m and b—for the data

p	\$ 2	\$3
x	10	5

Use the demand equation to find the demand x when the price is $p = \$4$.

(*b*) Suppose that x is the number of units of a commodity that a producer will make at price p per unit where $p = m'x + b'$ for some numbers m' and b'. In economics this equation is called a *supply equation*. Determine the supply equation—that is, find m' and b'—for the data

p	\$ 3	\$2
x	10	5

Use the supply equation to calculate the supply x if the price is set at $p = \$4$.

(*c*) Graph the demand and supply equations of (a) and (b) in the same coordinate system. Compute the intersection (x, p) of these lines. This point is called the *equilibrium point*. Interpret the equilibrium point in terms of supply and demand. *Hint:* If the price of the commodity is set at p dollars, then x represents....

27. The table below gives the population of the United States as determined by the Census Bureau.

YEAR	POPULATION
1910	91.9 million
1920	105.7 million
1930	122.7 million

(*a*) By finding a line $p = mx + b$ through (1920, 105.7) and (1930, 122.7) and then computing $p = m(1927) + b$, estimate the population in 1927.

(*b*) Using the same technique, estimate the population in 1912 (the census is taken once each decade, so exact figures for 1912 and 1927 are not known). The technique by which these estimates are obtained is called *linear interpolation*.

(Leontief input-output model) In order to study interdependent industries, such as manufacturing, agriculture, energy, etc., Wassily Leontief developed what is called an *input-output model* (see References). We will describe this model for a farm-labor relationship. A large farm produces more than enough goods to meet the needs of those living on the farm, whom we will call labor. Suppose that, to produce \$1 worth of goods, the farm requires \$0.30 worth of its own products (seed, etc.) and \$0.50 worth of labor (to work fields, etc.). Further, suppose that, to produce \$1 worth of labor (wages for workers, etc.), \$0.60 worth of farm products and \$0.20 worth of labor (for maintaining homes, etc.) are needed. The situation is depicted in Figure 1.5.

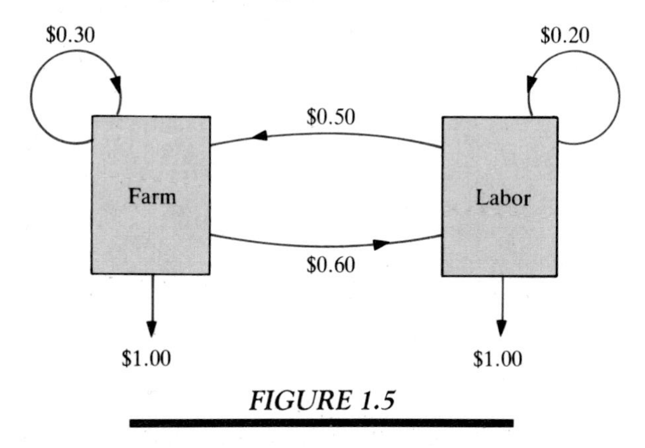

FIGURE 1.5

To produce $\$x_1$ worth of farm products, $\$0.50x_1$ of labor is required. Continuing with this line of reasoning, we can depict the requirements to produce $\$x_1$ worth of farm products and $\$x_2$ worth of labor as in Figure 1.6.

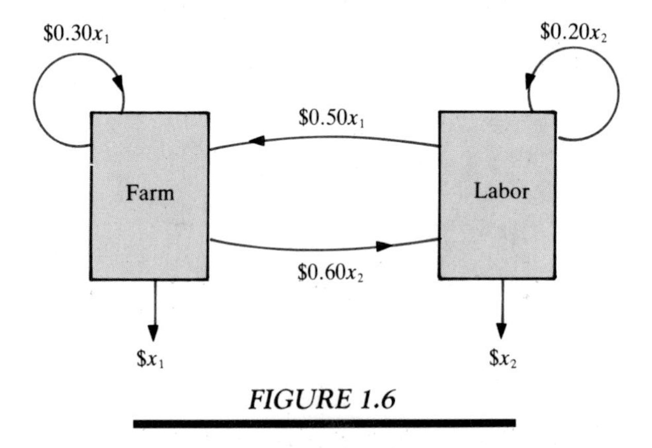

FIGURE 1.6

Suppose there is some demand (e.g., from town) for $\$d_1$ worth of farm products and $\$d_2$ worth of labor. To determine the amounts $\$x_1$ of farm products and $\$x_2$ of labor required in order to meet the farm's own needs as well as the demand, we must find x_1 and x_2 such that

$$\begin{aligned} \text{output} &= \text{farm needs} + \text{labor needs} + \text{demand} \\ x_1 &= 0.30x_1 + 0.60x_2 + d_1 \\ x_2 &= 0.50x_1 + 0.20x_2 + d_2 \end{aligned}$$

or

$$\begin{aligned} 0.70x_1 - 0.60x_2 &= d_1 \\ -0.50x_1 + 0.80x_2 &= d_2 \end{aligned}$$

28. Determine x_1 and x_2 if the demands are $d_1 = \$2{,}600$ and $d_2 = \$260$.

29. Determine x_1 and x_2 given $d_1 = \$4{,}800$ and $d_2 = \$3{,}200$ in the situation diagrammed in Figure 1.7.

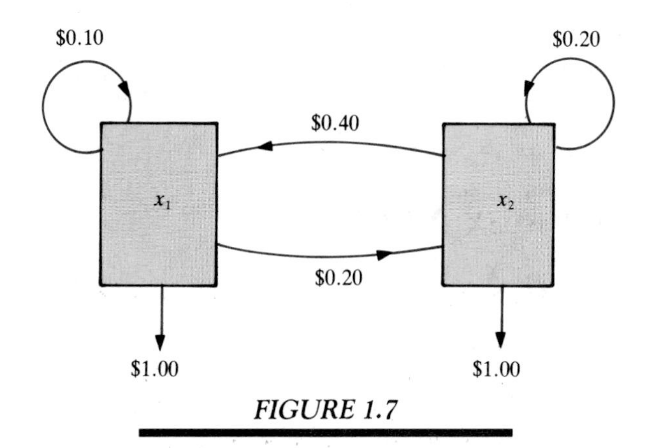

FIGURE 1.7

COMPUTER EXERCISES

30. (the k-digit computer) A k-digit number is any number that can be written as a number having k digits times a power of 10. For example,

(*i*) $125000 = 125 \times 10^3$, $0.00125 = 125 \times 10^{-5}$, and $1 = 100 \times 10^{-2}$ are three-digit numbers.

(*ii*) 125.6 is not a three-digit number

If x and y are two k-digit numbers, then a k-digit computer will compute $x + y$, $x - y$, xy, and x/y (if $y \neq 0$) and then round the result to a k-digit number before the next calculation.* We will refer to this form of calculation as *k-digit arithmetic*. For example, with $k = 2$,

(*i*) $2400 + 120 \approx 2500$ (2520 rounded)

(*ii*) $2900 + 32 \approx 2900$ (2932 rounded)

(*iii*) $25 \times 25 \approx 630$ (625 rounded)

(*iv*) $8900 \times 76 \approx 680000$ (676400 rounded)

Notice that the errors in the calculations are 20, 32, 5, and 3600, respectively. Any k-digit calculation can be in error and sometimes very much so. However, we can expect only that the calculated result is close to the exact value in the first k digits. A useful measurement of the accuracy of a computed result $\bar{x}$ compared to an exact result x is the

* We will always round up a number whose $(k + 1)$th digit is 5.

percentage error, given by

$$\frac{\text{error in the result}}{\text{exact result}} \times 100 = \frac{|x - \bar{x}|}{|x|} \times 100$$

(*a*) Show how a two-digit computer would solve

$$\begin{aligned} x + 125y &= 126 \\ 6x + \ \ 95y &= 101 \end{aligned}$$

Since the computer only accepts two-digit numbers, you must use two-digit arithmetic to solve

$$\begin{aligned} x + 130y &= 130 \\ 6x + \ \ 95y &= 100 \end{aligned}$$

(*b*) Compute the solution to the system in part (a) using a four-digit computer.
(*c*) Using percentage error, compare the x and y values obtained in parts (a) and (b) with the exact solution $x = y = 1$.

1.2 Solving Systems of Linear Algebraic Equations by the Gaussian Elimination Algorithm

In this section we describe the Gaussian elimination algorithm, an organized method for solving systems of linear algebraic equations. Such organization is critical when systems with a large number of equations and variables are to be solved (see exercise 32).

As can be observed in the last section, the idea of the substitution method for solving these systems is to algebraically manipulate the systems into ones that are more easily solved. In choosing manipulations, or operations, to change the system, however, we had to be sure not to change the solution set of the system.*

To apply this idea in developing an organized procedure for solving systems of linear algebraic equations, we need to know what operations on systems do not change their solution sets. Two such operations are obvious.

1. Interchange operation: Form a new system by interchanging equation i with equation j. To indicate that this operation has been performed on a system, we will write† $R_i \leftrightarrow R_j$.

Example 11

Applying $R_1 \leftrightarrow R_2$ to

$$\begin{aligned} a_{11}x_1 + a_{12}x_2 &= b_1 \\ a_{21}x_1 + a_{22}x_2 &= b_2 \end{aligned} \quad \text{yields} \quad \begin{aligned} a_{21}x_1 + a_{22}x_2 &= b_2 \\ a_{11}x_1 + a_{12}x_2 &= b_1 \end{aligned}$$

□

Example 12

Applying $R_1 \leftrightarrow R_2$ to

$$\begin{aligned} x_1 + 2x_2 &= 3 \\ 4x_1 + 5x_2 &= 6 \end{aligned} \quad \text{yields} \quad \begin{aligned} 4x_1 + 5x_2 &= 6 \\ x_1 + 2x_2 &= 3 \end{aligned}$$

□

* Systems that have the same solution set are often called equivalent systems.

† We use the symbol R_i, for row i, rather than the more natural E_i, for equation i, because we will shortly be writing systems of linear algebraic equations as arrays of numbers, and in these arrays each equation becomes a row. Further, R_i provides a natural couple with C_i, for column i, which will be used for companion operations on columns later.

EXAMPLE 13

Applying $R_2 \leftrightarrow R_3$ to

$$\begin{aligned} x_1 + x_2 + x_3 &= 1 \\ 2x_1 - x_2 + 3x_3 &= 2 \\ 3x_1 + 2x_2 - x_3 &= -1 \end{aligned} \quad \text{yields} \quad \begin{aligned} x_1 + x_2 + x_3 &= 1 \\ 3x_1 + 2x_2 - x_3 &= -1 \\ 2x_1 - x_2 + 3x_3 &= 2 \end{aligned}$$

□

2. Scaling* operation: Form a new system by multiplying any equation i by a nonzero constant c. We will denote this operation on a system by $cR_i \rightarrow R_i$.

EXAMPLE 14

Applying $cR_2 \rightarrow R_2$ to

$$\begin{aligned} a_{11}x_1 + a_{12}x_2 &= b_1 \\ a_{21}x_1 + a_{22}x_2 &= b_2 \end{aligned} \quad \text{yields} \quad \begin{aligned} a_{11}x_1 + a_{12}x_2 &= b_1 \\ ca_{21}x_1 + ca_{22}x_2 &= cb_2 \end{aligned}$$

Performing $cR_1 \rightarrow R_1$ on the original system yields

$$\begin{aligned} ca_{11}x_1 + ca_{12}x_2 &= cb_1 \\ a_{21}x_1 + a_{22}x_2 &= b_2 \end{aligned}$$

□

EXAMPLE 15

Applying $4R_1 \rightarrow R_1$ to

$$\begin{aligned} x_1 + 2x_2 &= 3 \\ -x_1 + 3x_2 &= 1 \end{aligned} \quad \text{yields} \quad \begin{aligned} 4x_1 + 8x_2 &= 12 \\ -x_1 + 3x_2 &= 1 \end{aligned}$$

□

EXAMPLE 16

Applying $-2R_2 \rightarrow R_2$ to

$$\begin{aligned} 3x_1 + 2x_2 + x_3 &= 1 \\ x_1 + 2x_2 + 3x_3 &= 4 \\ 2x_1 + x_2 - x_3 &= 2 \end{aligned} \quad \text{yields} \quad \begin{aligned} 3x_1 + 2x_2 + x_3 &= 1 \\ -2x_1 - 4x_2 - 6x_3 &= -8 \\ 2x_1 + x_2 - x_3 &= 2 \end{aligned}$$

□

The last operation we will list is basically that of substitution. Consider

$$\begin{aligned} a_{11}x_1 + a_{12}x_2 + \cdots + a_{1n}x_n &= b_1 \\ a_{21}x_1 + a_{22}x_2 + \cdots + a_{2n}x_n &= b_2 \end{aligned}$$

If $a_{11} \neq 0$, we can solve the first equation for x_1, getting

$$x_1 = \frac{1}{a_{11}}(b_1 - a_{12}x_2 - \cdots - a_{1n}x_n)$$

* Scalar is another word for number.

Substituting into the second equation to eliminate x_1 yields

$$\frac{a_{21}}{a_{11}}(b_1 - a_{12}x_2 - \cdots - a_{1n}x_n) + a_{22}x_2 + \cdots + a_{2n}x_n = b_2$$

Rearranging terms, we get the new system

$$\begin{aligned} a_{11}x_1 + \quad a_{12}x_2 + \cdots + \quad a_{1n}x_n &= b_1 \\ \left(a_{22} - a_{12}\frac{a_{21}}{a_{11}}\right)x_2 + \cdots + \left(a_{2n} - a_{1n}\frac{a_{21}}{a_{11}}\right)x_n &= b_2 - \frac{a_{21}}{a_{11}}b_1 \end{aligned}$$

This new system can be obtained more simply as follows. Beginning with the original system

$$\begin{aligned} a_{11}x_1 + a_{12}x_2 + \cdots + a_{1n}x_n &= b_1 \\ a_{21}x_1 + a_{22}x_2 + \cdots + a_{2n}x_n &= b_2 \end{aligned}$$

we replace the second equation by the second equation plus $-a_{21}/a_{11}$ times the first equation, which yields the system

$$\begin{aligned} a_{11}x_1 + \quad a_{12}x_2 + \cdots + \quad a_{1n}x_n &= b_1 \\ \left(a_{22} - a_{12}\frac{a_{21}}{a_{11}}\right)x_2 + \cdots + \left(a_{2n} - a_{1n}\frac{a_{21}}{a_{11}}\right)x_n &= b_2 - \frac{a_{21}}{a_{11}}b_1 \end{aligned}$$

This calculation suggests the third operation.

3. Substitution operation: Form a new system by replacing an equation j with the sum of equation j and a constant c times any other equation i in the system. Equation i is unchanged in the new system. Notationally we will indicate this operation by $\mathrm{R}_j + c\mathrm{R}_i \to \mathrm{R}_j$, making both end expressions R_j.

EXAMPLE 17

Applying $\mathrm{R}_1 + c\mathrm{R}_2 \to \mathrm{R}_1$ to

$$\begin{aligned} a_{11}x_1 + a_{12}x_2 &= b_1 \\ a_{21}x_1 + a_{22}x_2 &= b_2 \end{aligned} \quad \text{yields} \quad \begin{aligned} (a_{11} + ca_{21})x_1 + (a_{12} + ca_{22})x_2 &= b_1 + cb_2 \\ a_{21}x_1 + \quad a_{22}x_2 &= b_2 \end{aligned}$$

Applying $\mathrm{R}_2 + c\mathrm{R}_1 \to \mathrm{R}_2$ to the original system yields

$$\begin{aligned} a_{11}x_1 + \quad a_{12}x_2 &= b_1 \\ (a_{21} + ca_{11})x_1 + (a_{22} + ca_{12})x_2 &= b_2 + cb_1 \end{aligned}$$

□

EXAMPLE 18

Applying $\mathrm{R}_1 + 2\mathrm{R}_2 \to \mathrm{R}_1$ to

$$\begin{aligned} x_1 + 2x_2 &= 3 \\ 4x_1 + 5x_2 &= 6 \end{aligned} \quad \text{yields} \quad \begin{aligned} 9x_1 + 12x_2 &= 15 \\ 4x_1 + 5x_2 &= 6 \end{aligned}$$

□

EXAMPLE 19

Applying $R_2 - 3R_3 \rightarrow R_2$ to

$$\begin{aligned} 3x_1 - x_2 + x_3 &= 1 \\ x_1 + x_2 - x_3 &= 0 \\ 2x_1 - 2x_2 - x_3 &= 4 \end{aligned} \quad \text{yields} \quad \begin{aligned} 3x_1 - x_2 + x_3 &= 1 \\ -5x_1 + 7x_2 + 2x_3 &= -12 \\ 2x_1 - 2x_2 - x_3 &= 4 \end{aligned}$$ □

We summarize these operations, which we will call *Gaussian operations,** and their corresponding notations in Table 1.1.

TABLE 1.1

OPERATION	NOTATION
1. Interchange operation: Interchange equations i and j.	$R_i \leftrightarrow R_j$
2. Scaling operation: Multiply equation i by a constant $c \neq 0$.	$cR_i \rightarrow R_i$
3. Substitution operation: Replace equation j with the sum of equation j and a constant c times any other equation i.	$R_j + cR_i \rightarrow R_j$

As shown below, the application of these operations to any system does not change the solution set of that system.

*THEOREM 1.1 *

If any interchange, scaling, or substitution operations are applied to a system of linear algebraic equations to obtain a second system of linear algebraic equations, then the solution sets of the two systems are the same.

PROOF

The proof that the interchange and scaling operations do not change the solution set is left to you in exercise 25. For the substitution operation we demonstrate the proof for system I,

$$\begin{aligned} a_{11}x_1 + a_{12}x_2 + a_{13}x_3 &= b_1 \\ a_{21}x_1 + a_{22}x_2 + a_{23}x_3 &= b_2 \\ a_{31}x_1 + a_{32}x_2 + a_{33}x_3 &= b_3 \end{aligned}$$

and the operation $R_2 + cR_3 \rightarrow R_2$. The general proof is essentially the same except for size.

Applying the operation yields system II,

$$\begin{aligned} a_{11}x_1 + a_{12}x_2 + a_{13}x_3 &= b_1 \\ (a_{21} + ca_{31})x_1 + (a_{22} + ca_{32})x_2 + (a_{23} + ca_{33})x_3 &= b_2 + cb_3 \\ a_{31}x_1 + a_{32}x_2 + a_{33}x_3 &= b_3 \end{aligned}$$

To prove the theorem, we need to show that solution set to I = solution set to II.

Let (s_1, s_2, s_3) be in the solution set to I. Then

$$\begin{aligned} a_{11}s_1 + a_{12}s_2 + a_{13}s_3 &= b_1 \\ a_{21}s_1 + a_{22}s_2 + a_{23}s_3 &= b_2 \\ a_{31}s_1 + a_{32}s_2 + a_{33}s_3 &= b_3 \end{aligned}$$

* Some texts call these elementary operations.

Applying $R_2 + cR_3 \rightarrow R_2$ to these equations of numbers gives

$$\begin{aligned} a_{11}s_1 + \quad & a_{12}s_2 + \quad & a_{13}s_3 &= b_1 \\ (a_{21} + ca_{31})s_1 + (a_{22} + ca_{32})s_2 + (a_{23} + ca_{33})s_3 &= b_2 + cb_3 \\ a_{31}s_1 + \quad & a_{32}s_2 + \quad & a_{33}s_3 &= b_3 \end{aligned}$$

But this says that (s_1, s_2, s_3) is a solution to system II; i.e., $(s_1, s_2, s_3) \in$ solution set to II. Since (s_1, s_2, s_3) was arbitrary, solution set to I $\subseteq$ solution set to II.

Now, let (t_1, t_2, t_3) be in the solution set to II. Substituting these values into system II gives

$$\begin{aligned} a_{11}t_1 + \quad & a_{12}t_2 + \quad & a_{13}t_3 &= b_1 \\ (a_{21} + ca_{31})t_1 + (a_{22} + ca_{32})t_2 + (a_{23} + ca_{33})t_3 &= b_2 + cb_3 \\ a_{31}t_1 + \quad & a_{32}t_2 + \quad & a_{33}t_3 &= b_3 \end{aligned}$$

Applying $R_2 - cR_3 \rightarrow R_2$ to these equations of numbers yields

$$\begin{aligned} a_{11}t_1 + a_{12}t_2 + a_{13}t_3 &= b_1 \\ a_{21}t_1 + a_{22}t_2 + a_{23}t_3 &= b_2 \\ a_{31}t_1 + a_{32}t_2 + a_{33}t_3 &= b_3 \end{aligned}$$

This says that $(t_1, t_2, t_3) \in$ solution set to I and solution set to II $\subseteq$ solution set to I.

Thus solution set to I = solution set to II. ■

Given a system of linear algebraic equations, we intend to manipulate the system, by applying Gaussian operations, to obtain a system that is easily solved. Examples of easily solved systems that we might hope to obtain follow.

EXAMPLE 20

(i) A system such as

$$\begin{aligned} x_1 \qquad\qquad &= 1 \\ x_2 \qquad &= 2 \\ x_3 &= 3 \end{aligned}$$

specifies its solution and hence is easily solved.

(ii) A system such as

$$\begin{aligned} x_1 + x_2 + x_3 + x_4 &= 10 \\ x_2 + 2x_3 - x_4 &= 5 \end{aligned}$$

can be readily solved. Here, we let x_3 and x_4 be any numbers. Then we solve the second equation to get $x_2 = 5 - 2x_3 + x_4$. Substituting this expression into the first equation yields

$$x_1 + (5 - 2x_3 + x_4) + x_3 + x_4 = 10$$

or

$$x_1 = 5 + x_3 - 2x_4$$

Thus the solution set can be described in terms of x_3 and x_4 as

$$\{(5 + x_3 - 2x_4, 5 - 2x_3 + x_4, x_3, x_4) : x_3, x_4 \text{ are any numbers}\} \quad \square$$

By using the procedure of Example 20, we can easily solve any system that appears in a stair-step form such as

$$\begin{aligned} rx_1 + {*}x_2 + {*}x_3 &= * \\ rx_2 + {*}x_3 &= * \\ rx_3 &= * \end{aligned} \qquad \text{or} \qquad \begin{aligned} rx_1 + {*}x_2 + {*}x_3 + {*}x_4 &= * \\ rx_3 + {*}x_4 &= * \\ rx_4 &= * \end{aligned}$$

or the like, where r stands for any nonzero number and the symbol $*$ stands for any number. Notice that each equation has a different first variable. These variables are called *bound variables.** Any variable that is not bound is called *free*, since these variables can be assigned any values and the solution set described in terms of them.

EXAMPLE 21

In the system

$$\begin{aligned} x_1 + 2x_2 - x_3 + x_4 &= 1 \\ 3x_3 - x_4 &= 0 \end{aligned}$$

the variables x_1 and x_3 are bound. The free variables, x_2 and x_4, can be assigned arbitrary values and then the equations can be solved for x_1 and x_3 in terms of x_2 and x_4. Thus, let x_2 and x_4 be any numbers. Then, from the second equation, we have

$$x_3 = \tfrac{1}{3}x_4$$

Now substituting the values for x_2, x_3, and x_4 into the first equation yields

$$x_1 + 2x_2 - \tfrac{1}{3}x_4 + x_4 = 1$$

or

$$x_1 = 1 - 2x_2 - \tfrac{2}{3}x_4$$

Thus the solution set is

$$\{(1 - 2x_2 - \tfrac{2}{3}x_4, x_2, \tfrac{1}{3}x_4, x_4) : x_2, x_4 \text{ are any numbers}\}$$ □

We will now show through examples how Gaussian operations can be used in an organized way to manipulate a system into one having a stair-step form.

EXAMPLE 22

Solve

$$\begin{aligned} x + 2y - z &= 1 \\ 2x - y + 2z &= 3 \\ 4x - 7y + 8z &= 7 \end{aligned} \tag{1}$$

* We call the variables bound variables since they are tied down, or bound, by the equations that determine them. Free variables might also be described as arbitrary.

If x appeared only in the first equation, then any solution for y and z from the second and third equations would give an immediate value for x in the first equation. We can eliminate the term involving x in the second equation by applying $R_2 - 2R_1 \to R_2$, and we can eliminate the term involving x from the third equation by applying $R_3 - 4R_1 \to R_3$. These operations give

$$\begin{aligned} x + 2y - z &= 1 \\ -5y + 4z &= 1 \\ -15y + 12z &= 3 \end{aligned} \tag{2}$$

Similarly, we can eliminate the term involving y in the third equation by applying $R_3 - 3R_2 \to R_3$. We get

$$\begin{aligned} x + 2y - z &= 1 \\ -5y + 4z &= 1 \\ 0 &= 0 \end{aligned}$$

Here x and y are bound variables and z is a free variable. Thus, we let z be any number. Then $-5y + 4z = 1$ can be solved to give $y = -\frac{1}{5} + \frac{4}{5}z$. Substituting these values into the first equation yields

$$x + 2(-\tfrac{1}{5} + \tfrac{4}{5}z) - z = 1$$

or $x = \frac{7}{5} - \frac{3}{5}z$. Hence the solution set is

$$\{(\tfrac{7}{5} - \tfrac{3}{5}z, -\tfrac{1}{5} + \tfrac{4}{5}z, z) : z \text{ is any number}\}$$ □

Notice that, in going from system (1) to system (2) in the preceding example, we did not change the first equation. We used the variable x in this equation to eliminate x from all other equations. This required two operations of the type $R_i - cR_1 \to R_i$ for $i > 1$. To save labor, we rewrote the second and the third equation in one step. This is a practice we will continue.

EXAMPLE 23

Solve

$$\begin{aligned} y - 2z &= 3 \\ 2x + 5y - z &= 1 \\ x - 3y + z &= 2 \\ 4x + 19y - z &= -2 \end{aligned}$$

First, to obtain an equation involving x as the first equation of the system, we interchange the first and third equations $(R_1 \leftrightarrow R_3)$* to get

$$\begin{aligned} x - 3y + z &= 2 \\ 2x + 5y - z &= 1 \\ y - 2z &= 3 \\ 4x + 19y - z &= -2 \end{aligned}$$

* We could also apply $R_1 \leftrightarrow R_2$ or $R_1 \leftrightarrow R_4$.

Sequentially applying $R_2 - 2R_1 \to R_2$ and $R_4 - 4R_1 \to R_4$ yields

$$\begin{aligned} x - 3y + z &= 2 \\ 11y - 3z &= -3 \\ y - 2z &= 3 \\ 31y - 5z &= -10 \end{aligned}$$

To avoid fractions,* we apply $R_2 \leftrightarrow R_3$, which yields

$$\begin{aligned} x - 3y + z &= 2 \\ y - 2z &= 3 \\ 11y - 3z &= -3 \\ 31y - 5z &= -10 \end{aligned}$$

Applying $R_3 - 11R_2 \to R_3, R_4 - 31R_2 \to R_4$, and $R_4 - 3R_3 \to R_4$ produces

$$\begin{aligned} x - 3y + z &= 2 \\ y - 2z &= 3 \\ 19z &= -36 \\ 0 &= 5 \end{aligned}$$

There is no choice of x, y, and z that can satisfy this last equation. Thus, the solution set is $\varnothing$, the empty set. □

The method we used to solve the previous two problems is called *Gaussian elimination.* We formally describe the method below.

GAUSSIAN ELIMINATION ALGORITHM†

To solve a system of linear algebraic equations:

1. Let x be the left-most variable of the system with a nonzero coefficient. Select an equation i having variable x with a nonzero coefficient. If equation i is not equation 1, interchange equations 1 and i.

2. Use equation 1 and substitution operations to eliminate terms involving x from all remaining equations. Although it is not necessary, the scaling operation can also be used in this step. The variable x in equation 1 is now a bound variable.

3. If the first equation is the only nonzero equation, go to Step 4. If some equation has reached the form $0 = b$ for a real number $b \neq 0$, stop; the solution set is $\varnothing$. Otherwise, restrict your attention to the second through the last equations. Renumber these equations $1, 2, \ldots,$ and return to Step 1.

* We could use $R_3 - \frac{1}{11}R_2 \to R_3$, $R_4 - \frac{31}{11}R_2 \to R_4$ instead. These steps, however, would have introduced fractions into the calculations, which we wanted to avoid.

† The Chinese book *Nine Chapters on the Mathematical Art,* written about the time 206 B.C.–A.D. 220, included many methods for solving problems in agriculture, engineering, surveying, etc. Among these was a method for solving systems of linear algebraic equations. This method is essentially the same as that given by the Gaussian elimination algorithm.

4. Every variable that is not bound at this time is a free variable. Now, beginning with the last equation and working up, solve for the bound variables in terms of the free variables. This yields the solution.

Note that the first three steps in this algorithm replace the original system of linear algebraic equations with one that is easily solved. Step 4, called *back substitution,* then expresses the bound variables in terms of the free variables, which can be assigned arbitrary values.

EXAMPLE 24

Using the Gaussian elimination algorithm, solve

$$\begin{aligned} x - 3y + 2z &= 4 \\ 2x - 6y + 5z &= 2 \end{aligned}$$

To eliminate the term involving x from the second equation, we apply $R_2 - 2R_1 \to R_2$, obtaining

$$\begin{aligned} x - 3y + 2z &= 4 \\ z &= -6 \end{aligned}$$

Now, x and z are bound variables and y is a free variable. Thus, $z = -6$ and, from the first equation,

$$\begin{aligned} x &= 4 + 3y - 2z \\ &= 4 + 3y - 2(-6) \\ &= 16 + 3y \end{aligned}$$

Hence the solution set is $\{(16 + 3y, y, -6) : y \text{ is any number}\}$. □

In the next section we will revise the Gaussian elimination algorithm so that back substitution is not required for the solution.

EXERCISES FOR SECTION 1.2

COMPUTATIONAL EXERCISES

For each of the systems in exercises 1 through 4, perform the indicated operation or operations in the order specified. See exercise 31.

1. $\begin{aligned} 2x_1 + x_2 - x_3 &= 3 \\ x_1 - 3x_2 + x_3 &= 1 \end{aligned}$, $\quad R_1 \leftrightarrow R_2$

(This puts a $1x_1$ in the upper left position, allowing you to work with integers as long as possible.)

2. $\begin{aligned} \tfrac{1}{3}x_1 + \tfrac{2}{3}x_2 &= 1 \\ \tfrac{2}{3}x_1 - \tfrac{1}{2}x_2 &= 3 \end{aligned}$, $\quad 3R_1 \to R_1, 6R_2 \to R_2, R_2 - 4R_1 \to R_2$

(The first two operations temporarily eliminate fractions.)

3. $\begin{aligned} 3x + 6y - 12z &= 18 \\ 2x - 3y + z &= 5 \end{aligned}$, $\quad \tfrac{1}{3}R_1 \to R_1$

(This operation eliminates multiples in the first equation.)

4. $\begin{aligned} 5x + y - 3z + w &= -1 \\ 3x - 5y - w &= 2, \\ 8x + y - z + 3w &= 1 \end{aligned}$ $\quad R_1 - R_2 \to R_1, R_2 - R_1 \to R_2, R_3 - 4R_1 \to R_3, R_1 \leftrightarrow R_2$

(Note how a 1 is obtained as the coefficient of x in the first equation.)

In exercises 5 through 18, use the Gaussian elimination algorithm to solve the given system of linear algebraic equations. Use the notation for Gaussian operations to explain what you are doing at each step. In exercises 5 through 8, check your solutions by substituting them into the equations.

5.
$$\begin{aligned} 3x - 6y + 5z &= -11 \\ -6x + 12y - z &= 13 \end{aligned}$$

6.
$$\begin{aligned} 2x - 4y + z &= -3 \\ x - 2y + 2z &= 0 \\ -4x + 8y + 2z &= 4 \end{aligned}$$

7.
$$\begin{aligned} x + 3y - z &= -2 \\ -3x - 10y + 5z &= 11 \\ 2x - y + 3z &= 4 \end{aligned}$$

8.
$$\begin{aligned} 2x - 4y + z &= -1 \\ x - 2y + 2z &= 1 \end{aligned}$$

9.
$$\begin{aligned} 3x - 6y + 5z &= 4 \\ -6x + 12y - z &= 7 \\ x - 2y - 2z &= 2 \end{aligned}$$

10.
$$\begin{aligned} x - 4y + 2z &= -9 \\ -x + 5y + 3z &= 16 \\ 5x + y - 3z &= -16 \end{aligned}$$

11.
$$\begin{aligned} 4x + 5y - z &= 0 \\ 3x - y + z &= 0 \\ 19y - 7z &= 0 \end{aligned}$$

12.
$$\begin{aligned} 3x_1 - 5x_2 + x_3 &= 2 \\ 3x_1 - 5x_2 + x_3 &= -3 \end{aligned}$$

13.
$$\begin{aligned} \frac{x}{2} + 2y - \frac{z}{3} &= 22 \\ -\frac{3x}{4} + \frac{y}{5} + z &= 2 \\ 3x - y - 2z &= 2 \end{aligned}$$

14.
$$\begin{aligned} \frac{x}{5} - \frac{y}{2} - 4z &= -44 \\ -\frac{3x}{5} + y + z &= 2 \\ 2x - \frac{3y}{2} - \frac{z}{3} &= 22 \end{aligned}$$

15.
$$\begin{aligned} -2x + y - 3z &= 2 \\ 6x - 5y + 8z &= -4 \\ 4x - 6y + 4z &= 0 \\ -2y - z &= 2 \end{aligned}$$

16.
$$\begin{aligned} x + y + z &= 0 \\ 3x + 2y + 4z &= 0 \\ x + 2y &= 0 \end{aligned}$$

17.
$$\begin{aligned} -x_1 + 2x_2 + x_3 - 2x_4 &= 3 \\ 3x_1 - 5x_2 - 3x_3 + x_4 &= -1 \\ 2x_1 + x_2 - 2x_3 - 17x_4 &= 2 \\ 5x_1 + 4x_2 - 5x_3 - 48x_4 &= 1 \end{aligned}$$

18.
$$\begin{aligned} -x_1 + 2x_2 + x_3 - 2x_4 &= 3 \\ 3x_1 - 5x_2 - 3x_3 + x_4 &= -1 \\ 3x_1 - 5x_2 - 3x_3 + x_4 &= -1 \\ 3x_1 - 5x_2 - 3x_3 + x_4 &= -1 \end{aligned}$$

In exercises 19 and 20, find the polynomial $p(x) = a_2x^2 + a_1x + a_0$ passing through the specified points, and graph the curve.

19. (1, 1), (2, 2), (3, 1)

20. (1, 3), (2, 1), (3, 3)

In exercises 21 and 22, apply Gaussian operations to manipulate the system into one with the given properties. Solve that system. Also solve each system by the Gaussian elimination algorithm. Note that the solution sets for each system are the same, even though their descriptions are not.

21.
$$\begin{aligned} x_1 + x_2 - x_3 &= 2 \\ 2x_1 + 2x_2 + 2x_3 &= 4 \end{aligned}$$
, x_2 appearing only in the first equation

Then let x_1 be "free" and solve for x_2 in terms of x_1.

22.
$$\begin{aligned} x_1 + x_2 + x_3 + x_4 &= 1 \\ 2x_1 - x_2 + 3x_3 + 2x_4 &= -3 \end{aligned}$$
, x_3 appearing only in the first equation

Then let x_1 and x_2 be "free" and solve for x_3 and x_4 in terms of x_1 and x_2. (From these problems we see that other definitions of "bound" and "free" variables are possible.)

COMPLEX NUMBERS

In exercises 23 and 24, solve the given systems of linear algebraic equations by the Gaussian elimination algorithm.

23.
$$\begin{aligned} (2 - i)z + (1 - i)w &= 14 + i \\ (4 - 3i)z + (2 + 3i)w &= 38 + 30i \end{aligned}$$

24.
$$\begin{aligned} (3 - 2i)z + (2 - 3i)w &= 11 - 13i \\ (2 - 3i)z + (3 - 2i)w &= 9 - 7i \end{aligned}$$

THEORETICAL EXERCISES

25. Show, as in Theorem 1.1, that the interchange and scaling operations preserve the solution set of a system of linear algebraic equations.

26. In Step 3 of the Gaussian elimination algorithm, when an equation involves only constants, the left side is 0. Explain why.

APPLICATIONS EXERCISES

27. Let D be the demand for a certain product. Suppose D depends on the variables x and y according to $D = ax + by + c$ for some constants a, b, and c. Compute these constants for the data below. Then find the demand if $x = 8$ and $y = 5$.

x	8	9	10
y	4	5	6
D	19	22	25

28. In order to determine the elliptical orbit of a planet by finding the coefficients of $ax^2 + bxy + cy^2 + dx + ey + f = 0$, we need to determine and record the coordinates of the planet at at least how many points?

29. (calculus) A beam of length L, supported at its ends as shown in Figure 1.8, bends under a uniform load in such a way that its deviation at x is

$$y(x) = ax^4 + bx^3 + cx^2 + dx + e$$

where $y(0) = y(L) = 0$, and for $y''(x) = 12ax^2 + 6bx + 2c$, $y''(0) = y''(L) = 0$. If $L = 2$ and $y(1) = -1$, find and graph the deformation of the beam.

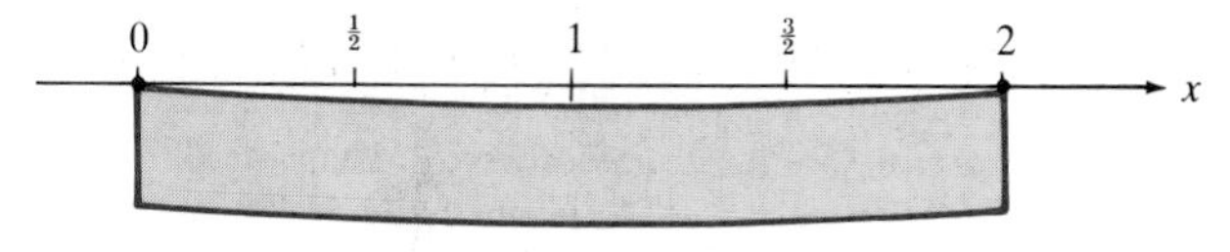

FIGURE 1.8

COMPUTER EXERCISES

30. Consider the system of linear algebraic equations

$$a_{11}x_1 + a_{12}x_2 = b_1$$
$$a_{21}x_1 + a_{22}x_2 = b_2$$

Assume the constants are such that the graphs of these equations are lines. Note that if these lines have about the same slope, then small changes in the constants can change the slopes and y-intercepts slightly and yet create rather large changes in the solution, as shown in Figure 1.9.

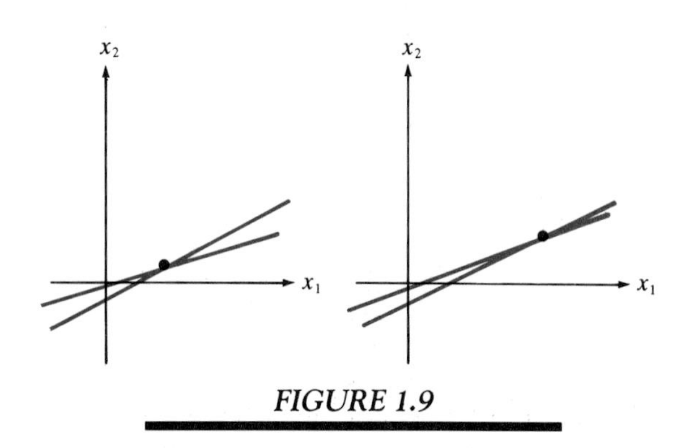

FIGURE 1.9

The following system has the property described above.

$$x_1 + x_2 = 2$$
$$1.15x_1 + x_2 = 3$$

Solve it using a two-digit computer and then a four-digit computer. By computing percentage error, compare your results to the exact answer. (See exercise 30 of Section 1.1 for percentage error.)

31. Using the Gaussian operations, develop a method for proceeding from a system of linear algebraic equations with integer coefficients and integer constants to a simple system using only integer arithmetic. See exercises 1 through 4 for ideas. Apply your method to the following systems.

(*a*) $\begin{aligned} 3x - y &= 2 \\ 2x + y &= 5 \end{aligned}$ (*b*) $\begin{aligned} 5x + 2y &= 7 \\ 3x - y &= 2 \\ 2x + 3y &= -1 \end{aligned}$ (*c*) $\begin{aligned} 4x + 2y - 3z &= 4 \\ 7x + 4y + 2z &= 1 \\ 9x + 3y - 5z &= 2 \end{aligned}$

(This technique does not involve decimal fractions until back substitution begins, so no round-off, a serious problem in computer calculations, is necessary until then.)

32. (large systems) You may come across systems involving hundreds of equations and unknowns when numerically solving differential and partial differential equations that arise in physical problems. Although such problems are common in practice, space allows us to show only one. For others, see any numerical analysis text.

If the temperature $u(x, y)$ at a point (x, y) on the square plate shown in Figure 1.10 does not change with time, then u satisfies the partial differential equation

$$\frac{\partial^2}{\partial x^2} u(x, y) + \frac{\partial^2}{\partial y^2} u(x, y) = 0$$

at all points (x, y) on the plate. If the temperature is known on the boundary, or edges, of the plate, then we can approximate the solution for u at specified points on the plate as follows.

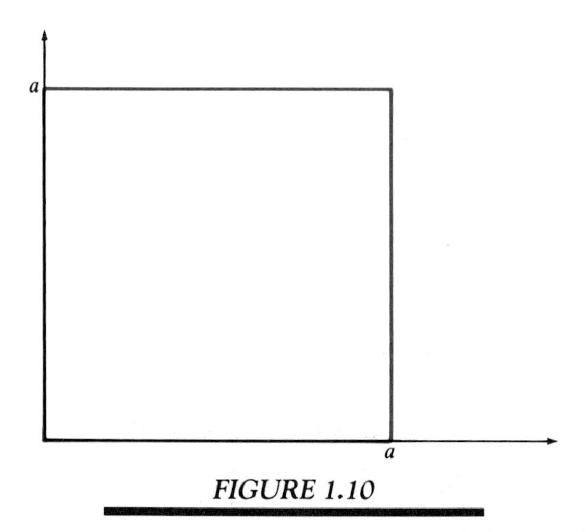

FIGURE 1.10

Let n be a positive integer used to determine the number of points at which u is to be approximated. Let (x_i, y_j) be the specified points on the plate where $i = 0, 1, \ldots, n$ and $j = 0, 1, \ldots, n$.

It can be shown (see finite difference techniques in any

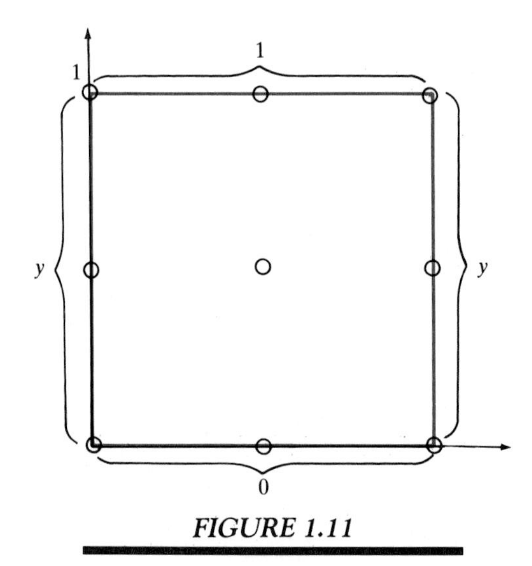

FIGURE 1.11

numerical analysis text) that, if we solve

$$u_{ij} = \frac{u_{i+1,j} + u_{i-1,j} + u_{i,j+1} + u_{i,j-1}}{4}$$

where $u_{i0} = u(x_i, 0)$ and $u_{0j} = u(0, y_j)$ for $0 < i < n$ and $0 < j < n$, then $u_{ij} \approx u(x_i, y_j)$. (The formula is easily remembered since it states that the value u_{ij} at (x_i, y_j) is the average of the values $u_{i-1,j}$, $u_{i+1,j}$, $u_{i,j-1}$, and $u_{i,j+1}$ at the four points closest to (x_i, y_j), that is, (x_{i-1}, y_j), (x_{i+1}, y_j), (x_i, y_{j-1}), and (x_i, y_{j+1}). Hence we could set up the equations without knowing anything about partial differential equations.)

Thus, as shown in Figure 1.11, if $a = 1$, $u(x, 0) = 0$, $u(x, 1) = 1$, $u(0, y) = u(1, y) = y$, and $n = 2$, then we have $u_{0,0} = 0$, $u_{1,0} = 0$, $u_{2,0} = 0$, $u_{0,1} = \frac{1}{2}$, $u_{0,2} = 1$, $u_{1,2} = 1$, $u_{2,1} = \frac{1}{2}$, $u_{2,2} = 1$. The equation involving the only unknown is

$$u_{1,1} = \tfrac{1}{4}(u_{2,1} + u_{0,1} + u_{1,2} + u_{1,0})$$

The solution to this equation is obvious; the resulting values of u are shown in Figure 1.12.

(*a*) Draw the diagram and write out the equations for $n = 3$. Solve these equations.

(*b*) For arbitrary n, sketch the diagram and give the number of equations involving unknowns, as obtained from your diagram.

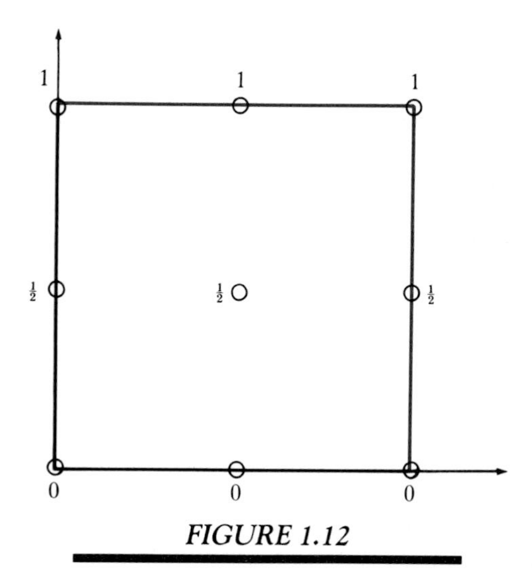

FIGURE 1.12

1.3 THE GAUSS-JORDAN ALGORITHM AND THE NATURE OF SOLUTIONS OF SYSTEMS OF LINEAR ALGEBRAIC EQUATIONS

In this section we show how Step 4 of the Gaussian elimination algorithm can be replaced by a step involving further applications of the Gaussian operations to obtain a still simpler system. Our approach here is to continue to manipulate the system until each bound variable appears, with a coefficient of 1, in only one equation. Then the bound variables are directly expressible in terms of the free variables, and consequently back substitution is not required for solution. We will use examples to show how this can be done.

EXAMPLE 25

Consider

$$\begin{aligned} x_1 + 2x_2 + x_3 + x_4 &= 7 \\ x_3 - x_4 &= 1 \end{aligned}$$

Here x_1 and x_3 are the bound variables, and x_2 and x_4 are free. To eliminate x_3 from the first equation, we apply $R_1 - R_2 \rightarrow R_1$. This yields

$$\begin{aligned} x_1 + 2x_2 \qquad + 2x_4 &= 6 \\ x_3 - \quad x_4 &= 1 \end{aligned}$$

Now x_1 and x_3 are directly expressible in terms of x_2 and x_4; that is, $x_3 = 1 + x_4$ and $x_1 = 6 - 2x_2 - 2x_4$. Thus, the solution set is

$$\{(6 - 2x_2 - 2x_4, x_2, 1 + x_4, x_4) : x_2, x_4 \text{ are any numbers}\} \qquad \square$$

EXAMPLE 26

Consider the system

$$
\begin{aligned}
-x_1 + \tfrac{3}{2}x_2 - x_3 + \tfrac{3}{2}x_4 &= \tfrac{1}{2} \\
x_2 - 6x_3 + 7x_4 &= -3 \\
2x_3 - 3x_4 &= \tfrac{1}{2}
\end{aligned}
$$

Here the bound variables are x_1, x_2, and x_3, and the free variable is x_4. To eliminate x_3 from the first and second equations, we apply $R_1 + \frac{1}{2}R_3 \to R_1$ and $R_2 + 3R_3 \to R_2$ to get

$$
\begin{aligned}
-x_1 + \tfrac{3}{2}x_2 &= \tfrac{3}{4} \\
x_2 - 2x_4 &= -\tfrac{3}{2} \\
2x_3 - 3x_4 &= \tfrac{1}{2}
\end{aligned}
$$

To eliminate x_2 from the first equation, we apply $R_1 - \frac{3}{2}R_2 \to R_1$ to get

$$
\begin{aligned}
-x_1 + 3x_4 &= 3 \\
x_2 - 2x_4 &= -\tfrac{3}{2} \\
2x_3 - 3x_4 &= \tfrac{1}{2}
\end{aligned}
$$

Applying $-R_1 \to R_1$ and $\frac{1}{2}R_3 \to R_3$ yields

$$
\begin{aligned}
x_1 - 3x_4 &= -3 \\
x_2 - 2x_4 &= -\tfrac{3}{2} \\
x_3 - \tfrac{3}{2}x_4 &= \tfrac{1}{4}
\end{aligned}
$$

The bound variables now can be expressed directly in terms of the free variable x_4. Hence, the solution set is

$$\{(-3 + 3x_4, -\tfrac{3}{2} + 2x_4, \tfrac{1}{4} + \tfrac{3}{2}x_4, x_4) : x_4 \text{ is any number}\} \quad \square$$

The above change in the Gaussian elimination algorithm yields the Gauss-Jordan algorithm. A description of this method follows.*

GAUSS-JORDAN ALGORITHM

To solve a system of linear algebraic equations:

1. Apply the Gaussian elimination algorithm to the system. Identify all bound variables.

* In the next chapter we will find that this algorithm not only is convenient but is important to an efficient method for matrix inversion.

2. Apply the following in either order.

   (a) For each bound variable, use the substitution operation to eliminate the bound variable from all equations above the one in which it is the first-appearing variable.

   (b) Multiply each equation by the reciprocal of the coefficient of its bound variable.

3. Express the bound variables in terms of the free variables.

EXAMPLE 27

Solve, indicating the operations applied.

$$\begin{aligned} 3x - y + 2z &= 1 \\ x + 2y - z &= 3 \\ 2x + 3y + 2z &= 1 \end{aligned} \xrightarrow{R_1 \leftrightarrow R_2} \begin{aligned} x + 2y - z &= 3 \\ 3x - y + 2z &= 1 \\ 2x + 3y + 2z &= 1 \end{aligned} \quad \begin{array}{l}\text{(a move of convenience}\\ \text{to avoid fractions for}\\ \text{a while)}\end{array}$$

$$\xrightarrow[R_3 - 2R_1 \to R_3]{R_2 - 3R_1 \to R_2} \begin{aligned} x + 2y - z &= 3 \\ -7y + 5z &= -8 \\ -y + 4z &= -5 \end{aligned}$$

$$\xrightarrow{R_2 \leftrightarrow R_3} \begin{aligned} x + 2y - z &= 3 \\ -y + 4z &= -5 \\ -7y + 5z &= -8 \end{aligned}$$

$$\xrightarrow{R_3 - 7R_2 \to R_3} \begin{aligned} x + 2y - z &= 3 \\ -y + 4z &= -5 \\ -23z &= 27 \end{aligned}$$

$$\xrightarrow[-\frac{1}{23}R_3 \to R_3]{-R_2 \to R_2} \begin{aligned} x + 2y - z &= 3 \\ y - 4z &= 5 \\ z &= -\tfrac{27}{23} \end{aligned} \quad \text{(for Step 2(b))}$$

$$\xrightarrow[R_2 + 4R_3 \to R_2]{R_1 + R_3 \to R_1} \begin{aligned} x + 2y &= \tfrac{42}{23} \\ y &= \tfrac{7}{23} \\ z &= -\tfrac{27}{23} \end{aligned} \quad \text{(for Step 2(a))}$$

$$\xrightarrow{R_1 - 2R_2 \to R_1} \begin{aligned} x &= \tfrac{28}{23} \\ y &= \tfrac{7}{23} \\ z &= -\tfrac{27}{23} \end{aligned}$$

Thus $(\frac{28}{23}, \frac{7}{23}, -\frac{27}{23})$ is the solution. □

EXAMPLE 28

Solve, indicating the operations applied.

$$\begin{aligned} 2x_1 + x_2 - 3x_3 &= 1 \\ x_1 - x_2 + 2x_3 &= 2 \end{aligned} \xrightarrow{R_2 - \frac{1}{2}R_1 \to R_2} \begin{aligned} 2x_1 + x_2 - 3x_3 &= 1 \\ -\tfrac{3}{2}x_2 + \tfrac{7}{2}x_3 &= \tfrac{3}{2} \end{aligned}$$

$$\xrightarrow{R_1 + \frac{2}{3}R_2 \to R_1} \begin{aligned} 2x_1 \qquad - \tfrac{2}{3}x_3 &= 2 \\ -\tfrac{3}{2}x_2 + \tfrac{7}{2}x_3 &= \tfrac{3}{2} \end{aligned} \quad \text{(for Step 2(a))}$$

$$\xrightarrow[-\frac{2}{3}R_2 \to R_2]{\frac{1}{2}R_1 \to R_1} \begin{aligned} x_1 \qquad - \tfrac{1}{3}x_3 &= 1 \\ x_2 - \tfrac{7}{3}x_3 &= -1 \end{aligned} \quad \text{(for Step 2(b))}$$

Thus $x_1 = 1 + \frac{1}{3}x_3$ and $x_2 = -1 + \frac{7}{3}x_3$, where x_3 is any number. Hence the solution set is

$$\{(1 + \tfrac{1}{3}x_3, -1 + \tfrac{7}{3}x_3, x_3) : x_3 \text{ is any number}\} \qquad \square$$

In Section 1.1 we showed various kinds of possible solution sets for systems of linear algebraic equations. Using the Gaussian elimination algorithm or the Gauss-Jordan algorithm, we can show that these are the only kinds of solution sets possible.

THEOREM 1.2

The system of linear algebraic equations

$$\begin{aligned} a_{11}x_1 + a_{12}x_2 + \cdots + a_{1n}x_n &= b_1 \\ a_{21}x_1 + a_{22}x_2 + \cdots + a_{2n}x_n &= b_2 \\ \vdots \qquad \vdots \qquad \vdots \qquad \vdots \quad &\quad \vdots \\ a_{m1}x_1 + a_{m2}x_2 + \cdots + a_{mn}x_n &= b_m \end{aligned}$$

has either no solutions, exactly one solution, or infinitely many solutions. Further, if $m < n$, then the system has either no solutions or infinitely many solutions.*

PROOF

Apply the Gaussian elimination or Gauss-Jordan algorithm to the system, obtaining bound variables that can be expressed as sums of free variables and perhaps some equations of the type $0 = b$, where b is a constant.

If the system includes an equation $0 = b$ but, in fact, $b \neq 0$, it follows that the system has no solution. If the system does not include such an equation, then two cases arise. If there are no free variables, then the solution provided by the algorithm is the only solution. On the other hand, if there are free variables, then the system has infinitely many solutions, one for each choice of value for each free variable. ■

The forms of the final systems of linear algebraic equations that lead to the various kinds of solution sets are demonstrated below.

EXAMPLE 29

Solve

$$\begin{aligned} 3x - 2y &= 4 \\ 5x + y &= 1 \\ 9x + 7y &= -5 \end{aligned}$$

Applying the Gauss-Jordan algorithm, we have the following:

$$\begin{aligned} 3x - 2y &= 4 \\ 5x + y &= 1 \\ 9x + 7y &= -5 \end{aligned} \quad \xrightarrow[R_3 - 3R_1 \to R_3]{R_2 - \frac{5}{3}R_1 \to R_2} \quad \begin{aligned} 3x - 2y &= 4 \\ \tfrac{13}{3}y &= -\tfrac{17}{3} \\ 13y &= -17 \end{aligned}$$

* The case $m \geqslant n$ is the same as the general case.

$$\xrightarrow{R_3 - 3R_2 \to R_3} \begin{aligned} 3x - 2y &= 4 \\ \tfrac{13}{3}y &= -\tfrac{17}{3} \\ 0 &= 0 \end{aligned}$$

$$\xrightarrow{R_1 + \frac{6}{13}R_2 \to R_1} \begin{aligned} 3x \quad &= \tfrac{18}{13} \\ \tfrac{13}{3}y &= -\tfrac{17}{3} \\ 0 &= 0 \end{aligned}$$

$$\xrightarrow[\frac{3}{13}R_2 \to R_2]{\frac{1}{3}R_1 \to R_1} \begin{aligned} x \quad &= \tfrac{6}{13} \\ y &= -\tfrac{17}{13} \\ 0 &= 0 \end{aligned}$$

Thus the solution set is $\{(\frac{6}{13}, -\frac{17}{13})\}$. □

The pattern that emerges when there is a unique solution is that, after the system is solved, every variable is bound and all equations not containing bound variables are of the type $0 = 0$.

EXAMPLE 30

Solve

$$\begin{aligned} 3x + 7y - 3z &= 2 \\ 2x + 5y + z &= -4 \\ 2x + 6y + 10z &= 3 \end{aligned}$$

Applying the Gaussian elimination algorithm, we have the following:

$$\begin{aligned} 3x + 7y - 3z &= 2 \\ 2x + 5y + z &= -4 \\ 2x + 6y + 10z &= 3 \end{aligned} \xrightarrow{R_1 - R_2 \to R_1} \begin{aligned} x + 2y - 4z &= 6 \\ 2x + 5y + z &= -4 \\ 2x + 6y + 10z &= 3 \end{aligned}$$

$$\xrightarrow[R_3 - 2R_1 \to R_3]{R_2 - 2R_1 \to R_2} \begin{aligned} x + 2y - 4z &= 6 \\ y + 9z &= -16 \\ 2y + 18z &= -9 \end{aligned}$$

$$\xrightarrow{R_3 - 2R_2 \to R_3} \begin{aligned} x + 2y - 4z &= 6 \\ y + 9z &= -16 \\ 0 &= 23 \end{aligned}$$

The algorithm stops since the equation $0 = 23$ has occurred, and thus there is no solution. □

The characteristic of any system having no solution is that, while the system is being solved, there appears an equation of the type $0 = b$ when, in fact, $b \neq 0$.

EXAMPLE 31

Solve

$$\begin{aligned} 3x + 7y - 3z &= 2 \\ 2x + 5y + z &= -4 \\ 2x + 6y + 10z &= -20 \end{aligned}$$

Applying the Gauss-Jordan algorithm, we have the following:

$$\begin{aligned} 3x + 7y - 3z &= 2 \\ 2x + 5y + z &= -4 \\ 2x + 6y + 10z &= -20 \end{aligned} \xrightarrow{R_1 - R_2 \to R_1} \begin{aligned} x + 2y - 4z &= 6 \\ 2x + 5y + z &= -4 \\ 2x + 6y + 10z &= -20 \end{aligned}$$

$$\xrightarrow[R_3 - 2R_1 \to R_3]{R_2 - 2R_1 \to R_2} \begin{aligned} x + 2y - 4z &= 6 \\ y + 9z &= -16 \\ 2y + 18z &= -32 \end{aligned}$$

$$\xrightarrow{R_3 - 2R_2 \to R_3} \begin{aligned} x + 2y - 4z &= 6 \\ y + 9z &= -16 \\ 0 &= 0 \end{aligned}$$

$$\xrightarrow{R_1 - 2R_2 \to R_1} \begin{aligned} x \quad - 22z &= 38 \\ y + 9z &= -16 \\ 0 &= 0 \end{aligned}$$

Thus z is free, $y = -16 - 9z$, and $x = 38 + 22z$. The solution set is

$$\{(38 + 22z, -16 - 9z, z) : z \text{ is any number}\}$$ □

The basic pattern for the infinite solution set is that after the system is solved, not every variable is a bound variable and all equations not containing bound variables are of the type $0 = 0$.

A special type of system of linear algebraic equations, called a *homogeneous system*, has all constant terms $b_1 = \cdots = b_m = 0$. This type of system always has $x_1 = \cdots = x_n = 0$ as a solution. In this case, the theorem yields the following corollary.

COROLLARY 1.2A

A homogeneous system of m linear algebraic equations with n unknowns has either one solution or infinitely many solutions. If $m < n$, then the system has infinitely many solutions.

PROOF

The proof is left to you in exercise 23. ■

EXAMPLE 32

Applying the Gauss-Jordan algorithm to the homogeneous system

$$\begin{aligned} x + y &= 0 \\ x - y &= 0 \\ 2x + 3y &= 0 \end{aligned}$$

yields

$$\begin{aligned} x + y &= 0 \\ x - y &= 0 \\ 2x + 3y &= 0 \end{aligned} \xrightarrow[R_3 - 2R_1 \to R_3]{R_2 - R_1 \to R_2} \begin{aligned} x + y &= 0 \\ -2y &= 0 \\ y &= 0 \end{aligned}$$

$$\xrightarrow{R_3+\frac{1}{2}R_2\to R_3} \begin{aligned} x + \ \ y &= 0 \\ -2y &= 0 \\ 0 &= 0 \end{aligned}$$

$$\xrightarrow{-\frac{1}{2}R_2\to R_2} \begin{aligned} x + y &= 0 \\ y &= 0 \\ 0 &= 0 \end{aligned}$$

$$\xrightarrow{R_1-R_2\to R_1} \begin{aligned} x \qquad &= 0 \\ y &= 0 \\ 0 &= 0 \end{aligned}$$

Thus the solution set is $\{(0,0)\}$, which has exactly one element. □

EXAMPLE 33

Applying the Gauss-Jordan algorithm to the homogeneous system

$$\begin{aligned} x_1 + x_2 - x_3 + 3x_4 &= 0 \\ 2x_1 - x_2 + 3x_3 - 4x_4 &= 0 \end{aligned}$$

yields

$$\begin{aligned} x_1 + x_2 - x_3 + 3x_4 &= 0 \\ 2x_1 - x_2 + 3x_3 - 4x_4 &= 0 \end{aligned} \xrightarrow{R_2-2R_1\to R_2} \begin{aligned} x_1 + x_2 - x_3 + 3x_4 &= 0 \\ -3x_2 + 5x_3 - 10x_4 &= 0 \end{aligned}$$

$$\xrightarrow{-\frac{1}{3}R_2\to R_2} \begin{aligned} x_1 + x_2 - x_3 + 3x_4 &= 0 \\ x_2 - \tfrac{5}{3}x_3 + \tfrac{10}{3}x_4 &= 0 \end{aligned}$$

$$\xrightarrow{R_1-R_2\to R_1} \begin{aligned} x_1 \qquad + \tfrac{2}{3}x_3 - \tfrac{1}{3}x_4 &= 0 \\ x_2 - \tfrac{5}{3}x_3 + \tfrac{10}{3}x_4 &= 0 \end{aligned}$$

Thus the solution set is

$$\{(-\tfrac{2}{3}x_3 + \tfrac{1}{3}x_4, \tfrac{5}{3}x_3 - \tfrac{10}{3}x_4, x_3, x_4) : x_3, x_4 \text{ are any numbers}\}$$

which is infinite. □

Our work in the next section will place the methods presented in the last two sections into a completely numerical framework, thus allowing the methods to be implemented on a computer.

EXERCISES FOR SECTION 1.3

COMPUTATIONAL EXERCISES

Recall that linear algebraic equations represent lines. In exercises 1 through 8, use geometry to show whether there are zero, one, or infinitely many solutions to the given system. Then solve the problem by the Gauss-Jordan algorithm. Note the form of the final system of linear algebraic equations. Compare your geometric and algebraic results.

1. $\begin{aligned} 3x + 5y &= 2 \\ 6x + 10y &= 4 \end{aligned}$

2. $\begin{aligned} x + 2y &= 3 \\ 4x + 7y &= 2 \\ 3x + 4y &= -11 \end{aligned}$

3. $3x + 2y = 4$
$x - 7y = 1$
$7x + 20y = 10$

4. $6x - 9y = 18$
$2x - 3y = 6$

5. $3x + 2y = 1$
$x - 7y = 1$
$7x + 20y = 10$

6. $5x - y = 0$
$2x + 3y = 0$

7. $3x + y = 0$
$5x - 2y = 0$
$6x + y = 0$

8. $x + 2y = 3$
$4x + 7y = 3$
$3x + 4y = -11$

In exercises 9 through 20, solve the system by the Gauss-Jordan algorithm. In exercises 9 through 12, check your answers by substituting them into the equations.

9. $y + 2z = -1$
$-2x + 3y - z = 2$
$4x - 7y + 2z = 5$

10. $2x - 4y = 3$
$-6x + 13y - z = -6$
$18x - 38y + 2z = 2$

11. $x_1 - 2x_2 + x_3 = 3$
$x_1 - 2x_2 + x_3 = 4$

12. $x_2 + x_3 = 1$
$x_1 - x_3 = 2$
$-2x_1 - x_2 = 3$

13. $3x - 5y + z = 1$
$5x - 8y + 2z = 3$

14. $3x - 4y + 2z = 3$
$-9x + 12y - 4z = 3$
$9x - 12y + 8z = 21$
$-3x + 4y = 9$

15. $x_2 - 3x_3 = 4$
$3x_1 - 2x_2 + x_3 = -2$
$6x_1 - 7x_2 + 11x_3 = 3$

16. $x_1 - 3x_2 + x_3 - x_4 = 4$
$3x_1 - 10x_2 + 5x_3 = -1$
$2x_1 - 5x_2 - 5x_4 = 21$

17. $x_1 - x_2 + x_3 = 0$
$3x_1 + x_2 - x_3 = 0$
$-x_1 - 3x_2 + 3x_3 = 0$
$9x_1 - x_2 + x_3 = 0$

18. $5x - 2y + z = 0$
$x + y + z = 0$
$3x - 11y + z = 0$
$4x + 2y - 3z = 0$

19. $3x - y + 2z = 0$
$2x + 5y - z = 0$
$-x + 2y + 11z = 0$

20. $2x_1 + 3x_2 - x_3 = 0$
$3x_1 - x_2 + 2x_3 = 0$
$x_1 + 7x_2 - 4x_3 = 0$
$11x_2 - 7x_3 = 0$

Complex Numbers

In exercises 21 and 22, solve by the Gauss-Jordan algorithm.

21. $iz + (1 - i)w = -4 - 3i$
$(1 - i)z + (2 - 3i)w = -1 - 4i$

22. $(1 - i)z + (2 - 3i)w = 1 - 5i$
$iz + (1 - i)w = -1 - i$

Theoretical Exercises

23. Prove Corollary 1.2A.

24. The quadratic equation $x^2 - 3x + 2 = 0$ has precisely two solutions. Can a system of linear algebraic equations have precisely two solutions? Explain your answer. (Note that Theorem 1.2 also tells you what you should not expect as the number of solutions.)

25. To find the coefficients of a polynomial $p(x) = a_n x^n + \cdots + a_0$, we need to know the values of p at at least how many points?

Applications Exercises

(electrical circuits) An electrical circuit is a network made up of resistors and voltage sources connected by wires. Diagrammatically we will represent a resistor by saw-tooth lines and a voltage source by unequal parallel bars, the larger bar being on the positive side.

A voltage source such as a battery, when connected in a loop, causes a flow of electrons through that loop. We measure the strength of the voltage source by a number V of volts.

A resistor resists the flow of electrons. The resistance R is measured in ohms. The current I measures the number of electrons passing through a piece of wire in a given amount of time. The sign of I indicates the direction of flow compared to an arbitrary preassigned direction of the line, with a positive number indicating that the current and the line have the same direction. The current flows from the positive to the negative side of the voltage source.

Kirchoff's laws govern the behavior of current and voltage:

(*i*) The current law states that the sum of all currents entering a connecting point of wires is 0.

(*ii*) The voltage law states that the sum of the voltage drops or rises around any loop is 0. The voltage drops can be calculated across a resistor by the formula $V = IR$, where I is the current flowing across the resistor and R is the resistance of the resistor.

In exercises 26 and 27, check the current equations for the given circuit. Then solve by applying the Gauss-Jordan algorithm.

26.

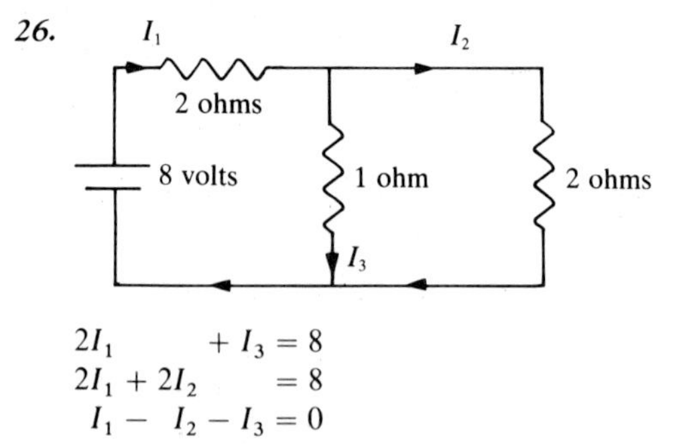

$$2I_1 + I_3 = 8$$
$$2I_1 + 2I_2 = 8$$
$$I_1 - I_2 - I_3 = 0$$

27.

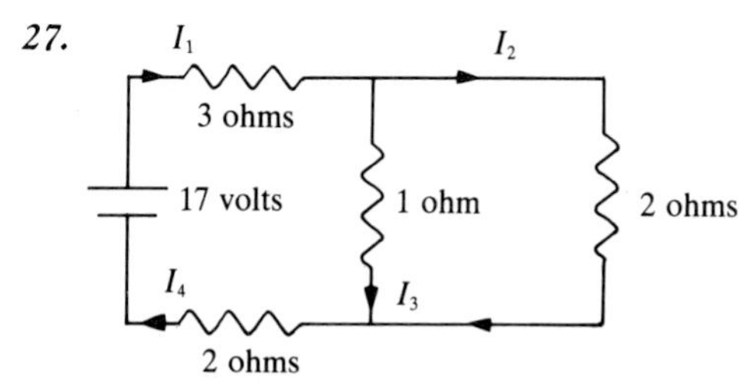

$$\begin{aligned} 3I_1 \quad\quad + I_3 + 2I_4 &= 17 \\ 3I_1 + 2I_2 \quad\quad + 2I_4 &= 17 \\ I_1 - I_2 - I_3 \quad\quad &= 0 \\ I_2 + I_3 - I_4 &= 0 \end{aligned}$$

In exercises 28 and 29, write out and solve the current equations for the given circuits.

28.

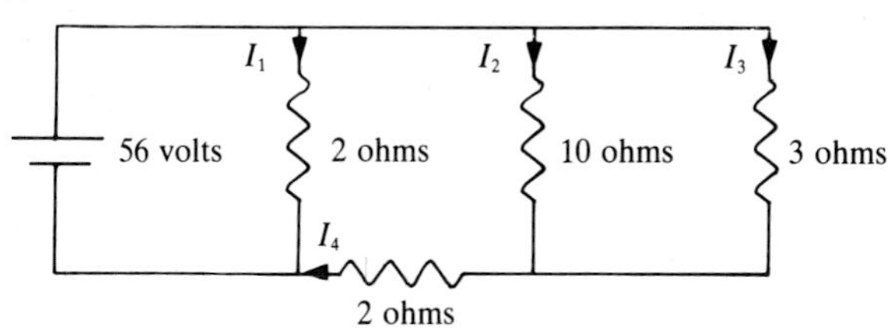

29.

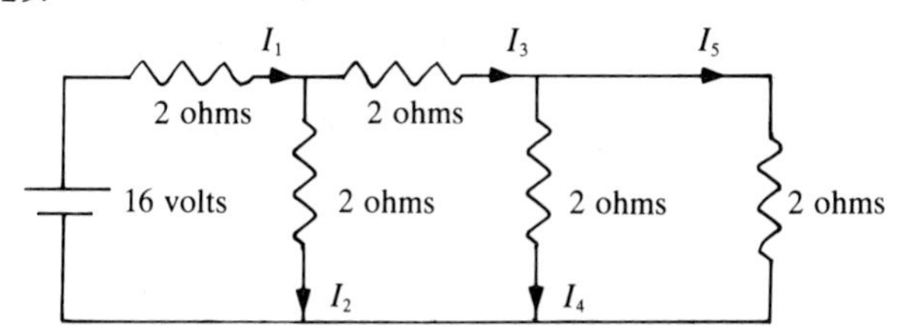

COMPUTER EXERCISES

30. Solve

$$\begin{aligned} x + y &= 2 \\ \tfrac{1}{3}x + y &= 1 \end{aligned}$$

by the Gauss-Jordan algorithm using a two-digit computer and then a four-digit computer. Compare your solutions with the exact solution $x = \frac{3}{2}$, $y = \frac{1}{2}$. Graph the equations and explain why round-off error has little effect.

31. Multiplication and division on a computer usually require more time than addition and subtraction. (To test this assertion, you can write small programs that compute the sum and the product of, say, 10,000 numbers and compare the two running times.) Thus, a comparative count of the multiplications and divisions in different algorithms is a good indicator of which is faster. Count the number of multiplications and divisions needed to solve a system of three linear algebraic equations with three unknowns when (a) the Gaussian elimination algorithm with back substitution is used and (b) the Gauss-Jordan algorithm is used. Explain why (a) is faster than (b). Remember, when you know a calculation will produce a 0, it need not be done. (Counts for the general case can be found in most numerical analysis texts.)

1.4 MATRICES AND SOLVING SYSTEMS USING THEM

In this section we introduce matrices and show how they can be used to solve systems of linear algebraic equations in a completely numerical format. This format provides a means for implementing the methods of the previous sections on a computer, thus providing a way to solve large systems.

A *matrix* (plural: *matrices*) is any rectangular array of numbers placed in *rows* and *columns*. For example,

$$[1 \quad 2 \quad 3], \quad \begin{bmatrix} \frac{1}{2} \\ \pi \\ -2 \\ 0 \end{bmatrix}, \quad \begin{bmatrix} 8 & -6 & \frac{2}{3} \\ -\frac{1}{2} & 2\frac{1}{6} & 1 \\ 3 & \sqrt{2} & -1 \end{bmatrix}, \quad \begin{bmatrix} \dfrac{\sqrt{2}}{2} & 1.36 & -0.1 & \frac{3}{2} \\ 0.107 & 0.2 & -\sqrt{10} & 3 \end{bmatrix}$$

are matrices.

Just as we often denote a number by a symbol such as a, b, c, etc., we will let

$$A = \begin{bmatrix} a_{11} & a_{12} & \cdots & a_{1n} \\ a_{21} & a_{22} & \cdots & a_{2n} \\ \vdots & \vdots & \vdots & \vdots \\ a_{m1} & a_{m2} & \cdots & a_{mn} \end{bmatrix}$$

denote a matrix. Sometimes it will be more convenient to write

$$A = [a_{ij}]$$

rather than exhibit the full array. Further, if we denote a matrix by a given uppercase letter, we will denote its entries by the corresponding lowercase letter. Since A has m rows and n columns, we say that the *size* of A is m by n, written $m \times n$.

We call the matrix $\mathbf{a}_i = [a_{i1} \quad a_{i2} \quad \cdots \quad a_{in}]$ the *i-row* of A, the matrix

$$\mathbf{a}^j = \begin{bmatrix} a_{1j} \\ a_{2j} \\ \vdots \\ a_{mj} \end{bmatrix}$$

the *j-column* of A, and the number a_{ij} the *i,j-entry* of A.

EXAMPLE 34

Let*

$$A = \begin{bmatrix} 1 & 2 & 3 \\ 4 & 5 & 6 \end{bmatrix}$$

The first row of A is $\mathbf{a}_1 = [1 \quad 2 \quad 3]$ and the second row is $\mathbf{a}_2 = [4 \quad 5 \quad 6]$.† The first, second, and third columns of A are

$$\mathbf{a}^1 = \begin{bmatrix} 1 \\ 4 \end{bmatrix}, \qquad \mathbf{a}^2 = \begin{bmatrix} 2 \\ 5 \end{bmatrix}, \qquad \mathbf{a}^3 = \begin{bmatrix} 3 \\ 6 \end{bmatrix}$$

respectively. The 1,1-entry of A is 1, the 1,2-entry of A is 2, the 1,3-entry of A is 3, the 2,1-entry of A is 4, the 2,2-entry of A is 5, and the 2,3-entry of A is 6. Finally, we note that A is 2×3. □

EXAMPLE 35

Let

$$B = \begin{bmatrix} 1 & -1 & 2 & -5 \\ 0 & -3 & 4 & 3 \end{bmatrix}$$

* As examples we often use a matrix whose entries are the first few integers, since this gives us a simple way of identifying entries by numbers.

† Matrices are sometimes written with commas to enhance clarity. Thus $\mathbf{a}_1 = [1 \quad 2 \quad 3]$ could also be written as $\mathbf{a}_1 = [1, 2, 3]$.

Then

$$\mathbf{b}_1 = [1 \quad -1 \quad 2 \quad -5], \qquad \mathbf{b}_2 = [0 \quad -3 \quad 4 \quad 3]$$

$$\mathbf{b}^1 = \begin{bmatrix} 1 \\ 0 \end{bmatrix}, \quad \mathbf{b}^2 = \begin{bmatrix} -1 \\ -3 \end{bmatrix}, \quad \mathbf{b}^3 = \begin{bmatrix} 2 \\ 4 \end{bmatrix}, \quad \mathbf{b}^4 = \begin{bmatrix} -5 \\ 3 \end{bmatrix}$$

Further, $b_{11} = 1$, $b_{12} = -1$, $b_{13} = 2$, $b_{14} = -5$, $b_{21} = 0$, $b_{22} = -3$, $b_{23} = 4$, and $b_{24} = 3$. Finally, we note that B is 2×4. □

Matrices are often used to store information or data. Some examples of this application are given in exercises 36 through 39. Another follows.

EXAMPLE 36

Markov Chain

OPTIONAL

A special type of applied problem can be described in general terms as an *exchange system** for two sets (see exercises 33, 40, and 41). Consider two sets of objects labeled Set 1 and Set 2. Suppose that the sets exchange objects and that a count of objects is made at times $t = 0, 1, \ldots, n, \ldots$. Suppose further that between any two of these times a fraction of the objects in each set move to the other set. Let

s_{ij} = the fractional part of Set i that goes to Set j during the time period

We assume that s_{ij} is constant for all time periods. This situation is depicted diagrammatically in Figure 1.13. The information on the exchange system shown in the diagram can be stored numerically by defining a *transition matrix* $S = [s_{ij}]$. Note that if S is known, then so is the exchange system, and vice versa. For example,

$$S = \begin{bmatrix} \frac{1}{3} & \frac{2}{3} \\ \frac{3}{4} & \frac{1}{4} \end{bmatrix}$$

yields the diagram in Figure 1.14, and the diagram in Figure 1.14 would yield S. □

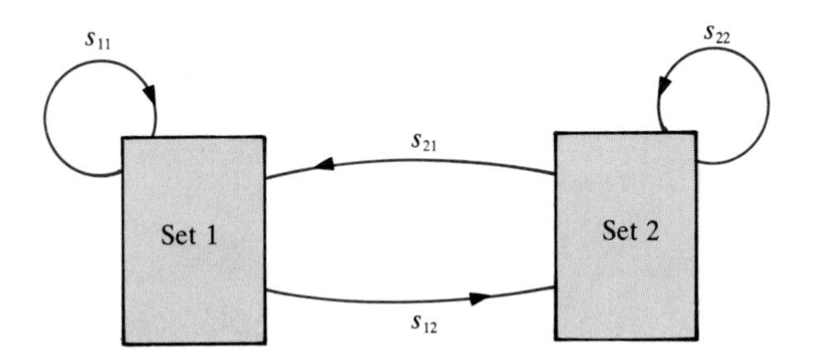

FIGURE 1.13

* The term *Markov chain* is also used. However, exchange system indicates more clearly the nature of these systems.

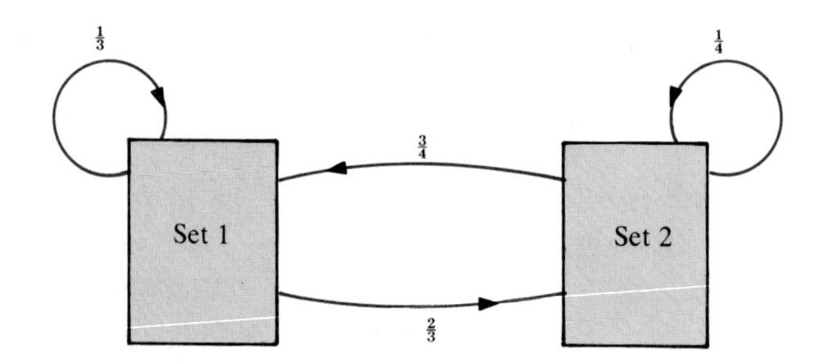

FIGURE 1.14

In the next chapter we will show how storing the diagram's information in a transition matrix makes various calculations involving exchange systems easier.

A system of linear algebraic equations

$$\begin{aligned} a_{11}x_1 + a_{12}x_2 + \cdots + a_{1n}x_n &= b_1 \\ a_{21}x_1 + a_{22}x_2 + \cdots + a_{2n}x_n &= b_2 \\ \vdots \qquad \vdots \qquad \vdots \qquad \vdots \quad &\quad \vdots \\ a_{m1}x_1 + a_{m2}x_2 + \cdots + a_{mn}x_n &= b_m \end{aligned}$$

can be stored numerically as an *augmented matrix*,

$$\begin{bmatrix} a_{11} & a_{12} & \cdots & a_{1n} & b_1 \\ a_{21} & a_{22} & \cdots & a_{2n} & b_2 \\ \vdots & \vdots & \vdots & \vdots & \vdots \\ a_{m1} & a_{m2} & \cdots & a_{mn} & b_m \end{bmatrix}$$

EXAMPLE 37

The augmented matrix of

$$\begin{aligned} 2x - 3y &= 6 \\ x - y &= 8 \\ 4x + y &= 3 \end{aligned}$$

is

$$\begin{bmatrix} 2 & -3 & 6 \\ 1 & -1 & 8 \\ 4 & 1 & 3 \end{bmatrix}$$

□

EXAMPLE 38

The system of linear algebraic equations for the augmented matrix

$$\begin{bmatrix} 1 & -2 & 1 & -1 & 1 \\ 1 & 0 & -1 & 7 & 0 \\ 0 & 3 & 0 & -2 & 0 \end{bmatrix}$$

is
$$\begin{aligned} x_1 - 2x_2 + x_3 - x_4 &= 1 \\ x_1 \qquad\quad - x_3 + 7x_4 &= 0 \\ 3x_2 \qquad - 2x_4 &= 0 \end{aligned}$$
□

In Sections 1.2 and 1.3 we saw how to solve systems of linear algebraic equations by the Gaussian elimination algorithm and the Gauss-Jordan algorithm. Reviewing solved examples in which these algorithms were used, we can see that the arithmetic operations take place on the constants in the system, and that the variables simply keep these constants in order during the computations. This ordering is also maintained in the augmented matrix. Hence we can solve a system of linear algebraic equations by applying the Gaussian operations directly to the augmented matrix.

ALGORITHM TO SOLVE A SYSTEM BY USING THE AUGMENTED MATRIX

To solve a system of linear algebraic equations:

1. Write out the augmented matrix for the given system.

2. Apply the Gaussian elimination algorithm, Steps 1 through 3, or the Gauss-Jordan algorithm, Steps 1 and 2, directly to the matrix.

3. Write out the equations from the augmented matrix obtained in Step 2.

4. Solve these equations.

The matrix obtained in Step 2 has a stair-step form. This form, called an *echelon* form*, has the property that each nonzero row begins with a sequence of zeros whose length is greater than that of the preceding row, and all zero rows follow all nonzero rows. Thus the format is

$$\begin{bmatrix} r & * & * & * & * \\ 0 & r & * & * & * \\ 0 & 0 & 0 & r & * \end{bmatrix}, \quad \begin{bmatrix} r & * & * \\ 0 & r & * \\ 0 & 0 & r \\ 0 & 0 & 0 \\ 0 & 0 & 0 \end{bmatrix}, \quad \begin{bmatrix} r & * & * & * & * & * & * \\ 0 & 0 & r & * & * & * & * \\ 0 & 0 & 0 & r & * & * & * \\ 0 & 0 & 0 & 0 & 0 & 0 & r \end{bmatrix}$$

or the like, where each r is nonzero and each * can be replaced by any number.

We now give some examples involving solving systems of linear algebraic equations by using the augmented matrix.

EXAMPLE 39

Using the Gaussian elimination algorithm, solve
$$\begin{aligned} x_1 + x_2 + x_3 &= 0 \\ 2x_1 \qquad + 2x_3 &= 0 \\ x_2 + x_3 &= -1 \end{aligned}$$

* The word "echelon" is taken from a formation of troops placed in parallel lines, each line positioned to the right of the line in front of it, so as to appear step-like.

Writing out the augmented matrix and solving yields the following:

$$\begin{bmatrix} 1 & 1 & 1 & 0 \\ 2 & 0 & 2 & 0 \\ 0 & 1 & 1 & -1 \end{bmatrix} \xrightarrow{R_2 - 2R_1 \to R_2} \begin{bmatrix} 1 & 1 & 1 & 0 \\ 0 & -2 & 0 & 0 \\ 0 & 1 & 1 & -1 \end{bmatrix}$$

$$\xrightarrow{R_3 + \frac{1}{2}R_2 \to R_3} \begin{bmatrix} 1 & 1 & 1 & 0 \\ 0 & -2 & 0 & 0 \\ 0 & 0 & 1 & -1 \end{bmatrix}$$

The corresponding equations are

$$\begin{aligned} x_1 + x_2 + x_3 &= 0 \\ -2x_2 \quad &= 0 \\ x_3 &= -1 \end{aligned}$$

Thus $x_3 = -1$, $x_2 = 0$, and $x_1 = 1$, and the solution set is $\{(1, 0, -1)\}$. □

EXAMPLE 40

Using the Gaussian elimination algorithm, solve

$$\begin{aligned} 2x_1 - x_2 \quad &= 0 \\ x_1 + x_2 + x_3 &= 2 \end{aligned}$$

Writing out the augmented matrix and solving yields the following:

$$\begin{bmatrix} 2 & -1 & 0 & 0 \\ 1 & 1 & 1 & 2 \end{bmatrix} \xrightarrow{R_2 - \frac{1}{2}R_1 \to R_2} \begin{bmatrix} 2 & -1 & 0 & 0 \\ 0 & \frac{3}{2} & 1 & 2 \end{bmatrix}$$

Writing out the corresponding equations gives

$$\begin{aligned} 2x_1 - x_2 \quad &= 0 \\ \tfrac{3}{2}x_2 + x_3 &= 2 \end{aligned}$$

Now x_1 and x_2 are the bound variables, and x_3 can be assigned any number since it is free. Solving the second equation for x_2 yields $x_2 = \frac{4}{3} - \frac{2}{3}x_3$. Substituting this value into the first equation gives

$$2x_1 - (\tfrac{4}{3} - \tfrac{2}{3}x_3) = 0 \qquad \text{or} \qquad x_1 = \tfrac{2}{3} - \tfrac{1}{3}x_3$$

Thus the solution set is

$$\{(\tfrac{2}{3} - \tfrac{1}{3}x_3, \tfrac{4}{3} - \tfrac{2}{3}x_3, x_3) : x_3 \text{ is any number}\}$$ □

EXAMPLE 41

Using the Gauss-Jordan algorithm, solve

$$\begin{aligned} x - y + 2z &= 3 \\ 3x - 2y + z &= 1 \\ 2x + y - 3z &= 6 \\ x + 6y - 9z &= 19 \end{aligned}$$

Writing out the augmented matrix and applying the algorithm yields

$$\begin{bmatrix} 1 & -1 & 2 & 3 \\ 3 & -2 & 1 & 1 \\ 2 & 1 & -3 & 6 \\ 1 & 6 & -9 & 19 \end{bmatrix} \xrightarrow[R_4 - R_1 \to R_4]{\substack{R_2 - 3R_1 \to R_2 \\ R_3 - 2R_1 \to R_3}} \begin{bmatrix} 1 & -1 & 2 & 3 \\ 0 & 1 & -5 & -8 \\ 0 & 3 & -7 & 0 \\ 0 & 7 & -11 & 16 \end{bmatrix}$$

$$\xrightarrow[R_4 - 7R_2 \to R_4]{R_3 - 3R_2 \to R_3} \begin{bmatrix} 1 & -1 & 2 & 3 \\ 0 & 1 & -5 & -8 \\ 0 & 0 & 8 & 24 \\ 0 & 0 & 24 & 72 \end{bmatrix}$$

$$\xrightarrow{\frac{1}{8}R_3 \to R_3} \begin{bmatrix} 1 & -1 & 2 & 3 \\ 0 & 1 & -5 & -8 \\ 0 & 0 & 1 & 3 \\ 0 & 0 & 24 & 72 \end{bmatrix}$$

$$\xrightarrow{R_4 - 24R_3 \to R_4} \begin{bmatrix} 1 & -1 & 2 & 3 \\ 0 & 1 & -5 & -8 \\ 0 & 0 & 1 & 3 \\ 0 & 0 & 0 & 0 \end{bmatrix}$$

$$\xrightarrow[R_2 + 5R_3 \to R_2]{R_1 - 2R_3 \to R_1} \begin{bmatrix} 1 & -1 & 0 & -3 \\ 0 & 1 & 0 & 7 \\ 0 & 0 & 1 & 3 \\ 0 & 0 & 0 & 0 \end{bmatrix}$$

$$\xrightarrow{R_1 + R_2 \to R_1} \begin{bmatrix} 1 & 0 & 0 & 4 \\ 0 & 1 & 0 & 7 \\ 0 & 0 & 1 & 3 \\ 0 & 0 & 0 & 0 \end{bmatrix}$$

The corresponding system of equations is

$$\begin{aligned} x_1 \qquad\qquad &= 4 \\ x_2 \qquad &= 7 \\ x_3 &= 3 \end{aligned}$$

Thus, $\{(4, 7, 3)\}$ is the solution set. □

EXERCISES FOR SECTION 1.4

COMPUTATIONAL EXERCISES

1. Let

$$A = \begin{bmatrix} 1 & 2 \\ 3 & 4 \\ 5 & 6 \end{bmatrix}$$

List the 1,2-entry, the 2,1-entry, the 3,2-entry, the 2-row, and the 1-column.

2. Let $A = [1 \quad 2 \quad 3 \quad 4]$. List the 1,2-entry, the 3-column, and the 1-row.

3. Let

$$A = \begin{bmatrix} 1 & 2 & 3 & 4 \\ 5 & 6 & 7 & 8 \\ 9 & 10 & 11 & 12 \\ 13 & 14 & 15 & 16 \end{bmatrix}$$

List $a_{12}, a_{23}, a_{41}, a_{33}, \mathbf{a}_1, \mathbf{a}_3, \mathbf{a}^2$, and $\mathbf{a}^4$.

4. Let

$$A = \begin{bmatrix} 1 & 0 & 3 \\ 2 & -2 & -1 \\ -1 & 3 & 0 \end{bmatrix}$$

List $a_{12}, a_{23}, a_{33}, \mathbf{a}_2, \mathbf{a}_3, \mathbf{a}^1$, and $\mathbf{a}^3$.

5. State the size of each matrix.

(*a*) $[-1 \quad 1 \quad 2]$

(*b*) $[3]$

(*c*) $\begin{bmatrix} 2 & -1 \\ 3 & 1 \end{bmatrix}$

(*d*) $\begin{bmatrix} 1 \\ 3 \\ -2 \end{bmatrix}$

(*e*) $\begin{bmatrix} 1 & 2 & 3 \\ 3 & 2 & 1 \end{bmatrix}$

6. State the size of each matrix.

(*a*) $[0]$

(*b*) $[0 \quad 0 \quad 0]$

(*c*) $\begin{bmatrix} 2 \\ 2 \\ -1 \end{bmatrix}$

(*d*) $\begin{bmatrix} 7 & 7 \\ 7 & 7 \end{bmatrix}$

(*e*) $\begin{bmatrix} 1 & 0 & 0 \\ 0 & 1 & 0 \\ 0 & 0 & 1 \end{bmatrix}$

7. Write the augmented matrix for each system.

(*a*)
$$\begin{aligned} 2x - 3y &= 6 \\ 5x + 6y &= -4 \end{aligned}$$

(*b*)
$$\begin{aligned} -x_1 + x_2 - 3x_3 &= 5 \\ 4x_1 - 3x_3 &= 6 \\ 3x_2 + 2x_4 &= 1 \end{aligned}$$

(*c*)
$$\begin{aligned} x_1 - 2x_2 - 6x_3 &= 5 \\ 2x_1 + x_3 &= -8 \\ -x_1 - 2x_3 &= 0 \\ -x_2 &= 6 \end{aligned}$$

8. Write out the system of linear algebraic equations for the given augmented matrix.

(*a*) $\begin{bmatrix} 1 & -1 & 0 & 2 & 1 \\ 4 & 2 & -3 & 0 & 6 \\ -1 & 5 & 2 & -1 & 1 \end{bmatrix}$

(*b*) $\begin{bmatrix} 2 & 1 & -1 \\ 1 & 0 & 5 \\ 1 & 1 & -2 \\ 3 & 2 & -1 \end{bmatrix}$

(*c*) $\begin{bmatrix} 0 & 1 & -1 & 1 \\ 1 & 0 & 1 & 0 \\ 1 & 1 & 0 & -1 \\ -1 & 0 & 1 & -1 \\ -1 & 0 & 1 & 1 \end{bmatrix}$

In exercises 9 through 16, redo the indicated exercises from Section 1.2, using the augmented matrix and the Gaussian elimination algorithm. Use the notation for Gaussian operations to indicate what you are doing at each step. (You can use grid or graph paper to help you keep track of row and column entries.)

9. #7 *10.* #6 *11.* #9 *12.* #10 *13.* #13 *14.* #14 *15.* #17 *16.* #18

In exercises 17 through 26, redo the indicated exercises from Section 1.3, using the augmented matrix and the Gauss-Jordan algorithm. Use the notation for Gaussian operations to indicate what you are doing at each step.

17. #9 *18.* #10 *19.* #11 *20.* #12 *21.* #13 *22.* #16 *23.* #17 *24.* #18 *25.* #19 *26.* #20

27. By applying the Gaussian elimination algorithm, find an echelon form for the following matrices. Then use the scaling operation to obtain a 1 as the leading entry in each nonzero row. (In some texts this special echelon form is called a *row echelon form.*)

(*a*) $A = \begin{bmatrix} 1 & 2 & -1 \\ 4 & 3 & 2 \\ -1 & 1 & 1 \\ 2 & -1 & 3 \end{bmatrix}$ (*b*) $A = \begin{bmatrix} 2 & -3 & 1 \\ 4 & -6 & 3 \\ -2 & 1 & 2 \\ -6 & -1 & 2 \end{bmatrix}$

28. By applying the Gauss-Jordan algorithm, find an echelon form for each of the following matrices. (In some texts this echelon form is given the special name *reduced row echelon form.*)

(*a*) $\begin{bmatrix} 1 & 0 & 1 & -1 & 0 \\ -2 & 1 & 2 & 0 & -1 \\ -1 & 4 & 0 & 1 & 2 \end{bmatrix}$

(*b*) $\begin{bmatrix} -3 & 0 & 1 & 1 & 0 \\ 3 & 1 & 2 & -2 & 2 \\ -6 & -2 & -4 & 1 & 2 \end{bmatrix}$

If a system of linear equations has a solution, it is called *consistent*; if it does not, it is called *inconsistent*. In exercises 29 and 30, determine which systems are consistent and which are inconsistent by computing the echelon forms.

29. (*a*)
$$\begin{aligned} x + y &= 3 \\ 2x + 2y &= 4 \end{aligned}$$

(*b*)
$$\begin{aligned} x - y + z &= 1 \\ x + y - z &= 1 \\ -x + y + z &= 1 \end{aligned}$$

(*c*)
$$\begin{aligned} x - y + 2z &= 1 \\ x + 2y - z &= 2 \end{aligned}$$

30. (a) $x - y = 1$
$x + y = 2$

(b) $x + y = 2$
$y + z = 1$
$x + 2y + z = 2$

(c) $2x - y + z = 1$
$-4x + 2y - 2z = 1$

COMPLEX NUMBERS

In exercises 31 and 32, solve by Gaussian elimination, using the augmented matrix.

31. $(2 - i)z + (3 + 2i)w = 11 + 2i$
$(1 - 3i)z + (2 - 4i)w = 4 - 12i$

32. $(1 - i)z + (4 + 3i)w = 8$
$(4 - 2i)z + (3 - 2i)w = 3 - i$

THEORETICAL EXERCISES

33. Define an exchange system for three sets, including the corresponding diagram and transition matrix. Write out the transition matrix for the diagram in Figure 1.15..

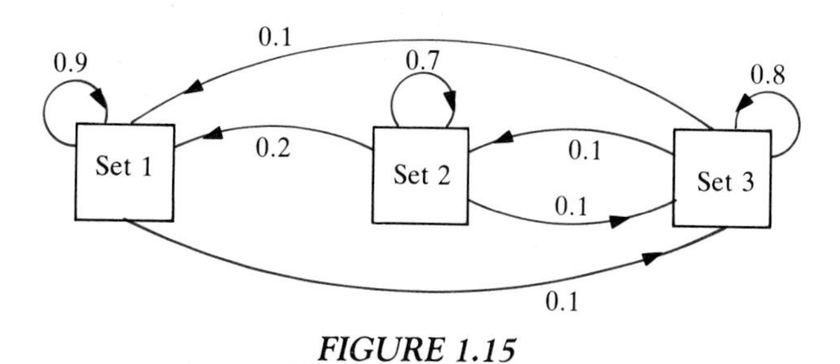

FIGURE 1.15

34. Let A be an $m \times n$ matrix with $m > n$. In terms of m and n, what is the maximum number of nonzero rows in an echelon form of A? (Look at some examples before deciding on your answer.)

35. (a) If we randomly write down m linear equations with n variables and $m > n$, should we expect a solution to exist? Based on intuition rather than mathematical results, explain your answer in terms of echelon forms. (When $m > n$, the system is called *overdetermined.*)
(b) Answer the same question for $m < n$. (When $m < n$, the system is called *underdetermined.*)

APPLICATIONS EXERCISES

36. Matrices are sometimes used to store inventory. Suppose a partial inventory of an auto dealership service department indicates that the number of parts for each type of auto is as follows:

	Carburetor	Generator	Alternator	
$A =$	2	3	2	Z-10
	4	3	3	Z-12
	3	4	2	Z-16

(a) How many Z-12 generators are in stock?
(b) How many Z-16 carburetors are in stock?
(c) How many Z-10 alternators are in stock?

37. Part of a personal computer store's sales sheet reads as follows:

	Computers	Printers	Disk Drives	
$A =$	2	1	2	May 5
	1	1	1	May 6
	3	2	1	May 7

(a) How many computers were sold on May 5?
(b) How many disk drives were sold on May 6?
(c) How many printers were sold on May 7?

38. The inventory for an auto dealership service department is given as follows:

3 carburetors for Z-16 cars
3 alternators for Z-10 cars
1 carburetor for Z-12 cars
2 generators for Z-16 cars
2 carburetors for Z-10 cars
1 generator for Z-12 cars

List this inventory in matrix form.

39. A directed graph can be stored as a matrix. Consider the directed graph in Figure 1.16, where the numbers are called *vertices* and a directed line segment from vertex i to vertex j is called an *arc*. Define the *adjacency matrix* $A = [a_{ij}]$ to be an $n \times n$ matrix, where n is the number of vertices, such that $a_{ij} = 1$ if there is an arc from vertex i to vertex j and $a_{ij} = 0$ otherwise. Find the matrix A for the graph of Figure 1.16.

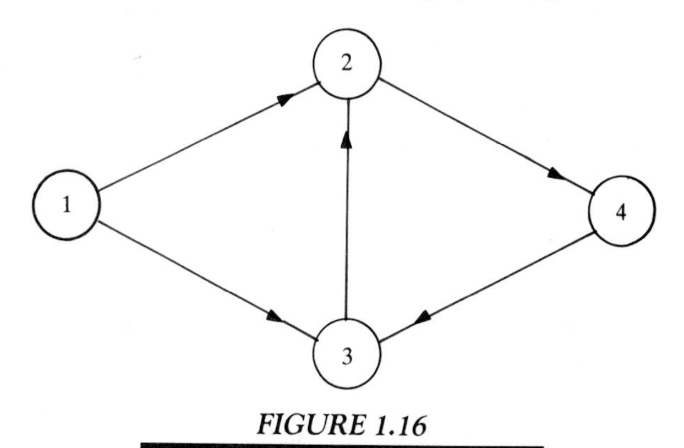

FIGURE 1.16

40. As in Example 36, write out the transition matrix corresponding to each of the following diagrams.

(*a*)

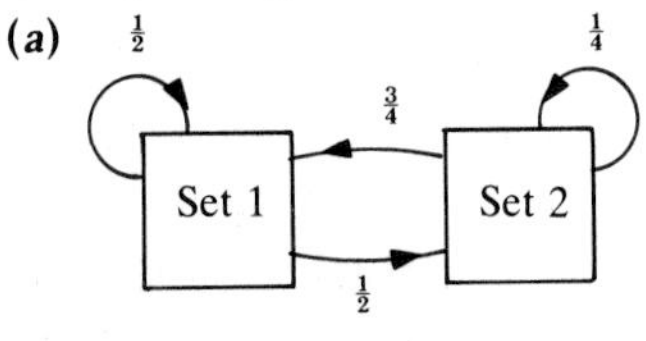

(*b*)

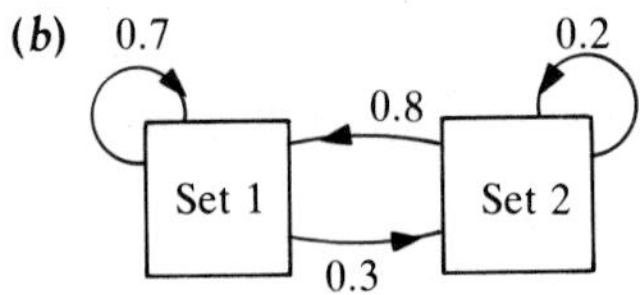

41. Each of the following is an exchange system. Draw the corresponding diagram and write out the transition matrix.
(*a*) Let C_1 and C_2 be neighboring countries. Suppose $\frac{1}{10}$ of the population of C_1 migrates to C_2 and $\frac{1}{20}$ of the population of C_2 migrates to C_1 each year.
(*b*) Let C be the set of drinkers of soft drink C and P be the set of drinkers of soft drink P in the United States. Suppose each year $\frac{1}{100}$ of the C drinkers switch to P and $\frac{1}{150}$ of the P drinkers to C.
(*c*) A set of rats were run sequentially through a maze in which they had a choice of running to the right or to the left. One-eighth of the rats that ran left on a particular run ran right on their next run, while one-sixth of those who ran right on a particular run ran left on their next run.

42. Write out the current equations for the circuit in Figure 1.17. Use Gaussian elimination to solve the system for I_2. For this circuit an unknown resistance y can be determined by using an ammeter to measure I_2 for various known resistances x. Why? (*Hint:* When does $I_2 = 0$?) This circuit is known as Wheatstone's bridge.

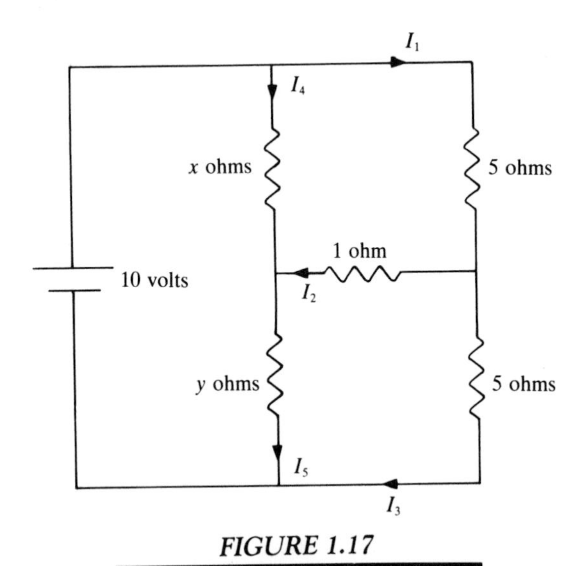

FIGURE 1.17

43. A city wants to know the traffic flow at a certain T-type intersection (see Figure 1.18). For a given day, it wants to know the number x_{NE} of cars traveling north that turn east, the number x_{WS} of cars traveling west that turn south, etc. There are six unknowns in all: $x_{\text{NE}}, x_{\text{NW}}, x_{\text{WS}}, x_{\text{WW}}, x_{\text{EE}}$, and x_{ES}. The city has available mechanical traffic counters that count the number of cars that run over them. An employee suggests putting six such counters as indicated in Figure 1.18. Assume that this is done, and that at the end of the day the totals for the six counters are C_1, C_2, C_3, C_4, C_5, and C_6.

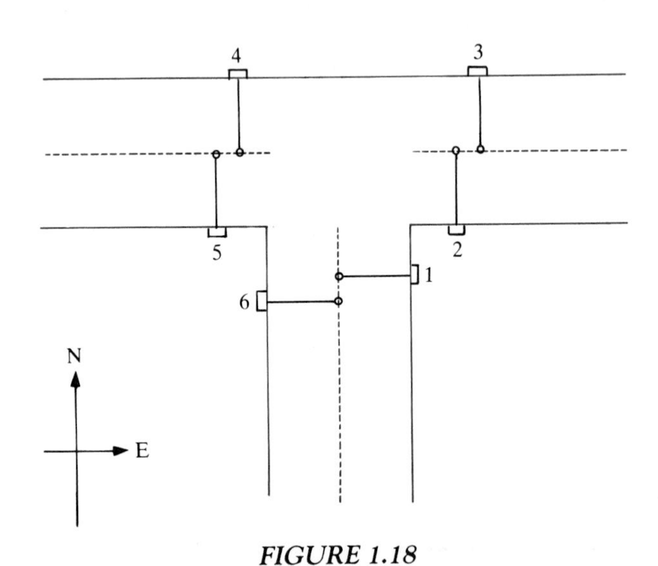

FIGURE 1.18

(*a*) Is this enough information to determine the traffic flow at the intersection? (*Hint:* Set up the equations $C_1 = x_{\text{NE}} + x_{\text{NW}}, \ldots$, and solve the resulting system of equations.)
(*b*) If this is not sufficient information, then the city will need to hire people to count part of the traffic flow. Each person can count one unknown. For example, one person might count x_{NE}, another x_{WS}, etc. In addition to the information obtained by the mechanical counters, how many people does the city need to determine the traffic flow? Are all of the mechanical counters needed? Explain. (*Hint:* How many free variables are there?)

COMPUTER EXERCISES

($m \times n$ memory computer) An $m \times n$ memory computer is a pedagogical tool that contains an $m \times n$ matrix in which entries can be stored. To store a number x in the i,j-entry, we write

$$\text{Store } a_{ij} = x.$$

If there is a number already in this entry, it is erased and x is written in its place.

44. Store the inventory of exercise 38 in a 3×3 memory computer as described below, with parts listed in the order (carburetor, generator, alternator) and cars in the order (Z-10, Z-12, Z-16).

(*a*) List cars along the rows and parts along the columns.

(*b*) List parts along the rows and cars along the columns.

(*c*) Suppose today's sales are 1 Z-10 carburetor and 1 Z-16 generator. Using the format of part (a), tell the computer how to update its inventory list.

45. Store the system

$$\begin{aligned} x - \tfrac{1}{3}y &= \tfrac{1}{7} \\ \tfrac{1}{4}x - \tfrac{1}{6}y &= \tfrac{3}{8} \end{aligned}$$

in a 2×3 memory computer. Assume that only two-digit numbers can be stored in each memory location.

46. Using a two-digit and then a four-digit computer, solve each of the following systems. Using percentage error, compare your results to the exact solutions rounded to integers. Graph the equations and explain the accuracy of your results geometrically.

(*a*) $\begin{aligned} 100x + 27y &= 300 \\ 114x + 27y &= 100 \end{aligned}$

(*b*) $\begin{aligned} 11x + 21y &= 42 \\ 21x + 10y &= 42 \end{aligned}$

CHAPTER TWO

MATRIX ARITHMETIC AND THE DETERMINANT

In this chapter we introduce the arithmetic of matrices and show how this arithmetic leads to a way of writing systems of linear algebraic equations as matrix equations. You will find that matrix equations can be arithmetically manipulated in much the same way as scalar equations.

2.1 MATRIX ADDITION, SUBTRACTION, AND TRANSPOSITION

An $m \times n$ matrix is a rectangular array of numbers. Since a function at each value of the variable is a number, function entries are also allowed. In this text, however, the work is given for numbers, and then, when appropriate, applied to the function case.

The notation of the previous chapter allows us to express a matrix in terms of its entries, its rows, or its columns by writing

$$A = [a_{ij}], \qquad A = \begin{bmatrix} \mathbf{a}_1 \\ \mathbf{a}_2 \\ \vdots \\ \mathbf{a}_m \end{bmatrix}, \qquad \text{or} \qquad A = [\mathbf{a}^1 \quad \mathbf{a}^2 \quad \cdots \quad \mathbf{a}^n]$$

respectively. We will use this notation throughout the text.

Several special matrices arise sufficiently often that they are given special names. A $1 \times n$ matrix is called a *row vector*, and an $m \times 1$ matrix is called a *column vector*. When the case is clear from context, we will refer to these two matrices simply as *vectors*. It should be noted that double subscripting is not necessary in a vector, and so we will write

$$\mathbf{x} = [x_1 \quad x_2 \quad \cdots \quad x_n] \qquad \text{and} \qquad \mathbf{x} = \begin{bmatrix} x_1 \\ x_2 \\ \vdots \\ x_m \end{bmatrix}$$

or simply $\mathbf{x} = [x_i]$. The special vectors

$$\mathbf{e}_i = [0 \quad \cdots \quad 0 \quad 1 \quad 0 \quad \cdots \quad 0] \qquad \text{and} \qquad \mathbf{e}^i = \begin{bmatrix} 0 \\ \vdots \\ 0 \\ 1 \\ 0 \\ \vdots \\ 0 \end{bmatrix}$$

have a 1 as the i entry and 0's elsewhere.

An $n \times n$ matrix is referred to as a *square matrix* for obvious reasons. In a square matrix $A = [a_{ij}]$, the entries $a_{11}, a_{22}, \ldots, a_{nn}$ are called the *main diagonal* of A. Using this notion, we define several special square matrices.

(*i*) A square matrix $T = [t_{ij}]$ is called *upper triangular* if the entries below the main diagonal are 0. Notationally, this means that $t_{ij} = 0$ for all $i > j$. *Lower triangular* is defined similarly. The word *triangular* is used to refer to either case.

(*ii*) A square matrix $D = [d_{ij}]$ is called a *diagonal matrix* if it is both upper and lower triangular. Notationally, this means that $d_{ij} = 0$ for all $i \neq j$.

(*iii*) An $n \times n$ matrix $I_n = [\delta_{ij}]$ is called an *identity matrix* if the only nonzero entries are on the main diagonal and these are all 1s. Notationally, we write that $\delta_{ij} = 0$ if $i \neq j$ and $\delta_{ii} = 1$ for all i.

EXAMPLE 1

Some examples of the above definitions follow.

(*i*) The matrix

$$\begin{bmatrix} 1 & 2 & 3 \\ 0 & 4 & 5 \\ 0 & 0 & 6 \end{bmatrix}$$

is upper triangular and is thus triangular. The main diagonal is 1, 4, 6.

(*ii*) The matrix

$$\begin{bmatrix} 1 & 0 & 0 \\ 0 & 2 & 0 \\ 0 & 0 & 3 \end{bmatrix}$$

is upper triangular and lower triangular and is thus triangular. It is also a diagonal matrix with main diagonal 1, 2, 3.

(*iii*) The matrices

$$I_2 = \begin{bmatrix} 1 & 0 \\ 0 & 1 \end{bmatrix} \qquad \text{and} \qquad I_3 = \begin{bmatrix} 1 & 0 & 0 \\ 0 & 1 & 0 \\ 0 & 0 & 1 \end{bmatrix}$$

are identity matrices. □

Calculations with matrices often involve equality. Two matrices $A = [a_{ij}]$ and $B = [b_{ij}]$ of the same size are *equal* if and only if their corresponding entries are equal; that is, $a_{ij} = b_{ij}$ for all i and j.

EXAMPLE 2

(*i*) By definition,

$$\begin{bmatrix} 1 & 1+1 \\ \frac{6}{2} & 2 \cdot 2 \end{bmatrix} = \begin{bmatrix} 1 & 2 \\ 3 & 4 \end{bmatrix} \quad \text{and} \quad [\sin \pi/2 \quad 8-2] = [1 \quad 6]$$

(*ii*) If we write

$$\begin{bmatrix} w & x \\ y & z \end{bmatrix} = \begin{bmatrix} 1 & 2 \\ 3 & 4 \end{bmatrix}$$

then $w = 1$, $x = 2$, $y = 3$, and $z = 4$.

(*iii*) Finally, we write

$$\begin{bmatrix} 1 & 1+2 \\ \frac{6}{2} & 2 \cdot 2 \end{bmatrix} \neq \begin{bmatrix} 1 & 2 \\ 3 & 4 \end{bmatrix} \quad \text{and} \quad [1 \quad 2 \quad 3 \quad 4] \neq \begin{bmatrix} 1 & 2 \\ 3 & 4 \end{bmatrix}$$

□

As shown in the first chapter, a matrix is often used to store data. As desired, the data stored in rows can be transposed, or interchanged, with the data stored in columns. Mathematically, this operation is defined as follows:

DEFINITION

If $A = [a_{ij}] = [\mathbf{a}^1 \cdots \mathbf{a}^n]$ is an $m \times n$ matrix, then the *transpose* of A is the $n \times m$ matrix

$$A^t = [a_{ij}^{(t)}] = \begin{bmatrix} \mathbf{a}_1^{(t)} \\ \vdots \\ \mathbf{a}_n^{(t)} \end{bmatrix}$$

where $a_{ij}^{(t)} = a_{ji}$ for all i and j. In terms of rows and columns, we have $\mathbf{a}_i^{(t)} = (\mathbf{a}^i)^t$ for all i.

EXAMPLE 3

Some examples of computing transposes follow.

(*i*) For

$$A = \begin{bmatrix} 1 & 2 \\ 3 & 4 \end{bmatrix}$$

the transpose is

$$A^t = \begin{bmatrix} 1 & 3 \\ 2 & 4 \end{bmatrix}$$

(*ii*) For

$$A = \begin{bmatrix} 2 & -1 \\ -4 & 0 \\ 8 & -7 \end{bmatrix}$$

the transpose is

$$A^t = \begin{bmatrix} 2 & -4 & 8 \\ -1 & 0 & -7 \end{bmatrix}$$

(*iii*) The matrix

$$A = \begin{array}{c} \\ \end{array}\begin{matrix} \text{Carburetor} & \text{Generator} & \text{Alternator} \\ \end{matrix}$$

$$A = \begin{bmatrix} 4 & 3 & 3 \\ 3 & 4 & 2 \end{bmatrix} \begin{matrix} \text{Z-2} \\ \text{Z-10} \end{matrix}$$

(columns: Carburetor, Generator, Alternator)

which stores inventory, can be transposed to give

$$A^t = \begin{bmatrix} 4 & 3 \\ 3 & 4 \\ 3 & 2 \end{bmatrix} \begin{matrix} \text{Carburetor} \\ \text{Generator} \\ \text{Alternator} \end{matrix}$$

(columns: Z-2, Z-10) □

As shown below, transposing a matrix twice results in the same matrix.

THEOREM 2.1

Let A be a matrix. Then $(A^t)^t = A$.

PROOF

To gain experience, we first prove the result for

$$A = \begin{bmatrix} a_{11} & a_{12} \\ a_{21} & a_{22} \end{bmatrix}$$

Here,

$$(A^t)^t = \begin{bmatrix} a_{11} & a_{21} \\ a_{12} & a_{22} \end{bmatrix}^t = \begin{bmatrix} a_{11} & a_{12} \\ a_{21} & a_{22} \end{bmatrix} = A$$

In general, we prove the result by viewing the i, j-entry of the matrix A through the above calculations. Thus,

$$(A^t)^t = [a_{ij}^{(t)}]^t = [a_{ji}]^t = [a_{ji}^{(t)}] = [a_{ij}] = A$$ ■

Having given this preliminary work on matrices, we can now begin the development of the arithmetic of matrices. This arithmetic has many similarities with the arithmetic of numbers, largely caused by the use of the arithmetic of numbers in developing the arithmetic of matrices. However, there are also some significant differences, as we will show.

Addition of matrices is defined as follows.

DEFINITION

If $A = [a_{ij}]$ and $B = [b_{ij}]$ are matrices of the same size, their *sum* is

$$A + B = [a_{ij} + b_{ij}]$$

In other words, addition is done entrywise.

EXAMPLE 4

(*i*) By definition, we write

$$\begin{bmatrix} 1 & 2 \\ 3 & 4 \end{bmatrix} + \begin{bmatrix} 5 & 6 \\ 7 & 8 \end{bmatrix} = \begin{bmatrix} 1+5 & 2+6 \\ 3+7 & 4+8 \end{bmatrix} = \begin{bmatrix} 6 & 8 \\ 10 & 12 \end{bmatrix}$$

and

$$\begin{bmatrix} 1 & 2 & 3 \\ 4 & 5 & 6 \end{bmatrix} + \begin{bmatrix} 7 & 8 & -9 \\ 10 & 11 & 12 \end{bmatrix} = \begin{bmatrix} 1+7 & 2+8 & 3+(-9) \\ 4+10 & 5+11 & 6+12 \end{bmatrix}$$
$$= \begin{bmatrix} 8 & 10 & -6 \\ 14 & 16 & 18 \end{bmatrix}$$

(*ii*) The matrices

$$\begin{bmatrix} 1 & 2 & 1 \\ 3 & 1 & -1 \end{bmatrix} \quad \text{and} \quad \begin{bmatrix} 1 & 2 & 1 & 2 \\ 3 & 1 & 1 & 2 \end{bmatrix}$$

do not have the same size, and hence they cannot be added. □

This addition satisfies the usual arithmetical properties of addition used in computing.

THEOREM 2.2

Let $A = [a_{ij}]$, $B = [b_{ij}]$, $C = [c_{ij}]$, and $\mathbf{0} = [0]$ (the matrix whose entries are all 0) be matrices of the same size. Then

(*i*) $A + B = B + A$ (commutative law)

(*ii*) $(A + B) + C = A + (B + C)$ (associative law)

(*iii*) $\mathbf{0} + A = A + \mathbf{0} = A$ (additive identity)

(*iv*) the matrix $-A = [-a_{ij}]$ has the property that $(-A) + A = A + (-A) = \mathbf{0}$ (additive inverse)

(*v*) $(A + B)^t = A^t + B^t$

PROOF

We show the proof of (i), leaving the remainder as exercise 43. Before proving (i) for the general case, we first gain some experience with the notation by proving it for the 2×2 case.

$$A + B = \begin{bmatrix} a_{11} & a_{12} \\ a_{21} & a_{22} \end{bmatrix} + \begin{bmatrix} b_{11} & b_{12} \\ b_{21} & b_{22} \end{bmatrix} = \begin{bmatrix} a_{11}+b_{11} & a_{12}+b_{12} \\ a_{21}+b_{21} & a_{22}+b_{22} \end{bmatrix}$$
$$= \begin{bmatrix} b_{11}+a_{11} & b_{12}+a_{12} \\ b_{21}+a_{21} & b_{22}+a_{22} \end{bmatrix} = \begin{bmatrix} b_{11} & b_{12} \\ b_{21} & b_{22} \end{bmatrix} + \begin{bmatrix} a_{11} & a_{12} \\ a_{21} & a_{22} \end{bmatrix}$$
$$= B + A$$

Now, the proof of (i) for the general case is the same except for size. Viewing the i,j-entry, we write

$$A + B = [a_{ij}] + [b_{ij}] = [a_{ij} + b_{ij}] = [b_{ij} + a_{ij}] = [b_{ij}] + [a_{ij}] = B + A \quad \blacksquare$$

As a consequence of (iv), we define *subtraction* of two matrices $A = [a_{ij}]$ and $B = [b_{ij}]$ of the same size in the natural way:

$$A - B = A + (-B) = [a_{ij} - b_{ij}]$$

Multiples of a matrix can also be defined. More generally, scalar multiplication is defined as follows.

DEFINITION

Let r be a scalar and $A = [a_{ij}]$ be any matrix. Then the *scalar product* is

$$rA = [ra_{ij}]$$

Thus rA is obtained by multiplying each entry of A by r.

Note that if $k > 0$ is an integer, then kA is the result of adding A to itself k times, and so scalar multiplication is a generalization of multiples of a matrix.

EXAMPLE 5

(*i*) By definition, we have

$$2\begin{bmatrix} 1 & 2 \\ 3 & 4 \end{bmatrix} = \begin{bmatrix} 2\cdot 1 & 2\cdot 2 \\ 2\cdot 3 & 2\cdot 4 \end{bmatrix} = \begin{bmatrix} 2 & 4 \\ 6 & 8 \end{bmatrix}$$

and

$$-8[1 \quad \tfrac{1}{2} \quad -3] = [-8\cdot 1 \quad -8\cdot\tfrac{1}{2} \quad -8\cdot(-3)] = [-8 \quad -4 \quad 24]$$

(*ii*) We can use scalar multiplication to record sums in a compact way. If A is a matrix, then

$$A + A + A + A + A + A + A = 7A$$

(*iii*) We can also factor as follows:

$$\begin{bmatrix} 2 & -6 & 8 \\ 4 & 0 & -2 \end{bmatrix} = \begin{bmatrix} 2\cdot 1 & 2\cdot(-3) & 2\cdot 4 \\ 2\cdot 2 & 2\cdot 0 & 2\cdot(-1) \end{bmatrix}$$
$$= 2\begin{bmatrix} 1 & -3 & 4 \\ 2 & 0 & -1 \end{bmatrix} \quad \square$$

The arithmetical properties of scalar multiplication used in computing follow from the corresponding properties of real numbers.

THEOREM 2.3

Let r and s be scalars and A and B be matrices of the same size. Then

(*i*) $r(A + B) = rA + rB$

(*ii*) $(r + s)A = rA + sA$

(*iii*) $(rs)A = r(sA)$

(*iv*) $(rA)^t = rA^t$

PROOF

We will prove (i) and leave the remaining properties for exercise 44. To gain experience with the notation, before proving (i) for the general case we will consider a special case. For 2×3 matrices A and B,

$$\begin{aligned}
r(A + B) &= r\left(\begin{bmatrix} a_{11} & a_{12} & a_{13} \\ a_{21} & a_{22} & a_{23} \end{bmatrix} + \begin{bmatrix} b_{11} & b_{12} & b_{13} \\ b_{21} & b_{22} & b_{23} \end{bmatrix}\right) \\
&= r\begin{bmatrix} a_{11} + b_{11} & a_{12} + b_{12} & a_{13} + b_{13} \\ a_{21} + b_{21} & a_{22} + b_{22} & a_{23} + b_{23} \end{bmatrix} \\
&= \begin{bmatrix} r(a_{11} + b_{11}) & r(a_{12} + b_{12}) & r(a_{13} + b_{13}) \\ r(a_{21} + b_{21}) & r(a_{22} + b_{22}) & r(a_{23} + b_{23}) \end{bmatrix} \\
&= \begin{bmatrix} ra_{11} + rb_{11} & ra_{12} + rb_{12} & ra_{13} + rb_{13} \\ ra_{21} + rb_{21} & ra_{22} + rb_{22} & ra_{23} + rb_{23} \end{bmatrix} \\
&= \begin{bmatrix} ra_{11} & ra_{12} & ra_{13} \\ ra_{21} & ra_{22} & ra_{23} \end{bmatrix} + \begin{bmatrix} rb_{11} & rb_{12} & rb_{13} \\ rb_{21} & rb_{22} & rb_{23} \end{bmatrix} \\
&= r\begin{bmatrix} a_{11} & a_{12} & a_{13} \\ a_{21} & a_{22} & a_{23} \end{bmatrix} + r\begin{bmatrix} b_{11} & b_{12} & b_{13} \\ b_{21} & b_{22} & b_{23} \end{bmatrix} = rA + rB
\end{aligned}$$

The proof of (i) for the general case follows the same pattern. Viewing the i,j-entry, we write

$$\begin{aligned}
r(A + B) &= r([a_{ij}] + [b_{ij}]) = r[a_{ij} + b_{ij}] \\
&= [r(a_{ij} + b_{ij})] = [ra_{ij} + rb_{ij}] \\
&= [ra_{ij}] + [rb_{ij}] = r[a_{ij}] + r[b_{ij}] \\
&= rA + rB
\end{aligned}$$

■

Theorems 2.2 and 2.3 mean that we can arrange sums for convenience of computation.

EXAMPLE 6

If A and B are $m \times n$ matrices, then $2A - B + 4A + 2B - 3A$ can be computed by combining all expressions involving A and then those involving B, to get

$$(2A + 4A - 3A) + (-B + 2B) = 3A + B$$

□

In the next section we will continue to develop the arithmetic of matrices by defining the product of two matrices.

EXERCISES FOR SECTION 2.1

COMPUTATIONAL EXERCISES

In exercises 1 through 4, state whether the given matrix is upper triangular, lower triangular, triangular, diagonal, and/or the identity matrix.

1. $\begin{bmatrix} 1 & -1 & 1 \\ 0 & 2 & 0 \\ 0 & 0 & 0 \end{bmatrix}$

2. $\begin{bmatrix} -1 & 0 & 0 \\ 3 & 2 & 0 \\ -3 & 1 & -2 \end{bmatrix}$

3. $\begin{bmatrix} 0 & 0 & 0 \\ 0 & 0 & 0 \\ 0 & 0 & 0 \end{bmatrix}$

4. $\begin{bmatrix} 1 & 0 & 0 \\ 0 & 1 & 0 \\ 0 & 0 & 1 \end{bmatrix}$

In exercises 5 through 8, find A^t for the given matrix A.

5. $A = \begin{bmatrix} 1 & 3 \\ 2 & 1 \end{bmatrix}$

6. $A = [1 \quad 2 \quad 1]$

7. $A = \begin{bmatrix} 1 & 1 & -2 \\ 2 & 1 & -3 \\ -1 & 2 & 0 \end{bmatrix}$

8. $A = \begin{bmatrix} 8 & -6 & 3 & 12 \\ 10 & -9 & 4 & -8 \\ 6 & 3 & -8 & 4 \\ 1 & 7 & -6 & 3 \end{bmatrix}$

9. If $A = \begin{bmatrix} 1 & 2 \\ 3 & 4 \end{bmatrix}$, find (a) $a^{(t)}_{11}$ (b) $a^{(t)}_{21}$ (c) $a^{(t)}_{12}$.

10. If $A = \begin{bmatrix} 1 & 2 & x \\ 5 & y & -3 \\ z & 3 & -2 \end{bmatrix}$, find (a) $a^{(t)}_{33}$ (b) $a^{(t)}_{31}$ (c) $a^{(t)}_{23}$.

11. Find $(A^t)^t$ for $A = \begin{bmatrix} 1 & 2 \\ 3 & 4 \end{bmatrix}$.

12. Find $(A^t)^t$ for $A = \begin{bmatrix} 3 & 1 \\ -2 & -4 \end{bmatrix}$.

In exercises 13 through 20, add or subtract as indicated, if possible, or say why the operation is not possible.

13. $\begin{bmatrix} 1 & 1 \\ 2 & 3 \end{bmatrix} + \begin{bmatrix} -1 & -4 \\ -6 & -8 \end{bmatrix}$

14. $\begin{bmatrix} 2 & 2 & 1 \\ 3 & -5 & 1 \end{bmatrix} + \begin{bmatrix} -5 & 3 & -6 \\ -1 & 1 & -2 \end{bmatrix}$

15. $[1 \quad 2 \quad 4] + \begin{bmatrix} 3 \\ 2 \\ 1 \end{bmatrix}$

16. $\begin{bmatrix} 1 & 2 \\ 4 & 6 \end{bmatrix} - \begin{bmatrix} 1 & 1 & -2 \\ 2 & 2 & 3 \end{bmatrix}$

17. $\begin{bmatrix} -1 & -4 \\ -6 & -8 \\ 1 & 4 \end{bmatrix} - \begin{bmatrix} 1 & 1 \\ 2 & 3 \\ -1 & 2 \end{bmatrix}$

18. $\begin{bmatrix} -2 & 0 & 1 \\ 3 & -4 & 0 \\ 1 & 7 & -2 \end{bmatrix} + \begin{bmatrix} -3 & 0 & -5 \\ -3 & 4 & 4 \\ 10 & 6 & -4 \end{bmatrix}$

19. $\begin{bmatrix} 1 & 5 & -2 & 1 \\ 2 & 3 & -7 & 2 \\ 8 & -4 & 6 & -1 \\ 2 & -3 & 7 & 4 \end{bmatrix} + \begin{bmatrix} 6 & 0 & -2 & 3 \\ 4 & -1 & 2 & 6 \\ 5 & 4 & -3 & 1 \\ 2 & -7 & 4 & 5 \end{bmatrix}$

20. $\begin{bmatrix} -5 & 1 & 3 & 2 \\ 6 & 0 & 10 & 14 \\ 8 & -21 & 3 & 4 \\ 7 & 16 & -3 & 27 \end{bmatrix} - \begin{bmatrix} 6 & 5 & -2 & 1 \\ 4 & -2 & 8 & 3 \\ -3 & 2 & 6 & -2 \\ -1 & 5 & 9 & 8 \end{bmatrix}$

In exercises 21 through 26, find the scalar product.

21. $5[-2]$

22. $5\begin{bmatrix} 2 & 6 \\ -3 & 2 \end{bmatrix}$

23. $-3\begin{bmatrix} 2 & 6 \\ -3 & 2 \end{bmatrix}$

24. $0\begin{bmatrix} 1 & -3 & 5 \\ -2 & 1 & 3 \\ 5 & 2 & 4 \end{bmatrix}$

25. $-2\begin{bmatrix} 1 & -3 & 5 & 2 \\ 2 & 1 & -1 & 0 \\ 6 & 0 & -4 & 8 \end{bmatrix}$

26. $-5\begin{bmatrix} 1 & 5 & 2 \\ -1 & 3 & 1 \end{bmatrix}$

Let $A = \begin{bmatrix} 1 & 0 & 2 \\ -1 & 3 & -4 \end{bmatrix}$ and $B = \begin{bmatrix} 2 & -1 & 0 \\ -4 & 3 & 5 \end{bmatrix}$.

27. Find $A + B$.

28. Find $A + 3B$.

In exercises 29 and 30, simplify the given expression.

29. $3A - B + 2A + 6B$

30. $C - 2A + 2C - 3A + C$

In exercises 31 through 34, solve the matrix equations for the unknowns.

31. $2\begin{bmatrix} 2 & -1 \\ -y & 2z \end{bmatrix} + 3\begin{bmatrix} x & 2 \\ y & -z \end{bmatrix} = \begin{bmatrix} 1 & 4 \\ 7 & -4 \end{bmatrix}$

32. $3\begin{bmatrix} x & y \\ -5 & 3 \end{bmatrix} + 5\begin{bmatrix} -2x & y+1 \\ 2z & -1 \end{bmatrix} = \begin{bmatrix} 14 & 21 \\ -15 & 4 \end{bmatrix}$

33. $5\begin{bmatrix} 4 & -1 \\ 3y & z+1 \end{bmatrix} - \begin{bmatrix} 2 & 2x \\ y & -3z \end{bmatrix} = \begin{bmatrix} 18 & 3x \\ 28 & -16 \end{bmatrix}$

34. $3\begin{bmatrix} 2x & 3 \\ y & z+3 \end{bmatrix} - \begin{bmatrix} 3x & 4 \\ -2y & 2z \end{bmatrix} = \begin{bmatrix} 6 & 5 \\ 10 & -4 \end{bmatrix}$

In exercises 35 and 36, factor out as much as possible from the matrices, leaving integer entries.

35. $\begin{bmatrix} 6 & -2 \\ 0 & 12 \end{bmatrix}$

36. $\begin{bmatrix} 14\pi & 16\pi \\ -8\pi & 6\pi \end{bmatrix}$

COMPLEX NUMBERS

In exercises 37 through 40, carry out the indicated calculations.

37. $(1-5i)\begin{bmatrix} 1+i & 2i \\ 3-6i & -2-3i \end{bmatrix} + \begin{bmatrix} -3+4i & 7-4i \\ 1-3i & 6 \end{bmatrix}$

38. $\begin{bmatrix} 2-3i & 7-6i \\ 8-i & 1+3i \end{bmatrix} - \begin{bmatrix} 4-i & 4 \\ 2 & 3+2i \end{bmatrix}$

39. $(2+3i)\begin{bmatrix} 2-3i & 7-6i \\ 8+i & 1+3i \end{bmatrix}$

40. $\begin{bmatrix} 2-i \\ 3+2i \end{bmatrix} - (4+i)\begin{bmatrix} 1+i \\ 3-2i \end{bmatrix}$

41. If a matrix A has complex entries, we define $A^* = [\bar{a}_{ij}]^t$, where $\bar{a}_{ij}$ is the conjugate of a_{ij}. Find A^* for

$$A = \begin{bmatrix} 1-i & 2-3i \\ 2+i & 3+2i \end{bmatrix}$$

42. Prove
(a) $(A^*)^* = A$
(b) $(cA)^* = \bar{c}A^*$
(c) $(A+B)^* = A^* + B^*$

THEORETICAL EXERCISES

43. Prove Theorem 2.2, parts (ii), (iii), (iv), and (v).

44. Prove Theorem 2.3, parts (ii), (iii), and (iv).

45. Prove that if A is an $m \times n$ matrix, r is a scalar, and $rA = \mathbf{0}$, then $r = 0$ or $A = \mathbf{0}$.

46. When is $A + A^t$ defined? *Hint*: Look at some examples.

47. A matrix $A = [a_{ij}]$ is *symmetric* if $A = A^t$. Show that, if A is symmetric, then
(a) A is square
(b) $a_{ij} = a_{ji}$ for all i and j
(c) $\mathbf{a}_i^{(t)} = \mathbf{a}^i$
(d) $(\mathbf{a}^j)^t = \mathbf{a}_j$
Hint: First write out some symmetric matrices to become familiar with them.

COMPUTER EXERCISES

48. Use a two-digit computer, and then a four-digit computer, to compute

$$\begin{bmatrix} \frac{1}{3} & \frac{1}{6} \\ \frac{1}{2} & 2 \end{bmatrix} + \begin{bmatrix} \frac{1}{8} & \frac{1}{3} \\ \frac{1}{4} & 3 \end{bmatrix}$$

Compare your calculations with the exact result,

$$\begin{bmatrix} \frac{11}{24} & \frac{1}{2} \\ \frac{3}{4} & 5 \end{bmatrix} \approx \begin{bmatrix} 0.458333 & 0.5 \\ 0.75 & 5 \end{bmatrix}$$

rounded to two digits and four digits, respectively.

49. Show, by counting, that there are fewer multiplications in computing $(rs)A$ than in computing $r(sA)$.

2.2 MATRIX MULTIPLICATION

In this section we develop the technically intricate operation of multiplication of two matrices. For this definition, we will proceed as did Arthur Cayley in his 1850s papers in which the arithmetic of matrices was introduced.* By using this approach we can show that multiplication of matrices arises in a somewhat natural manner.

Consider the pair of equations

$$\begin{aligned} b_{11}u_1 + b_{12}u_2 &= x_1 \\ b_{21}u_1 + b_{22}u_2 &= x_2 \end{aligned}$$

* In several research papers published in the late 1850s, Arthur Cayley developed the algebra of matrices. He defined the sum, scalar product, and multiplication as is done in this text. Other ideas appearing in these papers included the transposed matrix, powers of matrices, and the inverse of a matrix, all of which appear in this chapter. See References.

where $\begin{bmatrix} b_{11} & b_{12} \\ b_{21} & b_{22} \end{bmatrix}$ is a matrix of constants and x_1, x_2 and u_1, u_2 are variables. Consider also a second pair of equations which express the variables y_1 and y_2 in terms of variables x_1 and x_2:

$$\begin{aligned} a_{11}x_1 + a_{12}x_2 &= y_1 \\ a_{21}x_1 + a_{22}x_2 &= y_2 \end{aligned}$$

where $\begin{bmatrix} a_{11} & a_{12} \\ a_{21} & a_{22} \end{bmatrix}$ is a matrix of constants.

Suppose we want to express the variables y_1 and y_2 in terms of the variables u_1 and u_2.

EXAMPLE 7

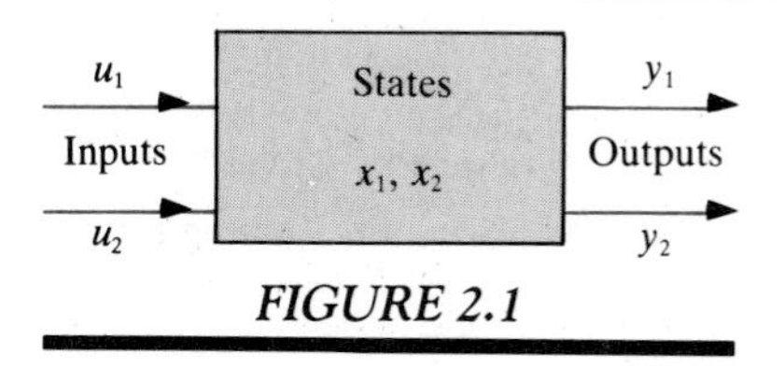

FIGURE 2.1

The equations of a control problem, under certain conditions, can be written as above, where u_1 and u_2 represent the inputs, x_1 and x_2 the states, and y_1 and y_2 the outputs (see exercise 48). Diagrammatically this situation can be viewed as shown in Figure 2.1. Here the problem is to express the outputs directly in terms of the inputs. □

Substituting the expressions for x_1 and x_2 given in the first pair of equations into the second pair of equations yields, after rearrangement,

$$\begin{aligned} (a_{11}b_{11} + a_{12}b_{21})u_1 + (a_{11}b_{12} + a_{12}b_{22})u_2 &= y_1 \\ (a_{21}b_{11} + a_{22}b_{21})u_1 + (a_{21}b_{12} + a_{22}b_{22})u_2 &= y_2 \end{aligned}$$

It is easy to see that the matrix of constants

$$\begin{bmatrix} a_{11}b_{11} + a_{12}b_{21} & a_{11}b_{12} + a_{12}b_{22} \\ a_{21}b_{11} + a_{22}b_{21} & a_{21}b_{12} + a_{22}b_{22} \end{bmatrix}$$

can be computed directly from the matrices

$$\begin{bmatrix} a_{11} & a_{12} \\ a_{21} & a_{22} \end{bmatrix} \quad \text{and} \quad \begin{bmatrix} b_{11} & b_{12} \\ b_{21} & b_{22} \end{bmatrix}$$

The rule by which this computation is performed is called *multiplication.* That is,*

$$\begin{bmatrix} a_{11} & a_{12} \\ a_{21} & a_{22} \end{bmatrix} \cdot \begin{bmatrix} b_{11} & b_{12} \\ b_{21} & b_{22} \end{bmatrix} = \begin{bmatrix} a_{11}b_{11} + a_{12}b_{21} & a_{11}b_{12} + a_{12}b_{22} \\ a_{21}b_{11} + a_{22}b_{21} & a_{21}b_{12} + a_{22}b_{22} \end{bmatrix}$$

Simply put, this rule states that the i,j-entry of the product is found by multiplying the corresponding entries of the i-row of $A = [a_{ij}]$ and the j-column of $B = [b_{ij}]$ and summing these products.

* Henceforth, we will omit the dot, which indicates multiplication, between the matrices.

$$(i)\ \begin{bmatrix} a_{11} & a_{12} \\ a_{21} & a_{22} \end{bmatrix} \begin{bmatrix} b_{11} & b_{12} \\ b_{21} & b_{22} \end{bmatrix} = \begin{bmatrix} a_{11}b_{11} + a_{12}b_{21} & a_{11}b_{12} + a_{12}b_{22} \\ a_{21}b_{11} + a_{22}b_{21} & a_{21}b_{12} + a_{22}b_{22} \end{bmatrix}$$

$$(ii)\ \begin{bmatrix} a_{11} & a_{12} \\ a_{21} & a_{22} \end{bmatrix} \begin{bmatrix} b_{11} & b_{12} \\ b_{21} & b_{22} \end{bmatrix} = \begin{bmatrix} a_{11}b_{11} + a_{12}b_{21} & a_{11}b_{12} + a_{12}b_{22} \\ a_{21}b_{11} + a_{22}b_{21} & a_{21}b_{12} + a_{22}b_{22} \end{bmatrix}$$

$$(iii)\ \begin{bmatrix} a_{11} & a_{12} \\ a_{21} & a_{22} \end{bmatrix} \begin{bmatrix} b_{11} & b_{12} \\ b_{21} & b_{22} \end{bmatrix} = \begin{bmatrix} a_{11}b_{11} + a_{12}b_{21} & a_{11}b_{12} + a_{12}b_{22} \\ a_{21}b_{11} + a_{22}b_{21} & a_{21}b_{12} + a_{22}b_{22} \end{bmatrix}$$

$$(iv)\ \begin{bmatrix} a_{11} & a_{12} \\ a_{21} & a_{22} \end{bmatrix} \begin{bmatrix} b_{11} & b_{12} \\ b_{21} & b_{22} \end{bmatrix} = \begin{bmatrix} a_{11}b_{11} + a_{12}b_{21} & a_{11}b_{12} + a_{12}b_{22} \\ a_{21}b_{11} + a_{22}b_{21} & a_{21}b_{12} + a_{22}b_{22} \end{bmatrix}$$

Note that, in order for this rule to apply to arbitrary size matrices A and B, the number of columns of A need only equal the number of rows of B.

DEFINITION

If $A = [a_{ij}]$ is $m \times r$ and $B = [b_{ij}]$ is $r \times n$, the *product matrix* $AB = C = [c_{ij}]$ is the $m \times n$ matrix with

$$c_{ij} = \sum_{k=1}^{r} a_{ik}b_{kj}$$

That is, in the above format, to find the i, j-entry in AB, we multiply the entries of the i-row of A by the corresponding entries of the j-column of B and sum these products. Diagrammatically,

$$\underbrace{\begin{bmatrix} a_{11} & a_{12} & \cdots & a_{1r} \\ \vdots & \vdots & \vdots & \vdots \\ a_{i1} & a_{i2} & \cdots & a_{ir} \\ \vdots & \vdots & \vdots & \vdots \\ a_{m1} & a_{m2} & \cdots & a_{mr} \end{bmatrix}}_{m \times r} \underbrace{\begin{bmatrix} b_{11} & \cdots & b_{1j} & \cdots & b_{1n} \\ b_{21} & \cdots & b_{2j} & \cdots & b_{2n} \\ \vdots & \vdots & \vdots & \vdots & \vdots \\ b_{r1} & \cdots & b_{rj} & \cdots & b_{rn} \end{bmatrix}}_{r \times n} = \underbrace{\begin{bmatrix} c_{11} & \cdots & c_{1j} & \cdots & c_{1n} \\ \vdots & \vdots & \vdots & \vdots & \vdots \\ c_{i1} & \cdots & c_{ij} & \cdots & c_{in} \\ \vdots & \vdots & \vdots & \vdots & \vdots \\ c_{m1} & \cdots & c_{mj} & \cdots & c_{mn} \end{bmatrix}}_{m \times n}$$

We provide some examples to demonstrate this rule of multiplication.

EXAMPLE 8

(*i*) A 2×3 matrix multiplied by a 3×3 matrix yields a 2×3 matrix.

$$\begin{bmatrix} 1 & 2 & 3 \\ 4 & 5 & 6 \end{bmatrix} \begin{bmatrix} 7 & 10 & 13 \\ 8 & 11 & 14 \\ 9 & 12 & 15 \end{bmatrix}$$

$$= \begin{bmatrix} 1 \cdot 7 + 2 \cdot 8 + 3 \cdot 9 & 1 \cdot 10 + 2 \cdot 11 + 3 \cdot 12 & 1 \cdot 13 + 2 \cdot 14 + 3 \cdot 15 \\ 4 \cdot 7 + 5 \cdot 8 + 6 \cdot 9 & 4 \cdot 10 + 5 \cdot 11 + 6 \cdot 12 & 4 \cdot 13 + 5 \cdot 14 + 6 \cdot 15 \end{bmatrix}$$

$$= \begin{bmatrix} 50 & 68 & 86 \\ 122 & 167 & 212 \end{bmatrix}$$

(*ii*) A 3×1 matrix multiplied by a 1×2 matrix yields a 3×2 matrix.

$$\begin{bmatrix} 1 \\ 2 \\ 3 \end{bmatrix} [4 \quad 5] = \begin{bmatrix} 1 \cdot 4 & 1 \cdot 5 \\ 2 \cdot 4 & 2 \cdot 5 \\ 3 \cdot 4 & 3 \cdot 5 \end{bmatrix} = \begin{bmatrix} 4 & 5 \\ 8 & 10 \\ 12 & 15 \end{bmatrix}$$

(*iii*) A 1×3 matrix multiplied by a 3×1 matrix yields a 1×1 matrix.

$$[1 \quad 2 \quad 3]\begin{bmatrix} 4 \\ 5 \\ 6 \end{bmatrix} = [1 \cdot 4 + 2 \cdot 5 + 3 \cdot 6] = [4 + 10 + 18] = [32]$$

(*iv*) A 2×2 matrix multiplied by a 2×1 matrix yields a 2×1 matrix.

$$\begin{bmatrix} 1 & 2 \\ 3 & 4 \end{bmatrix}\begin{bmatrix} 5 \\ 6 \end{bmatrix} = \begin{bmatrix} 1 \cdot 5 + 2 \cdot 6 \\ 3 \cdot 5 + 4 \cdot 6 \end{bmatrix} = \begin{bmatrix} 17 \\ 39 \end{bmatrix}$$

(*v*) The multiplication

$$\begin{bmatrix} 1 & 2 & 3 \\ 4 & 5 & 6 \end{bmatrix}\begin{bmatrix} 7 & 8 \\ 9 & 10 \end{bmatrix}$$

is not defined, since we cannot multiply a 2×3 matrix by a 2×2 matrix. □

The definition of multiplication of matrices shows how to compute the i,j-entry in the product matrix. On occasion, we need to calculate the rows and columns of the product matrix. These can be computed as follows.

LEMMA 2.4

Let

$$A = \begin{bmatrix} \mathbf{a}_1 \\ \vdots \\ \mathbf{a}_m \end{bmatrix}$$

be an $m \times r$ matrix and

$$B = [\mathbf{b}^1 \cdots \mathbf{b}^n]$$

be an $r \times n$ matrix. Then

$$AB = \begin{bmatrix} \mathbf{a}_1 B \\ \vdots \\ \mathbf{a}_m B \end{bmatrix} \quad \text{and} \quad AB = [A\mathbf{b}^1 \cdots A\mathbf{b}^n]$$

PROOF

The proof is left to you in exercise 38. ■

EXAMPLE 9

(*i*) By the lemma,

$$\begin{bmatrix} 1 & 2 \\ 3 & 4 \end{bmatrix}\begin{bmatrix} 5 & 6 & 7 \\ 8 & 9 & 10 \end{bmatrix} = \begin{bmatrix} [1 \quad 2]\begin{bmatrix} 5 & 6 & 7 \\ 8 & 9 & 10 \end{bmatrix} \\ [3 \quad 4]\begin{bmatrix} 5 & 6 & 7 \\ 8 & 9 & 10 \end{bmatrix} \end{bmatrix} = \begin{bmatrix} 21 & 24 & 27 \\ 47 & 54 & 61 \end{bmatrix}$$

(*ii*) Again by the lemma,

$$\begin{bmatrix}1&2\\3&4\end{bmatrix}\begin{bmatrix}5&6&7\\8&9&10\end{bmatrix}=\left[\begin{bmatrix}1&2\\3&4\end{bmatrix}\begin{bmatrix}5\\8\end{bmatrix}\ \begin{bmatrix}1&2\\3&4\end{bmatrix}\begin{bmatrix}6\\9\end{bmatrix}\ \begin{bmatrix}1&2\\3&4\end{bmatrix}\begin{bmatrix}7\\10\end{bmatrix}\right]$$
$$=\begin{bmatrix}21&24&27\\47&54&61\end{bmatrix}$$

(*iii*) If A is a 2×2 matrix, then

$$A\begin{bmatrix}\lambda_1&0\\0&\lambda_2\end{bmatrix}=\left[A\begin{bmatrix}\lambda_1\\0\end{bmatrix}\ A\begin{bmatrix}0\\\lambda_2\end{bmatrix}\right]=[\lambda_1\mathbf{a}^1\quad\lambda_2\mathbf{a}^2]$$

(Such calculations will occur in Chapter 6.) □

Although the rule for computing products of matrices is somewhat intricate, matrix multiplication shares some (but not all) of the arithmetical properties of multiplication of numbers. For example, we can show that the associative and distributive properties hold, while the commutative property does not. Further, we can show that the square matrix I_n is the multiplicative identity for multiplication.

THEOREM 2.5

Assuming that the sizes of the matrices A, B, and C are such that the operations are defined, then

(*i*) $(AB)C = A(BC)$ (associative law)
(*ii*) $A(B + C) = AB + AC$ (left distributive law)
(*iii*) $(B + C)A = BA + CA$ (right distributive law)
(*iv*) $I_mA = A$, $AI_n = A$ (multiplicative identity)*
(*v*) $(AB)^t = B^tA^t$

PROOF

To prove these properties, it is first advisable to gain some experience by proving them in the case where all matrices are 2×2 (exercise 39). We will provide the proofs for part (i) and the first equation of part (iv); the other equations can be proved similarly (exercise 40).

For part (i), let $A = [a_{ij}]$ be $m \times r$, $B = [b_{ij}]$ be $r \times s$, and $C = [c_{ij}]$ be $s \times n$. Then, expressing the various matrices in this product entrywise, we have

$$(AB)C=\left[\sum_{k=1}^{r}a_{ik}b_{kj}\right]C=\left[\sum_{l=1}^{s}\left(\sum_{k=1}^{r}a_{ik}b_{kl}\right)c_{lj}\right]=\left[\sum_{l=1}^{s}\sum_{k=1}^{r}a_{ik}b_{kl}c_{lj}\right]$$

Now, by exercise 46,

$$\left[\sum_{l=1}^{s}\sum_{k=1}^{r}a_{ik}b_{kl}c_{lj}\right]=\left[\sum_{k=1}^{r}\sum_{l=1}^{s}a_{ik}b_{kl}c_{lj}\right]=\left[\sum_{k=1}^{r}a_{ik}\left(\sum_{l=1}^{s}b_{kl}c_{lj}\right)\right]$$
$$=A\left[\sum_{l=1}^{s}b_{il}c_{lj}\right]=A(BC)$$

* The multiplicative identity for numbers is 1. From (iv), I_n plays the role of 1 in matrix arithmetic.

For the first equation of part (iv), computing columns yields

$$I_m A = [I_m\mathbf{a}^1 \quad I_m\mathbf{a}^2 \quad \cdots \quad I_m\mathbf{a}^n] = [\mathbf{a}^1 \quad \mathbf{a}^2 \quad \cdots \quad \mathbf{a}^n] = A \qquad \blacksquare$$

The following consequence of this theorem shows how to handle scalar matrix products in which the scalar is between two matrices.

COROLLARY 2.5A

Let r be a scalar and A and B be matrices such that their product AB is defined. Then

$$A(rB) = r(AB) = (rA)B$$

PROOF

The proof is left to you in exercise 41. ■

An example showing how scalars can occur between matrices follows.

EXAMPLE 10

Let A, B, and C be $n \times n$ matrices. By the corollary, we have

$$\begin{aligned} A(2B - 3C) + AB - 4AC &= A(2B) - A(3C) + AB - 4AC \\ &= 2AB - 3AC + AB - 4AC \\ &= 3AB - 7AC \end{aligned}$$

□

Computationally there are three important differences between the product for matrices and the product for numbers.

1. The commutative law does not hold for matrix multiplication.

Even if AB and BA are both defined, the commutative law will never hold if the matrices are not square. If A is $p \times q$ and B is $r \times s$, then for AB to be defined we must have $q = r$ and for BA to be defined we must have $p = s$. Under those conditions, AB must be $p \times p$ and BA must be $q \times q$. If $AB = BA$, then $p = q = r = s$ and A and B must be square and of the same size.

The following example shows that even square matrices do not necessarily commute under multiplication.

EXAMPLE 11

(*i*) If $A = \begin{bmatrix} 1 & 2 \\ 1 & 2 \end{bmatrix}$ and $B = \begin{bmatrix} 3 & 4 \\ 3 & 4 \end{bmatrix}$, then

$$AB = \begin{bmatrix} 9 & 12 \\ 9 & 12 \end{bmatrix} \quad \text{and} \quad BA = \begin{bmatrix} 7 & 14 \\ 7 & 14 \end{bmatrix}$$

so $AB \neq BA$.

(*ii*) However, if $A = \begin{bmatrix} 1 & 0 \\ 0 & 2 \end{bmatrix}$ and $B = \begin{bmatrix} 3 & 0 \\ 0 & 4 \end{bmatrix}$, then

$$AB = \begin{bmatrix} 3 & 0 \\ 0 & 8 \end{bmatrix} \quad \text{and} \quad BA = \begin{bmatrix} 3 & 0 \\ 0 & 8 \end{bmatrix}$$

so $AB = BA$. □

The second difference concerns matrices whose product is $\mathbf{0}$.

2. The $\mathbf{0}$-product property does not hold for matrix multiplication.

That is, when $AB = \mathbf{0}$, it may be that neither A nor B is $\mathbf{0}$.

EXAMPLE 12

Let $A = \begin{bmatrix} 1 & 1 \\ 1 & 1 \end{bmatrix}$ and $B = \begin{bmatrix} 1 & 1 \\ -1 & -1 \end{bmatrix}$. Then $AB = \begin{bmatrix} 0 & 0 \\ 0 & 0 \end{bmatrix}$, yet neither A nor B is $\mathbf{0}$. (See exercises 44 and 45.) □

The third difference concerns the cancellation law.

3. The cancellation law does not hold for matrix multiplication.

That is, if $AB = AC$ with $A \neq \mathbf{0}$, we cannot, in general, cancel A and conclude that $B = C$.

EXAMPLE 13

Let $A = \begin{bmatrix} 1 & 1 \\ 1 & 1 \end{bmatrix}$, $B = \begin{bmatrix} 1 & 2 \\ 3 & 4 \end{bmatrix}$, and $C = \begin{bmatrix} 3 & 4 \\ 1 & 2 \end{bmatrix}$. Then

$$AB = \begin{bmatrix} 4 & 6 \\ 4 & 6 \end{bmatrix} = AC$$

yet $B \neq C$. □

These differences must be kept in mind when one is calculating with matrices.

A convenient device for reducing the length of some products is to use integer exponents. To define exponents for matrices, let A be a square matrix. Define $A^0 = I_n$, and, for any positive integer k, define $A^k = AA \cdots A$, with k factors of A.

EXAMPLE 14

Some examples of powers of matrices follow.

(*i*) $\begin{bmatrix} 1 & 2 \\ 2 & 1 \end{bmatrix}^2 = \begin{bmatrix} 1 & 2 \\ 2 & 1 \end{bmatrix}\begin{bmatrix} 1 & 2 \\ 2 & 1 \end{bmatrix} = \begin{bmatrix} 5 & 4 \\ 4 & 5 \end{bmatrix}$

(*ii*) $\begin{bmatrix} d_1 & 0 & 0 \\ 0 & d_2 & 0 \\ 0 & 0 & d_3 \end{bmatrix}^3 = \begin{bmatrix} d_1^3 & 0 & 0 \\ 0 & d_2^3 & 0 \\ 0 & 0 & d_3^3 \end{bmatrix}$

(*iii*) $A \cdot A \cdot A \cdot A \cdot A \cdot A \cdot A = A^7$ □

The following arithmetical properties of exponents hold.

THEOREM 2.6

Let A and B be square matrices of the same size. Then for any nonnegative integers r and s,

(*i*) $A^r A^s = A^{r+s}$

(*ii*) $(A^r)^s = A^{rs}$

(*iii*) $(AB)^s = A^s B^s$ if $AB = BA$

PROOF

The proofs are the same as those for real numbers. ∎

EXAMPLE 15

If A and B are square matrices of the same size, then

$$AAB + 3AB + 2A(AB - B) = A^2B + 3AB + 2A^2B - 2AB = 3A^2B + AB$$

Also, $$(A + B)(A - B) = A^2 + BA - AB - B^2$$

which reduces to $(A + B)(A - B) = A^2 - B^2$ if A and B commute. □

An example of how products arise in practice follows. Others appear in exercises 29, 30, 48, 49, and 50.

EXAMPLE 16

OPTIONAL—EXCHANGE SYSTEMS

In Example 36 of Chapter 1 we defined exchange systems and showed how to find the transition matrix for such a system. For the system diagrammed in Figure 2.2, the transition matrix is

$$\begin{array}{cc} & \text{To} \\ & \begin{array}{cc} S_1 & S_2 \end{array} \\ \text{From} \begin{array}{c} S_1 \\ S_2 \end{array} & \begin{bmatrix} s_{11} & s_{12} \\ s_{21} & s_{22} \end{bmatrix} = S \end{array}$$

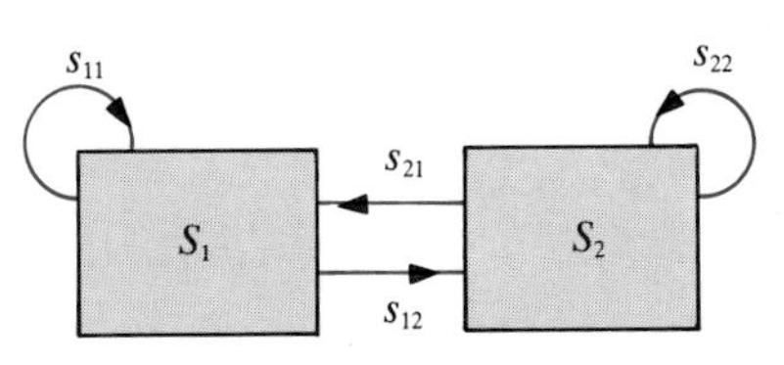

FIGURE 2.2

Let the initial number of objects in S_1 be x_1 and that in S_2 be x_2. The vector

$$\mathbf{x} = [x_1 \quad x_2]$$

is called the *initial distribution vector*.

To calculate the number of objects in each set at the next evaluation time $t = 1$, we have, for S_1,

(the number of objects in S_1) $\cdot$ (the portion that remains in S_1)
+ (the number of objects in S_2) $\cdot$ (the portion that moves to S_1)
$= x_1 s_{11} + x_2 s_{21}$

Similarly, the number of objects in S_2 at the next evaluation time $t = 1$ is $x_1 s_{12} + x_2 s_{22}$. These calculations can also be made by finding the first and second entries of $\mathbf{x}S$.

We can now find the number of objects in each set at time $t = 2$ by multiplying $\mathbf{x}S$ by S [that is, by evaluating $(\mathbf{x}S)S = \mathbf{x}S^2$] and the number at $t = n$ by computing $\mathbf{x}S^n$. (Some specific examples appear in exercise 50.) □

In the next section we will finish introducing the arithmetic of matrices by defining multiplicative inverses.

EXERCISES FOR SECTION 2.2

COMPUTATIONAL EXERCISES

In exercises 1 through 14, multiply as indicated or say why multiplication is impossible.

1. $\begin{bmatrix} 2 & -1 & 4 \end{bmatrix}\begin{bmatrix} 1 \\ 2 \\ 0 \end{bmatrix}$

2. $\begin{bmatrix} 1 & 3 & -2 \end{bmatrix}\begin{bmatrix} 0 \\ -1 \\ 2 \end{bmatrix}$

3. $\begin{bmatrix} 0 \\ -1 \\ 2 \end{bmatrix}\begin{bmatrix} 1 & 3 & -2 \end{bmatrix}$

4. $\begin{bmatrix} 1 \\ 2 \\ 0 \end{bmatrix}\begin{bmatrix} 2 & -1 & 4 \end{bmatrix}$

5. $\begin{bmatrix} -1 & 2 \\ 3 & -4 \end{bmatrix}\begin{bmatrix} 1 & -1 & 2 \\ 3 & 7 & -2 \end{bmatrix}$

6. $\begin{bmatrix} 2 & -3 \\ -4 & 5 \end{bmatrix}\begin{bmatrix} 2 & -4 & -6 \\ 3 & 1 & 0 \end{bmatrix}$

7. $\begin{bmatrix} 2 & -4 & -6 \\ 3 & 1 & 0 \end{bmatrix}\begin{bmatrix} 2 & -1 & 3 \\ -2 & 3 & -4 \\ 8 & -3 & 5 \end{bmatrix}$

8. $\begin{bmatrix} 1 & -1 \\ 2 & 1 \end{bmatrix}\begin{bmatrix} 1 & 6 \\ 5 & -3 \\ -2 & 0 \end{bmatrix}$

9. $\begin{bmatrix} 3 & 1 \\ -1 & 0 \\ 2 & 1 \end{bmatrix}\begin{bmatrix} 2 & -1 \\ 1 & -3 \\ -2 & 4 \end{bmatrix}$

10. $\begin{bmatrix} 1 & -1 & 2 \\ 3 & 7 & -2 \end{bmatrix}\begin{bmatrix} 4 & 0 & -7 \\ 6 & -1 & 0 \\ 0 & 2 & -3 \end{bmatrix}$

11. $\begin{bmatrix} -2 & -3 & 5 \\ 2 & 1 & -6 \\ 3 & -4 & 5 \end{bmatrix}\begin{bmatrix} 4 & 0 & 3 \\ -1 & 2 & 1 \\ 3 & 2 & 0 \end{bmatrix}$

12. $\begin{bmatrix} 0 & -2 & 1 \\ 2 & -3 & 4 \\ -1 & 3 & 2 \end{bmatrix}\begin{bmatrix} 4 & -5 & 3 \\ 6 & 2 & -4 \\ -3 & 2 & 1 \end{bmatrix}$

13. $\begin{bmatrix} 1 & 0 & -2 & 1 \\ 0 & 1 & 0 & 6 \\ 3 & 2 & -1 & 0 \\ 4 & 0 & 1 & 5 \end{bmatrix}\begin{bmatrix} 1 & 3 & -1 & 2 \\ 0 & -1 & 2 & 1 \\ 1 & 3 & 0 & -2 \\ 4 & 0 & -1 & 0 \end{bmatrix}$

14. $\begin{bmatrix} 2 & 1 & -1 & 1 \\ 3 & -1 & 2 & 2 \\ 2 & 0 & 3 & 4 \end{bmatrix}\begin{bmatrix} 1 & 0 & 2 & 1 & 4 \\ 2 & 1 & 7 & -2 & -1 \\ -1 & 6 & 0 & 3 & 6 \\ 1 & -3 & -1 & 0 & 3 \end{bmatrix}$

In exercises 15 through 18, use Lemma 2.4 with the given matrices to compute the indicated rows and columns of AB.

15. $A = \begin{bmatrix} 1 & 1 \\ -1 & 1 \end{bmatrix}$, $B = \begin{bmatrix} -1 & 2 & 3 \\ 1 & -2 & -3 \end{bmatrix}$; rows 1 and 2, columns 1 and 3

16. $A = \begin{bmatrix} 1 & 0 & -1 \\ 2 & -3 & 0 \\ 0 & 1 & -4 \end{bmatrix}$, $B = \begin{bmatrix} 2 & 4 \\ -1 & 1 \\ 3 & -2 \end{bmatrix}$; rows 1 and 3, columns 1 and 2

17. $A = \begin{bmatrix} 2 & 1 \\ 1 & 2 \\ 1 & -1 \end{bmatrix}$, $B = \begin{bmatrix} 3 & 1 \\ -2 & 4 \end{bmatrix}$; rows 1 and 2, columns 1 and 2

18. $A = [\mathbf{a}^1 \quad \mathbf{a}^2 \quad \mathbf{a}^3]$, $B = \begin{bmatrix} \lambda_1 & 0 & 0 \\ 0 & \lambda_2 & 0 \\ 0 & 0 & \lambda_3 \end{bmatrix}$; columns 1, 2, and 3

19. Let $A = \begin{bmatrix} 1 & 0 & 2 \\ -1 & 3 & -4 \end{bmatrix}$ and $B = \begin{bmatrix} 2 & -1 & 0 \\ -4 & 3 & 5 \end{bmatrix}$. Find

(*a*) $B^t(3A + B)$ (*b*) $A^t(A - 3B)$

20. Let $A = \begin{bmatrix} 1 & 2 \\ 3 & -1 \end{bmatrix}$ and $B = \begin{bmatrix} 4 & -1 \\ 2 & -3 \end{bmatrix}$. Find

(*a*) $AB - 2A$ (*b*) $A(B - 2A)$

In exercises 21 through 24, simplify the given expression.

21. $2AB - A(3B) + (4A)B$

22. $A - 2BC + 3A + B(4C)$

23. $BC - B(3A - 4C) + 2B(5C - 3A)$

24. $(A + B)C - 2A(3B + 4C) - A(-2B + C)$

In exercises 25 through 28, given $n \times n$ matrices A, B, and C, expand the given expressions.

25. $(A + B)^2$

26. $(A + B + C)^2$

27. $(A + B)^3$ if A and B do not commute

28. $(A + B)^3$ if A and B do commute

In exercises 29 and 30, the variables y_i are given in terms of the variables x_i and the variables z_i in terms of the variables y_i. Find the variables z_i in terms of the variables x_i by (a) substituting into the equations and (b) computing coefficients directly from the matrices using matrix products. Compare your results.

29. $\begin{aligned} 3x_1 + x_2 &= y_1 \\ x_1 - 2x_2 &= y_2 \end{aligned}$, $\begin{aligned} -2y_1 + 3y_2 &= z_1 \\ y_1 - y_2 &= z_2 \end{aligned}$

30. $\begin{aligned} x_1 - 2x_2 &= y_1 \\ 2x_1 + x_2 &= y_2 \end{aligned}$, $\begin{aligned} y_1 + 2y_2 &= z_1 \\ -3y_1 + 5y_2 &= z_2 \end{aligned}$

In exercises 31 through 34, calculate the given power of the matrix.

31. A^3 if $A = \begin{bmatrix} 1 & 1 \\ -1 & 1 \end{bmatrix}$

32. A^3 if $A = \begin{bmatrix} 2 & -1 \\ 3 & 1 \end{bmatrix}$

33. A^2 if $A = \begin{bmatrix} 1 & 0 & 1 \\ 0 & 1 & 0 \\ 1 & 1 & 0 \end{bmatrix}$

34. A^2 if $A = \begin{bmatrix} 2 & 0 & 1 \\ -1 & 1 & 0 \\ 3 & -2 & 4 \end{bmatrix}$

COMPLEX NUMBERS

In exercises 35 and 36, carry out the indicated calculations.

35. $\begin{bmatrix} 1-i & i \\ 2+i & 3-2i \end{bmatrix} \begin{bmatrix} 2i & 3+2i \\ 4+2i & 5-3i \end{bmatrix}$

36. $\begin{bmatrix} 2-i & 1+i \\ 3+2i & 3-2i \end{bmatrix} \begin{bmatrix} 5+i & 3i \\ -2i & 6 \end{bmatrix}$

37. Referring to exercise 41 of Section 2.1 for the definition of A^*, prove $(AB)^* = B^*A^*$.

THEORETICAL EXERCISES

38. Prove Lemma 2.4.

39. Prove Theorem 2.5 for the 2×2 case.

40. Prove Theorem 2.5, parts (ii), (iii), and (v) and the second equation in part (iv).

41. Prove Corollary 2.5A. *Hint*: Show that $(rI_m)A = A(rI_n) = rA$ and use Theorem 2.5.

42. Show that $(ABC)^t = C^tB^tA^t$. What is the rule for computing $(A_1A_2 \cdots A_k)^t$ in terms of $A_1^t, \ldots, A_k^t$?

43. Is there a 2×2 matrix X such that

$$X^2 = \begin{bmatrix} 0 & 1 \\ 0 & 0 \end{bmatrix}$$

Hint: Write out the entries in X^2 and equate them to the corresponding entries in $\begin{bmatrix} 0 & 1 \\ 0 & 0 \end{bmatrix}$. Then solve. (Some matrix equations have no solutions.)

44. If X is a 2×2 matrix and $X^2 = \mathbf{0}$, show that X need not be $\mathbf{0}$.

45. To find all 2×2 matrices X such that $X^2 = I_2$, we can write $X^2 - I_2 = \mathbf{0}$ and factor this equation into $(X - I_2) \cdot (X + I_2) = \mathbf{0}$, thus obtaining $X = I_2$ or $X = -I_2$. Let $A = \begin{bmatrix} 0 & 1 \\ 1 & 0 \end{bmatrix}$. Note that A also satisfies the equation $X^2 = I_2$. Why did we not find all solutions to the equation by the above procedure?

46. Prove:

(a) If a_{11}, a_{12}, a_{21}, and a_{22} are numbers, then

$$\sum_{i=1}^{2} \sum_{j=1}^{2} a_{ij} = \sum_{j=1}^{2} \sum_{i=1}^{2} a_{ij}$$

(b) If a_{ij} with $i \in \{1, 2, \ldots, m\}$ and $j \in \{1, 2, \ldots, n\}$ are numbers, then

$$\sum_{i=1}^{m} \sum_{j=1}^{n} a_{ij} = \sum_{j=1}^{n} \sum_{i=1}^{m} a_{ij}$$

Note that, if $A = [a_{ij}]$, the expressions give the sum of the entries in A. In the sum on the left-hand side, the row entries are summed first, whereas in the sum on the right-hand side, the column entries are summed first.

47. **(a)** Let A be an $m \times n$ matrix. Perform any Gaussian operation on the rows of A to obtain B. Perform the same Gaussian operation on I_m to obtain a matrix E. This matrix E is called an *elementary matrix*. Show that $EA = B$ for each of the three types of Gaussian operations. Thus, Gaussian operations on A can be recorded by premultiplying A by corresponding matrices. For example, if Gaussian operations are applied to a matrix A to obtain an echelon form E, then we can write $E_r \cdots E_1 A = E$, where each E_i is the elementary matrix representing the corresponding operation. (This representation is useful in various numerical studies on solving systems of linear algebraic equations.)

(b) Find an echelon form E for each of the given matrices A and determine the corresponding elementary matrices $E_1, \ldots, E_r$.

$$A = \begin{bmatrix} 1 & 0 & 1 \\ 1 & 1 & 0 \\ 1 & 1 & 1 \end{bmatrix}$$

and

$$A = \begin{bmatrix} -1 & 2 & 0 & 1 \\ 2 & 0 & 2 & -3 \\ 3 & 2 & 4 & -2 \end{bmatrix}$$

Check your work by showing that $E_r \cdots E_1 A = E$.

APPLICATIONS EXERCISES

48. A production process takes two raw materials, wood and paint, and produces two products, tables and chairs. The process has two stages, or states, construction and painting. Let

u_1 (input) = number of units of material 1 put into the process

u_2 (input) = number of units of material 2 put into the process

x_1 (state) = number of employees at stage 1

x_2 (state) = number of employees at stage 2

y_1 (output) = number of units of product 1 produced

y_2 (output) = number of units of product 2 produced

Suppose

$$\begin{aligned} x_1 &= 2u_1 + u_2 \\ x_2 &= 4u_1 + 3u_2 \end{aligned} \quad \text{and} \quad \begin{aligned} y_1 &= 5x_1 + 6x_2 \\ y_2 &= 7x_1 + 8x_2 \end{aligned}$$

As in Example 7, determine the outputs y_1 and y_2 directly from the inputs u_1 and u_2 by multiplying matrices.

49. Consider the directed graph shown in Figure 2.3. Let $A = [a_{ij}]$ be the 4×4 matrix such that $a_{ij} = 1$ if there is an arc (arrow) from vertex i to vertex j, and $a_{ij} = 0$ otherwise. Calculate $A^2 = [a_{ij}^{(2)}]$. Show that if $a_{ij}^{(2)} > 0$, then there is a directed walk of length 2 from i to j. (By a *directed walk* of length k from vertex i to vertex j we mean a set of arcs going from i to i_1, i_1 to $i_2, \ldots, i_{k-1}$ to $i_k = j$.) Calculate $A^3 = [a_{ij}^{(3)}]$. Show that if $a_{ij}^{(3)} > 0$, then there is a directed walk of length 3 from i to j.

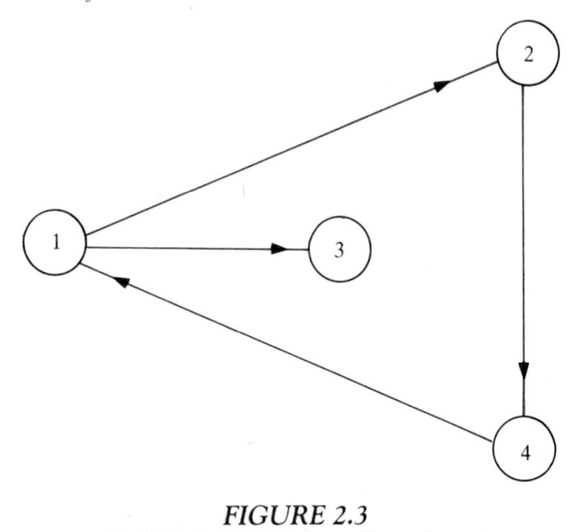

FIGURE 2.3

50. For each exchange system below, write the transition matrix and initial distribution (time $t = 0$). Then compute the distribution when $t = 2$.

(a) Let C_1 and C_2 be neighboring countries, with 2 million people in C_1 and 5 million people in C_2. Suppose $\frac{1}{8}$ of the population of C_1 migrates to C_2 and $\frac{1}{10}$ of the population of C_2 migrates to C_1 each year.

(b) Let C be the set of coffee drinkers and T the set of tea drinkers in the United States. Suppose C contains 20 million people and T contains 15 million people. Suppose $\frac{1}{4}$ of the coffee drinkers switch to tea and $\frac{1}{5}$ of the tea drinkers switch to coffee each decade.

(c) A set of rats were sequentially run through a maze in which a choice had to be made to run right or run left. Suppose on the initial trial 600 rats ran right and 400 ran left. When the rats were run again, $\frac{1}{4}$ of the rats that had run left in the previous trial ran right in the new trial, and $\frac{1}{5}$ of the rats that had run right in the previous trial ran left in the new trial.

COMPUTER EXERCISES

51. Let A, B, and C be 2×2 matrices. Note that $AB + AC = A(B + C)$. Which computation, $AB + AC$ or $A(B + C)$, requires fewer multiplications?

52. Let A be an $m \times r$ matrix and B be an $r \times n$ matrix. Show that the number of multiplications used to compute AB is mrn. Thus, if $m = r = n$, the number is n^3.

53. Use a two-digit computer, and then a four-digit computer, to compute

$$\begin{bmatrix} \frac{1}{3} & \frac{1}{6} \\ \frac{1}{2} & 2 \end{bmatrix} \begin{bmatrix} \frac{1}{8} & \frac{1}{3} \\ \frac{1}{4} & 3 \end{bmatrix}$$

Compare your calculations with the exact result,

$$\begin{bmatrix} \frac{1}{12} & \frac{11}{18} \\ \frac{9}{16} & \frac{37}{6} \end{bmatrix} \approx \begin{bmatrix} 0.083333 & 0.611111 \\ 0.5625 & 6.166667 \end{bmatrix}$$

by rounding to two digits and four digits, respectively.

2.3 INVERSES

In this section we complete the introduction of the arithmetic of matrices by defining multiplicative inverses.

By *multiplicative inverses* we mean a pair of matrices A and B such that*

$$AB = BA = I_n$$

* This mimics the definition of inverses of numbers.

In this case, we call B an *inverse* of A, and consequently A is an inverse of B. Matrices that have inverses are called *invertible* or nonsingular. Matrices that are not invertible are called *noninvertible* or singular.* In order for A and B to be multiplicative inverses, it is clear that A and B must be square and of the same size. Even so, some square matrices have inverses and some do not.

EXAMPLE 17

Since $I_n I_n = I_n I_n = I_n$, it follows that I_n is invertible and an inverse of I_n is I_n. □

EXAMPLE 18

Let $A = \begin{bmatrix} 1 & 1 \\ 1 & 1 \end{bmatrix}$. Since

$$AB = \begin{bmatrix} b_{11} + b_{21} & b_{12} + b_{22} \\ b_{11} + b_{21} & b_{12} + b_{22} \end{bmatrix} \neq I_2$$

for any choices of b_{11}, b_{12}, b_{21}, and b_{22}, it follows that A has no inverse. □

In order to decide if a square matrix A has an inverse, we can

(i) solve the defining equation $AB = I_n$ for B and then

(ii) check to see if B satisfies the defining equation $BA = I_n$.

To solve $AB = I_n$, we equate the corresponding columns of both sides to get the system of linear algebraic equations

$$A\mathbf{b}^1 = \mathbf{e}^1, A\mathbf{b}^2 = \mathbf{e}^2, \ldots, A\mathbf{b}^n = \mathbf{e}^n$$

We can solve this system by applying the Gauss-Jordan algorithm to the augmented matrices

$$[A \quad \mathbf{e}^1], [A \quad \mathbf{e}^2], \ldots, [A \quad \mathbf{e}^n]$$

EXAMPLE 19

Let $A = \begin{bmatrix} 1 & 1 \\ 1 & -1 \end{bmatrix}$. To see if A has an inverse, we first solve the defining equation $AB = I_2$, where $B = \begin{bmatrix} b_{11} & b_{12} \\ b_{21} & b_{22} \end{bmatrix}$. Finding the columns of the left side and equating them to the columns of the right side yields

$$\begin{bmatrix} 1 & 1 \\ 1 & -1 \end{bmatrix}\begin{bmatrix} b_{11} \\ b_{21} \end{bmatrix} = \begin{bmatrix} 1 \\ 0 \end{bmatrix} \quad \text{and} \quad \begin{bmatrix} 1 & 1 \\ 1 & -1 \end{bmatrix}\begin{bmatrix} b_{12} \\ b_{22} \end{bmatrix} = \begin{bmatrix} 0 \\ 1 \end{bmatrix}$$

Putting these systems in augmented matrix form and applying Gaussian

* The words *invertible* and *noninvertible* will be used in this text.

operations gives

$$\begin{bmatrix} 1 & 1 & 1 \\ 1 & -1 & 0 \end{bmatrix} \xrightarrow{R_2 - R_1 \to R_2} \begin{bmatrix} 1 & 1 & 1 \\ 0 & -2 & -1 \end{bmatrix} \xrightarrow{-\frac{1}{2}R_2 \to R_2} \begin{bmatrix} 1 & 1 & 1 \\ 0 & 1 & \frac{1}{2} \end{bmatrix} \xrightarrow{R_1 - R_2 \to R_1} \begin{bmatrix} 1 & 0 & \frac{1}{2} \\ 0 & 1 & \frac{1}{2} \end{bmatrix}$$

$$\begin{bmatrix} 1 & 1 & 0 \\ 1 & -1 & 1 \end{bmatrix} \xrightarrow{R_2 - R_1 \to R_2} \begin{bmatrix} 1 & 1 & 0 \\ 0 & -2 & 1 \end{bmatrix} \xrightarrow{-\frac{1}{2}R_2 \to R_2} \begin{bmatrix} 1 & 1 & 0 \\ 0 & 1 & -\frac{1}{2} \end{bmatrix} \xrightarrow{R_1 - R_2 \to R_1} \begin{bmatrix} 1 & 0 & \frac{1}{2} \\ 0 & 1 & -\frac{1}{2} \end{bmatrix}$$

Thus

$$\begin{bmatrix} b_{11} \\ b_{21} \end{bmatrix} = \begin{bmatrix} \frac{1}{2} \\ \frac{1}{2} \end{bmatrix} \quad \text{and} \quad \begin{bmatrix} b_{12} \\ b_{22} \end{bmatrix} = \begin{bmatrix} \frac{1}{2} \\ -\frac{1}{2} \end{bmatrix}$$

so $B = \begin{bmatrix} \frac{1}{2} & \frac{1}{2} \\ \frac{1}{2} & -\frac{1}{2} \end{bmatrix}$.

Now we need to check the second defining equation, $BA = I_2$. By direct calculation,

$$BA = \begin{bmatrix} \frac{1}{2} & \frac{1}{2} \\ \frac{1}{2} & -\frac{1}{2} \end{bmatrix} \begin{bmatrix} 1 & 1 \\ 1 & -1 \end{bmatrix} = \begin{bmatrix} 1 & 0 \\ 0 & 1 \end{bmatrix}$$

so $B = \begin{bmatrix} \frac{1}{2} & \frac{1}{2} \\ \frac{1}{2} & -\frac{1}{2} \end{bmatrix}$ is an inverse for $A = \begin{bmatrix} 1 & 1 \\ 1 & -1 \end{bmatrix}$. □

As can be observed in the example above, when the Gauss-Jordan algorithm is applied to the augmented matrices, the sequence of Gaussian operations used on the first system also yields the echelon form for the second augmented matrix, and in fact for all of the augmented matrices. This is due to the fact that the operations chosen are completely determined by the entries of A. Thus, the operations are repeated on each augmented matrix. This duplication of calculations can be eliminated by eliminating A from all but the first augmented matrix and considering

$$[A \quad \mathbf{e}^1], [\bullet \quad \mathbf{e}^2], \ldots, [\bullet \quad \mathbf{e}^n]$$

For efficiency, this can be contracted to

$$[A \quad \mathbf{e}^1 \quad \mathbf{e}^2 \quad \cdots \quad \mathbf{e}^n] = [A \quad I_n]$$

By applying the operations to $[A \quad I_n]$, we obtain the results for all augmented matrices simultaneously.

EXAMPLE 20

We will recompute the solution B to $AB = I_2$ for $A = \begin{bmatrix} 1 & 1 \\ 1 & -1 \end{bmatrix}$ so that the calculations of this example can be compared to the calculations of Example 19. This comparison will demonstrate the contraction described above. Here, we apply Gaussian operations directly to

$$[A \quad I_2] = \begin{bmatrix} 1 & 1 & 1 & 0 \\ 1 & -1 & 0 & 1 \end{bmatrix}$$

$$\xrightarrow{R_2 - R_1 \to R_2} \begin{bmatrix} 1 & 1 & 1 & 0 \\ 0 & -2 & -1 & 1 \end{bmatrix}$$

$$\xrightarrow{-\frac{1}{2}R_2 \to R_2} \begin{bmatrix} 1 & 1 & 1 & 0 \\ 0 & 1 & \frac{1}{2} & -\frac{1}{2} \end{bmatrix}$$

$$\xrightarrow{R_1 - R_2 \to R_1} \begin{bmatrix} 1 & 0 & \frac{1}{2} & \frac{1}{2} \\ 0 & 1 & \frac{1}{2} & -\frac{1}{2} \end{bmatrix}$$

Thus, $B = \begin{bmatrix} \frac{1}{2} & \frac{1}{2} \\ \frac{1}{2} & -\frac{1}{2} \end{bmatrix}$. □

We will now show that, to find an inverse for a square matrix A, we need only solve the single equation $AB = I_n$ for B. Thus, the defining equation $BA = I_n$ need not enter into the calculations.

THEOREM 2.7

Let A be an $n \times n$ matrix and E be an echelon form for A obtained by the Gaussian elimination algorithm.

(*i*) The matrix A has an inverse if and only if E has no row of zeros. (A check for inverses.)

(*ii*) If A has an inverse, it can be obtained by solving $AB = I_n$ for B. (Thus we do not need to check if $BA = I_n$.)

(*iii*) There is at most one inverse for A.

PROOF

This proof is given in general. Working through it using, say, $A = \begin{bmatrix} 1 & 2 \\ 1 & 1 \end{bmatrix}$ will help increase your understanding of the principles involved. For part (i), we apply the Gaussian elimination algorithm to $[A \quad I_n]$ to get $[E \quad C]$. Suppose E has a row of zeros. As the Gaussian elimination algorithm proceeds, $c\mathbf{R}_j \to \mathbf{R}_j$ and $\mathbf{R}_i + c\mathbf{R}_j \to \mathbf{R}_i$ are not applied unless row j contains a bound variable. Thus no row of zeros of E is used in this way. Given that E has a row of zeros, its last row must be a row of zeros and C must contain $\mathbf{e}^n$ as some column. Since in this case the system $E\mathbf{x} = \mathbf{e}^n$ has no solution, it follows that $AB = I_n$ has no solution and A has no inverse.

Conversely, suppose E has no row of zeros. Then, applying the Gauss-Jordan algorithm to $[A \quad I_n]$, we can get $[I_n \quad B]$, and $AB = I_n$. Now we need to show that $BA = I_n$. To do this, we solve $BX = I_n$ for X and show $X = A$. In order to do this, we define reverse Gaussian operations as shown in Table 2.1. If we obtained $[I_n \quad B]$ from $[A \quad I_n]$ by a sequence of operations, then we can obtain $[A \quad I_n]$ from $[I_n \quad B]$ by applying the corresponding reverse operations in reverse order. Thus we can use these operations to obtain $[I_n \quad A]$ from $[B \quad I_n]$, and so the solution to $BX = I_n$ is $X = A$. Hence part (i) follows.

TABLE 2.1

OPERATION	REVERSE OPERATION
$R_i \leftrightarrow R_j$	$R_i \leftrightarrow R_j$
$cR_i \to R_i$	$\frac{1}{c}R_i \to R_i$
$R_i + cR_j \to R_i$	$R_i - cR_j \to R_i$

Since B was obtained above by solving $AB = I_n$ and was then shown to be an inverse of A, part (ii) follows. For part (iii), suppose B and C are inverses of A. Then $AB = BA = I_n$ and $AC = CA = I_n$. Hence

$$B = BI_n = B(AC) = (BA)C = I_nC = C$$

Thus B and C are the same matrix. ■

From part (iii) of Theorem 2.7, we see that if A has an inverse, it is unique, and so we can denote it by A^{-1}. Also, this theorem gives us an algorithm for determining whether a matrix has an inverse and for finding it when it exists.

ALGORITHM TO COMPUTE THE INVERSE OF AN $n \times n$ MATRIX A

1. Form the matrix $[A \quad I_n]$.
2. Apply the Gauss-Jordan algorithm to $[A \quad I_n]$ to obtain $[E \quad B]$. If E has a row of zeros, then A has no inverse.
3. Otherwise, $E = I_n$ and $B = A^{-1}$.

EXAMPLE 21

Let $A = \begin{bmatrix} 1 & 2 \\ 2 & 4 \end{bmatrix}$. If possible, find the inverse of A.

Applying the above algorithm yields

$$[A \quad I_2] = \begin{bmatrix} 1 & 2 & 1 & 0 \\ 2 & 4 & 0 & 1 \end{bmatrix} \xrightarrow{R_2 - 2R_1 \to R_2} \begin{bmatrix} 1 & 2 & 1 & 0 \\ 0 & 0 & -2 & 1 \end{bmatrix}$$

Since E, $\begin{bmatrix} 1 & 2 \\ 0 & 0 \end{bmatrix}$, has a row of zeros, A has no inverse. □

EXAMPLE 22

Let

$$A = \begin{bmatrix} 1 & 0 & 1 \\ -1 & 2 & 1 \\ 2 & -1 & 0 \end{bmatrix}$$

If possible, find the inverse of A.

Here

$$[A \quad I_3] = \begin{bmatrix} 1 & 0 & 1 & 1 & 0 & 0 \\ -1 & 2 & 1 & 0 & 1 & 0 \\ 2 & -1 & 0 & 0 & 0 & 1 \end{bmatrix}$$

$$\xrightarrow{\substack{R_2 + R_1 \to R_2 \\ R_3 - 2R_1 \to R_3}} \begin{bmatrix} 1 & 0 & 1 & 1 & 0 & 0 \\ 0 & 2 & 2 & 1 & 1 & 0 \\ 0 & -1 & -2 & -2 & 0 & 1 \end{bmatrix}$$

$$\xrightarrow{\frac{1}{2}R_2 \to R_2} \begin{bmatrix} 1 & 0 & 1 & 1 & 0 & 0 \\ 0 & 1 & 1 & \frac{1}{2} & \frac{1}{2} & 0 \\ 0 & -1 & -2 & -2 & 0 & 1 \end{bmatrix}$$

$$\xrightarrow{R_3 + R_2 \to R_3} \begin{bmatrix} 1 & 0 & 1 & 1 & 0 & 0 \\ 0 & 1 & 1 & \frac{1}{2} & \frac{1}{2} & 0 \\ 0 & 0 & -1 & -\frac{3}{2} & \frac{1}{2} & 1 \end{bmatrix}$$

$$\xrightarrow{\substack{R_1 + R_3 \to R_1 \\ R_2 + R_3 \to R_2}} \begin{bmatrix} 1 & 0 & 0 & -\frac{1}{2} & \frac{1}{2} & 1 \\ 0 & 1 & 0 & -1 & 1 & 1 \\ 0 & 0 & -1 & -\frac{3}{2} & \frac{1}{2} & 1 \end{bmatrix}$$

$$\xrightarrow{-R_3 \to R_3} \begin{bmatrix} 1 & 0 & 0 & -\frac{1}{2} & \frac{1}{2} & 1 \\ 0 & 1 & 0 & -1 & 1 & 1 \\ 0 & 0 & 1 & \frac{3}{2} & -\frac{1}{2} & -1 \end{bmatrix}$$

Thus, $A^{-1} = \begin{bmatrix} -\frac{1}{2} & \frac{1}{2} & 1 \\ -1 & 1 & 1 \\ \frac{3}{2} & -\frac{1}{2} & -1 \end{bmatrix}$. □

Using part (iii) of Theorem 2.7, we can also show that inverses satisfy most of the standard arithmetical properties of inverses for numbers.

THEOREM 2.8

Let A and B be invertible matrices of the same size, and let r be a scalar with $r \neq 0$. Then

(*i*) $(A^{-1})^{-1} = A$

(*ii*) $(AB)^{-1} = B^{-1}A^{-1}$

(*iii*) $(rA)^{-1} = r^{-1}A^{-1}$

(*iv*) $(A^{-1})^t = (A^t)^{-1}$

Further, letting $A^{-k} = (A^{-1})^k$ for any positive integer k, we have for any integers s and t

(*v*) $A^s A^t = A^{s+t}$

(*vi*) $(A^s)^t = A^{st}$

PROOF

We will show the proof of part (i); parts (ii) through (vi) are left as exercise 42.

One method for showing that two matrices are equal is to show that the corresponding i,j-entries of the two matrices are equal. However, it would be difficult to find an expression for the i,j-entry of $(A^{-1})^{-1}$. Instead, we use Theorem 2.7, part (iii), the key result for developing properties of the inverse of a matrix. Let $C = A^{-1}$. Then $CA = AC = I_n$. By part (iii) of Theorem 2.7, this means that the inverse of C is A; that is, $A = C^{-1} = (A^{-1})^{-1}$. ■

Having established a multiplicative inverse, it is natural for us to discuss division of matrices. Recall that, for numbers a and b with $b \neq 0$, we can write $a/b = ab^{-1} = b^{-1}a$. Since not all matrices have inverses and matrix multiplication does not necessarily commute, we cannot, however, write $A/B = AB^{-1} = B^{-1}A$. Thus this definition cannot be extended to matrices.

One of the most important uses of matrix arithmetic is in writing systems of various equations as matrix equations so that they can be manipulated to a solution. We will now show how to write a system of linear algebraic equations as a matrix equation and how such a matrix equation can be manipulated. How other kinds of systems can be written as matrix equations and manipulated is shown in exercises 48 through 52. More will be shown in Chapters 6 and 7.

Consider the system of linear algebraic equations

$$\begin{array}{ccccccccc} a_{11}x_1 & + & a_{12}x_2 & + & \cdots & + & a_{1n}x_n & = & b_1 \\ a_{21}x_1 & + & a_{22}x_2 & + & \cdots & + & a_{2n}x_n & = & b_2 \\ \vdots & & \vdots & & \vdots & & \vdots & & \vdots \\ a_{m1}x_1 & + & a_{m2}x_2 & + & \cdots & + & a_{mn}x_n & = & b_m \end{array}$$

where $A = [a_{ij}]$, the *coefficient matrix*, is a matrix of constants, $\mathbf{b} = [b_i]$ is a vector of constants, and $\mathbf{x} = [x_i]$ is a vector of unknowns. Using the arithmetic of matrices, this system of linear algebraic equations can be written as the matrix equation

$$A\mathbf{x} = \mathbf{b}$$

Some general examples follow.

EXAMPLE 23

(*i*) We can write

$$\begin{aligned} a_{11}x_1 + a_{12}x_2 &= b_1 \\ a_{21}x_1 + a_{22}x_2 &= b_2 \end{aligned}$$

as

$$\begin{bmatrix} a_{11}x_1 + a_{12}x_2 \\ a_{21}x_1 + a_{22}x_2 \end{bmatrix} = \begin{bmatrix} b_1 \\ b_2 \end{bmatrix}$$

and thus as

$$\begin{bmatrix} a_{11} & a_{12} \\ a_{21} & a_{22} \end{bmatrix} \begin{bmatrix} x_1 \\ x_2 \end{bmatrix} = \begin{bmatrix} b_1 \\ b_2 \end{bmatrix}$$

(*ii*) For

$$a_{11}x_1 + a_{12}x_2 + a_{13}x_3 = b_1$$
$$a_{21}x_1 + a_{22}x_2 + a_{23}x_3 = b_2$$

we can write

$$\begin{bmatrix} a_{11} & a_{12} & a_{13} \\ a_{21} & a_{22} & a_{23} \end{bmatrix} \begin{bmatrix} x_1 \\ x_2 \\ x_3 \end{bmatrix} = \begin{bmatrix} b_1 \\ b_2 \end{bmatrix}$$

□

Some numerical examples are given below.

EXAMPLE 24

(*i*) For $x_1 - x_2 = 6$, we can write

$$\begin{bmatrix} 1 & -1 \end{bmatrix} \begin{bmatrix} x_1 \\ x_2 \end{bmatrix} = [6]$$

(*ii*) For

$$x_1 - 2x_2 + 3x_3 = 1$$
$$4x_2 + 5x_3 = 8$$

we have

$$\begin{bmatrix} 1 & -2 & 3 \\ 0 & 4 & 5 \end{bmatrix} \begin{bmatrix} x_1 \\ x_2 \\ x_3 \end{bmatrix} = \begin{bmatrix} 1 \\ 8 \end{bmatrix}$$

(*iii*) For

$$x_1 + x_2 = 1$$
$$x_1 - x_2 = 6$$
$$2x_1 - 3x_2 = 4$$

we write

$$\begin{bmatrix} 1 & 1 \\ 1 & -1 \\ 2 & -3 \end{bmatrix} \begin{bmatrix} x_1 \\ x_2 \end{bmatrix} = \begin{bmatrix} 1 \\ 6 \\ 4 \end{bmatrix}$$

□

Being able to write a system of linear algebraic equations as a matrix equation allows us to manipulate these equations in much the same way that equations involving one unknown are manipulated. For example,

(*i*) If A is an invertible matrix and $\mathbf{b}$ is a vector such that $A\mathbf{x} = \mathbf{b}$ is defined, then multiplying through by A^{-1} yields

$$\mathbf{x} = A^{-1}\mathbf{b}$$

(*ii*) If we have $\mathbf{y} = A\mathbf{x}$ and $\mathbf{x} = B\mathbf{u}$ as in Example 7 of Section 2.2, then we can express the variable $\mathbf{y}$ in terms of the variable $\mathbf{u}$ by direct substitution of $B\mathbf{u}$ for $\mathbf{x}$

in the first equation, obtaining

$$\mathbf{y} = A(B\mathbf{u}) \qquad \text{or} \qquad \mathbf{y} = (AB)\mathbf{u}$$

In the next section we start a study which will lead to a way of determining when a matrix is invertible by computing an expression involving only the entries of the matrix.

EXERCISES FOR SECTION 2.3

COMPUTATIONAL EXERCISES

In exercises 1 and 2, show that A is the inverse of B by showing that $AB = BA = I_n$.

1. $A = \begin{bmatrix} 4 & 9 \\ 3 & 7 \end{bmatrix}, B = \begin{bmatrix} 7 & -9 \\ -3 & 4 \end{bmatrix}$

2. $A = \begin{bmatrix} 2 & 3 & 5 \\ -1 & -1 & -2 \\ -1 & 1 & 1 \end{bmatrix}, B = \begin{bmatrix} 1 & 2 & -1 \\ 3 & 7 & -1 \\ -2 & -5 & 1 \end{bmatrix}$

In exercises 3 through 12, for the given matrix A, use the algorithm of this section to find A^{-1}, if possible. Using the defining equations, check your answers in 3, 4, 8, and 9.

3. $A = \begin{bmatrix} 1 & -1 \\ 2 & -1 \end{bmatrix}$

4. $A = \begin{bmatrix} 3 & 1 \\ 2 & 1 \end{bmatrix}$

5. $A = \begin{bmatrix} -2 & 4 \\ 4 & -8 \end{bmatrix}$

6. $A = \begin{bmatrix} -3 & 4 \\ 6 & -8 \end{bmatrix}$

7. $A = \begin{bmatrix} 2 & -1 & 3 \\ -5 & 1 & 2 \\ 9 & -3 & 4 \end{bmatrix}$

8. $A = \begin{bmatrix} 1 & -3 & 2 \\ -2 & 5 & -1 \\ 7 & 2 & 1 \end{bmatrix}$

9. $A = \begin{bmatrix} 1 & -2 & 5 \\ 3 & -7 & 2 \\ -1 & 1 & 2 \end{bmatrix}$

10. $A = \begin{bmatrix} 3 & -3 & -2 \\ -2 & 3 & 5 \\ 0 & 11 & 11 \end{bmatrix}$

11. $A = \begin{bmatrix} 1 & 1 & 1 & 1 \\ 0 & 1 & 1 & 0 \\ 1 & 1 & 0 & 1 \\ 0 & 0 & 0 & 1 \end{bmatrix}$

12. $A = \begin{bmatrix} 0 & 1 & 1 & 1 \\ 1 & 0 & 0 & 1 \\ 1 & 0 & 1 & 0 \\ 1 & 1 & 1 & 1 \end{bmatrix}$

In exercises 13 and 14, calculate the given power of the matrix.

13. A^{-2} if $A = \begin{bmatrix} 1 & 2 \\ 1 & 1 \end{bmatrix}$

14. A^{-3} if $A = \begin{bmatrix} 0 & 1 \\ -1 & 0 \end{bmatrix}$

In exercises 15 through 18, let A be an invertible matrix and let $\mathbf{x}$ and $\mathbf{b}$ be vectors. Solve the given equation for $\mathbf{x}$.

15. $A\mathbf{x} = 3A\mathbf{x}$

16. $A(\mathbf{x} + \mathbf{b}) = A\mathbf{x} - \mathbf{x}$

17. $\mathbf{x}A = \mathbf{b}$

18. $(\mathbf{x} + \mathbf{b})A = \mathbf{x} + \mathbf{x}A$

In exercises 19 through 22, given that A is invertible, solve the given equation for X.

19. $AX = BA$

20. $2AX - B = C$

21. $3XA + B = C$

22. $AXA = B$

In exercises 23 and 24, solve the given system by finding the inverse of the coefficient matrix. (To solve $A\mathbf{x} = \mathbf{d}$, it is easier to use the Gaussian elimination or the Gauss-Jordan algorithm than to compute A^{-1}, even when A is invertible. However, if the equation must be solved for several choices of $\mathbf{d}$ as below, then it can be advantageous to compute A^{-1} and then calculate $A^{-1}\mathbf{d}$ for the various $\mathbf{d}$'s.)

23. $\begin{bmatrix} 1 & 1 \\ 1 & 0 \end{bmatrix}\mathbf{x} = \mathbf{d}$ for $\mathbf{d} = \begin{bmatrix} 1 \\ 1 \end{bmatrix}, \begin{bmatrix} -1 \\ 1 \end{bmatrix}, \begin{bmatrix} 0 \\ 0 \end{bmatrix}$

24. $\begin{bmatrix} 1 & 2 \\ 2 & 1 \end{bmatrix}\mathbf{x} = \mathbf{d}$ for $\mathbf{d} = \begin{bmatrix} 2 \\ -2 \end{bmatrix}, \begin{bmatrix} -2 \\ 1 \end{bmatrix}, \begin{bmatrix} 0 \\ 0 \end{bmatrix}$

In exercises 25 through 28, write the matrix equation as a system of linear algebraic equations.

25. $\begin{bmatrix} 5 & 6 \\ 7 & 0 \end{bmatrix}\begin{bmatrix} x_1 \\ x_2 \end{bmatrix} = \begin{bmatrix} -1 \\ 3 \end{bmatrix}$

26. $\begin{bmatrix} -3 & 2 \\ 5 & 8 \end{bmatrix}\begin{bmatrix} x_1 \\ x_2 \end{bmatrix} = \begin{bmatrix} 4 \\ -1 \end{bmatrix}$

27. $\begin{bmatrix} 1 & 4 & 7 \\ 3 & 5 & 8 \\ -6 & -4 & -2 \end{bmatrix}\begin{bmatrix} x_1 \\ x_2 \\ x_3 \end{bmatrix} = \begin{bmatrix} -1 \\ 2 \\ 3 \end{bmatrix}$

28. $\begin{bmatrix} 0 & -1 & 3 \\ 2 & 5 & -2 \\ -6 & 3 & 1 \end{bmatrix}\begin{bmatrix} x_1 \\ x_2 \\ x_3 \end{bmatrix} = \begin{bmatrix} 4 \\ -5 \\ 2 \end{bmatrix}$

In exercises 29 through 32, write the system of linear algebraic equations from each exercise as a matrix equation.

29. Section 1.1, exercise 9

30. Section 1.1, exercise 10

31. Section 1.1, exercise 15

32. Section 1.1, exercise 14

In exercises 33 and 34, write the two systems of linear algebraic equations as matrix equations. Then, by direct substitution, find the matrix equation relating the y's and u's, and use a matrix inverse to find the u's in terms of the y's.

33. $\begin{aligned} x_1 &= u_1 + u_2 \\ x_2 &= u_1 - u_2 \end{aligned}$, $\begin{aligned} y_1 &= x_1 + 2x_2 \\ y_2 &= -x_1 + x_2 \end{aligned}$

34. $x_1 = u_1 - 2u_2$ $\quad y_1 = x_1 + x_2 + x_3$
$x_2 = 2u_1 + u_2,$ $\quad y_2 = x_1 - x_2 + 2x_3$
$x_3 = u_1 + u_2$

In exercises 35 through 38, solve the matrix equation $AX = B$, if possible. As with computing inverses, the work can be simplified by computing all of the columns of X simultaneously, using the augmented matrix $[A \quad B]$.

35. $\begin{bmatrix} 1 & 1 \\ 1 & 1 \end{bmatrix} X = \begin{bmatrix} 2 & 3 \\ 2 & 3 \end{bmatrix}$

36. $\begin{bmatrix} -1 & 1 \\ 1 & -1 \end{bmatrix} X = \begin{bmatrix} 2 & -3 \\ -2 & 3 \end{bmatrix}$

37. $\begin{bmatrix} 1 & 2 \\ 3 & 4 \end{bmatrix} X = \begin{bmatrix} 1 & 1 \\ 1 & 2 \end{bmatrix}$ $\quad$ *38.* $\begin{bmatrix} 1 & 2 \\ 2 & 4 \end{bmatrix} X = \begin{bmatrix} 3 & 1 \\ 6 & 1 \end{bmatrix}$

COMPLEX NUMBERS

In exercises 39 and 40, compute the inverse of the given matrix, if possible.

39. $\begin{bmatrix} i & 1-i \\ 2+i & -1-2i \end{bmatrix}$ $\quad$ *40.* $\begin{bmatrix} 1-i & 1+i \\ 1+i & -1+i \end{bmatrix}$

41. Referring to exercise 41 of Section 2.1 for the definition of A^*, prove that $(A^{-1})^* = (A^*)^{-1}$.

THEORETICAL EXERCISES

42. Prove Theorem 2.8, parts (ii) through (vi). Note that in part (ii), it is not feasible to compute the i,j-entries of $(AB)^{-1}$ and $B^{-1}A^{-1}$. You need to use Theorem 2.7, part (iii).

43. Prove that if A, B, and C are invertible and of the same size, then $(ABC)^{-1} = C^{-1}B^{-1}A^{-1}$. Write out the rule for a product of k invertible matrices.

44. Simplify $((((A^{-1})^t)^{-1})^t)^{-1}$.

45. Simplify $((((A^{-1})^t)^t)^{-1})^t$.

46. Let D be a diagonal matrix with no 0 entries on the main diagonal. Find D^{-1}. (Inverses of diagonal matrices are easily calculated.)

47. Let

$$T = \begin{bmatrix} t_{1,1} & t_{1,2} & t_{1,3} \\ 0 & t_{2,2} & t_{2,3} \\ 0 & 0 & t_{3,3} \end{bmatrix}$$

an upper triangular matrix with no 0 entries on the main diagonal. Find T^{-1}. *Hint:* Set $TX = I_3$. Then, by equating entries, you have $x_{31} = x_{32} = 0$ and $x_{33} = 1/t_{33}$. Now solve for the second row of X, and finally for the first row of X. (Inverses of triangular matrices are easily calculated.)

48. Let $\mathbf{x} = A\mathbf{y}$, $\mathbf{y} = B\mathbf{u}$, $\mathbf{u} = C\mathbf{z}$ for $n \times n$ matrices A, B, and C. Write the matrix equation linking $\mathbf{x}$ to $\mathbf{z}$. Assuming that A, B, and C are each invertible, write the equation that links $\mathbf{z}$ to $\mathbf{x}$.

49. Let A be an $n \times n$ matrix and $\mathbf{y} = A\mathbf{x} + \mathbf{u}$. Let P be an $n \times n$ invertible matrix. Substitute $\mathbf{y} = P\mathbf{y}'$ and $\mathbf{x} = P\mathbf{x}'$ into the equation and rewrite it as $\mathbf{y}' = B\mathbf{x}' + \mathbf{w}$. Find B and $\mathbf{w}$. (This kind of substitution occurs in various problems, some of which are shown in Chapters 6 and 7.)

50. Let A and B be $n \times n$ matrices and let P and Q be $n \times n$ invertible matrices. Substitute $X = PYQ$ into the equation $AX = XB$ and rearrange it to the form $RY = YS$. What are R and S in terms of A, B, P, and Q?

51. Let $x_1(k)$ and $x_2(k)$ be functions defined on the nonnegative integers. Then

$$x_1(k+1) = a_{11}x_1(k) + a_{12}x_2(k)$$
$$x_2(k+1) = a_{21}x_1(k) + a_{22}x_2(k)$$

where each a_{ij} is a constant, is called a *system of difference equations.* Letting $\mathbf{x}(k) = \begin{bmatrix} x_1(k) \\ x_2(k) \end{bmatrix}$ for all nonnegative integers k and letting $A = [a_{ij}]$, write this system as a matrix equation.

52. (calculus) Let $x_1(t)$ and $x_2(t)$ be functions of a variable t. Then

$$\frac{d}{dt}x_1(t) = a_{11}x_1(t) + a_{12}x_2(t)$$
$$\frac{d}{dt}x_2(t) = a_{21}x_1(t) + a_{22}x_2(t)$$

where each a_{ij} is a constant, is called a *system of differential equations.* Letting

$$\mathbf{x}(t) = \begin{bmatrix} x_1(t) \\ x_2(t) \end{bmatrix}, \qquad A = [a_{ij}], \qquad \frac{d}{dt}\begin{bmatrix} x_1(t) \\ x_2(t) \end{bmatrix} = \begin{bmatrix} \frac{d}{dt}x_1(t) \\ \frac{d}{dt}x_2(t) \end{bmatrix}$$

write this system as a matrix equation.

APPLICATIONS EXERCISES

53. Define a correspondence between the letters of the alphabet and numbers as follows:

A	B	C	D	...	X	Y	Z
1	2	3	4	...	24	25	26

Now messages can be sent by sending the corresponding numerical words. For example, for "SELL STOCK," we would send "19, 5, 12, 12, 19, 20, 15, 3, 11."

Even when the letters are numbered with a scrambled ordering a problem arises, since some letters in the English language are used more frequently than others. For example, e is used most frequently and z least frequently. Thus, by analyzing the frequency of numbers, a person may be able to decipher the message, especially if the text is long. This problem can be overcome as follows.

Let A be any invertible matrix having integer entries, say $A = \begin{bmatrix} 2 & 1 \\ 3 & 2 \end{bmatrix}$. Take the numerical message 19, 5, 12, 12, 19, 20, 15, 3, 11 and pair the letters to get

$$\begin{bmatrix} 19 \\ 5 \end{bmatrix}, \begin{bmatrix} 12 \\ 12 \end{bmatrix}, \begin{bmatrix} 19 \\ 20 \end{bmatrix}, \begin{bmatrix} 15 \\ 3 \end{bmatrix}, \begin{bmatrix} 11 \\ 0 \end{bmatrix}$$

where the 0 is used to fill in the last vector. Multiply each member of this list by A to get

$$\begin{bmatrix} 43 \\ 67 \end{bmatrix}, \begin{bmatrix} 36 \\ 60 \end{bmatrix}, \begin{bmatrix} 58 \\ 97 \end{bmatrix}, \begin{bmatrix} 33 \\ 51 \end{bmatrix}, \begin{bmatrix} 22 \\ 33 \end{bmatrix}$$

Now we can send the message 43, 67, 36, 60, 58, 97, 33, 51, 22, 33 in which the frequencies of the numbers have been changed.

The receiver of the numerical message can pair the numbers, write them in vector form, and then multiply the vectors by A^{-1} to obtain the original message.

Let $A = \begin{bmatrix} 2 & 2 \\ 1 & 2 \end{bmatrix}$.

(*a*) Code the word "MISSISSIPPI" using A.

(*b*) Decode the message 36, 33, 72, 54, 16, 11, 40, 20 using A^{-1}.

Computer Exercises

54. Let A be an invertible matrix. The system $A\mathbf{x} = \mathbf{b}$ can be solved by computing A^{-1} and forming $A^{-1}\mathbf{b}$. The Gauss-Jordan algorithm can also be used. Explain why the latter method requires less work. *Hint:* Note that the methods require forming $[A \quad I_n]$ or $[A \quad \mathbf{b}]$ and applying the Gauss-Jordan algorithm. Count the number of columns of these matrices.

55. Compute the inverse of $A = \begin{bmatrix} 1 & \frac{1}{6} \\ \frac{1}{3} & \frac{7}{8} \end{bmatrix}$ using a two-digit computer, and then a four-digit computer. Compare your solutions to the exact inverse,

$$A^{-1} = \frac{3}{59}\begin{bmatrix} 21 & -4 \\ -8 & 24 \end{bmatrix} \approx \begin{bmatrix} 1.067797 & -0.203390 \\ -0.406780 & 1.220339 \end{bmatrix}$$

by rounding to two digits and four digits, respectively.

2.4 The Determinant

In some work involving matrices, it is useful to be able to determine whether a matrix A is invertible directly from the entries of A. (See the results of Section 2.6.) In this section we show how this can be done. We introduce this work somewhat in the spirit in which it was done historically by Leibniz, Maclaurin, Cramer, and Bezout in the seventeenth and eighteenth centuries.* (Readers

* Determinants historically have been linked to systems of linear algebraic equations. Gottfried W. Leibniz, in a 1693 letter to G. F. A. l'Hôpital, wrote that

$$\begin{aligned} a_1 + b_1x + c_1y &= 0 \\ a_2 + b_2x + c_2y &= 0 \\ a_3 + b_3x + c_3y &= 0 \end{aligned}$$

where x and y are unknowns, has a nonzero solution if

$$\det \begin{bmatrix} a_1 & b_1 & c_1 \\ a_2 & b_2 & c_2 \\ a_3 & b_3 & c_3 \end{bmatrix} = 0$$

Colin Maclaurin (1729, 1748) and Cramer (1750) used the determinant to solve systems of linear algebraic equations by what is now called Cramer's Rule (to be shown later). Bezout showed that, if zero is the value of the determinant of the coefficient matrix for n linear algebraic equations with n unknowns, and if the constant terms of the system are all zero, then the system has a nonzero solution.

wishing to omit the introductory remarks and begin at the technical description of the determinant can proceed to the definition of permutation, ignoring the remarks on δ and Δ.)

We start by considering the 2×2 matrix $A = \begin{bmatrix} a_{11} & a_{12} \\ a_{21} & a_{22} \end{bmatrix}$. Suppose A is invertible. Since applying the Gauss-Jordan algorithm to A must yield I_2, it follows that one of a_{11} or a_{21} is not 0. We will assume that $a_{11} \neq 0$; the case $a_{21} \neq 0$ can be handled as described below by first applying $R_1 \leftrightarrow R_2$ to A. We apply $R_2 - (a_{21}/a_{11})\, R_1 \rightarrow R_2$ to A to get

$$\begin{bmatrix} a_{11} & a_{12} \\ 0 & a_{22} - \dfrac{a_{21}}{a_{11}} a_{12} \end{bmatrix}$$

Then we apply $a_{11} R_2 \rightarrow R_2$ to get

$$\begin{bmatrix} a_{11} & a_{12} \\ 0 & a_{11}a_{22} - a_{21}a_{12} \end{bmatrix}$$

Since this echelon form can have no zero rows, it follows that

$$\delta(A) = a_{11}a_{22} - a_{12}a_{21} \neq 0$$

On the other hand, if A is noninvertible, then either both a_{11} and a_{21} are 0, and so $\delta(A) = 0$, or the technique used above shows that $\delta(A) = 0$. Thus it follows that A is invertible if and only if the entries satisfy $\delta(A) \neq 0$.

EXAMPLE 25

Applying this result, we have that

$$A = \begin{bmatrix} 1 & 2 \\ 3 & 4 \end{bmatrix}$$

is invertible since $\delta(A) = 1 \cdot 4 - 2 \cdot 3 \neq 0$, and

$$A = \begin{bmatrix} 1 & 2 \\ 2 & 4 \end{bmatrix}$$

is noninvertible since $\delta(A) = 1 \cdot 4 - 2 \cdot 2 = 0$. Thus invertibility is determined directly from the entries of A. □

For a 3×3 matrix A, we will merely provide the result, since the calculations involved in this case are rather lengthy. Proceeding as in the 2×2 case, we can obtain the expression

$$\begin{aligned} \Delta(A) = {} & a_{11}a_{22}a_{33} + a_{12}a_{23}a_{31} + a_{13}a_{21}a_{32} \\ & - a_{13}a_{22}a_{31} - a_{12}a_{21}a_{33} - a_{11}a_{23}a_{32} \end{aligned}$$

and A is invertible if and only if the entries satisfy $\Delta(A) \neq 0$.

EXAMPLE 26

Applying the result, we see that

$$A = \begin{bmatrix} 1 & 2 & 1 \\ 2 & 1 & -1 \\ 3 & 4 & 0 \end{bmatrix}$$

is invertible because

$$\begin{aligned} \Delta(A) &= (1)(1)(0) + (2)(-1)(3) + (1)(2)(4) - (1)(1)(3) - (2)(2)(0) - (1)(-1)(4) \\ &= 3 \neq 0 \end{aligned}$$

and

$$A = \begin{bmatrix} 1 & 1 & 1 \\ 1 & 1 & 1 \\ 2 & 3 & 2 \end{bmatrix}$$

is noninvertible because

$$\Delta(A) = (1)(1)(2) + (1)(1)(2) + (1)(1)(3) - (1)(1)(2) - (1)(1)(2) - (1)(1)(3) = 0$$

Thus invertibility is determined by using the entries of A. □

The length of the calculations hinders use of this approach to find a similar expression for a general $n \times n$ matrix. However, the approach has given us expressions for $n = 2$ and 3, which can be used as patterns for extension to arbitrary n. Using these patterns, we now generalize the expressions $\delta(A)$ and $\Delta(A)$ to an expression for any $n \times n$ matrix A. Of course, we will have to show that our generalization has the desired property of detecting invertibility.

Note that both $\delta(A)$ and $\Delta(A)$ are sums of signed products, the products formed by taking exactly one entry from each row and column of A in every possible way. The choice of sign depends on the arrangement of the list of column indices for each product when the row indices are in increasing order. To indicate sign we will use the notation sign (r); for any real number r, sign (r) is $+1$ if $r \geqslant 0$ and -1 if $r < 0$.

Gathering information, we find that for $\delta(A)$ we have two terms:

PRODUCT	ARRANGEMENT OF COLUMN INDICES	SIGN
$a_{11}a_{22}$	1, 2	$1 = \text{sign}(2-1)$
$a_{12}a_{21}$	2, 1	$-1 = \text{sign}(1-2)$

In general,

$a_{1k_1}a_{2k_2}$	k_1, k_2	$\text{sign}(k_2 - k_1)$

For $\Delta(A)$, we have six terms:

PRODUCT	ARRANGEMENT OF COLUMN INDICES	SIGN
$a_{11}a_{22}a_{33}$	1, 2, 3	$1 = \text{sign}\,[(2-1)(3-1)(3-2)]$
$a_{13}a_{21}a_{32}$	3, 1, 2	$1 = \text{sign}\,[(1-3)(2-3)(2-1)]$
$a_{12}a_{23}a_{31}$	2, 3, 1	$1 = \text{sign}\,[(3-2)(1-2)(1-3)]$
$a_{11}a_{23}a_{32}$	1, 3, 2	$-1 = \text{sign}\,[(3-1)(2-1)(2-3)]$
$a_{12}a_{21}a_{33}$	2, 1, 3	$-1 = \text{sign}\,[(1-2)(3-2)(3-1)]$
$a_{13}a_{22}a_{31}$	3, 2, 1	$-1 = \text{sign}\,[(2-3)(1-3)(1-2)]$

In general, each number is subtracted from all numbers to its right:

$a_{1k_1}a_{2k_2}a_{3k_3}$	k_1, k_2, k_3	$\text{sign}\,([(k_2-k_1)(k_3-k_1)](k_3-k_2))$

These observations allow us to generalize the expressions $\delta(A)$ and $\Delta(A)$ to an arbitrary $n \times n$ matrix A. Some preliminary notations are needed for this generalization.

Let $N = \{1, 2, \ldots, n\}$ be a set of integers. A *permutation of* N is a sequence $(k_1, k_2, \ldots, k_n)$ such that each $k_i \in N$ and each element of N is some k_i. Thus a permutation of N is an ordered arrangement of the numbers in N.

Example 27

Let $N = \{1, 2\}$. The permutations of N are $(1, 2)$ and $(2, 1)$. For $N = \{1, 2, 3\}$, the permutations are $(1, 2, 3)$, $(1, 3, 2)$, $(2, 1, 3)$, $(2, 3, 1)$, $(3, 1, 2)$, and $(3, 2, 1)$. □

Let perm (N) denote the set of all permutations of N. It can be shown that perm (N) contains exactly $n!$ permutations (exercise 17). We define the *sign function* s as follows: For any sequence of distinct positive integers $(k) = (k_1, k_2, \ldots, k_n)$,

$$s(k) = \text{sign}\,([(k_2-k_1)(k_3-k_1)\cdots(k_n-k_1)][(k_3-k_2)\cdots(k_n-k_2)]\cdots[k_n-k_{n-1}])$$

Thus the function s is defined on every permutation in perm (N).

Example 28

For $N = \{1, 2\}$,

$$\text{perm}\,(N) = \{(1, 2), (2, 1)\}$$

and

$$s(1, 2) = \text{sign}\,(2-1) = 1$$
$$s(2, 1) = \text{sign}\,(1-2) = -1$$

For $N = \{1, 2, 3\}$,

$$\text{perm}(N) = \{(1,2,3), (1,3,2), (2,1,3), (2,3,1), (3,1,2), (3,2,1)\}$$

and

$$\begin{aligned} s(1,2,3) &= \text{sign}([(2-1)(3-1)](3-2)) = 1 \\ s(1,3,2) &= -1 \\ s(2,1,3) &= -1 \\ s(2,3,1) &= 1 \\ s(3,1,2) &= 1 \\ s(3,2,1) &= -1 \end{aligned}$$

□

EXAMPLE 29

We also have

$$s(2,5,3) = \text{sign}([(5-2)(3-2)](3-5)) = -1$$

and

$$s(3,1,6,5) = \text{sign}([(1-3)(6-3)(5-3)][(6-1)(5-1)](5-6)) = 1$$

□

We can now generalize $\delta(A)$ and $\Delta(A)$ as follows.

DEFINITION

Let $A = [a_{ij}]$ be an $n \times n$ matrix. If $n = 1$, then the *determinant* of A is

$$\det A = a_{11}$$

For $n > 1$, the determinant of A is defined by

$$\det A = \sum_{(k) \in \text{perm}(N)} s(k) a_{1k_1} a_{2k_2} \cdots a_{nk_n}$$

Thus the determinant of A is found by choosing one entry from each row and each column, say $a_{1k_1}, a_{2k_2}, \ldots, a_{nk_n}$, forming the product $a_{1k_1}a_{2k_2}\cdots a_{nk_n}$, multiplying this by $s(k_1, k_2, \ldots, k_n)$, and summing all such terms.

We first give some general examples.

EXAMPLE 30

(*i*) Let

$$A = \begin{bmatrix} a_{11} & a_{12} \\ a_{21} & a_{22} \end{bmatrix}$$

Then

$$\det A = s(1,2)a_{11}a_{22} + s(2,1)a_{12}a_{21} = a_{11}a_{22} - a_{12}a_{21}$$

This expression can be obtained by placing a cross in the matrix as depicted, multiplying a_{11} by a_{22} and a_{12} by a_{21} on the cross, and then subtracting.

(*ii*) Let

$$A = \begin{bmatrix} a_{11} & a_{12} & a_{13} \\ a_{21} & a_{22} & a_{23} \\ a_{31} & a_{32} & a_{33} \end{bmatrix}$$

Then

$$\begin{aligned} \det A &= s(1,2,3)a_{11}a_{22}a_{33} + s(3,1,2)a_{13}a_{21}a_{32} + s(2,3,1)a_{12}a_{23}a_{31} \\ &\quad + s(1,3,2)a_{11}a_{23}a_{32} + s(2,1,3)a_{12}a_{21}a_{33} + s(3,2,1)a_{13}a_{22}a_{31} \\ &= a_{11}a_{22}a_{33} + a_{13}a_{21}a_{32} + a_{12}a_{23}a_{31} \\ &\quad - a_{11}a_{23}a_{32} - a_{12}a_{21}a_{33} - a_{13}a_{22}a_{31} \end{aligned}$$

(See exercise 18 for a schematic for computing this determinant.)

Thus the determinant gives $\delta(A)$ when $n = 2$ and $\Delta(A)$ when $n = 3$. □

We now give some numerical examples.

EXAMPLE 31

(*i*) Let $A = [2]$. Then $\det A = 2$. If $A = [-3]$, then $\det A = -3$.

(*ii*) Let $A = \begin{bmatrix} 1 & -1 \\ 3 & \frac{1}{2} \end{bmatrix}$. Then

$$\det A = s(1,2) \cdot 1 \cdot (\tfrac{1}{2}) + s(2,1) \cdot (-1) \cdot 3 = \tfrac{1}{2} - (-1) \cdot 3 = \tfrac{1}{2} + 3 = \tfrac{7}{2}$$

(*iii*) Let $A = \begin{bmatrix} 1 & 2 \\ 3 & 4 \end{bmatrix}$. Then, by the cross method,

$$\det A = 1 \cdot 4 - 2 \cdot 3 = 4 - 6 = -2$$

(*iv*) Let $A = \begin{bmatrix} 1 & 2 & 3 \\ 4 & 5 & 6 \\ 7 & 8 & 9 \end{bmatrix}$. Then

$$\begin{aligned} \det A &= s(1,2,3) \cdot 1 \cdot 5 \cdot 9 + s(3,1,2) \cdot 3 \cdot 4 \cdot 8 + s(2,3,1) \cdot 2 \cdot 6 \cdot 7 \\ &\quad + s(1,3,2) \cdot 1 \cdot 6 \cdot 8 + s(2,1,3) \cdot 2 \cdot 4 \cdot 9 + s(3,2,1) \cdot 3 \cdot 5 \cdot 7 \\ &= 1 \cdot 5 \cdot 9 + 3 \cdot 4 \cdot 8 + 2 \cdot 6 \cdot 7 - 1 \cdot 6 \cdot 8 - 2 \cdot 4 \cdot 9 - 3 \cdot 5 \cdot 7 \\ &= 45 + 96 + 84 - 48 - 72 - 105 = 0 \end{aligned}$$

□

Occasionally (see Corollary 2.16A and Theorem 2.17 in Section 2.6, as well as Example 7 and exercise 39 of Section 7.1) the value of the determinant must be computed. Since perm (N) contains $n!$ permutations, it follows that $\det A$ contains $n!$ terms. For $n = 10$, $n! \approx 3.6 \times 10^6$; for $n = 20$, $n! \approx 2.4 \times 10^{18}$. Calculating and keeping track of so many terms makes the definition of the determinant impractical for computational purposes. Thus, our first task will be to develop a practical method for computing determinants. (Readers not interested in the proofs of determinant results can proceed to Section 2.5.)

Since the expression for det A arose by applying Gaussian operations to A, it is natural to try to develop an algorithm based on Gaussian operations for computing determinants. Since these operations also affect the sign function s, we first need to see how changes in entries of a permutation affect s. If the i- and j-entries of a permutation $(k) = (k_1, \ldots, k_n)$ are interchanged, we will denote this operation by $E_i \leftrightarrow E_j$. For numerical experience, we look at some examples of how interchanges affect the sign function.

EXAMPLE 32

Here we apply $E_i \leftrightarrow E_j$ to (k) to obtain (k') and compare $s(k)$ to $s(k')$.

$s(k)$	(k)	INTERCHANGE	(k')	$s(k')$
1	(1, 2, 3)	$E_1 \leftrightarrow E_3$	(3, 2, 1)	−1
1	(1, 2, 3)	$E_1 \leftrightarrow E_2$	(2, 1, 3)	−1
1	(1, 2, 3)	$E_2 \leftrightarrow E_3$	(1, 3, 2)	−1
−1	(2, 1, 3)	$E_2 \leftrightarrow E_3$	(2, 3, 1)	1
−1	(3, 2, 1)	$E_2 \leftrightarrow E_3$	(3, 1, 2)	1
−1	(2, 4, 1, 3)	$E_2 \leftrightarrow E_4$	(2, 3, 1, 4)	1

□

In the above example we see that an interchange between two entries in a permutation causes a sign change in s. The following theorem shows that this is always true.

THEOREM 2.9

Let $N = \{1, 2, \ldots, n\}$ and let (k) be a permutation of N. Let (k') be obtained from (k) by applying $E_i \leftrightarrow E_j$. Then $s(k') = -s(k)$.

PROOF

To provide experience, we prove this theorem for $N = \{1, 2, 3, 4\}$. The general proof of the theorem is essentially the same except for size.

For convenience in arranging factors in s, we first suppose that the interchange is done with two adjacent entries. Suppose (k') is obtained from $(k) = (k_1, k_2, k_3, k_4)$ by $E_2 \leftrightarrow E_3$. Then

$$s(k) = \text{sign}\,([(k_2 - k_1)(k_3 - k_1)(k_4 - k_1)][(k_3 - k_2)(k_4 - k_2)][(k_4 - k_3)])$$

and

$$s(k') = \text{sign}\,([(k_3 - k_1)(k_2 - k_1)(k_4 - k_1)][(k_2 - k_3)(k_4 - k_3)][(k_4 - k_2)])$$

Comparing factors, we see that the factors of $s(k)$ and the factors of $s(k')$ are the same except that $s(k)$ contains $(k_3 - k_2)$ while $s(k')$ contains $(k_2 - k_3)$. As a consequence, $s(k') = -s(k)$.

To complete the proof, we suppose the interchange is done with two arbitrary entries, say $E_i \leftrightarrow E_j$, where we can assume $j > i$. (Working through the general argument for the rest of the proof, with a particular example or two will help increase your understanding.) By a sequence of $j - i$ interchanges of adjacent entries, we put k_j in the ith position. Now k_i is in the $(i+1)$th position. By $j-i-1$ adjacent interchanges, we put k_i into the jth position. The above sequence of adjacent interchanges on (k) yields (k'). But now, by using the result on interchanging adjacent entires, we have

$$s(k') = (-1)^{(j-i)+(j-i-1)}s(k) = -s(k)$$ ■

As a consequence of this theorem, we can also link the sign function and the number of interchanges required to put a permutation into increasing order.

COROLLARY 2.9A

Let $N = \{1, 2, \ldots, n\}$. If t interchanges can return a permutation (k) of N to increasing order $(1, 2, \ldots, n)$, then

$$s(k) = (-1)^t$$

PROOF

Note that $s(k) = (-1)^t s(1, 2, \ldots, n) = (-1)^t$. ■

EXAMPLE 33

For each permutation (k) below, we compute the number of interchanges sufficient to put (k) into increasing order and thus we find $s(k)$.

(*i*) $(k) = (3, 1, 4, 2)$ $\quad E_1 \leftrightarrow E_2$
$(1, 3, 4, 2)$ $\quad E_2 \leftrightarrow E_4$
$(1, 2, 4, 3)$ $\quad E_3 \leftrightarrow E_4$
$(1, 2, 3, 4)$

Thus $s(k) = (-1)^3 = -1$.

(*ii*) $(k) = (4, 3, 1, 5, 2)$ $\quad E_1 \leftrightarrow E_3$
$(1, 3, 4, 5, 2)$ $\quad E_2 \leftrightarrow E_5$
$(1, 2, 4, 5, 3)$ $\quad E_3 \leftrightarrow E_5$
$(1, 2, 3, 5, 4)$ $\quad E_4 \leftrightarrow E_5$
$(1, 2, 3, 4, 5)$

Thus $s(k) = (-1)^4 = 1$. □

EXAMPLE 34

Let

$$A = \begin{bmatrix} 0 & 1 & 0 & 0 \\ 0 & 0 & 0 & 2 \\ 3 & 0 & 0 & 0 \\ 0 & 0 & 4 & 0 \end{bmatrix}$$

Then

$$\det A = s(2, 4, 1, 3)(1)(2)(3)(4)$$

Computing $s(2,4,1,3)$, we have

$$\begin{array}{ll} (2,4,1,3) & E_1 \leftrightarrow E_3 \\ (1,4,2,3) & E_2 \leftrightarrow E_3 \\ (1,2,4,3) & E_3 \leftrightarrow E_4 \\ (1,2,3,4) & \end{array}$$

Thus $s(2,4,1,3) = (-1)^3 = -1$, and so $\det A = (-1)(24) = -24$. □

In the next section we will continue the study of the determinant by giving a practical method, based on Gaussian operations, for calculating it.

EXERCISES FOR SECTION 2.4

COMPUTATIONAL EXERCISES

In exercises 1 and 2, for each of the permutations (k) given, find $s(k)$ by using the definition. Also perform interchanges on (k) to obtain the increasing order and compute $s(k)$ by Corollary 2.9A. Compare your results.

1. *(a)* $(1,2)$ *(b)* $(1,2,3)$ *(c)* $(1,3,2)$ *(d)* $(2,4,1,3)$ *(e)* $(4,3,5,2,1)$

2. *(a)* $(2,1)$ *(b)* $(3,1,2)$ *(c)* $(3,2,1)$ *(d)* $(4,3,2,1)$ *(e)* $(5,1,4,2,3)$

In exercises 3 through 8, evaluate the determinant as in Example 31.

3. $\det\begin{bmatrix} 3 & 1 \\ 7 & 2 \end{bmatrix}$

4. $\det\begin{bmatrix} 4 & 2 \\ 3 & 1 \end{bmatrix}$

5. $\det\begin{bmatrix} 3 & 1 & 2 \\ 4 & -1 & 0 \\ -2 & -4 & -3 \end{bmatrix}$

6. $\det\begin{bmatrix} 1 & -1 & 3 \\ 0 & 4 & -2 \\ 2 & -3 & -4 \end{bmatrix}$

7. $\det\begin{bmatrix} 1 & -1 & 3 \\ 5 & 4 & -2 \\ 2 & -3 & -4 \end{bmatrix}$

8. $\det\begin{bmatrix} 3 & 1 & 2 \\ 4 & -1 & 5 \\ -2 & -4 & -3 \end{bmatrix}$

In exercises 9 and 10, compute the determinants of the given matrices. Note that in each case there is only one nonzero term in the determinant.

9. *(a)* $\begin{bmatrix} 0 & 1 & 0 \\ 0 & 0 & 2 \\ 3 & 0 & 0 \end{bmatrix}$ *(b)* $\begin{bmatrix} 0 & 0 & 2 & 0 \\ -3 & 0 & 0 & 0 \\ 0 & 0 & 0 & -4 \\ 0 & 3 & 0 & 0 \end{bmatrix}$

10. *(a)* $\begin{bmatrix} 1 & 0 & 0 & 0 \\ 2 & 3 & 0 & 0 \\ 4 & 5 & 6 & 0 \\ 7 & 8 & 9 & 10 \end{bmatrix}$ *(b)* $\begin{bmatrix} 0 & 0 & 1 & 0 \\ 5 & 0 & 2 & -1 \\ 0 & 0 & 3 & 6 \\ 8 & 7 & 4 & -3 \end{bmatrix}$

In exercises 11 through 14, find λ such that $\det A = 0$.

11. $A = \begin{bmatrix} \lambda & 1 \\ 1 & 1 \end{bmatrix}$

12. $A = \begin{bmatrix} \lambda & 1 \\ \lambda & 1 \end{bmatrix}$

13. $A = \begin{bmatrix} \lambda - 1 & 2 \\ 2 & \lambda - 1 \end{bmatrix}$

14. $\begin{bmatrix} \lambda - 1 & 2 & 3 \\ 0 & \lambda - 4 & 5 \\ 0 & 0 & \lambda - 6 \end{bmatrix}$

COMPLEX NUMBERS

In exercises 15 and 16, calculate the determinant of the given matrix.

15. $\begin{bmatrix} 1 - i & 2 + 3i \\ -4 + 2i & 3 + i \end{bmatrix}$

16. $\begin{bmatrix} 2 & i & 1 - i \\ 2 - i & 1 - 2i & 3i \\ -2i & 4 & 2 - 3i \end{bmatrix}$

THEORETICAL EXERCISES

17. Let $N = \{1, 2, \ldots, n\}$. Show that there are $n!$ permutations in perm (N). *Hint:* Use induction. Fix the first entry and find all permutations with that first entry.

18. Note that $\det\begin{bmatrix} a_{11} & a_{12} \\ a_{21} & a_{22} \end{bmatrix}$ can be computed using the cross method. Show that the pattern illustrated in Figure 2.4 can be used similarly to obtain the determinant of a 3×3

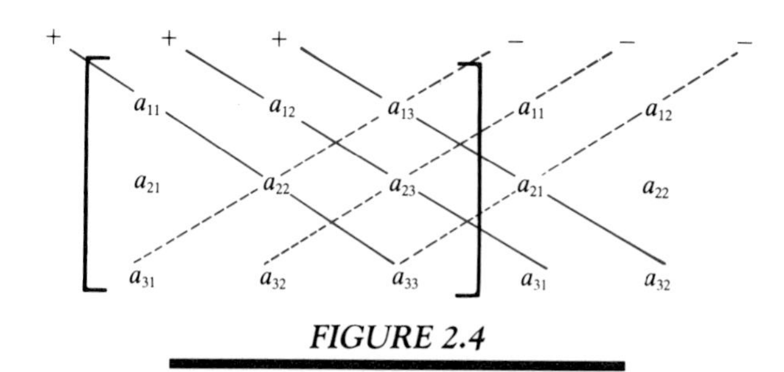

FIGURE 2.4

matrix, if multiplication is done as previously described along the lines marked, with + chosen on the solid lines and − on the dashed lines. Explain why this type of pattern fails for an $n \times n$ matrix with $n \geqslant 4$. *Hint:* Count the number of terms generated by this pattern if $n \geqslant 4$.

19. A permutation (k) is called *even* if $s(k) = 1$ and *odd* if $s(k) = -1$. Let $n \geqslant 2$. Prove that exactly half of the permutations in perm (N) are even and half are odd. *Hint:* Match the permutations by relating every even permutation with the odd permutation obtained by interchanging the first two entries.

20. Can the letters a, b, ..., y, z be rearranged to z, y, ..., b, a in exactly 37 interchanges? In 42 interchanges?

COMPUTER EXERCISES

21. Let A be an $n \times n$ matrix. Show that the number of multiplications used to compute $\det A$ is $(n-1)n!$.

22. Compute $\det\begin{bmatrix} 1 & \frac{1}{3} \\ \frac{1}{6} & -1 \end{bmatrix}$ using a two-digit computer and a four-digit computer. Compare your results with the exact value $-19/18 \approx -1.05555556$ by rounding to two digits and four digits, respectively.

23. The number $s(k_1, \ldots, k_n)$ can be calculated by definition or by Corollary 2.9A. Compare the two methods of calculation.

2.5 PROPERTIES OF THE DETERMINANT

In this section we develop some computational properties of the determinant and use them to obtain a practical way of computing it.

The determinant of a matrix A is the sum of signed products of entries of A chosen one from each row and each column. Of course, the determinant of A^t will contain these same products. In fact, the signs also agree, so we have the following theorem.

THEOREM 2.10

Let A be a square matrix. Then $\det A = \det A^t$.

PROOF

We prove the theorem for 3×3 matrices. The proof is easily generalized to $n \times n$ matrices. By definition,

$$\det A^t = \sum_{(k) \in \text{perm}(N)} s(k) a^{(t)}_{1k_1} a^{(t)}_{2k_2} a^{(t)}_{3k_3}$$

Recalling that $a^{(t)}_{ij} = a_{ji}$, we have

$$\det A^t = \sum_{(k) \in \text{perm}(N)} s(k) a_{k_1 1} a_{k_2 2} a_{k_3 3}$$

Note that the column indices are now in increasing order. Rearranging factors in the terms to get the row indices in increasing order, we have

$$a_{k_1 1} a_{k_2 2} a_{k_3 3} = a_{1 j_1} a_{2 j_2} a_{3 j_3}$$

for some permutation (j_1, j_2, j_3). Now, if it takes t interchanges to rearrange $a_{k_1 1} a_{k_2 2} a_{k_3 3}$ into $a_{1 j_1} a_{2 j_2} a_{3 j_3}$, we see by viewing indices that it takes t interchanges to put (k_1, k_2, k_3) into increasing order $(1, 2, 3)$. Thus $s(k) = (-1)^t$.

Further, since these same interchanges can be applied in reverse order to $a_{1j_1}a_{2j_2}a_{3j_3}$ to obtain $a_{k_1 1}a_{k_2 2}a_{k_3 3}$, it follows that $s(j) = (-1)^t$. Thus $s(k) = s(j)$.

Putting everything together, we have

$$\begin{aligned} \det A^t &= \sum_{(k) \in \text{perm}(N)} s(k)a_{k_1 1}a_{k_2 2}a_{k_3 3} \\ &= \sum_{(j) \in \text{perm}(N)} s(j)a_{1j_1}a_{2j_2}a_{3j_3} \\ &= \det A \end{aligned}$$

■

EXAMPLE 35

Let $A = \begin{bmatrix} 1 & 2 \\ 3 & 4 \end{bmatrix}$. Then $A^t = \begin{bmatrix} 1 & 3 \\ 2 & 4 \end{bmatrix}$. Note that $\det A = -2 = \det A^t$. □

Theorem 2.10 will let us convert determinantal results about row operations to corresponding results about column operations. The column operations are defined in Table 2.2.

TABLE 2.2

OPERATION	NOTATION
1. Interchange operation: Interchange column i and column j.	$C_i \leftrightarrow C_j$
2. Scaling operation: Multiply column i by a constant r, where $r \neq 0$.	$rC_i \rightarrow C_i$
3. Substitution operation: Add a constant r times column i to column j.	$C_j + rC_i \rightarrow C_j$

A theorem regarding the change in the determinant of a matrix due to application of a Gaussian operation to that matrix follows.

THEOREM 2.11

Let A be a square matrix. Suppose B is obtained from A by applying a Gaussian operation.

(*i*) For the interchange operation ($R_i \leftrightarrow R_j$ or $C_i \leftrightarrow C_j$ for $i \neq j$),

$$\det B = -\det A$$

Consequently, if two rows or two columns of A are identical, then

$$\det A = 0$$

(*ii*) For the scaling operation ($rR_i \rightarrow R_i$ or $rC_j \rightarrow C_j$),

$$\det B = r(\det A)$$

(*iii*) For the substitution operation ($R_i + rR_j \rightarrow R_i$ or $C_i + rC_j \rightarrow C_i$),

$$\det B = \det A$$

PROOF

To prove part (i), we will suppose that $i < j$ and that $R_i \leftrightarrow R_j$ is applied to A to obtain B. Let us take a term from $\det B$, say

$$s(k_1, \ldots, k_i, \ldots, k_j, \ldots, k_n)b_{1k_1} \cdots b_{ik_i} \cdots b_{jk_j} \cdots b_{nk_n}$$

Identifying the entries of B in terms of those of A, we have

$$s(k_1, \ldots, k_i, \ldots, k_j, \ldots, k_n) a_{1k_1} \cdots a_{jk_i} \cdots a_{ik_j} \cdots a_{nk_n}$$

By ordering the row indices of the factors as in the definition of $\det A$, we get

$$s(k_1, \ldots, k_i, \ldots, k_j, \ldots, k_n) a_{1k_1} \cdots a_{ik_j} \cdots a_{jk_i} \cdots a_{nk_n}$$

and by interchanging k_i and k_j in the permutation to match the arrangement of the column indices, we find that the preceding term equals

$$-s(k_1, \ldots, k_j, \ldots, k_i, \ldots, k_n) a_{1k_1} \cdots a_{ik_j} \cdots a_{jk_i} \cdots a_{nk_n}$$

Thus, if we pair off the terms of $\det A$ with those of $\det B$ as shown above, it follows that $\det B = -\det A$.

Suppose the operation $C_i \leftrightarrow C_j$ is applied to A to obtain B. Since B^t can be obtained from A^t by $R_i \leftrightarrow R_j$, it follows from the above result that $\det A = \det A^t = -\det B^t = -\det B$.

To prove the rest of part (i), if A has two identical rows or columns, we interchange them to obtain B. Then $\det B = -\det A$. But $B = A$, so $\det A = -\det A$ and thus $\det A = 0$.

The proofs of parts (ii) and (iii) are left as exercise 33. ■

Some examples follow.

EXAMPLE 36

(*i*) Let

$$A = \begin{bmatrix} 1 & 2 \\ 3 & 4 \end{bmatrix}$$

Applying $C_1 \leftrightarrow C_2$ yields

$$B = \begin{bmatrix} 2 & 1 \\ 4 & 3 \end{bmatrix}$$

and $\det B = 6 - 4 = -(4 - 6) = -\det A$.

(*ii*) Let

$$A = \begin{bmatrix} 1 & 0 & 2 \\ 0 & 1 & 3 \\ -1 & 2 & 0 \end{bmatrix}$$

Applying $R_1 \leftrightarrow R_3$ yields

$$B = \begin{bmatrix} -1 & 2 & 0 \\ 0 & 1 & 3 \\ 1 & 0 & 2 \end{bmatrix}$$

Now $\det B = 4 = -\det A$.

(*iii*) Let

$$A = \begin{bmatrix} 1 & 2 \\ 3 & 4 \end{bmatrix}$$

Form B by applying $5C_1 \to C_1$ to A. Then

$$B = \begin{bmatrix} 5 \cdot 1 & 2 \\ 5 \cdot 3 & 4 \end{bmatrix}$$

Now $\det B = (5 \cdot 1) \cdot 4 - 2 \cdot (5 \cdot 3) = 5 \cdot (4 - 6) = 5 \cdot \det A$

(*iv*) A constant can be factored out of a row or column in accord with part (ii) of Theorem 2.11. For example, let

$$A = \begin{bmatrix} 2 & 0 & 4 \\ 0 & -2 & 1 \\ 0 & 1 & 3 \end{bmatrix}$$

Factoring a 2 from the first row yields

$$\det A = 2 \det \begin{bmatrix} 1 & 0 & 2 \\ 0 & -2 & 1 \\ 0 & 1 & 3 \end{bmatrix} = 2(-7) = -14$$

(*v*) Let

$$A = \begin{bmatrix} 1 & 2 \\ 3 & 4 \end{bmatrix}$$

Apply $C_2 + 5C_1 \to C_2$ to form

$$B = \begin{bmatrix} 1 & 7 \\ 3 & 19 \end{bmatrix}$$

Then $\det B = 19 - 21 = -2 = \det A$.

(*vi*) Let

$$A = \begin{bmatrix} 1 & -1 & 2 \\ 1 & 2 & 4 \\ 0 & 1 & -1 \end{bmatrix}$$

Apply $R_2 - R_1 \to R_2$ to obtain

$$B = \begin{bmatrix} 1 & -1 & 2 \\ 1 - 1 & 2 + 1 & 4 - 2 \\ 0 & 1 & -1 \end{bmatrix} = \begin{bmatrix} 1 & -1 & 2 \\ 0 & 3 & 2 \\ 0 & 1 & -1 \end{bmatrix}$$

Then $\det B = -5 = \det A$. □

From the above results, we see how Gaussian operations applied to a matrix affect its determinant. Interchange operations and substitution operations may be applied to a square matrix to obtain a matrix in echelon form. In this case, the echelon form is an upper triangular matrix. The determinant of a triangular matrix is easily obtained.

THEOREM 2.12

Let $T = [t_{ij}]$ be a triangular matrix. Then

$$\det T = t_{11}t_{22}\cdots t_{nn}$$

PROOF

We first look at the 3×3 upper triangular matrix

$$T = \begin{bmatrix} t_{11} & t_{12} & t_{13} \\ 0 & t_{22} & t_{23} \\ 0 & 0 & t_{33} \end{bmatrix}$$

The terms of the determinant are formed by choosing exactly one entry from each row and column of T. To choose a term containing none of the 0's as factors, we must choose t_{33}, then t_{22} because t_{23} is no longer available, and then t_{11} because t_{12} and t_{13} are no longer available. Thus

$$\det T = s(1, 2, 3)t_{11}t_{22}t_{33} = t_{11}t_{22}t_{33}$$

For the general case, note that all terms contain a 0 factor, as above, with the possible exception of $t_{11}t_{22}\cdots t_{nn}$. Hence

$$\det T = t_{11}t_{22}\cdots t_{nn}$$ ■

EXAMPLE 37

By Theorem 2.12,

(*i*) $\det\begin{bmatrix} 1 & 2 \\ 0 & 3 \end{bmatrix} = 1 \cdot 3 = 3$

(*ii*) $\det\begin{bmatrix} 1 & 0 & 0 \\ 2 & 3 & 0 \\ 4 & 5 & 6 \end{bmatrix} = 1 \cdot 3 \cdot 6 = 18$

(*iii*) $\det\begin{bmatrix} 1 & 0 & 0 & 0 \\ 0 & 2 & 0 & 0 \\ 0 & 0 & 3 & 0 \\ 0 & 0 & 0 & 4 \end{bmatrix} = 1 \cdot 2 \cdot 3 \cdot 4 = 24$ □

Based on these results, we have the following method for computing determinants.

ALGORITHM TO COMPUTE THE DETERMINANT OF A SQUARE MATRIX A

1. Apply interchange ($R_i \leftrightarrow R_j$) and substitution ($R_i + rR_j \rightarrow R_i$) operations on A to obtain an upper triangular matrix $T = [t_{ij}]$.

2. Then

$$\det A = (-1)^t \det T = (-1)^t t_{11} t_{22} \cdots t_{nn}$$

where t is the number of row interchanges performed in obtaining T.

3. Of course, if desired, the scaling operation, as well as column operations, can be used in obtaining the triangular matrix T. In this case,

$$\det A = (-1)^t \frac{1}{r_1 r_2 \cdots r_k} \det T = (-1)^t \frac{t_{11} t_{22} \cdots t_{nn}}{r_1 r_2 \cdots r_k}$$

where $r_1, r_2, \ldots, r_k$ are the numbers used in the scaling operation and t is the number of row and column interchanges performed in obtaining T.

EXAMPLE 38

Let

$$A = \begin{bmatrix} 1 & 0 & 1 \\ 1 & -1 & 3 \\ 2 & 3 & 2 \end{bmatrix}$$

Applying the algorithm yields

$$\det \begin{bmatrix} 1 & 0 & 1 \\ 1 & -1 & 3 \\ 2 & 3 & 2 \end{bmatrix} \overset{\substack{R_2 - R_1 \to R_2 \\ R_3 - 2R_1 \to R_3}}{=} \det \begin{bmatrix} 1 & 0 & 1 \\ 0 & -1 & 2 \\ 0 & 3 & 0 \end{bmatrix}$$

$$\overset{R_3 + 3R_2 \to R_3}{=} \det \begin{bmatrix} 1 & 0 & 1 \\ 0 & -1 & 2 \\ 0 & 0 & 6 \end{bmatrix} = (1)(-1)(6) = -6$$

□

EXAMPLE 39

Let

$$A = \begin{bmatrix} 1 & 1 & 1 \\ \frac{2}{7} & \frac{2}{7} & \frac{1}{7} \\ 3 & 4 & 0 \end{bmatrix}$$

Applying Gaussian operations as described in the algorithm yields

$$\det A \overset{7R_2 \to R_2}{=} \tfrac{1}{7} \det \begin{bmatrix} 1 & 1 & 1 \\ 2 & 2 & 1 \\ 3 & 4 & 0 \end{bmatrix}$$

$$\overset{\substack{C_2 - C_1 \to C_2 \\ C_3 - C_1 \to C_3}}{=} \tfrac{1}{7} \det \begin{bmatrix} 1 & 0 & 0 \\ 2 & 0 & -1 \\ 2 & 1 & -3 \end{bmatrix}$$

$$\overset{C_2 \leftrightarrow C_3}{=} -\tfrac{1}{7} \det \begin{bmatrix} 1 & 0 & 0 \\ 2 & -1 & 0 \\ 3 & -3 & 1 \end{bmatrix} = -\tfrac{1}{7}(1)(-1)(1) = \tfrac{1}{7}$$

□

Using this algorithm to compute the determinant, we can see that the determinant, as intended, determines invertibility.

THEOREM 2.13

Let A be a square matrix. The following are equivalent.

(*i*) $\det A \neq 0$.

(*ii*) A is invertible.

(*iii*) $A\mathbf{x} = \mathbf{0}$ has only the solution $\mathbf{x} = \mathbf{0}$.

Similarly, the following are also equivalent.

(*i*) $\det A = 0$.

(*ii*) A is noninvertible.

(*iii*) $A\mathbf{x} = \mathbf{0}$ has a nonzero solution.

PROOF

Compute an echelon form E for A. If $\det A \neq 0$, then E has no row of zeros, so, by Theorem 2.7, A is invertible. Given that A is invertible, multiplying by A^{-1} shows that $A\mathbf{x} = \mathbf{0}$ has $\mathbf{x} = \mathbf{0}$ as its only solution. Finally, if $A\mathbf{x} = \mathbf{0}$ has only $\mathbf{x} = \mathbf{0}$ as a solution, then $E\mathbf{x} = \mathbf{0}$ has no free variables, and so E has no row of zeros. Thus, $\det A \neq 0$. ■

We conclude the section by showing how to calculate the determinant of matrices combined by the various arithmetic operations.

Let A and B be $n \times n$ matrices. For the sum, in general,

$$\det(A + B) \neq \det A + \det B$$

EXAMPLE 40

Note that

$$\det\left(\begin{bmatrix} 1 & 0 \\ 0 & 0 \end{bmatrix} + \begin{bmatrix} 0 & 0 \\ 0 & 1 \end{bmatrix}\right) = \det\begin{bmatrix} 1 & 0 \\ 0 & 1 \end{bmatrix} = 1$$

but

$$\det\begin{bmatrix} 1 & 0 \\ 0 & 0 \end{bmatrix} + \det\begin{bmatrix} 0 & 0 \\ 0 & 1 \end{bmatrix} = 0 + 0 = 0$$

However,

$$\det\left(\begin{bmatrix} 0 & 1 \\ 0 & 0 \end{bmatrix} + \begin{bmatrix} 1 & 0 \\ 0 & 0 \end{bmatrix}\right) = \det\begin{bmatrix} 1 & 1 \\ 0 & 0 \end{bmatrix} = 0$$

and

$$\det\begin{bmatrix} 0 & 1 \\ 0 & 0 \end{bmatrix} + \det\begin{bmatrix} 1 & 0 \\ 0 & 0 \end{bmatrix} = 0 + 0 = 0$$ □

For multiplication we have the following theorem.

THEOREM 2.14

Let r be a scalar and A and B be $n \times n$ matrices. Then

(*i*) $\det(rA) = r^n \det A$ (scalar product rule)

(*ii*) $\det(AB) = (\det A)(\det B)$ (product rule)

PROOF

See exercise 34. ■

EXAMPLE 41

Let $A = \begin{bmatrix} 1 & 2 \\ 3 & 4 \end{bmatrix}$ and $r = 5$. Then $rA = \begin{bmatrix} 5 & 10 \\ 15 & 20 \end{bmatrix}$. Since

$$\det(rA) = -50 \qquad \text{and} \qquad r^2 \det A = 25(-2) = -50$$

we see that $\det(rA) = r^2 \det A$. □

EXAMPLE 42

Let $A = \begin{bmatrix} 1 & 1 \\ -1 & 1 \end{bmatrix}$ and $B = \begin{bmatrix} 2 & 1 \\ 1 & 1 \end{bmatrix}$. Then $AB = \begin{bmatrix} 3 & 2 \\ -1 & 0 \end{bmatrix}$. Further, $\det(AB) = 2$, while $\det A = 2$ and $\det B = 1$. Thus, as the theorem states, $\det(AB) = (\det A)(\det B)$. □

EXAMPLE 43

If A is invertible, show that

$$\det A^{-1} = \frac{1}{\det A}$$

We involve A and A^{-1} in an equation by noting that $A^{-1}A = I_n$. Then we calculate the determinant and use the product rule to get $(\det A^{-1})(\det A) = 1$. Solving for $\det A^{-1}$, we have

$$\det A^{-1} = \frac{1}{\det A}$$

□

In the next section we show some uses for the expression $\det A$, where A is an invertible matrix, by giving formulas for the entries of A^{-1} and the entries of the solution $\mathbf{x}$ to $A\mathbf{x} = \mathbf{b}$, all in terms of the entries of A and $\mathbf{b}$.

EXERCISES FOR SECTION 2.5

COMPUTATIONAL EXERCISES

In each part of exercises 1 through 4, carry out the listed row and column operations on matrix A to get a matrix B. Calculate $\det A$, and predict $\det B$ from $\det A$ and the listed operations. Then check your prediction by independently calculating $\det B$.

1. $A = \begin{bmatrix} 3 & -1 \\ 2 & 2 \end{bmatrix}$ (*a*) $R_1 \leftrightarrow R_2$ (*b*) $3C_1 \to C_1$

(*c*) $R_1 - 2R_2 \to R_1, R_2 + 2R_1 \to R_2$

2. $A = \begin{bmatrix} 2 & -3 \\ 4 & 1 \end{bmatrix}$ (a) $C_1 \leftrightarrow C_2$ (b) $2R_1 \to R_1$

(c) $C_1 + C_2 \to C_1, C_2 - 3C_1 \to C_2$

3. $A = \begin{bmatrix} 3 & 0 & 4 \\ -2 & 2 & 6 \\ -1 & 0 & 1 \end{bmatrix}$ (a) $C_1 \leftrightarrow C_2, R_1 \leftrightarrow R_2$

(b) $R_1 + 2R_3 \to R_1, C_1 + C_2 \to C_1, R_3 + R_1 \to R_3$
(c) $\frac{1}{2}C_2 \to C_2, R_2 - 2R_3 \to R_2, R_1 + 3R_3 \to R_1$

4. $A = \begin{bmatrix} 2 & 6 & -4 \\ 0 & 3 & 0 \\ 1 & -3 & 2 \end{bmatrix}$ (a) $\frac{1}{2}R_1 \to R_1, C_3 \leftrightarrow C_2$

(b) $C_2 \leftrightarrow C_3, C_2 + 2C_1 \to C_2, R_2 \leftrightarrow R_3, R_1 \leftrightarrow R_2, C_2 \leftrightarrow C_1$
(c) $\frac{1}{2}R_1 \to R_1, R_3 - R_1 \to R_3, R_3 + 2R_2 \to R_3$

In exercises 5 through 8, let

$$A = \begin{bmatrix} \mathbf{a}_1 \\ \mathbf{a}_2 \\ \mathbf{a}_3 \end{bmatrix}$$

where the vectors $\mathbf{a}_i$ are 1×3, and let

$$B = [\mathbf{b}^1 \quad \mathbf{b}^2 \quad \mathbf{b}^3]$$

where the vectors $\mathbf{b}^i$ are 3×1. Find $\det C$, given $\det A$ or $\det B$.

5. $\det A = 2$, $C = \begin{bmatrix} \mathbf{a}_2 \\ \mathbf{a}_1 \\ \mathbf{a}_3 \end{bmatrix}$

6. $\det A = -3$, $C = \begin{bmatrix} \mathbf{a}_1 - 3\mathbf{a}_2 \\ 2\mathbf{a}_2 \\ \mathbf{a}_3 \end{bmatrix}$

7. $\det B = 3$, $C = [\mathbf{b}^1 \quad \mathbf{b}^3 \quad 4\mathbf{b}^2]$

8. $\det B = -7$, $C = [\mathbf{b}^1 + 4\mathbf{b}^3 \quad \mathbf{b}^2 \quad 2\mathbf{b}^3]$

In exercises 9 through 20, use the algorithm of this section to evaluate the determinant of the given matrix.

9. $\begin{bmatrix} 1 & 3 & 2 \\ 4 & -1 & 7 \\ 2 & 1 & -1 \end{bmatrix}$

10. $\begin{bmatrix} 3 & -4 & 5 \\ 5 & 4 & -11 \\ 2 & 1 & -3 \end{bmatrix}$

11. $\begin{bmatrix} 2 & -2 & 3 \\ -6 & 0 & 2 \\ 4 & 1 & -2 \end{bmatrix}$

12. $\begin{bmatrix} 2 & -3 & 1 \\ 4 & -6 & 2 \\ 1 & 3 & 1 \end{bmatrix}$

13. $\begin{bmatrix} 0 & 3 & -1 \\ \frac{5}{7} & -\frac{25}{7} & \frac{10}{7} \\ 3 & 0 & -2 \end{bmatrix}$

14. $\begin{bmatrix} 0 & 2 & -2 \\ -4 & -1 & 0 \\ \frac{7}{5} & -\frac{28}{5} & \frac{14}{5} \end{bmatrix}$

15. $\begin{bmatrix} 0 & \frac{2}{3} & 0 \\ -\frac{6}{7} & \frac{1}{3} & \frac{3}{8} \\ \frac{3}{5} & \frac{5}{7} & 0 \end{bmatrix}$

16. $\begin{bmatrix} 0 & -\frac{1}{5} & 0 \\ \frac{3}{5} & \frac{9}{7} & \frac{4}{7} \\ \frac{5}{8} & \frac{2}{5} & 0 \end{bmatrix}$

17. $\begin{bmatrix} 1 & 2 & -1 & 4 \\ 3 & 5 & 2 & 3 \\ -1 & 2 & 1 & 4 \\ 5 & -1 & 2 & 3 \end{bmatrix}$

18. $\begin{bmatrix} 1 & -2 & 3 & 5 \\ 4 & -9 & 6 & 12 \\ -2 & 5 & 1 & -7 \\ 3 & -4 & 2 & -17 \end{bmatrix}$

19. $\begin{bmatrix} \frac{1}{3} & \frac{12}{7} & 0 & -\frac{2}{5} \\ 0 & \frac{1}{9} & 0 & 0 \\ -\frac{1}{8} & \frac{7}{2} & \frac{3}{5} & \frac{2}{3} \\ 0 & \frac{6}{5} & 0 & -\frac{1}{5} \end{bmatrix}$

20. $\begin{bmatrix} \frac{7}{3} & 0 & \frac{9}{4} & 0 \\ \frac{6}{5} & 0 & \frac{3}{8} & -\frac{1}{3} \\ \frac{7}{2} & \frac{12}{5} & -\frac{1}{8} & \frac{2}{9} \\ 0 & 0 & \frac{1}{5} & 0 \end{bmatrix}$

In exercises 21 through 26, write a 2×2 matrix with the given properties, if possible.

21. Noninvertible with no 0 entries

22. Noninvertible with two 0 entries

23. Invertible with two 0 entries

24. Invertible with all entries the same

25. Invertible with three 0 entries

26. Invertible with a column of 0's

In exercises 27 and 28, verify the product rules, Theorem 2.14, for the products AB and rA, using the given matrices A and B and scalar r.

27. $A = \begin{bmatrix} 1 & 2 \\ 1 & 3 \end{bmatrix}$, $B = \begin{bmatrix} 2 & 1 \\ 1 & 2 \end{bmatrix}$, $r = 2$

28. $A = \begin{bmatrix} 0 & 1 & 3 \\ 2 & 1 & 1 \\ -1 & 2 & 0 \end{bmatrix}$, $B = \begin{bmatrix} 1 & 0 & 2 \\ -3 & 1 & -1 \\ 2 & -2 & 0 \end{bmatrix}$, $r = 3$

29. If $\det A = -5$, what is $\det A^2$?

30. If $\det A = 4$, what is $\det A^3$?

COMPLEX NUMBERS

In exercises 31 and 32, evaluate the given determinant.

31. $\det \begin{bmatrix} 2 - i & i & 3 \\ -2i & 1 - i & 2 + 3i \\ 3 - 2i & 4 & 1 + 5i \end{bmatrix}$

32. $\det \begin{bmatrix} i & 2 - i & 4 - 3i \\ 3 - 2i & 0 & 2 + 2i \\ 1 - i & 4 - 2i & 1 \end{bmatrix}$

THEORETICAL EXERCISES

33. Prove Theorem 2.11, parts (ii) and (iii).

34. Prove Theorem 2.14 for 3×3 matrices. [Franklin (see References) contains a proof of this theorem for general n.]

35. Prove that if A and B are $n \times n$ matrices, then $\det AB = \det BA$.

36. Prove that if P and A are $n \times n$ matrices with P invertible, then $\det(P^{-1}AP) = \det A$.

37. Prove that if A is an $n \times n$ matrix and $A^2 = A$, then $\det A = 0$ or $\det A = 1$.

38. Prove that

$$\det\begin{bmatrix} x & y & 1 \\ a_1 & a_2 & 1 \\ b_1 & b_2 & 1 \end{bmatrix} = 0$$

is the equation of the line through the points (a_1, a_2) and (b_1, b_2) in the coordinate plane.

39. Let x_1, x_2, and x_3 be distinct numbers, and let y_1, y_2, and y_3 be arbitrary numbers. Let

$$p(x) = a_2x^2 + a_1x + a_0$$

Write out the system of equations that would need to be solved to find the coefficients a_0, a_1, and a_2 such that $p(x_1) = y_1$, $p(x_2) = y_2$, and $p(x_3) = y_3$. Show that the coefficient matrix is

$$A = \begin{bmatrix} x_1^2 & x_1 & 1 \\ x_2^2 & x_2 & 1 \\ x_3^2 & x_3 & 1 \end{bmatrix}$$

Prove that $\det A \neq 0$. Explain why this means that there is only one polynomial of degree 2 or less whose graph passes through (x_1, y_1), (x_2, y_2), and (x_3, y_3). More generally, it can be shown that

$$\det\begin{bmatrix} x_1^k & x_1^{k-1} & \cdots & x_1 & 1 \\ x_2^k & x_2^{k-1} & \cdots & x_2 & 1 \\ \vdots & \vdots & \vdots & \vdots & \vdots \\ x_{k+1}^k & x_{k+1}^{k-1} & \cdots & x_{k+1} & 1 \end{bmatrix} \neq 0$$

if $x_1, x_2, \ldots, x_{k+1}$ are distinct. The matrix here is called the *Vandermonde matrix*.

40. Let A_1 be a $p \times p$ matrix, A_2 a $q \times q$ matrix, and X a $p \times q$ matrix. Prove that

$$\det\begin{bmatrix} A_1 & \mathbf{0} \\ X & A_2 \end{bmatrix} = (\det A_1)(\det A_2)$$

Hint: Use row operations to get $\begin{bmatrix} T_1 & \mathbf{0} \\ Y & T_2 \end{bmatrix}$, whose determinant is clearly $(\det T_1)(\det T_2)$. Now count the number of interchanges used and adjust accordingly.

41. Write out an $n \times n$ matrix A by randomly choosing the entries. Intuitively, do you expect A to be invertible? *Hint:* Consider $\det A$.

COMPUTER EXERCISES

42. Let $A = \begin{bmatrix} 2 & -1 \\ -1 & 1 \end{bmatrix}$. Suppose a calculation is made which says that

$$A^{150} = \begin{bmatrix} 56 & -2 \\ -2 & 51 \end{bmatrix}$$

Is this calculation correct? *Hint:* Check the determinants of A and the proposed A^{150}.

43. Let A be a 4×4 matrix. Find the number of multiplications and divisions needed to compute $\det A$ by the algorithm of this section. Compare your result to the number required to compute the determinant by definition. Which method is faster?

44. Let

$$A = \begin{bmatrix} -\frac{1}{4} & \frac{1}{6} & 0 \\ \frac{1}{2} & 0 & -\frac{1}{8} \\ 0 & -\frac{1}{4} & \frac{1}{7} \end{bmatrix}$$

Compute $\det A$ using a two-digit computer, and then a four-digit computer. Compare your results to the exact value, $\det A = -11/2688 \approx -0.00409226$, by computing percentage error.

2.6 COFACTOR EXPANSIONS FOR THE DETERMINANT

In this section we give an organized way of calculating the determinant by using its definition. The method presented is thus not practical for larger matrices. However, this method is of some use in calculating determinants of $n \times n$ matrices for $n = 3$, perhaps for $n = 4$, and for cases with larger n where many zeros are present in the matrix. Our basic use, however, of the method will be in obtaining determinantal formulas for inverses of matrices and solutions to systems of linear algebraic equations. In some problems (see exercises 35–38, 49, and 53–55) such formulas are of more use than numerical calculations.

We use the following:

DEFINITION

Let $n > 1$ and let A be an $n \times n$ matrix. Define A_{ij} as the matrix obtained from A by deleting row i and column j. Define*

$$c_{ij} = (-1)^{i+j} \det A_{ij}$$

The number c_{ij} is called the *i,j-cofactor* of A.

EXAMPLE 44

Let $A = \begin{bmatrix} 1 & 2 \\ 3 & 4 \end{bmatrix}$. Then

$$c_{11} = (-1)^{1+1} \det [4] = 4, \qquad c_{12} = (-1)^{1+2} \det [3] = -3,$$
$$c_{21} = (-1)^{2+1} \det [2] = -2, \qquad c_{22} = (-1)^{2+2} \det [1] = 1$$

□

EXAMPLE 45

Let $A = \begin{bmatrix} 1 & 2 & 3 \\ 4 & 5 & 6 \\ 7 & 8 & 9 \end{bmatrix}$. Then

$$c_{11} = (-1)^{1+1} \det \begin{bmatrix} 5 & 6 \\ 8 & 9 \end{bmatrix} = +(45 - 48) = -3$$
$$c_{12} = (-1)^{1+2} \det \begin{bmatrix} 4 & 6 \\ 7 & 9 \end{bmatrix} = -(36 - 42) = 6$$
$$c_{13} = (-1)^{1+3} \det \begin{bmatrix} 4 & 5 \\ 7 & 8 \end{bmatrix} = +(32 - 35) = -3$$
$$c_{21} = (-1)^{2+1} \det \begin{bmatrix} 2 & 3 \\ 8 & 9 \end{bmatrix} = -(18 - 24) = 6$$
$$c_{22} = (-1)^{2+2} \det \begin{bmatrix} 1 & 3 \\ 7 & 9 \end{bmatrix} = +(9 - 21) = -12$$
$$c_{23} = (-1)^{2+3} \det \begin{bmatrix} 1 & 2 \\ 7 & 8 \end{bmatrix} = -(8 - 14) = 6$$
$$c_{31} = (-1)^{3+1} \det \begin{bmatrix} 2 & 3 \\ 5 & 6 \end{bmatrix} = +(12 - 15) = -3$$
$$c_{32} = (-1)^{3+2} \det \begin{bmatrix} 1 & 3 \\ 4 & 6 \end{bmatrix} = -(6 - 12) = 6$$
$$c_{33} = (-1)^{3+3} \det \begin{bmatrix} 1 & 2 \\ 4 & 5 \end{bmatrix} = +(5 - 8) = -3$$

□

* $\det A_{ij}$ is called the *i,j-minor* of A.

The determinant can be computed by using cofactors as follows.

THEOREM 2.15

Let $n > 1$ and let A be an $n \times n$ matrix. Then

(*i*) $\det A = a_{i1}c_{i1} + \cdots + a_{in}c_{in}$, called the *ith row expansion* of $\det A$, and

(*ii*) $\det A = a_{1j}c_{1j} + \cdots + a_{nj}c_{nj}$, called the *jth column expansion* of $\det A$.

Thus, if you multiply the cofactors from one row or column by the entries from that row or column, respectively, the sum of the expressions yields $\det A$.

PROOF

We will prove part (i) for the first row expansion in a 3×3 matrix A. That is, we will prove

$$\det A = a_{11}c_{11} + a_{12}c_{12} + a_{13}c_{13}$$

Other row or column expansions can be obtained from this result by using row interchanges and transposes (exercise 45). This approach can be generalized to $n \times n$ matrices.

Manipulating, we have

$$\begin{aligned}\det A &= a_{11}a_{22}a_{33} + a_{12}a_{23}a_{31} + a_{13}a_{21}a_{32} \\ &\quad - a_{13}a_{22}a_{31} - a_{12}a_{21}a_{33} - a_{11}a_{23}a_{32} \\ &= a_{11}(a_{22}a_{33} - a_{23}a_{32}) + a_{12}(a_{23}a_{31} - a_{21}a_{33}) \\ &\quad + a_{13}(a_{21}a_{32} - a_{22}a_{31}) \\ &= a_{11}c_{11} + a_{12}c_{12} + a_{13}c_{13}\end{aligned}$$

■

EXAMPLE 46

Let $A = \begin{bmatrix} 1 & 2 & 3 \\ 4 & 5 & 6 \\ 7 & 8 & 9 \end{bmatrix}$, as in Example 45.

Computing by the first row expansion yields

$$\det A = 1c_{11} + 2c_{12} + 3c_{13} = 1(-3) + 2(6) + 3(-3) = 0$$

Computing by the third column expansion yields

$$\det A = 3c_{13} + 6c_{23} + 9c_{33} = 3(-3) + 6(6) + 9(-3) = 0$$

Computing by the second row expansion yields

$$\det A = 4c_{21} + 5c_{22} + 6c_{23} = 4(6) + 5(-12) + 6(6) = 0$$

□

EXAMPLE 47

Let $A = \begin{bmatrix} 1 & 0 & 2 \\ 3 & 2 & 5 \\ 4 & 0 & -1 \end{bmatrix}$. Since column two contains only one nonzero entry, we can simplify this calculation by computing the determinant by a second column expansion. Thus

$$\det A = 0c_{12} + 2c_{22} + 0c_{32} = 2\left((-1)^{2+2}\det\begin{bmatrix} 1 & 2 \\ 4 & -1 \end{bmatrix}\right) = 2(-9) = -18$$

□

The two identities below, used later, are consequences of the preceding theorem.

COROLLARY 2.15A

Let $A = [a_{ij}]$ be an $n \times n$ matrix. If $r \neq s$, then

$$\sum_{k=1}^{n} a_{rk}c_{sk} = 0 \qquad \text{and} \qquad \sum_{k=1}^{n} a_{ks}c_{kr} = 0$$

Thus, if you multiply the cofactors from one row or column by the entries from another row or column, respectively, the sum of the expressions is 0.

PROOF

Since the proofs of the two identities are essentially the same, we will only prove the first one. Further, this proof will only be given for a 3×3 matrix A, $r = 1$ and $s = 2$. This proof can be extended to the general case.

Here,

$$A = \begin{bmatrix} a_{11} & a_{12} & a_{13} \\ a_{21} & a_{22} & a_{23} \\ a_{31} & a_{32} & a_{33} \end{bmatrix}$$

Replacing row two of A by row one, we obtain

$$B = \begin{bmatrix} a_{11} & a_{12} & a_{13} \\ a_{11} & a_{12} & a_{13} \\ a_{31} & a_{32} & a_{33} \end{bmatrix}$$

Calculating the determinant of B by a second row expansion yields

$$\det B = a_{11}c_{21} + a_{12}c_{22} + a_{13}c_{23}$$

where these cofactors can be computed from either A or B. Since B has two identical rows, its determinant is 0, so

$$a_{11}c_{21} + a_{12}c_{22} + a_{13}c_{23} = 0$$

■

As a direct consequence of the above theorem and corollary, we have a way of expressing the inverse of a matrix in terms of determinants. For this result, we need the following definition.

DEFINITION

Let $n > 1$ and let A be an $n \times n$ matrix. Form the *cofactor matrix* $C = [c_{ij}]$ using the i,j-cofactor of A as the i,j-entry of C. Then the *adjoint* of A is

$$\text{adj}\, A = C^t$$

EXAMPLE 48

Let $A = \begin{bmatrix} 1 & 2 \\ 3 & 4 \end{bmatrix}$. Then

$$C = \begin{bmatrix} c_{11} & c_{12} \\ c_{21} & c_{22} \end{bmatrix} = \begin{bmatrix} 4 & -3 \\ -2 & 1 \end{bmatrix} \quad \text{and} \quad \text{adj } A = C^t = \begin{bmatrix} 4 & -2 \\ -3 & 1 \end{bmatrix} \quad \square$$

EXAMPLE 49

Let $A = \begin{bmatrix} 1 & 0 & 1 \\ 0 & -1 & 2 \\ -1 & 1 & 0 \end{bmatrix}$. Then

$$C = \begin{bmatrix} (-1)^{1+1}\det\begin{bmatrix} -1 & 2 \\ 1 & 0 \end{bmatrix} & (-1)^{1+2}\det\begin{bmatrix} 0 & 2 \\ -1 & 0 \end{bmatrix} & (-1)^{1+3}\det\begin{bmatrix} 0 & -1 \\ -1 & 1 \end{bmatrix} \\ (-1)^{2+1}\det\begin{bmatrix} 0 & 1 \\ 1 & 0 \end{bmatrix} & (-1)^{2+2}\det\begin{bmatrix} 1 & 1 \\ -1 & 0 \end{bmatrix} & (-1)^{2+3}\det\begin{bmatrix} 1 & 0 \\ -1 & 1 \end{bmatrix} \\ (-1)^{3+1}\det\begin{bmatrix} 0 & 1 \\ -1 & 2 \end{bmatrix} & (-1)^{3+2}\det\begin{bmatrix} 1 & 1 \\ 0 & 2 \end{bmatrix} & (-1)^{3+3}\det\begin{bmatrix} 1 & 0 \\ 0 & -1 \end{bmatrix} \end{bmatrix}$$

$$= \begin{bmatrix} -2 & -2 & -1 \\ 1 & 1 & -1 \\ 1 & -2 & -1 \end{bmatrix}$$

Thus

$$\text{adj } A = C^t = \begin{bmatrix} -2 & 1 & 1 \\ -2 & 1 & -2 \\ -1 & -1 & -1 \end{bmatrix} \quad \square$$

Since the adjoint was defined using cofactors, by Theorem 2.15 and Corollary 2.15A we have the following.

THEOREM 2.16

Let $n > 1$ and let A be an $n \times n$ matrix. Then

$$A(\text{adj } A) = (\text{adj } A)A = (\det A)I_n$$

PROOF

To see this result simply, we will prove the theorem for $n = 2$ and leave the generalization to an arbitrary n as exercise 46. With $n = 2$, using Theorem 2.15 and Corollary 2.15A we have

$$A(\text{adj } A) = \begin{bmatrix} a_{11} & a_{12} \\ a_{21} & a_{22} \end{bmatrix}\begin{bmatrix} c_{11} & c_{21} \\ c_{12} & c_{22} \end{bmatrix} = \begin{bmatrix} a_{11}c_{11} + a_{12}c_{12} & a_{11}c_{21} + a_{12}c_{22} \\ a_{21}c_{11} + a_{22}c_{12} & a_{21}c_{21} + a_{22}c_{22} \end{bmatrix}$$

$$= \begin{bmatrix} \det A & 0 \\ 0 & \det A \end{bmatrix} = (\det A)I_2$$

Further,

$$(\operatorname{adj} A)A = \begin{bmatrix} c_{11} & c_{21} \\ c_{12} & c_{22} \end{bmatrix} \begin{bmatrix} a_{11} & a_{12} \\ a_{21} & a_{22} \end{bmatrix} = \begin{bmatrix} a_{11}c_{11} + a_{21}c_{21} & a_{12}c_{11} + a_{22}c_{21} \\ a_{11}c_{12} + a_{21}c_{22} & a_{12}c_{12} + a_{22}c_{22} \end{bmatrix}$$

$$= \begin{bmatrix} \det A & 0 \\ 0 & \det A \end{bmatrix} = (\det A)I_2$$

■

As a corollary, we have a determinantal formula for inverses.

COROLLARY 2.16A

Let $n > 1$ and let A be an $n \times n$ matrix with $\det A \neq 0$. Then A has an inverse and

$$A^{-1} = \frac{1}{\det A}(\operatorname{adj} A)$$

PROOF

Using the arithmetic of matrices, we see that if $\det A \neq 0$, then

$$A\left(\frac{1}{\det A}(\operatorname{adj} A)\right) = \frac{1}{\det A}[A(\operatorname{adj} A)] = \frac{1}{\det A}[(\det A)I_n] = I_n$$

Thus

$$A^{-1} = \frac{1}{\det A}(\operatorname{adj} A)$$

■

EXAMPLE 50

Let $A = [a_{ij}]$ be a 2×2 invertible matrix. Then

$$A^{-1} = \frac{1}{\det A}\begin{bmatrix} c_{11} & c_{21} \\ c_{12} & c_{22} \end{bmatrix} = \begin{bmatrix} \dfrac{a_{22}}{a_{11}a_{22} - a_{12}a_{21}} & \dfrac{-a_{12}}{a_{11}a_{22} - a_{12}a_{21}} \\ \dfrac{-a_{21}}{a_{11}a_{22} - a_{12}a_{21}} & \dfrac{a_{11}}{a_{11}a_{22} - a_{12}a_{21}} \end{bmatrix}$$

Thus, the entries of A^{-1} are given by formulas involving only the entries of A.

□

EXAMPLE 51

Let $A = \begin{bmatrix} 1 & 2 \\ 3 & 4 \end{bmatrix}$. Then, from Example 48,

$$\operatorname{adj} A = \begin{bmatrix} 4 & -2 \\ -3 & 1 \end{bmatrix}$$

Hence

$$A^{-1} = \frac{1}{-2}\begin{bmatrix} 4 & -2 \\ -3 & 1 \end{bmatrix} = \begin{bmatrix} -2 & 1 \\ \frac{3}{2} & -\frac{1}{2} \end{bmatrix}$$

□

EXAMPLE 52

Let $A = \begin{bmatrix} 1 & 0 & 1 \\ 0 & -1 & 2 \\ -1 & 1 & 0 \end{bmatrix}$. Then, from Example 49,

$$\operatorname{adj} A = \begin{bmatrix} -2 & 1 & 1 \\ -2 & 1 & -2 \\ -1 & -1 & -1 \end{bmatrix}$$

so

$$A^{-1} = \frac{1}{-3}\begin{bmatrix} -2 & 1 & 1 \\ -2 & 1 & -2 \\ -1 & -1 & -1 \end{bmatrix} = \begin{bmatrix} \frac{2}{3} & -\frac{1}{3} & -\frac{1}{3} \\ \frac{2}{3} & -\frac{1}{3} & \frac{2}{3} \\ \frac{1}{3} & \frac{1}{3} & \frac{1}{3} \end{bmatrix}$$ □

EXAMPLE 53

OPTIONAL—COMPUTING

To see how a determinantal formula can be of use in studying the effects on the inverse of a matrix A of small changes in the entries of A, let $A = \begin{bmatrix} 1 & 1 \\ 1 & \lambda \end{bmatrix}$. Then

$$A^{-1} = \begin{bmatrix} \dfrac{\lambda}{\lambda - 1} & \dfrac{-1}{\lambda - 1} \\ \dfrac{-1}{\lambda - 1} & \dfrac{1}{\lambda - 1} \end{bmatrix}$$

The graphs of the entries of A^{-1} are shown in Figure 2.5. Note that when λ is

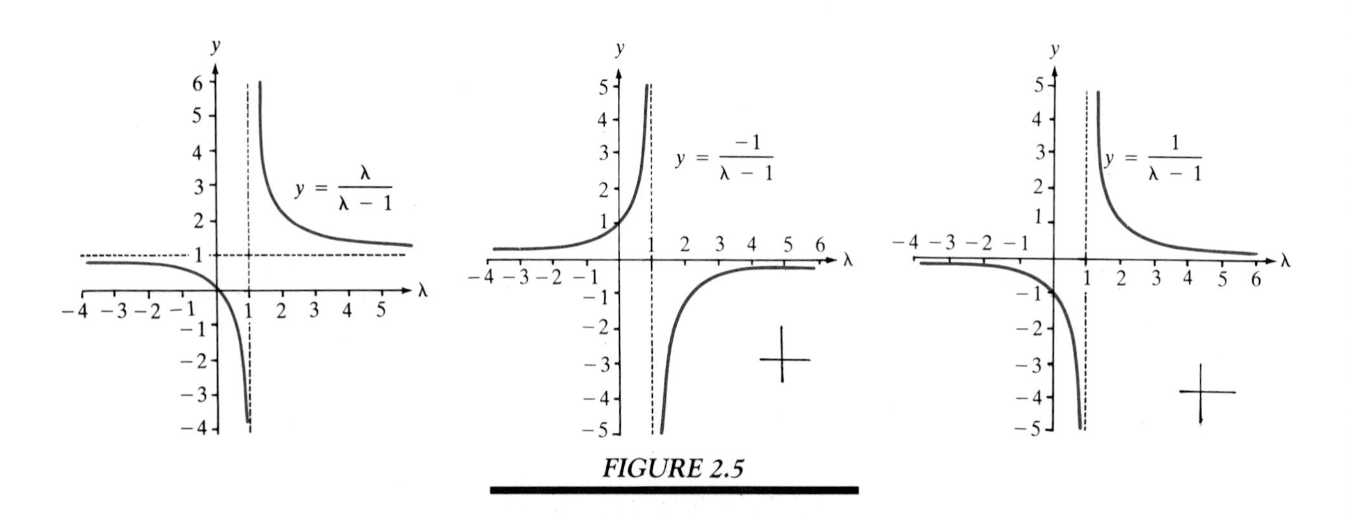

FIGURE 2.5

close to 1, small changes in λ give large changes in the entries of A^{-1}. For example, if $A = \begin{bmatrix} 1 & 1 \\ 1 & 1.01 \end{bmatrix}$, then

$$A^{-1} = \begin{bmatrix} \dfrac{1.01}{0.01} & -\dfrac{1}{0.01} \\ -\dfrac{1}{0.01} & \dfrac{1}{0.01} \end{bmatrix} = \begin{bmatrix} 101 & -100 \\ -100 & 100 \end{bmatrix}$$

and if $A = \begin{bmatrix} 1 & 1 \\ 1 & 1.02 \end{bmatrix}$, then

$$A^{-1} = \begin{bmatrix} \dfrac{1.02}{0.02} & -\dfrac{1}{0.02} \\ -\dfrac{1}{0.02} & \dfrac{1}{0.02} \end{bmatrix} = \begin{bmatrix} 51 & -50 \\ -50 & 50 \end{bmatrix}$$

The basic idea in this example is that when the determinant of A is small compared to one of its cofactors, say c_{ij}, then the entry $c_{ij}/\det A$ of A^{-1} is sensitive to small changes in A; that is, if $\det A$ changes a little, $c_{ij}/\det A$ can change a lot. Keep this example in mind when rounding off numbers during the computing of an inverse (see exercise 57). □

A determinantal formula can also be obtained for the solution of a system of linear algebraic equations whose coefficient matrix is square and invertible. This result, known as *Cramer's Rule*, is given next.

THEOREM 2.17

Let $n > 1$ and let A be an $n \times n$ invertible matrix. Then the system of linear equations

$$A\mathbf{x} = \mathbf{b}$$

has solution $\mathbf{x}$ with

$$x_i = \frac{\det A_i}{\det A}$$

for all i, where A_i is the matrix formed from A by replacing the i-column of A by $\mathbf{b}$.

PROOF

First, note that

$$\mathbf{x} = A^{-1}\mathbf{b} = \frac{1}{\det A}(\operatorname{adj} A)\mathbf{b}$$

Now, the result will follow if we can show that the i-entry of $(\operatorname{adj} A)\mathbf{b}$ is $\det A_i$. Note that the i-entry of $(\operatorname{adj} A)\mathbf{b}$ is $\sum_{k=1}^{n} b_k c_{ki}$. Expanding the determinant of A_i along column i yields the same result. Hence

$$x_i = \frac{\det A_i}{\det A}$$

■

EXAMPLE 54

Let $A = [a_{ij}]$ be a 2×2 invertible matrix and let $\mathbf{b} = [b_i]$ be a 2×1 vector. For $A\mathbf{x} = \mathbf{b}$,

$$x_1 = \frac{\det [\mathbf{b} \quad \mathbf{a}^2]}{\det A} = \frac{b_1 a_{22} - b_2 a_{12}}{a_{11}a_{22} - a_{12}a_{21}}$$

and

$$x_2 = \frac{\det [\mathbf{a}^1 \quad \mathbf{b}]}{\det A} = \frac{b_2 a_{11} - b_1 a_{21}}{a_{11}a_{22} - a_{12}a_{21}}$$

Thus the solutions are given by formulas involving the entries of A and $\mathbf{b}$. □

EXAMPLE 55

Let

$$A = \begin{bmatrix} -3 & 0 & 1 \\ 2 & 0 & 0 \\ -1 & 4 & 0 \end{bmatrix} \quad \text{and} \quad \mathbf{b} = \begin{bmatrix} 1 \\ -1 \\ 1 \end{bmatrix}$$

Then $A\mathbf{x} = \mathbf{b}$ has solution

$$x_1 = \frac{\det [\mathbf{b} \quad \mathbf{a}^2 \quad \mathbf{a}^3]}{\det A} = \frac{\det \begin{bmatrix} 1 & 0 & 1 \\ -1 & 0 & 0 \\ 1 & 4 & 0 \end{bmatrix}}{\det \begin{bmatrix} -3 & 0 & 1 \\ 2 & 0 & 0 \\ -1 & 4 & 0 \end{bmatrix}} = \frac{-4}{8} = -\frac{1}{2}$$

$$x_2 = \frac{\det [\mathbf{a}^1 \quad \mathbf{b} \quad \mathbf{a}^3]}{\det A} = \frac{\det \begin{bmatrix} -3 & 1 & 1 \\ 2 & -1 & 0 \\ -1 & 1 & 0 \end{bmatrix}}{8} = \frac{1}{8}$$

$$x_3 = \frac{\det [\mathbf{a}^1 \quad \mathbf{a}^2 \quad \mathbf{b}]}{\det A} = \frac{\det \begin{bmatrix} -3 & 0 & 1 \\ 2 & 0 & -1 \\ -1 & 4 & 1 \end{bmatrix}}{8} = \frac{-4}{8} = -\frac{1}{2}$$

□

EXAMPLE 56

OPTIONAL—ANALYSIS

Consider two plants that manufacture compact and mid-size cars as shown in the following table.

	PLANT A	PLANT B
Compact Car	2 per day	1 per day
Mid-size Car	3 per day	2 per day

Let x_1 = the number of days of operation for Plant A and x_2 = the number of days of operation for Plant B. Suppose d_1 = the number of mid-size cars demanded and d_2 = the number of compact cars demanded. By using formulas for x_1 and x_2, we can do an if-then analysis on this problem. In particular, if the demand for compact cars increases, then how does that affect the working time for each plant?

First, note that

$$\begin{bmatrix} 2 & 1 \\ 3 & 2 \end{bmatrix} \begin{bmatrix} x_1 \\ x_2 \end{bmatrix} = \begin{bmatrix} d_1 \\ d_2 \end{bmatrix}$$

Since $\det\begin{bmatrix} 2 & 1 \\ 3 & 2 \end{bmatrix} = 1$,

$$x_1 = \det\begin{bmatrix} d_1 & 1 \\ d_2 & 2 \end{bmatrix} = 2d_1 - d_2 \qquad \text{and} \qquad x_2 = \det\begin{bmatrix} 2 & d_1 \\ 3 & d_2 \end{bmatrix} = 2d_2 - 3d_1$$

Thus, if d_1 increases, x_1 increases and x_2 decreases. That is, Plant A operates more while Plant B operates less. □

EXAMPLE 57

OPTIONAL—COMPUTING

Consider the system of linear algebraic equations

$$\begin{aligned} x + y &= 1 \\ x + \varepsilon y &= 2 \end{aligned}$$

Using the determinantal formulas for x and y, we can show how changes in ε affect the solutions x and y to this system.

Note that

$$x = \frac{\det\begin{bmatrix} 1 & 1 \\ 2 & \varepsilon \end{bmatrix}}{\det\begin{bmatrix} 1 & 1 \\ 1 & \varepsilon \end{bmatrix}} = \frac{\varepsilon - 2}{\varepsilon - 1} = 1 - \frac{1}{\varepsilon - 1}$$

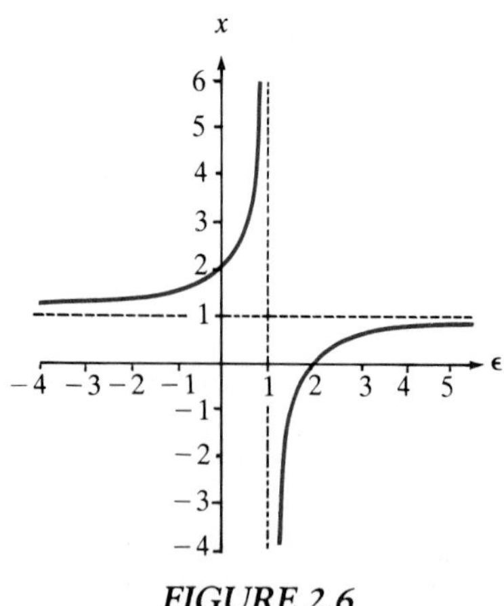

FIGURE 2.6

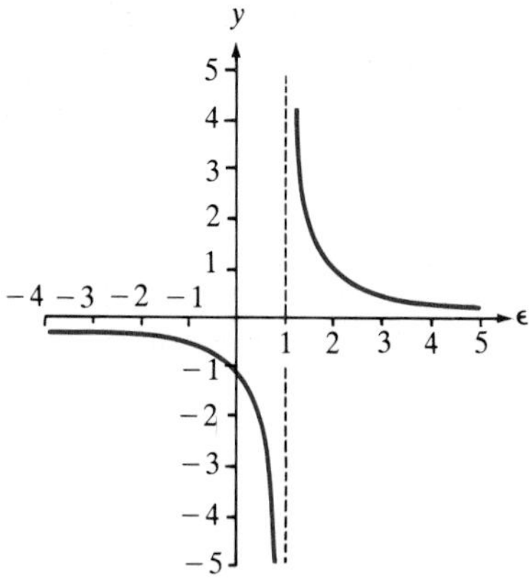

FIGURE 2.7

Graphing x yields Figure 2.6. If ε is close to 1, and thus $\det\begin{bmatrix}1 & 1\\ 1 & \varepsilon\end{bmatrix}$ is close to 0, then the solution x changes greatly for small changes in ε.

Similarly,

$$y = \frac{\det\begin{bmatrix}1 & 1\\ 1 & 2\end{bmatrix}}{\det\begin{bmatrix}1 & 1\\ 1 & \varepsilon\end{bmatrix}} = \frac{1}{\varepsilon - 1}$$

Graphing, we get Figure 2.7. Thus, when ε is close to 1, small changes in ε result in large changes in y.

In general, if $\det A$ is small compared to some $\det A_i$, then $x_i = \det A_i/\det A$ is sensitive to small changes in $\det A$. □

This concludes our work on the determinant. In the next chapter we start the study of vector spaces.

EXERCISES FOR SECTION 2.6

COMPUTATIONAL EXERCISES

In exercises 1 through 4, write the i,j-cofactor of A for the given matrix A and the given i and j.

1. $A = \begin{bmatrix} 1 & -3 & 2\\ 6 & 5 & 17\\ -8 & 3 & 2\end{bmatrix}$ (a) $i = j = 1$ (b) $i = 2, j = 3$

2. $A = \begin{bmatrix} -4 & 3 & -2\\ 4 & 6 & -7\\ 7 & 2 & 2\end{bmatrix}$ (a) $i = 1, j = 3$ (b) $i = 3, j = 2$

3. $A = \begin{bmatrix} -1 & 3 & 6 & 5\\ 17 & 1 & 8 & 2\\ 9 & 3 & 2 & 1\\ 2 & 1 & -1 & 4\end{bmatrix}$ (a) $i = 1, j = 3$ (b) $i = 3, j = 2$

4. $A = \begin{bmatrix} 17 & 8 & -2 & 6 \\ 5 & -1 & 2 & 3 \\ 8 & 2 & 3 & -5 \\ 3 & -1 & 5 & 1 \end{bmatrix}$ (a) $i = 1, j = 3$ (b) $i = 3, j = 2$

In exercises 5 through 8, evaluate the determinant of the given matrix by expansion on the given row, and then on the given column. Compare your answers.

5. $\begin{bmatrix} 6 & -1 & 2 \\ 0 & 0 & 3 \\ 4 & 5 & -2 \end{bmatrix}$, row 2, column 3

6. $\begin{bmatrix} -5 & 3 & -2 \\ 6 & 1 & -1 \\ 2 & 4 & 0 \end{bmatrix}$, row 3, column 3

7. $\begin{bmatrix} 3 & -2 & 4 & -1 \\ 5 & 6 & -1 & 2 \\ 0 & 2 & 1 & 0 \\ 3 & 3 & 0 & -4 \end{bmatrix}$, row 3, column 3

8. $\begin{bmatrix} -1 & 4 & 1 & 3 \\ 17 & 8 & -4 & -6 \\ 3 & 0 & 2 & 2 \\ -2 & 1 & 0 & 5 \end{bmatrix}$, row 3, column 3

In exercises 9 through 16, use expansion along the row or column of your choice to evaluate the determinant of the given matrix. Make use of the zeros to reduce the number of computations needed.

9. $\begin{bmatrix} -1 & 2 & 3 \\ 4 & 0 & 0 \\ -1 & 4 & 2 \end{bmatrix}$

10. $\begin{bmatrix} 3 & -2 & 0 \\ -1 & 2 & 0 \\ 4 & 1 & 5 \end{bmatrix}$

11. $\begin{bmatrix} 0 & \frac{1}{3} & 0 \\ -3 & \frac{2}{7} & \frac{1}{4} \\ 12 & \frac{1}{8} & -5 \end{bmatrix}$

12. $\begin{bmatrix} 5 & -\frac{4}{7} & \frac{1}{4} \\ 0 & \frac{1}{5} & 0 \\ -20 & \frac{6}{11} & 2 \end{bmatrix}$

13. $\begin{bmatrix} \frac{7}{5} & -\frac{12}{7} & -\frac{6}{11} \\ 0 & \frac{1}{5} & 0 \\ \frac{3}{7} & \frac{2}{9} & 0 \end{bmatrix}$

14. $\begin{bmatrix} \frac{6}{17} & 0 & -\frac{4}{7} \\ -\frac{9}{5} & \frac{3}{8} & \frac{12}{5} \\ \frac{3}{11} & 0 & 0 \end{bmatrix}$

15. $\begin{bmatrix} -1 & 0 & 5 & -2 \\ 4 & 2 & 3 & -4 \\ 0 & 0 & 3 & 0 \\ 5 & 1 & -1 & 2 \end{bmatrix}$

16. $\begin{bmatrix} -3 & 0 & 2 & -5 \\ 7 & -2 & 1 & -1 \\ 3 & 0 & 4 & 0 \\ 2 & 0 & -1 & 3 \end{bmatrix}$

In exercises 17 through 22, write the adjoint of the given matrix.

17. $\begin{bmatrix} -1 & 2 \\ -2 & 3 \end{bmatrix}$

18. $\begin{bmatrix} 3 & -1 \\ -4 & -1 \end{bmatrix}$

19. $\begin{bmatrix} 2 & 1 \\ 0 & -2 \end{bmatrix}$

20. $\begin{bmatrix} -3 & 0 \\ -1 & 2 \end{bmatrix}$

21. $\begin{bmatrix} 3 & 3 & -1 \\ -1 & 3 & 3 \\ 3 & -1 & 3 \end{bmatrix}$

22. $\begin{bmatrix} 2 & 1 & 2 \\ 1 & 2 & 2 \\ 2 & 2 & 1 \end{bmatrix}$

In exercises 23 through 28, use the adjoint and the determinant of the matrix to write its inverse.

23. The matrix of exercise 17
24. The matrix of exercise 18
25. The matrix of exercise 19
26. The matrix of exercise 20
27. The matrix of exercise 21
28. The matrix of exercise 22

In exercises 29 through 34, use Cramer's Rule to solve the given system of linear algebraic equations for the given variable or variables.

29. $\begin{aligned} x - 2y &= 3 \\ 2x + y &= 1 \end{aligned}$ for x

30. $\begin{aligned} 2x - 3y &= 4 \\ 7x + y &= 6 \end{aligned}$ for y

31. $\begin{aligned} 8x_1 + 3x_2 &= 9 \\ 2x_1 - 5x_2 &= 7 \end{aligned}$ for x_2

32. $\begin{aligned} 6x_1 - 12x_2 &= 7 \\ 5x_1 + x_2 &= 2 \end{aligned}$ for x_1

33. $\begin{aligned} 6x - y + 3z &= -3 \\ 9x + 5y - 2z &= 7 \\ -5x - y + 8z &= 2 \end{aligned}$ for y and z

34. $\begin{aligned} x + 5y - z &= 2 \\ 2x - y + 3z &= -1 \\ 9x + y - 5z &= 5 \end{aligned}$ for x and z

In exercises 35 and 36, given matrix A, graph the entries in A^{-1}. Where is the calculation of A^{-1} sensitive to λ?

35. $A = \begin{bmatrix} 1 & 2 \\ \lambda & 1 \end{bmatrix}$

36. $A = \begin{bmatrix} 1 & 2 \\ 0 & \lambda \end{bmatrix}$

In exercises 37 and 38, graph x_1 and x_2 as functions of λ. Where is the calculation of $\mathbf{x}$ sensitive to λ, and which variables are affected by variations in λ?

37. $\begin{bmatrix} 1 & \lambda \\ 1 & 1 \end{bmatrix}\mathbf{x} = \begin{bmatrix} 1 \\ 2 \end{bmatrix}$

38. $\begin{bmatrix} 2 & 4 \\ \lambda & 2 \end{bmatrix}\mathbf{x} = \begin{bmatrix} 1 \\ 2 \end{bmatrix}$

COMPLEX NUMBERS

In exercises 39 and 40, use a row or column expansion to find the determinant.

39. $\det\begin{bmatrix} 2-i & i & 3 \\ -2i & 1-i & 2+3i \\ 3-2i & 4 & 6+5i \end{bmatrix}$

40. $\det\begin{bmatrix} i & 2-i & 4-3i \\ 3-2i & 0 & 2+2i \\ 1-i & 4-2i & 1 \end{bmatrix}$

In exercises 41 and 42, compute A^{-1} using the adjoint formula.

41. $A = \begin{bmatrix} i & 1-i \\ 1+i & 3 \end{bmatrix}$

42. $A = \begin{bmatrix} 2-i & 3+2i \\ 1-2i & 2+3i \end{bmatrix}$

In exercises 43 and 44, solve by Cramer's Rule.

43. $\begin{bmatrix} i & 1-i \\ 1+i & 3 \end{bmatrix}\mathbf{x} = \begin{bmatrix} 1-i \\ i \end{bmatrix}$

44. $\begin{bmatrix} 2-i & 3+2i \\ 1-2i & 2+3i \end{bmatrix}\mathbf{x} = \begin{bmatrix} i \\ 2 \end{bmatrix}$

THEORETICAL EXERCISES

45. Prove the general case of Theorem 2.15 for 3×3 matrices.

46. Prove the general case of Theorem 2.16.

47. Let A be a 5×5 matrix. Using a row expansion, we can reduce the problem of computing $\det A$ to that of computing the determinant of five 4×4 matrices. Continuing, we can eventually reduce this to a problem of computing determinants of how many 2×2 matrices?

48. Prove that if A is an $n \times n$ matrix, then $\det(\operatorname{adj} A) = (\det A)^{n-1}$.

49. Let A be an $n \times n$ matrix with integer entries. Show that adj A has integer entries. Show that if $\det A \neq 0$, then the entries of A^{-1} are rational numbers. Show that if $\det A = \pm 1$, then the inverse has integer entries.

50. Let A be an $n \times n$ matrix with integer entries and $\det A \neq 0$. Let $\mathbf{b}$ be an $n \times 1$ vector with integer entries. Does the solution to $A\mathbf{x} = \mathbf{b}$ have integer entries? Under what conditions on A will it necessarily have integer entries?

51. (calculus) Let A be an $n \times n$ matrix. Show that when $\det A \neq 0$, then the entries in A^{-1} are continuous functions of the entries of A. (It can be shown that algorithms used in solving $A\mathbf{x} = \mathbf{b}$, because of round-off errors, actually give the solution to a system $\hat{A}\hat{\mathbf{x}} = \hat{\mathbf{b}}$, where $\hat{A}$ is close to A and $\hat{\mathbf{b}}$ is close to $\mathbf{b}$. Thus it follows, at least theoretically, that $\hat{\mathbf{x}}$ can be brought as close to $\mathbf{x}$ as desired by assuring that $\hat{A}$ is sufficiently close to A and $\hat{\mathbf{b}}$ is sufficiently close to $\mathbf{b}$.)

52. (calculus) Let A be an $n \times n$ matrix. Let $\mathbf{b}$ be an $n \times 1$ vector. If $A\mathbf{x} = \mathbf{b}$, show that when $\det A \neq 0$, $x_i = \det A_i/\det A$ is a continuous function of the entries of A and $\mathbf{b}$.

APPLICATIONS EXERCISES

53. Write out the equations for the network given in Figure 2.8. Perform an if-then analysis to determine how an increase in the resistance R_1 affects I_1, I_2, and I_3. Perform the same analysis for R_2 and R_3.

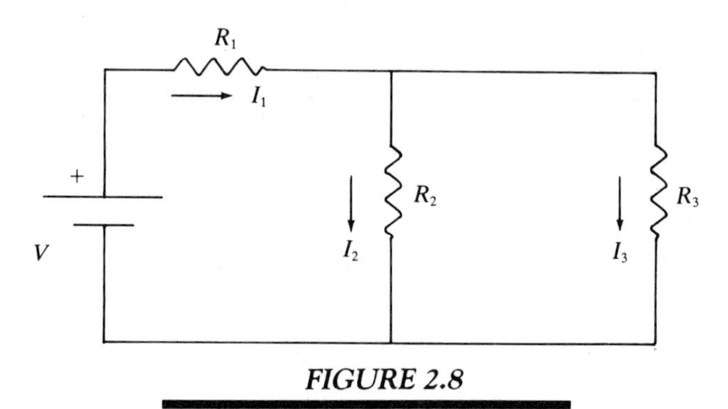

FIGURE 2.8

54. Suppose the supply and demand equations for a product are, respectively,

$$p - 2x = \varepsilon$$

and

$$p + 3x = 1$$

where ε depends on the cost of raw materials. Do an if-then analysis on this system of equations to determine how increases in ε affect price p and quantity x at equilibrium.

55. Consider springs having natural length l and spring constants k_1, k_2, and k_3, respectively, attached as shown in Figure 2.9. Let x_1 and x_2 be the positions of the particles of

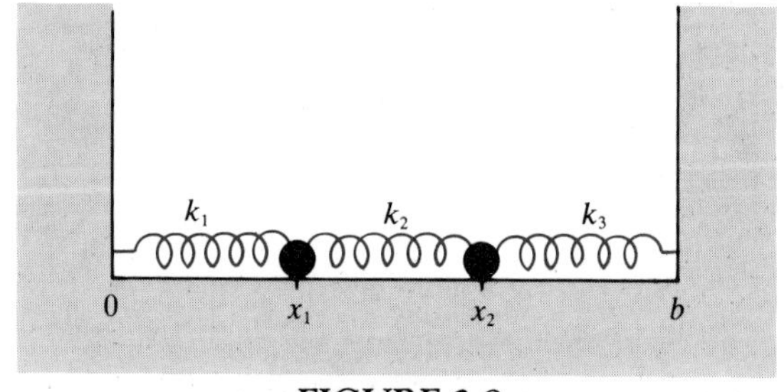

FIGURE 2.9

mass at equilibrium, with the springs attached to walls at 0 and b, as shown in the figure.

From Hooke's law, which states that the force exerted on a particle of mass by a spring is equal to the spring constant times the displacement of the spring from its equilibrium length, it follows that

$$k_1(x_1 - l) = k_2(x_2 - x_1 - l)$$

and $$k_2(x_2 - x_1 - l) = k_3(b - x_2 - l)$$

Find formulas for x_1 and x_2 in terms of the k's, b, and l. Using these, decide what happens to x_1 and x_2 if k_2 increases.

COMPUTER EXERCISES

56. Let A be a 4×4 invertible matrix. Find the number of multiplications and divisions needed to calculate A^{-1} using determinants. Count the number of multiplications and divisions required to calculate A^{-1} by the algorithm of Section 2.3. Compare these counts and decide which is the faster technique.

57. Let $A = \begin{bmatrix} 1 & 2 \\ 2 & 4.11 \end{bmatrix}$. Predict whether the calculations of the entries of A^{-1} are sensitive to roundoff. Check your prediction by using a two-digit computer to calculate A^{-1}.

CHAPTER THREE

VECTOR SPACES

Define $\mathbf{R}_n$ as the set of all $1 \times n$ matrices, or row vectors. From the previous chapter, we know that $\mathbf{R}_n$ satisfies the basic arithmetical properties listed below. In this list, for use later, we let $\mathbf{V} = \mathbf{R}_n$.

The set $\mathbf{V}$ of vectors satisfies each property listed below.

(*i*) An addition, written as $+$, is defined on $\mathbf{V}$. That is, for every pair of vectors $\mathbf{x}, \mathbf{y} \in \mathbf{V}$, there is a sum vector, denoted by $\mathbf{x} + \mathbf{y}$, in $\mathbf{V}$ (closure property of addition). Further, this addition satisfies the following properties used in computing.

1. For any vectors $\mathbf{x}$ and $\mathbf{y}$,

$$\mathbf{x} + \mathbf{y} = \mathbf{y} + \mathbf{x} \qquad \text{(commutative law)}$$

2. For any vectors $\mathbf{x}$, $\mathbf{y}$, and $\mathbf{z}$,

$$(\mathbf{x} + \mathbf{y}) + \mathbf{z} = \mathbf{x} + (\mathbf{y} + \mathbf{z}) \qquad \text{(associative law)}$$

3. There is a unique vector, written $\mathbf{0}$, such that for any vector $\mathbf{x}$,

$$\mathbf{x} + \mathbf{0} = \mathbf{0} + \mathbf{x} = \mathbf{x} \qquad \text{(additive identity)}$$

4. For any vector $\mathbf{x}$ there is a unique vector, written $-\mathbf{x}$, such that

$$\mathbf{x} + (-\mathbf{x}) = (-\mathbf{x}) + \mathbf{x} = \mathbf{0} \qquad \text{(additive inverse)}$$

(*ii*) A scalar multiplication, denoted by $\cdot$, is also defined on $\mathbf{V}$. That is, for every scalar $r \in \mathbf{R}$ and vector $\mathbf{x} \in \mathbf{V}$, there is a scalar product vector, written $r \cdot \mathbf{x}$, in $\mathbf{V}$* (closure property of scalar multiplication). Further, this scalar multiplication satisfies the following properties used in computing:

5. For any $r, s \in \mathbf{R}$ and any vector $\mathbf{x}$,

$$(r + s)\mathbf{x} = r\mathbf{x} + s\mathbf{x} \qquad \text{(distributive law)}$$

6. For any $r \in \mathbf{R}$ and any vectors $\mathbf{x}$ and $\mathbf{y}$,

$$r(\mathbf{x} + \mathbf{y}) = r\mathbf{x} + r\mathbf{y} \qquad \text{(distributive law)}$$

* For simplicity, we write $r \cdot \mathbf{x}$ as $r\mathbf{x}$ except when the dot is essential for clarity.

7. For any $r, s \in \mathbf{R}$ and any vector $\mathbf{x}$,

$$(rs)\mathbf{x} = r(s\mathbf{x}) \qquad \text{(associative law)}$$

8. For $1 \in \mathbf{R}$ and any vector $\mathbf{x}$,

$$1\mathbf{x} = \mathbf{x} \qquad \text{(scalar identity)}$$

Since the elements of $\mathbf{R}_n$ are called vectors, we will call $\mathbf{R}_n$ a vector space. More particularly, $\mathbf{R}_n$ is called the *Euclidean n-space* of row vectors.

Other sets also satisfy the arithmetical properties listed in (i) and (ii) and 1 through 8; that is, the same arithmetical properties hold for computations on these sets. For example, $\mathbf{R}^n$, the vector space of $n \times 1$ column vectors, called the *Euclidean n-space* of column vectors, and $\mathbf{R}_{m,n}$, the *matrix space* of $m \times n$ matrices, satisfy all these properties. In addition, the *polynomial space* of all polynomials and other such special sets of functions satisfy these properties. All of these spaces are important for the same reasons that the set of real numbers $\mathbf{R}$ and the set of ordered pairs of real numbers $\mathbf{R}_2$ are important. That is, when applied and geometric problems are formulated mathematically, the problems are described in these spaces. Thus, as we did with $\mathbf{R}$ and $\mathbf{R}_2$, we need to do a study of these spaces.

To study the various individual spaces, of course, would require a lot of time. However, we intend to eliminate that problem by doing the study for all of the spaces simultaneously. We will do this by defining a general space of objects, called vectors, having addition $(+)$ and scalar multiplication $(\cdot)$ such that all the arithmetical properties in (i), (ii), and 1 through 8 above are satisfied. We will study this general space, and then apply our results to any particular space that arises.

Although we intend to study spaces in general, it will often be helpful to view our work geometrically. Drawing sketches helps us obtain a geometrical feeling for our work. Therefore, in the first section we will show various ways of viewing $\mathbf{R}_3$ geometrically. This work will simultaneously provide us with geometric views for $\mathbf{R}_2$.

3.1 WAYS TO VIEW THE VECTOR SPACE R_3 GEOMETRICALLY

Since $\mathbf{R}^3$ and $\mathbf{R}_3$ are both sets containing triples of numbers, and since both have entrywise addition and scalar multiplication, we can see that these two vector spaces are the same except for the way in which the triples are recorded. Thus, it is clear that the results we obtain for $\mathbf{R}_3$ will also hold for $\mathbf{R}^3$.

In this section we give three ways to view $\mathbf{R}_3$ geometrically. These views will give us ways of envisioning, by sketches, the results obtained in the general study of vector spaces. We present this work in three parts, one for each geometric view. Parts I and III are used throughout the text; part II is optional.

I. $\mathbf{R}_3$ AS POINTS IN SPACE

In this part, we show how a vector in $\mathbf{R}_3$ can be viewed as a point in space. We assume that in previous studies you saw the development of a coordinate line and a coordinate plane. The development of a coordinate space extends that work.

Take three coordinate lines and place them in space so that they intersect at their origins and are mutually perpendicular, as in Figure 3.1. These lines, which we call the *x-, y-,* and *z-axes*, can be used to link points in space to triples of numbers as follows. Let P_1 be a point in space. Draw three planes through P_1 so that there is one plane perpendicular to each of the x-axis, the y-axis, and the z-axis. As shown in Figure 3.2, let these planes intersect the axes at the numbers x_1, y_1, z_1, respectively, which we call the *coordinates* of P_1. The set of all points in space together with their coordinates is called the *x,y,z-coordinate space.**

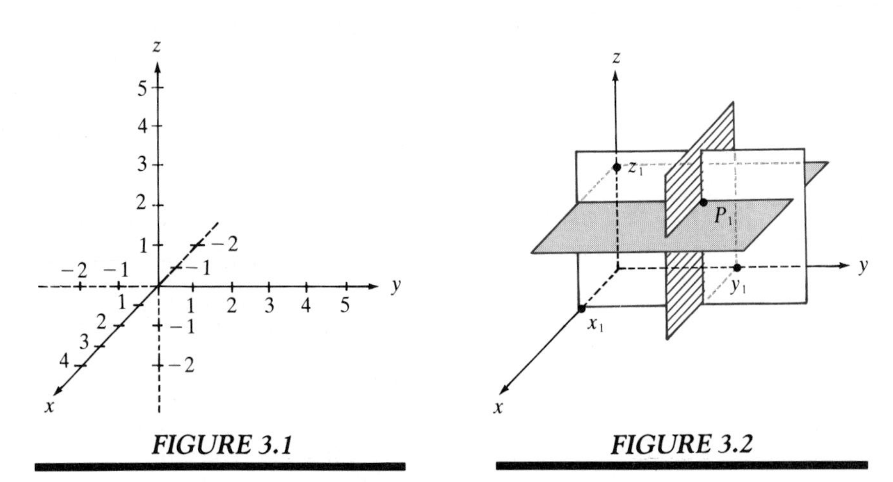

FIGURE 3.1

FIGURE 3.2

By writing the coordinates x_1, y_1, z_1 of P_1 as a row vector $[x_1, y_1, z_1]$, we can see that there is a one-to-one correspondence between vectors in $\mathbf{R}_3$ and points in space. Hence, we may identify the vector $[x_1, y_1, z_1] \in \mathbf{R}_3$ with the point P_1 in the x,y,z-coordinate space.

EXAMPLE 1

In Figure 3.3 we locate the points $[1, 2, 3]$, $[2, 2, 0]$, and $[1, 0, 2]$ in the given x,y,z-coordinate space. □

The plane containing the x- and y-axes in the x,y,z-coordinate space is an x,y-coordinate plane. Hence each point $[x, y, 0]$ in this plane can be written simply as $[x, y]$. Thus all of the work presented about the x,y,z-coordinate space and $\mathbf{R}_3$ can also be applied to the x,y-coordinate plane and $\mathbf{R}_2$.

* The letters x, y, and z indicate the axes in the space and thus identify the coordinate space.

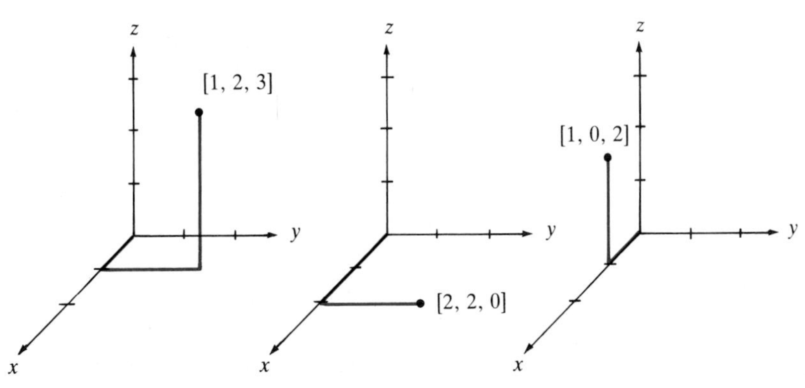

FIGURE 3.3

The study of coordinate spaces is like that of coordinate planes and can be found in most analytic geometry texts. For our work, it is sufficient in this part for us to show how to compute the distance from one point to another by using the corresponding coordinates. This can be accomplished by using the Pythagorean theorem.

The *distance* $d(\mathbf{x}, \mathbf{y})$ from a point $\mathbf{x} = [x_1, y_1, z_1]$ to a point $\mathbf{y} = [x_2, y_2, z_2]$ in the constructed x,y,z-coordinate space is

$$d(\mathbf{x}, \mathbf{y}) = \sqrt{(x_2 - x_1)^2 + (y_2 - y_1)^2 + (z_2 - z_1)^2}$$

EXAMPLE 2

The distance from the point $\mathbf{x} = [0, 2, 2]$ to the point $\mathbf{y} = [2, 0, 1]$ is

$$d(\mathbf{x}, \mathbf{y}) = \sqrt{(0 - 2)^2 + (2 - 0)^2 + (2 - 1)^2} = 3$$

as shown in Figure 3.4. □

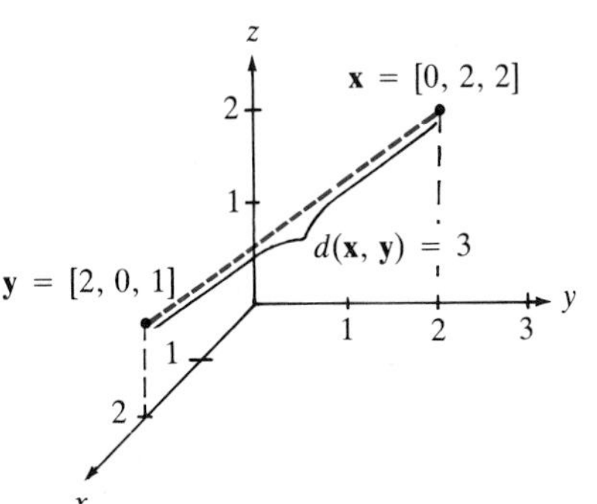

FIGURE 3.4

Since a vector in $\mathbf{R}_3$ can be identified with, and thus geometrically envisioned as, a point in a coordinate space, in the remainder of this text, when interpreting a vector as a point, we will use these two terms interchangeably.

II. $\mathbf{R}_3$ AS FREE ARROWS (OPTIONAL)

In this part, we show how to view a vector in $\mathbf{R}_3$ as a free arrow. By the term *free arrow* we mean a directed line segment with ends, called the *tail* and the

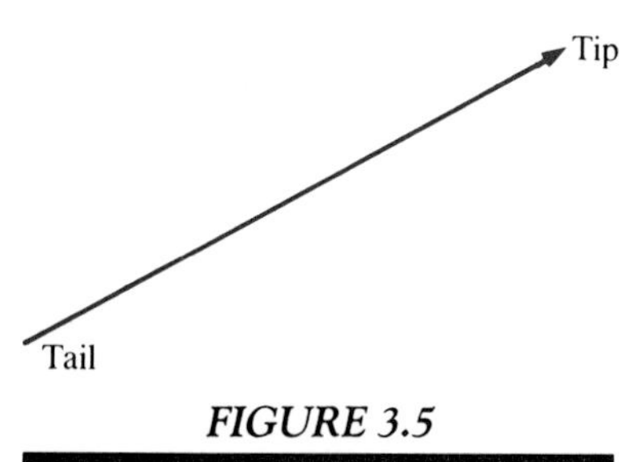

FIGURE 3.5

tip, such that the direction of the segment is from the tail to the tip, as shown in Figure 3.5. Any free arrow carries two pieces of information, its length and its direction. We allow a free arrow and duplicates of it to be placed anywhere in a given coordinate space, provided its direction and length are not changed. We will say that any two free arrows in the coordinate space are *equal* if they have the same length and direction.

Let us place a free arrow **a** in the given coordinate space so that its tail is at the origin and its tip is at the point $\mathbf{x} = [x_1, x_2, x_3] \in \mathbf{R}_3$. We will associate the vector $\mathbf{x} \in \mathbf{R}_3$ and the free arrow **a**, thus setting up a one-to-one correspondence between free arrows and vectors in $\mathbf{R}_3$. This allows us to visualize a vector as a free arrow.

EXAMPLE 3

The free arrows associated with [1, 2, 3], [1, 2, 0], and [1, 0, 2] can be viewed as shown in Figure 3.6. □

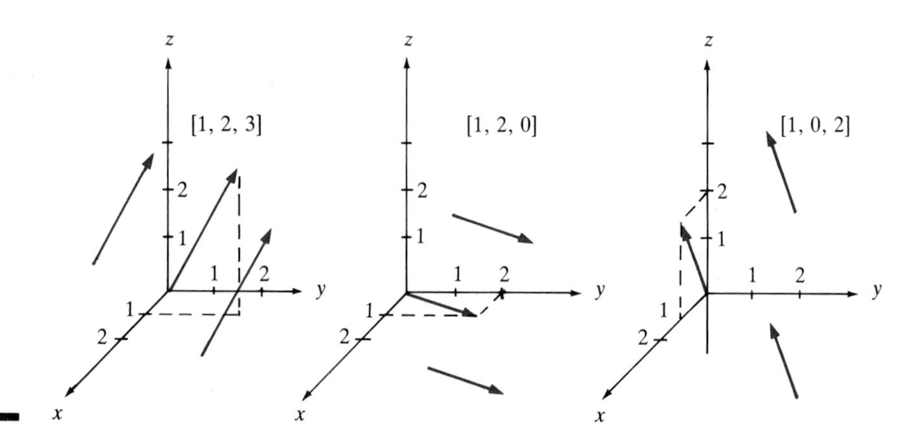

FIGURE 3.6

If the free arrow **a** is placed in the given coordinate space so that its tail and tip have coordinates $\mathbf{x} = [x_1, x_2, x_3]$ and $\mathbf{y} = [y_1, y_2, y_3]$, respectively, then it can be shown geometrically that the vector in $\mathbf{R}_3$ associated with **a** is

$$\mathbf{z} = \mathbf{y} - \mathbf{x} = [y_1 - x_1, y_2 - x_2, y_3 - x_3]$$

EXAMPLE 4

Suppose the free arrow **a** is placed in the coordinate space given in Figure 3.7 so that its tail has coordinates [1, 1, 0] and its tip has coordinates [1, 1, 2]. Then, as shown, the vector in $\mathbf{R}_3$ associated with **a** is

$$[1, 1, 2] - [1, 1, 0] = [1 - 1, 1 - 1, 2 - 0] = [0, 0, 2]$$ □

We now develop a geometric arithmetic for free arrows. To define the *sum* of two free arrows, say **a** and **b**, we place **a** anywhere in the given coordinate space. We then place **b** in the coordinate space so that its tail is at the tip of **a**, as seen in Figure 3.8. The sum of **a** and **b**, written **a** + **b**, is the free

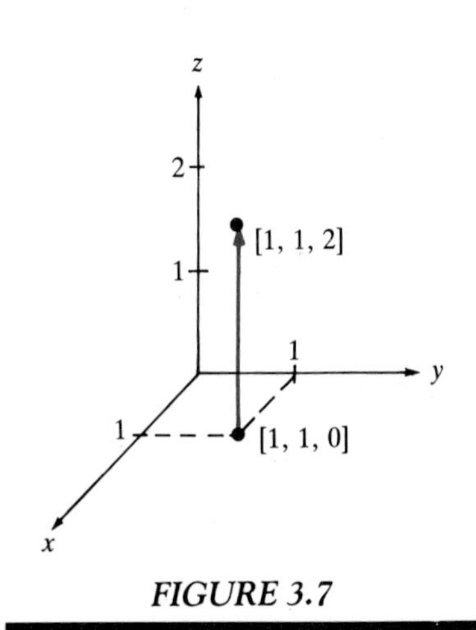

FIGURE 3.7

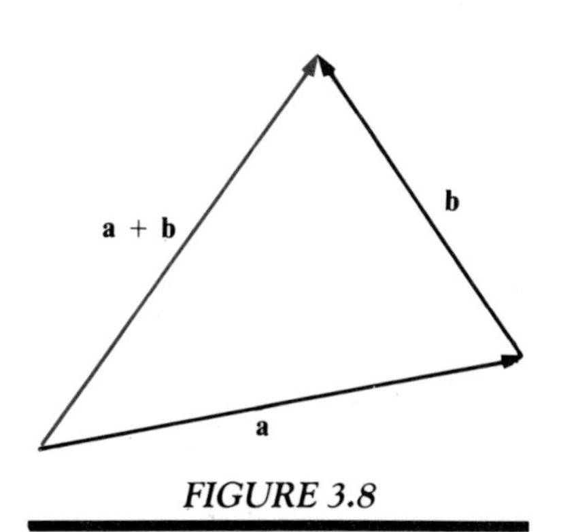

FIGURE 3.8

arrow whose tail is at the tail of **a** and whose tip is at the tip of **b**. It can be shown that this sum does not depend on the initial placement of **a** (exercise 49).

The sum of free arrows, though defined geometrically, can be calculated numerically by using the associated vectors as described below.

THEOREM 3.1

If free arrows **a** and **b** have associated vectors **x** and **y**, respectively, in $\mathbf{R}_3$, then **a** + **b** has associated vector **x** + **y**.

PROOF

Referring to Figure 3.9, place **a** with its tail at the origin and its tip at $\mathbf{x} = [x_1, x_2, x_3]$, and place **b** so that its tail is at the tip of **a**. Then the coordinates

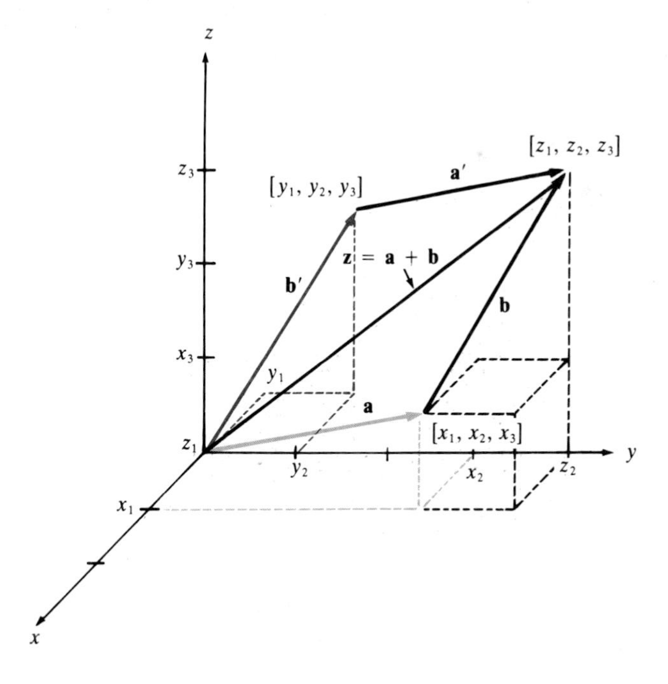

FIGURE 3.9

$[z_1, z_2, z_3]$ of the tip of **b** in this position are also the entries of the vector **z** of $\mathbf{R}_3$ associated with **a** + **b**. Now place a free arrow **b**′ equal to **b** with the tail of **b**′ at the origin. Then the tip of **b**′ is at the point $[y_1, y_2, y_3]$.

Drop a perpendicular from each of the points $[x_1, x_2, x_3]$, $[y_1, y_2, y_3]$, and $[z_1, z_2, z_3]$ to the x,y-plane, and drop a perpendicular from each of the bases of these perpendiculars to the x-axis. The resulting points have x-coordinates x_1, y_1, and z_1, respectively. But since **b** and **b**′ have the same length and direction, the directed distance along the x-axis from the origin to $[y_1, 0, 0]$ equals the directed distance from $[x_1, 0, 0]$ to $[z_1, 0, 0]$. Hence

$$y_1 = z_1 - x_1 \qquad \text{or} \qquad z_1 = x_1 + y_1$$

Similarly,

$$z_2 = x_2 + y_2 \qquad \text{and} \qquad z_3 = x_3 + y_3$$

■

A *scalar multiplication* can also be defined on free arrows. Let **a** be a free arrow and let r be a scalar. Then r**a** is the free arrow having length $|r| \cdot$ (length of **a**) and

(*i*) the same direction as **a** if $r \geqslant 0$, or

(*ii*) the direction opposite that of **a** if $r < 0$.

Examples of this definition are depicted in Figure 3.10.

FIGURE 3.10

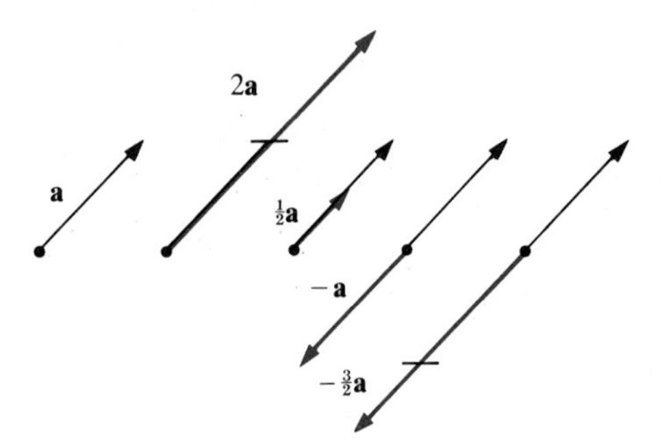

This multiplication, although defined geometrically, agrees with that of the associated vectors in $\mathbf{R}_3$.

THEOREM 3.2

Let **a** be a free arrow and let r be a scalar. Let **x** be the vector in $\mathbf{R}_3$ that is associated with **a**. Then $r\mathbf{x}$ is the vector in $\mathbf{R}_3$ associated with r**a**.

PROOF

See exercise 50. ■

As a consequence of these theorems, it follows that the set of free arrows, with + and · as defined above, satisfies the properties given in (i) and (ii) and 1 through 8 in the introduction to this chapter.

Some applications of the arithmetic of free arrows follow.

EXAMPLE 5

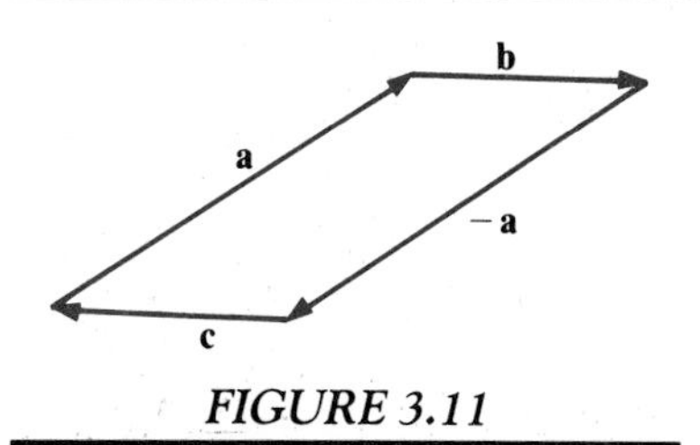

FIGURE 3.11

Show that, if two sides of a quadrilateral are parallel and have equal length, then the quadrilateral is a parallelogram.

Associate free arrows to the sides of the quadrilateral, as shown in Figure 3.11. Then $\mathbf{a} + \mathbf{b} - \mathbf{a} + \mathbf{c} = \mathbf{0}$. Using the arithmetic, it follows that $\mathbf{b} = -\mathbf{c}$. Hence these two sides are also parallel and are equal in length, and the quadrilateral is a parallelogram. □

By Theorems 3.1 and 3.2, we can see that the association between free arrows and vectors allows us to make numerical computations on free arrows by doing the computations on the associated vectors.

EXAMPLE 6

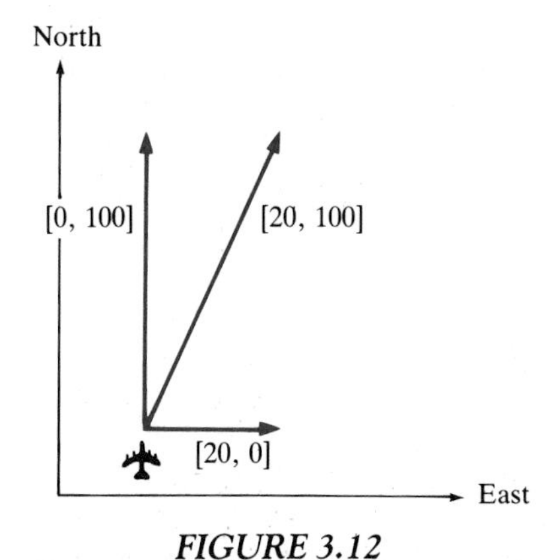

FIGURE 3.12

The velocity of an object moving in a given coordinate space may be specified with a free arrow by letting the free arrow point in the direction of motion and letting the length of the free arrow be the speed. It is known that velocities can be added as free arrows. Thus, if an airplane flies due north at 100 miles per hour in a crosswind blowing due east at 20 miles per hour, as shown in Figure 3.12, we can compute the resulting ground speed and direction of the airplane as follows. The vector $[0, 100]$ is associated with the free arrow representing the velocity of the airplane, and $[20, 0]$ with that representing the velocity of the wind. Hence

$$[0, 100] + [20, 0] = [20, 100]$$

is associated with the free arrow giving the ground direction of the airplane. The ground speed is the length of this vector; thus, by the Pythagorean theorem, the ground speed is

$$\text{the length of } [20, 100] = \sqrt{10400}$$

or approximately 102 miles per hour. □

EXAMPLE 7

EQUATION OF A LINE

In the coordinate space shown in Figure 3.13 is a line l through the point $[x_0, y_0, z_0]$ in the direction of the nonzero free arrow $[a, b, c]$. To find the equation of this line, let $[x, y, z]$ be any point on it. Then the free arrow

$$[x - x_0, y - y_0, z - z_0] = t[a, b, c]$$

for some scalar t. Conversely, if $[x, y, z]$ is a point that satisfies this equation for some t, it is on the line l. Hence, entrywise,

$$\begin{aligned} x - x_0 &= at \\ y - y_0 &= bt \\ z - z_0 &= ct \end{aligned}$$

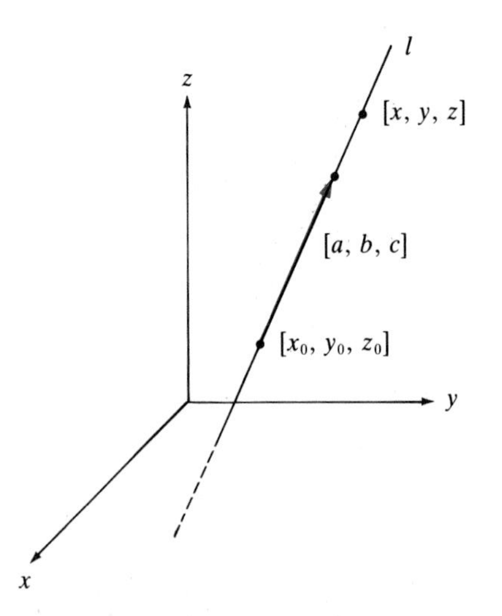

FIGURE 3.13

are the *parametric* equations of the line*. That is, if for each t we plot the corresponding point $[x, y, z]$, these points will yield the line l.

If none of $a, b,$ or c is 0, then by solving these equations for t and equating them we get

$$\frac{x - x_0}{a} = \frac{y - y_0}{b} = \frac{z - z_0}{c}$$

These equations are called the *symmetric equations* for the line. Of course, if $a = 0$ while neither of b or c is 0, the parametric equations reduce to

$$x = x_0 \qquad \text{and} \qquad \frac{y - y_0}{b} = \frac{z - z_0}{c}$$

The remaining cases, such as $b = 0$, yield similar sets of equations.

For example, to find the equation of a line through $[1, 2, 3]$ in the direction of $[1, -1, 0]$, we have

$$\begin{aligned} x - 1 &= t \\ y - 2 &= -t \\ z - 3 &= 0 \end{aligned}$$

as parametric equations. The symmetric equations are

$$\frac{x - 1}{1} = \frac{y - 2}{-1} \qquad \text{and} \qquad z = 3$$ □

Lengths of and angles between free arrows can also be computed by using the associated vectors. A concept useful in this regard is that of the norm of a vector. For any vector $\mathbf{x} \in \mathbf{R}_3$, we define the *norm* of $\mathbf{x}$, written

* "Parameter" is another word for variable. Since x, y, z are also variables, traditionally the independent variable t is called a parameter.

$||x||$, as

$$||\mathbf{x}|| = \sqrt{x_1^2 + x_2^2 + x_3^2}$$

where $\mathbf{x} = [x_1, x_2, x_3]$. Geometrically, the norm gives the length of the free arrow associated with $\mathbf{x}$ and is related to the distance between points, as given in Part I, since for any two points $\mathbf{x}$ and $\mathbf{y}$ we have

$$d(\mathbf{x}, \mathbf{y}) = ||\mathbf{x} - \mathbf{y}||$$

EXAMPLE 8

Let $\mathbf{a}$ be a free arrow, with tail having coordinates $[1, 0, -1]$ and tip having coordinates $[2, 1, 0]$. Then the associated vector is

$$\mathbf{x} = [2 - 1, 1 - 0, 0 - (-1)]$$

and

$$\text{length } \mathbf{a} = ||\mathbf{x}|| = \sqrt{(2-1)^2 + (1-0)^2 + [0-(-1)]^2} = \sqrt{3} \qquad \square$$

The norm satisfies the following properties, often used in computing. These properties are intuitive when the norm is considered as a length.

(*i*) $||\mathbf{x}|| \geqslant 0$ for all $\mathbf{x} \in \mathbf{R}_3$ with equality if and only if $\mathbf{x} = \mathbf{0}$

(*ii*) $||r\mathbf{x}|| = |r| \, ||\mathbf{x}||$ for all $r \in \mathbf{R}$ and $\mathbf{x} \in \mathbf{R}_3$

(*iii*) $||\mathbf{x} + \mathbf{y}|| \leqslant ||\mathbf{x}|| + ||\mathbf{y}||$ for all $\mathbf{x}, \mathbf{y} \in \mathbf{R}_3$

PROOF

Property (i) follows, since

$$||\mathbf{x}|| = \sqrt{x_1^2 + x_2^2 + x_3^2} \geqslant 0$$

for all $\mathbf{x}$ with equality if and only if $x_1 = x_2 = x_3 = 0$, or $\mathbf{x} = \mathbf{0}$.

For property (ii), let $\mathbf{x} = [x_1, x_2, x_3] \in \mathbf{R}_3$ and $r \in \mathbf{R}$. Then

$$||r\mathbf{x}|| = \sqrt{(rx_1)^2 + (rx_2)^2 + (rx_3)^2} = \sqrt{r^2(x_1^2 + x_2^2 + x_3^2)} = \sqrt{r^2}\sqrt{x_1^2 + x_2^2 + x_3^2} = |r| \, ||\mathbf{x}||$$

Property (iii) follows from the fact that

$$||\mathbf{x} + \mathbf{y}||^2 \leqslant (||\mathbf{x}|| + ||\mathbf{y}||)^2$$

which can be shown by expanding both sides into their entries and comparing. ∎

Angles between free arrows can be computed using the associated vectors as follows. Let $\mathbf{a}$ and $\mathbf{b}$ be two nonzero free arrows in the given coordinate space having their tails at the same point. Let

$$\mathbf{x} = [x_1, x_2, x_3] \qquad \text{and} \qquad \mathbf{y} = [y_1, y_2, y_3]$$

be the vectors associated with $\mathbf{a}$ and $\mathbf{b}$, respectively. Let θ, $0 \leqslant \theta \leqslant \pi$, be the angle between $\mathbf{a}$ and $\mathbf{b}$, as shown in Figure 3.14. Consider the triangle defined by these vectors, as given in Figure 3.15. Note that $\mathbf{a} + (-\mathbf{b}) = \mathbf{a} - \mathbf{b}$ is a free

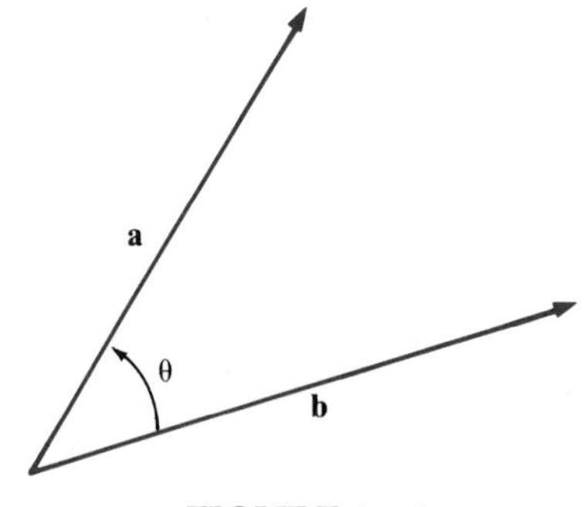

FIGURE 3.14

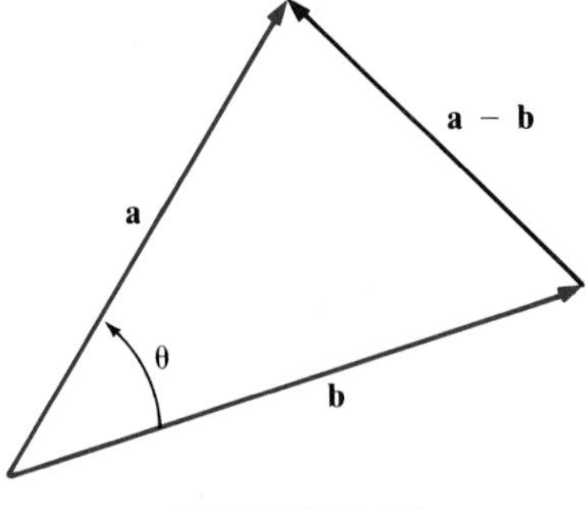

FIGURE 3.15

arrow which forms the third side of the triangle. Now applying the law of cosines* to this triangle yields

$$\|\mathbf{x} - \mathbf{y}\|^2 = \|\mathbf{x}\|^2 + \|\mathbf{y}\|^2 - 2\|\mathbf{x}\| \cdot \|\mathbf{y}\| \cos\theta$$

Thus,

$$(x_1 - y_1)^2 + (x_2 - y_2)^2 + (x_3 - y_3)^2$$
$$= x_1^2 + x_2^2 + x_3^2 + y_1^2 + y_2^2 + y_3^2 - 2\|\mathbf{x}\| \cdot \|\mathbf{y}\| \cos\theta$$

and solving for $\cos\theta$ yields

$$\cos\theta = \frac{x_1y_1 + x_2y_2 + x_3y_3}{\|\mathbf{x}\|\,\|\mathbf{y}\|}$$

To write the numerator of this fraction in compact notation, we define the *inner product*† of $\mathbf{x}$ and $\mathbf{y}$, written $(\mathbf{x}, \mathbf{y})$, as

$$(\mathbf{x}, \mathbf{y}) = x_1y_1 + x_2y_2 + x_3y_3$$

The properties for the inner product listed below, used in computing, can be obtained by direct calculation.

(*i*) $(\mathbf{x}, \mathbf{x}) \geqslant 0$ for all $\mathbf{x} \in \mathbf{R}_3$ with equality if and only if $\mathbf{x} = \mathbf{0}$ (positive property)

(*ii*) $(\mathbf{x}, \mathbf{y}) = (\mathbf{y}, \mathbf{x})$ for all $\mathbf{x}, \mathbf{y} \in \mathbf{R}_3$ (symmetric property)

(*iii*) $(r\mathbf{x}, \mathbf{y}) = (\mathbf{x}, r\mathbf{y}) = r(\mathbf{x}, \mathbf{y})$ for all $\mathbf{x}, \mathbf{y} \in \mathbf{R}_3$ and $r \in \mathbf{R}$ (homogeneous property)

(*iv*) $(\mathbf{x} + \mathbf{y}, \mathbf{z}) = (\mathbf{x}, \mathbf{z}) + (\mathbf{y}, \mathbf{z})$ and $(\mathbf{x}, \mathbf{y} + \mathbf{z}) = (\mathbf{x}, \mathbf{y}) + (\mathbf{x}, \mathbf{z})$ for all $\mathbf{x}, \mathbf{y}, \mathbf{z} \in \mathbf{R}_3$ (distributive property)

PROOF

Property (i) follows since

$$(\mathbf{x}, \mathbf{x}) = x_1^2 + x_2^2 + x_3^2 \geqslant 0$$

* The law of cosines, $\|\mathbf{a} - \mathbf{b}\|^2 = \|\mathbf{a}\|^2 + \|\mathbf{b}\|^2 - 2\|\mathbf{a}\|\,\|\mathbf{b}\|\cos\theta$, can be found in any trigonometry text.

† The inner product is sometimes called the *dot product* and is then written $\mathbf{x} \cdot \mathbf{y}$.

for all $\mathbf{x}$ with equality if and only if $\mathbf{x} = \mathbf{0}$. For property (ii), let $\mathbf{x}, \mathbf{y} \in \mathbf{R}_3$. Then, by definition,

$$(\mathbf{x}, \mathbf{y}) = x_1 y_1 + x_2 y_2 + x_3 y_3 = y_1 x_1 + y_2 x_2 + y_3 x_3 = (\mathbf{y}, \mathbf{x})$$

Property (iii) follows in the same way. Finally, property (iv) can be shown by expanding both sides and comparing terms. ■

Using the inner product notation, we can write

$$\cos \theta = \frac{(\mathbf{x}, \mathbf{y})}{\|\mathbf{x}\| \, \|\mathbf{y}\|}$$

Since $\theta = \pi/2$ if and only if $(\mathbf{x}, \mathbf{y}) = 0$, we note that **a** and **b** are orthogonal* if and only if the inner product of the associated nonzero vectors $\mathbf{x}$ and $\mathbf{y}$ is 0.

EXAMPLE 9

Place two free arrows in the coordinate space given in Figure 3.16 so that their tails are at [1, 1, 1] and their tips are at [1, 2, 2] and [1, 2, 1], respectively. Find the angle between these arrows.

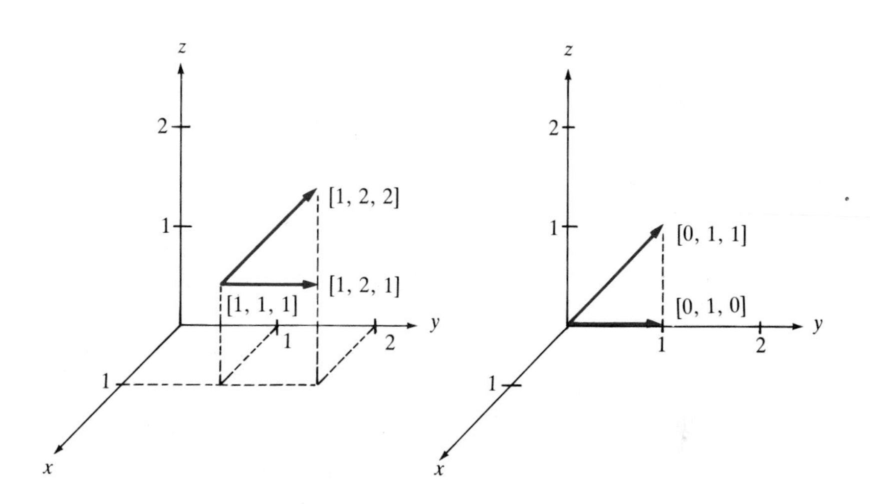

FIGURE 3.16

The free arrows have associated vectors

$$[1, 2, 2] - [1, 1, 1] = [0, 1, 1] \qquad \text{and} \qquad [1, 2, 1] - [1, 1, 1] = [0, 1, 0]$$

respectively, as shown. Now,

$$\cos \theta = \frac{([0, 1, 1], [0, 1, 0])}{\|[0, 1, 1]\| \, \|[0, 1, 0]\|} = \frac{\sqrt{2}}{2}$$

Thus $\theta = \pi/4$. □

* Another word for orthogonal is perpendicular.

EXAMPLE 10

Two free arrows are placed in the coordinate plane so that their tails are at $[1, 1]$ and their tips at $[2, 2]$ and $[2, 0]$, respectively, as shown in Figure 3.17. Decide if these two free arrows are orthogonal.

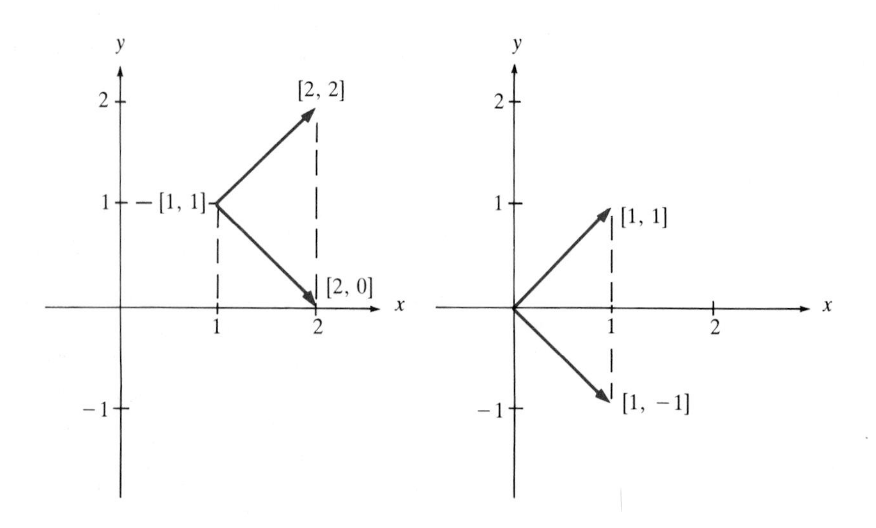

FIGURE 3.17

The associated vectors are

$$[2,2] - [1,1] = [1,1] \qquad \text{and} \qquad [2,0] - [1,1] = [1,-1]$$

respectively. Computing the inner product yields

$$([1,1],[1,-1]) = 1 - 1 = 0$$

Thus the free arrows are orthogonal. □

EXAMPLE 11

EQUATION OF A PLANE

The plane P through the point $[x_0, y_0, z_0]$ and orthogonal to the nonzero free arrow $[a, b, c]$ is shown in Figure 3.18. To find the equation of this plane, let $[x, y, z]$ be any point on it. Then the free arrow $[x - x_0, y - y_0, z - z_0]$ is orthogonal to $[a, b, c]$; that is,

$$([x - x_0, y - y_0, z - z_0], [a, b, c]) = 0$$

Conversely, if $[x, y, z]$ is a point that satisfies this equation, then it is on the plane. Thus, the equation of the plane in expanded form is

$$a(x - x_0) + b(y - y_0) + c(z - z_0) = 0$$

For example, the plane through the point $[1, 2, 3]$ and orthogonal to the free arrow [4, 5, 6] is

$$4(x - 1) + 5(y - 2) + 6(z - 3) = 0$$

or

$$4x + 5y + 6z = 32$$

□

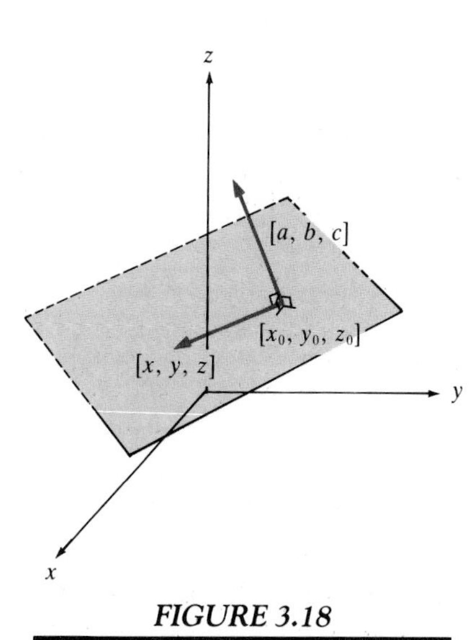

FIGURE 3.18

Since in this part we have shown that vectors in $\mathbf{R}_3$ can be visualized as free arrows, when interpreting vectors as free arrows, we will use these terms interchangeably, often referring to "angles between vectors" and "orthogonal vectors."

III. R_3 AS FIXED ARROWS

Throughout this chapter and the remainder of the text, we will use fixed arrows as a pedagogical tool to visualize the general results of vector spaces. The work in this part is most important for that reason.

Take any directed line segment (free arrow) and place it in a given coordinate space so that its tail is at the origin. A directed line segment in this position in the coordinate space is called a *fixed arrow*. If the tip of the fixed arrow is at the point $\mathbf{x} \in \mathbf{R}_3$, we will associate the vector $\mathbf{x}$ and the fixed arrow. Thus, there is a one-to-one correspondence between vectors in $\mathbf{R}_3$ and fixed arrows. This allows us to view vectors as fixed arrows and vice versa.

EXAMPLE 12

Some fixed arrows and associated vectors are shown in Figure 3.19. □

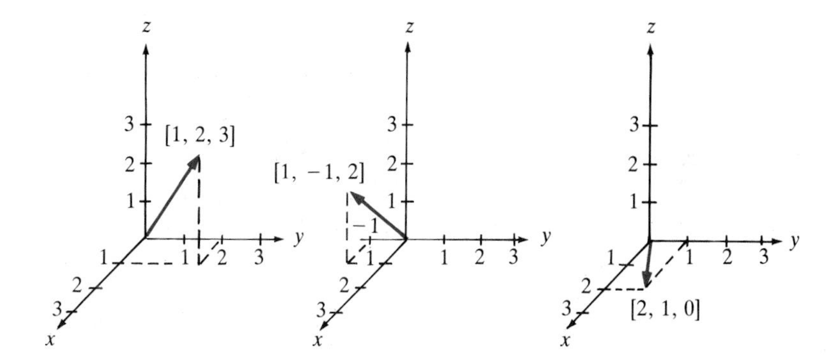

FIGURE 3.19

To develop the geometrical arithmetic of fixed arrows, we first define addition as follows. Let **a** and **b** be fixed arrows. Form the parallelogram determined by these fixed arrows as shown in Figure 3.20. Then $\mathbf{a} + \mathbf{b}$ is the fixed arrow along the diagonal of the parallelogram with its tip at the vertex opposite the origin, as shown in Figure 3.21.

The association between fixed arrows and vectors in $\mathbf{R}_3$ can be used to compute sums.

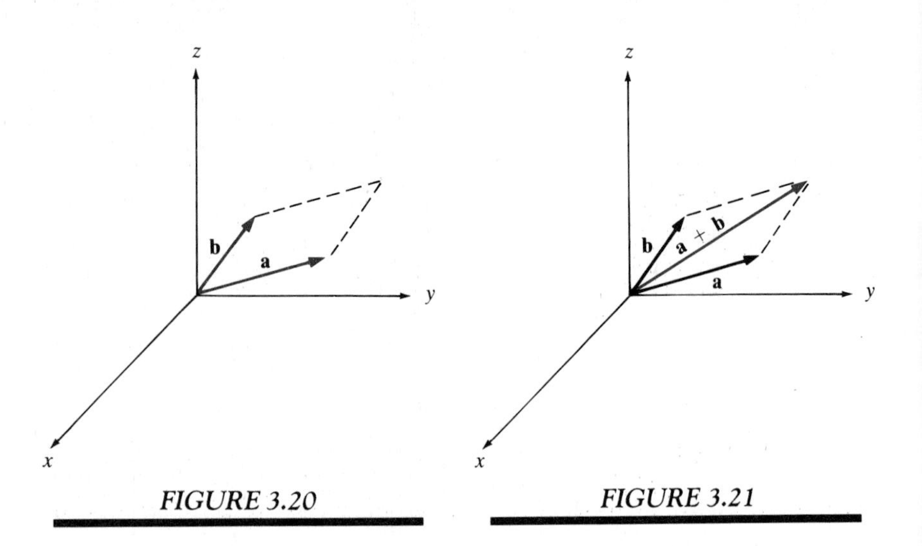

FIGURE 3.20

FIGURE 3.21

THEOREM 3.3

If **a** and **b** are fixed arrows with associated vectors **x** and **y**, respectively, then the fixed arrow **a** + **b** is associated with the vector **x** + **y**.

PROOF

As that of Theorem 3.1. ■

EXAMPLE 13

Fixed arrows and associated vectors demonstrating this theorem are shown in Figure 3.22. □

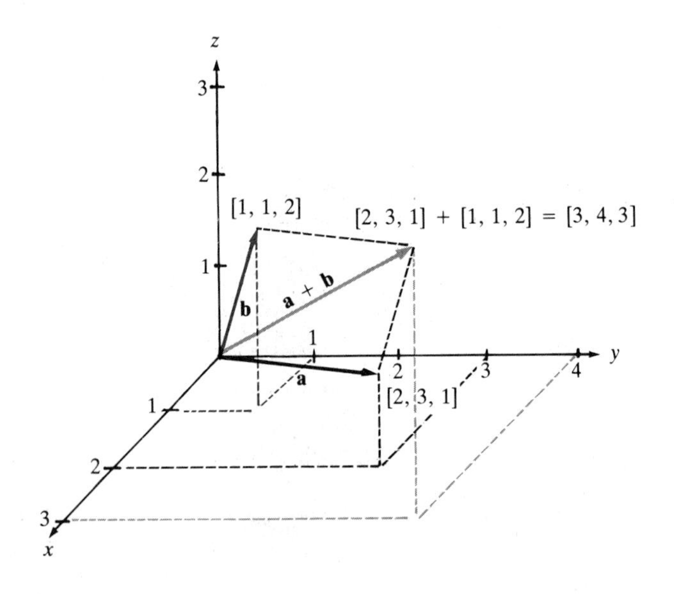

FIGURE 3.22

To define scalar multiplication for fixed arrows, let **a** be a fixed arrow and let r be a scalar. Define $r\mathbf{a}$ as the fixed arrow such that

(*i*) $r\mathbf{a}$ has length $|r| \cdot$ (length of **a**), and

(*ii*) $r\mathbf{a}$ has the same direction as **a** if $r > 0$ and the direction opposite that of **a** if $r < 0$.

As shown in Figure 3.23, this means that the scalar r in $r\mathbf{a}$ scales **a** by stretching, shrinking, and possibly reflecting **a**.

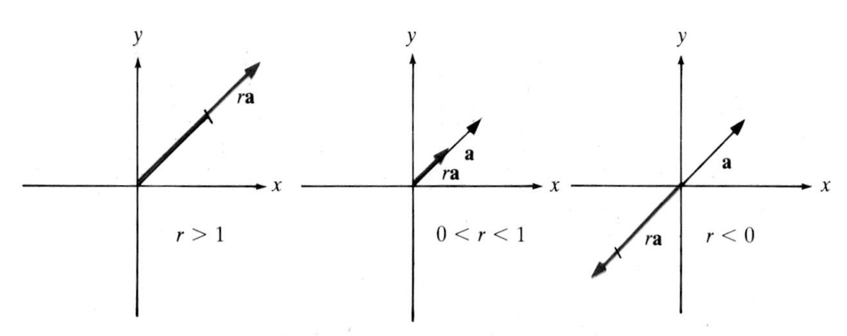

FIGURE 3.23

Again the association between fixed arrows and vectors in $\mathbf{R}_3$ can be used to compute scalar products.

THEOREM 3.4

If **a** is a fixed arrow with associated vector $\mathbf{x} \in \mathbf{R}_3$ and r is a scalar, then the fixed arrow $r\mathbf{a}$ has associated vector $r\mathbf{x}$.

PROOF

As that of Theorem 3.2 (exercise 50). ■

EXAMPLE 14

Fixed arrows and associated vectors demonstrating this theorem are shown in Figure 3.24. □

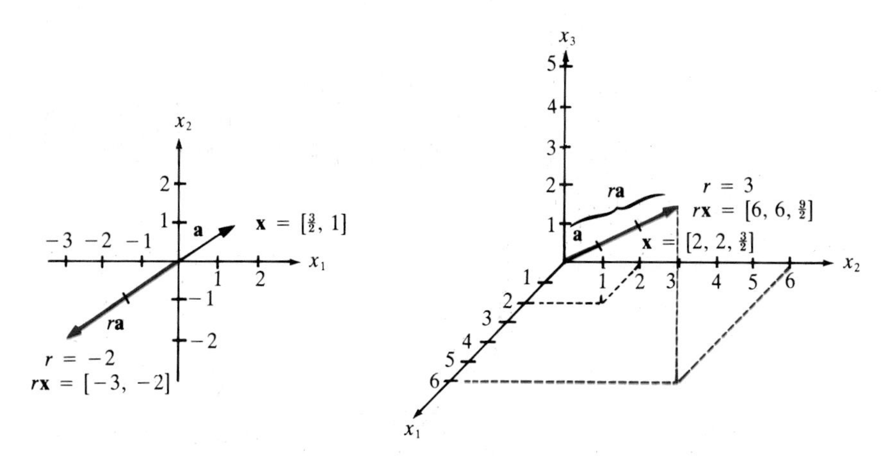

FIGURE 3.24

As a consequence of these Theorems 3.3 and 3.4, it follows that fixed arrows satisfy the arithmetic properties given in (i), (ii), and 1 through 8 in the introduction to this chapter and that computations involving fixed arrows can be done on the associated vectors. Further, computations involving vectors in $\mathbf{R}_3$ can be visualized by using the associated fixed arrows, as we will do throughout this chapter and the next.

EXAMPLE 15

In Figure 3.25, the sum

$$[0, 1, 2] + [0, 3, 1] = [0, 4, 3]$$

is depicted as the sum of associated fixed arrows using the parallelogram rule. □

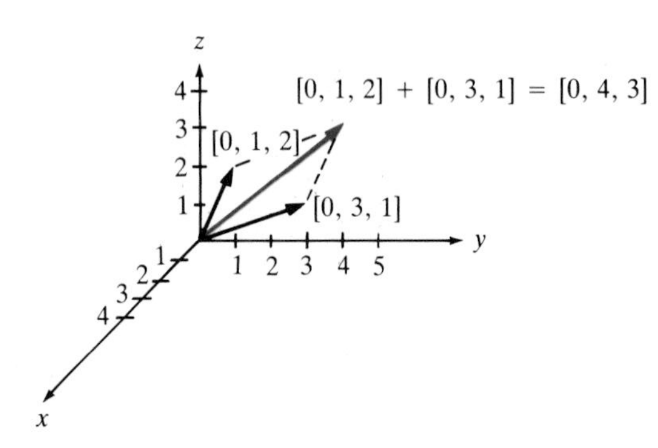

FIGURE 3.25

EXAMPLE 16

The product $2[1, 3, 2]$ can be visualized geometrically in terms of fixed arrows as a doubling of the fixed arrow $[1, 3, 2]$, as shown in Figure 3.26. □

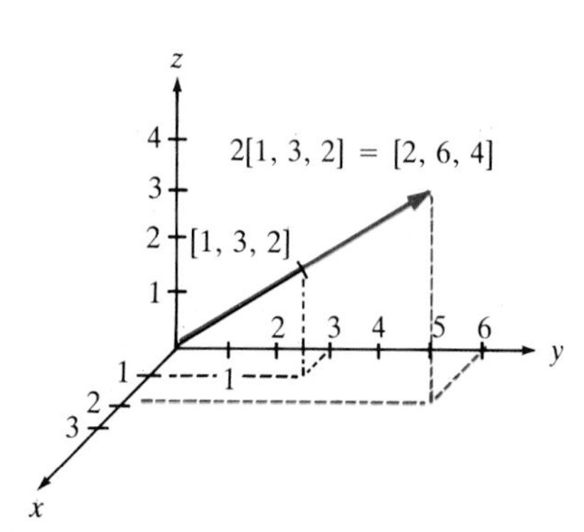

FIGURE 3.26

In a more general form, we have the following.

EXAMPLE 17

If $\mathbf{x}$ and $\mathbf{y}$ are vectors and r and s are scalars, the sum $r\mathbf{x} + s\mathbf{y}$ can be envisioned by scaling $\mathbf{x}$ by r, then scaling $\mathbf{y}$ by s, and finally using the parallelogram rule to obtain the sum, as shown in Figure 3.27. □

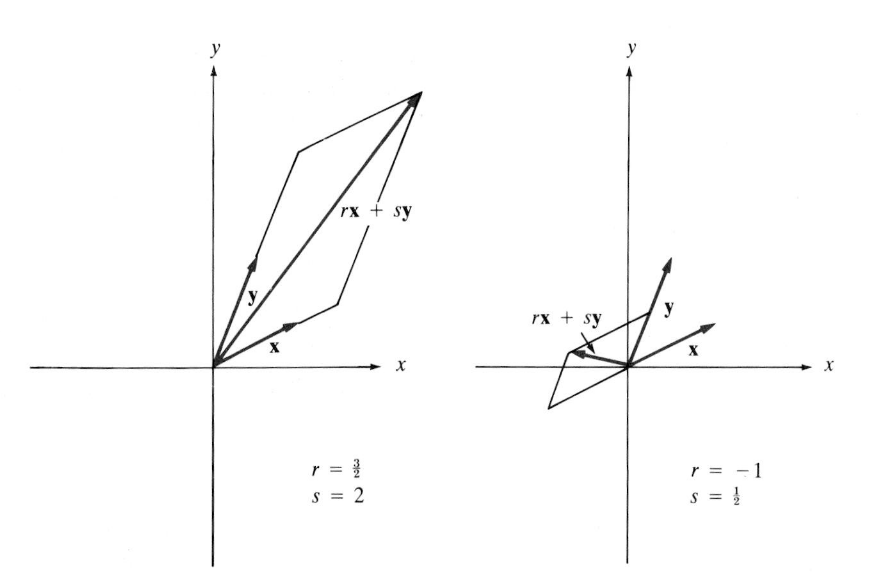

FIGURE 3.27

EXAMPLE 18

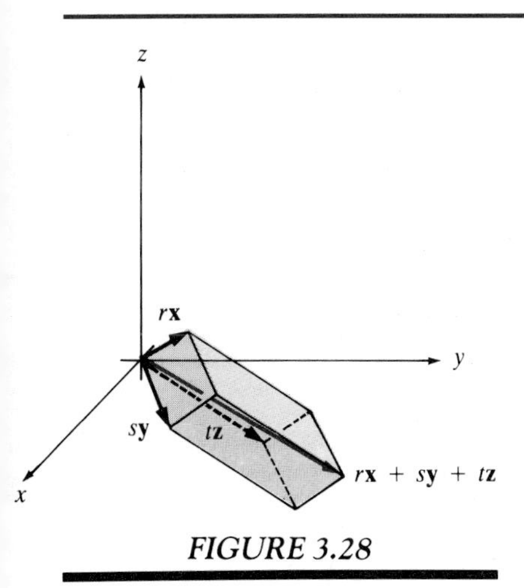

FIGURE 3.28

If $\mathbf{x}$, $\mathbf{y}$, and $\mathbf{z}$ are vectors and r, s, and t are scalars, the sum $r\mathbf{x} + s\mathbf{y} + t\mathbf{z}$ can be seen as follows. Scale $\mathbf{x}$ and $\mathbf{y}$ by r and s, respectively. Compute $r\mathbf{x} + s\mathbf{y}$ as shown in Example 17. Then scale $\mathbf{z}$ by t and add $t\mathbf{z}$ to $(r\mathbf{x} + s\mathbf{y})$ by the parallelogram rule, as shown in Figure 3.28. The net result is a fixed arrow that is a diagonal of the box formed by $r\mathbf{x}$, $s\mathbf{y}$, and $t\mathbf{z}$. □

For geometrical calculations with fixed arrows, we use, as in Part II, the *inner product*

$$(\mathbf{x}, \mathbf{y}) = x_1 y_1 + x_2 y_2 + x_3 y_3$$

for all $\mathbf{x}, \mathbf{y} \in \mathbf{R}_3$. Using the inner product, we have that the *norm* of a vector $\mathbf{x}$, or the length of the associated fixed arrow, is

$$\|\mathbf{x}\| = (\mathbf{x}, \mathbf{x})^{1/2} = \sqrt{x_1^2 + x_2^2 + x_3^2}$$

and that the angle θ, $0 \leqslant \theta \leqslant \pi$, between the nonzero fixed arrows having associated vectors $\mathbf{x}$ and $\mathbf{y}$ satisfies

$$\cos\theta = \frac{(\mathbf{x}, \mathbf{y})}{\|\mathbf{x}\|\,\|\mathbf{y}\|}$$

Thus, two such fixed arrows are orthogonal if and only if the inner product of their associated vectors is 0.

For the remainder of the text, we will use the set of fixed arrows as a model with which to geometrically view concepts and results about general vector spaces. This model provides us with a way of drawing sketches and picturing our general results, thus giving us a geometrical feeling for this study. When visualizing a vector as a fixed arrow, we will often use the two terms interchangeably.

EXERCISES FOR SECTION 3.1

COMPUTATIONAL EXERCISES

PART I

In exercises 1 through 6, sketch and find the distance between the given points.

1. [1, 1], [2, 3]

2. [−1, 2], [5, 4]

3. [1, 0, 1], [3, 2, −1]

4. [0, −1, 3], [1, 3, −2]

5. [0, 0, 0], [1, 1, 1]

6. [2, −2, 3], [1, 3, −4]

PART II

In exercises 7 through 12, sketch the described free arrows and give the vector associated with each.

7. From [−1, 2] to [3, −2]

8. From [4, −1] to [−1, −1]

9. From [5, 1] to [−2, −3]

10. From [3, 2] to [0, 0]

11. From [1, −1, 2] to [0, 0, 0]

12. From [2, 1, −1] to [3, 1, −1]

In exercises 13 through 16, a triangle in the coordinate plane is determined by three points. In each case, decide if the triangle is isosceles or equilateral, and/or if it is a right triangle. *Hint:* Use the vectors determined by these points.

13. [2, 3], [3, 3], [2, 4]

14. [−1, −3], [−1, −2], [2, −3]

15. [2, 2], [3, 2], $[5/2, 2 + \sqrt{3}/2]$

16. [−2, 4], [−3, 5], [−1, 5]

In exercises 17 and 18, decide whether the given points determine a square.

17. [0, 2], [2, 4], [4, 6], and [2, 0]

18. [2, 2], [2, 8], [5, 5], and [5, −1]

In exercises 19 and 20, write both parametric and symmetric equations of the line for the given data.

19. $[x_0, y_0, z_0] = [-1, 1, -1]$, $[a, b, c] = [2, -3, 1]$

20. $[x_0, y_0, z_0] = [1, 0, -1]$, $[a, b, c] = [1, 0, -1]$

In exercises 21 and 22, write out the equation of the plane for the given data.

21. $[x_0, y_0, z_0] = [1, 1, 1]$, $[a, b, c] = [1, 1, 1]$

22. $[x_0, y_0, z_0] = [2, -1, -3]$, $[a, b, c] = [1, 0, 2]$

PART III

In exercises 23 through 28, depict the given vectors and vector arithmetic geometrically using fixed arrows.

23. [1, 2], [2, 1], [1, 2] + [2, 1]

24. [1, 1], [−1, 1], 3[1, 1], −2[−1, 1], 3[1, 1] − 2[−1, 1]

25. [−1, 2], [2, 3], −1[−1, 2], −2[2, 3], −1[−1, 2] − 2[2, 3]

26. [1, 3], [3, 1], [1, 1], 2[1, 3] + [3, 1], 2[1, 3] + [3, 1] + 2[1, 1]. *Hint:* Compute from left to right.

27. [1, 1, 0], [0, 1, 1], 2[1, 1, 0], 3[0, 1, 1], 2[1, 1, 0] + 3[0, 1, 1]

28. [1, 0, 0], [0, 1, 0], [0, 0, 1], 2[1, 0, 0], 3[0, 1, 0], 4[0, 0, 1], 2[1, 0, 0] + 3[0, 1, 0] + 4[0, 0, 1]. *Hint:* Compute from left to right.

In exercises 29 through 32, shade the points representing the tips of fixed arrows $r_1\mathbf{x} + r_2\mathbf{y}$ for all

(*a*) r_1 and r_2 (*b*) $r_1 \geqslant 0, r_2 \geqslant 0$ (*c*) $r_1 \geqslant 0, r_2 \leqslant 0$
(*d*) $r_1 \leqslant 0, r_2 \geqslant 0$ (*e*) $r_1 \leqslant 0, r_2 \leqslant 0$

29.

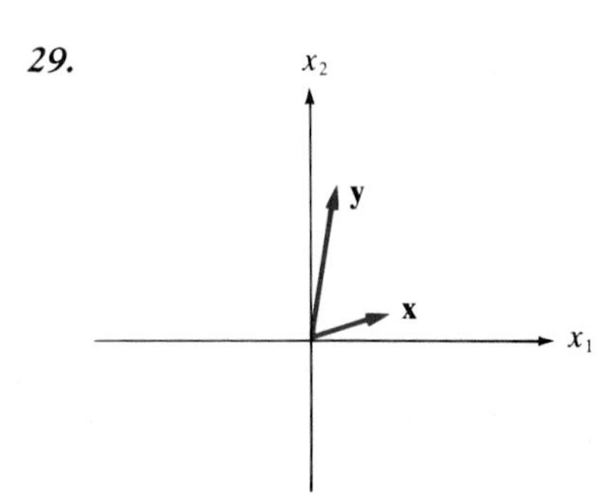

30.

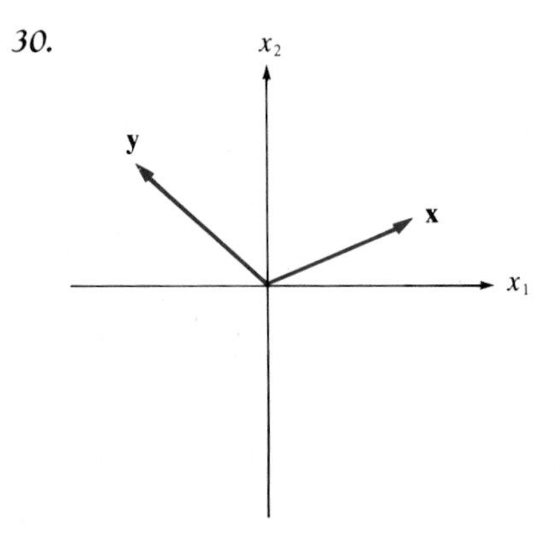

31.

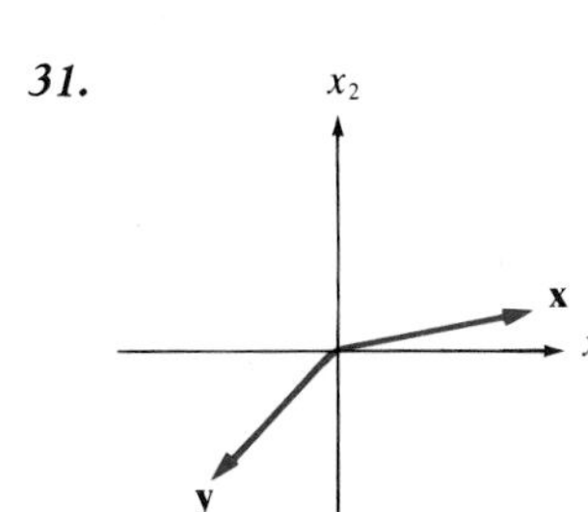

32.

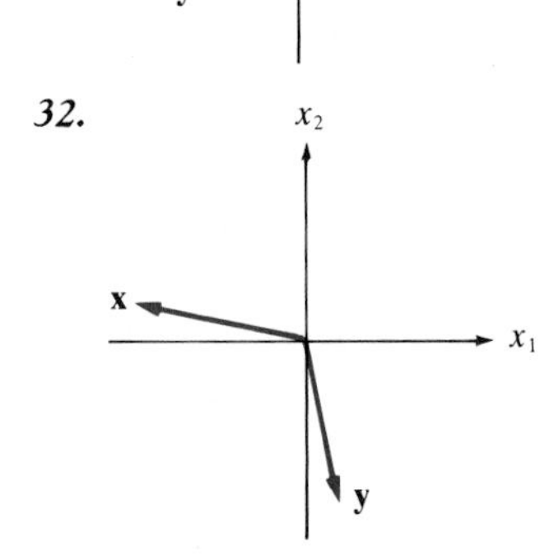

In exercises 33 and 34, compute the inner product of the vectors given.

33. (*a*) $[-1,2], [3,1]$
(*b*) $[1,-1,4], [-2,2,1]$
(*c*) $[-2,1,3], [0,0,0]$

34. (*a*) $[0,-1], [3,2]$
(*b*) $[3,1,-2], [-1,2,4]$
(*c*) $[0,0,0], [2,-1,3]$

In exercises 35 through 38, compute the norm of each vector given, graph the associated fixed arrow, and state its length.

35. (*a*) $[1,6]$
(*b*) $[-1,1]$

36. (*a*) $[-1,-2]$
(*b*) $[2,-1]$

37. (*a*) $[1,-1,2]$
(*b*) $[-2,0,5]$

38. (*a*) $[3,-1,1]$
(*b*) $[2,-2,1]$

In exercises 39 through 48, find the cosine of the angle θ between the given pair of vectors. Then use a calculator or table to find θ. Sketch the associated fixed arrows, indicating the angle calculated.

39. $[1,-3], [1,2]$

40. $[-4,1], [1,4]$

41. $[-2,1], [3,-2]$

42. $[-1,2], [3,5]$

43. $[\sqrt{2},\sqrt{3}], [\sqrt{2},\sqrt{5}]$

44. $[\sqrt{3},-\sqrt{7}], [\sqrt{12},2\sqrt{5}]$

45. $[1,3,2], [3,-5,6]$

46. $[-1,0,-1], [1,-1,2]$

47. $[2,0,-2], [0,3,3]$

48. $[0,3,-3], [-2,2,-1]$

THEORETICAL EXERCISES

49. Show geometrically that addition of the free arrows **a** and **b** does not depend on the initial placement of **a**.

50. Prove Theorem 3.2.

51. Sometimes in vector calculations it is important to find the component of a vector **x** along a second vector **y**, as shown in Figure 3.29. The *component* of **x** along **y** is the number $\|\mathbf{x}\| \cos \theta$. Show that the component can be calculated in terms of the coordinates of **x** and **y** by $(\mathbf{x}, \mathbf{y})/\|\mathbf{y}\|$, an easier calculation. Then, in the exercises below, compute the component of **x** along **y**. Sketch **x** and **y** and the component of **x** along **y**. Compare your computations with your sketch.
(*a*) $\mathbf{x} = [1,1], \mathbf{y} = [1,0]$
(*b*) $\mathbf{x} = [1,1], \mathbf{y} = [-1,0]$
(*c*) $\mathbf{x} = [1,2], \mathbf{y} = [1,1]$
(*d*) $\mathbf{x} = [1,2,1], \mathbf{y} = [0,2,5]$

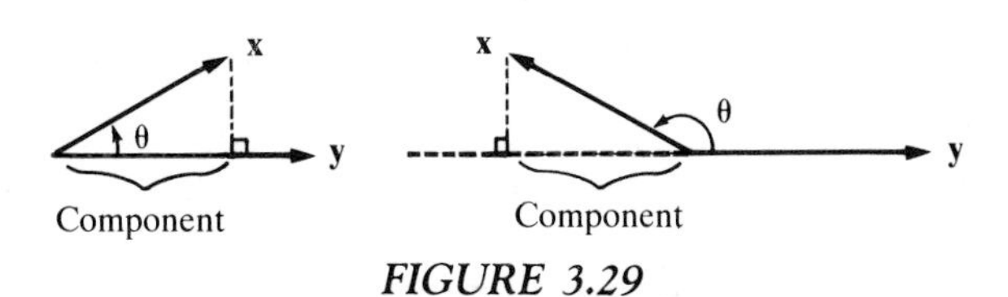

FIGURE 3.29

APPLICATIONS EXERCISES

52. Using free arrows, show that the diagonals of a parallelogram bisect each other.

53. Show that, in an equilateral triangle, a line drawn from a vertex perpendicular to the opposite side also bisects that side.

In exercises 54 through 56, the velocity of an object in the coordinate space may be specified with a free arrow by letting the free arrow point in the direction of motion and letting the length of the free arrow be the speed. It is known that velocities can be added as free arrows.

54. An airplane flies due north at 150 miles per hour in a wind blowing east at 10 miles per hour. Compute the direction and ground speed of the airplane.

55. An airplane flies in the direction $[1,1]$ at a speed of 140 miles per hour into a wind blowing in the direction $[-1,-2]$ at a speed of 10 miles per hour. Calculate the direction and ground speed of the airplane.

56. An airplane can fly at 250 miles per hour. The pilot wants to fly at this speed from point $[0,0]$ to point $[1,500]$, with coordinates measured in miles. For the flight there will be a crosswind blowing due east at 8 miles per hour. In what direction should the pilot fly? How long will it take to reach the destination? (This problem is a bit more realistic than exercises 54 and 55.)

For exercises 57 and 58, note that a force is a push or pull that can be measured in pounds. (If you weigh 150 pounds, the earth is pulling you toward it with a force of 150 pounds; if you push a car, the force is the number of pounds of push you apply to it.) A force can be represented with a free arrow whose direction is the direction of the force and whose length is the number of pounds of force applied. Like velocities, forces can be added as free vectors.

57. Apply to an object a force of 150 pounds in the direction $[1, 1]$ and another of 110 pounds in the direction $[1, -1]$. Find the resultant force (that is, the single force on the object that is equivalent to these two.)

58. Apply to an object a force of 11 pounds in the direction $[1, -1, 2]$ and another of 17 pounds in the direction $[2, -1, -1]$. Find the resultant force.

3.2 *Vector Spaces*

Placing a problem in a mathematical setting requires a space, such as one of those given in the introduction, and perhaps a function or an equation defined on that space. This space can be $\mathbf{R}^n$, $\mathbf{R}_n$, $\mathbf{R}_{m,n}$, a function space, etc. Our interest in this chapter is to study all of these spaces simultaneously; in following chapters, we will study functions and equations defined on them. To do this, we first define what we mean, in general, by a vector space.

Definition

A *vector space* is a nonempty set $\mathbf{V}$ of objects, called *vectors* or occasionally *points*, that satisfies the following description.*

(*i*) Defined on $\mathbf{V}$ there is an addition,† written $+$, such that each pair of vectors $\mathbf{x}, \mathbf{y} \in \mathbf{V}$ can be combined into a sum vector $\mathbf{x} + \mathbf{y} \in \mathbf{V}$ (closure property of addition). This addition satisfies the following properties.

1. For all $\mathbf{x}, \mathbf{y} \in \mathbf{V}$,

$$\mathbf{x} + \mathbf{y} = \mathbf{y} + \mathbf{x} \qquad \text{(commutative law)}$$

2. For all $\mathbf{x}, \mathbf{y}, \mathbf{z} \in \mathbf{V}$,

$$(\mathbf{x} + \mathbf{y}) + \mathbf{z} = \mathbf{x} + (\mathbf{y} + \mathbf{z}) \qquad \text{(associative law)}$$

3. There is a unique vector, written $\mathbf{0}$, in $\mathbf{V}$ such that for any $\mathbf{x} \in \mathbf{V}$,

$$\mathbf{x} + \mathbf{0} = \mathbf{0} + \mathbf{x} = \mathbf{x} \qquad \text{(additive identity)}$$

4. For any $\mathbf{x} \in \mathbf{V}$, there is a unique vector, written $-\mathbf{x}$, in $\mathbf{V}$ such that

$$\mathbf{x} + (-\mathbf{x}) = (-\mathbf{x}) + \mathbf{x} = \mathbf{0} \qquad \text{(additive inverse)}$$

* You may find these properties easier to remember if you associate them with the arithmetical properties for $+$ and $\cdot$ in $\mathbf{R}_3$, given in the introduction to this chapter.

† In cases where $+$ is traditionally used with a different meaning, we will use $\oplus$ for the vector space operation.

(*ii*) Also defined on $\mathbf{V}$ there is a scalar multiplication,* denoted by $\cdot$, such that each scalar $r \in \mathbf{R}$ can be combined with any vector $\mathbf{x} \in \mathbf{V}$ to form a scalar product vector $r \cdot \mathbf{x} \in \mathbf{V}$ (closure property of scalar multiplication). This scalar multiplication satisfies the following properties.

5. For all $r, s \in \mathbf{R}$ and all $\mathbf{x} \in \mathbf{V}$,

$$(r + s) \cdot \mathbf{x} = r \cdot \mathbf{x} + s \cdot \mathbf{x} \qquad \text{(distributive law)}$$

6. For all $r \in \mathbf{R}$ and all $\mathbf{x}, \mathbf{y} \in \mathbf{V}$,

$$r \cdot (\mathbf{x} + \mathbf{y}) = r \cdot \mathbf{x} + r \cdot \mathbf{y} \qquad \text{(distributive law)}$$

7. For all $r, s \in \mathbf{R}$ and all $\mathbf{x} \in \mathbf{V}$,

$$(rs) \cdot \mathbf{x} = r \cdot (s \cdot \mathbf{x}) \qquad \text{(associative law)}$$

8. For all $\mathbf{x} \in \mathbf{V}$,

$$1 \cdot \mathbf{x} = \mathbf{x} \qquad \text{(scalar identity)}$$

where $1 \in \mathbf{R}$.

This definition allows us to study vector spaces in general and then apply our general results to any particular vector space. To demonstrate how this can be done, we give a theorem describing some further basic arithmetical properties of any vector space.

THEOREM 3.5

Let $\mathbf{V}$ be a vector space. Then

(*i*) $0 \cdot \mathbf{x} = \mathbf{0}$ for all $\mathbf{x} \in \mathbf{V}$

(*ii*) $r \cdot \mathbf{0} = \mathbf{0}$ for all $r \in \mathbf{R}$

(*iii*) $(-1) \cdot \mathbf{x} = -\mathbf{x}$ for all $\mathbf{x} \in \mathbf{V}$

(*iv*) If $r \in \mathbf{R}$, $\mathbf{x} \in \mathbf{V}$, and $r \cdot \mathbf{x} = \mathbf{0}$, it follows that $r = 0$ or $\mathbf{x} = \mathbf{0}$.

(*v*) If $r_1, \ldots, r_m \in \mathbf{R}$ and $\mathbf{x}_1, \ldots, \mathbf{x}_m \in \mathbf{V}$, then†

$$\mathbf{x} = r_1 \cdot \mathbf{x}_1 + \cdots + r_m \cdot \mathbf{x}_m \in \mathbf{V}$$

(*vi*) The word "unique" in properties 3 and 4 of the definition of a vector space can be deleted, since uniqueness is assured by the remaining properties.

PROOF

For the proof of (i), note that if $\mathbf{x} \in \mathbf{V}$, then

$$0 \cdot \mathbf{x} = (0 + 0) \cdot \mathbf{x} = 0 \cdot \mathbf{x} + 0 \cdot \mathbf{x}$$

* In cases where $\cdot$ is traditionally used with a different meaning, we will use $\odot$ for the vector space operation. In cases where no confusion can arise, we will use juxtaposition in place of $\cdot$ for scalar multiplication.

† Recall that the closure properties only assure that two vectors or a scalar and a vector can be combined to form a vector in $\mathbf{V}$.

by the distributive property. Now adding $-(0 \cdot \mathbf{x})$ to both sides yields

$$\begin{aligned} 0 \cdot \mathbf{x} + [-(0 \cdot \mathbf{x})] &= (0 \cdot \mathbf{x} + 0 \cdot \mathbf{x}) + [-(0 \cdot \mathbf{x})] \\ &= 0 \cdot \mathbf{x} + (0 \cdot \mathbf{x} + [-(0 \cdot \mathbf{x})]) && \text{(associative property)} \\ &= 0 \cdot \mathbf{x} + \mathbf{0} && \text{(additive inverse)} \\ &= 0 \cdot \mathbf{x} && \text{(additive identity)} \end{aligned}$$

Thus, by the additive inverse property,

$$\mathbf{0} = 0 \cdot \mathbf{x}$$

We now prove (iii). Note that if $\mathbf{x} \in \mathbf{V}$, then

$$[1 + (-1)] \cdot \mathbf{x} = 0 \cdot \mathbf{x} = \mathbf{0}$$

Applying the distributive property and inverting the order of the equation, we have

$$\begin{aligned} \mathbf{0} &= 1 \cdot \mathbf{x} + (-1) \cdot \mathbf{x} \\ &= \mathbf{x} + (-1) \cdot \mathbf{x} && \text{(scalar identity property)} \end{aligned}$$

Now adding $-\mathbf{x}$ to both sides and using the additive identity property, we get

$$\begin{aligned} -\mathbf{x} &= (-\mathbf{x}) + [\mathbf{x} + (-1) \cdot \mathbf{x}] \\ &= [(-\mathbf{x}) + \mathbf{x}] + (-1) \cdot \mathbf{x} && \text{(associative property)} \\ &= \mathbf{0} + (-1) \cdot \mathbf{x} && \text{(additive inverse property)} \end{aligned}$$

Thus, using the additive identity property again yields

$$(-1) \cdot \mathbf{x} = -\mathbf{x}$$

Parts (ii), (iv), (v) and (vi) are left as exercise 28. ■

Notice that, in the statement of the theorem and its proof, no particular vector space is identified. The work was done in general and so it applies to all particular vector spaces.

We now begin a list of examples of particular vector spaces on which our general results can be applied. In these examples, note that the set **V** changes and that the rules for + and · change. However, the arithmetical properties used in computing do not change.

EXAMPLE 19

Let $\mathbf{V} = \{\mathbf{0}\}$ with $\mathbf{0} + \mathbf{0} = \mathbf{0}$ and $r\mathbf{0} = \mathbf{0}$ for all $r \in \mathbf{R}$. Then **V**, called the **0** *space*, is a vector space (exercise 29). □

EXAMPLE 20

The spaces $\mathbf{R}_n$, $\mathbf{R}^n$, and $\mathbf{R}_{m,n}$ were shown to be vector spaces in Chapter 2. The spaces of free arrows and fixed arrows were shown to be vector spaces in the previous section. □

Some vector spaces of functions follow.

EXAMPLE 21

Let $\mathbf{P}$ = the set of all polynomials

$$a_n t^n + a_{n-1} t^{n-1} + \cdots + a_1 t + a_0$$

such that n is any nonnegative integer, $a_i \in \mathbf{R}$ for all i, and t is a variable, with the usual $+$ and $\cdot$ for polynomials. We will show that $\mathbf{P}$ is a vector space by showing that it satisfies the definition of a vector space.

For the closure of addition property, let

$$p(t) = a_n t^n + \cdots + a_0 \qquad \text{and} \qquad q(t) = b_m t^m + \cdots + b_0$$

Suppose that $m \leqslant n$ and write

$$q(t) = b_n t^n + \cdots + b_0$$

where $b_n = \cdots = b_{m+1} = 0$. Then

$$(p + q)(t) = (a_n + b_n)t^n + \cdots + (a_0 + b_0) \in \mathbf{P}$$

For property 1, note that

$$p(t)+q(t)=(a_n+b_n)t^n+\cdots+(a_0+b_0)=(b_n+a_n)t^n+\cdots+(b_0+a_0)=q(t)+p(t)$$

Property 2 is shown similarly. Property 3 follows using the polynomial $z(t) = 0$, since

$$p(t) + z(t) = (a_n + 0)t^n + \cdots + (a_0 + 0) = p(t)$$

Finally, for property 4, let

$$-p(t) = -a_n t^n - \cdots - a_0 \in \mathbf{P}$$

Then

$$p(t) + [-p(t)] = z(t)$$

For the closure of scalar multiplication property, let $r \in \mathbf{R}$. Then

$$rp(t) = (ra_n)t^n + \cdots + (ra_0)$$

Hence $rp(t) \in \mathbf{P}$. For property 5, let $r, s \in \mathbf{R}$; then

$$\begin{aligned} r(sp)(t) &= r(sa_n t^n + \cdots + sa_0) = r(sa_n)t^n + \cdots + r(sa_0) \\ &= (rs)a_n t^n + \cdots + (rs)a_0 = (rs)(a_n t^n + \cdots + a_0) = (rs)p(t) \end{aligned}$$

Properties 6, 7, and 8 are proved similarly. Thus $\mathbf{P}$, called the *polynomial space*, is a vector space. □

EXAMPLE 22

Let $\mathbf{D} \subseteq \mathbf{R}$ and let $\mathbf{F(D)}$ be the set of all real-valued functions having domain $\mathbf{D}$, with the usual rules for $+$ and $\cdot$. For $f, g \in \mathbf{F(D)}$, define $f + g$ as the function that assigns to t the value $f(t) + g(t)$, that is,

$$(f + g)(t) = f(t) + g(t)$$

Further, for any $r \in \mathbf{R}$, define rf as

$$(rf)(t) = r[f(t)]$$

Then, using $z(t) = 0$ for all t as the zero for property 3 and $(-f)(t) = -f(t)$ for property 4, it can be shown that $\mathbf{F(D)}$, called the *function space*, is a vector space (exercise 30). The usual choices for $\mathbf{D}$ will be $\mathbf{R}$, the interval $[a, b]$, and the set of nonnegative integers $\mathbf{N} = \{0, 1, 2, \ldots\}$. □

EXAMPLE 23

Let

$$\mathbf{D} \subseteq \mathbf{R} \qquad \text{and} \qquad \mathbf{F}^n(\mathbf{D}) = \left\{ \begin{bmatrix} x_1(t) \\ \vdots \\ x_n(t) \end{bmatrix} : x_i(t) \in \mathbf{F(D)} \right\}$$

with addition and scalar multiplication defined entrywise as usual in matrices.* We will show that $\mathbf{F}^n(\mathbf{D})$ is a vector space.

To simplify notation, we write

$$\mathbf{x}(t) = \begin{bmatrix} x_1(t) \\ \vdots \\ x_n(t) \end{bmatrix}$$

For closure of addition, if $\mathbf{x}(t), \mathbf{y}(t) \in \mathbf{F}^n(\mathbf{D})$, then

$$\mathbf{x}(t) + \mathbf{y}(t) = \begin{bmatrix} x_1(t) + y_1(t) \\ \vdots \\ x_n(t) + y_n(t) \end{bmatrix} = \begin{bmatrix} (x_1 + y_1)(t) \\ \vdots \\ (x_n + y_n)(t) \end{bmatrix}$$

Since the sum of two functions is a function, it follows that

$$\mathbf{x}(t) + \mathbf{y}(t) \in \mathbf{F}^n(\mathbf{D})$$

* As is common in mathematics, in this text we often denote a function either by f or, when we feel it is helpful to indicate the variable, by $f(t)$.

For property 1, if $\mathbf{x}(t), \mathbf{y}(t) \in \mathbf{F}^n(\mathbf{D})$, then

$$\mathbf{x}(t) + \mathbf{y}(t) = \begin{bmatrix} x_1(t) + y_1(t) \\ \vdots \\ x_n(t) + y_n(t) \end{bmatrix} = \begin{bmatrix} y_1(t) + x_1(t) \\ \vdots \\ y_n(t) + x_n(t) \end{bmatrix} = \mathbf{y}(t) + \mathbf{x}(t)$$

Property 2 is shown similarly. For property 3, let

$$\mathbf{z}(t) = \begin{bmatrix} 0 \\ \vdots \\ 0 \end{bmatrix}$$

for all t. Then $\mathbf{z}(t) \in \mathbf{F}^n(\mathbf{D})$ and

$$\mathbf{x}(t) + \mathbf{z}(t) = \mathbf{x}(t)$$

for all $\mathbf{x}(t) \in \mathbf{F}^n(\mathbf{D})$. For property 4, if $\mathbf{x}(t) \in \mathbf{F}^n(\mathbf{D})$, define

$$-\mathbf{x}(t) = \begin{bmatrix} -x_1(t) \\ \vdots \\ -x_n(t) \end{bmatrix}$$

Then

$$-\mathbf{x}(t) \in \mathbf{F}^n(\mathbf{D}) \qquad \text{and} \qquad \mathbf{x}(t) + [-\mathbf{x}(t)] = \mathbf{z}(t)$$

For scalar multiplication, let $r \in \mathbf{R}$ and $\mathbf{x}(t) \in \mathbf{F}^n(\mathbf{D})$. Then

$$r\mathbf{x}(t) = \begin{bmatrix} rx_1(t) \\ \vdots \\ rx_n(t) \end{bmatrix} = \begin{bmatrix} (rx_1)(t) \\ \vdots \\ (rx_n)(t) \end{bmatrix}$$

Since a scalar multiple of a function is a function, it follows that $r\mathbf{x}(t) \in \mathbf{F}^n(\mathbf{D})$.

For property 5 we have, for $r, s \in \mathbf{R}$ and $\mathbf{x}(t) \in \mathbf{F}^n(\mathbf{D})$,

$$(r + s)\mathbf{x}(t) = \begin{bmatrix} (r + s)x_1(t) \\ \vdots \\ (r + s)x_n(t) \end{bmatrix} = \begin{bmatrix} rx_1(t) + sx_1(t) \\ \vdots \\ rx_n(t) + sx_n(t) \end{bmatrix}$$

$$= \begin{bmatrix} rx_1(t) \\ \vdots \\ rx_n(t) \end{bmatrix} + \begin{bmatrix} sx_1(t) \\ \vdots \\ sx_n(t) \end{bmatrix} = r\mathbf{x}(t) + s\mathbf{x}(t)$$

Properties 6, 7, and 8 can be shown similarly, so $\mathbf{F}^n(\mathbf{D})$, called the *n-function space*, is a vector space. The usual choices for $\mathbf{D}$ are $\mathbf{R}$, the interval $[a, b]$, and the set $\mathbf{N}$ of nonnegative integers. □

Some vector spaces arise inside of others. More specifically, let $\mathbf{V}$ be a vector space and $\mathbf{W}$ a nonempty subset of $\mathbf{V}$. We will say that $\mathbf{W}$ is a *subspace* of $\mathbf{V}$ if $\mathbf{W}$ satisfies the definition of a vector space using the same $+$ and $\cdot$ as for $\mathbf{V}$. Thus, a subspace is itself a vector space.

In deciding if a nonempty subset of a vector space is a subspace, you can make use of the fact that Properties 1, 2, 5, 6, 7, and 8 will always hold true since they are arithmetical properties for $+$ and $\cdot$ of the vector space and thus hold for any subset of the vector space. In fact, we can show that none of the arithmetical properties 1 through 8 need be checked.

THEOREM 3.6

(test for subspaces) Let $\mathbf{V}$ be a vector space. A subset $\mathbf{W} \subseteq \mathbf{V}$ is a subspace of $\mathbf{V}$ if and only if

(*i*) $\mathbf{W}$ is nonempty,

(*ii*) for all $\mathbf{x}, \mathbf{y} \in \mathbf{W}$, it follows that $\mathbf{x} + \mathbf{y} \in \mathbf{W}$ (closure of addition property), and

(*iii*) for all $r \in \mathbf{R}$ and $\mathbf{x} \in \mathbf{W}$, it follows that $r\mathbf{x} \in \mathbf{W}$ (closure of scalar multiplication property).

PROOF

If $\mathbf{W}$ is a subspace of $\mathbf{V}$, then since a subspace is a vector space and a vector space is nonempty and satisfies (ii) and (iii), it follows that $\mathbf{W}$ satisfies (i), (ii), and (iii). Thus we need only show that if $\mathbf{W}$ is a nonempty subset of $\mathbf{V}$ satisfying (ii) and (iii), then $\mathbf{W}$ is a subspace.

The closure of addition property of a vector space is satisfied, since it is (ii). Thus, we proceed to show that properties 1 through 4 hold. Properties 1 and 2 follow as usual, since these are arithmetical properties of $+$ for every set of vectors in the vector space $\mathbf{V}$. To show property 3, take any $\mathbf{x} \in \mathbf{W}$. By (iii), $0\mathbf{x} \in \mathbf{W}$. But by Theorem 3.5, $0\mathbf{x} = \mathbf{0}$, so $\mathbf{0} \in \mathbf{W}$. Finally, to show property 4, let $\mathbf{x} \in \mathbf{W}$. Then by (iii), $(-1)\mathbf{x} \in \mathbf{W}$. By Theorem 3.5, $(-1)\mathbf{x} = -\mathbf{x}$, so $-\mathbf{x} \in \mathbf{W}$ and property 4 is satisfied.

The closure of scalar multiplication property of a vector space holds for $\mathbf{W}$ by (iii). Thus we need only check properties 5 through 8. But these properties are arithmetical properties of $+$ and $\cdot$ for all subsets of the vector space $\mathbf{V}$. Hence, $\mathbf{W}$ is a vector space, and so it is a subspace of $\mathbf{V}$. ■

This theorem gives us an easy way to check if a subset of a vector space is a subspace and consequently a vector space. (See exercise 27.)

EXAMPLE 24

Let

$$\mathbf{W} = \left\{ \begin{bmatrix} r \\ s \\ t \end{bmatrix} : r, s, t \in \mathbf{R} \text{ and } r + s + t = 0 \right\}$$

We will show that $\mathbf{W}$ is a vector space by showing that $\mathbf{W}$ is a subspace of $\mathbf{R}^3$.

First note that $\begin{bmatrix} 0 \\ 0 \\ 0 \end{bmatrix} \in \mathbf{W}$, so $\mathbf{W}$ is not empty. Let

$$\mathbf{x} = \begin{bmatrix} x_1 \\ x_2 \\ x_3 \end{bmatrix} \quad \text{and} \quad \mathbf{y} = \begin{bmatrix} y_1 \\ y_2 \\ y_3 \end{bmatrix}$$

be in $\mathbf{W}$. To check the closure of addition property, note that

$$\mathbf{x} + \mathbf{y} = \begin{bmatrix} x_1 + y_1 \\ x_2 + y_2 \\ x_3 + y_3 \end{bmatrix}$$

Since

$$(x_1 + y_1) + (x_2 + y_2) + (x_3 + y_3) = (x_1 + x_2 + x_3) + (y_1 + y_2 + y_3) = 0 + 0 = 0$$

we know that $\mathbf{x} + \mathbf{y} \in \mathbf{W}$.

Finally, we check the closure of scalar multiplication property. Let $r \in \mathbf{R}$; then

$$r\mathbf{x} = \begin{bmatrix} rx_1 \\ rx_2 \\ rx_3 \end{bmatrix}$$

Since

$$(rx_1) + (rx_2) + (rx_3) = r(x_1 + x_2 + x_3) = r0 = 0$$

we see that $r\mathbf{x} \in \mathbf{W}$. By the theorem, $\mathbf{W}$ is a subspace of $\mathbf{R}^3$. □

EXAMPLE 25

Let n be a positive integer and let

$$\mathbf{P}_n = \{a_n t^n + \cdots + a_0 : a_i \in \mathbf{R} \text{ for all } i \text{ and } t \text{ is a variable}\}$$

Then $\mathbf{P}_n \subseteq \mathbf{P}$. To show that $\mathbf{P}_n$ is a vector space, we will show that it is a subspace of $\mathbf{P}$.

Since $0 \in \mathbf{P}_n$, $\mathbf{P}_n \neq \varnothing$. Further, since the sum of two polynomials of degree n or less is a polynomial of degree n or less, the closure of addition property holds. Finally, since a scalar times a polynomial in $\mathbf{P}_n$ yields a polynomial in $\mathbf{P}_n$, the closure of scalar multiplication property holds. Thus, by the test for subspaces, $\mathbf{P}_n$, called the *n-degree polynomial space*, is a vector space. □

EXAMPLE 26

CALCULUS

Let $\mathbf{D} \subseteq \mathbf{R}$ and let $\mathbf{C}(\mathbf{D}) \subseteq \mathbf{F}(\mathbf{D})$ be the set of all continuous functions having domain $\mathbf{D}$, with the usual $+$ and $\cdot$ for functions. Since the sum of continuous functions is continuous and a scalar times a continuous function is continuous, it follows from the test for subspaces that $\mathbf{C}(\mathbf{D})$, called the *continuous function space*, is a vector space. Similarly, $\mathbf{C}_1(\mathbf{D})$, the set of all differentiable functions having domain $\mathbf{D}$, with the usual $+$ and $\cdot$ for functions—called the *differentiable function space*—is a vector space. The usual choices for $\mathbf{D}$ are $\mathbf{R}$ and intervals (a, b). □

In the next section we show some special vector spaces.

EXERCISES FOR SECTION 3.2

COMPUTATIONAL EXERCISES

Let $\mathbf{V}$ be a vector space and let $\mathbf{x}, \mathbf{y}, \mathbf{z} \in \mathbf{V}$. In exercises 1 through 4, simplify the given expression, stating the arithmetical properties that you use. (Recall that multiplication is done before addition and subtraction. Otherwise you work from left to right.)

1. $3\mathbf{x} - 2\mathbf{y} + 5(\mathbf{y} - 2\mathbf{x})$

2. $\mathbf{x} + \mathbf{y} - 2\mathbf{x} + 4(\mathbf{x} - 3\mathbf{y})$

3. $\mathbf{x} + \mathbf{y} - \mathbf{z} - 4\mathbf{x} + 2(\mathbf{z} - \mathbf{y})$

4. $-(-\mathbf{x})$. *Hint:* $-\mathbf{x} = (-1)\mathbf{x}$

Let $\mathbf{V}$ be a vector space and $\mathbf{x}, \mathbf{a}, \mathbf{b} \in \mathbf{V}$. In exercises 5 through 8, solve the given equation for $\mathbf{x}$, stating the arithmetical properties you use.

5. $3(\mathbf{x} + \mathbf{a}) - 2(\mathbf{b} - 2\mathbf{a}) = \mathbf{a}$

6. $4[\mathbf{a} - (\mathbf{x} + \mathbf{b})] + 3\mathbf{a} - \mathbf{x} = \mathbf{0}$

7. $\mathbf{a} - 3(\mathbf{x} - \mathbf{b}) = \mathbf{b} - 4(\mathbf{a} - \mathbf{x})$

8. $2[(\mathbf{x} - 2\mathbf{a}) - 3(\mathbf{x} - 2\mathbf{b})] + 2(\mathbf{a} + \mathbf{b})$
$= 4(\mathbf{a} + \mathbf{b} + \mathbf{x}) - 2[\mathbf{a} + 3(\mathbf{x} - \mathbf{b})]$

In exercises 9 through 16, show which vector space properties fail for the given set and show that all other vector space properties are satisfied.

9. $\mathbf{V} = \{[x, y, x + y - 1] : x, y \in \mathbf{R}\}$ with the usual operations in $\mathbf{R}_3$

10. $\mathbf{V} = \{r[1, 2] + s[2, 1] : r \geqslant 0, s \geqslant 0\} \subseteq \mathbf{R}_2$ with the usual operations in $\mathbf{R}_2$

11. $\mathbf{V} = \{\text{all integers}\}$ with the usual $+$ and $\cdot$ for $\mathbf{R}$

12. $\mathbf{V} \subseteq \mathbf{R}_3$ with $[x_1, x_2, x_3] \in \mathbf{V}$ if and only if $x_1^2 + x_2^2 + x_3^2 = 1$. Sketch the set.

13. $\{$the 0 polynomial together with all polynomials of degree exactly $n\} \subseteq \mathbf{P}_n$

14. The subset of all $n \times n$ invertible matrices in $\mathbf{R}_{n,n}$

15. $\mathbf{V} = \{[x_1, x_2] : x_1, x_2 \in \mathbf{R}\}$. For $[x_1, x_2]$, $[y_1, y_2]$ in $\mathbf{V}$, let

$$[x_1, x_2] + [y_1, y_2] = [x_1 + y_1, x_2 + y_2]$$

and for $r \in \mathbf{R}$, let

$$r[x_1, x_2] = [rx_1, x_2]$$

16. $\mathbf{V} = \mathbf{R}$, $\mathbf{x} \oplus \mathbf{y} = \mathbf{x} - \mathbf{y}$, and $r \odot \mathbf{x} = r\mathbf{x}$ for all $\mathbf{x}$ and $\mathbf{y}$ in $\mathbf{V}$ and $r \in \mathbf{R}$.

In exercises 17 through 24, decide whether or not the given set with the usual operations $+$ and $\cdot$ is a vector space, and prove your answer. Use the test for subspaces.

17. $\mathbf{V} = \{[x, y, 0] : x, y \in \mathbf{R}\} \subseteq \mathbf{R}_3$. Sketch the set.

18. $\mathbf{V} = \{[x, y, 2x - 3y] : x, y \in \mathbf{R}\} \subseteq \mathbf{R}_3$

19. $\mathbf{V} = \{f(x) : f \in \mathbf{F}(\mathbf{R}) \text{ and } f \geqslant 0\} \subseteq \mathbf{F}(\mathbf{R})$

20. $\mathbf{V} = \{r[1, 1] + s[-1, 1] : r, s \in \mathbf{R}\} \subseteq \mathbf{R}_2$. Sketch the set.

21. $\mathbf{V} = \left\{\begin{bmatrix} a & b \\ -b & a \end{bmatrix} : a, b \in \mathbf{R}\right\} \subseteq \mathbf{R}_{2,2}$

22. $\mathbf{V}$ is the subset of all noninvertible $n \times n$ matrices in $\mathbf{R}_{n,n}$.

23. $\mathbf{V} = \{[x_{ij}] : [x_{ij}] \in \mathbf{R}_{m,n}$ and $x_{ij} \neq 0$ if and only if $i = j\} \subseteq \mathbf{R}_{m,n}$

24. $\mathbf{V} = \{f(x) : f \in \mathbf{F}(\mathbf{R}) \text{ and } f(1) = 0\} \subseteq \mathbf{F}(\mathbf{R})$

COMPLEX NUMBERS

Let $\mathbf{V}$ be a nonempty set of objects, called vectors, on which there is defined an addition $+$. Let $\mathbf{C}$ be the set of complex numbers. Suppose there is a scalar multiplication $\cdot$ between numbers in $\mathbf{C}$ and vectors in $\mathbf{V}$. If the operations $+$ and $\cdot$ satisfy the properties of a vector space listed in (i) and (ii) and 1 through 8 on pages 122 and 123, using $\mathbf{C}$ for $\mathbf{R}$, we call $\mathbf{V}$ a *vector space over the complex numbers.*

In exercises 25 and 26, show that the given set is a vector space over the complex numbers.

25. $\mathbf{V} = \mathbf{C}_n = \{[z_1, \ldots, z_n] : z_i \in \mathbf{C}$ for all $i\}$ with componentwise $+$ and $\cdot$

26. $\mathbf{V} = \mathbf{C}_{m,n} = \{m \times n$ matrices $A = [a_{ij}] : a_{ij} \in \mathbf{C}$ for all $i, j\}$ with the usual $+$ and $\cdot$ for matrices

THEORETICAL EXERCISES

27. Explain the computational difference between verifying a vector space and verifying a subspace. Also make clear when each verification can be used.

28. Prove parts (ii), (iv), (v), and (vi) of Theorem 3.5. *Hint for (ii):* Note that $r\mathbf{0} = r(\mathbf{0} + \mathbf{0})$. *Hint for (iv):* Suppose $r \neq 0$. Then show that $\mathbf{x} = \mathbf{0}$. *Hint for (vi):* Suppose $\mathbf{0}$ and $\mathbf{0}'$ are additive identities. Calculate $\mathbf{0} + \mathbf{0}'$, first using that $\mathbf{0}$ is an identity and then again using that $\mathbf{0}'$ is an identity.

29. Prove that $\{\mathbf{0}\}$, as described in Example 19, is a vector space.

30. Prove that $\mathbf{F}(\mathbf{D})$, as given in Example 22, is a vector space.

31. (space of linear equations) Let $\mathbf{V}$ be the set of all linear equations of the form

$$a_1x_1 + \cdots + a_nx_n = b$$

where $a_1, \ldots, a_n, b \in \mathbf{R}$. Define

$$(a_1x_1 + \cdots + a_nx_n = b) + (c_1x_1 + \cdots + c_nx_n = d) \\ = [(a_1 + c_1)x_1 + \cdots + (a_n + c_n)x_n = b + d]$$

and

$$c \cdot (a_1x_1 + \cdots + a_nx_n = b) = (ca_1x_1 + \cdots + ca_nx_n = cb)$$

Show that $\mathbf{V}$ is a vector space.

32. (space of equations) Let $\mathbf{V}$ be the set of all equations of the form

$$a_1x_1^{k_1} + \cdots + a_nx_n^{k_n} = b$$

where $a_1, \ldots, a_n, b \in \mathbf{R}$ and $k_1, \ldots, k_n$ are n chosen integers. Define $+$ and $\cdot$ as in exercise 31. Show that $\mathbf{V}$ is a vector space.

33. (function space of two variables) Let

$$F(x, y) = \{f(x, y) : x, y \in \mathbf{R}\}.$$

Defining addition and scalar multiplication in the usual way for functions, prove that $F(x, y)$ is a vector space.

In exercises 34 and 35, show that the given $\mathbf{V}$ is a vector space by showing that it is a subspace.

34. (Fourier sum space) **(*a*)** For a positive integer n, the set

$$\mathbf{V} = \{r_0 + r_1 \cos t + r_2 \cos 2t + \cdots + r_n \cos nt : r_0, \ldots, r_n \in \mathbf{R}\}$$

with the usual $+$ and $\cdot$ for functions.

(*b*) For a positive integer n, the set

$$\mathbf{V} = \{r_0 + r_1 \sin t + r_2 \sin 2t + \cdots + r_n \sin nt : r_0, \ldots, r_n \in \mathbf{R}\}$$

with the usual $+$ and $\cdot$ for functions.

35. (special polynomials space) For fixed polynomials $p_1, \ldots, p_n$, the set

$$\mathbf{V} = \{r_0 + r_1p_1(t) + \cdots + r_np_n(t) : r_0, \ldots, r_n \in \mathbf{R}\}$$

with the usual $+$ and $\cdot$ for functions. (Such polynomial sums occur in approximations with Legendre, Lagrange, Chebyshev, and other polynomials.)

36. Prove that if a vector space $\mathbf{V} \neq \{\mathbf{0}\}$, then $\mathbf{V}$ contains infinitely many vectors. *Hint:* Pick $\mathbf{x} \in \mathbf{V}$ with $\mathbf{x} \neq \mathbf{0}$, and consider $r\mathbf{x}$ for all $r \in \mathbf{R}$.

37. (logarithm space) The set $\mathbf{V} = \{\text{positive real numbers}\}$ with

$$\mathbf{x} \oplus \mathbf{y} = \mathbf{xy} \qquad \text{and} \qquad c \odot \mathbf{x} = \mathbf{x}^c$$

for all $\mathbf{x}$ and $\mathbf{y}$ in $\mathbf{V}$ and $c \in \mathbf{R}$. Show that $\mathbf{V}$ is a vector space.

38. Let $\mathbf{V}$ be a vector space. Prove that $\{\mathbf{0}\}$ and $\mathbf{V}$ are subspaces of $\mathbf{V}$.

39. "Linear algebra students do not know what they are talking about." Interpret this statement in two different ways. *Hint:* Linear algebra students speak of "general" vector spaces.

40. In algebra, you write "Let x be a number" without stating which number is intended. In a sentence or two, compare this statement with the statement "Let $\mathbf{V}$ be a vector space."

41. Explain why exchange systems are general systems. Give some particular exchange systems. Explain why studying exchange systems in general is useful.

3.3 SOME SPECIAL VECTOR SPACES

In this section we show some special vector spaces. Our first example shows geometrically all vector spaces in $\mathbf{R}_3$.

EXAMPLE 27

GEOMETRICALLY DETERMINING ALL SUBSPACES IN $\mathbf{R}_3$

In this example, we find all subsets of $\mathbf{R}_3$ that are subspaces of $\mathbf{R}_3$. To understand the problem, consider

$$B = \{[x, y, z] : x \geq 0, y \geq 0, z = 0\}$$

This subset is shown in Figure 3.30. Note that $\mathbf{x} = [1, 1, 0] \in B$. However, if

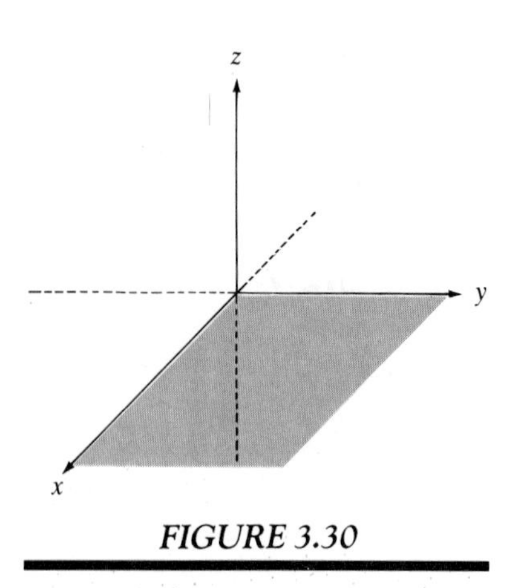

FIGURE 3.30

$r = -1$, then

$$r\mathbf{x} = -1[1, 1, 0] = [-1, -1, 0] \notin B$$

Thus, not all scalar multiples of $\mathbf{x}$ are in B and so B does not satisfy the closure of scalar multiplication property. Thus, B is not a subspace.

The subsets of $\mathbf{R}_3$ that are subspaces in $\mathbf{R}_3$ are of four types:

(*i*) the origin,
(*ii*) any line through the origin,
(*iii*) any plane through the origin, and
(*iv*) $\mathbf{R}_3$ itself.

That the geometric sets described in (i) through (iv) are subspaces of $\mathbf{R}_3$ follows by viewing vectors as fixed arrows as in Part III of Section 3.1 and applying the test for subspaces directly to them. Thus we need only show that any subspace of $\mathbf{R}_3$ is of one of the types described in (i) through (iv).

Let $\mathbf{W}$ be a subspace of $\mathbf{R}_3$. If $\mathbf{W} = \{\mathbf{0}\}$, it is the origin. Thus suppose $\mathbf{W}$ contains a nonzero vector $\mathbf{x}$. Then, by the test for subspaces,

$$\{r\mathbf{x} : r \in \mathbf{R}\} \subseteq \mathbf{W}$$

If equality holds, as shown in Figure 3.31, $\mathbf{W}$ is described by (ii). If $\mathbf{W}$ contains a vector

$$\mathbf{y} \notin \{r\mathbf{x} : r \in \mathbf{R}\}$$

then

$$\{r\mathbf{x} + s\mathbf{y} : r, s \in \mathbf{R}\} \subseteq \mathbf{W}$$

If equality holds, then $\mathbf{W}$ is described by (iii), since, as depicted in Figure 3.32, the region labeled 1 can be obtained by making all choices of $r \geqslant 0$, $s \geqslant 0$; the region labeled 2 can be obtained by making all choices of $r \geqslant 0$, $s \leqslant 0$; etc. Finally, if $\mathbf{W}$ contains a vector

$$\mathbf{z} \notin \{r\mathbf{x} + s\mathbf{y} : r, s \in \mathbf{R}\}$$

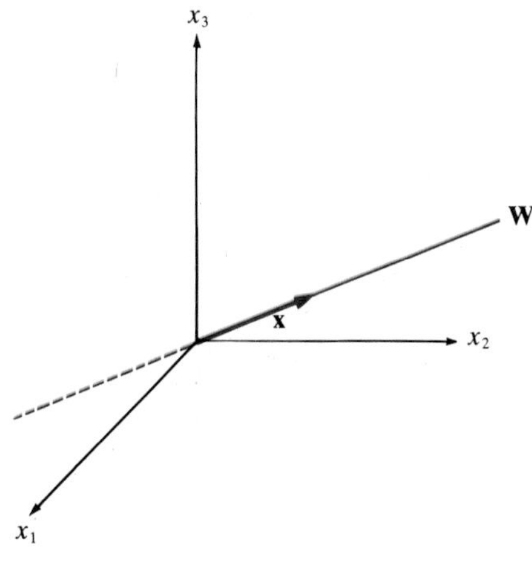

FIGURE 3.31

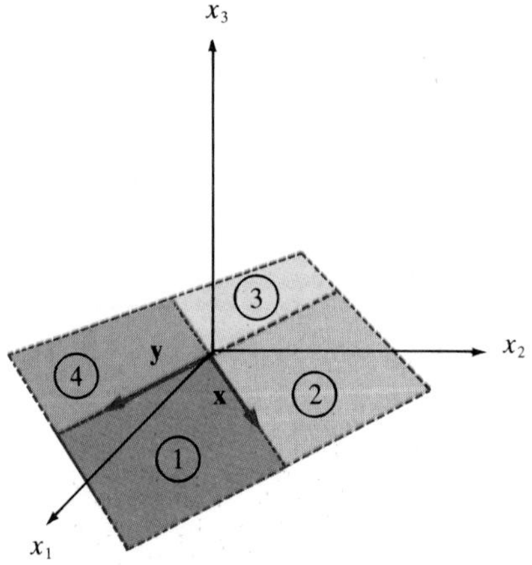

FIGURE 3.32

then $$\mathbf{Y} = \{r\mathbf{x} + s\mathbf{y} + t\mathbf{z} : r, s, t \in \mathbf{R}\} \subseteq \mathbf{W}$$

Applying the work of Example 18 in Section 3.1, we can show geometrically that $\mathbf{Y} = \mathbf{R}_3$. Thus it follows that $\mathbf{W} = \mathbf{R}_3$, which is (iv). □

Any matrix determines two vector spaces.

EXAMPLE 28

Let A be an $m \times n$ matrix. Define the *column space* of A as

$$\mathbf{V} = \{x_1\mathbf{a}^1 + \cdots + x_n\mathbf{a}^n : x_1, \ldots, x_n \in \mathbf{R}\}$$

This space can be written more compactly as

$$\mathbf{V} = \{A\mathbf{x} : \mathbf{x} \in \mathbf{R}^n\}$$

For example, since

$$x_1\begin{bmatrix} 1 \\ 3 \end{bmatrix} + x_2\begin{bmatrix} 2 \\ 4 \end{bmatrix} = \begin{bmatrix} 1x_1 + 2x_2 \\ 3x_1 + 4x_2 \end{bmatrix} = \begin{bmatrix} 1 & 2 \\ 3 & 4 \end{bmatrix}\begin{bmatrix} x_1 \\ x_2 \end{bmatrix}$$

the column space of $\begin{bmatrix} 1 & 2 \\ 3 & 4 \end{bmatrix}$ is

$$\left\{x_1\begin{bmatrix} 1 \\ 3 \end{bmatrix} + x_2\begin{bmatrix} 2 \\ 4 \end{bmatrix} : x_1, x_2 \in \mathbf{R}\right\} = \left\{\begin{bmatrix} 1 & 2 \\ 3 & 4 \end{bmatrix}\mathbf{x} : \mathbf{x} \in \mathbf{R}^2\right\}$$

This space arises in studies of solutions to $A\mathbf{x} = \mathbf{b}$. Note, for example, that $A\mathbf{x} = \mathbf{b}$ has a solution if and only if $\mathbf{b} \in$ column space of A. (See exercises 11 and 12 and the work of Section 4.3.)

Since this set is a subset of $\mathbf{R}^m$, we can show that it is a vector space by showing that it is a subspace of $\mathbf{R}^m$. Applying the test for subspaces, first note that the column space is not empty. To verify the closure of addition property, take $A\mathbf{x}$ and $A\mathbf{y}$ in $\mathbf{V}$, where $\mathbf{x}, \mathbf{y} \in \mathbf{R}^n$. Then

$$A\mathbf{x} + A\mathbf{y} = A(\mathbf{x} + \mathbf{y})$$

Since $\mathbf{x} + \mathbf{y} \in \mathbf{R}^n$, it follows that $A\mathbf{x} + A\mathbf{y} \in \mathbf{V}$. For the closure of scalar multiplication property, let $r \in \mathbf{R}$ and take $A\mathbf{x} \in \mathbf{V}$, where $\mathbf{x} \in \mathbf{R}^n$. Then $r(A\mathbf{x}) = A(r\mathbf{x})$. Since $r\mathbf{x} \in \mathbf{R}^n$, it follows that $r(A\mathbf{x}) \in \mathbf{V}$. Hence the column space of A is a vector space.

Similarly, the *row space* of A,

$$\{x_1\mathbf{a}_1 + \cdots + x_m\mathbf{a}_m : x_1, \ldots, x_m \in \mathbf{R}\}$$

can be written as $\{\mathbf{x}A : \mathbf{x} \in \mathbf{R}_m\}$ and is a vector space because it is a subspace of $\mathbf{R}_n$. □

A vector space arising as the solution of an equation follows.

EXAMPLE 29

Let A be an $m \times n$ matrix. The equation $A\mathbf{x} = \mathbf{0}$ is called *homogeneous*, in reference to the fact that the right side of this equation is $\mathbf{0}$.* Define the *solution space* of A as

$$\mathbf{V} = \{\mathbf{x} : A\mathbf{x} = \mathbf{0}\} \subseteq \mathbf{R}^n$$

with the usual $+$ and $\cdot$ for $\mathbf{R}^n$. We show that $\mathbf{V}$ is a vector space by applying the test for subspaces to show that $\mathbf{V}$ is a subspace of $\mathbf{R}^n$.

Since $A\mathbf{0} = \mathbf{0}$, it follows that $\mathbf{0} \in \mathbf{V}$, so $\mathbf{V} \neq \varnothing$. Next, we show the closure of addition property. For this, if $\mathbf{x}, \mathbf{y} \in \mathbf{V}$, it follows that $A\mathbf{x} = \mathbf{0}$ and $A\mathbf{y} = \mathbf{0}$. Thus,

$$A(\mathbf{x} + \mathbf{y}) = A\mathbf{x} + A\mathbf{y} = \mathbf{0} + \mathbf{0} = \mathbf{0}$$

so $\mathbf{x} + \mathbf{y} \in \mathbf{V}$.

Finally, we show the closure of scalar multiplication property. Let $r \in \mathbf{R}$ and $\mathbf{x} \in \mathbf{V}$. Then, from $A\mathbf{x} = \mathbf{0}$, we have

$$A(r\mathbf{x}) = rA\mathbf{x} = r\mathbf{0} = \mathbf{0}$$

so $r\mathbf{x} \in \mathbf{V}$. Thus, we have shown that $\mathbf{V}$ is a vector space. □

The role of vector spaces in equations applies to difference equations, differential equations, partial differential equations, and integral equations, as well as other types of equations. Some of these are clearly beyond the level of this text. However, it is helpful to study at least one type of equation other than the algebraic one. Thus we introduce difference equations, which require no additional mathematical background. How such equations arise can be seen, for example, in exercises 35, 36, 37, and 38. Other examples follow throughout the text.

Let $x(k)$ be a function, to be determined, defined on the set of nonnegative integers $\mathbf{N} = \{0, 1, \ldots\}$, and let $a \in \mathbf{R}$ and $g(k) \in \mathbf{F}(\mathbf{N})$ be given. Any equation of the form

$$x(k + 1) + ax(k) = g(k)$$

is called a *first-order linear difference equation.*† For $a, b, c \in \mathbf{R}$ given, with $a \neq 0$, we call the equation

$$ax(k + 2) + bx(k + 1) + cx(k) = g(k)$$

a *second-order linear difference equation.* If $g(k) = 0$ for all $k \in \mathbf{N}$, we call these equations *homogeneous.**

* To decide if an equation is homogeneous, place all expressions involving unknowns to the left side of the equation and all constants to the right side. Then see if the right side is 0.

† Difference equations are also called *recursion relations.*

By a *solution* to either equation we mean any function defined on $\mathbf{N}$ that, when substituted for $x(k)$, satisfies the equation.

EXAMPLE 30

The equation

$$x(k+1) - 2x(k) = 0$$

has $x(k) = 2^k$ as a solution since, by substitution,

$$x(k+1) - 2x(k) = 2^{k+1} - 2(2^k) = 0$$

for all $k \in \mathbf{N}$.

The equation

$$x(k+1) - 3x(k) = 4^k$$

has $x(k) = 4^k$ as a solution since

$$x(k+1) - 3x(k) = 4^{k+1} - 3(4^k) = 4^k$$

for all $k \in \mathbf{N}$.

The equation

$$x(k+2) - 3x(k+1) + 2x(k) = 0$$

has $x(k) = 2^k$ as a solution since

$$x(k+2) - 3x(k+1) + 2x(k) = 2^{k+2} - 3(2^{k+1}) + 2(2^k) = 2^k(4 - 6 + 2) = 0$$

for all $k \in \mathbf{N}$. Also, $x(k) = 1$ is a solution since

$$x(k+2) - 3x(k+1) + 2x(k) = 1 - 3(1) + 2(1) = 0$$

for all $k \in \mathbf{N}$.

The equation

$$x(k+2) - x(k+1) + x(k) = 3(\tfrac{1}{2})^{k+2}$$

has $x(k) = (\frac{1}{2})^k$ as a solution since

$$\begin{aligned} x(k+2) - x(k+1) + x(k) &= (\tfrac{1}{2})^{k+2} - (\tfrac{1}{2})^{k+1} + (\tfrac{1}{2})^k = (\tfrac{1}{2})^k(\tfrac{1}{4} - \tfrac{1}{2} + 1) \\ &= \tfrac{3}{4}(\tfrac{1}{2})^k = 3(\tfrac{1}{2})^{k+2} \end{aligned}$$

for all $k \in \mathbf{N}$. □

To solve a difference equation, we need to find all solutions to that equation. The first-order homogeneous equation

$$x(k+1) + ax(k) = 0$$

can be solved as follows. Let $x(0) = r$ where r is any number. Then

$$\begin{aligned}
x(0) &= r \\
x(1) &= -ax(0) = -ar \\
x(2) &= -ax(1) = (-a)^2 r \\
&\vdots \\
x(k) &= -ax(k-1) = (-a)^k r
\end{aligned}$$

for all $k \in \mathbf{N}$, where, in the latter expression, we use $0^0 = 1$ when $a = 0$.*

EXAMPLE 31

The equation

$$x(k+1) + 3x(k) = 0$$

has solution $x(k) = r(-3)^k$ for any $r \in \mathbf{R}$.

The equation

$$x(k+1) - 2x(k) = 0$$

has solution $x(k) = r(2^k)$ for all $r \in \mathbf{R}$.

The equation

$$x(k+1) = 0$$

has solution $x(k) = r(0^k)$ for all $r \in \mathbf{R}$; that is, $x(0) = r$, $x(1) = x(2) = \cdots = 0$.

□

The solution to the second-order homogeneous equation

$$ax(k+2) + bx(k+1) + cx(k) = 0$$

is much more complicated to compute than that to the first-order homogeneous difference equation. If $x(0) = r$ and $x(1) = s$ for some chosen $r, s \in \mathbf{R}$, then

$$\begin{aligned}
x(0) &= r \\
x(1) &= s \\
x(2) &= -\frac{b}{a}x(1) - \frac{c}{a}x(0) = -\frac{b}{a}s - \frac{c}{a}r \\
x(3) &= -\frac{b}{a}x(2) - \frac{c}{a}x(1) = -\frac{b}{a}\left(-\frac{b}{a}s - \frac{c}{a}r\right) - \frac{c}{a}s \\
&= \frac{b^2}{a^2}s + \frac{bc}{a^2}r - \frac{c}{a}s \\
&= \left(\frac{b^2}{a^2} - \frac{c}{a}\right)s + \frac{bc}{a^2}r
\end{aligned}$$

* That $0^0 = 1$ is used only in this formula.

From this, it is clear that $x(k)$ is completely determined once $x(0) = r$ and $x(1) = s$ are specified. However, a formula or simple expression for it is not obvious.

Throughout this chapter we will show how various parts of the study of vector spaces apply to these equations. The cumulative result will be a method for solving all homogeneous difference equations. In Chapter 5 we will show how vector spaces are used to solve the nonhomogeneous difference equations as well. For now, we show how vector spaces arise when these equations are solved.

EXAMPLE 32

Let $a \in \mathbf{R}$ and let $\mathbf{V}$ be the set of all solutions to the homogeneous equation

$$x(k+1) + ax(k) = 0$$

with the usual $+$ and $\cdot$ for functions. To show that $\mathbf{V}$ is a vector space, we apply the test for subspaces to show that it is a subspace of the vector space $\mathbf{F}(\mathbf{N})$. For this, define $z(k) = 0$ for all $k \in \mathbf{N}$. Then

$$z(k+1) + az(k) = 0 + a \cdot 0 = 0$$

so $z(k) \in \mathbf{V}$ and $\mathbf{V} \neq \varnothing$. For the closure of addition property, let $x(k), y(k) \in \mathbf{V}$. Then

$$\begin{aligned}
&[x(k+1) + y(k+1)] + a[x(k) + y(k)] \\
&\quad = x(k+1) + ax(k) + y(k+1) + ay(k) \\
&\quad = 0 + 0 = 0
\end{aligned}$$

so $x(k) + y(k) \in \mathbf{V}$. Finally, for $r \in \mathbf{R}$ and $x(k) \in \mathbf{V}$, we have that

$$[rx(k+1)] + a[rx(k)] = r[x(k+1) + ax(k)] = r \cdot 0 = 0$$

so $rx(k) \in \mathbf{V}$. Hence the closure of scalar multiplication property holds and $\mathbf{V}$ is a vector space, which we call the *solution space* to the equation.

Similarly, if $a, b, c \in \mathbf{R}$ with $a \neq 0$, then the set $\mathbf{V}$ of all solutions to

$$ax(k+2) + bx(k+1) + cx(k) = 0$$

is also a vector space (exercise 26). □

EXAMPLE 33

Let A be an $n \times n$ matrix. Consider the system of linear difference equations written as the matrix equation

$$\mathbf{x}(k+1) = A\mathbf{x}(k)$$

where $\mathbf{x}(k) \in \mathbf{F}^n(\mathbf{N})$. This equation is called homogeneous since it can be written as*

$$\mathbf{x}(k+1) - A\mathbf{x}(k) = \mathbf{0}$$

Let $\mathbf{V}$ be the set of all solutions to this equation. Then, mimicking the proof in the previous example, we can show that $\mathbf{V}$ is a vector space (exercise 28). □

In the next section, we continue to find examples of vector spaces by showing a general way in which most subspaces, and consequently vector spaces, arise.

EXERCISES FOR SECTION 3.3

COMPUTATIONAL EXERCISES

In exercises 1 through 4, write out the column space and row space of the given matrix A. Then write them in compact form as in Example 28.

1. $A = \begin{bmatrix} 1 & 1 \\ 1 & -1 \end{bmatrix}$

2. $A = \begin{bmatrix} 1 & -1 & 2 \\ 2 & 1 & 1 \end{bmatrix}$

3. $A = \begin{bmatrix} 1 & 0 & -1 \\ 2 & 1 & 3 \\ -1 & 4 & 2 \end{bmatrix}$

4. $A = \begin{bmatrix} 2 & 1 \\ -1 & -2 \\ 2 & 1 \end{bmatrix}$

In exercises 5 through 10, write the system as a matrix equation. Decide if its solution set is a vector space. Use Example 29 when possible.

5. $\begin{aligned} 2x + y &= 0 \\ x - y &= 0 \\ -3x + 2y &= 0 \end{aligned}$

6. $\begin{aligned} 2x - 3y &= 3 \\ 4x - 6y &= 0 \end{aligned}$

7. $\begin{aligned} x_1 - 2x_2 + x_3 &= 1 \\ 2x_1 \qquad - x_3 &= 2 \end{aligned}$

8. $\begin{aligned} x_1 + 3x_2 - x_3 &= 0 \\ -x_1 + x_2 - 3x_3 &= 0 \end{aligned}$

9. $x_1 + x_2 + x_3 = 0$

10. $\begin{aligned} 2x_1 + x_2 - x_3 + x_4 &= 0 \\ x_1 - 2x_2 \qquad + x_4 &= 0 \\ 3x_1 \qquad + x_3 \qquad &= 0 \end{aligned}$

In exercises 11 and 12, graph the column space of A, using Example 28. Graph the given vector $\mathbf{b}$. From this graph, state if $A\mathbf{x} = \mathbf{b}$ has a solution.

11. $A = \begin{bmatrix} -1 & 1 \\ 1 & -1 \end{bmatrix}, \mathbf{b} = \begin{bmatrix} 2 \\ 3 \end{bmatrix}$

12. $A = \begin{bmatrix} -1 & 2 \\ 2 & -4 \end{bmatrix}, \mathbf{b} = \begin{bmatrix} 3 \\ -6 \end{bmatrix}$

In exercises 13 and 14, calculate $x(0)$, $x(1)$, $x(k+1)$, and $x(k+1) - x(k)$ for each of the given functions in $\mathbf{F}(\mathbf{N})$.

13. (*a*) $x(k) = 2^k$ (*b*) $x(k) = 3k + 7$
(*c*) $x(k) = \dfrac{1}{k+1}$

14. (*a*) $x(k) = k - 3^k$ (*b*) $x(k) = -k + 2$
(*c*) $x(k) = 2k^2 - k + 3$

In exercises 15 and 16, sketch the graph of each function $x(k) \in \mathbf{F}(\mathbf{N})$. Use the function $x(t)$ as a guide.

15. (*a*) $x(k) = k$ (*b*) $x(k) = k^3$ (*c*) $x(k) = 2^k$
(*d*) $x(k) = \cos k\pi$

16. (*a*) $x(k) = 2k + 3$ (*b*) $x(k) = \dfrac{1}{k}$ (*c*) $x(k) = (-1)^k$
(*d*) $x(k) = (-\frac{1}{2})^k$

In exercises 17 through 20, decide whether or not the given function is a solution to the given equation.

17. $x(k+1) - x(k) = 0$, $x(k) = 3$

18. $x(k+2) - 3x(k+1) + 2x(k) = 0$, $x(k) = 3^k$

19. $x(k+2) - 5x(k+1) + 3x(k) = 2^k$, $x(k) = 2^k$

20. $x(k+1) - 2x(k) = 3^k$, $x(k) = 3^k$

In exercises 21 through 24, solve the given equation.

21. $x(k+1) - x(k) = 0$

22. $x(k+1) + 3x(k) = 0$

23. $2x(k+1) + x(k) = 0$

24. $3x(k+1) - 2x(k) = 0$

COMPLEX NUMBERS

25. Let $A \in \mathbf{C}_{m,n}$ and let $\mathbf{V} = \{\mathbf{x} : A\mathbf{x} = \mathbf{0}\}$. Show that $\mathbf{V}$ is a vector space.

* All unknowns are placed on the left side of the equation. The right side is then 0.

THEORETICAL EXERCISES

26. Show that the set of all solutions to

$$ax(k+2) + bx(k+1) + cx(k) = 0$$

as described in Example 32 is a vector space.

27. (nonhomogeneous system of linear algebraic equations) Let

$$\mathbf{V} = \{\text{all solutions to } A\mathbf{x} = \mathbf{b} \text{ where } A \text{ is an } m \times n \text{ matrix and } \mathbf{b} \neq \mathbf{0}\}$$

Prove that **V** is not a vector space. (Thus the solution set to a system of linear algebraic equations is a vector space if and only if the system is homogeneous.)

28. (homogeneous system of linear difference equations) Prove that the set of all solutions to the homogeneous difference equation given in Example 33 is a vector space.

29. (nonhomogeneous difference equation) Consider the difference equation

$$x(k+1) - rx(k) = b(k)$$

where b is not identically 0. Prove that the solution set of this equation is not a subspace of $\mathbf{F}(\mathbf{N})$.

30. (homogeneous differential equation) (calculus) Let **V** be the set of all functions $y(t)$ that satisfy the first-order differential equation

$$y'(t) - ay(t) = 0$$

where a is a real number. Prove that **V** is a vector space.

31. (nonhomogeneous differential equation) (calculus) Consider the differential equation

$$y'(t) - ry(t) = b(t)$$

where b is not identically 0. Prove that the solution set of this equation is not a subspace of $\mathbf{F}(\mathbf{R})$.

32. (homogeneous differential equation) (calculus) Prove that the solution set of the homogeneous linear differential equation

$$ay''(t) + by'(t) + cy(t) = 0$$

where $a \neq 0$, is a subspace of $\mathbf{F}(\mathbf{R})$.

33. (homogeneous integral equation) (calculus) For $a < b$ in **R**, let **V** = the set of all functions continuous over $[a, b]$ such that

$$\int_a^b f(t)\,dt = 0$$

with the usual $+$ and $\cdot$ for functions. Prove that **V** is a vector space.

34. (homogeneous system of linear differential equations) (calculus) Consider the homogeneous differential equation $\mathbf{x}'(t) = A\mathbf{x}(t)$, where A is an $n \times n$ matrix,

$$\mathbf{x}(t) = [x_1(t), \ldots, x_n(t)]^t \quad \text{and} \quad \mathbf{x}'(t) = [x_1'(t), \ldots, x_n'(t)]^t$$

Prove that the set of all solutions to this equation is a subspace of $\mathbf{F}^n(\mathbf{R})$.

APPLICATIONS EXERCISES

35. Let $x(k)$ be the number of smokers in the United States in the kth year. Suppose 1% of all smokers stop smoking each year. Then

$$x(k+1) = 0.99x(k)$$

Given that there are 10 million smokers in the United States now, find the expression for the number of smokers k years from now.

36. The growth rate of a culture of bacteria is 0.2; that is, if there are $x(k)$ bacteria at time k, then there will be an additional $0.2x(k)$ bacteria in the next time period $k + 1$. Thus

$$x(k+1) = x(k) + 0.2x(k) \quad \text{or} \quad x(k+1) = 1.2x(k)$$

If there are $x(0) = 1000$ bacteria in the culture now, find an expression for the number of bacteria at time k.

37. A junior college has freshman and sophomore classes. There are 1000 new students each year, all of whom are freshmen. In order to raise standards, the administration decides that grading should be done such that $\frac{1}{10}$ of each class fails each year. This situation is depicted diagrammatically in Figure 3.33. If there are presently 1000 freshmen and 900 sophomores, write the difference equations that yield the number of freshmen $x_1(k)$ and the number of sophomores $x_2(k)$ in k years. Write these as a matrix equation. If $\mathbf{x}(0) = [1000, 900]$, find $\mathbf{x}(1)$, $\mathbf{x}(2)$, and $\mathbf{x}(3)$, rounding your answer to the nearest integer. Can you tell what is happening to the sizes of these classes?

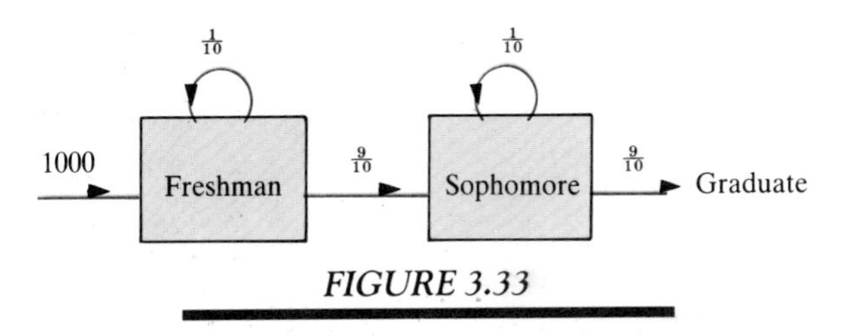

FIGURE 3.33

38. A small tobacco pipe manufacturer makes pipes in two processing stages, carving and painting, each stage taking one day for completion. Each pipe is made from a wooden block; 100 such blocks are supplied to the carving stage each day. At the end of the day, the painters return $\frac{1}{10}$ of the pipes they have to carving for some touch-up work, and carving scraps $\frac{1}{10}$ of all the pipes they have worked on. This situation is shown in Figure 3.34. Determine the equations for the numbers $x_1(k)$ and $x_2(k)$ of pipes in the carving and painting stages, respectively, on day k. Write these as a matrix equation.

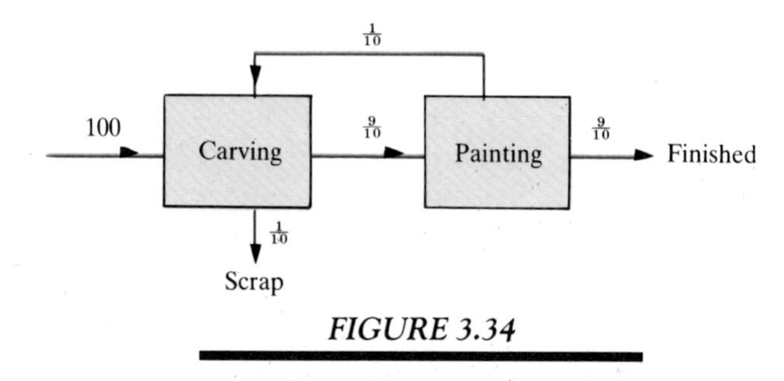

FIGURE 3.34

3.4 SPANNING SETS AND FINITE-DIMENSIONAL VECTOR SPACES

In this section we continue to find examples of vector spaces by showing a general way in which subspaces, and consequently vector spaces, arise. This description will be based on two preliminary definitions, the first of which follows.

DEFINITION

Let $\mathbf{V}$ be a vector space and let

$$S = \{\mathbf{x}_1, \ldots, \mathbf{x}_m\} \subseteq \mathbf{V}$$

where S is a nonempty set. A vector $\mathbf{y} \in \mathbf{V}$ is a *linear combination** of the vectors in S if

$$y = r_1\mathbf{x}_1 + \cdots + r_m\mathbf{x}_m$$

for some $r_1, \ldots, r_m \in \mathbf{R}$. As a mathematical convenience allowing simplicity in the statements of our results, we will say that $\mathbf{0}$, and only $\mathbf{0}$, is a linear combination of the vectors in $\varnothing$.†

The first example gives a geometrical view of linear combinations.**

* The word "linear" is used because $\mathbf{y}$ is constructed in a manner similar to that used for a linear function.

† An arithmetical reason for this is that we would want

$$\sum_{k=1}^{n} r_k\mathbf{x}_k + \sum_{k=1}^{m} s_k\mathbf{y}_k = \sum_{k=1}^{n} r_k\mathbf{x}_k$$

if $m = 0$. Thus we need

$$\sum_{k=1}^{0} s_k\mathbf{y}_k = \mathbf{0}$$

by definition.

** This geometrical view and those that follow are intended to provide feeling for and insight into this material.

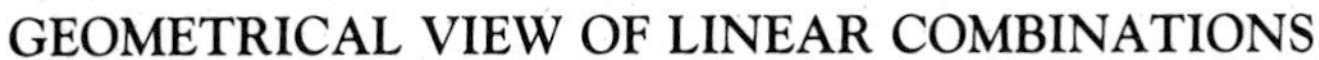

EXAMPLE 34

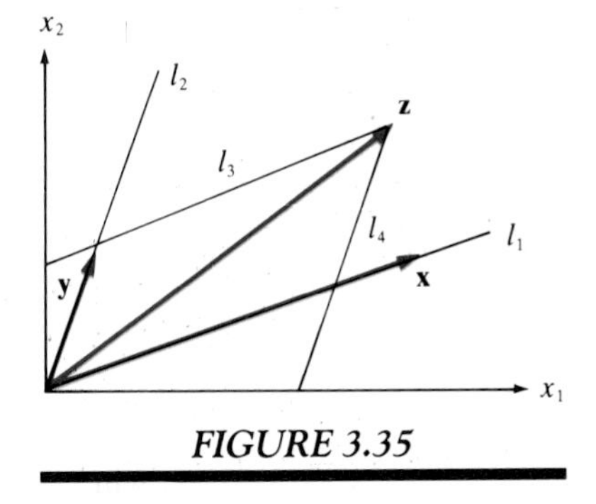

FIGURE 3.35

GEOMETRICAL VIEW OF LINEAR COMBINATIONS

Using the work of Part III of Section 3.1, consider vectors **x**, **y**, and **z**, depicted as fixed arrows in Figure 3.35.

To show that **z** is a linear combination of **x** and **y**, draw the lines l_1 and l_2 containing **x** and **y**. Now draw a line l_3 through the tip of **z** parallel to **x** and a line l_4 through the tip of **z** parallel to **y**, as shown in Figure 3.35. The intersections of these lines with l_1 and l_2 form the vertices of a parallelogram. Since the addition of vectors is by the parallelogram rule, **x** may be shrunk by $r \approx \frac{4}{5}$ and **y** stretched by $s \approx \frac{11}{10}$ to obtain

$$\mathbf{z} = r\mathbf{x} + s\mathbf{y}$$

In fact, using this scheme, we can show that any vector in $\mathbf{R}^2$ is a linear combination of **x** and **y** (see exercises 1 and 2). □

Some numerical examples follow.

EXAMPLE 35

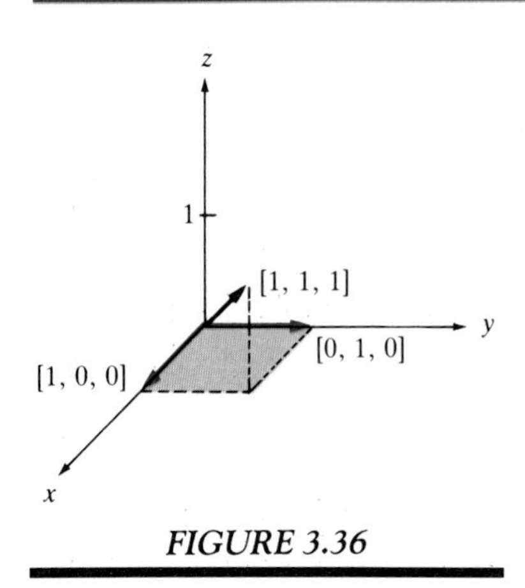

FIGURE 3.36

Is $[1, 1, 1]$ a linear combination of $[1, 0, 0]$ and $[0, 1, 0]$?

First we look at the vectors geometrically as fixed arrows, as in Figure 3.36. Any parallelogram formed from the vectors $r[1, 0, 0]$ and $s[0, 1, 0]$, for any scalars r and s, must lie in the x,y-plane. Thus linear combinations of $[1, 0, 0]$ and $[0, 1, 0]$ are vectors in the x,y-plane. Since $[1, 1, 1]$ is not in this plane, it is not a linear combination of $[1, 0, 0]$ and $[0, 1, 0]$.

This conclusion can also be reached algebraically by trying to compute the scalars such that

$$[1, 1, 1] = r[1, 0, 0] + s[0, 1, 0] = [r, s, 0]$$

This equation requires that $r = 1$, $s = 1$, and $0 = 1$. Since this is impossible, there are no $r, s \in \mathbf{R}$ that satisfy the equation. Hence $[1, 1, 1]$ is not a linear combination of $[1, 0, 0]$ and $[0, 1, 0]$. □

EXAMPLE 36

Is $[1, 0, 1]$ a linear combination of the vectors in

$$S = \{[1, 1, 1], [0, 1, 0], [1, 0, 0], [0, 1, 1]\}$$

To answer this question, we will see if there are r_1, r_2, r_3, and r_4 in $\mathbf{R}$ that satisfy

$$\begin{aligned}[1, 0, 1] &= r_1[1, 1, 1] + r_2[0, 1, 0] + r_3[1, 0, 0] + r_4[0, 1, 1] \\ &= [r_1 + r_3, r_1 + r_2 + r_4, r_1 + r_4]\end{aligned}$$

Thus we need to solve the system

$$\begin{aligned} r_1 \quad + r_3 \quad &= 1 \\ r_1 + r_2 \quad + r_4 &= 0 \\ r_1 \quad\quad + r_4 &= 1 \end{aligned}$$

Using the Gaussian elimination algorithm, we have

$$\begin{bmatrix} 1 & 0 & 1 & 0 & 1 \\ 1 & 1 & 0 & 1 & 0 \\ 1 & 0 & 0 & 1 & 1 \end{bmatrix} \xrightarrow[\mathbf{R}_3 - \mathbf{R}_1 \to \mathbf{R}_3]{\mathbf{R}_2 - \mathbf{R}_1 \to \mathbf{R}_2} \begin{bmatrix} 1 & 0 & 1 & 0 & 1 \\ 0 & 1 & -1 & 1 & -1 \\ 0 & 0 & -1 & 1 & 0 \end{bmatrix}$$

Writing this matrix as a system of equations yields

$$\begin{aligned} r_1 \quad + r_3 \quad &= 1 \\ r_2 - r_3 + r_4 &= -1 \\ -r_3 + r_4 &= 0 \end{aligned}$$

Now r_4 is free and $r_3 = r_4$, $r_2 = -1$, and $r_1 = 1 - r_4$. Thus there are infinitely many choices of r_1, r_2, r_3, and r_4 that give $[1, 0, 1]$ as a linear combination of the vectors in S. For example, we can choose $r_1 = 1$, $r_2 = -1$, $r_3 = 0$, and $r_4 = 0$, showing that

$$[1, 0, 1] = 1[1, 1, 1] - 1[0, 1, 0]$$

This same result could also be shown geometrically, as in the previous examples.* □

EXAMPLE 37

Is $q(t) = t^2 + 1$ a linear combination of

$$p_1(t) = 1 \qquad p_2(t) = t - 1 \qquad p_3(t) = t^2 + t$$

Here we need to see if there are scalars r_1, r_2. and r_3 that satisfy

$$q(t) = r_1 p_1(t) + r_2 p_2(t) + r_3 p_3(t)$$

or

$$\begin{aligned} t^2 + 1 &= r_1(1) + r_2(t - 1) + r_3(t^2 + t) \\ &= r_1 + r_2 t - r_2 + r_3 t^2 + r_3 t \\ &= r_3 t^2 + (r_2 + r_3)t + (r_1 - r_2) \end{aligned}$$

Now, equating coefficients of like terms, we have

$$\begin{aligned} r_3 &= 1 \\ r_2 + r_3 &= 0 \\ r_1 - r_2 \quad &= 1 \end{aligned}$$

* Comparisons of calculations to geometry, when possible, can also be used to check calculations.

Solving these equations yields $r_1 = 0$, $r_2 = -1$, and $r_3 = 1$. Thus $q(t)$ is a linear combination of $p_1(t)$, $p_2(t)$, and $p_3(t)$. □

The second of our preliminary definitions is as follows.

DEFINITION

Let **V** be a vector space and let

$$S = \{\mathbf{x}_1, \ldots, \mathbf{x}_k\} \subseteq \mathbf{V}$$

The set of all linear combinations of the vectors in S is called the *span** of that set S. Algebraically, we write

$$\begin{aligned} \text{span } S &= \text{span } \{\mathbf{x}_1, \mathbf{x}_2, \ldots, \mathbf{x}_k\} \\ &= \{r_1\mathbf{x}_1 + \cdots + r_k\mathbf{x}_k : r_1, \ldots, r_k \in \mathbf{R}\} \end{aligned}$$

Note that it follows by definition that span $\varnothing = \{\mathbf{0}\}$. Since S completely determines span S, we say that S is a *spanning set* of the set span S or that S *spans* the set span S.

A geometrical view of span follows.

EXAMPLE 38

GEOMETRICAL VIEW OF SPAN

For each set S in Figure 3.37, span S can be visualized, using fixed arrows, by finding the lines generated by all scalar multiples of each vector in S and then

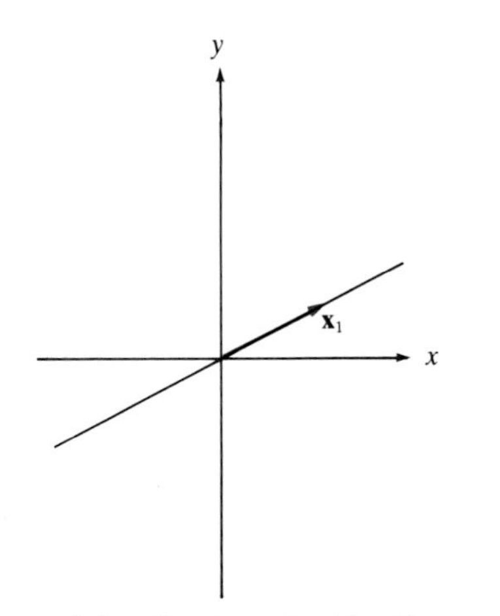

span $\{\mathbf{x}_1\}$ = line determined by this vector

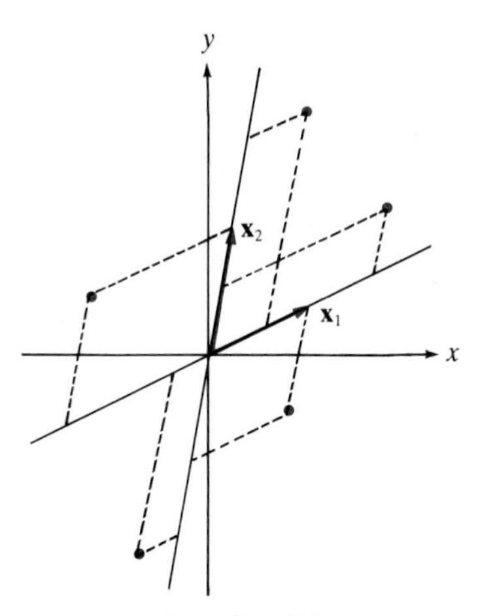

span $\{\mathbf{x}_1, \mathbf{x}_2\} = \mathbf{R}^2$

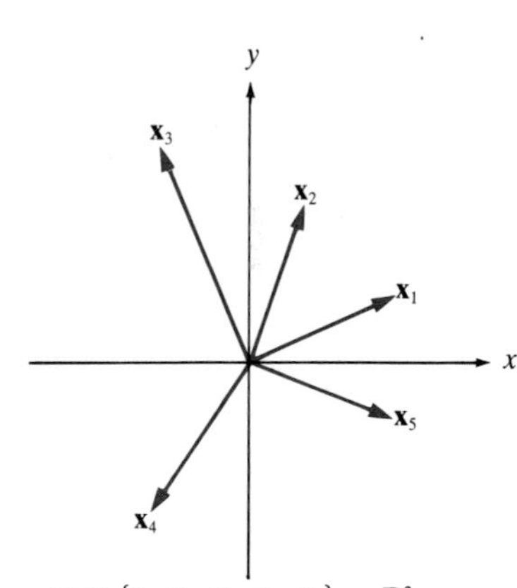

span $\{\mathbf{x}_1, \mathbf{x}_2, \mathbf{x}_3, \mathbf{x}_4, \mathbf{x}_5\} = \mathbf{R}^2$

FIGURE 3.37

(cont.)

* "Span" means reach, breadth, or extent.

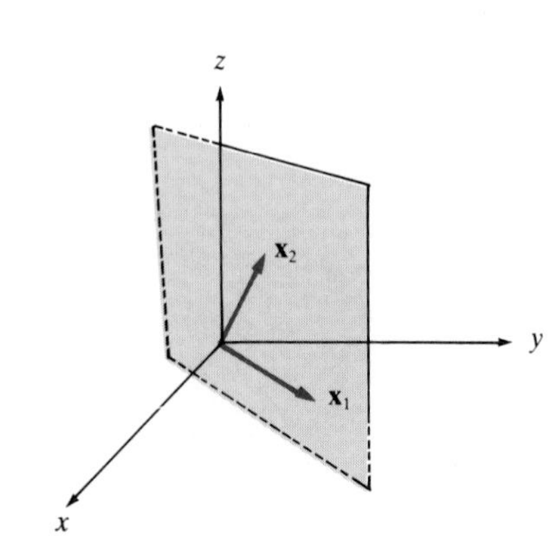

span $\{\mathbf{x}_1, \mathbf{x}_2\}$ = plane determined by these vectors

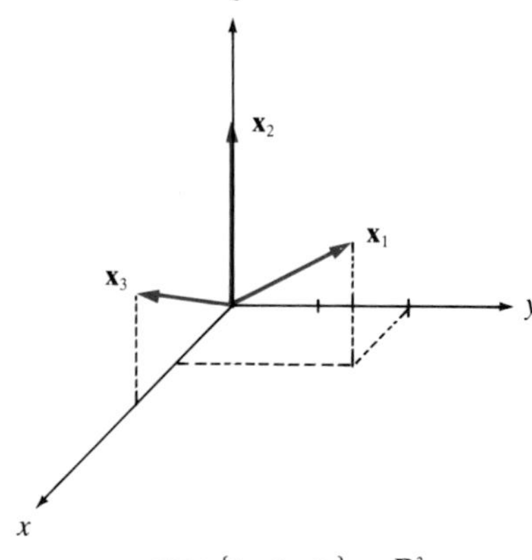

span $\{\mathbf{x}_1, \mathbf{x}_2, \mathbf{x}_3\} = \mathbf{R}^3$

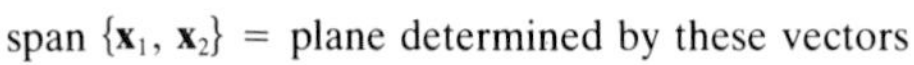

FIGURE 3.37 (cont.)

applying the parallelogram rule to these vectors to obtain all other linear combinations. Note that scalars can be chosen to be 0. □

We now provide some numerical examples.

EXAMPLE 39

In $\mathbf{R}^2$, show that

$$\text{span}\left\{\begin{bmatrix}1\\2\end{bmatrix}, \begin{bmatrix}2\\1\end{bmatrix}\right\} = \mathbf{R}^2$$

Geometrically, as in Example 38, we can see from Figure 3.38 that any vector in $\mathbf{R}^2$ is a linear combination of $\begin{bmatrix}1\\2\end{bmatrix}$ and $\begin{bmatrix}2\\1\end{bmatrix}$.

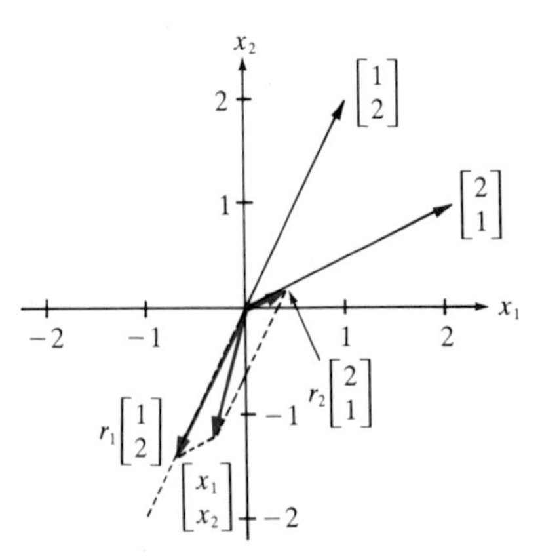

FIGURE 3.38

Algebraically, we need to show that, for any $\begin{bmatrix}x_1\\x_2\end{bmatrix} \in \mathbf{R}^2$, there are scalars r_1 and r_2 that satisfy

$$\begin{bmatrix}x_1\\x_2\end{bmatrix} = r_1\begin{bmatrix}1\\2\end{bmatrix} + r_2\begin{bmatrix}2\\1\end{bmatrix}$$
$$= \begin{bmatrix}r_1 + 2r_2\\2r_1 + r_2\end{bmatrix}$$

Writing the individual equations yields

$$\begin{aligned} r_1 + 2r_2 &= x_1 \\ 2r_1 + r_2 &= x_2 \end{aligned}$$

Solving by the Gauss-Jordan algorithm yields

$$\begin{bmatrix} 1 & 2 & x_1 \\ 2 & 1 & x_2 \end{bmatrix} \xrightarrow{R_2 - 2R_1 \to R_2} \begin{bmatrix} 1 & 2 & x_1 \\ 0 & -3 & x_2 - 2x_1 \end{bmatrix}$$

$$\xrightarrow{-\frac{1}{3}R_2 \to R_2} \begin{bmatrix} 1 & 2 & x_1 \\ 0 & 1 & \dfrac{x_2 - 2x_1}{-3} \end{bmatrix}$$

$$\xrightarrow{R_1 - 2R_2 \to R_1} \begin{bmatrix} 1 & 0 & \dfrac{2x_2 - x_1}{3} \\ 0 & 1 & \dfrac{x_2 - 2x_1}{-3} \end{bmatrix}$$

Thus, $r_1 = (2x_2 - x_1)/3$ and $r_2 = (x_2 - 2x_1)/(-3)$. So, regardless of the choice of x_1 and x_2, there are scalars r_1 and r_2, computable by the above formulas, that satisfy

$$\begin{bmatrix} x_1 \\ x_2 \end{bmatrix} = r_1 \begin{bmatrix} 1 \\ 2 \end{bmatrix} + r_2 \begin{bmatrix} 2 \\ 1 \end{bmatrix}$$

For example, if $\mathbf{x} = \begin{bmatrix} 5 \\ 5 \end{bmatrix}$, then by the formulas $r_1 = \frac{5}{3}$ and $r_2 = \frac{5}{3}$; if $\mathbf{x} = \begin{bmatrix} -1 \\ 3 \end{bmatrix}$, then $r_1 = \frac{7}{3}$ and $r_2 = -\frac{5}{3}$. These calculations, of course, can be checked geometrically. Hence,

$$\operatorname{span}\left\{ \begin{bmatrix} 1 \\ 2 \end{bmatrix}, \begin{bmatrix} 2 \\ 1 \end{bmatrix} \right\} = \mathbf{R}^2$$

□

EXAMPLE 40

Show that

$$\operatorname{span}\left\{ \begin{bmatrix} 1 \\ 2 \end{bmatrix}, \begin{bmatrix} 2 \\ 1 \end{bmatrix}, \begin{bmatrix} -1 \\ 1 \end{bmatrix} \right\} = \mathbf{R}^2$$

This follows from the previous example if we note that the coefficient of $\begin{bmatrix} -1 \\ 1 \end{bmatrix}$ can be taken to be 0, so

$$\mathbf{R}^2 = \operatorname{span}\left\{ \begin{bmatrix} 1 \\ 2 \end{bmatrix}, \begin{bmatrix} 2 \\ 1 \end{bmatrix} \right\} \subseteq \operatorname{span}\left\{ \begin{bmatrix} 1 \\ 2 \end{bmatrix}, \begin{bmatrix} 2 \\ 1 \end{bmatrix}, \begin{bmatrix} -1 \\ 1 \end{bmatrix} \right\} \subseteq \mathbf{R}^2$$

□

Some basic results on span, all of which can be envisioned geometrically, follow.

THEOREM 3.7

Let $\mathbf{V}$ be a vector space with S and T finite subsets of $\mathbf{V}$. Then

(*i*) $S \subseteq \text{span } S$;

(*ii*) if $S \subseteq T$ then $\text{span } S \subseteq \text{span } T$; and

(*iii*) if $S \subseteq \text{span } T$ then $\text{span } S \subseteq \text{span } T$.

PROOF

If $S = \varnothing$ or $T = \varnothing$, the corresponding results are obvious. Thus, we need only consider $S \neq \varnothing$ and $T \neq \varnothing$. For (i), let $S = \{\mathbf{x}_1, \ldots, \mathbf{x}_m\}$. Then

$$\mathbf{x}_i = 0\mathbf{x}_1 + \cdots + 0\mathbf{x}_{i-1} + 1\mathbf{x}_i + 0\mathbf{x}_{i+1} + \cdots + 0\mathbf{x}_m \in \text{span } S$$

for each i, so $S \subseteq \text{span } S$. For (ii), let S be as defined above and let

$$T = \{\mathbf{x}_1, \ldots, \mathbf{x}_m, \ldots, \mathbf{x}_n\}$$

Then, if $\mathbf{x} \in \text{span } S$,

$$\mathbf{x} = r_1\mathbf{x}_1 + \cdots + r_m\mathbf{x}_m = r_1\mathbf{x}_1 + \cdots + r_m\mathbf{x}_m + 0\mathbf{x}_{m+1} + \cdots + 0\mathbf{x}_n \in \text{span } T$$

Since $\mathbf{x}$ was chosen arbitrarily, $\text{span } S \subseteq \text{span } T$. We leave (iii) as exercise 38. ■

We now show that the span of a finite set of vectors is always a subspace and thus span gives a general way in which subspaces, and consequently vector spaces, arise.

THEOREM 3.8

Let $\mathbf{V}$ be a vector space and let S be a finite subset of $\mathbf{V}$. Then span S is a subspace of $\mathbf{V}$.

PROOF

If $S = \varnothing$, then $\text{span } S = \{\mathbf{0}\}$, which is a subspace of $\mathbf{V}$. Thus, let $S = \{\mathbf{x}_1, \ldots, \mathbf{x}_k\}$ with $k \geqslant 1$, and let $\mathbf{W} = \text{span } S$. Using the test for subspaces, we need to show that $\mathbf{W}$ is nonempty and that $\mathbf{W}$ satisfies the closure of addition and scalar multiplication properties. Since $k \geqslant 1$, it follows that $\mathbf{W}$ is not empty. To show the closure of addition property, let

$$\mathbf{x} = r_1\mathbf{x}_1 + \cdots + r_k\mathbf{x}_k \qquad \text{and} \qquad \mathbf{y} = s_1\mathbf{x}_1 + \cdots + s_k\mathbf{x}_k$$

be in $\mathbf{W}$, where $r_1, \ldots, r_k, s_1, \ldots, s_k \in \mathbf{R}$. Then

$$\mathbf{x} + \mathbf{y} = (r_1 + s_1)\mathbf{x}_1 + \cdots + (r_k + s_k)\mathbf{x}_k \in \mathbf{W}$$

Now, to show the closure of scalar multiplication property, let

$$s \in \mathbf{R} \qquad \text{and} \qquad \mathbf{x} = r_1\mathbf{x}_1 + \cdots + r_k\mathbf{x}_k \in \mathbf{W}$$

Then

$$s\mathbf{x} = (sr_1)\mathbf{x}_1 + \cdots + (sr_k)\mathbf{x}_k$$

which is in $\mathbf{W}$. Thus the theorem follows. ■

From this result we see that the span of any finite set of vectors in a vector space gives a vector space.

Example 41

For the vector space $\mathbf{P}$, we can construct subspaces as follows.

(*i*) Let $S = \{\frac{3}{2}(t^2 - \frac{1}{3}), t, 1\}$, the set of the first three Legendre polynomials. Then

$$\text{span } S = \{a_2[\tfrac{3}{2}(t^2 - \tfrac{1}{3})] + a_1 t + a_0 1 : a_0, a_1, a_2 \in \mathbf{R}\}$$

is a subspace of $\mathbf{P}$.

(*ii*) Let $S = \{8t^3 - 12t, 4t^2 - 2, 2t, 1\}$, the set of the first four Hermite polynomials. Then

$$\text{span } S = \{a_3(8t^3 - 12t) + a_2(4t^2 - 2) + a_1(2t) + a_0 1 : a_0, a_1, a_2, a_3 \in \mathbf{R}\}$$

is a subspace of $\mathbf{P}$. (These polynomials are used in approximation work.) □

Any vector space obtained as a span is determined by a finite number of vectors. Thus it is called a *finite-dimensional vector space*. Most vector spaces of interest, at least in this text, are finite dimensional. A partial list of finite-dimensional vector spaces follows.

Example 42

(*i*) $\mathbf{V} = \{\mathbf{0}\} = \text{span } \varnothing$

(*ii*) $\mathbf{R} = \text{span } \{1\}$

(*iii*) $\mathbf{R}_2 = \text{span } \{[1,0],[0,1]\}$ since

$$[x, y] = x[1,0] + y[0,1]$$

(*iv*) $\mathbf{R}_3 = \text{span } \{[1,0,0],[0,1,0],[0,0,1]\}$

(*v*) $\mathbf{R}_n = \text{span } \{[1,0,\ldots,0],[0,1,0,\ldots,0],\ldots,[0,\ldots,0,1]\}$

(*vi*) $\mathbf{R}_{2,2} = \text{span } \left\{\begin{bmatrix}1 & 0\\0 & 0\end{bmatrix}, \begin{bmatrix}0 & 1\\0 & 0\end{bmatrix}, \begin{bmatrix}0 & 0\\1 & 0\end{bmatrix}, \begin{bmatrix}0 & 0\\0 & 1\end{bmatrix}\right\}$ since

$$\begin{bmatrix}a & b\\c & d\end{bmatrix} = a\begin{bmatrix}1 & 0\\0 & 0\end{bmatrix} + b\begin{bmatrix}0 & 1\\0 & 0\end{bmatrix} + c\begin{bmatrix}0 & 0\\1 & 0\end{bmatrix} + d\begin{bmatrix}0 & 0\\0 & 1\end{bmatrix}$$

(*vii*) $\mathbf{R}_{m,n} = \text{span } S$, where S is the set of all possible different $m \times n$ matrices such that each has exactly one entry equal to 1 and all other entries equal to 0.

(*viii*) Column space $A = \text{span } \{\mathbf{a}^1, \ldots, \mathbf{a}^n\}$ and row space $A = \text{span } \{\mathbf{a}_1, \ldots, \mathbf{a}_m\}$

(*ix*) $\mathbf{P}_n = \text{span}\,\{t^n, t^{n-1}, \ldots, 1\}$ since

$$a_n t^n + \cdots + a_0 = a_n(t^n) + \cdots + a_0(1)$$

(*x*) Lines and planes through the origin in $\mathbf{R}_3$ (exercise 36) □

However, not all vector spaces are finite dimensional, as can be seen in the following example.

EXAMPLE 43

Let $\mathbf{P}$ be the vector space of all polynomials. Let

$$S = \{p_1(t), \ldots, p_m(t)\}$$

be a set of polynomials. Let k be greater than the degree of $p_i(t)$ for all i. Now, every linear combination of $p_1(t), \ldots, p_m(t)$ is a polynomial of degree less than k. Hence span $S \neq \mathbf{P}$. Since S was arbitrarily chosen, $\mathbf{P}$ is not finite dimensional. □

The concepts of linear combination, span, and finite-dimensional vector spaces arise in practice in various ways. One application follows.

EXAMPLE 44

OPTIONAL—CONTROL PROBLEMS*

Consider two columns T_1 and T_2 of water and an infinite water source S, as diagrammed in Figure 3.39. Suppose the numbers on the columns are such that one unit on each line corresponds to one gallon of water. Let

$$x_1(k) = \text{the water level in } T_1 \text{ on day } k$$
$$x_2(k) = \text{the water level in } T_2 \text{ on day } k$$

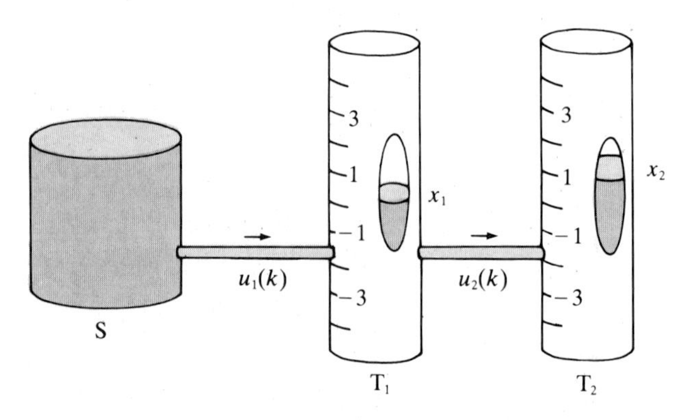

FIGURE 3.39

* Such problems arise, for example, in controlling chemical plants, electrical systems, and mechanical devices. Others can be found in books concerning control theory. We have chosen to describe a water tower system since it requires no technical background.

Suppose we can control, by valves and pumps, the daily flow of water between S, T_1, and T_2. Let

$u_1(k) =$ the number of gallons of water we decide to move from S to T_1 on day k

$u_2(k) =$ the number of gallons of water we decide to move from T_1 to T_2 on day k

Thus, if $u_i(k) > 0$, the flow is along the arrow, and if $u_i(k) < 0$, the flow is in the opposite direction. Given the water levels on day k, we can compute the levels on day $k + 1$ as

$$x_1(k + 1) = x_1(k) + u_1(k) - u_2(k)$$
$$x_2(k + 1) = x_2(k) + u_2(k)$$

or, in matrix form

$$\mathbf{x}(k + 1) = \mathbf{x}(k) + [u_1(k)]\begin{bmatrix} 1 \\ 0 \end{bmatrix} + [u_2(k)]\begin{bmatrix} -1 \\ 1 \end{bmatrix}$$

where $\mathbf{x}(k) = \begin{bmatrix} x_1(k) \\ x_2(k) \end{bmatrix}$.

Suppose the levels on day 0 are $\mathbf{x}(0) = \begin{bmatrix} 0 \\ 0 \end{bmatrix}$. We will decide if it is possible to control the flow between S, T_1, and T_2 so that, on day 1, $\mathbf{x}(1) = \begin{bmatrix} 2 \\ 0 \end{bmatrix}$. By the above equation, we set

$$\mathbf{x}(1) = \mathbf{x}(0) + [u_1(0)]\begin{bmatrix} 1 \\ 0 \end{bmatrix} + [u_2(0)]\begin{bmatrix} -1 \\ 1 \end{bmatrix}$$

or

$$\begin{bmatrix} 2 \\ 0 \end{bmatrix} = [u_1(0)]\begin{bmatrix} 1 \\ 0 \end{bmatrix} + [u_2(0)]\begin{bmatrix} -1 \\ 1 \end{bmatrix}$$

Thus, we need to see if $\begin{bmatrix} 2 \\ 0 \end{bmatrix}$ is in the span of $\left\{\begin{bmatrix} 1 \\ 0 \end{bmatrix}, \begin{bmatrix} -1 \\ 1 \end{bmatrix}\right\}$. Since $u_1(0) = 2$, $u_2(0) = 0$ solves this equation, we see that sending two gallons of water from S to T_1 provides the solution.

In fact, the above calculations show that the water levels that can be obtained are precisely those in span $\left\{\begin{bmatrix} 1 \\ 0 \end{bmatrix}, \begin{bmatrix} -1 \\ 1 \end{bmatrix}\right\}$. □

In this and the preceding two sections we have given numerous examples of vector spaces. In the next section we begin the general study of vector spaces, obtaining results that apply to all vector spaces.

Exercises for Section 3.4

Computational Exercises

For each of the diagrams in exercises 1 and 2, sketch a parallelogram as we did in Example 34 to show geometrically if **z** is a linear combination of **x**, **y**, and, when given, **w**. Also give an estimate of the scalings of **x**, **y**, and **w** used. A colored letter means that more than one solution is possible.

1. (*a*)

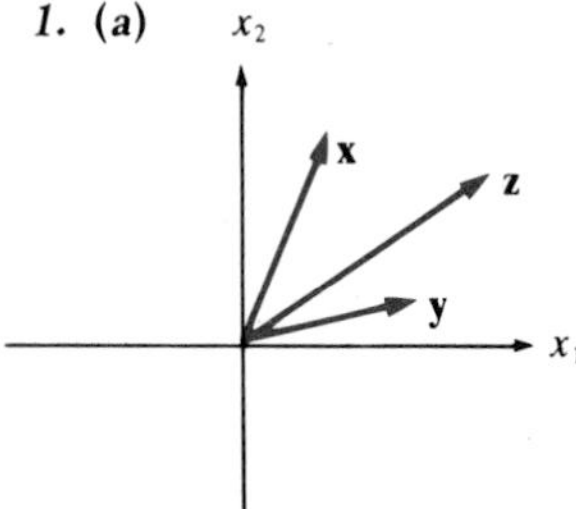

(*c*)

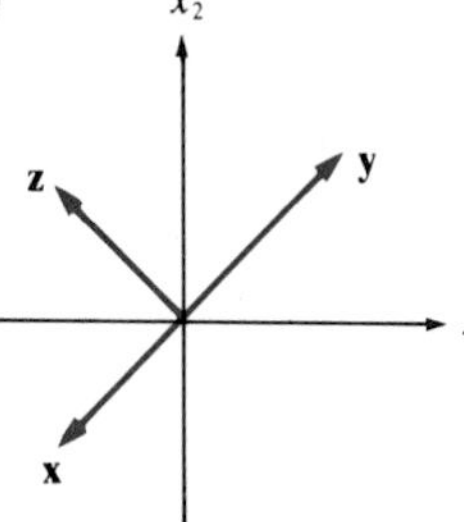

(*b*)

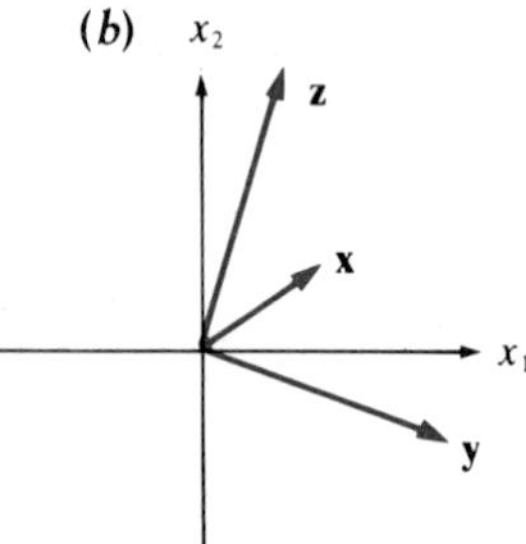

(*d*)

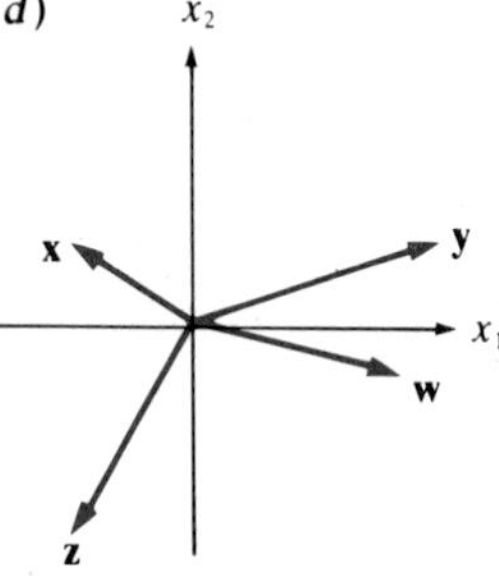

2. (*a*)

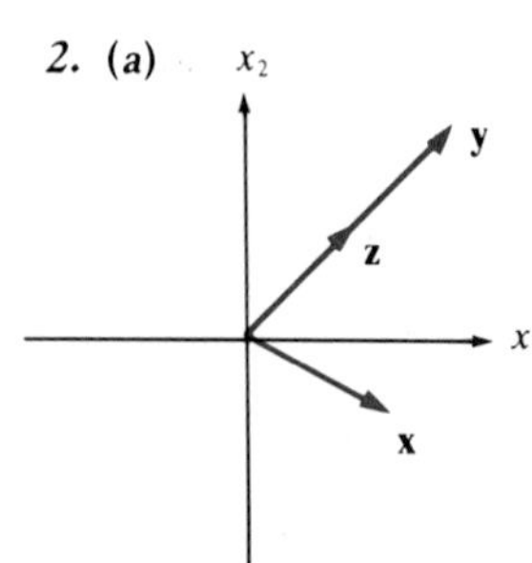

(*b*)

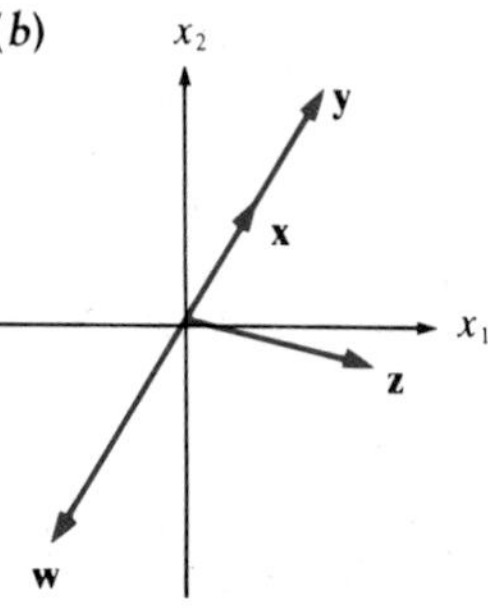

(*c*)

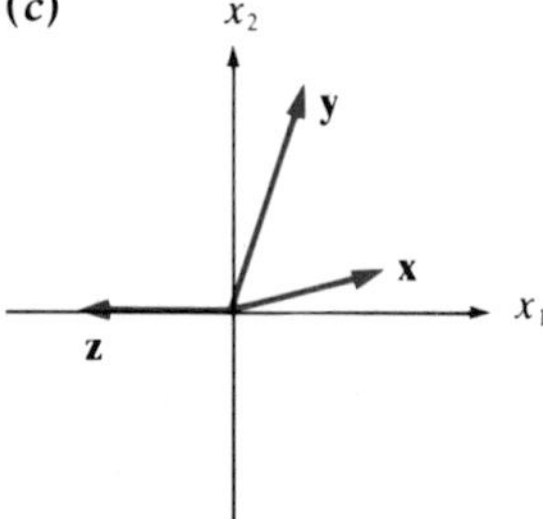

(*d*)

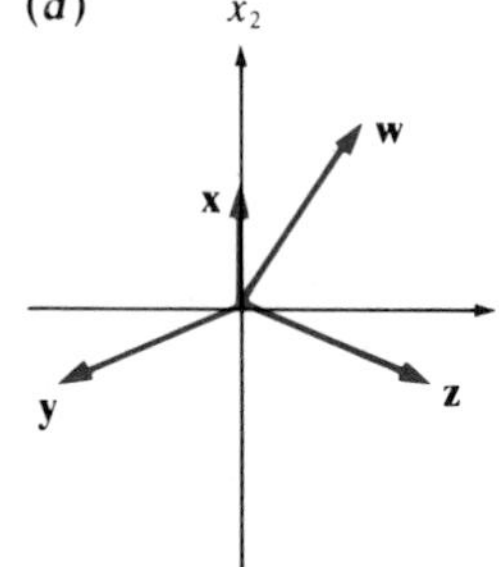

In exercises 3 and 4, show that the vectors given in (a), (b), and (c) are linear combinations of the vectors in the set S by computing the scalars in the linear combinations. Demonstrate geometrically that these scalars are also the scalings needed on the vectors of S in order to apply the parallelogram rule.

3. $S = \left\{\begin{bmatrix}1\\0\end{bmatrix}, \begin{bmatrix}3\\2\end{bmatrix}\right\}$ (*a*) $\begin{bmatrix}5\\2\end{bmatrix}$ (*b*) $\begin{bmatrix}-7\\-6\end{bmatrix}$ (*c*) $\begin{bmatrix}1\\-4\end{bmatrix}$

4. $S = \left\{\begin{bmatrix}2\\2\end{bmatrix}, \begin{bmatrix}2\\1\end{bmatrix}\right\}$ (*a*) $\begin{bmatrix}-2\\-4\end{bmatrix}$ (*b*) $\begin{bmatrix}6\\7\end{bmatrix}$ (*c*) $\begin{bmatrix}1\\-2\end{bmatrix}$

In exercises 5 through 10, determine algebraically which of the given vectors is a linear combination of the vectors in the given set S.

5. $S = \{[1,1,1], [1,1,-1], [1,-1,-1]\}$
(*a*) $[4,-2,0]$ (*b*) $[-1,-5,-1]$

6. $S = \{[-1,1,0], [1,0,-1], [1,-1,1]\}$
(*a*) $[-1,-1,6]$ (*b*) $[2,-1,-5]$

7. $S = \{[1,1,1,0], [0,1,1,1], [1,0,1,1]\}$
(*a*) $[2,2,3,2]$ (*b*) $[0,2,3,2]$

8. $S = \{[1,1,1,0], [1,1,0,1], [1,0,1,1], [0,1,1,1]\}$
(*a*) $[0,3,0,2]$ (*b*) $[2,1,2,1]$

9. $S = \{3t^2 - 2t - 1, -6t^2 + 4t + 2, t^2 - 4t - 2\}$
(*a*) $t^2 + 6t + 3$ (*b*) $t^2 - 2t + 1$

10. $S = \{6t^3 - 2t + 4, -3t^3 + t - 2, t^2 - 3, 3t^3 + 2t^2 - t - 4\}$
(*a*) $3t^3 - 3t^2 + t + 7$ (*b*) $6t^3 - t^2 - 2t + 7$

In exercises 11 through 14, shade the span of each of the given sets of vectors. Find all subsets of the vectors having the same span.

11. *13.*

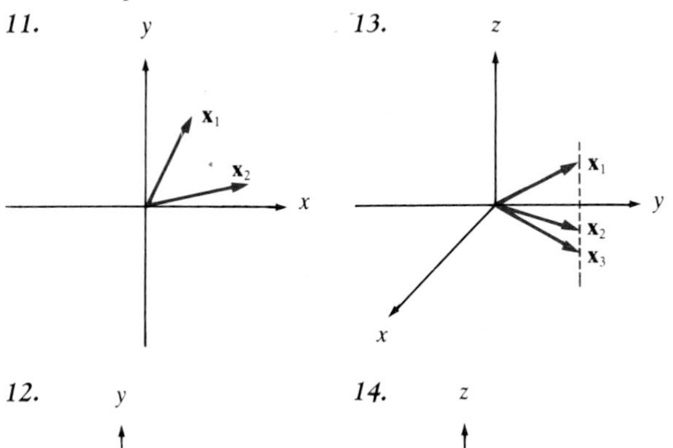

12. *14.*

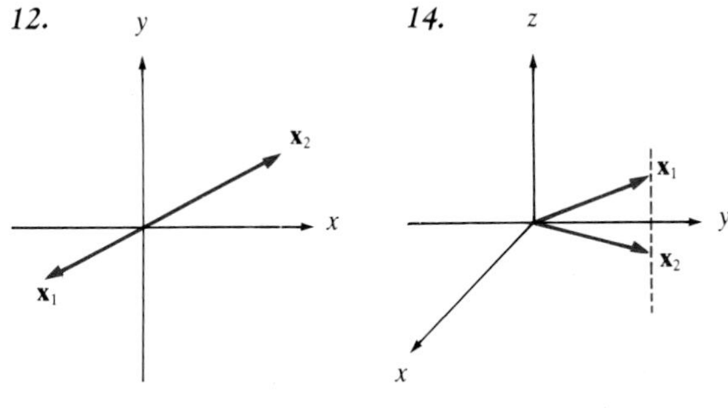

In exercises 15 through 18, geometrically describe the given set.

15. span $\{[0,0],[1,-2],[-3,6]\}$

16. span $\{[1,1],[-1,1]\}$

17. span $\{[1,2,1],[1,3,1]\}$

18. span $\{[0,0,0],[-1,1,1],[2,1,-1]\}$

In exercises 19 through 24, decide algebraically whether or not the given vector $\mathbf{x}$ is in span S.

19. $S=\{[2,1,3],[4,7,-1]\}, \mathbf{x}=[0,5,-7]$

20. $S=\{[1,5,-2],[-4,6,2]\}, \mathbf{x}=[7,5,-8]$

21. $S=\{[1,-1,0,1],[0,1,-1,1]\}, \mathbf{x}=[1,0,-1,2]$

22. $S=\{[1,1,0,1],[1,0,1,1],[0,1,1,1]\}, \mathbf{x}=[0,2,0,1]$

23. $S=\{t^2-t+1, 2t^2+3t-5\}, \mathbf{x}=5t^2+15t-28$

24. $S=\{-2t^2+3t-4, 3t^2+t+6\}, \mathbf{x}=11t$

In exercises 25 through 32, determine algebraically if the given set S spans the given space, and prove your answer. In exercises 25 and 26, show your result geometrically.

25. $S=\{[1,1],[1,2]\}, \mathbf{R}_2$

26. $S=\{[-1,1],[1,-1]\}, \mathbf{R}_2$

27. $S=\{[1,2,3],[1,2,2],[-1,1,1]\}, \mathbf{R}_3$

28. $S=\{[1,1,1],[1,1,2],[1,2,1]\}, \mathbf{R}_3$

29. $S=\left\{\begin{bmatrix}2 & 1\\ -1 & 0\end{bmatrix}, \begin{bmatrix}3 & 1\\ 1 & 1\end{bmatrix}, \begin{bmatrix}2 & 2\\ 1 & 1\end{bmatrix}, \begin{bmatrix}2 & 3\\ -4 & -1\end{bmatrix}\right\}, \mathbf{R}_{2,2}$

30. $S=\left\{\begin{bmatrix}-1 & 1\\ 1 & 1\end{bmatrix}, \begin{bmatrix}-1 & -1\\ 1 & 1\end{bmatrix}, \begin{bmatrix}-1 & -1\\ -1 & 1\end{bmatrix}, \begin{bmatrix}1 & 1\\ 1 & 1\end{bmatrix}\right\},$ $\mathbf{R}_{2,2}$

31. $S=\{2, -t+1, t^2+t+1\}, \mathbf{P}_2$

32. $S=\{t^2-t+1, 2t^2+t-3, t^2+5t-9\}, \mathbf{P}_2$

33. Let $p_0(t)=1$, $p_1(t)=t$, $p_2(t)=\frac{1}{2}(3t^2-1)$, and $p_3(t)=\frac{1}{2}(5t^3-3t)$, the first four Legendre polynomials. Show that t^3+t^2+t+1 is a linear combination of these polynomials. (Legendre polynomials are used in approximation.)

34. Let $T_0(t)=1$, $T_1(t)=t$, and $T_2(t)=2t^2-1$. Show that

$$\mathbf{P}_2 = \text{span}\,\{T_0, T_1, T_2\}$$

(T_0, T_1, and T_2 are known as the first three Chebyshev polynomials. These polynomials are used in approximation work.)

COMPLEX NUMBERS

35. Show that $\mathbf{C}_n=$ span $\{\mathbf{e}_1,\ldots,\mathbf{e}_n\}$, and hence it is finite dimensional.

THEORETICAL EXERCISES

36. Explain geometrically, using sketches, why the vector spaces consisting of the following are finite dimensional:

(*a*) a line through the origin in $\mathbf{R}^3$

(*b*) a plane through the origin in $\mathbf{R}^3$

37. Explain why the vector space of all fixed arrows is finite dimensional.

38. Prove part (iii) of Theorem 3.7. Give a few sketches in $\mathbf{R}^3$.

39. (geometry in $\mathbf{R}_n$) A line in $\mathbf{R}_3$ through $\mathbf{x}_0$ in the direction of $\mathbf{a}\neq\mathbf{0}$ is the set of those $\mathbf{x}\in\mathbf{R}_3$ that are obtained as $\mathbf{x}=\mathbf{x}_0+t\mathbf{a}$ for some $t\in\mathbf{R}$. The equation can also be used to define a line in $\mathbf{R}_n$ through $\mathbf{x}_0\in\mathbf{R}_n$ and in the direction of $\mathbf{a}\in\mathbf{R}_n$. Write the equation of the line for the following $\mathbf{x}_0$, $\mathbf{a}\in\mathbf{R}_n$. Graph the line in (a) and (b).

(*a*) $\mathbf{x}_0=[1,1], \mathbf{a}=[1,-1]$

(*b*) $\mathbf{x}_0=[1,2,0], \mathbf{a}=[0,0,1]$

(*c*) $\mathbf{x}_0=[1,-1,1,1], \mathbf{a}=[2,1,-2,3]$

(*d*) $\mathbf{x}_0=[1,0,1,-1,2], \mathbf{a}=[0,1,-1,3,2]$

40. (geometry in $\mathbf{R}_n$) The word "between" makes sense in $\mathbf{R}$, $\mathbf{R}_2$, and $\mathbf{R}_3$. For $\mathbf{R}_n$, we say that $\mathbf{x}\in\mathbf{R}_n$ is *between* $\mathbf{y}$ and $\mathbf{z}$ in

$\mathbf{R}_n$ if there is a scalar r, where $0 \leqslant r \leqslant 1$, such that

$$\mathbf{x} = r\mathbf{y} + (1 - r)\mathbf{z}$$

For the given **x**, **y**, and **z** below, decide if **x** is between **y** and **z**. Make a sketch in (a) and (b).

(*a*) $\mathbf{x} = [0, 3]$, $\mathbf{y} = [-1, 4]$, $\mathbf{z} = [1, 2]$

(*b*) $\mathbf{x} = [-1, 1, 2]$, $\mathbf{y} = [1, 1, 0]$, $\mathbf{z} = [0, 1, 1]$

(*c*) $\mathbf{x} = [2, 1, 2, 8]$, $\mathbf{y} = [1, -1, 3, 5]$, $\mathbf{z} = [3, 3, 1, 11]$

41. (difference equations) Consider the difference equation

$$x(k+2) - 5x(k+1) + 6x(k) = 0$$

Show that $x_1(k) = 2^k$ and $x_2(k) = 3^k$ are solutions to this equation, and thus

$$x(k) = r_1 x_1(k) + r_2 x_2(k)$$

is also a solution for any choice of r_1 and r_2. Among these solutions, can you find one such that $x(0) = 1$ and $x(1) = -1$? (Such problems arise when one is looking for a particular solution that describes a specific physical system.)

42. (differential equations) (calculus) Consider

$$y''(t) - 5y'(t) + 6y(t) = 0$$

Show that $y_1(t) = e^{2t}$ and $y_2(t) = e^{3t}$ are solutions to this equation, and thus

$$y(t) = r_1 y_1(t) + r_2 y_2(t)$$

is a solution for any choices of r_1 and r_2. Are there numbers r_1 and r_2 such that $y(0) = 0$ and $y'(0) = 1$? (See the comment in exercise 41 above.)

43. (systems of difference equations) Consider the equation

$$\mathbf{x}(k+1) = \begin{bmatrix} 1 & 2 \\ 2 & 1 \end{bmatrix} \mathbf{x}(k)$$

Show that

$$\mathbf{x}_1(k) = 3^k \begin{bmatrix} 1 \\ 1 \end{bmatrix} \quad \text{and} \quad \mathbf{x}_2(k) = (-1)^k \begin{bmatrix} 1 \\ -1 \end{bmatrix}$$

are solutions to this equation, and thus so is

$$\mathbf{x}(k) = r_1 \mathbf{x}_1(k) + r_2 \mathbf{x}_2(k)$$

for any real numbers r_1 and r_2. Are there numbers r_1 and r_2 such that $\mathbf{x}(0) = \begin{bmatrix} 1 \\ -2 \end{bmatrix}$? (See the comment in exercise 41.)

44. Let $\mathbf{a}^1, \ldots, \mathbf{a}^n \in \mathbf{R}^n$ and $A = [\mathbf{a}^1, \cdots, \mathbf{a}^n]$. Prove that, if $\det A \neq 0$, then

$$\text{span}\,\{\mathbf{a}^1, \ldots, \mathbf{a}^n\} = \mathbf{R}^n$$

Hint: Consider $\mathbf{b} = x_1\mathbf{a}^1 + \cdots + x_n\mathbf{a}^n = A\mathbf{x}$ and apply Cramer's Rule. (This gives an easy way to decide if n vectors in $\mathbf{R}^n$ span $\mathbf{R}^n$. Similar results apply in $\mathbf{R}_n$, $\mathbf{R}_{m,n}$, and $\mathbf{P}_n$.)

Applications Exercises

45. (*a*) Consider the water system shown in Figure 3.40. Write the equations for this system. If $\mathbf{x}(0) = \begin{bmatrix} 0 \\ 0 \\ 0 \end{bmatrix}$, decide if the levels $\begin{bmatrix} -1 \\ -\frac{1}{2} \\ \frac{3}{2} \end{bmatrix}$ can be achieved at $\mathbf{x}(1)$.

(*b*) What about $\begin{bmatrix} 1 \\ 0 \\ -1 \end{bmatrix}$?

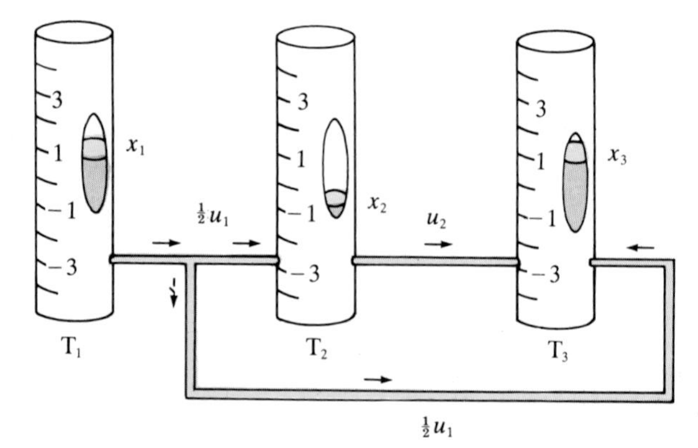

FIGURE 3.40

46. Consider the water tower system shown in Figure 3.41. Write the equations for the system. Let $\mathbf{x}(0) = \begin{bmatrix} 0 \\ 0 \\ 0 \end{bmatrix}$. Use

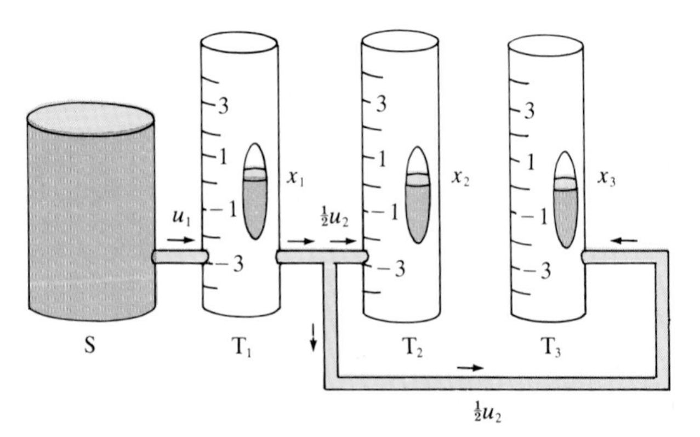

FIGURE 3.41

the span of a set of vectors to describe $\mathbf{W}(1)$, the set of all vectors $\mathbf{x}(1)$ that can be obtained from choices of $u_1(0)$ and $u_2(0)$. If $\mathbf{x} \in \mathbf{W}(1)$, what is the connection between the water levels and the components of $\mathbf{x}$?

COMPUTER EXERCISE

47. Plot the fixed arrows $\mathbf{x} = \begin{bmatrix} 1 \\ 0 \\ 0 \end{bmatrix}$, $\mathbf{y} = \begin{bmatrix} 0 \\ 1 \\ 0 \end{bmatrix}$, and $\mathbf{z} = \begin{bmatrix} 1 \\ 1 \\ 0 \end{bmatrix}$. Show that $\mathbf{z}$ is a linear combination of $\mathbf{x}$ and $\mathbf{y}$. Draw a sphere, of any radius, about the tip of $\mathbf{z}$. Show that there is a fixed arrow $\mathbf{z}'$ with tip in this sphere such that $\mathbf{z}'$ is not a linear combination of $\mathbf{x}$ and $\mathbf{y}$. Thus, if small errors are made in $\mathbf{x}$, $\mathbf{y}$, and $\mathbf{z}$ to obtain $\mathbf{x}'$, $\mathbf{y}'$, and $\mathbf{z}'$, it may be that $\mathbf{z}'$ is not a linear combination of $\mathbf{x}'$ and $\mathbf{y}'$ even though $\mathbf{z}$ is a linear combination of $\mathbf{x}$ and $\mathbf{y}$. (Round-off affects the determination of whether $\mathbf{z}$ is a linear combination of $\mathbf{x}$ and $\mathbf{y}$.)

3.5 LINEAR INDEPENDENCE AND BASIS

In the last three sections we showed numerous examples of vector spaces. We now start a general study of them, obtaining results that can be used in any vector space. Basically, this study generalizes, to any vector space $\mathbf{V}$, those properties of $\mathbf{R}_1$, $\mathbf{R}_2$, and $\mathbf{R}_3$ which are used in mathematically formulating and solving problems. We have already seen how to compute using the arithmetical properties of $\mathbf{V}$. Geometrically, we now develop the concepts of dimension, distance, and angle. This section and the next introduce the concept of dimension. Chapter 4 concerns the remaining two concepts.

We know $\mathbf{R}_2$ has dimension 2—length and width—and that $\mathbf{R}_3$ has dimension 3—length, width, and height. To extend the concept of dimension to all finite-dimensional vector spaces, we intend to find the minimum number of vectors required to span it. We will do this by finding a spanning set that contains no proper spanning set. Geometrically in $\mathbf{R}_3$, in such sets each vector points in a direction away from the span of the remaining vectors, as is shown in the example below. Thus, each vector gives a new dimension to the vector space.

EXAMPLE 45

Figure 3.42 shows sets in which each vector points away from the span of the remaining vectors. As a consequence, these sets can have no subsets with the same spanning capabilities. □

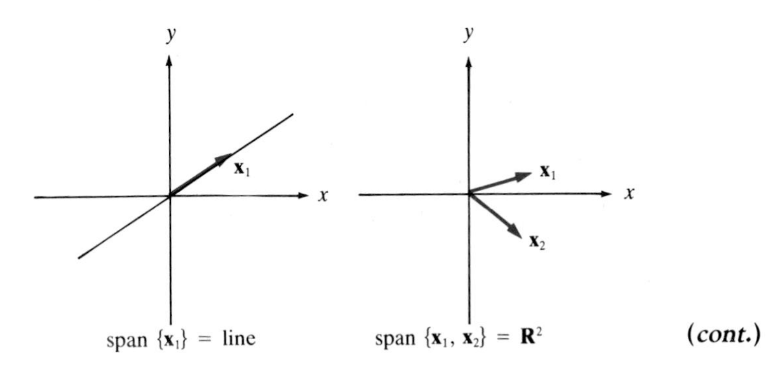

FIGURE 3.42

(cont.)

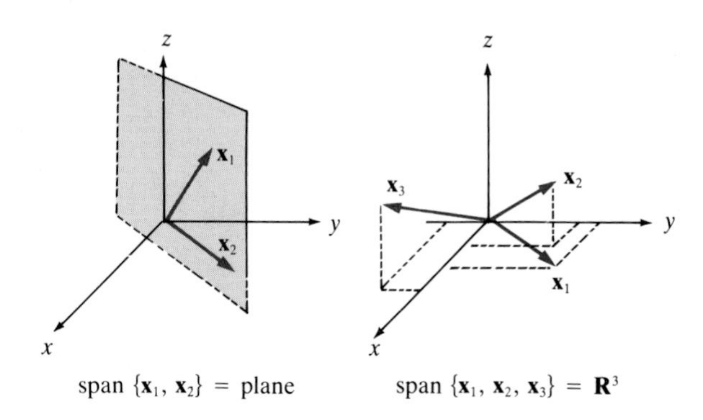

FIGURE 3.42 (cont.)

span $\{\mathbf{x}_1, \mathbf{x}_2\}$ = plane span $\{\mathbf{x}_1, \mathbf{x}_2, \mathbf{x}_3\} = \mathbf{R}^3$

We now describe this concept for general vector spaces.

DEFINITION

Let $\mathbf{V}$ be a vector space and let

$$S = \{\mathbf{x}_1, \ldots, \mathbf{x}_k\} \subseteq \mathbf{V}$$

We say that S is *linearly independent* if and only if

(*i*) $k = 0$ (so $S = \varnothing$), or

(*ii*) $k = 1$ and $\mathbf{x}_1 \neq \mathbf{0}$, or

(*iii*) $k > 1$ and no $\mathbf{x}_i$ is in the span of the remaining vectors $\mathbf{x}_1, \ldots, \mathbf{x}_{i-1}, \mathbf{x}_{i+1}, \ldots, \mathbf{x}_k$.

If S is not linearly independent, we call it *linearly dependent*.*

The theorem stating that linearly independent sets are minimal spanning sets follows.

THEOREM 3.9

Let $\mathbf{V}$ be a vector space and let $S = \{\mathbf{x}_1, \ldots, \mathbf{x}_n\}$ be a linearly independent set in $\mathbf{V}$. If S' is a proper subset of S, then span S' is a proper subset of span S.

PROOF

Suppose $\mathbf{x}_i \notin S'$. Then, by the definition of linear independence, $\mathbf{x}_i \notin \text{span } S'$. Since span $S' \subseteq$ span S, it follows that span S' must be a proper subset of span S. ■

The definitions of linear independence and linear dependence are general. We apply them in various particular vector spaces below.

EXAMPLE 46

The set $\varnothing$ is linearly independent by part (i) of the definition. There is no geometry associated with this part of the definition. It is given as a mathematical convenience so that we need not consider various special cases in the discussions that follow. □

* We use this definition for linear independence because it is geometrically appealing.

EXAMPLE 47

The set $\{[0,0,0]\}$ is linearly dependent by part (ii) of the definition. Geometrically, the vector $[0,0,0]$ does not provide a new direction or point away from span $\varnothing = \{\mathbf{0}\}$. □

EXAMPLE 48

Since $p(t) = t^2 - 1 \neq 0$, the set $\{t^2 - 1\}$ in the vector space $\mathbf{P}$ of polynomials is linearly independent. □

EXAMPLE 49

Decide whether the set $\{[1,1,0],[0,1,1]\}$ is linearly independent.

Applying the definition, we will first see if $[1,1,0]$ is in the span of the remaining vector $[0,1,1]$. For this, we need to see if we can solve

$$r[0,1,1] = [1,1,0]$$

for $r \in \mathbf{R}$. Equating entries, we have

$$\begin{aligned} 0 &= 1 \\ r &= 1 \\ r &= 0 \end{aligned}$$

This system clearly has no solution, and so $[1,1,0]$ is not in span $\{[0,1,1]\}$.

Now, to see if $[0,1,1]$ is in span $\{[1,1,0]\}$, we write

$$r[1,1,0] = [0,1,1]$$

Again we have a system with no solution:

$$\begin{aligned} r &= 0 \\ r &= 1 \\ 0 &= 1 \end{aligned}$$

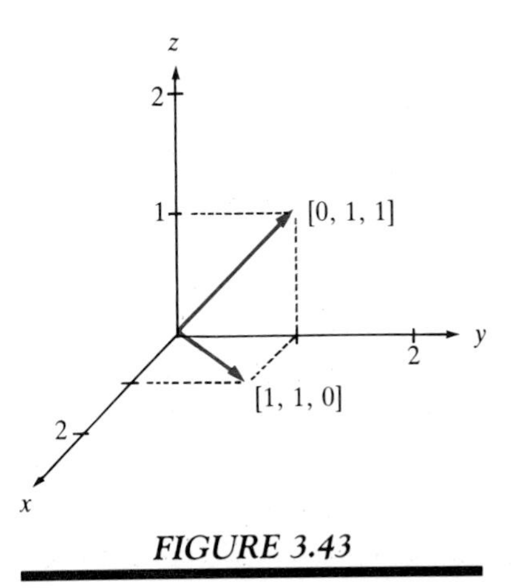

FIGURE 3.43

Thus $[0,1,1]$ is not in span $\{[1,1,0]\}$.

It follows from these calculations that $\{[1,1,0],[0,1,1]\}$ is linearly independent.

These calculations can be compared to the sketch in Figure 3.43, where we can see geometrically that this set is linearly independent. □

To decide if a set $\{\mathbf{x}_1, \ldots, \mathbf{x}_k\}$ of vectors is linearly independent we look at k equations, one for each time we check to see if $\mathbf{x}_i$ is in the span of the remaining vectors. Fortunately, the following theorem will let us make this decision by considering only one equation.

THEOREM 3.10

(test for linear independence) Let $\mathbf{V}$ be a vector space and let

$$S = \{\mathbf{x}_1, \ldots, \mathbf{x}_k\} \subseteq \mathbf{V}$$

with $k \geqslant 1$. Then S is linearly independent if and only if the equation

$$a_1\mathbf{x}_1 + \cdots + a_k\mathbf{x}_k = \mathbf{0}$$

has $a_1 = \cdots = a_k = 0$ as its only solution.

PROOF

We need to consider two cases.

CASE 1: Suppose $k = 1$. In this case, if S is linearly independent, then $\mathbf{x}_1 \neq \mathbf{0}$. Hence $a_1\mathbf{x}_1 = \mathbf{0}$ has $a_1 = 0$ as its only solution. On the other hand, if $a_1\mathbf{x}_1 = \mathbf{0}$ has $a_1 = 0$ as its only solution, then it is clear that $\mathbf{x}_1 \neq \mathbf{0}$, so $\{\mathbf{x}_1\}$ is linearly independent.

CASE 2: Suppose $k > 1$. In this case, first suppose S is linearly independent. Consider the equation

$$a_1\mathbf{x}_1 + \cdots + a_k\mathbf{x}_k = \mathbf{0}$$

If this equation has as its only solution $a_1 = \cdots = a_k = 0$, then we have the result. Thus, suppose there is another solution, say $b_1, \ldots, b_k$, in which $b_i \neq 0$ for some i. Then, solving for $b_i\mathbf{x}_i$, we get

$$b_i\mathbf{x}_i = -b_1\mathbf{x}_1 - \cdots - b_{i-1}\mathbf{x}_{i-1} - b_{i+1}\mathbf{x}_{i+1} - \cdots - b_k\mathbf{x}_k$$

so

$$\mathbf{x}_i = -\frac{b_1}{b_i}\mathbf{x}_1 - \cdots - \frac{b_{i-1}}{b_i}\mathbf{x}_{i-1} - \frac{b_{i+1}}{b_i}\mathbf{x}_{i+1} - \cdots - \frac{b_k}{b_i}\mathbf{x}_k$$

But this equation says that $\mathbf{x}_i$ is in the span of the remaining vectors in S, which is a contradiction. Hence we can conclude that the only solution to the equation is $a_1 = \cdots = a_k = 0$.

On the other hand, suppose the only solution to

$$a_1\mathbf{x}_1 + \cdots + a_k\mathbf{x}_k = \mathbf{0}$$

is $a_1 = \cdots = a_k = 0$. Now, either S is linearly independent or some $\mathbf{x}_i$ is in the span of the remaining vectors in S. If the latter were true, we would have

$$\mathbf{x}_i = b_1\mathbf{x}_1 + \cdots + b_{i-1}\mathbf{x}_{i-1} + b_{i+1}\mathbf{x}_{i+1} + \cdots + b_k\mathbf{x}_k$$

for some $b_1, \ldots, b_{i-1}, b_{i+1}, \ldots, b_k \in \mathbf{R}$. But then setting $a_1 = b_1, \ldots, a_{i-1} = b_{i-1}$,

$a_i = -1, a_{i+1} = b_{i+1}, \ldots, a_k = b_k$, we would have another solution to the equation

$$a_1\mathbf{x}_1 + \cdots + a_k\mathbf{x}_k = \mathbf{0}$$

Since this is a contradiction, S is linearly independent. ■

EXAMPLE 50

Decide if $S = \left\{ \begin{bmatrix} 1 \\ -1 \\ 0 \end{bmatrix}, \begin{bmatrix} 1 \\ 1 \\ -1 \end{bmatrix}, \begin{bmatrix} 0 \\ 1 \\ -1 \end{bmatrix} \right\}$ is linearly independent. By the above theorem, to make this decision we need to see if the equation

$$r_1 \begin{bmatrix} 1 \\ -1 \\ 0 \end{bmatrix} + r_2 \begin{bmatrix} 1 \\ 1 \\ -1 \end{bmatrix} + r_3 \begin{bmatrix} 0 \\ 1 \\ -1 \end{bmatrix} = \begin{bmatrix} 0 \\ 0 \\ 0 \end{bmatrix}$$

has as its only solution $r_1 = r_2 = r_3 = 0$. Rewriting this equation yields

$$\begin{aligned} r_1 + r_2 &= 0 \\ -r_1 + r_2 + r_3 &= 0 \\ -r_2 - r_3 &= 0 \end{aligned}$$

Solving this equation by the Gaussian elimination algorithm gives

$$\begin{bmatrix} 1 & 1 & 0 & 0 \\ -1 & 1 & 1 & 0 \\ 0 & -1 & -1 & 0 \end{bmatrix} \xrightarrow{R_2 + R_1 \to R_2} \begin{bmatrix} 1 & 1 & 0 & 0 \\ 0 & 2 & 1 & 0 \\ 0 & -1 & -1 & 0 \end{bmatrix}$$

$$\xrightarrow{R_3 + \frac{1}{2}R_2 \to R_3} \begin{bmatrix} 1 & 1 & 0 & 0 \\ 0 & 2 & 1 & 0 \\ 0 & 0 & -\frac{1}{2} & 0 \end{bmatrix}$$

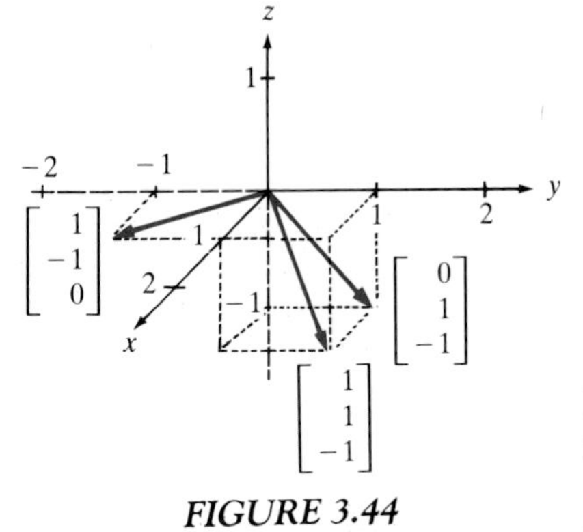

FIGURE 3.44

Thus

$$\begin{aligned} r_1 + r_2 &= 0 \\ 2r_2 + r_3 &= 0 \\ -\tfrac{1}{2}r_3 &= 0 \end{aligned}$$

Back substitution yields $r_1 = r_2 = r_3 = 0$. Since this is the only solution, it follows that S is linearly independent.

Depicting the vectors as fixed arrows, we can visualize the linear independence as shown in Figure 3.44. □

EXAMPLE 51

Decide if $S = \{3, t - 1, 2t + 1\}$ is linearly independent in the vector space **P** of polynomials.

To make this decision, we need to see if the equation

$$r_1(3) + r_2(t - 1) + r_3(2t + 1) = 0$$

has as its only solution $r_1 = r_2 = r_3 = 0$. Rewriting the equation gives

$$(r_2 + 2r_3)t + (3r_1 - r_2 + r_3) = 0$$

Equating coefficients of like terms produces the system

$$\begin{aligned} r_2 + 2r_3 &= 0 \\ 3r_1 - r_2 + r_3 &= 0 \end{aligned}$$

But since r_3 is free, this system has infinitely many solutions, so $r_1 = r_2 = r_3 = 0$ cannot be the only solution. Thus S is linearly dependent. □

EXAMPLE 52

Let $S = \{2^k, 2^{k+1}\} \subseteq \mathbf{F}(\mathbf{N})$. To decide if S is linearly independent, we write

$$r_1 2^k + r_2 2^{k+1} = 0 \qquad \text{or} \qquad 2^k(r_1 + 2r_2) = 0$$

Since $2^k \neq 0$,

$$r_1 + 2r_2 = 0$$

Since r_2 is free, there are infinitely many solutions to this equation, and so S is linearly dependent. □

Using linear independence, we can describe minimal spanning sets as follows.

DEFINITION

Let $\mathbf{V}$ be a vector space. A set

$$S = \{\mathbf{x}_1, \ldots, \mathbf{x}_s\} \subseteq \mathbf{V}$$

is a *basis** (plural: *bases*) for $\mathbf{V}$ if and only if

(*i*) S is linearly independent, and

(*ii*) span $S = \mathbf{V}$.

EXAMPLE 53

The set $\varnothing$ is a basis for $\mathbf{V} = \{\mathbf{0}\}$. □

EXAMPLE 54

Let $S = \{[1, 0, 0], [0, 1, 0], [0, 0, 1]\}$. We will show that S is a basis for $\mathbf{R}_3$.

Geometrically it can be seen that S is linearly independent and spans $\mathbf{R}_3$, as shown in Figure 3.45. Thus S is a basis for $\mathbf{R}_3$.

* "Basis" means the foundation or the fundamental part.

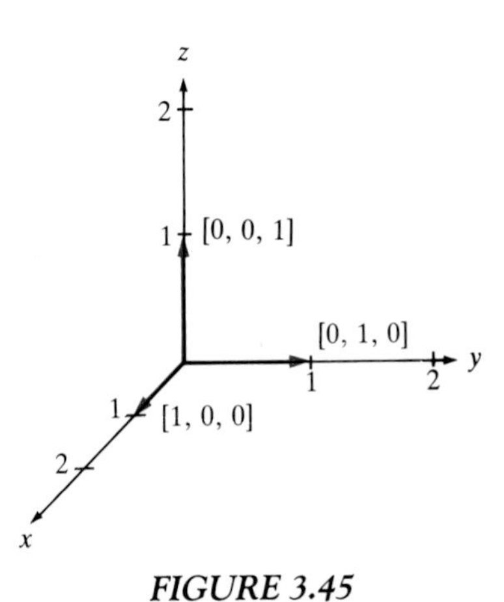

FIGURE 3.45

Algebraically, to show that S is linearly independent, we write the equation

$$a[1,0,0] + b[0,1,0] + c[0,0,1] = [0,0,0]$$

Solving this equation yields $a = b = c = 0$. Thus S is linearly independent.

Now, to show that span $S = \mathbf{R}_3$, let

$$\mathbf{x} = [x_1, x_2, x_3] \in \mathbf{R}_3$$

To show that $\mathbf{x} \in$ span S, we write the equation

$$a[1,0,0] + b[0,1,0] + c[0,0,1] = [x_1, x_2, x_3]$$

where the unknowns are a, b, and c. This equation has the obvious solution $a = x_1$, $b = x_2$, and $c = x_3$. Thus, since $\mathbf{x}$ can be chosen arbitrarily, it follows that every $\mathbf{x} \in \mathbf{R}_3$ is in span S, and so span $S = \mathbf{R}_3$. □

EXAMPLE 55

Using the methods of Example 54, we can show some rather natural bases for $\mathbf{R}_n$, $\mathbf{R}^n$, $\mathbf{R}_{m,n}$, and $\mathbf{P}_n$.

(*i*) The natural basis for $\mathbf{R}_n$ is the set $\{\mathbf{e}_1, \mathbf{e}_2, \ldots, \mathbf{e}_n\}$.

(*ii*) The natural basis for $\mathbf{R}^n$ is the set $\{\mathbf{e}^1, \mathbf{e}^2, \ldots, \mathbf{e}^n\}$.

(*iii*) The natural basis for $\mathbf{R}_{m,n}$ is the set of all distinct $m \times n$ matrices having exactly one entry equal to 1 and all other entries equal to 0.

(*iv*) The natural basis for $\mathbf{P}_n$ is the set $\{t^n, t^{n-1}, t^{n-2}, \ldots, t, 1\}$ □

EXAMPLE 56

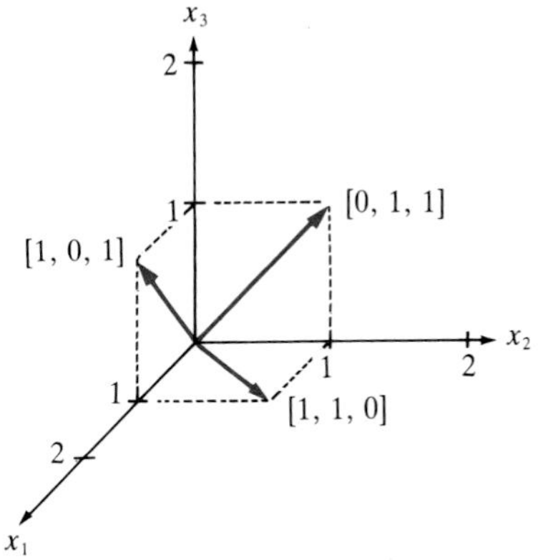

Show that $S = \{[1,0,1],[1,1,0],[0,1,1]\}$ is a basis for $\mathbf{R}_3$.

Geometrically, we see from Figure 3.46 that S is a basis for $\mathbf{R}_3$. To show this result algebraically, we check to see if S satisfies the definition of a basis for this space. Consider an arbitrary vector

$$\mathbf{x} = [x_1, x_2, x_3] \in \mathbf{R}_3$$

We need to see if $\mathbf{x}$ can be written as a linear combination of the vectors in S. We write

$$a[1,0,1] + b[1,1,0] + c[0,1,1] = [x_1, x_2, x_3]$$

FIGURE 3.46

Rewriting yields

$$\begin{aligned} a + b \quad\;\; &= x_1 \\ b + c &= x_2 \\ a \quad\;\; + c &= x_3 \end{aligned}$$

Solving by the Gaussian elimination algorithm gives

$$\begin{bmatrix} 1 & 1 & 0 & x_1 \\ 0 & 1 & 1 & x_2 \\ 1 & 0 & 1 & x_3 \end{bmatrix} \xrightarrow{R_3 - R_1 \to R_3} \begin{bmatrix} 1 & 1 & 0 & x_1 \\ 0 & 1 & 1 & x_2 \\ 0 & -1 & 1 & x_3 - x_1 \end{bmatrix}$$

$$\xrightarrow{R_3 + R_2 \to R_3} \begin{bmatrix} 1 & 1 & 0 & x_1 \\ 0 & 1 & 1 & x_2 \\ 0 & 0 & 2 & x_2 + x_3 - x_1 \end{bmatrix}$$

or

$$\begin{aligned} a + b \quad &= x_1 \\ b + c &= x_2 \\ c &= \frac{x_2 + x_3 - x_1}{2} \end{aligned}$$

Thus,

$$c = \frac{x_2 + x_3 - x_1}{2}$$

$$b = x_2 - c = \frac{x_1 + x_2 - x_3}{2}$$

$$a = x_1 - b = \frac{x_1 - x_2 + x_3}{2}$$

and so the equation has a solution. Hence $\mathbf{x} \in \operatorname{span} S$. Since $\mathbf{x}$ is arbitrary, it follows that $\operatorname{span} S = \mathbf{R}_3$.

Further, the above computation shows that, for each $\mathbf{x}$, the values of a, b, and c are unique. Thus, if we choose $\mathbf{x} = [0, 0, 0]$, the only possible solution to the equation

$$a[1, 0, 1] + b[1, 1, 0] + c[0, 1, 1] = \mathbf{x}$$

is $a = b = c = 0$. Hence S is also linearly independent and so is a basis of $\mathbf{R}_3$.

□

EXAMPLE 57

Decide if $S = \left\{ \begin{bmatrix} 1 \\ -1 \\ 1 \end{bmatrix}, \begin{bmatrix} 2 \\ 0 \\ 1 \end{bmatrix} \right\}$ is a basis for $\mathbf{R}^3$.

It is clear from Figure 3.47 that all linear combinations of the two vectors in S lie in the plane determined by the fixed arrows associated with these vectors. Hence no vector outside of this plane can be in span S, and so S is not a basis for $\mathbf{R}^3$.

To make this decision algebraically, we need to see if S satisfies the definition of a basis.

To see if $\operatorname{span} S = \mathbf{R}^3$, take any $\mathbf{x} = \begin{bmatrix} x_1 \\ x_2 \\ x_3 \end{bmatrix} \in \mathbf{R}^3$. We need to see if $\mathbf{x}$ is a

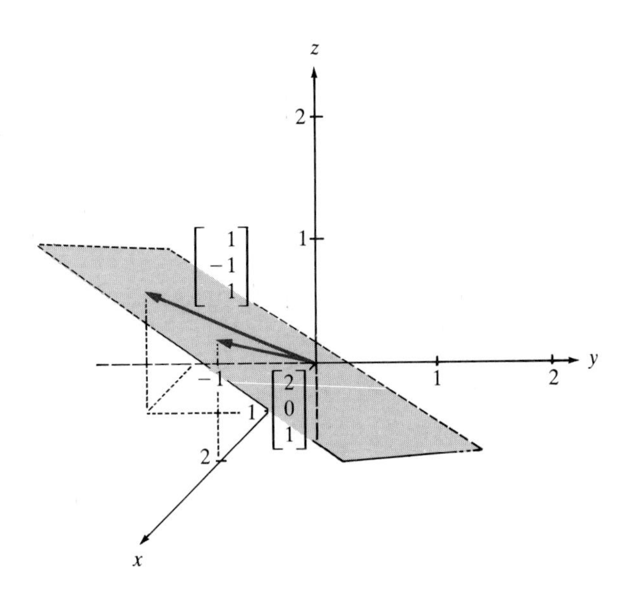

FIGURE 3.47

linear combination of the vectors in S. We write

$$a\begin{bmatrix} 1 \\ -1 \\ 1 \end{bmatrix} + b\begin{bmatrix} 2 \\ 0 \\ 1 \end{bmatrix} = \begin{bmatrix} x_1 \\ x_2 \\ x_3 \end{bmatrix}$$

or

$$\begin{aligned} a + 2b &= x_1 \\ -a \qquad &= x_2 \\ a + \ b &= x_3 \end{aligned}$$

Solving by the Gaussian elimination algorithm yields

$$\begin{bmatrix} 1 & 2 & x_1 \\ -1 & 0 & x_2 \\ 1 & 1 & x_3 \end{bmatrix} \xrightarrow[\mathbf{R}_3 - \mathbf{R}_1 \to \mathbf{R}_1]{\mathbf{R}_2 + \mathbf{R}_1 \to \mathbf{R}_2} \begin{bmatrix} 1 & 2 & x_1 \\ 0 & 2 & x_1 + x_2 \\ 0 & -1 & x_3 - x_1 \end{bmatrix}$$

$$\xrightarrow{\mathbf{R}_3 + \frac{1}{2}\mathbf{R}_2 \to \mathbf{R}_3} \begin{bmatrix} 1 & 2 & x_1 \\ 0 & 2 & x_1 + x_2 \\ 0 & 0 & \dfrac{-x_1 + x_2 + 2x_3}{2} \end{bmatrix}$$

But the last equation given by this matrix is

$$0 = \tfrac{1}{2}(2x_3 - x_1 + x_2)$$

so if we choose $x_1 = x_2 = x_3 = 1$, corresponding to $\mathbf{x} = \begin{bmatrix} 1 \\ 1 \\ 1 \end{bmatrix}$, it follows that

there is no solution to the system. Thus $\begin{bmatrix}1\\1\\1\end{bmatrix} \notin \text{span } S$, and so span $S \neq \mathbf{R}^3$.
Hence we conclude that S is not a basis for $\mathbf{R}^3$. □

The previous examples concerned linear independence and bases. Next we show how such concepts can arise in practice.

EXAMPLE 58

OPTIONAL—CONTROL PROBLEMS

Extend the water tower system of Example 44 of Section 3.4 as depicted in Figure 3.48. Then we have the equations

$$x_1(k+1) = x_1(k) + u_1(k) - u_2(k)$$
$$x_2(k+1) = x_2(k) + u_2(k) + u_3(k)$$

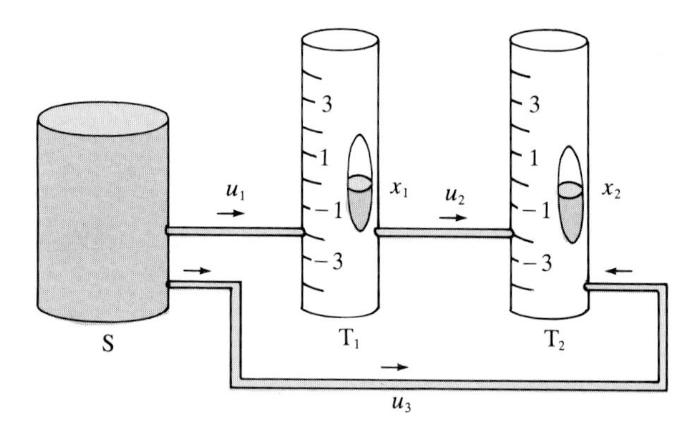

FIGURE 3.48

or $$\mathbf{x}(k+1) = \mathbf{x}(k) + u_1(k)\begin{bmatrix}1\\0\end{bmatrix} + u_2(k)\begin{bmatrix}-1\\1\end{bmatrix} + u_3(k)\begin{bmatrix}0\\1\end{bmatrix}$$

Note that $\left\{\begin{bmatrix}1\\0\end{bmatrix}, \begin{bmatrix}-1\\1\end{bmatrix}, \begin{bmatrix}0\\1\end{bmatrix}\right\}$ is linearly dependent. Hence we can delete one of the directed supply lines and still control the water levels as before. For example, since

$$\text{span}\left\{\begin{bmatrix}1\\0\end{bmatrix}, \begin{bmatrix}-1\\1\end{bmatrix}, \begin{bmatrix}0\\1\end{bmatrix}\right\} = \text{span}\left\{\begin{bmatrix}-1\\1\end{bmatrix}, \begin{bmatrix}0\\1\end{bmatrix}\right\}$$

we can delete u_1, with the resulting equation

$$\mathbf{x}(k+1) = \mathbf{x}(k) + u_2(k)\begin{bmatrix}-1\\1\end{bmatrix} + u_3(k)\begin{bmatrix}0\\1\end{bmatrix}$$

and the resulting diagram shown in Figure 3.49.

The calculations above show that the water tower system could be designed more efficiently to eliminate redundant lines. Alternatively, they

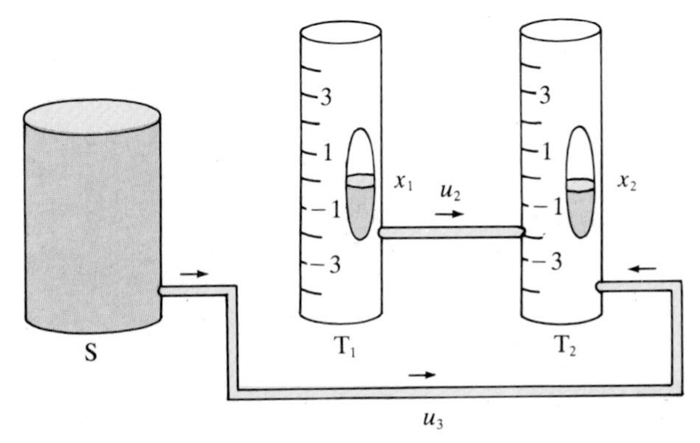

FIGURE 3.49

show that this system as presently constituted can still be controlled if a line fails. □

In the next section we continue the study of bases, using them to obtain dimensions for vector spaces.

EXERCISES FOR SECTION 3.5

COMPUTATIONAL EXERCISES

In exercises 1 and 2, decide geometrically whether the given sets of vectors are linearly independent, and explain why.

1. (*a*)

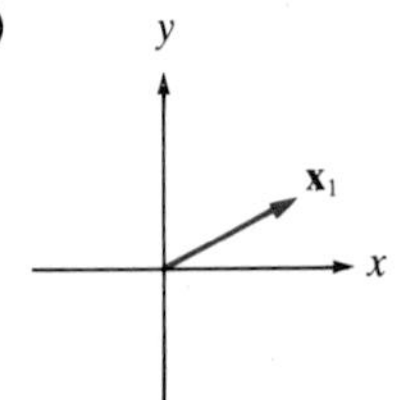

(*b*)

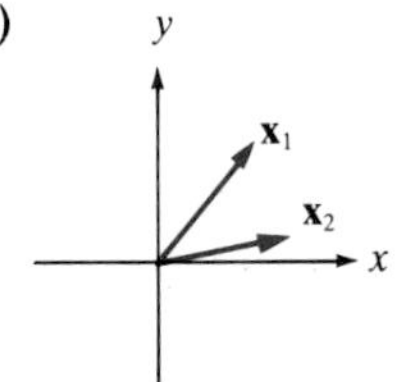

(*c*)

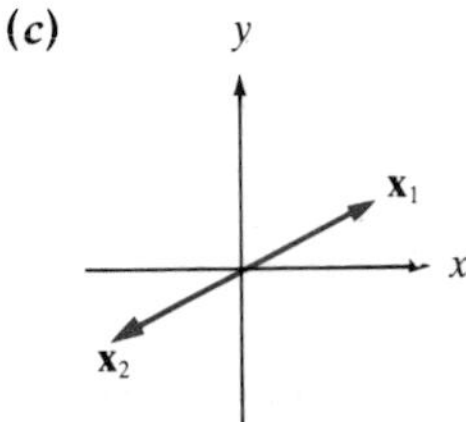

2. (*a*)

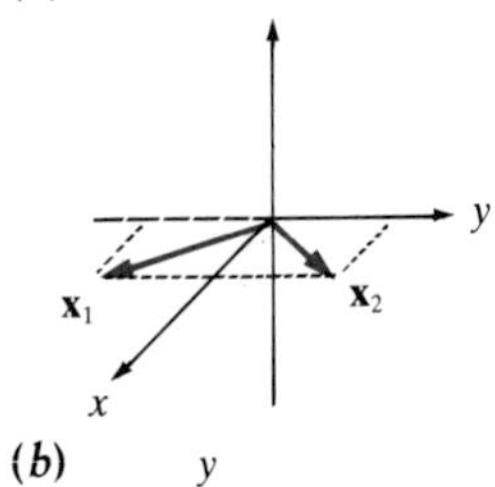

(*b*)

(*c*)

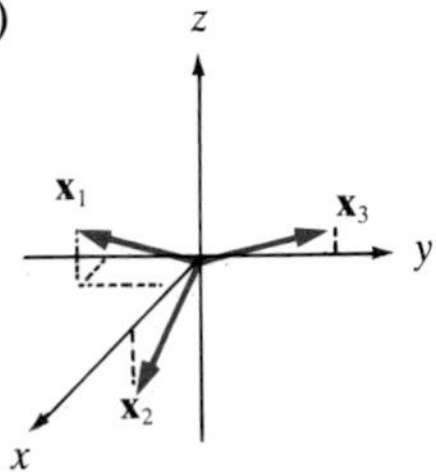

In exercises 3 and 4, graph the span of each of the given sets of vectors. Find all linearly independent subsets that span the same space.

3. (*a*)

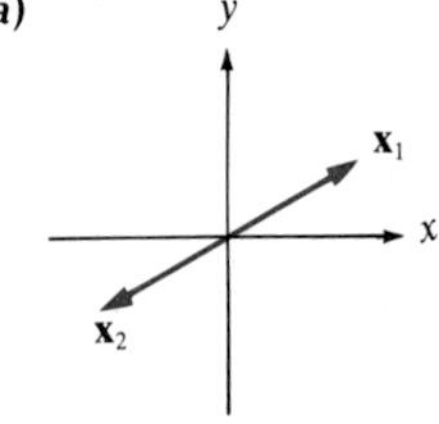

(*b*)

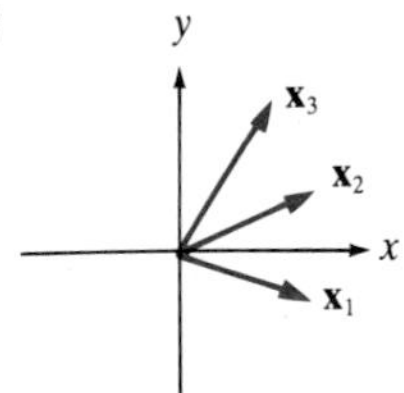

(*c*)

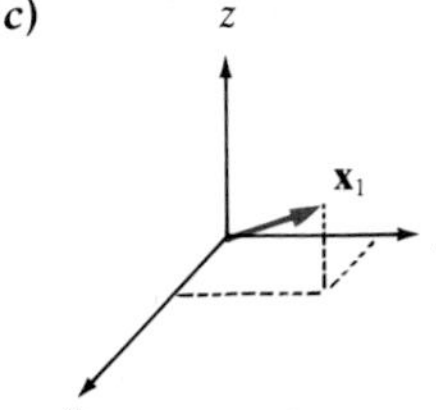

4. (*a*)

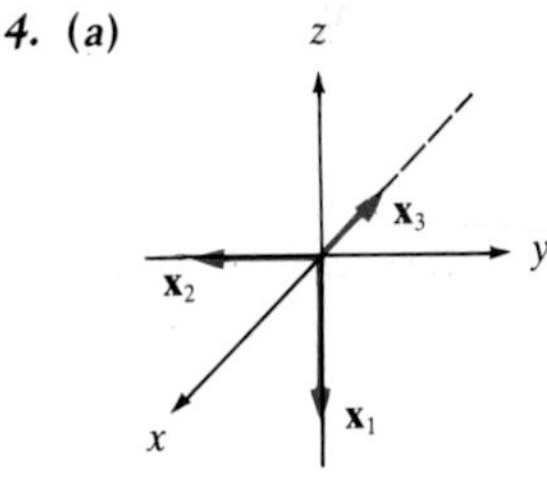

(*b*)

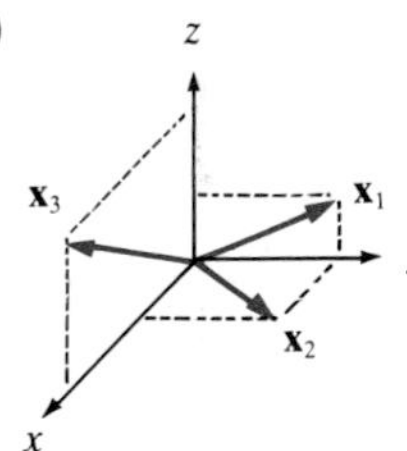

(*c*)

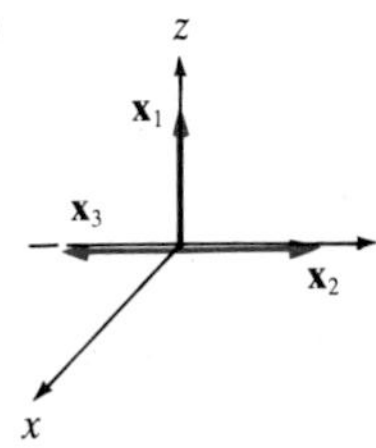

In exercises 5 through 10, show algebraically which of the sets given are linearly independent. Then graph and view each conclusion geometrically.

5. $\{[1,0],[-3,0]\}$

6. $\{[-3,2],[6,-3]\}$

7. $\{[2,1],[1,3]\}$

8. $\{[4,-2],[-2,1]\}$

9. $\{[1,-1],[2,3],[-5,8]\}$

10. $\{[1,1],[-1,1],[1,3]\}$

In exercises 11 through 24, determine algebraically if the given set is linearly independent or linearly dependent.

11. $\{[1,2,1],[-1,1,2],[3,3,1]\}$

12. $\{[-1,2,0],[1,2,1],[4,2,2]\}$

13. $\{[1,1,0,0],[0,1,1,0],[0,0,1,1],[1,0,0,1]\}$

14. $\{[1,-1,1,1],[1,1,-1,1],[-1,1,1,1],[1,1,1,-1]\}$

15. $\left\{\begin{bmatrix}1 & 1\\ 0 & 0\end{bmatrix},\begin{bmatrix}1 & 0\\ 1 & 1\end{bmatrix},\begin{bmatrix}1 & 1\\ 1 & 1\end{bmatrix}\right\}$

16. $\left\{\begin{bmatrix}1 & 0\\ 0 & 1\end{bmatrix},\begin{bmatrix}0 & 1\\ 1 & 0\end{bmatrix},\begin{bmatrix}-1 & 0\\ 0 & 1\end{bmatrix},\begin{bmatrix}0 & -1\\ 1 & 0\end{bmatrix}\right\}$

17. $\{3^k, 3^{k+1}\}$

18. $\{1, 2^k, 4^k\}$

19. $\{t+1, t-1\}$

20. $\{t-2, 2-t\}$

21. $\{t-2, t-1\}$

22. $\{t, t^2-1, t+2\}$

23. $\{t^2-t+1, 2t^2+2t-1, 4t-3\}$

24. $\{\sin^2 x, \cos^2 x, \cos 2x\}$. *Hint:* Use the double-angle formula for $\cos 2x$.

In exercises 25 through 28, decide geometrically if the given set S of vectors is a basis of the given vector space $\mathbf{V}$, and verify your result algebraically.

25. $S=\{[1,2],[2,1]\}$, $\mathbf{V}=\mathbf{R}_2$

26. $S=\left\{\begin{bmatrix}-1\\ 1\end{bmatrix},\begin{bmatrix}1\\ 1\end{bmatrix}\right\}$, $\mathbf{V}=\mathbf{R}^2$

27. $S=\left\{\begin{bmatrix}1\\ 1\\ -1\end{bmatrix},\begin{bmatrix}1\\ -1\\ 1\end{bmatrix},\begin{bmatrix}-1\\ 1\\ 1\end{bmatrix}\right\}$, $\mathbf{V}=\mathbf{R}^3$

28. $S=\{[1,0,1],[1,1,0],[0,1,1]\}$, $\mathbf{V}=\mathbf{R}_3$

In exercises 29 through 42, decide algebraically if the set is a basis for the given vector space $\mathbf{V}$. In exercises 29 through 34, graph the set and its span.

29. $S=\{[1,1],[-1,1]\}$, $\mathbf{V}=\mathbf{R}_2$

30. $S=\{[6,-12],[-2,4]\}$, $\mathbf{V}=\mathbf{R}_2$

31. $S=\{[-2,1,1],[1,-2,1],[1,1,-2]\}$, $\mathbf{V}=\mathbf{R}_3$

32. $S=\{[1,1,0],[0,1,1],[1,0,1]\}$, $\mathbf{V}=\mathbf{R}_3$

33. $S=\{[-1,1,1],[1,-1,1],[1,1,-1]\}$, $\mathbf{V}=\mathbf{R}_3$

34. $S=\{[2,1,1],[1,2,1],[-1,1,0]\}$, $\mathbf{V}=\mathbf{R}_3$

35. $S=\{[1,0,1,0],[0,1,0,1],[1,0,0,1],[0,1,1,0]\}$, $\mathbf{V}=\mathbf{R}_4$

36. $S=\{[1,0,-1,0],[0,-1,1,0],[0,0,1,-1],[0,0,1,0]\}$, $\mathbf{V}=\mathbf{R}_4$

37. $S=\left\{\begin{bmatrix}1 & 0\\ 0 & 1\end{bmatrix},\begin{bmatrix}0 & 1\\ 1 & 0\end{bmatrix},\begin{bmatrix}1 & 1\\ 0 & 0\end{bmatrix},\begin{bmatrix}0 & 0\\ 1 & 1\end{bmatrix}\right\}$, $\mathbf{V}=\mathbf{R}_{2,2}$

38. $S=\left\{\begin{bmatrix}1 & -1\\ 1 & 1\end{bmatrix},\begin{bmatrix}1 & 1\\ 1 & -1\end{bmatrix},\begin{bmatrix}-1 & 1\\ 1 & 1\end{bmatrix},\begin{bmatrix}1 & 1\\ -1 & 1\end{bmatrix}\right\}$, $\mathbf{V}=\mathbf{R}_{2,2}$

39. $S=\{3t-1, -3t-1\}$, $\mathbf{V}=\mathbf{P}_1$

40. $S=\{4t-2, -2t+1\}$, $\mathbf{V}=\mathbf{P}_1$

41. $S=\{t^2+t+1, t^2+t-1, t^2-t-1\}$, $\mathbf{V}=\mathbf{P}_2$

42. $S=\{t^2-3t+1, -t^2+2t+1, t^2-5t+5\}$, $\mathbf{V}=\mathbf{P}_2$

43. The equations for a line through the origin in the direction of the fixed arrow $[1,1,1]$ are $x=y=z$. There are three equations here. Show that the set of these equations is linearly dependent.

Complex Numbers

In exercises 44 and 45, decide if the sets are linearly independent.

44. $\{[1,i,0],[0,i,1+i],[1-i,3i,0]\}$

45. $\{[i,-i,0],[1,0,i],[1+i,-i,i]\}$

Theoretical Exercises

46. Let $\mathbf{V}$ be a vector space. Show that $\{\mathbf{x}_1,\mathbf{x}_2\}\subseteq\mathbf{V}$ is linearly dependent if and only if one of $\mathbf{x}_1$ or $\mathbf{x}_2$ is a scalar multiple of the other. (This is a useful tool for deciding if a set of two vectors is linearly independent or linearly dependent.) Test this result on exercises 5, 6, 7, 8, 17, 19, 20, and 21.

47. Give examples to show that a subset S of a linearly dependent set T in a vector space $\mathbf{V}$ may be linearly dependent or linearly independent. *Hint:* Look at some sketches in $\mathbf{R}_2$.

48. Prove that every subset of a linearly independent set is linearly independent. Make some sketches showing this in $\mathbf{R}_3$.

49. Prove that if S is linearly dependent and if $S\subseteq T$, then T is linearly dependent. Make some sketches showing this in $\mathbf{R}_2$.

50. Let $\mathbf{V}$ be a vector space and S_1 and S_2 be linearly independent subsets of $\mathbf{V}$. Prove that $S_1\cap S_2$ is also linearly independent. Make some sketches showing this in $\mathbf{R}_3$.

51. Prove that if S is a set of vectors and if $\mathbf{0}\in S$, then S is linearly dependent.

52. (calculus) Prove that if r_1 and r_2 are distinct real numbers, then $\{e^{r_1 t}, e^{r_2 t}\}$ is linearly independent. *Hint:* Let $a_1 e^{r_1 t} + a_2 e^{r_2 t} = 0$. Let $t = 0$ to get the equation $a_1 + a_2 = 0$. Now differentiate to get $r_1 a_1 e^{r_1 t} + r_2 a_2 e^{r_2 t} = 0$. Again let $t = 0$ to get the equation $r_1 a_1 + r_2 a_2 = 0$. Solve the system. This method can be extended to arbitrarily large sets with distinct real numbers $r_1, \ldots, r_n$.

53. Suppose $\{\mathbf{x}_1, \ldots, \mathbf{x}_n\}$ is a linearly independent subset of a vector space $\mathbf{V}$.
(*a*) Prove that if $r_1, \ldots, r_n$ in $\mathbf{R}$ are all nonzero, then $\{r_1\mathbf{x}_1, \ldots, r_n x_n\}$ is also linearly independent. Look at some sketches in $\mathbf{R}_3$.
(*b*) Prove that if $r \in \mathbf{R}$, then $\{\mathbf{x}_1, \ldots, \mathbf{x}_i + r\mathbf{x}_j, \ldots, \mathbf{x}_n\}$ is also linearly independent.
(*c*) Note that (a) and (b) show that the Gaussian operations, applied to a linearly independent set of vectors, preserve linear independence. Thus, if $\mathbf{x}_1, \ldots, \mathbf{x}_m \in \mathbf{R}_n$, we can decide if this set is linearly independent by forming $A = \begin{bmatrix} \mathbf{x}_1 \\ \vdots \\ \mathbf{x}_m \end{bmatrix}$ and obtaining an echelon form $\begin{bmatrix} \varepsilon_1 \\ \vdots \\ \varepsilon_m \end{bmatrix}$ for A. Then $\{\mathbf{x}_1, \ldots, \mathbf{x}_m\}$ is linearly independent if and only if $\{\varepsilon_1, \ldots, \varepsilon_m\}$ is linearly independent. Explain this.

54. (geometry in $\mathbf{R}_n$) A triangle in $\mathbf{R}_2$ and a tetrahedron in $\mathbf{R}_3$ can be generalized to an object in $\mathbf{R}_n$. To see how this is done, let $\mathbf{x}_0, \mathbf{x}_1, \ldots, \mathbf{x}_n$ be points in $\mathbf{R}_n$ such that

$$\{\mathbf{x}_1 - \mathbf{x}_0, \ldots, \mathbf{x}_n - \mathbf{x}_0\}$$

is linearly independent. By a *pyramid* in $\mathbf{R}_n$ we mean the points $\mathbf{x}_0, \ldots, \mathbf{x}_n$ and all points between them (see exercise 40 of Section 3.4). Decide whether the following sets of points describe pyramids. Sketch (a) and (b).
(*a*) $\{[0,1],[1,0],[1,1]\}$
(*b*) $\{[0,3,3],[3,3,0],[3,0,3],[2,2,2]\}$
(*c*) $\{[0,0,0,0],[1,0,1,0],[1,0,0,0],[0,1,1,0],[0,1,0,1]\}$

APPLICATIONS EXERCISE

55. Write the equations for the water tower system shown in Figure 3.50. As in Example 58, decide if the system could be designed more efficiently by deleting lines.

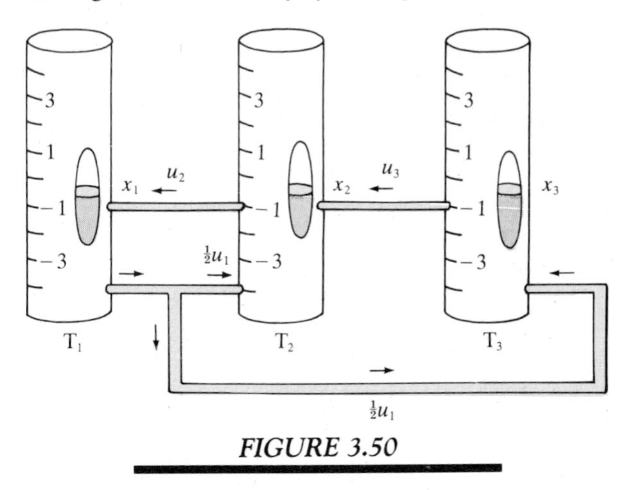

FIGURE 3.50

COMPUTER EXERCISE

56. Plot fixed arrows $\mathbf{x}_1 = \begin{bmatrix} 2 \\ -1 \end{bmatrix}$ and $\mathbf{x}_2 = \begin{bmatrix} 4 \\ -2 \end{bmatrix}$. Note that the set consisting of these vectors is linearly dependent. Draw circles of any radii about the tips of $\mathbf{x}_1$ and $\mathbf{x}_2$. Show that there are fixed arrows $\mathbf{y}_1$ and $\mathbf{y}_2$ with tips in these circles such that the set $\{\mathbf{y}_1, \mathbf{y}_2\}$ is linearly independent. Thus, if small errors are made in $\mathbf{x}_1$ and $\mathbf{x}_2$, the resulting set can be linearly independent. This result can be generalized to $\mathbf{R}^n$; that is, small errors made in a linearly dependent set can result in a linearly independent set. (Thus, in deciding on linear dependence, round-off error is not acceptable.)

3.6 DIMENSION

In this section we show that every finite-dimensional vector space $\mathbf{V}$ has a basis and how this basis can be used to determine a dimension for $\mathbf{V}$.

Bases for finite-dimensional vector spaces can be found using the following theorem.

THEOREM 3.11

(deletion theorem) Let $\mathbf{V}$ be a vector space and let

$$S = \{\mathbf{x}_1, \ldots, \mathbf{x}_m\} \subseteq \mathbf{V}$$

If $\mathbf{x}_i$ is a linear combination of the remaining vectors in S and

$$S' = \{\mathbf{x}_1, \dots, \mathbf{x}_{i-1}, \mathbf{x}_{i+1}, \dots, \mathbf{x}_m\}$$

then

$$\operatorname{span} S' = \operatorname{span} S$$

PROOF

Since $S' \subseteq S$, $\operatorname{span} S' \subseteq \operatorname{span} S$. Thus we only need to show that $\operatorname{span} S \subseteq \operatorname{span} S'$. For this, let $\mathbf{x} \in \operatorname{span} S$. Then

$$\mathbf{x} = r_1\mathbf{x}_1 + \cdots + r_m\mathbf{x}_m$$

for some $r_1, \dots, r_m \in \mathbf{R}$. Since $\mathbf{x}_i$ is a linear combination of $\mathbf{x}_1, \dots, \mathbf{x}_{i-1}, \mathbf{x}_{i+1}, \dots, \mathbf{x}_m$, we can write

$$\mathbf{x}_i = s_1\mathbf{x}_1 + \cdots + s_{i-1}\mathbf{x}_{i-1} + s_{i+1}\mathbf{x}_{i+1} + \cdots + s_m\mathbf{x}_m$$

By substitution,

$$\begin{aligned}
\mathbf{x} &= r_1\mathbf{x}_1 + \cdots + r_i(s_1\mathbf{x}_1 + \cdots + s_{i-1}\mathbf{x}_{i-1} + s_{i+1}\mathbf{x}_{i+1} + \cdots + s_m\mathbf{x}_m) \\
&\quad + r_{i+1}\mathbf{x}_{i+1} + \cdots + r_m\mathbf{x}_m \\
&= (r_1 + r_i s_1)\mathbf{x}_1 + \cdots + (r_{i-1} + r_i s_{i-1})\mathbf{x}_{i-1} + (r_{i+1} + r_i s_{i+1})\mathbf{x}_{i+1} \\
&\quad + \cdots + (r_m + r_i s_m)\mathbf{x}_m
\end{aligned}$$

Thus $\mathbf{x} \in \operatorname{span} S'$. Since $\mathbf{x}$ was arbitrarily chosen, $\operatorname{span} S \subseteq \operatorname{span} S'$. ∎

The algorithm for finding a basis contained in a spanning set of a finite-dimensional vector space follows.

DELETION ALGORITHM

Let $S = \{\mathbf{x}_1, \dots, \mathbf{x}_m\}$ be a spanning set of a vector space $\mathbf{V}$, and let $S_0 = S$.

1. Write the equation

$$a_1\mathbf{x}_1 + \cdots + a_m\mathbf{x}_m = \mathbf{0}$$

If $a_1 = \cdots = a_m = 0$ is the only solution to this equation, then S_0 is linearly independent. Otherwise, find a nonzero solution $b_1, \dots, b_m$ to this equation. Let b_i be the last nonzero number in this list.* Then solving for $\mathbf{x}_i$ shows that $\mathbf{x}_i$ is a linear combination of the remaining vectors in S_0.

2. Form S_1 by deleting $\mathbf{x}_i$ from S_0. Then $\operatorname{span} S_1 = \operatorname{span} S_0$.

3. Repeat this process until a set S_r is formed that is linearly independent. Then S_r is a basis for $\mathbf{V}$.

* Later in the text we will see that our decision in Step 1 to take the last nonzero number, thus protecting the initial vectors, can be very useful.

Applying the algorithm, we can show that every spanning set contains a basis.

COROLLARY 3.11A

Every finite-dimensional vector space has a basis.

PROOF

Let $S = \{\mathbf{x}_1, \ldots, \mathbf{x}_s\} \subseteq \mathbf{V}$, where span $S = \mathbf{V}$. By applying the deletion algorithm to S, we obtain a basis for $\mathbf{V}$. ■

EXAMPLE 59

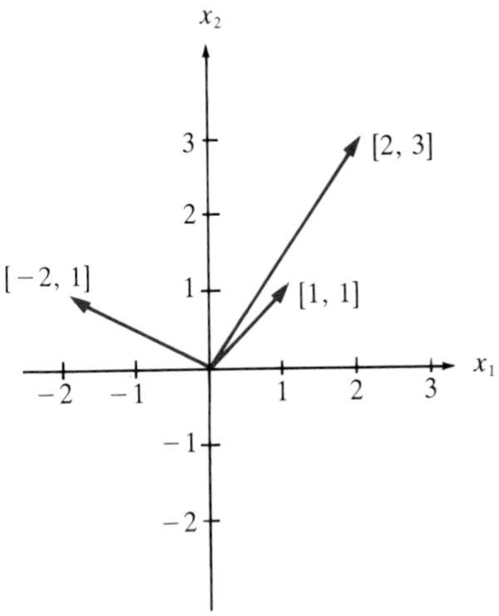

Let $S_1 = \{[1,1], [2,3], [-2,1]\}$, as shown in Figure 3.51. Note that span $S_1 = \mathbf{R}_2$. Applying the deletion algorithm, we have

$$r_1[1,1] + r_2[2,3] + r_3[-2,1] = \mathbf{0}$$

or

$$\begin{aligned} r_1 + 2r_2 - 2r_3 &= 0 \\ r_1 + 3r_2 + r_3 &= 0 \end{aligned}$$

Solving by the Gaussian elimination algorithm yields

$$\begin{bmatrix} 1 & 2 & -2 & 0 \\ 1 & 3 & 1 & 0 \end{bmatrix} \xrightarrow{\mathbf{R}_2 - \mathbf{R}_1 \to \mathbf{R}_2} \begin{bmatrix} 1 & 2 & -2 & 0 \\ 0 & 1 & 3 & 0 \end{bmatrix}$$

or

$$\begin{aligned} r_1 + 2r_2 - 2r_3 &= 0 \\ r_2 + 3r_3 &= 0 \end{aligned}$$

FIGURE 3.51

Thus r_3 is free, and we eliminate $[-2,1]$ from S_1 to obtain

$$S_2 = \{[1,1], [2,3]\}$$

Since S_2 is linearly independent, S_2 is a basis for $\mathbf{R}_2$, as can be seen by viewing Figure 3.51. □

EXAMPLE 60

Let

$$S = \{[1,1,1], [1,0,0], [1,0,1], [0,1,0], [0,0,1]\} \subseteq \mathbf{R}_3$$

as shown in Figure 3.52. Since the natural basis for $\mathbf{R}_3$ is contained in S, it is clear that span $S = \mathbf{R}_3$. We now find a basis in S, using the deletion algorithm. The equation

$$a_1[1,1,1] + a_2[1,0,0] + a_3[1,0,1] + a_4[0,1,0] + a_5[0,0,1] = \mathbf{0}$$

Using this result, we can show that the dimensional relationship between a finite-dimensional vector space and any of its subspaces is as you might expect from sketches.

COROLLARY 3.12C

If $\mathbf{V}$ is a finite-dimensional vector space and $\mathbf{W}$ is a subspace of $\mathbf{V}$, then $\mathbf{W}$ is finite dimensional and $\dim \mathbf{W} \leqslant \dim \mathbf{V}$.

PROOF

See exercise 37. ■

Also from sketches we can see that if the dimension of a vector space is known, then the following theorem can be used to show when a set is a basis.

THEOREM 3.13

Let $\mathbf{V}$ be a finite-dimensional vector space with $\dim \mathbf{V} = n$. Let

$$S = \{\mathbf{x}_1, \ldots, \mathbf{x}_n\} \subseteq \mathbf{V}$$

(*i*) If S is linearly independent, then S is a basis for $\mathbf{V}$.
(*ii*) If span $S = \mathbf{V}$, then S is a basis for $\mathbf{V}$.

PROOF

For (i), see exercise 38. For (ii), if S is not linearly independent, then by applying the deletion theorem we can delete some vectors from S to obtain a linearly independent set S' with span $S' = \mathbf{V}$. But then S' is a basis with fewer than n vectors. Since this cannot happen, S must be linearly independent and a basis for $\mathbf{V}$. ■

EXAMPLE 65

Show that the set of Hermite polynomials

$$S = \{4t^2 - 2, 2t, 1\}$$

is a basis for $\mathbf{P}_2$. Here we need only show that S is linearly independent or that it spans $\mathbf{P}_2$. We show the former. For this, we write

$$r_1(4t^2 - 2) + r_2(2t) + r_3(1) = 0$$

or

$$4r_1t^2 + 2r_2t + r_3 - 2r_1 = 0$$

Then

$$\begin{aligned} 4r_1 \qquad\qquad &= 0 \\ 2r_2 \qquad &= 0 \\ -2r_1 \qquad\quad + r_3 &= 0 \end{aligned}$$

so $r_1 = r_2 = r_3 = 0$. Thus S is linearly independent, and consequently S is a basis for $\mathbf{P}_2$. □

In the next section we continue the study of dimension, computing the dimension of several commonly encountered vector spaces.

EXERCISES FOR SECTION 3.6

COMPUTATIONAL EXERCISES

In exercises 1 through 4, find all bases for the given set of vectors geometrically. Count the number of vectors in each basis. State the property exhibited.

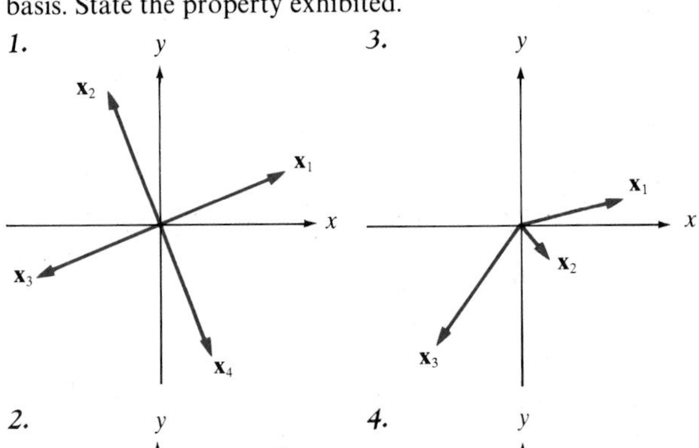

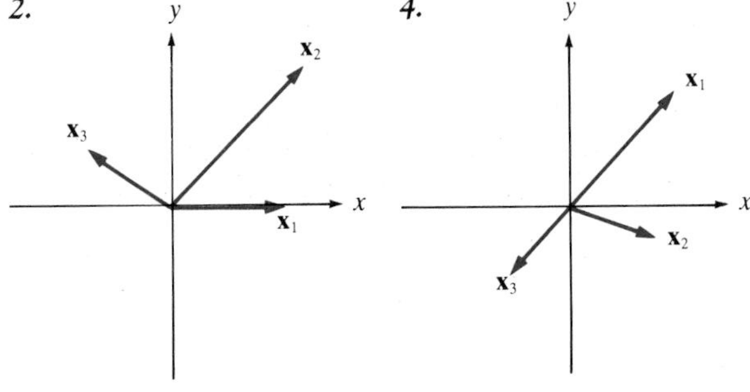

In exercises 5 and 6, sketch a basis with the stated properties for each given vector space. Count the number of vectors in each basis.

5. In $\mathbf{R}_2$,
(*a*) basis in the first quadrant
(*b*) basis in the fourth quadrant
(*c*) basis with a 0 entry in each vector

6. In $\mathbf{R}_3$,
(*a*) basis in which each vector has positive entries
(*b*) basis in which each vector has exactly one negative entry
(*c*) basis in which each vector has at least one 0 entry

In exercises 7 through 10, each given set is linearly independent. By counting, decide whether or not it is a basis for $\mathbf{R}_3$ or $\mathbf{P}_3$.

7. $\{[1,2,1],[1,3,1],[2,3,1]\}$

8. $\{[2,1,3],[-1,2,1],[3,0,0]\}$.

9. $\{t^3+t^2-t+3, t+2, t^2-5\}$

10. $\{t^2-t+2, t^3, t+5, t^3-t^2\}$

In exercises 11 through 14, the given set spans $\mathbf{R}_3$ or $\mathbf{P}_2$. By counting, decide if the set is a basis for its vector space.

11. $\{[-1,1,-1],[2,-1,3],[-3,2,-1],[4,-2,3]\}$

12. $\{[1,1,3],[3,1,1],[1,3,1]\}$

13. $\{t^2-1, t+2, 1\}$

14. $\{t, t^2+3, t-1, t^2+2\}$

In exercises 15 through 22, compute a basis from the given spanning set of $\mathbf{V}$ by the deletion algorithm. In exercises 15 through 18, demonstrate your work with sketches.

15. $\{[1,2],[1,3],[-2,3],[4,1]\}$

16. $\{[-1,3],[2,-6],[1,4],[2,3]\}$

17. $\{[-1,1,0],[1,0,-1],[-3,2,1],[0,1,1]\}$

18. $\{[1,2,3],[0,1,2],[3,4,5],[-1,2,5]\}$

19. $\{[1,1,0,0],[0,1,1,0],[0,0,1,1],[1,0,0,1],[1,0,1,0]\}$

20. $\{[1,0,0,-1],[0,-1,0,1],[1,-1,0,0],[0,0,1,-1],[-1,0,1,0]\}$

21. $\{t^2+t-1, 2t^2-3, 2t+1, t^2-t-2\}$

22. $\{-4t^2+2, 2t^2-1, 2t^2+t-1\}$

In exercises 23 through 26, show that the given functions are solutions to the given equation and then decide if the set of functions is linearly independent. If it is linearly dependent, use the deletion theorem to eliminate enough solutions to obtain a linearly independent set, thus describing the span of the set more efficiently. *Hint for 25 and 26:* See exercise 52 of Section 3.5.

23. $x(k+2)-3x(k+1)+2x(k)=0$, $\{1,2,2^k\}$

24. $x(k+2)-x(k+1)-6x(k)=0$, $\{3^k, 3^{k+1}\}$

25. $y''(t)-3y'(t)+2y(t)=0$, $\{e^t, e^{2t}\}$

26. $y''(t)-y'(t)-6y(t)=0$, $\{e^{3t}, e^{-2t}\}$

The set of all equations in three variables gives a vector space (see exercises 31 and 32 of Section 3.2 for subspaces of this space). Any set of such equations describes a set of points in $\mathbf{R}_3$. If one equation in the set is a linear combination of the other equations, that equation is not needed for the description of the set of points. In exercises 27 through 30, use the deletion algorithm to decide if any of the equations can be deleted without affecting the description of the surface. (Precise graphs of these equations, although hard to draw, will geometrically show linear dependence or independence in the equations.)

27.
$$\begin{aligned} x_1 - 2x_2 + 3x_3 &= 1 \\ -x_1 + 4x_2 - x_3 &= 2 \\ -x_1 + 8x_2 + 3x_3 &= 8 \end{aligned}$$

28.
$$\begin{aligned} 2x_1 + x_2 - 3x_3 &= 1 \\ -4x_1 - 3x_2 + 2x_3 &= 3 \\ 2x_1 + 3x_2 + 5x_3 &= -9 \end{aligned}$$

29.
$$\begin{aligned} x^2 + y^2 + z^2 &= 4 \\ x + y + z &= 1 \\ (x-1)^2 + (y-1)^2 + (z-1)^2 &= 5 \end{aligned}$$

30.
$$\begin{aligned} x^2 + y^2 - z &= 0 \\ x^2 + y^2 + z &= 2 \\ z &= 1 \end{aligned}$$

In exercises 31 through 34, identify the given subspace as a line, a plane, or $\mathbf{R}_3$. If the spanning set is linearly dependent, use the deletion algorithm to find a basis for the space contained in the given set. Graph the span and the spanning set, indicating the deleted vectors.

31. span $\{[1,-1,1],[-1,2,1],[1,0,2]\}$

32. span $\{[1,-2,-1],[2,-4,-2],[-3,6,3],[4,-8,-4]\}$

33. span $\{[-2,1,-1],[1,0,-1],[0,-1,1],[1,-1,2]\}$

34. span $\{[2,-2,3],[1,-2,3],[0,-2,3],[1,2,-3]\}$

COMPLEX NUMBERS

35. Show that $\{\mathbf{e}_1,\ldots,\mathbf{e}_n\}$ is a basis for $\mathbf{C}_n$ and thus that $\dim \mathbf{C}_n = n$. (See exercise 35 of Section 3.4.)

THEORETICAL EXERCISES

36. Draw several sketches in $\mathbf{R}_3$ demonstrating Theorem 3.12.

37. Prove Corollary 3.12C. *Hint:* Pick a largest linearly independent set $\{\mathbf{x}_1,\ldots,\mathbf{x}_n\}$ in $\mathbf{W}$ and show that it is a basis for $\mathbf{W}$. For this, note that if $\mathbf{x} \in \mathbf{W}$, then $\{\mathbf{x}_1,\ldots,\mathbf{x}_n,\mathbf{x}\}$ is linearly dependent. Note that this means that $\mathbf{x}$ is a linear combination of $\mathbf{x}_1,\ldots,\mathbf{x}_n$.

38. Prove part (i) of Theorem 3.13.

39. Let $\mathbf{V}$ be an infinite-dimensional vector space. Show that, for any positive integer n, $\mathbf{V}$ contains a linearly independent set of n vectors. (Thus, in this sense, $\mathbf{V}$ has infinitely many dimensions.) *Hint:* Show that, if $S = \{\mathbf{x}_1,\ldots,\mathbf{x}_m\} \subseteq \mathbf{V}$ is linearly independent and $\mathbf{x} \notin \operatorname{span} S$, then $\{\mathbf{x}_1,\ldots,\mathbf{x}_m,\mathbf{x}\}$ is linearly independent.

40. (geometry in $\mathbf{R}_n$) We define the *dimension* of a pyramid, as given in exercise 54 of Section 3.5, as n. Does this dimension agree with the geometric dimensions when $n = 2$ and 3?

41. Prove that if $\mathbf{V}$ is an n-dimensional vector space, then every set of $n+1$ or more vectors in $\mathbf{V}$ is linearly dependent.

42. Prove that $\dim \mathbf{F}(\mathbf{R}) = \infty$. *Hint:* Note that $\mathbf{P} \subseteq \mathbf{F}(\mathbf{R})$. Then recall Example 64, and use Corollary 3.12C.

43. Prove that if $\mathbf{W}$ is a subspace of a vector space $\mathbf{V}$ and if $\dim \mathbf{W} = \dim \mathbf{V}$, then $\mathbf{W} = \mathbf{V}$. *Hint:* Use Theorem 3.13.

APPLICATIONS EXERCISES

44. Suppose an object at point $[x_0, y_0, z_0]$ can apply forces (say, by using rockets) of arbitrary size for any required time in the directions $\pm[1,1,1]$, $\pm[1,-2,1]$, $\pm[2,-1,2]$. Can this object move to any point in space by applying these forces?

45. (economics) Divide the economy of a country into two disjoint sectors, the "private sector" (denoted by the number 1) and the "public sector" (denoted by the number 2). Let

$$[P(0), G(0)] \in \mathbf{R}_2$$

give the amount $P(0)$ of money in the private sector and the amount $G(0)$ of money in the public sector. Let matrix

$$A = \begin{bmatrix} 0.5 & 0.5 \\ 0.3 & 0.7 \end{bmatrix}$$

where a_{ij} = the proportion of money in sector i that is spent in sector j each month. Let $u_i(k)$ be the amount that a control agency gives or takes, depending on sign, from sector i in month k.

(*a*) Show mathematically that the agency can completely control the amount of money in each sector after two months.

(*b*) If we insist that any money given to one sector must be taken from the other sector, then we are requiring that

$$u_1(k) + u_2(k) = 0$$

for all k. Can the agency completely control the amount of money after three months in this case? Develop your answer mathematically. *Hint:* Write the problem as

$$[P(k+1), G(k+1)] = [P(k), G(k)]A + [u_1(k), u_2(k)]$$

where $u_1(k)$ and $u_2(k)$ are inputs in the kth month.

3.7 Bases and Dimensions of Some Special Vector Spaces

In this section we show how to compute bases and dimensions of several commonly encountered vector spaces.

I. Row and Column Spaces

Let A be an $m \times n$ matrix. Note that

$$\text{row space of } A = \text{span}\{\mathbf{a}_1, \ldots, \mathbf{a}_m\}$$

and

$$\text{column space of } A = \text{span}\{\mathbf{a}^1, \ldots, \mathbf{a}^n\}$$

Define

$$\text{rank } A = \dim(\text{row space of } A)$$

To compute rank A, we show how to find a special basis for the row space of A.

Theorem 3.14

Let A be an $m \times n$ matrix. Let $E = \begin{bmatrix} \varepsilon_1 \\ \vdots \\ \varepsilon_m \end{bmatrix}$ be any echelon form obtained from A by applying row operations. Then

(*i*) row space of A = row space of E and

(*ii*) the nonzero rows of E form a basis for the row space of A.

Further,

(*iii*) rank A = the number of nonzero rows in E.

Proof

For (i), we need to show that if we perform a scaling, interchange, or substitution operation on A to obtain a matrix B, then row space of A = row space of B. Applying this result for each Gaussian operation used to obtain E will then yield (i). We will show the result for the substitution operation. The remaining operations are done in the same way.

Suppose B is obtained from A by the substitution operation $R_i + sR_j \to R_i$. Then

$$\mathbf{b}_1 = \mathbf{a}_1, \ldots, \mathbf{b}_{i-1} = \mathbf{a}_{i-1},$$
$$\mathbf{b}_i = \mathbf{a}_i + s\mathbf{a}_j, \mathbf{b}_{i+1} = \mathbf{a}_{i+1}, \ldots, \mathbf{b}_m = \mathbf{a}_m$$

Thus we need to show that if

$$S_1 = \text{span}\{\mathbf{a}_1, \ldots, \mathbf{a}_i, \ldots, \mathbf{a}_m\}$$

and

$$S_2 = \text{span}\{\mathbf{a}_1, \ldots, \mathbf{a}_i + s\mathbf{a}_j, \ldots, \mathbf{a}_m\}$$

then $S_1 = S_2$. Let

$$\mathbf{x} = r_1\mathbf{a}_1 + \cdots + r_i\mathbf{a}_i + \cdots + r_m\mathbf{a}_m \in S_1$$

Then, by rearranging, we have

$$\mathbf{x} = r_1\mathbf{a}_1 + \cdots + r_i(\mathbf{a}_i + s\mathbf{a}_j) + \cdots + (r_j - r_i s)\mathbf{a}_j + \cdots + r_m\mathbf{a}_m \in S_2$$

Thus $S_1 \subseteq S_2$. Conversely, let

$$\mathbf{x} = t_1\mathbf{a}_1 + \cdots + t_i(\mathbf{a}_i + s\mathbf{a}_j) + \cdots + t_m\mathbf{a}_m \in S_2$$

Again by rearranging, we have

$$\mathbf{x} = t_1\mathbf{a}_1 + \cdots + t_i\mathbf{a}_i + \cdots + (t_j + t_i s)\mathbf{a}_j + \cdots + t_m\mathbf{a}_m \in S_1$$

Thus $S_2 \subseteq S_1$, so $S_1 = S_2$, and the result follows.

For (ii), let $\boldsymbol{\varepsilon}_1, \ldots, \boldsymbol{\varepsilon}_k$ be the nonzero rows of E. Write

$$r_1\boldsymbol{\varepsilon}_1 + \cdots + r_k\boldsymbol{\varepsilon}_k = \mathbf{0}$$

Since $\boldsymbol{\varepsilon}_1$ has its first nonzero entry where $\boldsymbol{\varepsilon}_2, \ldots, \boldsymbol{\varepsilon}_k$ have a zero, it follows that $r_1 = 0$. Thus the equation becomes

$$r_2\boldsymbol{\varepsilon}_2 + \cdots + r_k\boldsymbol{\varepsilon}_k = \mathbf{0}$$

Similarly, $\boldsymbol{\varepsilon}_2$ has its first nonzero entry where $\boldsymbol{\varepsilon}_3, \ldots, \boldsymbol{\varepsilon}_k$ have a zero, and so $r_2 = 0$. Continuing, we have $r_3 = \cdots = r_k = 0$, and so $\{\boldsymbol{\varepsilon}_1, \ldots, \boldsymbol{\varepsilon}_k\}$ is a basis for the row space of E. But by (i), row space of A = row space of E. Thus $\{\boldsymbol{\varepsilon}_1, \ldots, \boldsymbol{\varepsilon}_k\}$ is a basis for the row space of A.

Finally, (iii) follows from (ii) by definition. ■

EXAMPLE 66

Let $A = \begin{bmatrix} 1 & -1 & 0 \\ 0 & 2 & 2 \\ 1 & 1 & 2 \end{bmatrix}$. Applying $R_3 - R_1 \to R_3$, $R_3 - R_2 \to R_3$ yields

$$E = \begin{bmatrix} 1 & -1 & 0 \\ 0 & 2 & 2 \\ 0 & 0 & 0 \end{bmatrix}$$

Thus

$$\text{row space of } A = \text{span}\{[1, -1, 0]\ [0, 2, 2]\}$$

and

$$\dim(\text{row space of } A) = \text{rank } A = 2$$

This space is shown in Figure 3.57. □

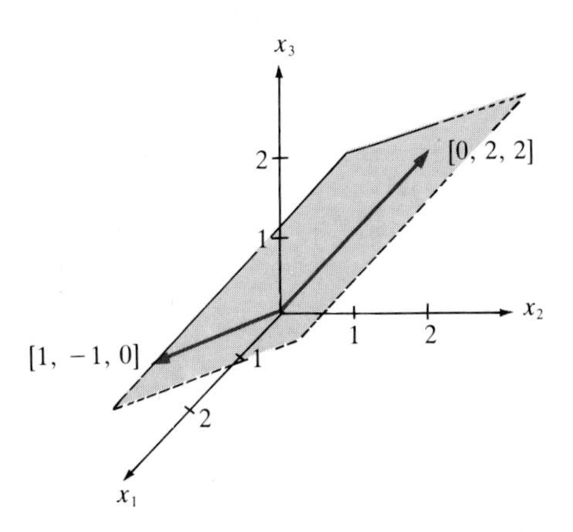

FIGURE 3.57

By this method of computing rank we also have the following.

COROLLARY 3.14A

The system of linear algebraic equations $A\mathbf{x} = \mathbf{b}$ has a solution if and only if

$$\text{rank } [A \quad \mathbf{b}] = \text{rank } A$$

PROOF

If we compute an echelon form for $[A \quad \mathbf{b}]$, such as $[E \quad \mathbf{e}]$, we see that the system has a solution if and only if $[E \quad \mathbf{e}]$ and E have the same number of nonzero rows. But by Theorem 3.14 this holds if and only if rank $[A \quad \mathbf{b}]$ = rank A. ■

EXAMPLE 67

For the system

$$\begin{aligned} x \quad\; + z &= 1 \\ x + y \quad\; &= 2 \\ -y + z &= 1 \end{aligned}$$

we have

$$[A \quad \mathbf{b}] = \begin{bmatrix} 1 & 0 & 1 & 1 \\ 1 & 1 & 0 & 2 \\ 0 & -1 & 1 & 1 \end{bmatrix}$$

Applying $R_2 - R_1 \to R_2$ yields

$$\begin{bmatrix} 1 & 0 & 1 & 1 \\ 0 & 1 & -1 & 1 \\ 0 & -1 & 1 & 1 \end{bmatrix}$$

Next, applying $R_3 + R_2 \to R_3$ gives

$$\begin{bmatrix} 1 & 0 & 1 & 1 \\ 0 & 1 & -1 & 1 \\ 0 & 0 & 0 & 2 \end{bmatrix}$$

Thus rank $[A \quad \mathbf{b}] = 3$ while rank $A = 2$, and so the system has no solution. □

By transposing A, thereby interchanging columns and rows, we also see that

$$\dim(\text{column space of } A) = \text{rank } A^t$$

EXAMPLE 68

Let $A = \begin{bmatrix} 1 & -1 & 0 \\ 0 & 2 & 2 \\ 1 & 1 & 2 \end{bmatrix}$. Exchanging rows and columns yields

$$A^t = \begin{bmatrix} 1 & 0 & 1 \\ -1 & 2 & 1 \\ 0 & 2 & 2 \end{bmatrix}$$

Applying $R_2 + R_1 \to R_2$ and $R_3 - R_2 \to R_3$ gives

$$E = \begin{bmatrix} 1 & 0 & 1 \\ 0 & 2 & 2 \\ 0 & 0 & 0 \end{bmatrix}$$

Exchanging rows and columns again yields

$$E^t = \begin{bmatrix} 1 & 0 & 0 \\ 0 & 2 & 0 \\ 1 & 2 & 0 \end{bmatrix}$$

and thus

$$\text{column space of } A = \text{span}\left\{\begin{bmatrix} 1 \\ 0 \\ 1 \end{bmatrix}, \begin{bmatrix} 0 \\ 2 \\ 2 \end{bmatrix}\right\}$$

and

$$\dim(\text{column space of A}) = \text{rank } A^t = 2$$

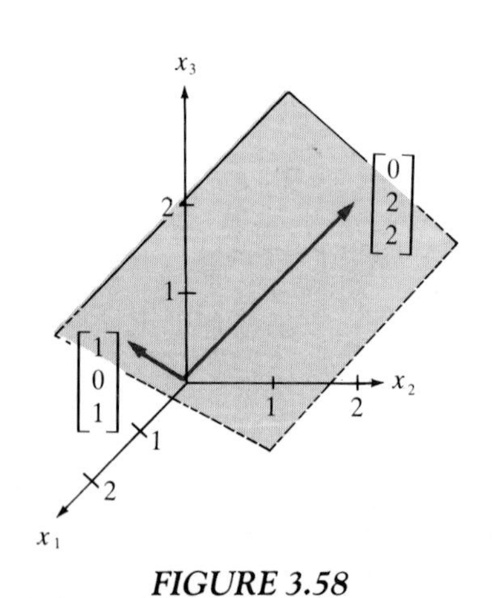

FIGURE 3.58

This column space is shown in Figure 3.58. □

From the two previous examples, we see that, although the row space and the column space of A are different,

$$\dim(\text{row space of } A) = \dim(\text{column space of } A) = 2$$

In fact, this is true for all matrices A.

COROLLARY 3.14B

Let A be an $m \times n$ matrix. Then

$$\text{rank } A = \text{rank } A^t$$

PROOF

Let E be an echelon form of A obtained by Gaussian elimination. Then, by Theorem 3.14, rank A = rank E = number of bound variables in $E\mathbf{x} = 0$. Take any solution $\mathbf{y}$ to this equation. Then

$$A\mathbf{y} = y_1\mathbf{a}^1 + \cdots + y_n\mathbf{a}^n = \mathbf{0}$$

Note that if any column r corresponding to a free variable x_r is deleted from A and E, forming A' and E', respectively, then, for $i \neq r$, a variable x_i is free in $E\mathbf{x} = \mathbf{0}$ if and only if the corresponding variable is free in $E'\mathbf{x}' = \mathbf{0}$. It follows by the deletion theorem that every column of A corresponding to a free variable in $E\mathbf{y} = \mathbf{0}$ can be deleted, forming S from the set of columns of $\mathbf{A}$, and span S = column space of A. Thus

$$\begin{aligned} \text{rank } A^t &\leqslant \text{number of bound variables} \\ &\leqslant \text{rank } A \end{aligned}$$

Now since this inequality holds for any matrix A, it also holds for A^t. Thus rank $(A^t)^t \leqslant$ rank A^t. But $(A^t)^t = A$. Putting all of these results together yields

$$\text{rank } A = \text{rank } A^t$$ ■

This result allows us to write

$$\dim(\text{column space of } A) = \dim(\text{row space of } A) = \text{rank } A$$

As a consequence, we have the following often-used corollary.

COROLLARY 3.14C

Let A be an $n \times n$ matrix. The following are equivalent.

(*i*) rank $A = n$

(*ii*) The set of columns and the set of rows of A are linearly independent.

(*iii*) A is invertible.

PROOF

We show the equivalence only for the columns of A. If (i) holds, then $\{\mathbf{a}^1, \ldots, \mathbf{a}^n\}$ must contain a basis for the column space of A. Since dim(column space A) = n, it follows that $\{\mathbf{a}^1, \ldots, \mathbf{a}^n\}$ must be the basis, and (ii) follows. If (ii) holds, then rank A = rank $A^t = n$. Hence det $A \neq 0$ and A is invertible. Finally, if (iii) holds, then det $A \neq 0$, so rank $A = n$. Hence (i) follows. ∎

II. SOLUTIONS TO $AX = 0$

Let A be an $m \times n$ matrix and **W** the vector space of solutions to $A\mathbf{x} = \mathbf{0}$. To see how to compute a basis and the dimension of **W**, we look at an example.

EXAMPLE 69

To find the solution set **W** of

$$\begin{aligned} x_1 + x_2 \qquad - x_4 + x_5 &= 0 \\ x_1 \qquad - x_3 \qquad + x_5 &= 0 \\ x_2 + x_3 - x_4 \qquad &= 0 \end{aligned}$$

we apply the Gaussian elimination algorithm.

$$\begin{bmatrix} 1 & 1 & 0 & -1 & 1 & 0 \\ 1 & 0 & -1 & 0 & 1 & 0 \\ 0 & 1 & 1 & -1 & 0 & 0 \end{bmatrix} \xrightarrow{R_2 - R_1 \to R_2} \begin{bmatrix} 1 & 1 & 0 & -1 & 1 & 0 \\ 0 & -1 & -1 & 1 & 0 & 0 \\ 0 & 1 & 1 & -1 & 0 & 0 \end{bmatrix}$$

$$\xrightarrow{R_3 + R_2 \to R_3} \begin{bmatrix} 1 & 1 & 0 & -1 & 1 & 0 \\ 0 & -1 & -1 & 1 & 0 & 0 \\ 0 & 0 & 0 & 0 & 0 & 0 \end{bmatrix}$$

Thus we have

$$\begin{aligned} x_1 + x_2 \qquad - x_4 + x_5 &= 0 \\ -x_2 - x_3 + x_4 \qquad &= 0 \end{aligned}$$

Writing the solution and separating it into vectors corresponding to each free variable yields

$$\mathbf{x} = \begin{bmatrix} x_1 \\ x_2 \\ x_3 \\ x_4 \\ x_5 \end{bmatrix} = \begin{bmatrix} x_3 \qquad - x_5 \\ -x_3 + x_4 \\ x_3 \\ x_4 \\ x_5 \end{bmatrix} = x_3 \begin{bmatrix} 1 \\ -1 \\ 1 \\ 0 \\ 0 \end{bmatrix} + x_4 \begin{bmatrix} 0 \\ 1 \\ 0 \\ 1 \\ 0 \end{bmatrix} + x_5 \begin{bmatrix} -1 \\ 0 \\ 0 \\ 0 \\ 1 \end{bmatrix}$$

Thus the solution set is

$$\text{span} \left\{ \begin{bmatrix} 1 \\ -1 \\ 1 \\ 0 \\ 0 \end{bmatrix}, \begin{bmatrix} 0 \\ 1 \\ 0 \\ 1 \\ 0 \end{bmatrix}, \begin{bmatrix} -1 \\ 0 \\ 0 \\ 0 \\ 1 \end{bmatrix} \right\}$$

Notice that the free variable x_3 is the only free variable in the third entry of $\mathbf{x}$, x_4 is the only free variable in the fourth entry of $\mathbf{x}$, and x_5 is the only free variable in the fifth entry of $\mathbf{x}$. This shows that if we set

$$x_3\begin{bmatrix}1\\-1\\1\\0\\0\end{bmatrix}+x_4\begin{bmatrix}0\\1\\0\\1\\0\end{bmatrix}+x_5\begin{bmatrix}-1\\0\\0\\0\\1\end{bmatrix}=\mathbf{0}$$

it follows that $x_3 = x_4 = x_5 = 0$. Hence

$$\left\{\begin{bmatrix}1\\-1\\1\\0\\0\end{bmatrix},\begin{bmatrix}0\\1\\0\\1\\0\end{bmatrix},\begin{bmatrix}-1\\0\\0\\0\\1\end{bmatrix}\right\}$$

is linearly independent and thus a basis for $\mathbf{W}$ with $\dim \mathbf{W} = 3$, the number of free variables. □

Using the technique given in this example, we can prove the following.

THEOREM 3.15

Let A be an $m \times n$ matrix. Let $[A \quad \mathbf{0}]$ be the augmented matrix for the system $A\mathbf{x} = \mathbf{0}$ and let $[E \quad \mathbf{0}]$ be the matrix obtained by applying the Gaussian elimination algorithm to $[A \quad \mathbf{0}]$. If $\mathbf{W}$ is the solution set for this system, then

(*i*) a basis for $\mathbf{W}$ can be obtained from the solution to $E\mathbf{x} = \mathbf{0}$ by using the set of coefficient vectors of the free variables, and

(*ii*) $\dim \mathbf{W}$ = the number of free variables for $[E \quad \mathbf{0}]$
= n − the number of bound variables for $[E \quad \mathbf{0}]$
= $n - \operatorname{rank} A$

To conclude this part, we will show how this theorem is of use in describing solution sets to homogeneous systems of linear algebraic equations.

EXAMPLE 70

Let $\mathbf{W}$ be the solution space for the system of equations

$$\begin{aligned} x - y + z &= 0 \\ -x + y - z &= 0 \end{aligned}$$

The coefficient matrix is

$$A = \begin{bmatrix} 1 & -1 & 1 \\ -1 & 1 & -1 \end{bmatrix}$$

Applying $R_2 + R_1 \to R_2$ yields

$$\begin{bmatrix} 1 & -1 & 1 \\ 0 & 0 & 0 \end{bmatrix}$$

Thus rank $A = 1$ and $\dim \mathbf{W} = 3 - 1 = 2$.

Solving the system by the Gaussian elimination algorithm yields

$$\begin{bmatrix} 1 & -1 & 1 & 0 \\ -1 & 1 & -1 & 0 \end{bmatrix} \xrightarrow{R_2 + R_1 \to R_2} \begin{bmatrix} 1 & -1 & 1 & 0 \\ 0 & 0 & 0 & 0 \end{bmatrix}$$

or

$$x - y + z = 0$$

Now y and z are free, so we have $x = y - z$ and

$$\begin{bmatrix} x \\ y \\ z \end{bmatrix} = \begin{bmatrix} y - z \\ y \\ z \end{bmatrix}$$

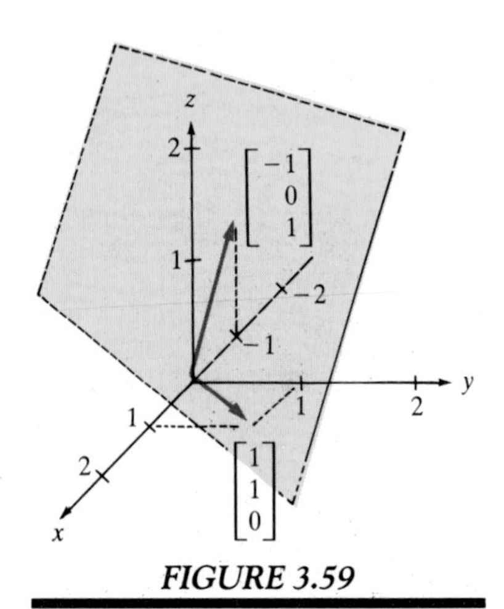

FIGURE 3.59

is the solution to the system. Separating this vector in terms of y and z yields

$$\begin{bmatrix} x \\ y \\ z \end{bmatrix} = y\begin{bmatrix} 1 \\ 1 \\ 0 \end{bmatrix} + z\begin{bmatrix} -1 \\ 0 \\ 1 \end{bmatrix}$$

Thus, $\mathbf{W}$ is the subspace with basis $\left\{\begin{bmatrix} 1 \\ 1 \\ 0 \end{bmatrix}, \begin{bmatrix} -1 \\ 0 \\ 1 \end{bmatrix}\right\}$. Geometrically, this subspace can be identified as the plane shown in Figure 3.59. □

III. *SOLUTIONS TO LINEAR DIFFERENCE EQUATIONS*

Let $\mathbf{W}$ be the solution set of a homogeneous linear difference equation. Recall from Example 32 of Section 3.3 that $\mathbf{W}$ is a vector space. We will compute the dimension of $\mathbf{W}$ and find a basis for it.

Let $a \in \mathbf{R}$ and consider the homogeneous difference equation

$$x(k + 1) + ax(k) = 0$$

Then, from Section 3.3,

$$x(k) = r(-a)^k$$

for any $r \in \mathbf{R}$. Letting $x_1(k) = (-a)^k$, we have the solution set

$$\mathbf{W} = \text{span}\,\{x_1(k)\}$$

and so $\dim \mathbf{W} = 1$.

Let a, b, $c \in \mathbf{R}$ with $a \neq 0$, and consider the homogeneous difference equation

$$ax(k+2) + bx(k+1) + cx(k) = 0$$

From Section 3.3, we know that there is a unique solution to this equation such that $x(0) = r$ and $x(1) = s$ for any $r, s \in \mathbf{R}$. Let $x_1(k)$ be the solution with $x_1(0) = 1$ and $x_1(1) = 0$, and let $x_2(k)$ be the solution with $x_2(0) = 0$ and $x_2(1) = 1$. We show that the solution set is

$$\mathbf{W} = \text{span}\,\{x_1(k), x_2(k)\}$$

Of course, $\text{span}\,\{x_1(k), x_2(k)\} \subseteq \mathbf{W}$. Thus we need only prove the reverse inclusion. For this, let $y(k) \in \mathbf{W}$. Suppose $y(0) = r$ and $y(1) = s$. Then

$$y(0) = rx_1(0) + sx_2(0) \qquad \text{and} \qquad y(1) = rx_1(1) + sx_2(1)$$

Now, since

$$x(k) = rx_1(k) + sx_2(k)$$

is a solution to the equation such that $x(0) = r = y(0)$ and $x(1) = s = y(1)$, and since there is only one solution to the equation with $x(0) = r$ and $x(1) = s$, it follows that $y(k) = x(k)$. Thus, we have

$$\mathbf{W} \subseteq \text{span}\,\{x_1(k), x_2(k)\}$$

Hence

$$\mathbf{W} = \text{span}\,\{x_1(k), x_2(k)\}$$

Finally, if $r_1x_1(k) + r_2x_2(k) = 0$, taking $k = 0$ shows that $r_1 = 0$ and taking $k = 1$ shows that $r_2 = 0$. Thus the set $\{x_1(k), x_2(k)\}$ is linearly independent. Consequently, this set is a basis for $\mathbf{W}$, and $\dim \mathbf{W} = 2$.

We can use these results to construct solution sets for these difference equations. The solution to the first-order homogeneous linear difference equation

$$x(k+1) + ax(k) = 0$$

is $x(k) = r(-a)^k$ for all $r \in \mathbf{R}$. Solving the second-order homogeneous difference equation

$$ax(k+2) + bx(k+1) + cx(k) = 0$$

with $a \neq 0$ is more complex.

Since the solution set **W** of this equation is of dimension 2, we know that we can find **W** by finding any two solutions that form a linearly independent set. We intend to find such solutions.

The form of this equation suggests that we try to find a solution of the type

$$x(k) = r^k$$

for some $r \in \mathbf{R}$. To compute the appropriate r, we substitute $x(k) = r^k$ into the equation to get

$$ar^{k+2} + br^{k+1} + cr^k = 0 \qquad \text{or} \qquad ar^2 + br + c = 0$$

Now solving the equation for r using the quadratic formula yields

$$r = \frac{-b \pm \sqrt{b^2 - 4ac}}{2a}$$

We need to look at cases.

CASE 1: Suppose $b^2 - 4ac > 0$. In this case there are two real solutions to this equation, say r_1 and r_2. Then

$$x_1(k) = r_1^k \qquad \text{and} \qquad x_2(k) = r_2^k$$

are solutions to the difference equation, and it can be shown that the set of these solutions is linearly independent (exercise 48). Thus, the set of solutions can be written as

$$x(k) = a_1 x_1(k) + a_2 x_2(k)$$

for all $a_1, a_2 \in \mathbf{R}$.

As an example, we solve

$$x(k + 2) - 5x(k + 1) + 6x(k) = 0$$

Here,

$$r = \frac{-b \pm \sqrt{b^2 - 4ac}}{2a} = \frac{5 \pm 1}{2}$$

so $r_1 = 2$ and $r_2 = 3$. Thus

$$x_1(k) = 2^k \qquad \text{and} \qquad x_2(k) = 3^k$$

and the solution is

$$x(k) = a_1 2^k + a_2 3^k$$

for all $a_1, a_2 \in \mathbf{R}$.

CASE 2: Suppose $b^2 - 4ac = 0$. In this case the solution to the quadratic equation is a real number r of multiplicity 2. Thus, $x_1(k) = r^k$ is a solution to the difference equation. We need to obtain another solution $x_2(k)$ such that $\{x_1(k), x_2(k)\}$ is linearly independent. By substitution, we can show that $x_2(k) = kr^k$ is a solution* (exercise 49). Further, the set $\{x_1(k), x_2(k)\}$ is linearly independent (exercise 50). Thus the set of solutions can be written as

$$x(k) = a_1 x_1(k) + a_2 x_2(k)$$

for all $a_1, a_2 \in \mathbf{R}$.

As an example, we solve

$$x(k + 2) - 2x(k + 1) + x(k) = 0$$

Here $r = 1$ with multiplicity 2. Thus,

$$x_1(k) = 1^k \qquad \text{and} \qquad x_2(k) = k(1^k)$$

and the solutions are

$$x(k) = a_1(1^k) + a_2 k(1^k)$$

for all $a_1, a_2 \in \mathbf{R}$.

CASE 3: (optional) Suppose $b^2 - 4ac < 0$. In this case the roots of the quadratic equation are complex numbers, and we will just show two functions that satisfy the equation.† Let

$$\alpha = \frac{-b}{2a}, \qquad \beta = \frac{\sqrt{4ac - b^2}}{2a}, \qquad \gamma = \sqrt{\alpha^2 + \beta^2}$$

Choose θ by $\theta = \text{Arccos}\,(\alpha/\gamma)$. Then

$$x_1(k) = \gamma^k \cos k\theta \qquad \text{and} \qquad x_2(k) = \gamma^k \sin k\theta$$

are solutions to the difference equation and the set of these solutions is linearly independent (exercise 51). Thus the solution can be written as

$$x(k) = a_1 x_1(k) + a_2 x_2(k)$$

for all $a_1, a_2 \in \mathbf{R}$.

* There is a method for computing this second solution to the difference equation. However, it is beyond the level of this text. For our work, we may suppose this solution was guessed by trial and error.

† The methods for converting complex number solutions of the quadratic equation into real solutions of the difference equation require a knowledge of complex functions. Hence they are not given here. Again, for our purposes, we may suppose these solutions were obtained by trial and error.

As an example, we solve

$$x(k + 2) - 2x(k + 1) + 2x(k) = 0$$

Here $\alpha = 1$ and $\beta = 1$, so $\gamma = \sqrt{2}$ and $\theta = \pi/4$. Thus

$$x_1(k) = (\sqrt{2})^k \cos k\pi/4$$
$$x_2(k) = (\sqrt{2})^k \sin k\pi/4$$

and $$x(k) = a_1(\sqrt{2})^k \cos k\pi/4 + a_2(\sqrt{2})^k \sin k\pi/4$$

for all $a_1, a_2 \in \mathbf{R}$.

IV. SOLUTIONS TO SYSTEMS OF LINEAR DIFFERENCE EQUATIONS (OPTIONAL)

Let A be an $n \times n$ matrix. The solution set to

$$\mathbf{x}(k + 1) = A\mathbf{x}(k)$$

where $\mathbf{x}(k) \in \mathbf{F}^n(\mathbf{N})$, is a vector space (Example 33 of Section 3.3). Further, for any $\mathbf{b} \in \mathbf{R}^n$, if we set $\mathbf{x}(0) = \mathbf{b}$, then

$$\mathbf{x}(1) = A\mathbf{x}(0) = A\mathbf{b}, \mathbf{x}(2) = A\mathbf{x}(1) = A^2\mathbf{b}, \ldots, \mathbf{x}(k) = A^k\mathbf{b}, \ldots$$

and it follows that $\mathbf{x}(k) = A^k\mathbf{b}$ is the only solution to the difference equation for which $\mathbf{x}(0) = \mathbf{b}$.

Let $\mathbf{x}_i(k)$ be the solution to the system for which $\mathbf{x}_i(0) = \mathbf{e}^i$, $i = 1, \ldots, n$. Set

$$S = \{\mathbf{x}_1(k), \ldots, \mathbf{x}_n(k)\}$$

We will show that S is a basis for $\mathbf{W}$. To see this, write

$$r_1\mathbf{x}_1(k) + \cdots + r_n\mathbf{x}_n(k) = \mathbf{0}$$

At $k = 0$ this yields

$$r_1\mathbf{e}^i + \cdots + r_n\mathbf{e}^n = \mathbf{0}$$

and so $r_1 = \cdots = r_n = 0$. Thus S is linearly independent.

Now take any $\mathbf{x}(k) \in \mathbf{W}$ and suppose $\mathbf{x}(0) = \mathbf{b}$. Since

$$b_1\mathbf{x}_1(k) + \cdots + b_n\mathbf{x}_n(k) \in \mathbf{W}$$

and at $k = 0$ is $\mathbf{b}$, it follows that

$$\mathbf{x}(k) = b_1\mathbf{x}_1(k) + \cdots + b_n\mathbf{x}_n(k)$$

But this means that $\mathbf{W} = \text{span } S$ and so S is a basis for $\mathbf{W}$. Thus

$$\dim \mathbf{W} = n$$

Although we have shown that S is a basis for $\mathbf{W}$, we have not found each $\mathbf{x}_i(k)$ in functional form. This type of work will be given in Section 6.3 of Chapter 6.

This concludes our work on dimension for vector spaces. In the next chapter we develop the geometrical concepts of distance and angle for all vector spaces.

EXERCISES FOR SECTION 3.7

COMPUTATIONAL EXERCISES

In exercises 1 through 4, find a basis for the row space and the column space of A. Give the dimension of each. Graph both vector spaces.

1. $A = \begin{bmatrix} 1 & 2 \\ 2 & 4 \end{bmatrix}$

2. $A = \begin{bmatrix} 1 & 0 \\ -1 & 1 \\ 0 & 1 \end{bmatrix}$

3. $A = \begin{bmatrix} -1 & 3 \\ 2 & 5 \\ 4 & -1 \end{bmatrix}$

4. $A = \begin{bmatrix} 1 & 1 & 2 \\ 1 & -1 & 3 \end{bmatrix}$

In exercises 5 through 10, use rank to compute the dimensions of the row space and column space of A.

5. $A = \begin{bmatrix} 2 & 2 \\ 1 & 1 \end{bmatrix}$

6. $A = \begin{bmatrix} 3 & 2 \\ 1 & 1 \end{bmatrix}$

7. $A = \begin{bmatrix} 2 & 3 & 5 \\ 1 & 2 & 4 \\ 1 & -1 & -5 \\ 3 & 2 & 0 \end{bmatrix}$

8. $A = \begin{bmatrix} 1 & 3 & 1 \\ 2 & -1 & -5 \\ -2 & 4 & 8 \end{bmatrix}$

9. $A = \begin{bmatrix} 2 & -1 & 10 \\ 1 & 4 & 4 \\ 3 & 2 & 8 \\ -1 & 1 & -6 \end{bmatrix}$

10. $A = \begin{bmatrix} -2 & 3 & 4 \\ 3 & 2 & -19 \\ -1 & -1 & 7 \end{bmatrix}$

In exercises 11 through 14, use Corollary 3.14A to decide whether or not $A\mathbf{x} = \mathbf{b}$ has a solution for the given matrix A and column vector $\mathbf{b}$.

11. $A = \begin{bmatrix} 1 & -2 \\ -2 & 4 \end{bmatrix}$ (a) $\mathbf{b} = \begin{bmatrix} -3 \\ 6 \end{bmatrix}$ (*b*) $\mathbf{b} = \begin{bmatrix} 4 \\ 8 \end{bmatrix}$

12. $A = \begin{bmatrix} -4 & 2 \\ 6 & -3 \end{bmatrix}$ (a) $\mathbf{b} = \begin{bmatrix} 12 \\ 18 \end{bmatrix}$ (*b*) $\mathbf{b} = \begin{bmatrix} 4 \\ -6 \end{bmatrix}$

13. $A = \begin{bmatrix} 1 & 3 \\ 2 & -5 \\ -1 & 2 \end{bmatrix}$ (a) $\mathbf{b} = \begin{bmatrix} 1 \\ -9 \\ 4 \end{bmatrix}$ (*b*) $\mathbf{b} = \begin{bmatrix} 0 \\ 11 \\ 5 \end{bmatrix}$

14. $A = \begin{bmatrix} 2 & 0 & 3 \\ -1 & -5 & 1 \\ 4 & 26 & -7 \end{bmatrix}$ (a) $\mathbf{b} = \begin{bmatrix} -6 \\ 2 \\ 14 \end{bmatrix}$ (*b*) $\mathbf{b} = \begin{bmatrix} 7 \\ -1 \\ 1 \end{bmatrix}$

In each of exercises 15 through 18, an equation $A\mathbf{x} = \mathbf{b}$ is given. For each problem, graph the column space of A. Graph $\mathbf{b}$ and note why $A\mathbf{x} = \mathbf{b}$ has no solution.

15. $\begin{bmatrix} 1 & 1 \\ 1 & 1 \end{bmatrix}\mathbf{x} = \begin{bmatrix} 1 \\ 2 \end{bmatrix}$

16. $\begin{bmatrix} 1 & -3 \\ -2 & 6 \end{bmatrix}\mathbf{x} = \begin{bmatrix} 1 \\ 3 \end{bmatrix}$

17. $\begin{bmatrix} -2 & 4 \\ 1 & -2 \end{bmatrix}\mathbf{x} = \begin{bmatrix} 6 \\ -4 \end{bmatrix}$

18. $\begin{bmatrix} 1 & 2 \\ 1 & 2 \end{bmatrix}\mathbf{x} = \begin{bmatrix} 5 \\ 3 \end{bmatrix}$

In exercises 19 through 22, find the rank geometrically by graphing the rows of the given matrix.

19. $\begin{bmatrix} 1 & -1 \\ 1 & 1 \end{bmatrix}$

20. $\begin{bmatrix} 3 & -6 \\ -1 & 2 \end{bmatrix}$

21. $\begin{bmatrix} 1 & -1 \\ -2 & 2 \\ 0 & 0 \end{bmatrix}$

22. $\begin{bmatrix} 1 & 2 \\ 2 & 1 \\ 1 & 1 \end{bmatrix}$

In exercises 23 through 26, find the rank of the given matrix A by computing an echelon form. State the dimension of the solution set to $A\mathbf{x} = \mathbf{0}$.

23. $A = \begin{bmatrix} 1 & 2 & 1 \\ 3 & 1 & -2 \\ 1 & 7 & 6 \\ 0 & 1 & 1 \end{bmatrix}$

24. $A = \begin{bmatrix} -1 & 4 & 6 \\ 7 & 2 & -3 \\ -4 & 14 & 15 \\ 0 & 0 & 1 \end{bmatrix}$

25. $A = \begin{bmatrix} 4 & -5 & 2 & 3 & 6 \\ 3 & 1 & -2 & 3 & 1 \\ 1 & 13 & -10 & 3 & -9 \\ 2 & 0 & 2 & 2 & 4 \end{bmatrix}$

26. $A = \begin{bmatrix} 1 & 0 & 1 & 3 & 1 & 0 \\ 0 & 1 & -1 & 2 & 0 & 1 \\ 2 & -1 & 0 & 1 & -1 & 0 \\ 3 & -1 & 1 & 4 & 0 & 0 \end{bmatrix}$

In exercises 27 through 32, find a basis of the solution set of the given system of linear algebraic equations, as in Example 70, and state the dimension of the solution set. In exercises 27 through 30, graph the basis and the solution set.

27. $\begin{aligned} 4x - 6y &= 0 \\ -2x + 3y &= 0 \end{aligned}$

28. $\begin{aligned} 2x + y &= 0 \\ 3x - y &= 0 \end{aligned}$

29. $\begin{aligned} x - 3y + 7z &= 0 \\ -2x + 6y - 14z &= 0 \end{aligned}$

30. $\begin{aligned} 4x - 6y + 2z &= 0 \\ 2x - 3y + z &= 0 \end{aligned}$

31. $\begin{aligned} 2x - y + z &= 0 \\ 6x + 5y - 3z &= 0 \\ 6x + 13y - 9z &= 0 \end{aligned}$

32. $\begin{aligned} 3x + 2y - z &= 0 \\ 5x + y - 2z &= 0 \\ x + 3y &= 0 \end{aligned}$

In exercises 33 and 34, use the given set of vectors to build as many invertible matrices as possible. Use geometry as a guide. Recall that an $n \times n$ matrix A is invertible if and only if its set of columns is linearly independent.

33. $\left\{ \begin{bmatrix} 1 \\ -1 \end{bmatrix}, \begin{bmatrix} 1 \\ 1 \end{bmatrix}, \begin{bmatrix} -1 \\ 1 \end{bmatrix}, \begin{bmatrix} 2 \\ 1 \end{bmatrix} \right\}$

34. $\left\{ \begin{bmatrix} 2 \\ 1 \end{bmatrix}, \begin{bmatrix} 1 \\ 2 \end{bmatrix}, \begin{bmatrix} 1 \\ 1 \end{bmatrix}, \begin{bmatrix} -1 \\ -2 \end{bmatrix} \right\}$

In exercises 35 through 38, solve the equation. Check your answers by substituting them into the equations.

35. $x(k+2) + x(k+1) - 12x(k) = 0$

36. $x(k+2) + 4x(k+1) + 3x(k) = 0$

37. $x(k+2) - 4x(k+1) + 4x(k) = 0$

38. (complex roots) $x(k+2) + 2x(k+1) + 17x(k) = 0$

In exercises 39 through 44, state the dimension of the solution set to $\mathbf{x}(k+1) = A\mathbf{x}(k)$. Decide if the given expression is the complete solution to the equation.

39. $\mathbf{x}(k+1) = \begin{bmatrix} 1 & 2 \\ 2 & 1 \end{bmatrix}\mathbf{x}(k)$, $\mathbf{x}(k) = r_1 3^k \begin{bmatrix} 1 \\ 1 \end{bmatrix} + r_2(-1)^k \begin{bmatrix} 1 \\ -1 \end{bmatrix}$

40. $\mathbf{x}(k+1) = \begin{bmatrix} 3 & 1 \\ 1 & 3 \end{bmatrix}\mathbf{x}(k)$, $\mathbf{x}(k) = r_1 4^k \begin{bmatrix} 1 \\ 1 \end{bmatrix} + r_2 2^k \begin{bmatrix} 1 \\ -1 \end{bmatrix}$

41. $\mathbf{x}(k+1) = \begin{bmatrix} 3 & 2 \\ 2 & 3 \end{bmatrix}\mathbf{x}(k)$, $\mathbf{x}(k) = r_1 5^k \begin{bmatrix} 1 \\ 1 \end{bmatrix} + r_2 5^{k+1} \begin{bmatrix} 1 \\ 1 \end{bmatrix}$

42. $\mathbf{x}(k+1) = \frac{1}{3}\begin{bmatrix} 8 & 2 \\ 1 & 7 \end{bmatrix}\mathbf{x}(k)$, $\mathbf{x}(k) = r_1 2^k \begin{bmatrix} 1 \\ -1 \end{bmatrix} + r_2 3^k \begin{bmatrix} 2 \\ 1 \end{bmatrix}$

43. $\mathbf{x}(k+1) = \begin{bmatrix} 1 & 0 & 0 \\ 0 & 2 & 3 \\ 0 & 3 & 2 \end{bmatrix}\mathbf{x}(k)$,

$$\mathbf{x}(k) = r_1 1^k \begin{bmatrix} 1 \\ 0 \\ 0 \end{bmatrix} + r_2 5^k \begin{bmatrix} 0 \\ 1 \\ 1 \end{bmatrix} + r_3(-1)^k \begin{bmatrix} 0 \\ 1 \\ -1 \end{bmatrix}$$

44. $\mathbf{x}(k+1) = \begin{bmatrix} 2 & 0 & 0 \\ 0 & 3 & 1 \\ 0 & 1 & 3 \end{bmatrix}\mathbf{x}(k)$,

$$\mathbf{x}(k) = r_1 2^k \begin{bmatrix} 1 \\ 0 \\ 0 \end{bmatrix} + r_2 4^k \begin{bmatrix} 0 \\ 1 \\ 1 \end{bmatrix} + r_3 2^k \begin{bmatrix} 0 \\ 1 \\ -1 \end{bmatrix}$$

COMPLEX NUMBERS

45. Find rank $\begin{bmatrix} i & -1 & 1+i & 0 \\ 1+i & i & -1 & 3+2i \\ 1+2i & -1+i & i & 3+2i \end{bmatrix}$

46. Find a basis for the row space of

$$A = \begin{bmatrix} i & -1 & 1+i \\ 1+2i & -1+i & i \\ i & -1+i & 1+i \end{bmatrix}$$

47. Decide if $\begin{bmatrix} i & -1 & \\ 1 & & i \\ 1+i & -1+i & \end{bmatrix}\mathbf{z} = \begin{bmatrix} 1-i \\ -1-i \\ 1+i \end{bmatrix}$ has a solution.

THEORETICAL EXERCISES

48. Let $r_1, r_2 \in \mathbf{R}$ with $r_1 \neq r_2$. Show that $\{r_1^k, r_2^k\}$ is linearly independent in $\mathbf{F}(\mathbf{N})$.

49. Let $a, b, c \in \mathbf{R}$ with $a \neq 0$. If $b^2 - 4ac = 0$ and if r solves $ax^2 + bx + c = 0$, then show that kr^k is a solution of

$$ax(k+2) + bx(k+1) + cx(k) = 0$$

50. Let $r \in \mathbf{R}$. Show that $\{r^k, kr^k\}$ is linearly independent in $\mathbf{F}(\mathbf{N})$.

51. If γ and θ are as defined in Case 3 of Part III of this section, using that

$$x_1(k) = \gamma^k \cos k\theta \qquad \text{and} \qquad x_2(k) = \gamma^k \sin k\theta$$

are solutions of the difference equation, show that the set of these two solutions is linearly independent.

52. Let $S = \{\mathbf{x}^1, \ldots, \mathbf{x}^n\} \subseteq \mathbf{R}^n$. Writing

$$r_1\mathbf{x}^1 + \cdots + r_n\mathbf{x}^n = \mathbf{0}$$

as a system of linear equations, show that a basis for span S can be found by simultaneously eliminating all vectors among $\mathbf{x}^1, \ldots, \mathbf{x}^n$ that correspond to free variables at the end of the Gaussian elimination algorithm. Apply this technique to the sets below.

(a) $\left\{ \begin{bmatrix} 1 \\ 0 \\ 1 \end{bmatrix}, \begin{bmatrix} -1 \\ 1 \\ 0 \end{bmatrix}, \begin{bmatrix} 0 \\ 1 \\ 1 \end{bmatrix}, \begin{bmatrix} 1 \\ 1 \\ 1 \end{bmatrix}, \begin{bmatrix} 1 \\ -1 \\ 1 \end{bmatrix} \right\}$

(b) $\left\{ \begin{bmatrix} 2 \\ 4 \\ -2 \end{bmatrix}, \begin{bmatrix} 1 \\ 3 \\ 1 \end{bmatrix}, \begin{bmatrix} -1 \\ 1 \\ 7 \end{bmatrix}, \begin{bmatrix} 1 \\ -2 \\ -9 \end{bmatrix}, \begin{bmatrix} 3 \\ 1 \\ -2 \end{bmatrix} \right\}$

53. Let A be an $n \times n$ matrix. Prove that the set of the rows of A and the set of the columns of A are linearly independent if and only if $\det A \neq 0$.

54. Let A be an $m \times n$ matrix. Prove

$$\text{rank } A \leqslant \min\{m, n\}$$

Hint: Use rank A = rank A^t.

55. Prove the following computational rules for rank.

(a) In general,

$$\text{rank}(A + B) \neq \text{rank } A + \text{rank } B$$

and $$\text{rank}(AB) \neq (\text{rank } A)(\text{rank } B)$$

(b) rank $AB \leqslant \min\{\text{rank } A, \text{rank } B\}$. *Hint:* Note that every row of AB is a linear combination of the rows of B. Thus rank $AB \leqslant$ rank B. Replacing A with B^t and B with A^t, we have rank $B^tA^t \leqslant$ rank A^t. Now use that rank C = rank C^t for any matrix C.

56. Prove that if P is an invertible $m \times m$ matrix, Q is an invertible $n \times n$ matrix, and A is an $m \times n$ matrix, then

$$\text{rank } PA = \text{rank } AQ = \text{rank } A$$

Hint: Use exercise 55(b) and that rank B = rank$(A^{-1}AB)$.

57. (calculus) Consider the equation

$$\frac{d}{dt} p(t) = 0$$

for all $p \in \mathbf{P}$. Find the dimension and a basis for the solution set.

58. (calculus) Consider the equation

$$\frac{d^2}{dt^2} p(t) = 0$$

for all $p \in \mathbf{P}$. Find the dimension and a basis for the solution set.

59. (calculus) Consider the equation

$$\int_0^1 p(t)\, dt = 0$$

for all $p \in \mathbf{P}_n$. Find the dimension and a basis for the solution set.

APPLICATIONS EXERCISE

60. Consider the water tower system depicted in Figure 3.60. Write the equations for this system. If $\mathbf{x}(0) = \mathbf{0}$, find $\mathbf{W}(1)$, the set of all $\mathbf{x}(1)$ for all choices of $u_1(0)$, and $u_2(0)$ in terms of span. Specify the dimension of $\mathbf{W}(1)$.

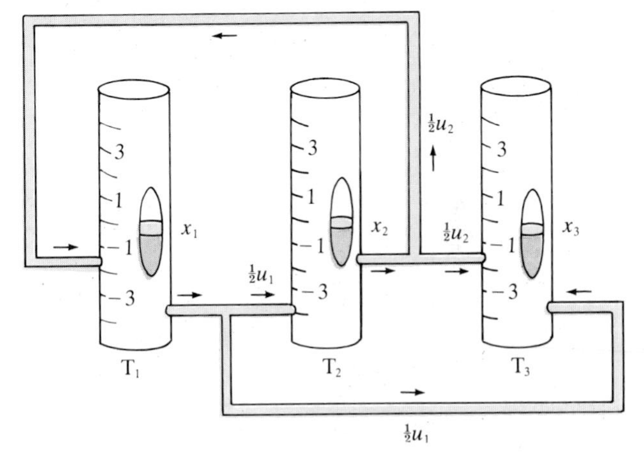

FIGURE 3.60

COMPUTER EXERCISE

61. Let the function $r(t)$ = rank A for the following matrices A. Graph this function, and notice that small changes in t can sometimes change the rank. Thus round-off is not acceptable in computing rank.

(a) $A = \begin{bmatrix} 1 & 1 \\ 1 & t \end{bmatrix}$ (b) $A = \begin{bmatrix} t & 2 \\ 2 & t \end{bmatrix}$

CHAPTER FOUR

INNER PRODUCT SPACES

In the last chapter we defined a vector space and introduced the geometrical concept of dimension for all vector spaces. In this chapter we introduce the remaining geometrical concepts—distance and angle. To see how it is possible to define these additional geometric concepts for all vector spaces, we recall how they were defined in $\mathbf{R}_3$. For vectors $\mathbf{x}$, $\mathbf{y} \in \mathbf{R}_3$, the inner product was defined as

$$(\mathbf{x}, \mathbf{y}) = x_1 y_1 + x_2 y_2 + x_3 y_3$$

and the norm of $\mathbf{x}$ as

$$\|\mathbf{x}\| = (\mathbf{x}, \mathbf{x})^{1/2}$$

Then,

1. Considering $\mathbf{x}$, $\mathbf{y} \in \mathbf{R}_3$ as points in a coordinate space, we defined the distance from $\mathbf{x}$ to $\mathbf{y}$ as

$$d(\mathbf{x}, \mathbf{y}) = \|\mathbf{x} - \mathbf{y}\|$$

2. Considering $\mathbf{x}$, $\mathbf{y} \in \mathbf{R}_3$ as fixed arrows in a coordinate space, we defined
(*a*) $\|\mathbf{x}\|$ as the length of $\mathbf{x}$ and
(*b*) $\cos\theta = \dfrac{(\mathbf{x}, \mathbf{y})}{\|\mathbf{x}\| \|\mathbf{y}\|}$ as the equation determining the angle θ between $\mathbf{x} \neq \mathbf{0}$ and $\mathbf{y} \neq \mathbf{0}$.

Note that all of these concepts are defined in terms of the inner product. Thus, to obtain these geometric concepts for any vector space, we need to start by defining an inner product on that vector space. To do so, we recall what an inner product is in $\mathbf{R}_3$. In $\mathbf{R}_3$, $(\mathbf{x}, \mathbf{y})$ is a number for each pair of vectors $\mathbf{x}$, $\mathbf{y}$ such that $(\bullet, \bullet)$ satisfies the following arithmetical properties:

1. For any vector $\mathbf{x}$,

$$(\mathbf{x}, \mathbf{x}) \geqslant 0$$

with equality if and only if $\mathbf{x} = \mathbf{0}$ (positive property).

2. For any vectors $\mathbf{x}$, $\mathbf{y}$,

$$(\mathbf{x}, \mathbf{y}) = (\mathbf{y}, \mathbf{x})$$

(symmetry property)

3. For any vectors $\mathbf{x}$ and $\mathbf{y}$ and for any $r \in \mathbf{R}$,

$$(r\mathbf{x}, \mathbf{y}) = (\mathbf{x}, r\mathbf{y}) = r(\mathbf{x}, \mathbf{y})$$

(homogeneous property)

4. For all vectors $\mathbf{x}$, $\mathbf{y}$, and $\mathbf{z}$,

$$(\mathbf{x} + \mathbf{y}, \mathbf{z}) = (\mathbf{x}, \mathbf{z}) + (\mathbf{y}, \mathbf{z}) \qquad \text{and} \qquad (\mathbf{x}, \mathbf{y} + \mathbf{z}) = (\mathbf{x}, \mathbf{y}) + (\mathbf{x}, \mathbf{z})$$

(distributive property)

In this chapter we use these four properties to define an inner product for any vector space and then use the inner product to define distance and angle.

4.1 Inner Product Spaces, Length, and Distance

In this section we define an inner product and, from it, distance and length on any vector space. These definitions will allow us to make geometrical calculations in any vector space just as we do in $\mathbf{R}_3$.

We define an inner product as follows.

Definition

Let $\mathbf{V}$ be a vector space. By an *inner product* on $\mathbf{V}$ we mean a function, written $(\bullet, \bullet)$, that assigns to each pair of vectors $\mathbf{x}$, $\mathbf{y}$ in $\mathbf{V}$ a number, written $(\mathbf{x}, \mathbf{y})$. This function satisfies the following properties:

1. For any $\mathbf{x} \in \mathbf{V}$,

$$(\mathbf{x}, \mathbf{x}) \geqslant 0$$

with equality if and only if $\mathbf{x} = \mathbf{0}$ (positive property)

2. For any $\mathbf{x}$, $\mathbf{y} \in \mathbf{V}$,

$$(\mathbf{x}, \mathbf{y}) = (\mathbf{y}, \mathbf{x})$$

(symmetry property)

3. For any $\mathbf{x}$, $\mathbf{y} \in \mathbf{V}$ and any $r \in \mathbf{R}$,

$$(r\mathbf{x}, \mathbf{y}) = (\mathbf{x}, r\mathbf{y}) = r(\mathbf{x}, \mathbf{y})$$

(homogeneous property)

4. For any $\mathbf{x}$, $\mathbf{y}$, $\mathbf{z} \in \mathbf{V}$,

$$(\mathbf{x} + \mathbf{y}, \mathbf{z}) = (\mathbf{x}, \mathbf{z}) + (\mathbf{y}, \mathbf{z}) \qquad \text{and} \qquad (\mathbf{x}, \mathbf{y} + \mathbf{z}) = (\mathbf{x}, \mathbf{y}) + (\mathbf{x}, \mathbf{z})$$

(distributive property)

Vector spaces that have inner products defined on them are called *inner product spaces*.

Some particular examples of inner product spaces follow.

EXAMPLE 1

For the vector space $\mathbf{R}_n$, let

$$(\mathbf{x}, \mathbf{y}) = x_1 y_1 + \cdots + x_n y_n$$

for all $\mathbf{x}, \mathbf{y} \in \mathbf{R}_n$. To show that $(\bullet, \bullet)$ is an inner product on $\mathbf{R}_n$, we need to show that it satisfies properties 1 through 4 as given in the definition. For property 1, note that

$$(\mathbf{x}, \mathbf{x}) = x_1^2 + \cdots + x_n^2 \geqslant 0$$

Further, the only way equality can hold is if $x_1 = \cdots = x_n = 0$; that is, $\mathbf{x} = \mathbf{0}$. Thus $(\mathbf{x}, \mathbf{x}) \geqslant 0$ with equality if and only if $\mathbf{x} = \mathbf{0}$.

For property 2,

$$(\mathbf{x}, \mathbf{y}) = x_1 y_1 + \cdots + x_n y_n = y_1 x_1 + \cdots + y_n x_n = (\mathbf{y}, \mathbf{x})$$

Now, let $r \in \mathbf{R}$. For property 3,

$$(r\mathbf{x}, \mathbf{y}) = (rx_1)y_1 + \cdots + (rx_n)y_n = r(x_1 y_1 + \cdots + x_n y_n) = r(\mathbf{x}, \mathbf{y})$$

That $(\mathbf{x}, r\mathbf{y}) = r(\mathbf{x}, \mathbf{y})$ follows similarly.

Finally, for property 4, if $\mathbf{x}, \mathbf{y}, \mathbf{z} \in \mathbf{R}_n$, then

$$\begin{aligned}(\mathbf{x}+\mathbf{y}, \mathbf{z}) &= (x_1+y_1)z_1 + \cdots + (x_n+y_n)z_n = x_1 z_1 + y_1 z_1 + \cdots + x_n z_n + y_n z_n \\ &= x_1 z_1 + \cdots + x_n z_n + y_1 z_1 + \cdots + y_n z_n = (\mathbf{x}, \mathbf{z}) + (\mathbf{y}, \mathbf{z})\end{aligned}$$

That $(\mathbf{x}, \mathbf{y} + \mathbf{z}) = (\mathbf{x}, \mathbf{y}) + (\mathbf{x}, \mathbf{z})$ follows similarly.

Thus $(\bullet, \bullet)$ is an inner product and $\mathbf{R}_n$ is an inner product space. With the same definition for $(\bullet, \bullet)$, the vector space $\mathbf{R}^n$ is also an inner product space. In both cases, we call this inner product the *standard inner product** for the space. □

EXAMPLE 2

For another inner product on $\mathbf{R}_n$, let $w_1 > 0, \ldots, w_n > 0$ be real numbers, which are called *weights* for the inner product. Define

$$(\mathbf{x}, \mathbf{y}) = w_1 x_1 y_1 + w_2 x_2 y_2 + \cdots + w_n x_n y_n$$

for all $\mathbf{x}, \mathbf{y} \in \mathbf{R}_n$. Then, as in Example 1, we can show that $(\bullet, \bullet)$ is an inner product, and so $\mathbf{R}_n$ is an inner product space in another way. Similarly, $\mathbf{R}^n$, with the same definition for $(\bullet, \bullet)$, is an inner product space in another way. □

* The standard inner product is a natural generalization of that given in Chapter 3 for $\mathbf{R}_3$.

EXAMPLE 3

For $\mathbf{R}_{m,n}$, let

$$(A, B) = \sum_{i=1}^{m} \sum_{j=1}^{n} a_{ij} b_{ij}$$

for all $A, B \in \mathbf{R}_{m,n}$. For example,

$$\left(\begin{bmatrix} 1 & 2 \\ 3 & 4 \end{bmatrix}, \begin{bmatrix} 5 & 6 \\ 7 & 8 \end{bmatrix} \right) = 1 \cdot 5 + 2 \cdot 6 + 3 \cdot 7 + 4 \cdot 8 = 70$$

Then $(\bullet, \bullet)$ is an inner product, called the *standard inner product* for $\mathbf{R}_{m,n}$ (exercise 41). Thus $\mathbf{R}_{m,n}$ is an inner product space. □

EXAMPLE 4

CALCULUS
For $\mathbf{C}[a, b]$, let

$$(f, g) = \int_a^b f(t)g(t)\,dt$$

for all $f, g \in \mathbf{C}[a, b]$. We will show that $(\bullet, \bullet)$ is an inner product, called the *standard inner product*, for $\mathbf{C}[a, b]$.

For property 1,

$$(f, f) = \int_a^b [f(t)]^2\,dt \geqslant 0$$

If $f = 0$, then clearly $(f, f) = 0$. Further, if, for some t_0, $a \leqslant t_0 \leqslant b$, we have $f(t_0) \neq 0$. Then, since f is continuous, there is a positive area under f^2, and so

$$\int_a^b [f(t)]^2\,dt > 0$$

Thus $(f, f) = 0$ implies that $f = 0$ and conversely.

Property 2 is shown by

$$(f, g) = \int_a^b f(t)g(t)\,dt = \int_a^b g(t)f(t)\,dt = (g, f)$$

For property 3, let $r \in \mathbf{R}$. Then

$$(rf, g) = \int_a^b rf(t)g(t)\,dt = \int_a^b f(t)[rg(t)]\,dt = (f, rg)$$

Similarly, $(rf, g) = r(f, g)$.

Finally, for property 4,

$$(f, g + h) = \int_a^b f(t)[g(t) + h(t)]\,dt = \int_a^b f(t)g(t) + f(t)h(t)\,dt$$
$$= \int_a^b f(t)g(t)\,dt + \int_a^b f(t)h(t)\,dt = (f, g) + (f, h)$$

Similarly, $(f + g, h) = (f, h) + (g, h)$. Thus $(\bullet, \bullet)$ is an inner product for $\mathbf{C}[a, b]$ and $\mathbf{C}[a, b]$ is an inner product space. □

EXAMPLE 5

CALCULUS
For another inner product on $\mathbf{C}[a, b]$, let $w \in \mathbf{C}[a, b]$ where $w(t) > 0$ for all $t \in [a, b]$. This function is called a *weight* for the inner product. Define

$$(f, g) = \int_a^b w(t)f(t)g(t)\,dt$$

Then, as in Example 4, we can show that $(\bullet, \bullet)$ is an inner product and thus $\mathbf{C}[a, b]$ is an inner product space with this inner product as well. □

Other examples are also possible. In fact, it can be shown that an inner product can be defined on any vector space. (See exercise 42 for the finite case.)

Following the outline in the introduction, we now show how to define a norm on any inner product space. This norm will be used to define distances between points and angles between vectors.

DEFINITION

Let $\mathbf{V}$ be an inner product space and let $\mathbf{x} \in \mathbf{V}$. We define the *norm*, or *length*, of $\mathbf{x}$, written $\|\mathbf{x}\|$, as*

$$\|\mathbf{x}\| = (\mathbf{x}, \mathbf{x})^{1/2}$$

EXAMPLE 6

Let $\mathbf{R}_n$ be the inner product space with the standard inner product. Then

$$\|\mathbf{x}\| = (x_1^2 + x_2^2 + \cdots + x_n^2)^{1/2}$$

Thus the length of $\mathbf{x}$ depends on the sizes of the entries in $\mathbf{x}$.

For $n = 4$ and $\mathbf{x} = [1, 2, 0, 1]$,

$$\|\mathbf{x}\| = (1^2 + 2^2 + 0^2 + 1^2)^{1/2} = \sqrt{6}$$ □

* Geometrically, $\mathbf{x}$ is thought of as a fixed arrow.

EXAMPLE 7

CALCULUS
For fixed a and b in $\mathbf{R}$, let $\mathbf{C}[a, b]$ be the inner product space with the standard inner product. Then, for any $f \in \mathbf{C}[a, b]$,

$$\|f\| = \left[\int_a^b [f(t)]^2 \, dt \right]^{1/2}$$

Thus the norm of f depends on the size and shape of the region from a to b between f and the t-axis, as illustrated in Figure 4.1.

For $a = -1$, $b = 1$, and $f(t) = t^2 - t$,

$$\|f\| = \left[\int_{-1}^1 (t^2 - t)^2 \, dt \right]^{1/2} = \left[\int_{-1}^1 t^4 - 2t^3 + t^2 \, dt \right]^{1/2} = \frac{4}{15}\sqrt{15}$$

See Figure 4.2. □

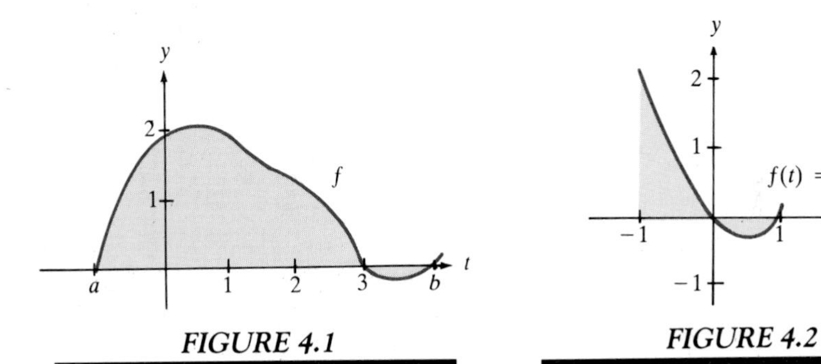

FIGURE 4.1

FIGURE 4.2

A vector of length 1 is called a *unit vector*. Since, for any nonzero vector $\mathbf{x}$,

$$\|\mathbf{x}\| = (\mathbf{x}, \mathbf{x}) > 0$$

and

$$\left\| \frac{1}{\|\mathbf{x}\|}\mathbf{x} \right\|^2 = \left(\frac{1}{\|\mathbf{x}\|}\mathbf{x}, \frac{1}{\|\mathbf{x}\|}\mathbf{x} \right) = \frac{1}{\|\mathbf{x}\|^2}(\mathbf{x}, \mathbf{x}) = \frac{\|\mathbf{x}\|^2}{\|\mathbf{x}\|^2} = 1$$

it follows that $\mathbf{x}$ can be divided by its length $\|\mathbf{x}\|$ to obtain the unit vector $\dfrac{1}{\|\mathbf{x}\|}\mathbf{x}$.

EXAMPLE 8

Let $\mathbf{x} = [-1, 1, 2, 3, 1]$. Using the standard inner product, we have

$$\|\mathbf{x}\| = \sqrt{(-1)^2 + (1)^2 + (2)^2 + (3)^2 + (1)^2} = 4$$

Thus $\dfrac{1}{\|\mathbf{x}\|}\mathbf{x} = [-\frac{1}{4}, \frac{1}{4}, \frac{2}{4}, \frac{3}{4}, \frac{1}{4}]$ is a unit vector. □

Before we give additional properties of norms, we need the following inequality.

THEOREM 4.1

(Cauchy-Schwarz inequality) Let $\mathbf{V}$ be an inner product space. For all $\mathbf{x}, \mathbf{y} \in \mathbf{V}$, we have

$$|(\mathbf{x}, \mathbf{y})| \leqslant \|\mathbf{x}\| \, \|\mathbf{y}\|$$

PROOF

For the proof, we consider two cases.

CASE 1: If $\mathbf{x} = \mathbf{0}$, then by the definition of the inner product we have

$$|(\mathbf{x}, \mathbf{y})| = |(\mathbf{0}, \mathbf{y})| = |(0\mathbf{y}, \mathbf{y})| = 0|(\mathbf{y}, \mathbf{y})| = 0$$

Thus $|(\mathbf{x}, \mathbf{y})| \leqslant \|\mathbf{x}\| \, \|\mathbf{y}\|$. In the same way, the inequality holds if $\mathbf{y} = \mathbf{0}$.

CASE 2: If $\mathbf{x} \neq \mathbf{0}$ and $\mathbf{y} \neq \mathbf{0}$, set

$$\mathbf{u} = \frac{1}{\|\mathbf{x}\|}\mathbf{x} \qquad \text{and} \qquad \mathbf{v} = \frac{1}{\|\mathbf{y}\|}\mathbf{y}$$

Then, using the arithmetical properties of the inner product, we have

$$\begin{aligned} 0 \leqslant (\mathbf{u} - \mathbf{v}, \mathbf{u} - \mathbf{v}) &= (\mathbf{u}, \mathbf{u}) - 2(\mathbf{u}, \mathbf{v}) + (\mathbf{v}, \mathbf{v}) \\ &= \|\mathbf{u}\|^2 - 2(\mathbf{u}, \mathbf{v}) + \|\mathbf{v}\|^2 \\ &= 2 - 2(\mathbf{u}, \mathbf{v}) \end{aligned}$$

Hence, $(\mathbf{u}, \mathbf{v}) \leqslant 1$. Substituting $\mathbf{u} = \frac{1}{\|\mathbf{x}\|}\mathbf{x}$ and $\mathbf{v} = \frac{1}{\|\mathbf{y}\|}\mathbf{y}$, it follows that

$$\left(\frac{1}{\|\mathbf{x}\|}\mathbf{x}, \frac{1}{\|\mathbf{y}\|}\mathbf{y}\right) \leqslant 1$$

Now, again using the properties of the inner product, we have

$$\frac{1}{\|\mathbf{x}\|}\frac{1}{\|\mathbf{y}\|}(\mathbf{x}, \mathbf{y}) \leqslant 1 \qquad \text{or} \qquad (\mathbf{x}, \mathbf{y}) \leqslant \|\mathbf{x}\| \, \|\mathbf{y}\|$$

Applying this result, we also have

$$-(\mathbf{x}, \mathbf{y}) = (-\mathbf{x}, \mathbf{y}) \leqslant \|-\mathbf{x}\| \, \|\mathbf{y}\| = \|\mathbf{x}\| \, \|\mathbf{y}\|$$

Hence,

$$|(\mathbf{x}, \mathbf{y})| \leqslant \|\mathbf{x}\| \, \|\mathbf{y}\| \qquad \blacksquare$$

We can now give the additional properties of norms often used in computing.

THEOREM 4.2

Let $\mathbf{V}$ be an inner product space. Then

(*i*) For all $\mathbf{x} \in \mathbf{V}$,

$$\|\mathbf{x}\| \geqslant 0$$

with equality if and only if $\mathbf{x} = \mathbf{0}$.

(*ii*) For all $r \in \mathbf{R}$ and $\mathbf{x} \in \mathbf{V}$,

$$\|r\mathbf{x}\| = |r|\,\|\mathbf{x}\|$$

(*iii*) For all $\mathbf{x}, \mathbf{y} \in \mathbf{V}$,

$$\|\mathbf{x} + \mathbf{y}\| \leqslant \|\mathbf{x}\| + \|\mathbf{y}\|$$

(triangle inequality)*

PROOF

For (i), using the definition of inner product, we have

$$\|\mathbf{x}\| = \sqrt{(\mathbf{x}, \mathbf{x})} \geqslant 0$$

with equality if and only if $\mathbf{x} = \mathbf{0}$. For (ii),

$$\|r\mathbf{x}\| = \sqrt{(r\mathbf{x}, r\mathbf{x})} = \sqrt{r^2(\mathbf{x}, \mathbf{x})} = |r|\sqrt{(\mathbf{x}, \mathbf{x})} = |r|\,\|\mathbf{x}\|$$

Finally, for (iii),

$$\|\mathbf{x} + \mathbf{y}\|^2 = (\mathbf{x} + \mathbf{y}, \mathbf{x} + \mathbf{y}) = (\mathbf{x}, \mathbf{x}) + 2(\mathbf{x}, \mathbf{y}) + (\mathbf{y}, \mathbf{y}) = \|\mathbf{x}\|^2 + 2(\mathbf{x}, \mathbf{y}) + \|\mathbf{y}\|^2$$

Now, the Cauchy-Schwarz inequality gives

$$\|\mathbf{x} + \mathbf{y}\|^2 \leqslant \|\mathbf{x}\|^2 + 2\|\mathbf{x}\|\,\|\mathbf{y}\| + \|\mathbf{y}\|^2$$

or

$$\|\mathbf{x} + \mathbf{y}\|^2 \leqslant (\|\mathbf{x}\| + \|\mathbf{y}\|)^2$$

Thus, taking square roots yields the result. ■

Continuing to use the outline in the introduction, we define distance as follows.

DEFINITION

Let $\mathbf{V}$ be an inner product space and $\mathbf{x}, \mathbf{y} \in \mathbf{V}$. Then the *distance* from $\mathbf{x}$ to $\mathbf{y}$, written $d(\mathbf{x}, \mathbf{y})$, is†

$$d(\mathbf{x}, \mathbf{y}) = \|\mathbf{x} - \mathbf{y}\|$$

* These results are intuitive when $\mathbf{x}$ and $\mathbf{y}$ are thought of as fixed arrows.

† Geometrically, $\mathbf{x}$ and $\mathbf{y}$ are thought of as points.

EXAMPLE 9

Let $\mathbf{R}_n$ be the inner product space with the standard inner product. Then

$$d(\mathbf{x},\mathbf{y}) = \|\mathbf{x}-\mathbf{y}\| = [(x_1-y_1)^2+(x_2-y_2)^2+\cdots+(x_n-y_n)^2]^{1/2}$$

Thus the distance from $\mathbf{x}$ to $\mathbf{y}$ depends on the distance between the corresponding entries of $\mathbf{x}$ and $\mathbf{y}$.

For $\mathbf{x} = [2,0,1,3]$ and $\mathbf{y} = [1,2,2,1]$, we have

$$\begin{aligned} d(\mathbf{x},\mathbf{y}) &= \|[2,0,1,3]-[1,2,2,1]\| \\ &= [(2-1)^2+(0-2)^2+(1-2)^2+(3-1)^2]^{1/2} \\ &= \sqrt{10} \end{aligned}$$

□

EXAMPLE 10

CALCULUS

Let $\mathbf{C}[a,b]$ be the inner product space with the standard inner product. Then

$$d(f,g) = \|f-g\| = \left(\int_a^b [f(t)-g(t)]^2\,dt\right)^{1/2}$$

Thus the distance from f to g depends on the size and shape of the region from a to b between f and g.

For $a=0$, $b=1$, $f(t)=t^2-2$, and $g(t)=t-2$, we have

$$d(f,g) = \|f-g\| = \left(\int_0^1 [(t^2-2)-(t-2)]^2\,dt\right)^{1/2}$$

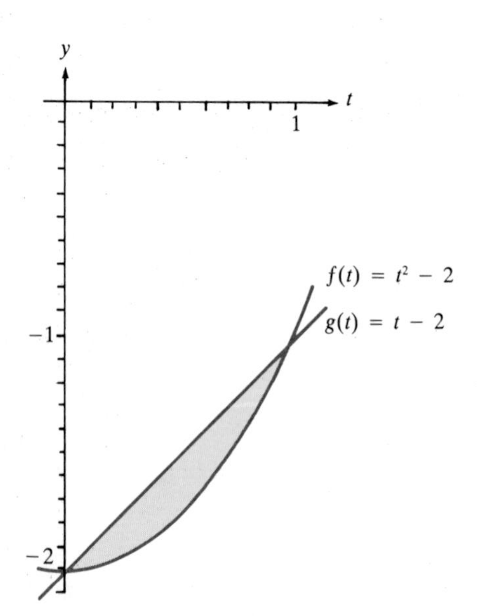

FIGURE 4.3

$$= \left(\int_0^1 (t^2 - t)^2 \, dt \right)^{1/2}$$
$$= \left(\int_0^1 t^4 - 2t^3 + t^2 \, dt \right)^{1/2}$$
$$= \left(\frac{t^5}{5} - \frac{2t^4}{4} + \frac{t^3}{3} \Big|_0^1 \right)^{1/2} = \sqrt{\frac{1}{30}}$$

See Figure 4.3. □

The following natural properties are used in distance computations.

THEOREM 4.3

Let **V** be an inner product space. Then

(*i*) For **x**, **y** ∈ **V**,

$$d(\mathbf{x}, \mathbf{y}) = 0$$

if and only if **x** = **y**.

(*ii*) For all **x**, **y** ∈ **V**,

$$d(\mathbf{x}, \mathbf{y}) = d(\mathbf{y}, \mathbf{x})$$

(*iii*) For all **x**, **y**, **z** ∈ **V**,

$$d(\mathbf{x}, \mathbf{y}) \leqslant d(\mathbf{x}, \mathbf{z}) + d(\mathbf{z}, \mathbf{y})$$

(triangle inequality)*

PROOF

For property (i),

$$d(\mathbf{x}, \mathbf{x}) = \|\mathbf{x} - \mathbf{x}\| = \|\mathbf{0}\| = 0$$

Further, if $0 = d(\mathbf{x}, \mathbf{y}) = \|\mathbf{x} - \mathbf{y}\|$, then, from the properties of norms, $\mathbf{x} - \mathbf{y} = \mathbf{0}$ or $\mathbf{x} = \mathbf{y}$. For property (ii),

$$d(\mathbf{x}, \mathbf{y}) = \|\mathbf{x} - \mathbf{y}\| = \|-(\mathbf{y} - \mathbf{x})\| = \|\mathbf{y} - \mathbf{x}\| = d(\mathbf{y}, \mathbf{x})$$

Finally, for (iii),

$$\begin{aligned} d(\mathbf{x}, \mathbf{y}) &= \|(\mathbf{x} - \mathbf{z}) + (\mathbf{z} - \mathbf{y})\| \leqslant \|\mathbf{x} - \mathbf{z}\| + \|\mathbf{z} - \mathbf{y}\| \\ &= d(\mathbf{x}, \mathbf{z}) + d(\mathbf{z}, \mathbf{y}) \end{aligned}$$ ■

We will use this distance for calculating in vector spaces in the same way we use it in $\mathbf{R}_3$.

* These properties are intuitive when **x** and **y** are thought of as points.

EXAMPLE 11

Let $\mathbf{R}_4$ be the inner product space with the standard inner product. Let

$$\mathbf{x} = [-1, 1, -1, 1], \qquad \mathbf{y} = [1, -1, 1, 0], \qquad \mathbf{z} = [1, 1, -1, 1]$$

in $\mathbf{R}_4$. We will decide which of $\mathbf{x}$ or $\mathbf{y}$ is closer to $\mathbf{z}$. To do this, we calculate

$$d(\mathbf{x}, \mathbf{z}) = [(-1-1)^2 + (1-1)^2 + (-1+1)^2 + (1-1)^2]^{1/2} = 2$$

and

$$d(\mathbf{y}, \mathbf{z}) = [(1-1)^2 + (-1-1)^2 + (1+1)^2 + (0-1)^2]^{1/2} = 3$$

Hence $\mathbf{x}$ is closer to $\mathbf{z}$ than is $\mathbf{y}$. □

EXAMPLE 12

CALCULUS

Let $\mathbf{C}[0, 1]$ be the inner product space with the standard inner product. Let

$$f(t) = t^2, \qquad p(t) = t - \tfrac{1}{16}, \qquad q(t) = t - \tfrac{1}{4}$$

The graphs of these functions are shown in Figure 4.4. Checking the regions between the pairs p and f and the pairs q and f, we see that it is difficult to decide which of p or q is closer to f. Thus, to make the decision, we need to calculate $d(f, p)$ and $d(f, q)$.

$$\begin{aligned} d(f, p) &= \left(\int_0^1 \left[t^2 - \left(t - \frac{1}{16} \right) \right]^2 dt \right)^{1/2} \\ &= \left(\int_0^1 t^4 - 2t^3 + \frac{9}{8}t^2 - \frac{1}{8}t + \frac{1}{256}\, dt \right)^{1/2} \\ &= \left(\frac{t^5}{5} - \frac{2t^4}{4} + \frac{3t^3}{8} - \frac{t^2}{16} + \frac{t}{256} \Bigg|_0^1 \right)^{1/2} = \left(\frac{21}{1280} \right)^{1/2} \approx 0.1281 \end{aligned}$$

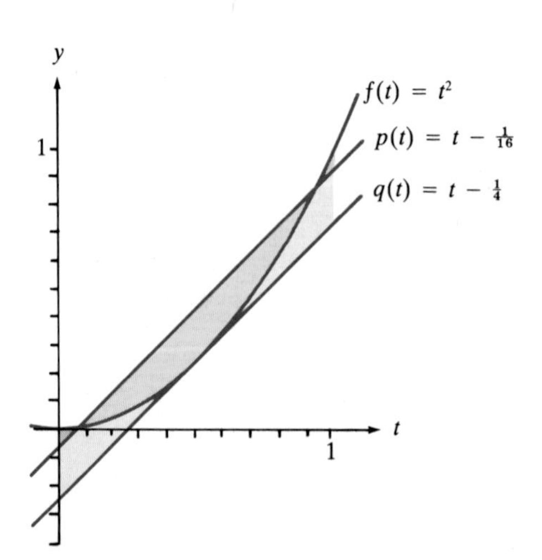

FIGURE 4.4

$$d(f,q) = \left(\int_0^1 \left[t^2 - \left(t - \frac{1}{4}\right)\right]^2 dt\right)^{1/2}$$
$$= \left(\int_0^1 t^4 - 2t^3 + \frac{3}{2}t^2 - \frac{1}{2}t + \frac{1}{16}\, dt\right)^{1/2}$$
$$= \left(\frac{t^5}{5} - \frac{2t^4}{4} + \frac{3t^3}{6} - \frac{t^2}{4} + \frac{t}{16}\bigg|_0^1\right)^{1/2} = \left(\frac{1}{80}\right)^{1/2} \approx 0.11180$$

Thus, $q(t) = t - \frac{1}{4}$ is the closer function. (Distance calculations such as these are used in determining which polynomial, of specified degree, is the closest approximation to a given function. The polynomial is then used as a simple expression for the function.) □

EXAMPLE 13

Let $\mathbf{R}_{2,2}$ be the inner product space with the standard inner product. Let

$$A = \begin{bmatrix} 1 & 2 \\ 3 & 4 \end{bmatrix} \quad \text{and} \quad B = \begin{bmatrix} 1.1 & 1.9 \\ 3.2 & 3.8 \end{bmatrix}$$

Thus, A is a one-digit approximation of B. To measure how close A and B are, calculate

$$d(A,B) = \|A - B\| = \left\| \begin{bmatrix} -0.1 & 0.1 \\ -0.2 & 0.2 \end{bmatrix} \right\|$$
$$= [(-0.1)^2 + (0.1)^2 + (-0.2)^2 + (0.2)^2]^{1/2}$$
$$= \sqrt{0.1}$$

(Distance measurements such as these are useful in studying round-off errors in computer calculations.) □

In the next section we introduce the geometrical concept of angle.

EXERCISES FOR SECTION 4.1

COMPUTATIONAL EXERCISES

In exercises 1 through 8, decide whether or not the given function is an inner product for the given vector space $\mathbf{V}$. If it is, prove it. If it is not, show which properties of the definition of inner product fail.

1. $\mathbf{V} = \mathbf{R}_2$, $([x_1, x_2], [y_1, y_2]) = 3x_1y_1 + x_1y_2 + x_2y_1 + 3x_2y_2$

2. $\mathbf{V} = \mathbf{R}_2$, $([x_1, x_2], [y_1, y_2]) = x_1y_1 + 2x_1y_2 + 2x_2y_1 + x_2y_2$

3. $\mathbf{V} = \mathbf{R}_2$, $([x_1, x_2], [y_1, y_2]) = x_1y_2 + x_2y_1$

4. $\mathbf{V} = \mathbf{R}_2$, $([x_1, x_2], [y_1, y_2]) = 2x_1y_1 + x_1y_2 + x_2y_1 + 3x_2y_2$

5. $\mathbf{V} = \mathbf{R}_4$, $([x_1, x_2, x_3, x_4], [y_1, y_2, y_3, y_4]) = x_1y_1 + 2x_2y_2 + 5x_3y_3 + x_4y_4$

6. $\mathbf{V} = \mathbf{R}_4$, $([x_1, x_2, x_3, x_4], [y_1, y_2, y_3, y_4]) = x_1y_1 + x_2y_2 + x_3y_3 - x_4y_4$
(This function occurs in the study of relativity.)

7. $\mathbf{V} = \mathbf{R}_{2,3}$, $\left(\begin{bmatrix} x_{11} & x_{12} & x_{13} \\ x_{21} & x_{22} & x_{23} \end{bmatrix}, \begin{bmatrix} y_{11} & y_{12} & y_{13} \\ y_{21} & y_{22} & y_{23} \end{bmatrix}\right) = x_{11}y_{11} + x_{12}y_{12} + x_{13}y_{13} - x_{21}y_{21} - x_{22}y_{22} - x_{13}y_{23}$

8. $\mathbf{V} = \mathbf{R}_{2,2}$, $\left(\begin{bmatrix} x_{11} & x_{12} \\ x_{21} & x_{22} \end{bmatrix}, \begin{bmatrix} y_{11} & y_{12} \\ y_{21} & y_{22} \end{bmatrix}\right) = x_{11}y_{11} + 2x_{12}y_{12} + 2x_{21}y_{21} + x_{22}y_{22}$

In exercises 9 through 12, expand the inner product.

9. $(\mathbf{x} + \mathbf{y}, \mathbf{x} + \mathbf{y})$

10. $(\mathbf{x} - \mathbf{y}, \mathbf{x} - \mathbf{y})$

11. $(\mathbf{x} + \mathbf{y} + \mathbf{z}, \mathbf{x} - \mathbf{y})$

12. $(2\mathbf{x} + \mathbf{y} - \mathbf{z}, \mathbf{x} - 2\mathbf{y} + 3\mathbf{z})$

In exercises 13 through 20, find $\|\mathbf{x}\|$ using the standard inner product. In exercises 13 and 14, graph the vector and visually check your answer.

13. $\mathbf{x} = [1, 1, 1]$

14. $\mathbf{x} = [1, 1, -1]$

15. $\mathbf{x} = [-1, 3, -2, 5]$

16. $\mathbf{x} = [1, -1, 2, -1, 3]$

17. $\mathbf{x} = \begin{bmatrix} 2 & -1 \\ 1 & 3 \end{bmatrix}$

18. $\mathbf{x} = \begin{bmatrix} 1 & -1 \\ 5 & 2 \end{bmatrix}$

19. In $\mathbf{C}[0, 1]$, $\mathbf{x} = t - 1$

20. In $\mathbf{C}[-1, 1]$, $\mathbf{x} = t^2$

In exercises 21 and 22, the function given is an inner product. Find $\|\mathbf{x}\|$.

21. $(\mathbf{x}, \mathbf{y}) = x_1y_1 + 2x_2y_2 + x_3y_3$, $\mathbf{x} = [0, 1, 0]$

22. $(\mathbf{x}, \mathbf{y}) = 3x_1y_1 + 2x_2y_2 + x_3y_3$, $\mathbf{x} = [1, 1, 1]$

In exercises 23 through 26, find the unit vector $\mathbf{x}/\|\mathbf{x}\|$ from $\mathbf{x}$ using the standard inner product. Graph the vector in exercise 23 and visually check your answer.

23. $\mathbf{x} = [1, 1, 1]$

24. $\mathbf{x} = [3, -1, 0, 2, 4]$

25. $\mathbf{x} = \begin{bmatrix} 1 & -2 \\ 4 & 0 \end{bmatrix}$

26. In $\mathbf{C}[-1, 1]$, $\mathbf{x} = t + 3$

In exercises 27 through 32, use the standard inner product to find $d(\mathbf{x}, \mathbf{y})$ for the given $\mathbf{x}, \mathbf{y}$. In exercises 27 and 28, graph the vectors and visually check your answer.

27. $\mathbf{x} = [1, 1, 0]$, $\mathbf{y} = [0, 1, 1]$

28. $\mathbf{x} = [1, 0, 0]$, $\mathbf{y} = [0, 1, 0]$

29. $\mathbf{x} = [1, -2, 3, 2]$, $\mathbf{y} = [2, 1, -2, 3]$

30. $\mathbf{x} = [1, 1, 1, 1]$, $\mathbf{y} = [2, -1, 2, -1]$

31. $\mathbf{x} = \begin{bmatrix} 1 & -1 \\ 2 & 1 \end{bmatrix}$, $\mathbf{y} = \begin{bmatrix} 2 & 1 \\ -1 & 3 \end{bmatrix}$

32. In $\mathbf{C}[-1, 1]$, $\mathbf{x} = t + 1$, $\mathbf{y} = t - 1$

In exercises 33 and 34, show that, for any inner product space, the given calculations are correct.

33. (*a*) $\|-\mathbf{x}\| = \|\mathbf{x}\|$ for all vectors $\mathbf{x}$. *Hint:* $\|-\mathbf{x}\| = \|(-1)\mathbf{x}\|$.
(*b*) $\|\mathbf{0}\| = 0$.

34. (*a*) $\|\mathbf{x} + \mathbf{y}\|^2 = \|\mathbf{x}\|^2 + 2(\mathbf{x}, \mathbf{y}) + \|\mathbf{y}\|^2$
(*b*) $\|\mathbf{x} - \mathbf{y}\|^2 = \|\mathbf{x}\|^2 - 2(\mathbf{x}, \mathbf{y}) + \|\mathbf{y}\|^2$

In exercises 35 through 38, use the standard inner product on the appropriate space to decide which of $\mathbf{x}$ or $\mathbf{y}$ is closer to $\mathbf{z}$.

35. $\mathbf{x} = [1, -1, 1, 1]$, $\mathbf{y} = [-1, 1, -1, 1]$, $\mathbf{z} = [1, 1, 1, 1]$

36. $\mathbf{x} = [0, 1, 0, -1]$, $\mathbf{y} = [1, 0, -1, 0]$, $\mathbf{z} = [1, 0, 1, 1]$

37. $\mathbf{x} = \begin{bmatrix} 1 & 1 \\ 1 & 1 \end{bmatrix}$, $\mathbf{y} = \begin{bmatrix} 1 & -1 \\ 1 & 1 \end{bmatrix}$, $\mathbf{z} = \begin{bmatrix} 1 & 1 \\ -1 & 1 \end{bmatrix}$

38. $\mathbf{x} = \begin{bmatrix} 2 & 1 \\ 1 & 1 \end{bmatrix}$, $\mathbf{y} = \begin{bmatrix} 1 & 2 \\ 1 & 1 \end{bmatrix}$, $\mathbf{z} = \begin{bmatrix} 1 & 1 \\ 1 & 2 \end{bmatrix}$

Complex Numbers

39. An inner product on a vector space over complex numbers is slightly different from that for real numbers. Specifically, a function $(\bullet, \bullet)$ from $\mathbf{V} \times \mathbf{V}$ to $\mathbf{C}$ is an inner product on a vector space $\mathbf{V}$ using the complex numbers $\mathbf{C}$ if and only if

1. For all $\mathbf{z} \in \mathbf{V}$, $(\mathbf{z}, \mathbf{z})$ is a real number and

$$(\mathbf{z}, \mathbf{z}) \geqslant 0$$

with equality if and only if $\mathbf{z} = \mathbf{0}$.

2. For all $\mathbf{z}, \mathbf{w} \in \mathbf{V}$,

$$(\mathbf{z}, \mathbf{w}) = \overline{(\mathbf{w}, \mathbf{z})}$$

3. For all $\mathbf{z}, \mathbf{w}_1, \mathbf{w}_2 \in \mathbf{V}$,

$$(\mathbf{z}, \mathbf{w}_1 + \mathbf{w}_2) = (\mathbf{z}, \mathbf{w}_1) + (\mathbf{z}, \mathbf{w}_2)$$

4. For all $\lambda \in \mathbf{C}$ and $\mathbf{z}, \mathbf{w} \in \mathbf{V}$,

$$(\lambda\mathbf{z}, \mathbf{w}) = \lambda(\mathbf{z}, \mathbf{w}) = (\mathbf{z}, \bar{\lambda}\mathbf{w})$$

Show that $(\mathbf{z}, \mathbf{w}) = z_1\overline{w_1} + \cdots + z_n\overline{w_n}$ is an inner product on $\mathbf{C}_n$. This inner product is called the *standard inner product* on $\mathbf{C}_n$.

40. Using the standard inner product on $\mathbf{C}_n$, find
(*a*) $\|[1 - i, 2 + 3i, -2 + i]\|$
(*b*) $d([1 + i, 2 - 3i, 2i], [2, 2 - i, 3 - 2i])$

Theoretical Exercises

41. Prove that the function given in Example 3 is an inner product.

42. Let $\mathbf{V}$ be any finite-dimensional vector space, and let $\{\mathbf{u}_1, \ldots, \mathbf{u}_n\}$ be a basis of $\mathbf{V}$. Define $(\mathbf{u}_i, \mathbf{u}_j) = \delta_{ij}$, where $\delta_{ij} = 1$ if $i = j$ and 0 otherwise. For

$$\mathbf{u} = r_1\mathbf{u}_1 + \cdots + r_n\mathbf{u}_n \quad \text{and} \quad \mathbf{v} = s_1\mathbf{u}_1 + \cdots + s_n\mathbf{u}_n$$

in $\mathbf{V}$, define

$$(\mathbf{u}, \mathbf{v}) = (r_1\mathbf{u}_1 + \cdots + r_n\mathbf{u}_n, s_1\mathbf{u}_1 + \cdots + s_n\mathbf{u}_n)$$
$$= r_1 s_1 + \cdots + r_n s_n$$

Show that $(\bullet, \bullet)$ is an inner product on $\mathbf{V}$. (Hence you can show that every finite-dimensional vector space has an inner product. Using this same idea and a more general idea of basis, it can also be shown that every infinite-dimensional vector space has an inner product.)

43. Prove that if $\mathbf{V}$ is any inner product space, then

$$(\mathbf{0}, \mathbf{y}) = (\mathbf{y}, \mathbf{0}) = 0$$

for all $\mathbf{y} \in \mathbf{V}$.

44. Show that $(\mathbf{x}, \mathbf{y}) = (\ln x)(\ln y)$ is an inner product for the logarithm space described in exercise 37 of Section 3.2.

45. Graph $w(t) = 1/\sqrt{1 - t^2}$ over the interval $(-1, 1)$. Explain why using this weight function in the inner product

$$\int_{-1}^{1} w(t)p(t)q(t)\,dt$$

puts more emphasis on the points near -1 and 1. (This inner product arises in approximation work involving orthogonal polynomials. See exercise 42 of Section 4.2.)

46. Let $(\bullet, \bullet)$ be an inner product on a vector space $\mathbf{V}$. Let $\mathbf{x}_0 \in \mathbf{V}$. Show that

$$d(\mathbf{x} + \mathbf{x}_0, \mathbf{y} + \mathbf{x}_0) = d(\mathbf{x}, \mathbf{y})$$

for all $\mathbf{x}$ and $\mathbf{y}$ in $\mathbf{V}$. Thus distance does not change under translation (that is, adding $\mathbf{x}_0$ to both vectors). Show by a sketch what this means geometrically in $\mathbf{R}_2$. Also, for the nonstandard inner product $(\mathbf{x}, \mathbf{y}) = x_1y_1 + 4x_2y_2$, graph

$$D_{[0,0]} = \{\mathbf{x} : d(\mathbf{x}, [0, 0]) \leqslant 1\}$$

and

$$D_{[1,1]} = \{\mathbf{x} : d(\mathbf{x}, [1, 1]) \leqslant 1\}$$

The sketches of $D_{[0,0]}$ and $D_{[1,1]}$ illustrate that the distance used will not change under translation.

47. The inner product $(\mathbf{x}, \mathbf{y}) = x_1y_1 + 9x_2y_2$ is a nonstandard inner product on $\mathbf{R}_2$. Graph the points

$$\mathbf{x}_1 = \left[\frac{1}{\sqrt{2}}, \frac{1}{\sqrt{2}}\right] \quad \text{and} \quad \mathbf{x}_2 = [1, 0]$$

Note that $\mathbf{x}_1$ can be obtained from $\mathbf{x}_2$ by rotating the plane $\pi/4$ radians counterclockwise. Find $d(\mathbf{x}_1, \mathbf{0})$ and $d(\mathbf{x}_2, \mathbf{0})$. Notice that distance for this inner product changes under rotation. Graph

$$D = \{\mathbf{x} : d(\mathbf{x}, \mathbf{0}) = 1\}$$

for a geometrical understanding of this. (For standard inner products it can be shown that distance does not change under rotation.)

APPLICATIONS EXERCISES

48. The polynomial $p(t) = 1 + t$ is a two-term approximation to $q(t) = 1 + t + t^2$. Using

$$(f, g) = \int_0^1 f(t)g(t)\,dt$$

as the inner product, find the error $d(p, q)$ of the approximation.

49. In estimating a point in $\mathbf{R}^2$, suppose the estimate for the abscissa is twice as reliable as that for the ordinate. Explain why

$$(\mathbf{x}, \mathbf{y}) = 4x_1y_1 + x_2y_2$$

is a good choice for the inner product when the accuracy of the estimate is being measured.

50. (transmission of data through a channel in which error can occur) Let P and Q denote two positions in space. Suppose it is possible to communicate a 1 or a 0 from P to Q by some means (holding up fingers; flashing a light; telegraphing, using long for 1 and short for 0; or using more sophisticated equipment). Then it is possible to send a sequence of such 0's and 1's and thereby send coded messages from P to Q.

We will further suppose that the channel over which we send the message is noisy—that is, it can cause some 1's to be received as 0's and vice versa. Of course, some errors can be corrected by inspection. For example, if you decoded a received message and got "Your grade if an F," you would know that "if" should be "is." On the other hand, you may have doubts about the "F." Thus, if the message is sent over a noisy channel, a code that can be corrected is of use.

To produce such a code, we will assign a vector of 0's and 1's in $\mathbf{R}_n$ to each letter of the alphabet. We will suppose that if any such vector is transmitted over the channel, the channel will change at most one entry in the vector. Thus, in building a code, we want to assign these vectors in $\mathbf{R}_n$ to letters of the alphabet such that, if $\mathbf{x}$ and $\mathbf{y}$ are vectors assigned to different letters, then

$$d(\mathbf{x}, \mathbf{y}) = \|\mathbf{x} - \mathbf{y}\| \geqslant 3$$

Then, if a vector $\mathbf{x}$ is sent and a vector $\mathbf{y} \neq \mathbf{x}$ is received, we know it has exactly one entry different from $\mathbf{x}$. Hence, if we calculate the distances between $\mathbf{y}$ and those vectors which are assigned to letters, we see that $d(\mathbf{x}, \mathbf{y}) = 1$ and $d(\mathbf{z}, \mathbf{y}) \geqslant 2$ for all other vectors $\mathbf{z}$ assigned to letters. Thus we can use distance to correct messages as we receive them.

An example of such a code follows:

A—1, 0, 1, 1, 1, 1 B—1, 1, 1, 0, 0, 0
C—0, 1, 1, 1, 1, 0 T—0, 0, 0, 0, 0, 0

(*a*) Decode the message 1, 1, 1, 1, 0, 0, 1, 0, 1, 1, 1, 1, 0, 0, 0, 0, 0, 0.

(*b*) Decode the message 1, 1, 1, 1, 1, 0, 1, 0, 1, 1, 1, 1, 0, 0, 0, 0, 0, 0.

51. Suppose a certain kind of fish finds it twice as hard to swim up and down as it does to swim horizontally. Explain why

$$d(\mathbf{x}, \mathbf{y}) = (\mathbf{x} - \mathbf{y}, \mathbf{x} - \mathbf{y})^{1/2}$$

where $(\mathbf{w}, \mathbf{z}) = w_1 z_1 + w_2 z_2 + 4w_3 z_3$ might be a useful way of measuring distance for this fish.

Computer Exercises

52. Consider $A\mathbf{x} = \mathbf{b}$, whose exact solution $\mathbf{x}$ is given. Suppose that by some algorithm we calculate a solution $\bar{\mathbf{x}}$, encountering some round-off error. Using the standard inner product, compute the error $\|\mathbf{x} - \bar{\mathbf{x}}\|$ and the relative error $\|\mathbf{x} - \bar{\mathbf{x}}\|/\|\bar{\mathbf{x}}\|$.

(*a*) $\mathbf{x} = [1, 1, 1]$, $\bar{\mathbf{x}} = [0.9, 1, 1.1]$

(*b*) $\mathbf{x} = [1, 2, -1]$, $\bar{\mathbf{x}} = [1.01, 1.99, -0.98]$

53. Let A be a 2×2 matrix. Suppose that by some algorithm we evaluate the inverse of A, encountering some round-off error, and obtain B. Calculate A^{-1} exactly, and, using the standard inner product, compute the error $\|B - A^{-1}\|$ and the relative error $\|B - A^{-1}\|/\|B\|$.

(*a*) $A = \begin{bmatrix} 1 & 1 \\ 0 & 1 \end{bmatrix}$, $B = \begin{bmatrix} 0.9 & -1.1 \\ 0 & 1.1 \end{bmatrix}$

(*b*) $A = \begin{bmatrix} 1 & 1 \\ 1 & 2 \end{bmatrix}$, $B = \begin{bmatrix} 1.9 & -0.9 \\ -0.9 & 1.1 \end{bmatrix}$

54. Two algorithms, say A1 and A2, for solving systems of linear algebraic equations are tested on a sequence of example problems. The calculated solutions and the exact solutions are given in Table 4.1. Using distance defined from the standard inner product, decide which algorithm is providing better results.

TABLE 4.1

A1	$[0.8, 1.1]^t$	$[1.3, -1.2]^t$	$[0.31, -2.1]^t$
A2	$[0.9, 0.9]^t$	$[1.3, -1.1]^t$	$[0.35, -2.4]^t$
Exact	$[1, 1]^t$	$[1.2, -1.1]^t$	$[0.34, -2.3]^t$

4.2 Angles, Orthogonality, and Orthogonal Bases

In this section we introduce the last geometrical concept, the angle between two vectors. Using this concept, we will show when vectors are orthogonal and how to obtain a basis of pairwise orthogonal vectors for any finite-dimensional inner product space.

Definition

Let $\mathbf{V}$ be an inner product space. Following the outline given in the introduction, for any nonzero vectors $\mathbf{x}, \mathbf{y} \in \mathbf{V}$, we define the *angle between* $\mathbf{x}$ *and* $\mathbf{y}$ as the number θ that satisfies

$$\cos\theta = \frac{(\mathbf{x}, \mathbf{y})}{\|\mathbf{x}\|\,\|\mathbf{y}\|}$$

where $0 \leqslant \theta \leqslant \pi$. By the Cauchy-Schwarz inequality,

$$\left| \frac{(\mathbf{x}, \mathbf{y})}{\|\mathbf{x}\|\,\|\mathbf{y}\|} \right| \leqslant 1$$

so there is precisely one θ that will satisfy this equation.

Thus we calculate angles in any vector space as we do in $\mathbf{R}_3$.

EXAMPLE 14

Let $\mathbf{R}_4$ be the inner product space with the standard inner product. Let

$$\mathbf{x} = [-3, 17, 1, -1] \qquad \text{and} \qquad \mathbf{y} = [0, 1, -1, 1]$$

Then the angle θ between $\mathbf{x}$ and $\mathbf{y}$ satisfies

$$\cos\theta = \frac{(\mathbf{x}, \mathbf{y})}{\|\mathbf{x}\|\,\|\mathbf{y}\|} = \frac{(-3)(0) + (17)(1) + (1)(-1) + (-1)(1)}{\sqrt{(-3)^2 + (17)^2 + 1^2 + (-1)^2}\sqrt{0^2 + 1^2 + (-1)^2 + 1^2}}$$
$$= \frac{15}{\sqrt{300}\sqrt{3}} = \frac{1}{2}$$

Solving this equation for θ yields

$$\theta = \frac{\pi}{3} \text{ radians}$$

□

EXAMPLE 15

CALCULUS

Let $\mathbf{C}[-1, 1]$ be the inner product space with the standard inner product. Let

$$f(t) = t^2 - t \qquad \text{and} \qquad g(t) = 1$$

Using the calculations of Example 7, Section 4.1, we can find the angle between f and g by solving

$$\cos\theta = \frac{(f, g)}{\|f\|\,\|g\|} = \frac{\left(\int_{-1}^{1} f(t)g(t)\,dt\right)}{\left(\int_{-1}^{1} [f(t)]^2\,dt\right)^{1/2}\left(\int_{-1}^{1} [g(t)]^2\,dt\right)^{1/2}}$$
$$= \frac{\frac{2}{3}}{\frac{4}{15}\sqrt{15}\sqrt{2}} = \frac{\sqrt{30}}{12} \approx 0.4564$$

Thus, from tables, the angle between f and g is approximately 1.10 radians.

□

As in $\mathbf{R}_3$, if the angle between nonzero vectors $\mathbf{x}$ and $\mathbf{y}$ in $\mathbf{V}$ is $\pi/2$, we say that $\mathbf{x}$ and $\mathbf{y}$ are *orthogonal*. We can check to see if two nonzero vectors are orthogonal by calculating their inner product as described below.

THEOREM 4.4

Let $\mathbf{V}$ be an inner product space and let $\mathbf{x}$ and $\mathbf{y}$ be nonzero vectors in $\mathbf{V}$. Then $\mathbf{x}$ and $\mathbf{y}$ are orthogonal if and only if $(\mathbf{x}, \mathbf{y}) = 0$.

PROOF

If θ is the angle between $\mathbf{x}$ and $\mathbf{y}$, then

$$\cos\theta = \frac{(\mathbf{x}, \mathbf{y})}{\|\mathbf{x}\|\,\|\mathbf{y}\|}$$

Now, $\theta = \pi/2$ if and only if $\cos\theta = 0$, which happens if and only if $(\mathbf{x}, \mathbf{y}) = 0$. ■

EXAMPLE 16

Let $\mathbf{R}_4$ be the inner product space with the standard inner product, and let

$$\mathbf{x} = [1, 1, 1, -1] \qquad \text{and} \qquad \mathbf{y} = [4, -1, 3, 6]$$

Then

$$(\mathbf{x}, \mathbf{y}) = (1)(4) + (1)(-1) + (1)(3) + (-1)(6) = 0$$

so $\mathbf{x}$ and $\mathbf{y}$ are orthogonal. □

EXAMPLE 17

Let $\mathbf{R}_4$ be the inner product space with

$$(\mathbf{x}, \mathbf{y}) = x_1 y_1 + x_2 y_2 + 8x_3 y_3 + 9x_4 y_4$$

This inner product, of course, is not standard.

Let $\mathbf{x} = [-3, 17, 1, -1]$ and $\mathbf{y} = [0, 1, -1, 1]$. Then, since

$$(\mathbf{x}, \mathbf{y}) = (-3)(0) + (17)(1) + 8(1)(-1) + 9(-1)(1) = 0$$

it follows that $\mathbf{x}$ and $\mathbf{y}$ are orthogonal with this inner product, although they are not orthogonal with the standard inner product. (Also see exercise 46.) □

EXAMPLE 18

CALCULUS

Let $\mathbf{C}[-1, 1]$ be the inner product space with the standard inner product. Let $f(t) = t$ and $g(t) = t - 1$. To decide if f and g are orthogonal, we need to calculate

$$(f, g) = \int_{-1}^{1} t(t-1)\,dt = \int_{-1}^{1} t^2 - t\,dt = \frac{2}{3}$$

Thus $f(t) = t$ and $g(t) = t - 1$ are not orthogonal. □

Orthogonality in vector spaces is often used to simplify calculations.

EXAMPLE 19

Let $\mathbf{V}$ be a finite-dimensional inner product space, and let $S = \{\mathbf{x}_1, \ldots, \mathbf{x}_n\}$ be a basis for $\mathbf{V}$. If $\mathbf{x} \in \mathbf{V}$, we can write

$$\mathbf{x} = a_1\mathbf{x}_1 + a_2\mathbf{x}_2 + \cdots + a_n\mathbf{x}_n$$

for suitable $a_1, \ldots, a_n \in \mathbf{R}$. If $\mathbf{V} = \mathbf{R}_n$ or $\mathbf{R}^n$, the coefficients $a_1, \ldots, a_n$ can be calculated by solving a system of n linear equations with n unknowns. If $\mathbf{V}$ is a subspace of $\mathbf{F(D)}$, calculating these coefficients can be more difficult.

If, in addition to being a basis of $\mathbf{V}$, S is also a set of pairwise orthogonal vectors, then the coefficients $a_1, \ldots, a_n$ can be calculated more simply. To calculate a_i, we compute the inner product of both sides of the equation with $\mathbf{x}_i$. This leads to

$$(\mathbf{x}, \mathbf{x}_i) = a_1(\mathbf{x}_1, \mathbf{x}_i) + \cdots + a_i(\mathbf{x}_i, \mathbf{x}_i) + \cdots + a_n(\mathbf{x}_n, \mathbf{x}_i)$$

Since $(\mathbf{x}_j, \mathbf{x}_i) = 0$ if $i \neq j$, we have

$$(\mathbf{x}, \mathbf{x}_i) = a_i(\mathbf{x}_i, \mathbf{x}_i)$$

and thus

$$a_i = \frac{(\mathbf{x}, \mathbf{x}_i)}{(\mathbf{x}_i, \mathbf{x}_i)}$$

for $i = 1, \ldots, n$, a rather easy calculation. (This is essentially the technique used to calculate coefficients of Fourier series which arise in the solution of problems concerning heat, vibration, etc.) □

In the last chapter we showed that every finite-dimensional vector space has a basis. Having defined orthogonality, we can now show that every finite-dimensional vector space has an *orthogonal basis*, a basis of pairwise orthogonal vectors, as well as an *orthonormal basis*, an orthogonal basis of unit vectors. The next theorem provides a procedure, called the Gram-Schmidt algorithm, for taking any basis and computing an orthogonal or orthonormal basis from it.

THEOREM 4.5

Let $\mathbf{V}$ be a finite-dimensional inner product space, and let $\{\mathbf{x}_1, \ldots, \mathbf{x}_n\}$ be a basis of $\mathbf{V}$. Then there is an orthogonal basis

$$\{\mathbf{y}_1, \ldots, \mathbf{y}_n\}$$

of $\mathbf{V}$. Further, letting $\mathbf{u}_i = \mathbf{y}_i/\|\mathbf{y}_i\|$ for all i yields an orthonormal basis $\{\mathbf{u}_1, \ldots, \mathbf{u}_n\}$ of $\mathbf{V}$.

PROOF

Although our proof is general, we will provide some sketches in $\mathbf{R}_3$. For example, in Figure 4.5, we depict $\{\mathbf{x}_1, \mathbf{x}_2, \mathbf{x}_3\}$.

We will show how to construct $\mathbf{y}_1, \ldots, \mathbf{y}_n$. Set

$$\mathbf{y}_1 = \mathbf{x}_1$$

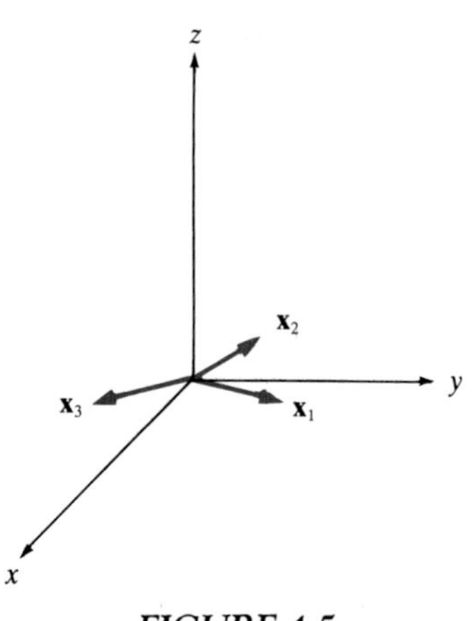

FIGURE 4.5

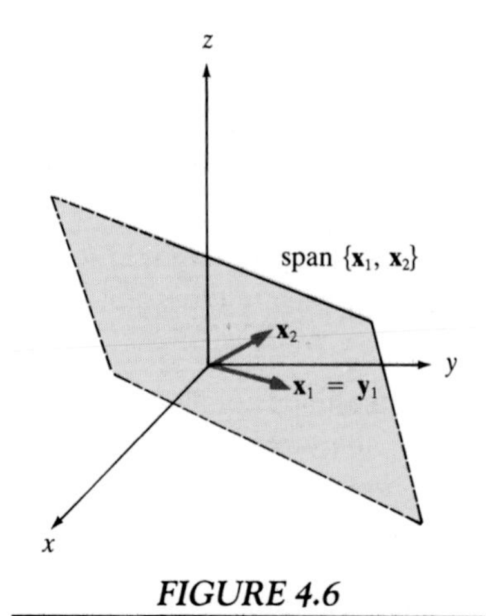

FIGURE 4.6

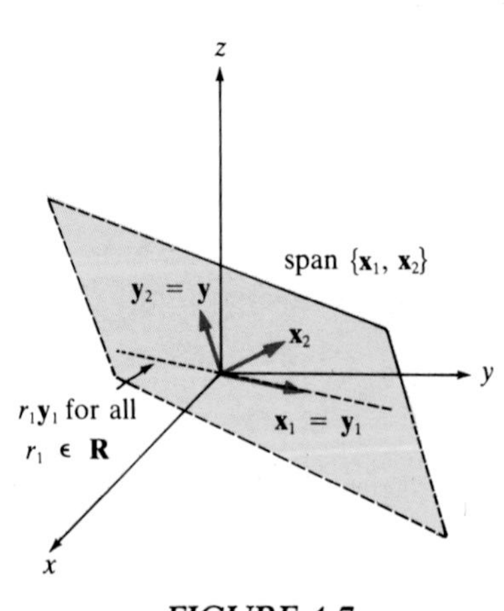

FIGURE 4.7

Then span $\{\mathbf{x}_1\}$ = span $\{\mathbf{y}_1\}$. To construct $\mathbf{y}_2$, set

$$\mathbf{y} = \mathbf{x}_2 - r_1\mathbf{y}_1$$

Note that $\mathbf{y} \in \text{span}\{\mathbf{y}_1, \mathbf{x}_2\} = \text{span}\{\mathbf{x}_1, \mathbf{x}_2\}$ as shown in the region shaded in Figure 4.6. We determine r_1 so that $(\mathbf{y}, \mathbf{y}_1) = 0$. This means that

$$(\mathbf{x}_2, \mathbf{y}_1) - r_1(\mathbf{y}_1, \mathbf{y}_1) = 0$$

Hence $r_1 = \dfrac{(\mathbf{x}_2, \mathbf{y}_1)}{(\mathbf{y}_1, \mathbf{y}_1)}$, and so

$$\mathbf{y} = \mathbf{x}_2 - \frac{(\mathbf{x}_2, \mathbf{y}_1)}{(\mathbf{y}_1, \mathbf{y}_1)}\mathbf{y}_1$$

Now, since $\mathbf{y}_1 = \mathbf{x}_1$ and $\{\mathbf{x}_1, \mathbf{x}_2\}$ is linearly independent, $\mathbf{y} \neq \mathbf{0}$. Thus, we define

$$\mathbf{y}_2 = \mathbf{y}$$

See Figure 4.7. By this choice, $(\mathbf{y}_1, \mathbf{y}_2) = (\mathbf{y}_1, \mathbf{y}) = 0$, so $\mathbf{y}_1$ and $\mathbf{y}_2$ are orthogonal vectors.

Also, since $\mathbf{x}_2 = \mathbf{y}_2 + \dfrac{(\mathbf{x}_2, \mathbf{y}_1)}{(\mathbf{y}_1, \mathbf{y}_1)}\mathbf{y}_1$, it follows that

$$\text{span}\{\mathbf{x}_1, \mathbf{x}_2\} \subseteq \text{span}\{\mathbf{y}_1, \mathbf{y}_2\}$$

Since dim(span $\{\mathbf{y}_1, \mathbf{y}_2\}$) = 2, we have that

$$\text{span}\{\mathbf{x}_1, \mathbf{x}_2\} = \text{span}\{\mathbf{y}_1, \mathbf{y}_2\}$$

Now, suppose we have constructed $\mathbf{y}_1, \mathbf{y}_2, \ldots, \mathbf{y}_k$ such that $\{\mathbf{y}_1, \mathbf{y}_2, \ldots, \mathbf{y}_k\}$ is an orthogonal set and

$$\text{span}\{\mathbf{x}_1, \mathbf{x}_2, \ldots, \mathbf{x}_k\} = \text{span}\{\mathbf{y}_1, \mathbf{y}_2, \ldots, \mathbf{y}_k\}$$

Using this, we will show how to construct $\mathbf{y}_{k+1}$. Set

$$\mathbf{y} = \mathbf{x}_{k+1} - r_1\mathbf{y}_1 - \cdots - r_k\mathbf{y}_k$$

We determine $r_1, r_2, \ldots, r_k$ so that $(\mathbf{y}, \mathbf{y}_1) = \cdots = (\mathbf{y}, \mathbf{y}_k) = 0$. This means that

$$(\mathbf{x}_{k+1}, \mathbf{y}_i) - r_1(\mathbf{y}_1, \mathbf{y}_i) - \cdots - r_k(\mathbf{y}_k, \mathbf{y}_i) = 0$$

Hence $r_i = \dfrac{(\mathbf{x}_{k+1}, \mathbf{y}_i)}{(\mathbf{y}_i, \mathbf{y}_i)}$ for all i and

$$\mathbf{y} = \mathbf{x}_{k+1} - \frac{(\mathbf{x}_{k+1}, \mathbf{y}_1)}{(\mathbf{y}_1, \mathbf{y}_1)}\mathbf{y}_1 - \cdots - \frac{(\mathbf{x}_{k+1}, \mathbf{y}_k)}{(\mathbf{y}_k, \mathbf{y}_k)}\mathbf{y}_k$$

Now, since span $\{\mathbf{y}_1, \ldots, \mathbf{y}_k\}$ = span $\{\mathbf{x}_1, \ldots, \mathbf{x}_k\}$, we can rewrite this equation as

$$\mathbf{y} = \mathbf{x}_{k+1} + s_1\mathbf{x}_1 + \cdots + s_k\mathbf{x}_k$$

for some $s_1, \ldots, s_k \in \mathbf{R}$. Since $\{\mathbf{x}_1, \ldots, \mathbf{x}_{k+1}\}$ is linearly independent, $\mathbf{y} \neq \mathbf{0}$. Thus, we define

$$\mathbf{y}_{k+1} = \mathbf{y}$$

Finally, since $\mathbf{x}_{k+1} = \mathbf{y} + r_1\mathbf{y}_1 + \cdots + r_k\mathbf{y}_k$, it follows that

$$\operatorname{span}\{\mathbf{x}_1, \ldots, \mathbf{x}_{k+1}\} \subseteq \operatorname{span}\{\mathbf{y}_1, \ldots, \mathbf{y}_{k+1}\}$$

Since $\dim(\operatorname{span}\{\mathbf{x}_1, \ldots, \mathbf{x}_{k+1}\}) = k + 1$,

$$\operatorname{span}\{\mathbf{y}_1, \ldots, \mathbf{y}_{k+1}\} = \operatorname{span}\{\mathbf{x}_1, \ldots, \mathbf{x}_{k+1}\}$$

This proves the first part of the theorem. The second part of the theorem is routine. ■

The proof contains the following algorithm.

GRAM-SCHMIDT ALGORITHM

Given a finite-dimensional inner product space $\mathbf{V}$ with basis $B = \{\mathbf{x}_1, \ldots, \mathbf{x}_n\}$, we construct an orthogonal and an orthonormal basis as follows.

1. Let $\mathbf{y}_1 = \mathbf{x}_1$.

2. Given $\mathbf{y}_1, \ldots, \mathbf{y}_k$, let

$$\mathbf{y}_{k+1} = \mathbf{x}_{k+1} - r_{1k}\mathbf{y}_1 - \cdots - r_{kk}\mathbf{y}_k$$

where
$$r_{ik} = \frac{(\mathbf{x}_{k+1}, \mathbf{y}_i)}{(\mathbf{y}_i, \mathbf{y}_i)}$$

for each i.

3. Continue until $k = n - 1$. Then $\{\mathbf{y}_1, \ldots, \mathbf{y}_n\}$ is an orthogonal basis.

4. For an orthonormal basis, set $\mathbf{u}_i = \dfrac{1}{\|\mathbf{y}_i\|}\mathbf{y}_i$ for $i = 1, \ldots, n$. Then $\{\mathbf{u}_1, \ldots, \mathbf{u}_n\}$ is an orthonormal basis.

EXAMPLE 20

The inner product space $\mathbf{R}_3$ with the standard inner product has

$$\{[1,1,0], [0,1,1], [1,0,1]\}$$

as a basis. To obtain an orthonormal basis, we apply the Gram-Schmidt algorithm to the set of these vectors.

Since $\mathbf{x}_1 = [1,1,0]$,

$$\mathbf{y}_1 = \mathbf{x}_1 = [1,1,0]$$

For the second vector, set

$$\mathbf{y}_2 = \mathbf{x}_2 - \frac{(\mathbf{x}_2, \mathbf{y}_1)}{(\mathbf{y}_1, \mathbf{y}_1)}\mathbf{y}_1 = [0,1,1] - [\tfrac{1}{2},\tfrac{1}{2},0] = [-\tfrac{1}{2},\tfrac{1}{2},1]$$

For the last vector, set

$$\begin{aligned}\mathbf{y}_3 &= \mathbf{x}_3 - \frac{(\mathbf{x}_3, \mathbf{y}_1)}{(\mathbf{y}_1, \mathbf{y}_1)}\mathbf{y}_1 - \frac{(\mathbf{x}_3, \mathbf{y}_2)}{(\mathbf{y}_2, \mathbf{y}_2)}\mathbf{y}_2 \\ &= [1,0,1] - \tfrac{1}{2}[1,1,0] - \frac{(\frac{1}{2})}{(\frac{3}{2})}[-\tfrac{1}{2},\tfrac{1}{2},1] = [\tfrac{2}{3},-\tfrac{2}{3},\tfrac{2}{3}]\end{aligned}$$

It follows that the required orthonormal basis is

$$\begin{aligned}\{\mathbf{u}_1,\mathbf{u}_2,\mathbf{u}_3\} &= \left\{\frac{1}{\|\mathbf{y}_1\|}\mathbf{y}_1, \frac{1}{\|\mathbf{y}_2\|}\mathbf{y}_2, \frac{1}{\|\mathbf{y}_3\|}\mathbf{y}_3\right\} \\ &= \left\{\left[\frac{\sqrt{2}}{2},\frac{\sqrt{2}}{2},0\right],\left[-\frac{\sqrt{6}}{6},\frac{\sqrt{6}}{6},\frac{\sqrt{6}}{3}\right],\left[\frac{\sqrt{3}}{3},-\frac{\sqrt{3}}{3},\frac{\sqrt{3}}{3}\right]\right\}\end{aligned}$$ □

EXAMPLE 21

CALCULUS

The inner product space $\mathbf{C}[-1,1]$ with the standard inner product has $\{1,t,t^2\}$ as a linearly independent set. This set is a basis for $\mathbf{P}_2$. We will apply the Gram-Schmidt algorithm to this set.

For the first vector, set $\mathbf{y}_1 = 1$.

For the second vector, set

$$\mathbf{y}_2 = \mathbf{x}_2 - \frac{(\mathbf{x}_2, \mathbf{y}_1)}{(\mathbf{y}_1, \mathbf{y}_1)}\mathbf{y}_1 = t - \frac{\int_{-1}^{1} t\,dt}{\int_{-1}^{1} dt}1 = t$$

For the last vector, set

$$\mathbf{y}_3 = \mathbf{x}_3 - \frac{(\mathbf{x}_3, \mathbf{y}_1)}{(\mathbf{y}_1, \mathbf{y}_1)}\mathbf{y}_1 - \frac{(\mathbf{x}_3, \mathbf{y}_2)}{(\mathbf{y}_2, \mathbf{y}_2)}\mathbf{y}_2 = t^2 - \frac{\int_{-1}^{1} t^2\,dt}{\int_{-1}^{1} dt}1 - \frac{\int_{-1}^{1} t^3\,dt}{\int_{-1}^{1} t^2\,dt}t = t^2 - \tfrac{1}{3}$$

Computing an orthonormal set, we have the Legendre polynomials

$$\mathbf{u}_1 = \frac{\mathbf{y}_1}{\|\mathbf{y}_1\|} = \frac{\sqrt{2}}{2}, \qquad \mathbf{u}_2 = \frac{\mathbf{y}_2}{\|\mathbf{y}_2\|} = \frac{\sqrt{6}}{2}t, \qquad \mathbf{u}_3 = \frac{\mathbf{y}_3}{\|\mathbf{y}_3\|} = \frac{3\sqrt{10}}{4}\left(t^2 - \frac{1}{3}\right)$$

□

This concludes our introduction of geometrical concepts for inner product spaces. In the next section we will show how these concepts can be used in finding "best" solutions to linear equations.

EXERCISES FOR SECTION 4.2

COMPUTATIONAL EXERCISES

In exercises 1 through 8, find the angle between the given vectors using the standard inner product. Graph the vectors in exercises 1 through 4 and visually check your answer.

1. $[3, -1], [2, 2]$

2. $[1, 2], [0, 1]$

3. $[0, 1, 1], [1, 0, 1]$

4. $[2, 1, 0], [1, 1, 2]$

5. $[1, -1, 1, -1], [0, 1, -1, 1]$

6. $[-1, 1, 1, -1], [-1, 0, 1, -1]$

7. t, t^2, in $\mathbf{C}[-1, 1]$

8. $t + 1, t - 1$, in $\mathbf{C}[-1, 1]$

In exercises 9 and 10, let $(\bullet, \bullet)$ be an inner product on a vector space $\mathbf{R}_n$. Find the angle between vectors $\mathbf{x}$ and $\mathbf{y}$.

9. $(\mathbf{x}, \mathbf{y}) = x_1 y_1 + 2x_2 y_2 + x_3 y_3$, $\mathbf{x} = [1, 1, 1]$, $\mathbf{y} = [-1, 1, -1]$

10. $(\mathbf{x}, \mathbf{y}) = x_1 y_1 + 2x_2 y_2 + 2x_3 y_3 + x_4 y_4$, $\mathbf{x} = [2, 1, -1, -2]$ $\mathbf{y} = [-1, 1, 1, -1]$

In exercises 11 through 20, decide which pairs of vectors are orthogonal using the standard inner product. For exercises 11 through 14, check your answer geometrically.

11. $[1, -1], [1, 1], [0, 0]$

12. $[3, 2], [-4, 6], [2, -3]$

13. $[1, 1, 1], [3, 2, 0], [2, -3, 1]$

14. $[-1, 1, 2], [0, 0, 0], [2, 4, -1]$

15. $[1, -1, 1, -1], [1, 1, 1, 1], [3, 1, -2, 0]$

16. $[-1, 2, -1, 3], [1, 2, 1, -1], [2, 1, 3, 1]$

17. $\begin{bmatrix} 1 & -1 \\ -1 & 1 \end{bmatrix}, \begin{bmatrix} 1 & 1 \\ 1 & 1 \end{bmatrix}, \begin{bmatrix} 1 & 1 \\ -1 & -1 \end{bmatrix}$

18. $\begin{bmatrix} 2 & -5 \\ 3 & 4 \end{bmatrix}, \begin{bmatrix} 1 & 1 \\ 1 & 0 \end{bmatrix}, \begin{bmatrix} -1 & 2 \\ 0 & 3 \end{bmatrix}$

19. $t, 1 - \frac{3t}{2}, t - \frac{2t}{3}$, in $\mathbf{C}[0, 1]$

20. $1, \sin t, \cos t$, in $\mathbf{C}[0, \pi]$

In exercises 21 through 24, using the standard inner product for the vector space named, apply the Gram-Schmidt algorithm to the vectors of S. Graph S and your result.

21. $S = \{[1, 1], [0, 1]\}$ in $\mathbf{R}_2$

22. $S = \{[1, 2], [2, 1]\}$ in $\mathbf{R}_2$

23. $S = \{[1, 1, 0], [1, 0, 1], [0, 1, 1]\}$ in $\mathbf{R}_3$

24. $S = \{[1, 1, 1], [1, 1, 0], [0, 1, 1]\}$ in $\mathbf{R}_3$

In exercises 25 and 26, use the Gram-Schmidt algorithm to find an orthonormal basis for span S, using the given inner product.

25. $S = \{[3, -1, 1], [-1, 2, 1]\}$, $(\mathbf{x}, \mathbf{y}) = x_1 y_1 + 2x_2 y_2 + 3x_3 y_3$

26. $S = \{[2, 1, 1], [-1, 1, 1]\}$, $(\mathbf{x}, \mathbf{y}) = x_1 y_1 + 2x_2 y_2 + 3x_3 y_3$

In exercises 27 through 32, use the standard inner product and the Gram-Schmidt algorithm to find an orthonormal basis for span S, where the given set S is linearly independent.

27. $S = \{[1, 0, -1, 1], [-1, 1, 0, 1], [0, -1, 1, -1]\}$

28. $S = \{[-2, 1, -1, 0], [1, 2, 0, 1], [1, 0, -2, 1]\}$

29. $S = \left\{ \begin{bmatrix} 1 & 1 \\ 1 & 0 \end{bmatrix}, \begin{bmatrix} 1 & 1 \\ 0 & 1 \end{bmatrix} \right\}$

30. $S = \left\{ \begin{bmatrix} 1 & 1 \\ 0 & 0 \end{bmatrix}, \begin{bmatrix} 1 & 0 \\ 1 & 0 \end{bmatrix}, \begin{bmatrix} 0 & 0 \\ 1 & 1 \end{bmatrix} \right\}$

31. $S = \{1, t, t^2\}$ in $\mathbf{C}[0, 1]$. (Compare your result with the one in Example 21.)

32. $S = \{1, \sin t, \cos t\}$ in $\mathbf{C}[0, \pi]$.

THEORETICAL EXERCISES

33. Let $\mathbf{V}$ be an inner product space and $S = \{\mathbf{x}_1, \ldots, \mathbf{x}_n\}$ an orthogonal set in $\mathbf{V}$. Prove that S is linearly independent.

34. (geometry in $\mathbf{R}_n$—Pythagorean theorem) Prove that, if $\mathbf{x} \neq \mathbf{0}$ and $\mathbf{y} \neq \mathbf{0}$ are in $\mathbf{R}_n$ and $(\mathbf{x}, \mathbf{y}) = 0$, then

$$\|\mathbf{x} - \mathbf{y}\|^2 = \|\mathbf{x}\|^2 + \|\mathbf{y}\|^2$$

Draw a sketch of this result in $\mathbf{R}_3$.

35. Let $\mathbf{x}$ be in inner product space $\mathbf{V}$. Let

$$\mathbf{W} = \{\mathbf{y}:\mathbf{y} \in \mathbf{V} \text{ and } (\mathbf{x},\mathbf{y}) = 0\}$$

Prove that $\mathbf{W}$ is a subspace of $\mathbf{V}$. Make some sketches of this result in $\mathbf{R}_3$.

36. What will happen if a linearly dependent set of vectors is used in the Gram-Schmidt algorithm? Prove your claim.

37. Let $\mathbf{V}$ be an inner product space with $\mathbf{x}, \mathbf{y}, \mathbf{z} \in \mathbf{V}$. Suppose $\mathbf{x}$ is orthogonal to each of $\mathbf{y}$ and $\mathbf{z}$. Prove that $\mathbf{x}$ is orthogonal to any nonzero linear combination $r\mathbf{y} + s\mathbf{z}$ of $\mathbf{y}$ and $\mathbf{z}$. Demonstrate this result geometrically in $\mathbf{R}_3$.

38. Let $\mathbf{x}, \mathbf{y} \in \mathbf{R}_n$. Show that, to find all $\mathbf{z} \in \mathbf{R}_n$ orthogonal to $\mathbf{x}$ and $\mathbf{y}$, we can solve

$$\begin{bmatrix} \mathbf{x} \\ \mathbf{y} \end{bmatrix} \mathbf{z}^t = \mathbf{0}$$

Use this procedure to find all vectors orthogonal to $[1, 1, 0]$ and $[0, 1, 1]$, and graph the result.

39. For $a > 0$ and $b > 0$, define

$$(\mathbf{x}, \mathbf{y}) = ax_1y_1 + bx_2y_2$$

for all $\mathbf{x}, \mathbf{y} \in \mathbf{R}^2$. Prove that there are choices of a and b for which a given pair $\mathbf{x}$ and $\mathbf{y}$ are orthogonal if and only if $x_2y_2/x_1y_1 < 0$. (Thus, viewing the vectors as fixed arrows, we have such a choice of weights a and b if and only if the slopes of the lines determined by the arrows have a negative product. Compare this result with the usual theorem about the slopes of perpendicular lines in the plane.)

In exercises 40 through 42, let $B = \{1, t, t^2, \ldots\} \subseteq \mathbf{P}$. Apply the Gram-Schmidt algorithm to the first three polynomials in B, using the stated inner product. *Hint:* In exercises 40 and 41, use a table of integrals.

40. $(p, q) = \int_0^\infty e^{-t} p(t)q(t)\,dt$. The polynomials obtained are called the *Laguerre polynomials*, used in approximations.

41. $(p, q) = \int_{-\infty}^{\infty} e^{-t^2} p(t)q(t)\,dt$. The polynomials obtained are called the *Hermite polynomials*, used in approximations.

42. $(p, q) = \int_{-1}^{1} (1 - t^2)^{-1/2} p(t)q(t)\,dt$. The polynomials obtained are called the *Chebyshev polynomials*, used in approximations.

(*Note:* The polynomials obtained in exercises 40, 41, and 42 are actually scalar multiples of the classically defined polynomials.)

43. Let $\mathbf{V}$ be an inner product space. Show that, if $\mathbf{x}$ and $\mathbf{y}$ are orthogonal, then so are $r\mathbf{x}$ and $s\mathbf{y}$ for any scalars $r \neq 0$ and $s \neq 0$. Thus orthogonality does not change under scaling.

44. Let $\mathbf{V}$ be an inner product space. Show that, if $\mathbf{x}, \mathbf{y} \in \mathbf{V}$ and $a > 0$ and $b > 0$ are in $\mathbf{R}$, then the angle between $\mathbf{x}$ and $\mathbf{y}$ is the same as that between $a\mathbf{x}$ and $b\mathbf{y}$. Thus, angles do not change under scaling by positive numbers.

45. For the inner product $(\mathbf{x}, \mathbf{y}) = x_1y_1 + 4x_2y_2$, graph

$$\mathbf{x}_1 = \left[\frac{1}{\sqrt{2}}, \frac{1}{\sqrt{2}}\right], \qquad \mathbf{y}_1 = \left[\frac{1}{\sqrt{2}}, -\frac{1}{\sqrt{2}}\right]$$

and $$\mathbf{x}_2 = [1, 0], \qquad \mathbf{y}_2 = [0, 1]$$

Find the angle between $\mathbf{x}_1$ and $\mathbf{y}_1$ and the angle between $\mathbf{x}_2$ and $\mathbf{y}_2$. (Notice that, with this inner product, angles change under rotation. This is not true with the standard inner product.)

APPLICATIONS EXERCISES

46. (a use for a nonstandard inner product) Prove that $(\mathbf{x}, \mathbf{y}) = x_1y_1 + 4x_2y_2$ is an inner product on $\mathbf{R}_2$. The graph of

$$D = \{\mathbf{x} : \|\mathbf{x}\| = 1\}$$

an ellipse, is shown in Figure 4.8. Note that $P = [1/2, \sqrt{3}/4]$ is on this ellipse. We will accept that the free arrow $\mathbf{v}_1 = [1, -\sqrt{3}/6]$ is tangent to the curve at P. Show that the free arrow $\mathbf{v}_2 = [-1/2, -\sqrt{3}/4]$ is orthogonal to $\mathbf{v}_1$. Graph $\mathbf{v}_2$ so that its tail is at P and note that it points to the origin. Thus this nonstandard inner product points directly to the origin, whereas the standard inner product does not. (In fact, it is true that, if $\mathbf{v}$ is any free arrow tangent to this curve, then a

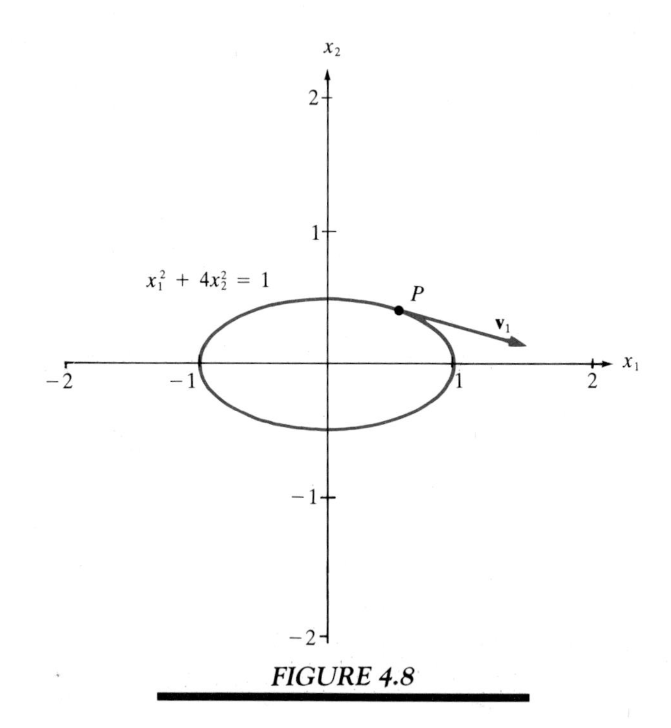

FIGURE 4.8

free arrow orthogonal to $\mathbf{v}$ will point to the origin. This property is useful in optimization—that is, finding maximum and minimum values of a function.) See also exercise 45.

47. (geometry in $\mathbf{R}_n$) A plane in $\mathbf{R}_3$ through $\mathbf{p}$ and orthogonal to $\mathbf{a} \neq \mathbf{0}$ is defined as those $\mathbf{x}$'s which satisfy $((\mathbf{x} - \mathbf{p}), \mathbf{a}) = 0$, or

$$a_1(x_1 - p_1) + a_2(x_2 - p_2) + a_3(x_3 - p_3) = 0$$

This equation can be extended to define a *hyperplane* in $\mathbf{R}_n$ through $\mathbf{p} \in \mathbf{R}_n$ and orthogonal to $\mathbf{a} \in \mathbf{R}_n$ as

$$a_1(x_1 - p_1) + a_2(x_2 - p_2) + \cdots + a_n(x_n - p_n) = 0$$

Write the equations for the hyperplanes with the given $\mathbf{p}$ and $\mathbf{a}$. Graph those of (a) and (b).

(*a*) $\mathbf{p} = [1, 1, 1]$, $\mathbf{a} = [1, 1, 1]$
(*b*) $\mathbf{p} = [1, 0, 1]$, $\mathbf{a} = [0, 1, 0]$
(*c*) $\mathbf{p} = [1, -1, 2, 1]$, $\mathbf{a} = [2, -1, 3, 1]$
(*d*) $\mathbf{p} = [-1, 2, 4, 3, 1]$, $\mathbf{a} = [3, -1, 2, -4, 5]$

4.3 LEAST SQUARES SOLUTIONS (OPTIONAL)

In previous chapters we saw equations that have no solution; that is, their solution set is $\varnothing$. In such cases, as we will see, it is often useful to obtain the "best" solution we can. In this section we will show how some of the work on inner product spaces can be used in finding such "best" solutions.

The work of this section is divided into three parts. Part I shows how to find "best" solutions to systems of linear algebraic equations. Part II shows how that kind of solution is useful in fitting polynomial curves, within a specified degree, to given data. Finally, Part III considers the general problem of finding the vector in a finite-dimensional subspace that is closest to a given vector. The result obtained is then used in fitting a polynomial, within a specified degree, to a given function.

I. LEAST SQUARES SOLUTIONS TO SYSTEMS OF LINEAR ALGEBRAIC EQUATIONS

Let A be an $m \times n$ matrix. By Corollary 3.14A, a system

$$A\mathbf{x} = \mathbf{b}$$

of linear algebraic equations has a solution if and only if rank A = rank $[A \quad \mathbf{b}]$. Thus, some such equations have no solution. In this case, we want to find an $\mathbf{x} \in \mathbf{R}^n$ such that the left side $A\mathbf{x}$ and the right side $\mathbf{b}$ are as close as possible. We can do this as follows. Let $(\bullet, \bullet)$ be the standard inner product on $\mathbf{R}^n$. For each $\mathbf{x} \in \mathbf{R}^n$, define an *error vector*

$$\boldsymbol{\varepsilon} = \begin{bmatrix} \varepsilon_1 \\ \vdots \\ \varepsilon_m \end{bmatrix}$$

by $\boldsymbol{\varepsilon} = A\mathbf{x} - \mathbf{b}$, or, in terms of entries,

$$\begin{aligned} \varepsilon_1 &= a_{11}x_1 + \cdots + a_{1n}x_n - b_1 \\ &\vdots \\ \varepsilon_m &= a_{m1}x_1 + \cdots + a_{mn}x_n - b_m \end{aligned}$$

Thus the entries of $\boldsymbol{\varepsilon}$ give the size of the error between the corresponding entries of $A\mathbf{x}$ and $\mathbf{b}$. A *least squares solution* to the system is defined as any vector $\mathbf{x} \in \mathbf{R}^n$ for which

$$(\varepsilon_1^2 + \cdots + \varepsilon_m^2)^{1/2} = \|\boldsymbol{\varepsilon}\| = \|A\mathbf{x} - \mathbf{b}\|$$

is smallest. In other words, $\mathbf{x}$ is a least squares solution if

$$\|A\mathbf{x} - \mathbf{b}\| \leqslant \|A\mathbf{y} - \mathbf{b}\|$$

for all $\mathbf{y} \in \mathbf{R}^n$. In this sense, $\mathbf{x}$ is the best solution we can get for the equation.

The basic geometric idea involved in finding a least squares solution to $A\mathbf{x} = \mathbf{b}$ is as follows. Recall that

$$\mathbf{W} = \{A\mathbf{x} : \mathbf{x} \in \mathbf{R}^n\}$$

is the column space of A. We need to find a vector $\mathbf{x}$ such that $A\mathbf{x} \in \mathbf{W}$ is closest to $\mathbf{b}$. As shown in Figure 4.9, for such an $\mathbf{x}$, $\mathbf{b} - A\mathbf{x}$ must be orthogonal to all vectors $A\mathbf{x} - A\mathbf{y}$ in $\mathbf{W}$.

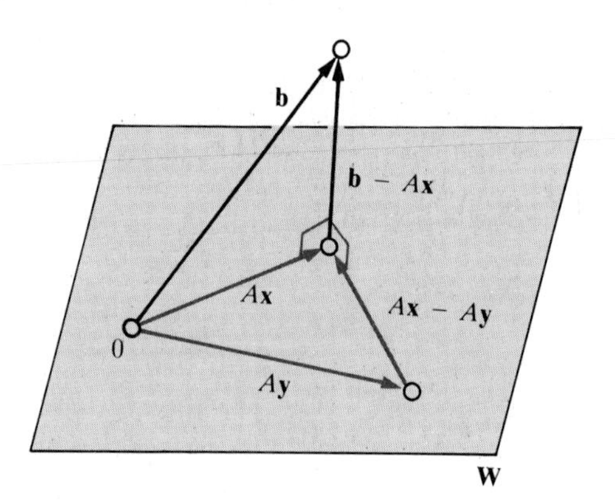

FIGURE 4.9

Setting $(A\mathbf{x} - A\mathbf{y}, A\mathbf{x} - \mathbf{b}) = 0$ and using that $(\mathbf{z}, \mathbf{w}) = \mathbf{z}^t\mathbf{w}$ yields

$$(\mathbf{x} - \mathbf{y})^t A^t(A\mathbf{x} - \mathbf{b}) = 0$$

for all $\mathbf{y}$. Since this equation must hold for all $\mathbf{y}$, we see by choosing $\mathbf{y} = \mathbf{x} - \mathbf{e}^i$ for $i = 1, \ldots, n$ that $\mathbf{x}$ is a least squares solution if and only if $A^t(A\mathbf{x} - \mathbf{b}) = \mathbf{0}$, or

$$A^tA\mathbf{x} = A^t\mathbf{b}$$

The equations of the system $A^tA\mathbf{x} = A^t\mathbf{b}$ are called the *normal equations* for $A\mathbf{x} = \mathbf{b}$.*

More formally, we have the following theorem.

* Note that geometrically there can be many least squares solutions $\mathbf{y}$ to $A\mathbf{x} = \mathbf{b}$, although all of these yield the same point $A\mathbf{y} \in \mathbf{W}$ closest to $\mathbf{b}$.

THEOREM 4.6

Let A be an $m \times n$ matrix. The least squares solutions to

$$A\mathbf{x} = \mathbf{b}$$

are precisely the solutions to the system of normal equations

$$A^t A\mathbf{x} = A^t\mathbf{b}$$

Further, this system always has at least one solution.

PROOF

It can be shown that

$$\text{rank}\, A^t A = \text{rank}\,[A^t A \quad A^t\mathbf{b}]$$

(exercise 21). Thus, $A^t A\mathbf{x} = A^t\mathbf{b}$ always has a solution, and so we only need to show that the set of least squares solutions is identical with the set of solutions to the normal equations.

The remainder of our proof is based on the Pythagorean theorem for inner product spaces. As depicted in Figure 4.10, if $(\mathbf{z}, \mathbf{w}) = 0$, then

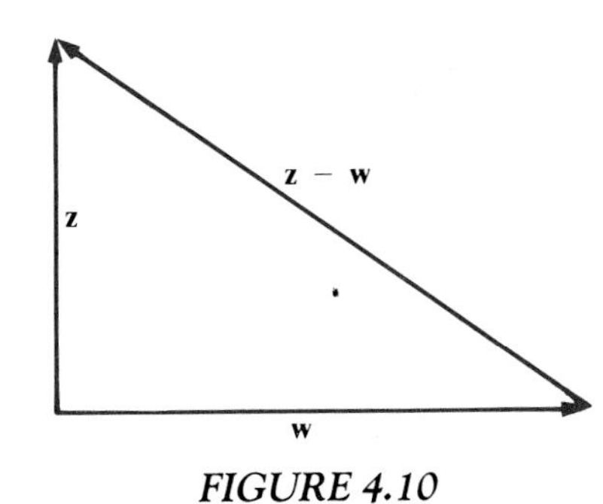

FIGURE 4.10

$$\|\mathbf{z} - \mathbf{w}\|^2 = \|\mathbf{z}\|^2 + \|\mathbf{w}\|^2$$

(exercise 22).

Let $\mathbf{u}$ be a solution to $A^t A\mathbf{x} = A^t\mathbf{b}$. Let $\mathbf{y}$ be any vector in $\mathbf{R}^n$. Set

$$\mathbf{z} = A\mathbf{y} - A\mathbf{u} \qquad \text{and} \qquad \mathbf{w} = \mathbf{b} - A\mathbf{u}$$

Then

$$\begin{aligned}(\mathbf{z}, \mathbf{w}) &= (A\mathbf{y} - A\mathbf{u}, \mathbf{b} - A\mathbf{u}) = (A\mathbf{y} - A\mathbf{u})^t(\mathbf{b} - A\mathbf{u}) = (\mathbf{y} - \mathbf{u})^t A^t(\mathbf{b} - A\mathbf{u}) \\ &= (\mathbf{y} - \mathbf{u})^t\mathbf{0} = 0\end{aligned}$$

Hence, by the Pythagorean theorem,

$$\|\mathbf{z} - \mathbf{w}\|^2 = \|\mathbf{z}\|^2 + \|\mathbf{w}\|^2$$

or

$$\|A\mathbf{y} - \mathbf{b}\|^2 = \|A\mathbf{y} - A\mathbf{u}\|^2 + \|\mathbf{b} - A\mathbf{u}\|^2 \geqslant \|\mathbf{b} - A\mathbf{u}\|^2$$

Since this result holds for any $\mathbf{y} \in \mathbf{R}^n$, $\mathbf{u}$ is a least squares solution to $A\mathbf{x} = \mathbf{b}$.

On the other hand, let $\mathbf{x}_0$ be a least squares solution to $A\mathbf{x} = \mathbf{b}$ and let $\mathbf{v}$ be any solution to $A^t A\mathbf{x} = A^t\mathbf{b}$. Then, as above,

$$\|A\mathbf{x}_0 - \mathbf{b}\|^2 = \|A\mathbf{x}_0 - A\mathbf{v}\|^2 + \|\mathbf{b} - A\mathbf{v}\|^2$$

This means that $\|A\mathbf{x}_0 - A\mathbf{v}\|^2 = 0$, or $A\mathbf{x}_0 = A\mathbf{v}$. Hence,

$$A^t A\mathbf{x}_0 = A^t A\mathbf{v} = A^t\mathbf{b}$$

and $\mathbf{x}_0$ is a solution to the normal equations. ■

EXAMPLE 22

To find the least squares solutions to

$$\begin{bmatrix} 1 & 0 \\ 0 & 0 \\ 0 & 1 \end{bmatrix} \mathbf{x} = \begin{bmatrix} 1 \\ 1 \\ 1 \end{bmatrix}$$

we form the system of normal equations

$$\begin{bmatrix} 1 & 0 & 0 \\ 0 & 0 & 1 \end{bmatrix} \begin{bmatrix} 1 & 0 \\ 0 & 0 \\ 0 & 1 \end{bmatrix} \mathbf{x} = \begin{bmatrix} 1 & 0 & 0 \\ 0 & 0 & 1 \end{bmatrix} \begin{bmatrix} 1 \\ 1 \\ 1 \end{bmatrix}$$

or

$$\begin{bmatrix} 1 & 0 \\ 0 & 1 \end{bmatrix} \mathbf{x} = \begin{bmatrix} 1 \\ 1 \end{bmatrix}$$

Solving this equation yields $\mathbf{x} = \begin{bmatrix} 1 \\ 1 \end{bmatrix}$ as the least squares solution. Further, as depicted in Figure 4.11,

$$\mathbf{W} = \{A\mathbf{x} : \mathbf{x} \in \mathbf{R}^2\}$$

is the x_1, x_3-plane and

$$A\mathbf{x} = \begin{bmatrix} 1 & 0 \\ 0 & 0 \\ 0 & 1 \end{bmatrix} \begin{bmatrix} 1 \\ 1 \end{bmatrix} = \begin{bmatrix} 1 \\ 0 \\ 1 \end{bmatrix}$$

is the point of $\mathbf{W}$ closest to $\begin{bmatrix} 1 \\ 1 \\ 1 \end{bmatrix}$. □

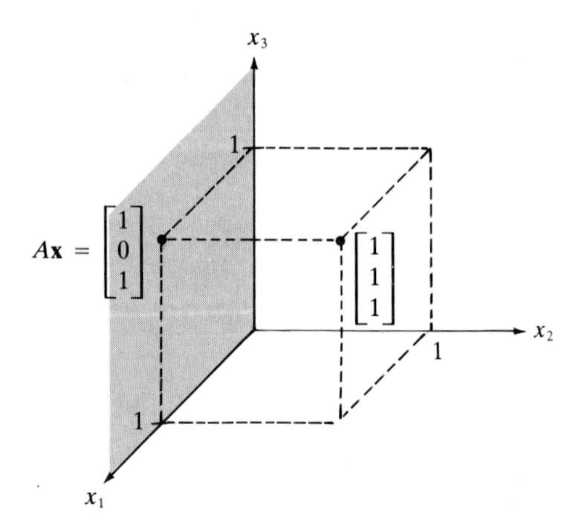

FIGURE 4.11

EXAMPLE 23

Find the least squares solutions to

$$\begin{bmatrix} 1 & 0 \\ 1 & 1 \\ 0 & 1 \end{bmatrix} \mathbf{x} = \begin{bmatrix} 0 \\ 1 \\ 0 \end{bmatrix}$$

For this system, the system of normal equations is

$$\begin{bmatrix} 2 & 1 \\ 1 & 2 \end{bmatrix} \mathbf{x} = \begin{bmatrix} 1 \\ 1 \end{bmatrix}$$

which has $\mathbf{x} = \begin{bmatrix} \frac{1}{3} \\ \frac{1}{3} \end{bmatrix}$ as its solution. As in the previous example, a sketch can be drawn showing that $A\mathbf{x} = \begin{bmatrix} \frac{1}{3} \\ \frac{2}{3} \\ \frac{1}{3} \end{bmatrix}$ is closest to $\mathbf{b} = \begin{bmatrix} 0 \\ 1 \\ 0 \end{bmatrix}$. □

II. LEAST SQUARES POLYNOMIAL APPROXIMATIONS TO GIVEN DATA

Equations requiring least squares solutions arise in problems of approximation, some of which we will now show.

Let f be a function of one variable and let $[x_1, y_1], \ldots, [x_n, y_n]$ be a set of points on the graph of f. If f is unknown, except at these points, we can find, as in Chapter 1, a polynomial p that passes through these points and approximate $f(x)$ by $p(x)$ for all x. This approximation process is called *interpolation.* If the values y_i are only estimates of the functional values $f(x_i)$ for all i, it may be advisable not to approximate f by fitting a polynomial of too high a degree through these points. For example, if the graph of the estimated values is that shown in Figure 4.12, then we might expect the function to be $f(x) = x$, rather than an oscillating polynomial actually passing through the plotted points. In these cases, we need to look at the graph of the given set of points, decide on

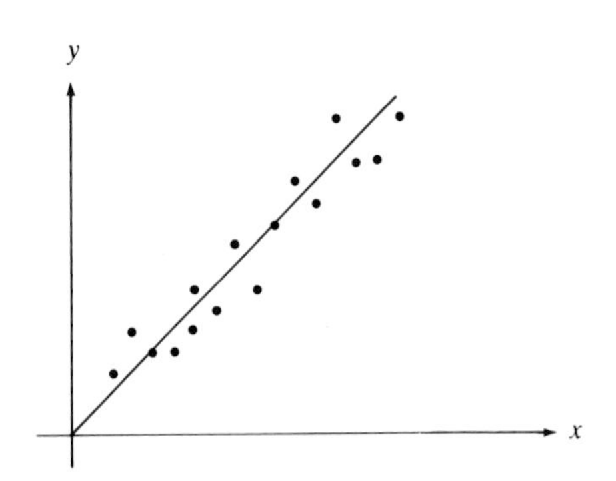

FIGURE 4.12

the type of function described by these points, and then find the "best" such function.

Suppose we decide by looking at the data (see the applications exercises) that the function is a polynomial p and that the degree of p should not exceed n. Then

$$p(x) = a_n x^n + \cdots + a_1 x + a_0$$

for some choice of $a_0, a_1, \ldots, a_n$. We define the errors between polynomial values $p(x_1), \ldots, p(x_m)$ and data values $y_1, \ldots, y_m$ as

$$\begin{array}{ccccccccc} \varepsilon_1 = & p(x_1) & - & y_1 = & a_n x_1^n + \cdots + & a_1 x_1 & + & a_0 & - y_1 \\ \vdots & \vdots & & \vdots & \vdots & \vdots & & \vdots & \vdots \\ \varepsilon_m = & p(x_m) & - & y_m = & a_n x_m^n + \cdots + & a_1 x_m & + & a_0 & - y_m \end{array}$$

Such errors are illustrated in Figure 4.13.

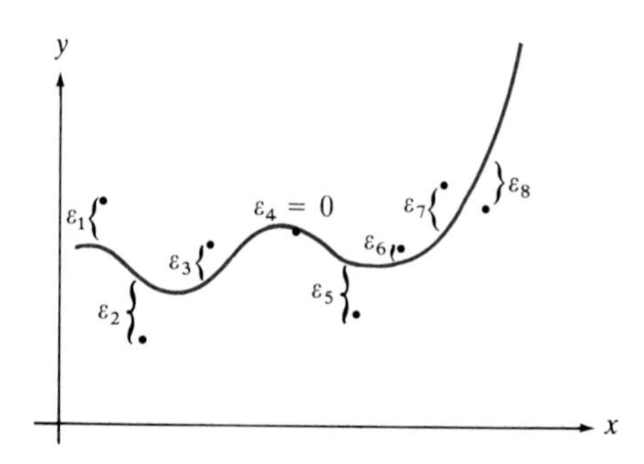

FIGURE 4.13

For

$$\boldsymbol{\varepsilon} = \begin{bmatrix} \varepsilon_1 \\ \vdots \\ \varepsilon_m \end{bmatrix}, \qquad A = \begin{bmatrix} 1 & x_1 & \cdots & x_1^n \\ \vdots & \vdots & & \vdots \\ 1 & x_m & \cdots & x_m^n \end{bmatrix}, \qquad \mathbf{x} = \begin{bmatrix} a_0 \\ \vdots \\ a_n \end{bmatrix}, \qquad \mathbf{b} = \begin{bmatrix} y_1 \\ \vdots \\ y_m \end{bmatrix}$$

these equations can be written as

$$\boldsymbol{\varepsilon} = A\mathbf{x} - \mathbf{b}$$

A *least squares polynomial approximation* to the points will be a polynomial for which the error

$$(\varepsilon_1^2 + \cdots + \varepsilon_m^2)^{1/2} = \|\boldsymbol{\varepsilon}\|$$

is smallest. Such a polynomial can be found by finding a least squares solution to

$$A\mathbf{x} = \mathbf{b}$$

EXAMPLE 24

Find the least squares polynomial approximation of degree 1 or less to the points [0, 1], [1, 1.4], and [2, 1.9].

We need to find $p(x) = a_1 x + a_0$ using least squares. To find this polynomial, we set

$$\begin{bmatrix} 1 & 0 \\ 1 & 1 \\ 1 & 2 \end{bmatrix} \begin{bmatrix} a_0 \\ a_1 \end{bmatrix} = \begin{bmatrix} 1 \\ 1.4 \\ 1.9 \end{bmatrix}$$

To find a least squares solution to this equation, we need to solve the system of normal equations

$$\begin{bmatrix} 1 & 1 & 1 \\ 0 & 1 & 2 \end{bmatrix} \begin{bmatrix} 1 & 0 \\ 1 & 1 \\ 1 & 2 \end{bmatrix} \begin{bmatrix} a_0 \\ a_1 \end{bmatrix} = \begin{bmatrix} 1 & 1 & 1 \\ 0 & 1 & 2 \end{bmatrix} \begin{bmatrix} 1 \\ 1.4 \\ 1.9 \end{bmatrix}$$

or

$$\begin{bmatrix} 3 & 3 \\ 3 & 5 \end{bmatrix} \begin{bmatrix} a_0 \\ a_1 \end{bmatrix} = \begin{bmatrix} 4.3 \\ 5.2 \end{bmatrix}$$

Thus, to two decimal places, $a_0 = 0.98$ and $a_1 = 0.45$, so $p(x) = 0.45x + 0.98$. We show the graph of p and the points in Figure 4.14. □

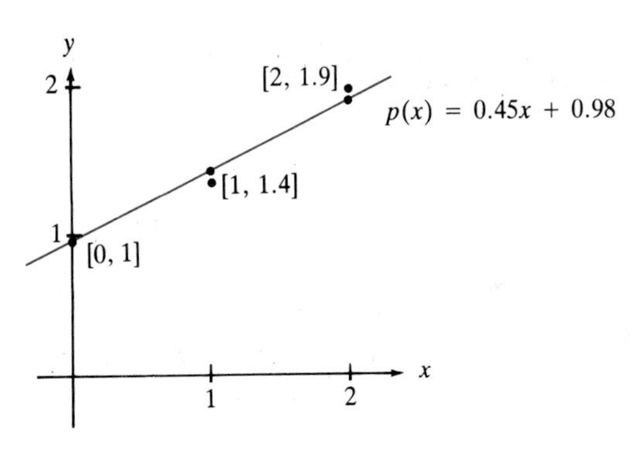

FIGURE 4.14

EXAMPLE 25

Find the least squares polynomial $p(x) = a_2 x^2 + a_1 x + a_0$ that approximates the data [−1, 1], [0, 0], [1, 1], and [2, 5].

To find this polynomial, we set

$$\begin{bmatrix} 1 & -1 & 1 \\ 1 & 0 & 0 \\ 1 & 1 & 1 \\ 1 & 2 & 4 \end{bmatrix} \begin{bmatrix} a_0 \\ a_1 \\ a_2 \end{bmatrix} = \begin{bmatrix} 1 \\ 0 \\ 1 \\ 5 \end{bmatrix}$$

The system of normal equations is

$$\begin{bmatrix} 4 & 2 & 6 \\ 2 & 6 & 8 \\ 6 & 8 & 18 \end{bmatrix} \begin{bmatrix} a_0 \\ a_1 \\ a_2 \end{bmatrix} = \begin{bmatrix} 7 \\ 10 \\ 22 \end{bmatrix}$$

which yields $a_0 = -\frac{3}{20}$, $a_1 = \frac{1}{20}$, and $a_2 = \frac{5}{4}$. Thus $p(x) = \frac{5}{4}x^2 + \frac{1}{20}x - \frac{3}{20}$. □
The polynomial and data are depicted in Figure 4.15.

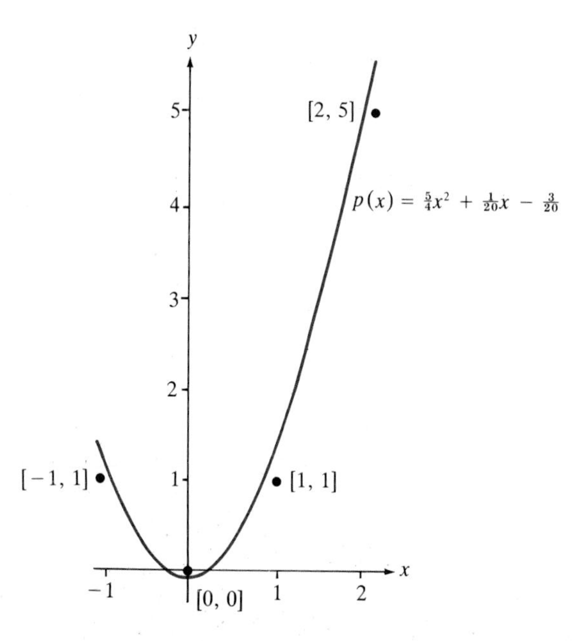

FIGURE 4.15

This work can be extended to approximations by functions of several variables. We will show how this can be done for a linear function

$$z = a_0 + a_1x + a_2y$$

Let the points $[x_1, y_1, z_1], \ldots, [x_m, y_m, z_m]$ be the given estimates to a function $f(x, y)$. Define

$$\begin{aligned} \varepsilon_1 &= a_0 + a_1x_1 + a_2y_1 - z_1 \\ &\vdots \\ \varepsilon_m &= a_0 + a_1x_m + a_2y_m - z_m \end{aligned}$$

The *least squares linear approximation* to these points will be a linear function for which $(\varepsilon_1^2 + \cdots + \varepsilon_m^2)^{1/2}$ is smallest. The coefficients of the linear function z can be found by finding a least squares solution to

$$\begin{bmatrix} 1 & x_1 & y_1 \\ \vdots & \vdots & \vdots \\ 1 & x_m & y_m \end{bmatrix} \begin{bmatrix} a_0 \\ a_1 \\ a_2 \end{bmatrix} = \begin{bmatrix} z_1 \\ \vdots \\ z_m \end{bmatrix}$$

EXAMPLE 26

Find the linear function $z = a_0 + a_1x + a_2y$ that is the least squares linear approximation to the points $[1, 1, 2]$, $[1, 2, 1]$, $[2, 1, 3]$, and $[1, 0, 2]$.

To find a_0, a_1, and a_2 we need to find a least squares solution

$$\begin{bmatrix} 1 & 1 & 1 \\ 1 & 1 & 2 \\ 1 & 2 & 1 \\ 1 & 1 & 0 \end{bmatrix} \begin{bmatrix} a_0 \\ a_1 \\ a_2 \end{bmatrix} = \begin{bmatrix} 2 \\ 1 \\ 3 \\ 2 \end{bmatrix}$$

The system of normal equations is

$$\begin{bmatrix} 1 & 1 & 1 & 1 \\ 1 & 1 & 2 & 1 \\ 1 & 2 & 1 & 0 \end{bmatrix} \begin{bmatrix} 1 & 1 & 1 \\ 1 & 1 & 2 \\ 1 & 2 & 1 \\ 1 & 1 & 0 \end{bmatrix} \begin{bmatrix} a_0 \\ a_1 \\ a_2 \end{bmatrix} = \begin{bmatrix} 1 & 1 & 1 & 1 \\ 1 & 1 & 2 & 1 \\ 1 & 2 & 1 & 0 \end{bmatrix} \begin{bmatrix} 2 \\ 1 \\ 3 \\ 2 \end{bmatrix}$$

or

$$\begin{bmatrix} 4 & 5 & 4 \\ 5 & 7 & 5 \\ 4 & 5 & 6 \end{bmatrix} \begin{bmatrix} a_0 \\ a_1 \\ a_2 \end{bmatrix} = \begin{bmatrix} 8 \\ 11 \\ 7 \end{bmatrix}$$

Solving this system yields $a_0 = \frac{5}{6}$, $a_1 = \frac{4}{3}$, and $a_2 = -\frac{1}{2}$; thus $z = \frac{5}{6} + \frac{4}{3}x - \frac{1}{2}y$. This solution is graphed in Figure 4.16. □

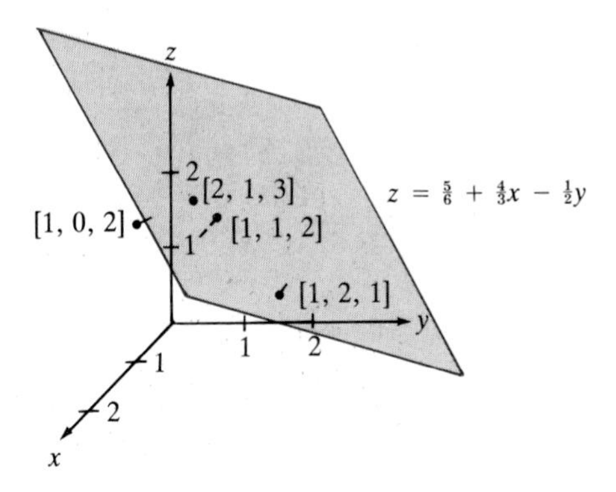

FIGURE 4.16

III. LEAST SQUARES APPROXIMATION IN A SUBSPACE TO A VECTOR

For the least squares work in Part II we were concerned with finding a polynomial, within a specified degree, to approximate a given function using a finite set of values of the function. But we can also fit a polynomial, within a specified degree, to a given function over its whole domain. We will conclude this section by showing how this can be done.

Let $\mathbf{V}$ be an inner product space with $\mathbf{W}$ a finite-dimensional subspace of $\mathbf{V}$. Let $\{\mathbf{u}_1,\ldots,\mathbf{u}_n\}$ be an orthogonal basis of $\mathbf{W}$, and let $\mathbf{v} \in \mathbf{V}$. Consider the equation

$$a_1\mathbf{u}_1 + \cdots + a_n\mathbf{u}_n = \mathbf{v}$$

If $\mathbf{v}$ is not a linear combination of $\mathbf{u}_1,\ldots,\mathbf{u}_n$, then the equation has no solution. However, as in the preceding work, we intend to find the "best" solution we can. That is, we want to find $a_1,\ldots,a_n$ so that the left side, $a_1\mathbf{u}_1 + \cdots + a_n\mathbf{u}_n$, is as close as possible to the right side, $\mathbf{v}$. When these conditions are satisfied, the numbers $a_1,\ldots,a_n$ are called *Fourier coefficients*, and $a_1\mathbf{u}_1 + \cdots + a_n\mathbf{u}_n$ is called a *Fourier sum of* $\mathbf{v}$.

To find such a solution, we define the error vector

$$\boldsymbol{\varepsilon} = a_1\mathbf{u}_1 + \cdots + a_n\mathbf{u}_n - \mathbf{v}$$

Any $\mathbf{u} = a_1\mathbf{u}_1 + \cdots + a_n\mathbf{u}_n$ for which $\|\boldsymbol{\varepsilon}\|$ is smallest is called a *least squares approximation of* $\mathbf{v}$ *in* $\mathbf{W}$. This approximation can be calculated as follows.

THEOREM 4.7

Let $\mathbf{V}$ be an inner product space and let $\mathbf{W}$ be a finite-dimensional subspace of $\mathbf{V}$. Let $\{\mathbf{u}_1,\ldots,\mathbf{u}_n\}$ be an orthogonal basis of $\mathbf{W}$. Then, for any $\mathbf{v} \in \mathbf{V}$, the least squares approximation of $\mathbf{v}$ in $\mathbf{W}$ is the Fourier sum

$$\mathbf{u} = \frac{(\mathbf{v},\mathbf{u}_1)}{(\mathbf{u}_1,\mathbf{u}_1)}\mathbf{u}_1 + \cdots + \frac{(\mathbf{v},\mathbf{u}_n)}{(\mathbf{u}_n,\mathbf{u}_n)}\mathbf{u}_n$$

PROOF

Note that, because $\{\mathbf{u}_1,\ldots,\mathbf{u}_n\}$ is orthogonal,

$$\|\boldsymbol{\varepsilon}\|^2 = (\boldsymbol{\varepsilon},\boldsymbol{\varepsilon}) = a_1^2(\mathbf{u}_1,\mathbf{u}_1) - 2a_1(\mathbf{v},\mathbf{u}_1) + a_2^2(\mathbf{u}_2,\mathbf{u}_2) - 2a_2(\mathbf{v},\mathbf{u}_2) + \cdots + a_n^2(\mathbf{u}_n,\mathbf{u}_n) - 2a_n(\mathbf{v},\mathbf{u}_n) + (\mathbf{v},\mathbf{v})$$

It can be shown that a quadratic polynomial $a^2b - 2ac$ with variable a has its smallest value at $a = c/b$. Since the value of a_i is independent of that of a_j if $i \neq j$, it follows that $\|\boldsymbol{\varepsilon}\|$ is smallest when

$$a_1 = \frac{(\mathbf{v},\mathbf{u}_1)}{(\mathbf{u}_1,\mathbf{u}_1)},\ldots,a_n = \frac{(\mathbf{v},\mathbf{u}_n)}{(\mathbf{u}_n,\mathbf{u}_n)}$$

■

EXAMPLE 27

In $\mathbf{R}_3$ with the standard inner product,

$$\{\mathbf{u}_1,\mathbf{u}_2\} = \{[-1,0,0],[0,2,0]\}$$

is an orthogonal basis for

$$\mathbf{W} = \{[x,y,z]:z=0\}$$

Let $\mathbf{x} = [1,1,1]$.

To find the least squares approximation of $\mathbf{x}$ in $\mathbf{W}$, we need to calculate

$$\frac{(\mathbf{x},\mathbf{u}_1)}{(\mathbf{u}_1,\mathbf{u}_1)} = -1 \qquad \text{and} \qquad \frac{(\mathbf{x},\mathbf{u}_2)}{(\mathbf{u}_2,\mathbf{u}_2)} = \frac{1}{2}$$

Hence, the approximation is

$$\mathbf{u} = \frac{(\mathbf{x},\mathbf{u}_1)}{(\mathbf{u}_1,\mathbf{u}_1)}\mathbf{u}_1 + \frac{(\mathbf{x},\mathbf{u}_2)}{(\mathbf{u}_2,\mathbf{u}_2)}\mathbf{u}_2 = [1,0,0] + [0,1,0] = [1,1,0]$$

The geometry of the problem is shown in Figure 4.17. □

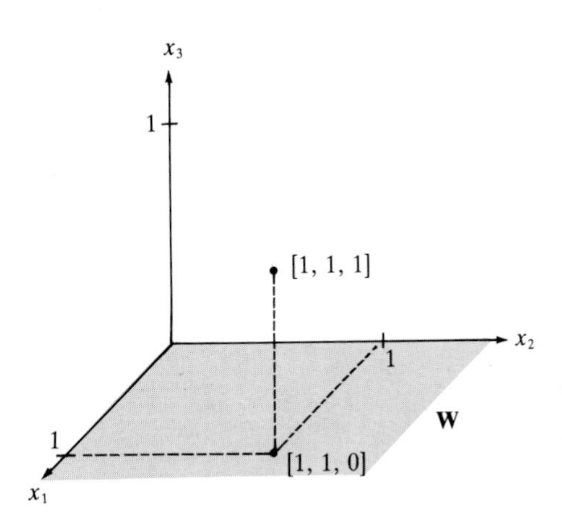

FIGURE 4.17

How this theorem can be used to find a polynomial, within a given degree, which least-squares-approximates a given function is shown in the following example.

EXAMPLE 28

CALCULUS

Let $\mathbf{C}[-1,1]$ be the inner product space with the standard inner product. Find $p(t) \in \mathbf{P}_2$ that least-squares-approximates $f(t) = e^t$.

From Example 21 of Section 4.2, an orthonormal basis for $\mathbf{P}_2$ consists of Legendre polynomials

$$\mathbf{u}_1 = \frac{\sqrt{2}}{2}, \qquad \mathbf{u}_2 = \frac{\sqrt{6}}{2}t, \qquad \mathbf{u}_3 = \frac{3\sqrt{10}}{4}\left(t^2 - \frac{1}{3}\right)$$

Because this basis is orthonormal, the inner products $(\mathbf{u}_i, \mathbf{u}_i) = 1$ and we do not need to compute these in finding the approximation.

Now, calculating the Fourier coefficients yields

$$(f,\mathbf{u}_1) = \int_{-1}^{1} e^t \frac{\sqrt{2}}{2}\,dt = \frac{\sqrt{2}}{2}(e^1 - e^{-1}) \approx 1.66$$

$$(f, \mathbf{u}_2) = \int_{-1}^{1} e^t\left(\frac{\sqrt{6}}{2}t\right)dt = \frac{\sqrt{6}}{2}(2e^{-1}) = \sqrt{6}\,e^{-1} \approx 0.90$$

$$(f, \mathbf{u}_3) = \int_{-1}^{1} e^t\left[\frac{3\sqrt{10}}{4}\left(t^2 - \frac{1}{3}\right)\right]dt = \frac{\sqrt{10}}{2}(e - 7e^{-1}) \approx 0.23$$

Thus

$$\begin{aligned}
\mathbf{u} &= (f, \mathbf{u}_1)\mathbf{u}_1 + (f, \mathbf{u}_2)\mathbf{u}_2 + (f, \mathbf{u}_3)\mathbf{u}_3 \\
&\approx 1.66\left(\frac{\sqrt{2}}{2}\right) + 0.90\left(\frac{\sqrt{6}}{2}t\right) + 0.23\left[\frac{3\sqrt{10}}{4}\left(t^2 - \frac{1}{3}\right)\right] \\
&\approx 1.17 + 1.10t + 0.55(t^2 - 0.33) \\
&\approx 0.99 + 1.10t + 0.55t^2
\end{aligned}$$

See Figure 4.18. □

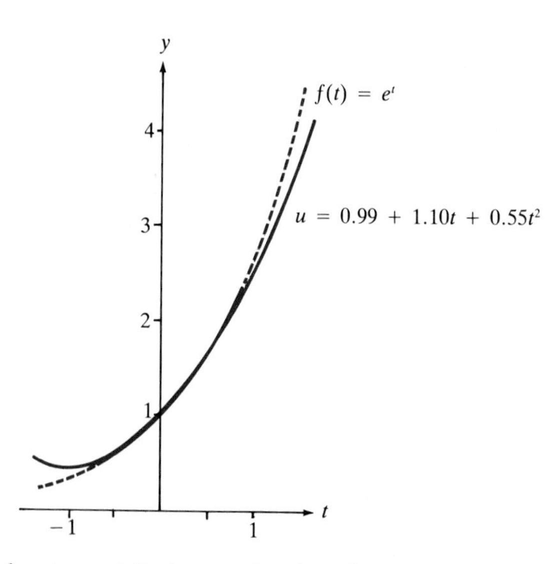

FIGURE 4.18

In this chapter and the last we developed a general study of vector spaces and showed some of its uses. In the next chapter we will study functions and equations defined on vector spaces.

EXERCISES FOR SECTION 4.3

COMPUTATIONAL EXERCISES

In exercises 1 and 2, find the error $\|A\mathbf{x} - \mathbf{b}\|$ for the given A, $\mathbf{x}$, and $\mathbf{b}$.

1. $A = \begin{bmatrix} 1 & 2 \\ 2 & 3 \end{bmatrix}$, $\mathbf{b} = \begin{bmatrix} 1 \\ 2 \end{bmatrix}$

(a) $\mathbf{x} = \begin{bmatrix} 1 \\ 0 \end{bmatrix}$ (b) $\mathbf{x} = \begin{bmatrix} 3 \\ -1 \end{bmatrix}$ (c) $\mathbf{x} = \begin{bmatrix} 1 \\ 1 \end{bmatrix}$

2. $A = \begin{bmatrix} 1 & 3 & 2 \\ -3 & -8 & 1 \\ 2 & 5 & -2 \end{bmatrix}$, $\mathbf{b} = \begin{bmatrix} 4 \\ 4 \\ -5 \end{bmatrix}$

$(a)\ \mathbf{x} = \begin{bmatrix} 13 \\ -4 \\ 3 \end{bmatrix}$ $(b)\ \mathbf{x} = \begin{bmatrix} 13 \\ -5 \\ 3 \end{bmatrix}$ $(c)\ \mathbf{x} = \begin{bmatrix} -3 \\ 1 \\ 2 \end{bmatrix}$

In exercises 3 through 8, find the least squares solution of the given equation.

3. $\begin{bmatrix} 0 & 1 \\ -1 & 2 \\ -2 & 1 \end{bmatrix} \begin{bmatrix} x \\ y \end{bmatrix} = \begin{bmatrix} 3 \\ 2 \\ 2 \end{bmatrix}$

4. $\begin{bmatrix} 1 & 1 \\ 1 & 2 \\ 1 & 3 \end{bmatrix} \begin{bmatrix} x \\ y \end{bmatrix} = \begin{bmatrix} 1 \\ 2 \\ 1 \end{bmatrix}$

5. $\begin{bmatrix} 2 & 1 \\ 3 & 1 \\ 5 & 1 \end{bmatrix} \begin{bmatrix} x \\ y \end{bmatrix} = \begin{bmatrix} 2 \\ 7 \\ 1 \end{bmatrix}$

6. $\begin{bmatrix} 1 & 2 & 1 \\ 1 & 3 & 2 \\ 1 & -1 & 2 \\ 1 & 2 & -1 \end{bmatrix} \begin{bmatrix} x \\ y \\ z \end{bmatrix} = \begin{bmatrix} 1 \\ -1 \\ 3 \\ 2 \end{bmatrix}$

7. $[1 \quad 2 \quad 3] \begin{bmatrix} x \\ y \\ z \end{bmatrix} = [2]$

8. $\begin{bmatrix} 1 & 1 \\ 1 & 0 \\ 1 & 2 \\ 1 & -1 \end{bmatrix} \begin{bmatrix} x \\ y \end{bmatrix} = \begin{bmatrix} 1 \\ 0 \\ -1 \\ 0 \end{bmatrix}$

In exercises 9 through 14, find the least squares polynomial approximation for the given points, where the polynomial should have no more than the specified degree. Graph the polynomial and the points.

9. [1, 0.1], [3, 0.8], [7, 3.2], degree 1

10. [0, 2.8], [1, 5.1], [2, 7], degree 1

11. [−2, 0], [−1, 1], [0, 4], [1, 8], degree 2

12. [0, 2], [1, 0], [2, 2], [3, 4], degree 2

13. [0, 1, 4], [1, 0, 0], [1, 1, 2], [1, 2, 5], degree 1

14. [1, 1, 2], [1, 2, 1], [2, 0, 3], [2, 1, 3], degree 1

In exercises 15 through 18, find the least squares approximation of $\mathbf{v}$ in $\mathbf{W} \subseteq \mathbf{R}_n$, where $\mathbf{W}$ has the given orthonormal basis S. Graph your results.

15. $S = \left\{\left[\frac{1}{\sqrt{2}}, \frac{1}{\sqrt{2}}\right]\right\}$, $\mathbf{v} = [1, 0]$

16. $S = \left\{\left[-\frac{1}{\sqrt{2}}, \frac{1}{\sqrt{2}}\right]\right\}$, $\mathbf{v} = [1, 1]$

17. $S = \left\{\left[-\frac{1}{\sqrt{2}}, \frac{1}{\sqrt{2}}, 0\right], \left[\frac{1}{\sqrt{3}}, \frac{1}{\sqrt{3}}, \frac{1}{\sqrt{3}}\right]\right\}$, $\mathbf{v} = [0, 0, 1]$

18. $S = \left\{\frac{1}{\sqrt{5}}[1, 0, -2], \frac{1}{\sqrt{30}}[2, 5, 1]\right\}$, $\mathbf{v} = [3, 2, 1]$

In exercises 19 and 20, find the least squares approximation of $\mathbf{v}$ in $\mathbf{P}_2 \subseteq \mathbf{C}[-1, 1]$ with the standard inner product and the basis the set

$$\left\{\frac{\sqrt{2}}{2}, \frac{\sqrt{6}}{2}t, \frac{3\sqrt{10}}{4}\left(t^2 - \frac{1}{3}\right)\right\}$$

of Legendre polynomials. Graph $\mathbf{v}$ and your approximation $\mathbf{p}$.

19. $\mathbf{v} = t^3$ *20.* $\mathbf{v} = t^3 - 1$

THEORETICAL EXERCISES

21. Let A be an $m \times n$ matrix and $\mathbf{b} \in \mathbf{R}^m$. Prove that

$$\text{rank } A^tA = \text{rank } [A^tA \quad A^t\mathbf{b}]$$

(*Hint:* Observe that rank $A^tA \leqslant$ rank $[A^tA \quad A^t\mathbf{b}] \leqslant$ rank A^t by noting that, for any $n \times k$ matrix B, every column in A^tB is a linear combination of the columns of A^t. Then show that rank A^tA = rank A by noting that, if $A^tA\mathbf{x} = \mathbf{0}$, then $\mathbf{x}^tA^tA\mathbf{x} = \mathbf{0}$, or $\|A\mathbf{x}\| = 0$, so $A\mathbf{x} = \mathbf{0}$, and conversely. Thus the sizes of the solution sets of $A^tA\mathbf{x} = \mathbf{0}$ and $A\mathbf{x} = \mathbf{0}$ are the same. Since these sizes are $m -$ rank A^tA and $m -$ rank A, respectively, rank A^tA = rank A. Then recall that rank A = rank A^t.)

22. Prove the Pythagorean theorem: If $(\mathbf{z}, \mathbf{w}) = 0$, then

$$\|\mathbf{z} - \mathbf{w}\|^2 = \|\mathbf{z}\|^2 + \|\mathbf{w}\|^2$$

23. If $A\mathbf{x} = \mathbf{b}$ has a solution, then any such solution is a least squares solution. Explain why this is true.

24. (calculus) Let

$$\mathbf{V}_n = \{r_1 \sin t + r_2 \sin 2t + \cdots + r_n \sin nt : r_1, \ldots, r_n \in \mathbf{R}\}$$

Let
$$(f, g) = \frac{1}{\pi}\int_0^{2\pi} f(t)g(t)\,dt$$

be the inner product on $\mathbf{V}_n$. You may assume that for integers r and s,

$$\int_0^{2\pi} (\sin rt)(\sin st)\,dt = \begin{cases} 0 & \text{if} \quad r \neq s \\ \pi & \text{if} \quad r = s \end{cases}$$

Thus $\{\sin t, \sin 2t, \ldots, \sin nt\}$ is an orthogonal basis for $\mathbf{V}_n$. Show that, for any continuous function f, the function

$$r_1 \sin t + r_2 \sin 2t + \cdots + r_n \sin nt$$

with $r_k = \dfrac{(f(t), \sin kt)}{\pi}$, is the least squares approximation of f from $\mathbf{V}_n$. For $f(t) = t$, find the least squares approximation of f from $\mathbf{V}_2$. (This is a basic Fourier series problem.)

25. Find all lines that least-squares-approximate the data $[1, 1]$ and $[1, 2]$. Graph the points and three of the lines. (This exercise shows that there can be many least squares solutions.)

26. Explain why the system $A\mathbf{x} = \mathbf{b}$, where A is $m \times n$, has infinitely many least squares solutions if $m < n$. (This system is called *underdetermined*.)

27. Suppose we find the line that least-squares-approximates a given set of data. Interchanging the roles of x and y, suppose we then find a line of the form $x = m'y + b'$ that least-squares-approximates the data. Show that these two lines need not be the same.

APPLICATIONS EXERCISES

28. Suppose a barber shop changes its prices and counts its numbers of customers at the various prices, obtaining the data in the table below.

Price	\$3.50	\$3.75	\$4.00
Number of Customers	70	65	61

Letting p = price and x = number of customers, find the least squares approximation $p = mx + b$ to these data. Graph the data and the line. Use the line to predict the number of customers at price \$5.00.

29. The price p of a stock on day k is given by the table below.

Price (p)	\$10	\$5	\$2	\$1
Day (k)	1	2	3	4

Find the least squares polynomial approximation of degree 2 or less for these data. Graph the data and the polynomial. Predict the price of the stock on the sixth day. Explain your answer. At this time, would you buy or sell?

30. A plant's growth is recorded in the table below.

Height (h)	2	2.7	3.3	3.6	inches
Day (d)	0	2	4	6	

Describe this growth with a formula by finding the least squares polynomial approximation of degree 2 or less to these data. Graph the data and the polynomial. (Exact calculation of the function h in terms of d is impossible.)

31. The United States census is taken every ten years. In Table 4.2 is a listing of the populations counted during various censuses. Find the least squares linear approximation to these data. Graph the line and the data. Use this line to estimate the population for 1940 and 1950. The actual population in 1940 was 131.7 million and in 1950 was 150.7 million. (Such techniques can be used for futuristic predictions.)

TABLE 4.2

YEAR	POPULATION IN MILLIONS
1890	62.9
1900	75.9
1910	91.9
1920	105.7
1930	122.7

CHAPTER FIVE

LINEAR TRANSFORMATIONS AND LINEAR EQUATIONS

In the previous two chapters, we gave a general study of vector spaces, introducing geometric concepts such as dimension, distance, and angles. We now consider functions and equations defined on vector spaces.

You have already studied linear algebraic equations, linear difference equations, and, perhaps in homework problems, linear differential equations and linear integral equations. In future work, you will encounter other such equations. Thus, instead of studying these linear equations individually as they arise, it is more reasonable to study all linear equations simultaneously. This general study, given in Section 5.1, will include results for finding and describing solution sets to any such equation.

Functions and equations usually arise when a problem is placed in a mathematical setting (see applied exercises). As you have no doubt seen in previous courses, the first step is to provide the space in which the problem arises with a grid determined from axes, called a coordinate system. Points in the space are quantified* using coordinates determined from this system. Functions and equations are then quantified in terms of the coordinates. A judicious choice of coordinate system can yield simple expressions for functions and equations, which in turn simplify observations or calculations that need to be made.

EXAMPLE 1

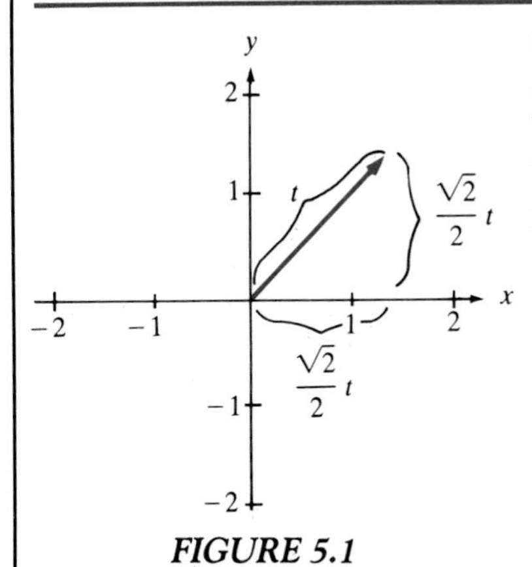

FIGURE 5.1

Suppose an object moves from a point 0 along a straight line making an angle of $\pi/4$ radian with the horizontal such that, at time t, it has traveled t units. We can place an x,y-coordinate system in the plane as shown in Figure 5.1. Then, in terms of the coordinates of this coordinate system, we have the position of the object at time t as

$$p(t) = \begin{bmatrix} \dfrac{\sqrt{2}t}{2} \\ \dfrac{\sqrt{2}t}{2} \end{bmatrix}$$

* "Quantify" means to express completely in terms of numbers.

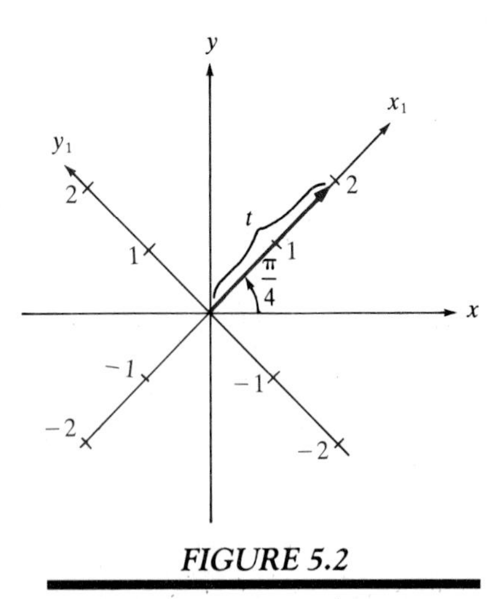

FIGURE 5.2

Of course, we could place other coordinate systems in this plane. For example, we could use the x_1,y_1-coordinate system shown in Figure 5.2, which is a $\pi/4$ rotation of the x,y-coordinate system. The position of the object at time t in terms of these coordinates is

$$p(t) = \begin{bmatrix} t \\ 0 \end{bmatrix}$$

Note that the function giving the position of the object has different coordinate descriptions in the two coordinate systems, since the coordinates of the points involved are different. Further, the second coordinate description is simpler than the first. □

In Sections 5.2 and 5.3 we will define various coordinate systems for any given finite-dimensional vector space. This process can be thought of geometrically as placing in the space various axes which determine a grid for that space. In each coordinate system, we will show how to express vectors in terms of coordinates and how to write linear functions in terms of those coordinates. Then in Chapters 6 and 7 we will develop some important problem-solving techniques by showing how to choose, for a given problem, a coordinate system that will allow the problem to be described in terms of relatively simple expressions.

5.1 LINEAR TRANSFORMATIONS AND LINEAR EQUATIONS

In this section we define and develop a general study of linear transformations and linear equations. To define a linear equation on a vector space, we first consider the linear algebraic equation $A\mathbf{x} = \mathbf{b}$, where A is an $m \times n$ matrix. Setting $L(\mathbf{x}) = A\mathbf{x}$, we have a function that associates to each vector $\mathbf{x} \in \mathbf{R}^n$ a vector $L(\mathbf{x}) = A\mathbf{x} \in \mathbf{R}^m$.

EXAMPLE 2

Let

$$L\left(\begin{bmatrix} x_1 \\ x_2 \end{bmatrix}\right) = \begin{bmatrix} 2 & -1 \\ 1 & 1 \end{bmatrix}\begin{bmatrix} x_1 \\ x_2 \end{bmatrix}$$

Then L is a function from $\mathbf{R}^2$ to $\mathbf{R}^2$. Note that

$$L\left(\begin{bmatrix} 1 \\ 1 \end{bmatrix}\right) = \begin{bmatrix} 1 \\ 2 \end{bmatrix} \quad \text{and} \quad L\left(\begin{bmatrix} 1 \\ -2 \end{bmatrix}\right) = \begin{bmatrix} 4 \\ -1 \end{bmatrix}$$

as shown geometrically in Figure 5.3. □

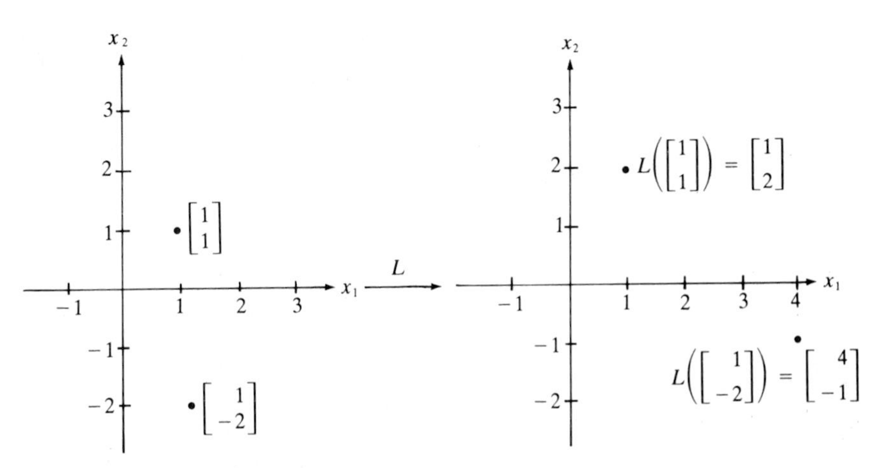

FIGURE 5.3

The arithmetical properties* of the function $L(\mathbf{x}) = A\mathbf{x}$ are as follows.

(*i*) $L(\mathbf{x} + \mathbf{y}) = L(\mathbf{x}) + L(\mathbf{y})$

(*ii*) $L(r\mathbf{x}) = rL(\mathbf{x})$

for all vectors $\mathbf{x}$ and $\mathbf{y}$ and scalars r. In terms of this function, the linear algebraic equation can be written as $L(\mathbf{x}) = \mathbf{b}$. Guided by these remarks, we will produce a general definition of a linear function, and consequently of a linear equation.

Let $\mathbf{V}$ and $\mathbf{W}$ be vector spaces. Let L be a function from $\mathbf{V}$ to $\mathbf{W}$; that is, for every vector $\mathbf{x} \in \mathbf{V}$, L assigns a vector $L(\mathbf{x}) \in \mathbf{W}$. We call L a *transformation*† in this vector-space setting. When thinking of L geometrically, we will often call the transformation a *mapping*. The transformation of interest in this course follows.

DEFINITION

Let $\mathbf{V}$ and $\mathbf{W}$ be vector spaces. Let L be a transformation from $\mathbf{V}$ to $\mathbf{W}$. If

(*i*) $L(\mathbf{x} + \mathbf{y}) = L(\mathbf{x}) + L(\mathbf{y})$ for all vectors $\mathbf{x}, \mathbf{y} \in \mathbf{V}$ and

(*ii*) $L(r\mathbf{x}) = rL(\mathbf{x})$ for all vectors $\mathbf{x} \in \mathbf{V}$ and scalars $r \in \mathbf{R}$,

then we call L a *linear transformation.***

EXAMPLE 3

Let A be an $m \times n$ matrix. Then $L(\mathbf{x}) = A\mathbf{x}$ is a linear transformation from $\mathbf{R}^n$ to $\mathbf{R}^m$. □

EXAMPLE 4

Let L be the mapping from the fixed arrows in $\mathbf{R}^2$ to the fixed arrows in $\mathbf{R}^2$ that takes any fixed arrow $\mathbf{x}$ into the fixed arrow $L(\mathbf{x})$ obtained by rotating the fixed

* These are the properties used in Example 29 of Chapter 3 to show that the solution set of $A\mathbf{x} = \mathbf{0}$ is a vector space.

† Sometimes the word "homomorphism" or "operator" is used in place of "transformation."

** These arithmetical properties for L can be remembered by recalling the properties for $L(\mathbf{x}) = A\mathbf{x}$.

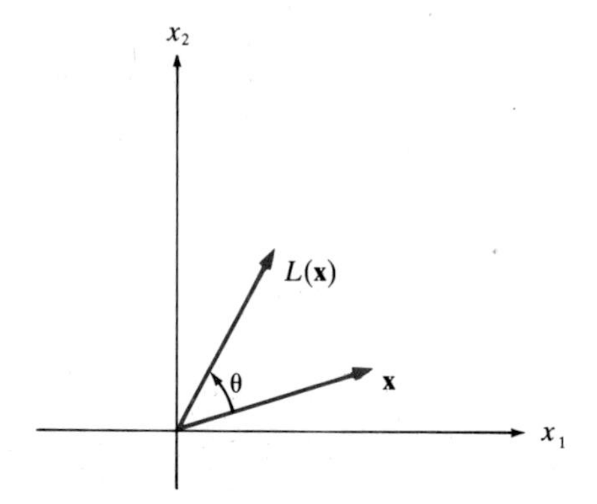

FIGURE 5.4

arrow $\mathbf{x}$ by θ radians, as shown in Figure 5.4. Then L, called the *θ-rotation mapping*, is a linear mapping (exercise 52). □

EXAMPLE 5

Let L be the transformation from $\mathbf{P}_2$ to $\mathbf{P}_1$ defined by

$$L(a_2t^2 + a_1t + a_0) = 2a_2t + a_1$$

To show that L is a linear transformation, we show that the arithmetical properties (i) and (ii) of the definition are satisfied.

For (i), let $p(t) = a_2t^2 + a_1t + a_0$ and $q(t) = b_2t^2 + b_1t + b_0$. Then

$$p(t) + q(t) = (a_2 + b_2)t^2 + (a_1 + b_1)t + (a_0 + b_0)$$

Thus

$$L(p(t) + q(t)) = 2(a_2 + b_2)t + a_1 + b_1 = 2a_2t + a_1 + 2b_2t + b_1$$

Further,

$$L(p(t)) + L(q(t)) = (2a_2t + a_1) + (2b_2t + b_1)$$

Thus $L(p(t) + q(t)) = L(p(t)) + L(q(t))$.

For (ii), let $r \in \mathbf{R}$. Then $rp(t) = ra_2t^2 + ra_1t + ra_0$. Thus

$$L(rp(t)) = 2ra_2t + ra_1 = r(2a_2t + a_1)$$

Further,

$$rL(p(t)) = r(2a_2t + a_1)$$

Hence $L(rp(t)) = rL(p(t))$, and consequently L is a linear transformation.* □

Example 6

Let $E(x(k)) = x(k+1)$ for all $x(k) \in \mathbf{F}(\mathbf{N})$. For example,

$$E(3k+2) = 3(k+1) + 2 = 3k + 5$$

and

$$E(k^2 + k) = (k+1)^2 + (k+1) = k^2 + 3k + 2$$

Thus E is a transformation, called the *shift transformation*,* from $\mathbf{F}(\mathbf{N})$ to $\mathbf{F}(\mathbf{N})$.

To show that E is a linear transformation, we need to verify properties (i) and (ii).

For (i), note that if $x(k), y(k) \in \mathbf{F}(\mathbf{N})$, then

$$E(x(k) + y(k)) = x(k+1) + y(k+1)$$

Also,

$$E(x(k)) + E(y(k)) = x(k+1) + y(k+1)$$

Thus $E(x(k) + y(k)) = E(x(k)) + E(y(k))$.

For (ii), if $r \in \mathbf{R}$ and $x(k) \in \mathbf{F}(\mathbf{N})$, then

$$E(rx(k)) = rx(k+1) \qquad \text{and} \qquad rE(x(k)) = rx(k+1)$$

Thus $E(rx(k)) = rE(x(k))$, and so E is a linear transformation. □

Example 7

For any numbers a, b, and c, the transformations

$$L(x(k)) = x(k+1) + ax(k)$$

and

$$L(x(k)) = ax(k+2) + bx(k+1) + cx(k)$$

from $\mathbf{F}(\mathbf{N})$ to $\mathbf{F}(\mathbf{N})$ are linear transformations (exercise 53). □

Example 8

Let $E(\mathbf{x}(k)) = \mathbf{x}(k+1)$ where $\mathbf{x}(k) = \begin{bmatrix} x_1(k) \\ \vdots \\ x_n(k) \end{bmatrix} \in \mathbf{F}^n(\mathbf{N})$. Then E is a linear transformation from $\mathbf{F}^n(\mathbf{N})$ to $\mathbf{F}^n(\mathbf{N})$ (exercise 54). □

* The graph of $w(k) = x(k+1)$ can be obtained by shifting the graph of $x(k)$ horizontally one unit to the left, hence the name "shift transformation."

EXAMPLE 9

CALCULUS

Let $L(f(t)) = \dfrac{df(t)}{dt}$ for all $f(t) \in \mathbf{C}_1(\mathbf{R})$. Then L is a linear transformation from $\mathbf{C}_1(\mathbf{R})$ to $\mathbf{F}(\mathbf{R})$ (exercise 55). □

EXAMPLE 10

CALCULUS

Let $L(f(t)) = \int_a^b f(t)\,dt$ for all $f(t) \in \mathbf{C}[a,b]$. Then L is a linear transformation from $\mathbf{C}[a,b]$ to $\mathbf{R}$ (exercise 56). □

It is also helpful to see an example of a transformation that is not linear.

EXAMPLE 11

Let $L(\mathbf{x}) = (\mathbf{x}, \mathbf{x})$ for all $\mathbf{x} \in \mathbf{R}_n$, where $(\bullet, \bullet)$ is the standard inner product. Then L is a transformation from $\mathbf{R}_n$ to $\mathbf{R}$. Note, however, that

$$L(\mathbf{x} + \mathbf{y}) = (\mathbf{x} + \mathbf{y}, \mathbf{x} + \mathbf{y}) = (\mathbf{x}, \mathbf{x}) + 2(\mathbf{x}, \mathbf{y}) + (\mathbf{y}, \mathbf{y})$$

whereas

$$L(\mathbf{x}) + L(\mathbf{y}) = (\mathbf{x}, \mathbf{x}) + (\mathbf{y}, \mathbf{y})$$

Hence $L(\mathbf{x} + \mathbf{y}) \neq L(\mathbf{x}) + L(\mathbf{y})$ for some $\mathbf{x}, \mathbf{y} \in \mathbf{R}_n$, such as $\mathbf{x} = [1, 0, \ldots, 0]$ and $\mathbf{y} = [1, 1, 0, \ldots, 0]$, and so L is not linear. □

Some general arithmetical results for all linear transformations follow as a consequence of the definition.

THEOREM 5.1

Let $\mathbf{V}$ and $\mathbf{W}$ be vector spaces with L a linear transformation from $\mathbf{V}$ to $\mathbf{W}$. Then

(*i*) $L(\mathbf{0}) = \mathbf{0}$

(*ii*) $L(r_1\mathbf{x}_1 + \cdots + r_m\mathbf{x}_m) = r_1L(\mathbf{x}_1) + \cdots + r_mL(\mathbf{x}_m)$ for all $\mathbf{x}_1, \ldots, \mathbf{x}_m \in \mathbf{V}$ and $r_1, \ldots, r_m \in \mathbf{R}$

PROOF

For (i), note that

$$L(\mathbf{0}) = L(\mathbf{0} + \mathbf{0}) = L(\mathbf{0}) + L(\mathbf{0})$$

Adding $-L(\mathbf{0})$ to both sides yields $\mathbf{0} = L(\mathbf{0})$.

For (ii), we can sequentially apply the definition of linear transformation to obtain

$$\begin{aligned} L(r_1\mathbf{x}_1 + \cdots + r_m\mathbf{x}_m) &= L(r_1\mathbf{x}_1) + L(r_2\mathbf{x}_2 + \cdots + r_m\mathbf{x}_m) = \cdots = L(r_1\mathbf{x}_1) + \cdots + L(r_m\mathbf{x}_m) \\ &= r_1L(\mathbf{x}_1) + \cdots + r_mL(\mathbf{x}_m) \end{aligned}$$

■

From this theorem, part (ii), we see that if the values of a linear transformation are known for vectors $\mathbf{x}_1, \ldots, \mathbf{x}_m$, then the values are known for all $\mathbf{x} \in \text{span}\{\mathbf{x}_1, \ldots, \mathbf{x}_m\}$, that is, for all linear combinations of $\mathbf{x}_1, \ldots, \mathbf{x}_m$.

EXAMPLE 12

Suppose L is a linear transformation from $\mathbf{R}_2$ to $\mathbf{R}_2$ and it is known that

$$L([1,0]) = [1,1] \qquad \text{and} \qquad L([0,1]) = [-1,1]$$

Since $\{[1,0], [0,1]\}$ is a basis for $\mathbf{R}_2$, the value of L can be computed for any $\mathbf{x} \in \mathbf{R}_2$. For example,

$$\begin{aligned} L([5,-2]) &= L(5[1,0] - 2[0,1]) = 5L([1,0]) - 2L([0,1]) \\ &= 5[1,1] - 2[-1,1] = [7,3] \end{aligned}$$

□

EXAMPLE 13

OPTIONAL

Let $L(\mathbf{x}) = \begin{bmatrix} 1 & 1 \\ 0 & 1 \end{bmatrix}\mathbf{x}$ for all $\mathbf{x} \in \mathbf{R}^2$. Then L is a linear mapping of points in $\mathbf{R}^2$ to points in $\mathbf{R}^2$. Using the theorem, we will give a geometrical view of how L maps points from $\mathbf{R}^2$ to $\mathbf{R}^2$.

First, note that $L\left(\begin{bmatrix} r \\ 0 \end{bmatrix}\right) = \begin{bmatrix} r \\ 0 \end{bmatrix}$, so L is the identity on the x-axis. Also, since $L\left(\begin{bmatrix} 0 \\ s \end{bmatrix}\right) = \begin{bmatrix} s \\ s \end{bmatrix}$, we see that L sends the points $\begin{bmatrix} 0 \\ s \end{bmatrix}$ on the y-axis horizontally to the points $\begin{bmatrix} s \\ s \end{bmatrix}$ on the line l, as shown in Figure 5.5.

Since

$$L\left(\begin{bmatrix} r \\ s \end{bmatrix}\right) = L\left(\begin{bmatrix} r \\ 0 \end{bmatrix} + \begin{bmatrix} 0 \\ s \end{bmatrix}\right) = L\left(\begin{bmatrix} r \\ 0 \end{bmatrix}\right) + L\left(\begin{bmatrix} 0 \\ s \end{bmatrix}\right) = \begin{bmatrix} r \\ 0 \end{bmatrix} + \begin{bmatrix} s \\ s \end{bmatrix}$$

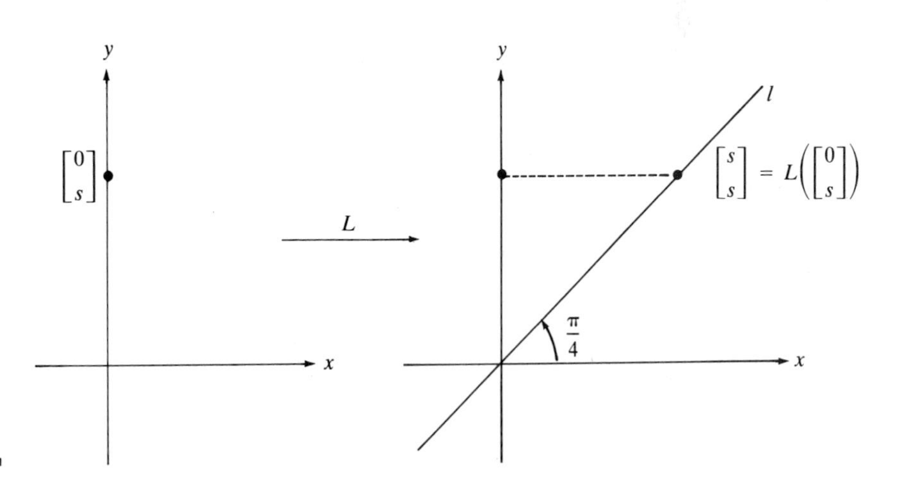

FIGURE 5.5

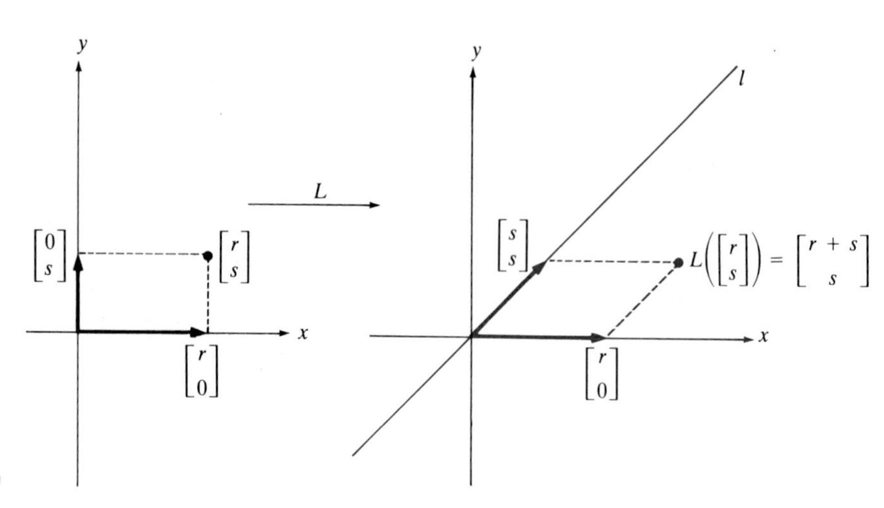

FIGURE 5.6

it follows that $L\left(\begin{bmatrix} r \\ s \end{bmatrix}\right)$ can be found from $\begin{bmatrix} r \\ s \end{bmatrix}$ geometrically by the parallelogram law applied to $\begin{bmatrix} r \\ 0 \end{bmatrix}$ and $\begin{bmatrix} s \\ s \end{bmatrix}$. See Figure 5.6. Thus, this transformation can be viewed as horizontally pushing the points on the y-axis to those on the line l, leaving the points on the x-axis fixed, with the remaining points mapped by the parallelogram law, as shown in Figure 5.7. □

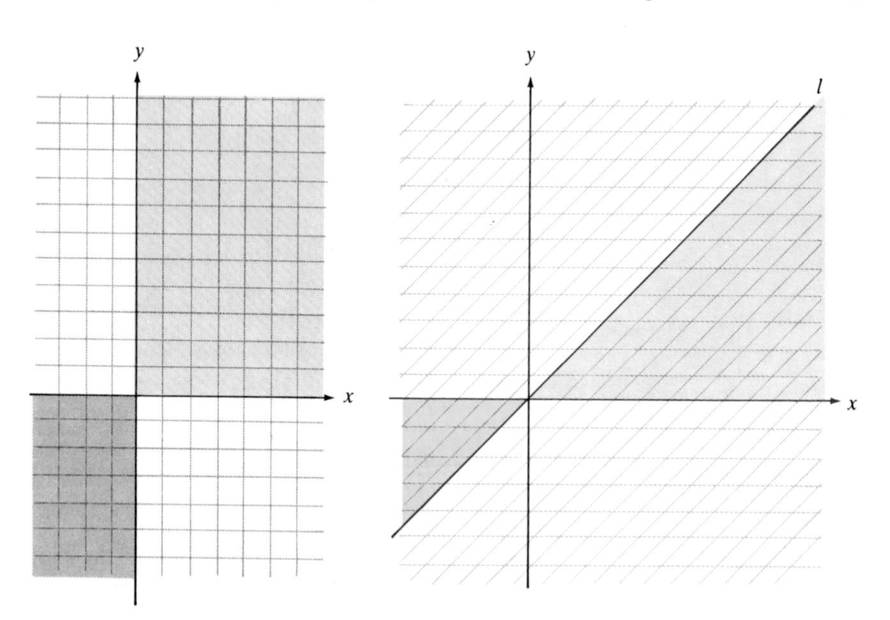

FIGURE 5.7

EXAMPLE 14

CALCULUS

Let $L(p(t)) = \dfrac{dp(t)}{dt}$ for all $p(t) \in \mathbf{P}$. We know that $\dfrac{d}{dt}t^n = nt^{n-1}$ for all $n \geqslant 0$.

Since L is a linear transformation, we can calculate the derivative of any polynomial. For example,

$$\begin{aligned} L(3t^5 - 2t^2 + t) &= 3L(t^5) - 2L(t^2) + L(t) \\ &= 3(5t^4) - 2(2t) + 1 \\ &= 15t^4 - 4t + 1 \end{aligned}$$

(Knowing that d/dt is linear means we need to know how to calculate it only on rather basic functions such as t^n, $\sin t$, and e^t. Then the derivative of any linear combination of these functions can be computed. This helps keep derivative tables small.) □

EXAMPLE 15

CALCULUS
Let $a, b \in \mathbf{R}$ and let

$$L(p(t)) = \int_a^b p(t)\,dt$$

for all $p(t) \in \mathbf{P}$. We know that

$$\int_a^b t^n\,dt = \left.\frac{t^{n+1}}{n+1}\right|_a^b = \frac{b^{n+1}}{n+1} - \frac{a^{n+1}}{n+1}$$

Since L is a linear transformation, we can calculate the integral of any polynomial. For example, if $a = 0$ and $b = 1$, then

$$\begin{aligned} L(3t^2 + 2t - 5) &= 3L(t^2) + 2L(t) - 5L(1) \\ &= 3\left(\frac{1^3}{3} - \frac{0^3}{3}\right) + 2\left(\frac{1^2}{2} - \frac{0^2}{2}\right) - 5(1 - 0) \\ &= 1 + 1 - 5 = -3 \end{aligned}$$

(Knowing that the integral is linear helps keep integral tables small.) □

Using the definition of a linear transformation and the outline given on page 228, we now give a general definition of a linear equation. Let L be a linear transformation from a vector space $\mathbf{V}$ to a vector space $\mathbf{W}$ and let $\mathbf{b} \in \mathbf{W}$. Any equation that can be written in the form

$$L(\mathbf{x}) = \mathbf{b}$$

is called a *linear equation*. Thus, to decide if an equation is linear, we need to place all constant terms on the right side of the equation and all terms involving $\mathbf{x}$ on the left side of the equation as $L(\mathbf{x})$. Then we can determine if L is linear. If $\mathbf{b} = \mathbf{0}$, we call the linear equation *homogeneous*.

EXAMPLE 16

Using the results of previous examples, we have the following linear equations.

(*i*) $A\mathbf{x} = \mathbf{b}$. Here

$$L(\mathbf{x}) = A\mathbf{x}$$

which is linear by Example 3.

(*ii*) $x(k + 1) + ax(k) = g(k)$. For this equation,

$$L(x(k)) = x(k + 1) + ax(k)$$

which is linear by Example 7. Similarly,

$$ax(k + 2) + bx(k + 1) + cx(k) = g(k)$$

is a linear equation.

(*iii*) $y'(t) + ay(t) = g(t)$. In this equation,

$$L(y(t)) = y'(t) + ay(t)$$

which is linear by exercise 57. Similarly,

$$ay''(t) + by'(t) + cy(t) = g(t)$$

is a linear equation.

(*iv*) $\int_a^b f(t)\,dt = c$. Here

$$L(f(t)) = \int_a^b f(t)\,dt$$

which is linear by Example 10. □

EXAMPLE 17

Let $A, B \in \mathbf{R}_{n,n}$. Consider the equation

$$AX = XB$$

with unknown $X \in \mathbf{R}_{n,n}$. To decide if this equation is linear, we place all terms involving X on the left side to obtain

$$AX - XB = \mathbf{0}$$

Then $L(X) = AX - XB$. To decide if L is linear, we note that if $X, Y \in \mathbf{R}_{n,n}$, then

$$L(X + Y) = A(X + Y) - (X + Y)B = AX - XB + AY - YB = L(X) + L(Y)$$

and if $r \in \mathbf{R}$, then

$$L(rX) = A(rX) - (rX)B = r(AX - XB) = rL(X)$$

Thus L is linear and $AX = XB$ is a linear equation. □

Linear equations often arise in practice (see the applied exercises). When they occur, we need to know how to find and describe their solution sets. Although methods for solving particular linear equations can vary from one type of equation to another, the vector space descriptions of the solution sets are all the same.

THEOREM 5.2

Let L be a linear transformation from a vector space $\mathbf{V}$ to a vector space $\mathbf{W}$.

(*i*) The solution set $\mathbf{K}$ of the homogeneous equation $L(\mathbf{x}) = \mathbf{0}$, called the *kernel** of L, is a subspace of $\mathbf{V}$.

(*ii*) For any $\mathbf{b} \in \mathbf{W}$, if the equation $L(\mathbf{x}) = \mathbf{b}$ has a solution $\mathbf{x}_0$, then

$$\mathbf{x}_0 + \mathbf{K} = \{\mathbf{x}_0 + \mathbf{w} : \mathbf{w} \in \mathbf{K}\}$$

is the complete solution to the equation.†

PROOF

To prove (i), applying the test for subspaces, we first show that $\mathbf{0} \in \mathbf{K}$ and so $\mathbf{K} \neq \varnothing$. For this, note from Theorem 5.1 that $L(\mathbf{0}) = \mathbf{0}$; thus $\mathbf{0} \in \mathbf{K}$. Now let $\mathbf{x}, \mathbf{y} \in \mathbf{K}$. Then $L(\mathbf{x}) = L(\mathbf{y}) = \mathbf{0}$. Thus

$$L(\mathbf{x} + \mathbf{y}) = L(\mathbf{x}) + L(\mathbf{y}) = \mathbf{0}$$

Hence $\mathbf{x} + \mathbf{y} \in \mathbf{K}$. Finally, let $\mathbf{x} \in \mathbf{K}$ and $r \in \mathbf{R}$. Then

$$L(r\mathbf{x}) = rL(\mathbf{x}) = r\mathbf{0} = \mathbf{0}$$

Hence $r\mathbf{x} \in \mathbf{K}$ and so $\mathbf{K}$ is a subspace of $\mathbf{V}$.

For the proof of (ii), we show that if T is the solution set to $L(\mathbf{x}) = \mathbf{b}$, then $T = \mathbf{x}_0 + \mathbf{K}$. To do this, let $\mathbf{x}_0 + \mathbf{w} \in \mathbf{x}_0 + \mathbf{K}$. Then

$$L(\mathbf{x}_0 + \mathbf{w}) = L(\mathbf{x}_0) + L(\mathbf{w}) = \mathbf{b} + \mathbf{0} = \mathbf{b}$$

so $\mathbf{x}_0 + \mathbf{w} \in T$. Hence $\mathbf{x}_0 + \mathbf{K} \subseteq T$. Now let $\mathbf{x} \in T$. Set $\mathbf{w} = \mathbf{x} - \mathbf{x}_0$. Then

$$L(\mathbf{w}) = L(\mathbf{x}) - L(\mathbf{x}_0) = \mathbf{b} - \mathbf{b} = \mathbf{0}$$

Thus $\mathbf{w} \in \mathbf{K}$ and $\mathbf{x} = \mathbf{x}_0 + \mathbf{w} \in \mathbf{x}_0 + \mathbf{K}$. Hence $T \subseteq \mathbf{x}_0 + \mathbf{K}$ and the result follows. ■

Using this theorem we can describe the solution sets to systems of linear algebraic equations in terms of vector spaces.

* The set $\mathbf{K}$ is also called the *null space* of L.

† There is no such theorem for nonlinear equations. Thus, nonlinear equations can be difficult to solve. See exercise 58.

EXAMPLE 18

OPTIONAL

Consider the equation

$$\begin{bmatrix} 1 & 0 & 1 \\ 1 & 1 & 0 \end{bmatrix} \mathbf{x} = \begin{bmatrix} 1 \\ 1 \end{bmatrix}$$

We solve this equation using the Gaussian elimination algorithm:

$$\begin{bmatrix} 1 & 0 & 1 & 1 \\ 1 & 1 & 0 & 1 \end{bmatrix} \xrightarrow{\mathbf{R}_2 - \mathbf{R}_1 \to \mathbf{R}_2} \begin{bmatrix} 1 & 0 & 1 & 1 \\ 0 & 1 & -1 & 0 \end{bmatrix}$$

or

$$\begin{aligned} x_1 \qquad + x_3 &= 1 \\ x_2 - x_3 &= 0 \end{aligned}$$

Thus x_3 is free and the solution is

$$\begin{bmatrix} x_1 \\ x_2 \\ x_3 \end{bmatrix} = \begin{bmatrix} 1 - x_3 \\ x_3 \\ x_3 \end{bmatrix}$$

for all choices of x_3. Separating constants and variables in the solution yields

$$\begin{bmatrix} x_1 \\ x_2 \\ x_3 \end{bmatrix} = \begin{bmatrix} 1 \\ 0 \\ 0 \end{bmatrix} + x_3 \begin{bmatrix} -1 \\ 1 \\ 1 \end{bmatrix}$$

Thus, taking $x_3 = 0$ gives

$$\mathbf{x} = \mathbf{x}_0 = \begin{bmatrix} 1 \\ 0 \\ 0 \end{bmatrix} \quad \text{and} \quad \mathbf{K} = \left\{ x_3 \begin{bmatrix} -1 \\ 1 \\ 1 \end{bmatrix} : x_3 \text{ is any number} \right\}$$

Geometrically, this solution is depicted in Figure 5.8 as a translation by $\mathbf{x}_0$ of the vector space $\mathbf{K}$. □

Theorem 5.2 also shows how we can solve any particular linear equation $L(\mathbf{x}) = \mathbf{b}$. This can be done by solving the homogeneous equation $L(\mathbf{x}) = \mathbf{0}$ to get $\mathbf{K}$, finding a solution $\mathbf{x}_0$ to $L(\mathbf{x}) = \mathbf{b}$, and then writing the solution $\mathbf{x} = \mathbf{x}_0 + \mathbf{w}$ for all $\mathbf{w} \in \mathbf{K}$. Methods for solving particular linear equations, such as linear difference equations and linear differential equations, usually consist of a method for finding $\mathbf{K}$ and a method for finding an $\mathbf{x}_0$. As described in Chapter 3, if $\mathbf{K}$ is finite dimensional, it can be found by finding finitely many vectors that form a basis for it.

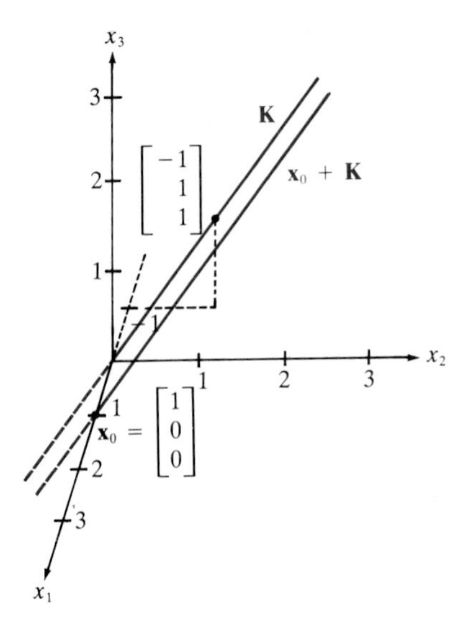

FIGURE 5.8

EXAMPLE 19

To solve the linear difference equation

$$x(k+2) - 5x(k+1) + 6x(k) = 4$$

we first solve the homogeneous equation

$$x(k+2) - 5x(k+1) + 6x(k) = 0$$

From the example of Case I of Part III of Section 3.7 we know that

$$x(k) = r_1 2^k + r_2 3^k$$

for all $r_1, r_2 \in \mathbf{R}$ is this solution.

Now we need a solution $x_0(k)$ to

$$x(k+2) - 5x(k+1) + 6x(k) = 4$$

Since $x(k+2) - 5x(k+1) + 6x(k)$ must be a constant, a reasonable choice for $x_0(k)$ is $x_0 = c$ (see exercises 45 and 46). Substituting into the equation yields $c - 5c + 6c = 4$, or $c = 2$. Hence we define $x_0(k) = 2$ for all $k \in \mathbf{N}$. Then, by Theorem 5.2, the solution set to $x(k+2) - 5x(k+1) + 6x(k) = 4$ is

$$x(k) = x_0(k) + r_1 2^k + r_2 3^k = 2 + r_1 2^k + r_2 3^k$$

for all $r_1, r_2 \in \mathbf{R}$. □

Since the solution set to a homogeneous linear equation is a vector space, it has a dimension. Knowing this dimension, of course, aids in finding the solution, since this number tells us the size of the solution set and consequently, at least if it is finite, how many solutions we need to find to determine it. However, we cannot give a useful general theorem on the dimension of the solution set of a linear equation. This number, when needed, must be calculated for the particular linear equation of interest. Instances of this sort of calculation occurred in Chapter 3, where we calculated the dimensions of the solution sets of first- and second-order linear difference equations.

EXERCISES FOR SECTION 5.1

COMPUTATIONAL EXERCISES

In exercises 1 and 2, compute $L(\mathbf{x})$ for the given transformation L and vectors $\mathbf{x}$. Show the work graphically, as in Figure 5.3, by plotting the vectors $\mathbf{x}$ and $L(\mathbf{x})$ as points in $\mathbf{R}^2$.

1. $L\left(\begin{bmatrix} x_1 \\ x_2 \end{bmatrix}\right) = \begin{bmatrix} x_1 + x_2 \\ x_1 + x_2 \end{bmatrix}$;

$\mathbf{x} = \begin{bmatrix} 0 \\ 0 \end{bmatrix}, \begin{bmatrix} 1 \\ 2 \end{bmatrix}, 2\begin{bmatrix} 1 \\ 2 \end{bmatrix}, -2\begin{bmatrix} 1 \\ 2 \end{bmatrix}, \begin{bmatrix} 2 \\ 1 \end{bmatrix}, 2\begin{bmatrix} 2 \\ 1 \end{bmatrix}, -2\begin{bmatrix} 2 \\ 1 \end{bmatrix}$

2. $L\left(\begin{bmatrix} x_1 \\ x_2 \end{bmatrix}\right) = \begin{bmatrix} 1 & 2 \\ 2 & 1 \end{bmatrix}\begin{bmatrix} x_1 \\ x_2 \end{bmatrix}$;

$\mathbf{x} = \begin{bmatrix} 0 \\ 0 \end{bmatrix}, \begin{bmatrix} 1 \\ 1 \end{bmatrix}, 2\begin{bmatrix} 1 \\ 1 \end{bmatrix}, \begin{bmatrix} 1 \\ -1 \end{bmatrix}, 3\begin{bmatrix} 1 \\ -1 \end{bmatrix}, 2\begin{bmatrix} 1 \\ 1 \end{bmatrix} + 3\begin{bmatrix} 1 \\ -1 \end{bmatrix}$

In exercises 3 through 18, decide whether or not the given transformation is a linear transformation, and prove your answer.

3. $L([x_1, x_2]) = [3x_1 + 2, x_2 - 3]$

4. $L([x_1, x_2]) = [2x_1 + x_2, 3x_2 - x_1 + 4]$

5. $L([x_1, x_2, x_3]) = [x_1 - x_2, x_2 - x_3, x_3 - x_1]$

6. $L([x_1, x_2, x_3]) = [x_1, x_1 + x_2, x_1 + x_2 + x_3]$

7. $L(\mathbf{x}) = \mathbf{x}A$, $\mathbf{x} \in \mathbf{R}_m$, $A \in \mathbf{R}_{m,n}$

8. $L(\mathbf{x}) = \mathbf{x}A\mathbf{x}^t$, $\mathbf{x} \in \mathbf{R}_n$, $A \in \mathbf{R}_{n,n}$

9. $L(\mathbf{x}) = (\mathbf{b}, \mathbf{x})$, $\mathbf{x} \in \mathbf{R}_n$, $\mathbf{b} \in \mathbf{R}_n$

10. $L(X) = X^t$, $X \in \mathbf{R}_{m,n}$

11. $L(X) = XA + A^tX$, A, $X \in \mathbf{R}_{n,n}$. (This is the Lyapunov transformation used in studies of stability.)

12. $L(X) = XA + BX$, A, B, $X \in \mathbf{R}_{n,n}$

13. $L(x(k)) = (x(k))^2$ for $x(k) \in \mathbf{F}(\mathbf{N})$

14. $L(x(k)) = x(k + 2) - 2x(k)$ for $x(k) \in \mathbf{F}(\mathbf{N})$

15. $L(y(t)) = y'(t) - y(t)$ for $y(t) \in \mathbf{C}_1(\mathbf{R})$

16. $L(y(t)) = y'(t) - [y(t)]^2$ for $y(t) \in \mathbf{C}_1(\mathbf{R})$

17. $L(y(t)) = y(t)\int_0^1 y(t)\,dt$ for $y(t) \in \mathbf{C}(\mathbf{R})$

18. $L(ax = b) = [a, b]$, $\mathbf{V} = \{ax = b : a, b \in \mathbf{R}\}$, $\mathbf{W} = \mathbf{R}_2$

In exercises 19 and 20, decide which of the given functions is a linear transformation from $\mathbf{R}$ to $\mathbf{R}$. State which computational properties are valid.

19. (*a*) $f(x) = \sqrt{x}$ (*b*) $f(x) = 4x - 6$

20. (*a*) $f(x) = 3x$ (*b*) $f(x) = \sin x$

In exercises 21 and 22, write each transformation in the form $L(\mathbf{x}) = A\mathbf{x}$, where A is a matrix, thus showing that each transformation is linear by Example 3.

21. (*a*) $L\left(\begin{bmatrix} x_1 \\ x_2 \end{bmatrix}\right) = \begin{bmatrix} x_1 + x_2 \\ x_1 - x_2 \end{bmatrix}$

(*b*) $L\left(\begin{bmatrix} x_1 \\ x_2 \\ x_3 \end{bmatrix}\right) = \begin{bmatrix} 3x_1 - x_2 + x_3 \\ 2x_1 + x_2 + 3x_3 \\ 5x_2 + 7x_3 \end{bmatrix}$

22. (*a*) $L\left(\begin{bmatrix} x_1 \\ x_2 \end{bmatrix}\right) = \begin{bmatrix} x_1 - 2x_2 \\ -2x_1 + x_2 \end{bmatrix}$

(*b*) $L\left(\begin{bmatrix} x_1 \\ x_2 \\ x_3 \end{bmatrix}\right) = \begin{bmatrix} 4x_1 + 3x_2 - x_3 \\ -2x_2 + 8x_3 \\ 3x_1 - 2x_2 - 4x_3 \end{bmatrix}$

In exercises 23 through 26, L is a linear transformation defined on span S for each given S. Find $L(\mathbf{x})$ and $L(\mathbf{y})$ for the specified vectors $\mathbf{x}$ and $\mathbf{y}$.

23. $S = \{[1, 2], [2, 1]\}$, $L([1, 2]) = [-1, 3]$, $L([2, 1]) = [1, 1]$, $\mathbf{x} = [0, 3]$, $\mathbf{y} = [1, -4]$

24. $S = \{[1, 1], [1, 0]\}$, $L([1, 1]) = [-1, 1]$, $L([1, 0]) = [1, 1]$, $\mathbf{x} = [5, 2]$, $\mathbf{y} = [2, 3]$

25. $S = \{2^k, (-3)^k\}$, $L(2^k) = 2^{k+1}$, $L((-3)^k) = (-3)^{k+1}$, $\mathbf{x} = 3(2^k) - 2(-3)^k$, $\mathbf{y} = (-1)(2^k) + 4(-3)^k$

26. $S = \{e^{3t}, e^{-t}\}$, $L(e^{3t}) = 3e^t$, $L(e^{-t}) = -e^{\ t}$, $\mathbf{x} = 2e^{3t} - e^{-t}$, $\mathbf{y} = 5e^{3t} + 2e^{-t}$

27. Let L be a linear transformation whose values, at selected vectors, are given in Table 5.1. Calculate L for the given expressions.

TABLE 5.1

x	$L(\mathbf{x})$
Ξ	Å
Γ	Λ
∇	Ψ
£	Π

(*a*) $3\Xi - 2\Gamma + 2\nabla$
(*b*) $2\Gamma - \nabla + 3£$
(*c*) $5\Xi - 3\Gamma + 2\nabla + £$

28. Assume that the linear transformation L of exercise 27 is one-to-one, and so L^{-1} is linear. Find L^{-1} of the following expressions.
(*a*) $2\text{Å} - 3\Lambda + 2\Pi$
(*b*) $4\text{Å} - \Psi + 3\Pi$
(*c*) $\text{Å} - 4\Lambda + 2\Psi - 3\Pi$
(Many calculations involving linear transformations are done by referring to tables as in exercises 27 and 28. Examples are the computations of derivatives, integrals, Laplace transforms, and Fourier transforms.)

In exercises 29 and 30, give a geometrical view of the given linear transformation, as in Example 13.

29. $L(\mathbf{x}) = \begin{bmatrix} 1 & 0 \\ 1 & 1 \end{bmatrix}\mathbf{x}$ *30.* $L(\mathbf{x}) = \begin{bmatrix} 1 & 1 \\ 1 & 1 \end{bmatrix}\mathbf{x}$

In exercises 31 through 40, decide if the given equation is linear. For the linear equations, state whether the equation is homogeneous or not. Assuming that a solution $\mathbf{x}_0 \neq \mathbf{0}$ exists when needed, state whether the solution has the form $\mathbf{K}$ or $\mathbf{x}_0 + \mathbf{K}$, where $\mathbf{K}$ is the vector space of Theorem 5.2.

31. $\begin{bmatrix} 2x_1 + x_2 - x_3 \\ x_1 - 2x_2 + 4x_3 \end{bmatrix} = \begin{bmatrix} 2 \\ 3 \end{bmatrix}$

32. $[3x_1 - x_2 + x_3, 2x_1 + x_2 + 3x_3, 5x_2 + 7x_3] = [2, 2, 2]$

33. $x(k+2) - 2x(k) = 0$

34. $x(k)x(k+2) - 2x(k) = 0$

35. $AX - XB = C$ with $A, B, C \in \mathbf{R}_{n,n}$ and $C \neq 0$

36. $\mathbf{x}A = \mathbf{c}$ with $A \in \mathbf{R}_{m,n}$, $\mathbf{c} \in \mathbf{R}_n$, and $\mathbf{c} \neq \mathbf{0}$.

37. $y''(t) = y^2(t)$

38. $y''(t) = 0$

39. $\int_{-1}^{1} f(t)\,dt = 0$

40. $\int_0^t e^{-s} f(s)\,ds = 0$

In exercises 41 through 44, solve the given linear equations. Write the solution set in the form $\mathbf{x}_0 + \mathbf{K}$ as in Example 18, and graph it.

41. $\begin{bmatrix} 1 & 1 \\ 2 & 2 \end{bmatrix}\begin{bmatrix} x \\ y \end{bmatrix} = \begin{bmatrix} 2 \\ 4 \end{bmatrix}$

42. $\begin{bmatrix} 2 & -3 \\ -4 & 6 \end{bmatrix}\begin{bmatrix} x \\ y \end{bmatrix} = \begin{bmatrix} -4 \\ 8 \end{bmatrix}$

43. $\begin{bmatrix} 1 & 1 & 1 \\ 2 & 0 & 2 \\ 1 & -1 & 1 \end{bmatrix}\begin{bmatrix} x \\ y \\ z \end{bmatrix} = \begin{bmatrix} 1 \\ 3 \\ 2 \end{bmatrix}$

44. $\begin{bmatrix} 2 & -4 & 10 \\ -4 & 8 & -20 \\ -1 & 2 & -5 \end{bmatrix}\begin{bmatrix} x \\ y \\ z \end{bmatrix} = \begin{bmatrix} -8 \\ 16 \\ 4 \end{bmatrix}$

In exercises 45 through 48, find the complete solution set of the given linear equation. Use the fact that the corresponding homogeneous equation has a solution set of dimension 2. In exercise 46, try ck for the particular solution.

45. $x(k+2) - 4x(k) = 3$

46. $x(k+2) - 3x(k+1) + 2x(k) = 4$

47. $y''(t) - 4y(t) = 3$ *48.* $y''(t) - 3y'(t) + 2y(t) = 4$

COMPLEX NUMBERS

49. Decide whether each of the following is a linear transformation. State which computational properties are valid.
(*a*) $L([z_1, z_2]) = [iz_1 + z_2, z_1 - iz_2]$
(*b*) $L([z_1, z_2]) = [(1+i)z_1 + (2-i)z_2, z_1 + i]$
(*c*) $L(z) = \bar{z}$, the conjugate of z

50. Decide whether each of the following is a system of linear equations.
(*a*) $[i \quad 1-i]\begin{bmatrix} z_1 \\ z_2 \end{bmatrix} + [z_1 \quad z_2]\begin{bmatrix} 1+i \\ -i \end{bmatrix} = 2 + 3i$
(*b*)
$$\begin{aligned} iz_1 + (2+i)z_2^2 &= 1 + i \\ (1-i)z_1 - (3-2i)z_2 &= 4 - 5i \end{aligned}$$

51. Write the solution to $\begin{bmatrix} 1 & 1+i & i \\ i & -1 & 1+i \end{bmatrix}\mathbf{z} = \begin{bmatrix} i \\ -i \end{bmatrix}$ in the form $\mathbf{z}_0 + \mathbf{K}$, as in Example 18.

THEORETICAL EXERCISES

52. Prove, making use of sketches, that the θ-rotation transformation is a linear transformation. *Hint:* To satisfy part (ii) of the definition, break the work into the cases $r < 0$, $0 \leqslant r \leqslant 1$, and $r > 1$.

53. Prove that
(*a*) $L(x(k)) = x(k+1) + ax(k)$ and
(*b*) $L(x(k)) = ax(k+2) + bx(k+1) + cx(k)$
are linear transformations from $\mathbf{F}(\mathbf{N})$ to $\mathbf{F}(\mathbf{N})$.

54. Prove that

$$E(\mathbf{x}(k)) = \begin{bmatrix} x_1(k+1) \\ \vdots \\ x_n(k+1) \end{bmatrix}$$

for all $\mathbf{x} \in \mathbf{F}^n(\mathbf{N})$ is a linear transformation.

55. (calculus) Prove that the transformation

$$L(f(t)) = \frac{df(t)}{dt}$$

from $\mathbf{C}_1(\mathbf{R})$ to $\mathbf{F}(\mathbf{R})$ is a linear transformation.

56. (calculus) Prove that the transformation

$$L(f(t)) = \int_a^b f(t)\,dt$$

from $\mathbf{C}[a, b]$ to $\mathbf{R}$ is a linear transformation.

57. (calculus) Prove that
(a) $L(y(t)) = y'(t) + ay(t)$ and
(b) $L(y(t)) = ay''(t) + by'(t) + cy(t)$
are linear transformations. Thus $y'(t) + ay(t) = g(t)$ and $ay''(t) + by'(t) + cy(t) = g(t)$ are linear equations.

58. Form a list of all
(a) numbers n for which you can solve polynomial equations of the form $a_n x^n + \cdots + a_0 = 0$ and
(b) differential equations of order 2 that you can solve.
How many nonlinear equations are there in your list?

59. Let $\mathbf{V}$ = the logarithm space of exercise 37, Section 3.2. Prove that $L(x) = \ln x$ is a linear transformation from $\mathbf{V}$ to $\mathbf{R}$.

60. Let L be the transformation that reflects each point in $\mathbf{R}^2$ in the x-axis. Show, making use of sketches, that L is linear.

61. For each $A \in \mathbf{R}_{m,n}$, define $L(A) = A^t$. Show that L is a linear transformation from $\mathbf{R}_{m,n}$ to $\mathbf{R}_{n,m}$.

62. Let $\mathbf{V}$ be an inner product space. Let $\mathbf{b} \in \mathbf{V}$. Show that $L(\mathbf{x}) = (\mathbf{b}, \mathbf{x})$ is a linear transformation from $\mathbf{V}$ to $\mathbf{R}$.

63. Let $1 \leqslant i < j \leqslant m$. For each $A \in \mathbf{R}_{m,n}$, apply the operation $\mathrm{R}_i \leftrightarrow \mathrm{R}_j$ to obtain B. Define $L(A) = B$. Show that L is linear. Also show that transformations defined from the other two Gaussian operations are linear.

64. Let L be a linear transformation from $\mathbf{R}^3$ to $\mathbf{R}^3$. Show that lines in $\mathbf{R}^3$ are mapped to lines or single points in $\mathbf{R}^3$. *Hint:* Write out the parametric expression for a line and compute L of it. (In fact, this result can be obtained for lines in any vector space. See exercise 39 of Section 3.4. Thus "linear" appropriately describes this transformation.)

65. Let $\mathbf{V}$ and $\mathbf{W}$ be vector spaces, and let L be a linear transformation from $\mathbf{V}$ to $\mathbf{W}$. The set

$$\text{range } L = \{\mathbf{y} : L(\mathbf{x}) = \mathbf{y} \text{ for some } \mathbf{x} \in \mathbf{V}\}$$

is called the *range* of L. Show that range L is a subspace of $\mathbf{W}$.

66. Show the subspaces kernel L and range L geometrically for the given L by finding and sketching a basis for each.

(a) $L(\mathbf{x}) = \begin{bmatrix} 1 & 1 \\ 1 & 1 \end{bmatrix}\begin{bmatrix} x_1 \\ x_2 \end{bmatrix}$ (b) $L(\mathbf{x}) = \begin{bmatrix} 1 & 0 \\ 1 & 1 \\ 0 & 1 \end{bmatrix}\begin{bmatrix} x_1 \\ x_2 \end{bmatrix}$

67. Let A be an $m \times n$ matrix. Then $L(\mathbf{x}) = A\mathbf{x}$ is a linear transformation from $\mathbf{R}^n$ to $\mathbf{R}^m$. Prove, as demonstrated in exercise 66, that

$$\dim(\text{range } L) + \dim(\text{kernel } L) = \dim \mathbf{R}^n$$

(This result can be proved for any finite-dimensional vector space.)

APPLICATIONS EXERCISES

68. A population triples its size each year. Write the difference equation for this phenomenon. Is the equation linear?

69. (calculus) Newton's law states that, for a falling object, $ma = -32$, where m = mass of the object and a = acceleration of the object as it falls. Letting $v(t)$ = velocity of the object and $x(t)$ = height of the object above the earth, we have by substitution that

$$mv'(t) = -32 \qquad \text{and} \qquad mx''(t) = -32$$

Are these differential equations linear?

70. (calculus) Not every problem, when formulated mathematically, leads to a linear equation. However, for many such problems, an approximating linear equation can be used. For example, if a pendulum of length l during its motion makes an angle of $\theta(t)$ with a vertical line at time t, as shown in Figure 5.9, then $\theta(t)$ satisfies

$$\frac{d^2\theta(t)}{dt^2} + \frac{g}{l}\sin\theta(t) = 0$$

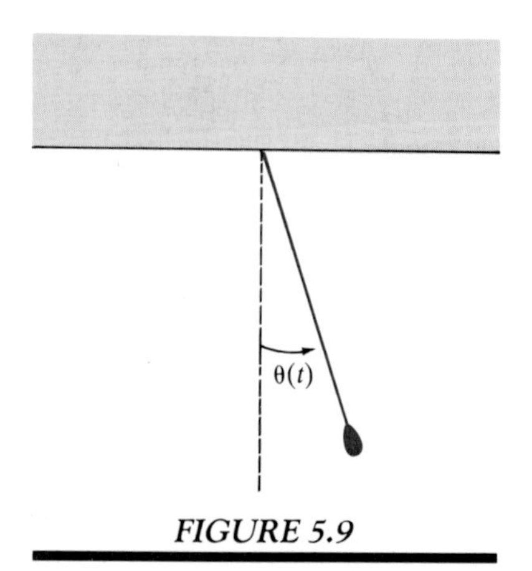

FIGURE 5.9

where g is a constant. Explain why this equation is not linear. For this problem, however, we note that $\sin \theta(t) \approx \theta(t)$ when $\theta(t)$ is small and use the approximating equation

$$\frac{d^2\theta(t)}{dt^2} + \frac{g}{l}\theta(t) = 0$$

Show that this equation is linear.

COMPUTER EXERCISE

71. A program is written that computes the square of any 3×3 matrix. Viewed as a function, is this program a linear transformation?

5.2 COORDINATE SYSTEMS AND COORDINATE SPACES

In this section, by introducing the concept of a coordinate system, we show how to associate a grid to any given finite-dimensional vector space $\mathbf{V}$.

EXAMPLE 20

Geometrically for $\mathbf{R}^2$, a grid can be obtained from two axes by sketching lines parallel to these axes, as shown in Figure 5.10. The appearance of the grid is that of graph paper. □

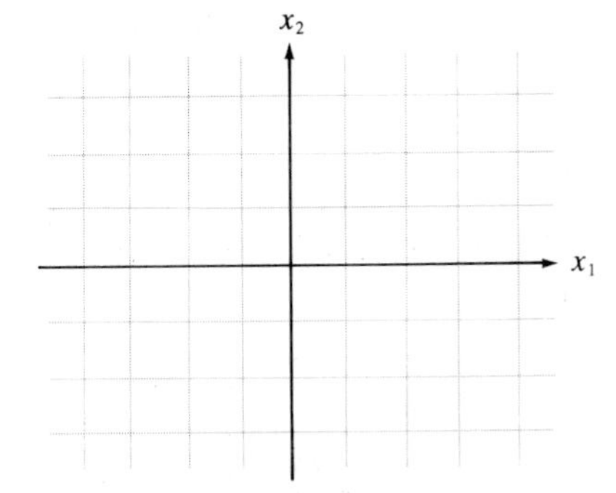

FIGURE 5.10

We show how this grid can be used to provide coordinates for each vector $\mathbf{x} \in \mathbf{V}$. Then, in Section 5.3, we show how to quantify linear transformations, and thus linear equations, defined on $\mathbf{V}$ by writing them in terms of these coordinates. As the introduction to this chapter shows, a judicious choice of coordinate system can lead to simple expressions for linear transformations and equations. Finding such a coordinate system can make a given linear transformation easier to understand or a given linear equation easier to solve.

To see how a coordinate system can be obtained for any finite-dimensional vector space, we first look at $\mathbf{R}^2$. A coordinate system for $\mathbf{R}^2$ is viewed as a grid determined from two axes labeled x_1, x_2 whose inclinations, orientations, and unit lengths are determined by $\mathbf{e}^1, \mathbf{e}^2$, as shown in Figure 5.11. Writing any $\mathbf{x} \in \mathbf{R}^2$ as a linear combination $\mathbf{x} = x_1\mathbf{e}^1 + x_2\mathbf{e}^2$, we can see by the parallelogram rule that the coefficients of $\mathbf{x}$ are exactly the usual coordinates $\begin{bmatrix} x_1 \\ x_2 \end{bmatrix}$ of $\mathbf{x}$.

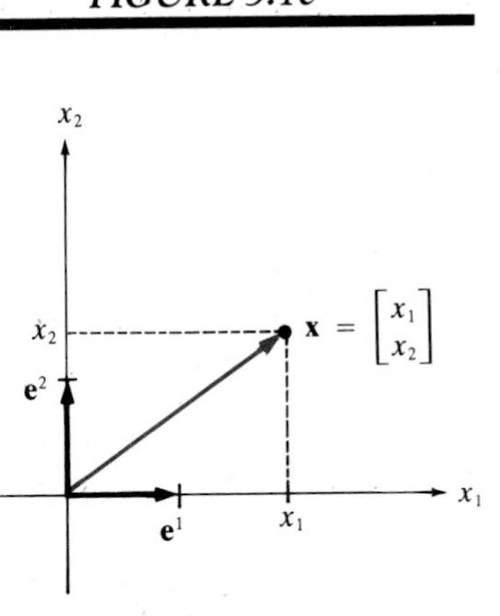

FIGURE 5.11

This technique of assigning coordinates can be done with any basis. For example, using $\mathbf{f}^1 = \begin{bmatrix} 1 \\ 1 \end{bmatrix}$ and $\mathbf{f}^2 = \begin{bmatrix} -1 \\ 1 \end{bmatrix}$, we can form a different grid for the plane by letting $\mathbf{f}^1$ and $\mathbf{f}^2$ determine inclinations, orientations, and unit lengths for the axes labeled y_1 and y_2, respectively, as shown in Figure 5.12. Further, if $\mathbf{x} \in \mathbf{R}^2$ with $\mathbf{x} = r_1\mathbf{f}^1 + r_2\mathbf{f}^2$, we see by the parallelogram rule that $\begin{bmatrix} r_1 \\ r_2 \end{bmatrix}$ gives the coordinates of the point $\mathbf{x}$ in terms of the y_1, y_2-axes. Thus, the coordinates of a point in this system are exactly the coefficients used in writing $\mathbf{x}$ as a linear combination of $\mathbf{f}^1$ and $\mathbf{f}^2$.

To define a coordinate system for any finite-dimensional vector space, we will generalize the above.

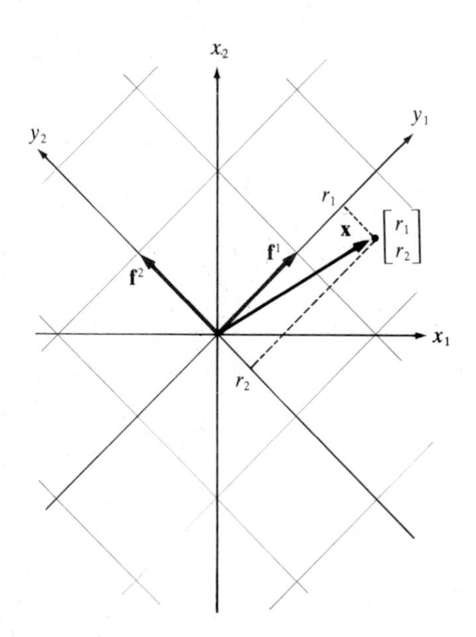

FIGURE 5.12

DEFINITION

Let **V** be a finite-dimensional vector space. Let $S = \{\mathbf{x}_1, \ldots, \mathbf{x}_n\}$ be a basis for **V**. The ordered set $\# = (\mathbf{x}_1, \ldots, \mathbf{x}_n)$ is called a *coordinate system* for **V**.

Geometrically, you can imagine that each of the vectors $\mathbf{x}_1, \ldots, \mathbf{x}_n$ determines an axis, including orientation and unit length, for the coordinate system. Axes can be numbered so that $|\mathbf{x}_i|$ is the unit length on axis x_i. These axes can be used to "grid" the space* **V**.

Write $\mathbf{x} = r_1\mathbf{x}_1 + \cdots + r_n\mathbf{x}_n$ for some $r_1, \ldots, r_n \in \mathbf{R}$. We will now show that the sequence $r_1, \ldots, r_n$ is completely determined by the vector **x**.

THEOREM 5.3

Let **V** be a finite-dimensional vector space and let $\# = (\mathbf{x}_1, \ldots, \mathbf{x}_n)$ be a coordinate system for **V**. For each $\mathbf{x} \in \mathbf{V}$, there is exactly one sequence $r_1, \ldots, r_n$ of coefficients of $\mathbf{x}_1, \ldots, \mathbf{x}_n$, respectively, for which $\mathbf{x} = r_1\mathbf{x}_1 + \cdots + r_n\mathbf{x}_n$.

PROOF

Suppose $\mathbf{x} = r_1\mathbf{x}_1 + \cdots + r_n\mathbf{x}_n$ and $\mathbf{x} = s_1\mathbf{x}_1 + \cdots + s_n\mathbf{x}_n$. Subtraction yields

$$\mathbf{0} = \mathbf{x} - \mathbf{x} = (r_1 - s_1)\mathbf{x}_1 + \cdots + (r_n - s_n)\mathbf{x}_n$$

Since $\{\mathbf{x}_1, \ldots, \mathbf{x}_n\}$ is linearly independent,

$$r_1 - s_1 = \cdots = r_n - s_n = 0$$

Thus $r_1 = s_1, \ldots, r_n = s_n$. ■

* We chose the notation # since it symbolizes a grid. We use the word "grid" when reading the symbol #. A coordinate system is sometimes called an *ordered basis* or *reference frame*.

Using this theorem, we can show how to assign coordinates to any vector in a finite-dimensional vector space.

DEFINITION

Let $\mathbf{V}$ be a finite-dimensional vector space with $\# = (\mathbf{x}_1, \ldots, \mathbf{x}_n)$ a coordinate system for $\mathbf{V}$. For any $\mathbf{x} \in \mathbf{V}$, write

$$\mathbf{x} = r_1\mathbf{x}_1 + \cdots + r_n\mathbf{x}_n$$

The numbers in the sequence $r_1, \ldots, r_n$ of coefficients are called the *coordinates of* $\mathbf{x}$ *with respect to* $\#$. As with $\mathbf{R}^3$, these numbers indicate how $\mathbf{x}$ is found using the coordinate system $\#$. Notationally, we will write*

$$[\mathbf{x}]_\# = \begin{bmatrix} r_1 \\ \vdots \\ r_n \end{bmatrix}$$

as the *coordinate vector of* $\mathbf{x}$. The set of coordinate vectors attached to points in $\mathbf{V}$ by $\#$ is called the *coordinate space of* $\mathbf{V}$ *with respect to* $\#$, written $\mathbf{V}_\#$.† In the case of $\mathbf{V} = \mathbf{R}^n$ and $\# = (\mathbf{e}^1, \ldots, \mathbf{e}^n)$, since the vectors and the attached coordinate vectors are identical, we often simply write $\mathbf{R}^n$ for $(\mathbf{R}^n)_\#$ and $\mathbf{x}$ for $[\mathbf{x}]_\#$.**

EXAMPLE 21

For $\mathbf{R}^2$, let $\# = (\mathbf{e}^1, \mathbf{e}^2)$. Since

$$\begin{bmatrix} 2 \\ 1 \end{bmatrix} = 2\mathbf{e}^1 + 1\mathbf{e}^2$$

it follows that

$$\left[\begin{bmatrix} 2 \\ 1 \end{bmatrix}\right]_\# = \begin{bmatrix} 2 \\ 1 \end{bmatrix}$$

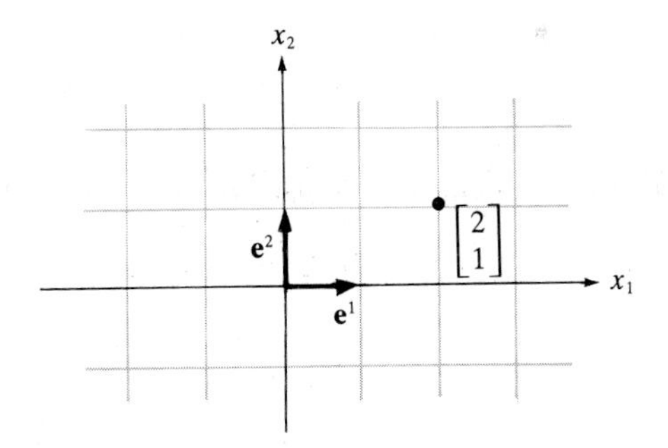

FIGURE 5.13

* The symbol $[\mathbf{x}]_\#$ is read "**x**-grid" or "the coordinates of **x** in grid."
† The symbol $\mathbf{V}_\#$ is read "**V** grid" or "the grid for **V**."
** As was done previously, the coordinate space $\mathbf{R}^3$ will always be depicted by drawing an x_1,x_2,x_3-coordinate system as described in Section 3.1. For $\mathbf{R}^2$, we draw an x_1,x_2-coordinate system.

as illustrated in Figure 5.13. Note that for this coordinate system, the vector and the coordinates of the vector are identical. □

EXAMPLE 22

For $\mathbf{R}^2$, let $\mathbf{f}^1 = \begin{bmatrix} 1 \\ 1 \end{bmatrix}$, $\mathbf{f}^2 = \begin{bmatrix} -1 \\ 1 \end{bmatrix}$, and $\# = (\mathbf{f}^1, \mathbf{f}^2)$. This coordinate system with axes determined by $\mathbf{f}^1, \mathbf{f}^2$ and labeled y_1, y_2, respectively, grids the space as depicted in Figure 5.14.

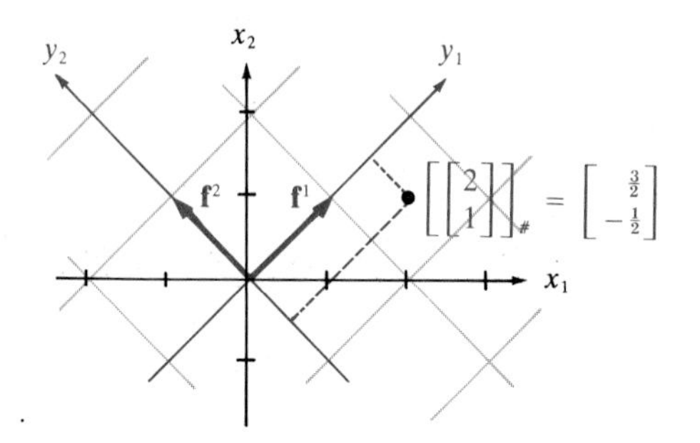

FIGURE 5.14

Let $\mathbf{x} = \begin{bmatrix} 2 \\ 1 \end{bmatrix}$. Then solving

$$\begin{bmatrix} 2 \\ 1 \end{bmatrix} = r_1 \mathbf{f}^1 + r_2 \mathbf{f}^2$$

yields $r_1 = \frac{3}{2}$ and $r_2 = -\frac{1}{2}$. Thus

$$[\mathbf{x}]_\# = \begin{bmatrix} \frac{3}{2} \\ -\frac{1}{2} \end{bmatrix}$$

is the coordinate vector of $\mathbf{x}$ in the coordinate space $(\mathbf{R}^2)_\#$. Note that the coordinates show how to find the point $\mathbf{x}$ in the plane by using the grid. □

Some coordinate systems for vector spaces that cannot be depicted with sketches follow.

EXAMPLE 23

For $\mathbf{R}_4$,

$$\# = ([1,1,0,0],[0,1,1,0],[0,0,1,1],[1,0,0,1])$$

provides a coordinate system. Let $\mathbf{x} = [3,0,-1,2]$. Since

$$\mathbf{x} = 1[1,1,0,0] - 1[0,1,1,0] + 0[0,0,1,1] + 2[1,0,0,1]$$

it follows that

$$[\mathbf{x}]_{\#} = \begin{bmatrix} 1 \\ -1 \\ 0 \\ 2 \end{bmatrix}$$

Note that this coordinate vector specifies how to find $\mathbf{x}$ in $\mathbf{R}_4$ by using $\#$. □

EXAMPLE 24

For $\mathbf{P}_2$,

$$\# = (t + 1, -t + 1, t^2 + t)$$

provides a coordinate system. Let $p(t) \in \mathbf{P}_2$ and suppose $[p(t)]_{\#} = \begin{bmatrix} 1 \\ 2 \\ 1 \end{bmatrix}$. Then $p(t)$ can be found from these coordinates by setting

$$p(t) = 1(t + 1) + 2(-t + 1) + 1(t^2 + t) = t^2 + 3$$

Note that the coordinates show how to find $p(t) = t^2 + 3$ by using $\#$. □

With entrywise $+$ and $\cdot$, each coordinate space is a vector space.* As the following lemma shows, calculations involving vectors in $\mathbf{V}$, which we use later, agree with those involving the associated coordinate vectors in $\mathbf{V}_{\#}$. (See exercise 34.)

LEMMA 5.4

Let $\mathbf{V}$ be a finite-dimensional vector space, and let $\# = (\mathbf{x}_1, \ldots, \mathbf{x}_n)$ be a coordinate system for $\mathbf{V}$. Then, for all $r \in \mathbf{R}$ and $\mathbf{x}, \mathbf{y} \in \mathbf{V}$,

(*i*) $[\mathbf{x} + \mathbf{y}]_{\#} = [\mathbf{x}]_{\#} + [\mathbf{y}]_{\#}$ and

(*ii*) $[r\mathbf{x}]_{\#} = r[\mathbf{x}]_{\#}$

Consequently,

(*iii*) $[r_1\mathbf{y}_1 + \cdots + r_m\mathbf{y}_m]_{\#} = r_1[\mathbf{y}_1]_{\#} + \cdots + r_m[\mathbf{y}_m]_{\#}$ for all $\mathbf{y}_1, \ldots, \mathbf{y}_m \in \mathbf{V}$ and all $r_1, \ldots, r_m \in \mathbf{R}$

PROOF

We will prove (i), leaving (ii) and (iii) as exercise 31. For (i), let

$$\mathbf{x} = r_1\mathbf{x}_1 + \cdots + r_n\mathbf{x}_n \qquad \text{and} \qquad \mathbf{y} = s_1\mathbf{x}_1 + \cdots + s_n\mathbf{x}_n$$

Then

$$\mathbf{x} + \mathbf{y} = (r_1 + s_1)\mathbf{x}_1 + \cdots + (r_n + s_n)\mathbf{x}_n$$

* The only difference between $\mathbf{V}_{\#}$ and the vector space $\mathbf{R}^n$ is that each coordinate vector $[\mathbf{x}]_{\#} \in \mathbf{V}_{\#}$ is attached to a vector $\mathbf{x} \in \mathbf{V}$.

so

$$[\mathbf{x} + \mathbf{y}]_\# = \begin{bmatrix} r_1 + s_1 \\ \vdots \\ r_n + s_n \end{bmatrix} = \begin{bmatrix} r_1 \\ \vdots \\ r_n \end{bmatrix} + \begin{bmatrix} s_1 \\ \vdots \\ s_n \end{bmatrix} = [\mathbf{x}]_\# + [\mathbf{y}]_\#$$

■

From the previous examples, we see that, if $\#_1$ and $\#_2$ are coordinate systems for $\mathbf{V}$, then the coordinates of a vector in $\#_1$ can differ from those in $\#_2$. See Figure 5.14. If we desire, however, we can convert coordinates of a vector with respect to $\#_2$ to those with respect to $\#_1$. A straightforward calculation, given below, shows that this conversion can be done by a single matrix, called a *transition matrix from* $\mathbf{V}_{\#_2}$ *to* $\mathbf{V}_{\#_1}$, defined in the following theorem.

THEOREM 5.5

(finding the transition matrix from $\mathbf{V}_{\#_2}$ to $\mathbf{V}_{\#_1}$) Let $\mathbf{V}$ be a finite-dimensional vector space having

$$\#_1 = (\mathbf{x}_1, \ldots, \mathbf{x}_n) \qquad \text{and} \qquad \#_2 = (\mathbf{y}_1, \ldots, \mathbf{y}_n)$$

as coordinate systems. Define the $n \times n$ matrix P by

$$P = [[\mathbf{y}_1]_{\#_1} \quad \cdots \quad [\mathbf{y}_n]_{\#_1}]$$

Then P is invertible and $P[\mathbf{x}]_{\#_2} = [\mathbf{x}]_{\#_1}$ for all $\mathbf{x} \in \mathbf{V}$.

Thus, if $\mathbf{V} = \mathbf{R}^n$ and $\#_1 = (\mathbf{e}^1, \ldots, \mathbf{e}^n)$, then $P = [\mathbf{y}_1 \quad \cdots \quad \mathbf{y}_n]$.*

PROOF

Write

$$\mathbf{x} = r_1\mathbf{x}_1 + \cdots + r_n\mathbf{x}_n = s_1\mathbf{y}_1 + \cdots + s_n\mathbf{y}_n$$

Then, by Lemma 5.4,

$$[\mathbf{x}]_{\#_1} = s_1[\mathbf{y}_1]_{\#_1} + \cdots + s_n[\mathbf{y}_n]_{\#_1} = P[\mathbf{x}]_{\#_2}$$

Now, to show that P is invertible, consider

$$\begin{aligned} P[\mathbf{x}_1]_{\#_2} &= [\mathbf{x}_1]_{\#_1} = \mathbf{e}^1 \\ P[\mathbf{x}_2]_{\#_2} &= [\mathbf{x}_2]_{\#_1} = \mathbf{e}^2 \\ &\vdots \\ P[\mathbf{x}_n]_{\#_2} &= [\mathbf{x}_n]_{\#_1} = \mathbf{e}^n \end{aligned}$$

Define a matrix

$$Q = [[\mathbf{x}_1]_{\#_2} \quad [\mathbf{x}_2]_{\#_2} \quad \cdots \quad [\mathbf{x}_n]_{\#_2}]$$

Then, computing columnwise, we have $PQ = I_n$, and so P is invertible. ■

* This case can be recalled by noting that $P[\mathbf{y}_i]_{\#_2} = [\mathbf{y}_i]_{\#_1}$, and, since $[\mathbf{y}_i]_{\#_2} = \mathbf{e}^i$ and $[\mathbf{y}_i]_{\#_1} = \mathbf{y}_i$, we have $\mathbf{p}^i = \mathbf{y}_i$.

EXAMPLE 25

For $\mathbf{R}^2$, let

$$\#_1 = \left(\begin{bmatrix}1\\0\end{bmatrix}, \begin{bmatrix}0\\1\end{bmatrix}\right) \quad \text{and} \quad \#_2 = \left(\begin{bmatrix}1\\1\end{bmatrix}, \begin{bmatrix}-1\\1\end{bmatrix}\right)$$

Since $(\mathbf{R}^2)_{\#_1} = \mathbf{R}^2$, the transition matrix from $(\mathbf{R}^2)_{\#_2}$ to $\mathbf{R}^2$ is $P = \begin{bmatrix}1 & -1\\1 & 1\end{bmatrix}$.

Now, to demonstrate the theorem, we use the result of Example 22 that $\left[\begin{bmatrix}2\\1\end{bmatrix}\right]_{\#_2} = \begin{bmatrix}\frac{3}{2}\\-\frac{1}{2}\end{bmatrix}$. And, of course, $\left[\begin{bmatrix}2\\1\end{bmatrix}\right]_{\#_1} = \begin{bmatrix}2\\1\end{bmatrix}$. Calculating, we have

$$P\left[\begin{bmatrix}2\\1\end{bmatrix}\right]_{\#_2} = \begin{bmatrix}1 & -1\\1 & 1\end{bmatrix}\begin{bmatrix}\frac{3}{2}\\-\frac{1}{2}\end{bmatrix} = \begin{bmatrix}2\\1\end{bmatrix} = \left[\begin{bmatrix}2\\1\end{bmatrix}\right]_{\#_1}$$

Note, as in Figure 5.15, that only the coordinates of the point change; the point itself is fixed. As described in the theorem, the transition matrix links the different coordinate vectors associated to the fixed point. □

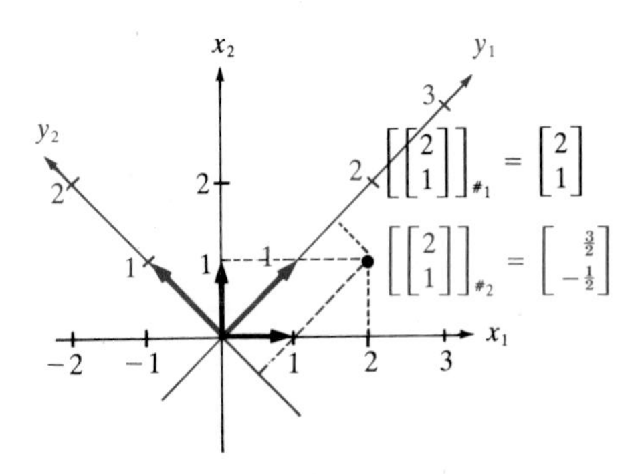

FIGURE 5.15

The next example is different from the previous one in that neither $\#_1$ nor $\#_2$ is the usual coordinate system for $\mathbf{R}_2$.

EXAMPLE 26

OPTIONAL

For $\mathbf{R}_2$, let

$$\#_1 = ([1,1],[-1,1]) \quad \text{and} \quad \#_2 = ([3,1],[1,3])$$

To find the transition matrix from $(\mathbf{R}_2)_{\#_2}$ to $(\mathbf{R}_2)_{\#_1}$ we need to find

$$P = [[[3,1]]_{\#_1} \quad [[1,3]]_{\#_1}]$$

Set

$$[3,1] = r_1[1,1] + r_2[-1,1]$$
$$[1,3] = s_1[1,1] + s_2[-1,1]$$

Solving these equations yields

$$[[3,1]]_{\#_1} = \begin{bmatrix} 2 \\ -1 \end{bmatrix} \quad \text{and} \quad [[1,3]]_{\#_1} = \begin{bmatrix} 2 \\ 1 \end{bmatrix}$$

Hence $P = \begin{bmatrix} 2 & 2 \\ -1 & 1 \end{bmatrix}$.

Now, adapting the result of Example 22 to $\mathbf{R}_2$, we know that $[[2,1]]_{\#_1} = \begin{bmatrix} \frac{3}{2} \\ -\frac{1}{2} \end{bmatrix}$. Further, $[[2,1]]_{\#_2} = \begin{bmatrix} \frac{5}{8} \\ \frac{1}{8} \end{bmatrix}$. Demonstrating the property of P given in Theorem 5.5 yields

$$P[[2,1]]_{\#_2} = \begin{bmatrix} 2 & 2 \\ -1 & 1 \end{bmatrix} \begin{bmatrix} \frac{5}{8} \\ \frac{1}{8} \end{bmatrix} = \begin{bmatrix} \frac{3}{2} \\ -\frac{1}{2} \end{bmatrix} = [[2,1]]_{\#_1}$$

The point $[2, 1]$ is shown in both coordinate spaces in Figure 5.16.

Again note that only the coordinates of the point are different, and P links these different coordinate vectors. □

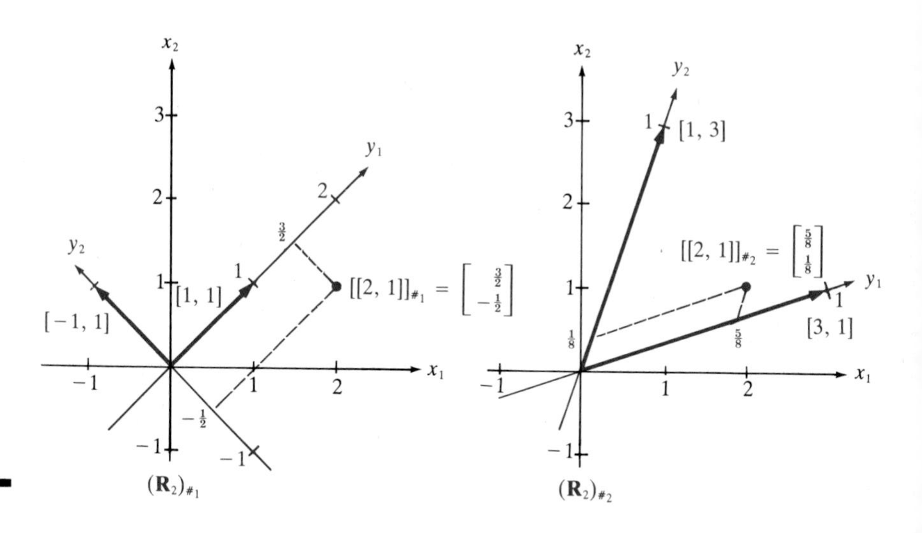

FIGURE 5.16

EXAMPLE 27

For $\mathbf{V} = P_2$, let

$$\#_1 = (1, t, t^2) \quad \text{and} \quad \#_2 = (t+1, -t+1, t^2+t)$$

To determine the transition matrix from $\mathbf{V}_{\#_2}$ to $\mathbf{V}_{\#_1}$, we need to find

$$P = [[t+1]_{\#_1} \quad [-t+1]_{\#_1} \quad [t^2+t]_{\#_1}]$$

Write

$$\begin{aligned} t + 1 &= q_1(t^2) + q_2(t) + q_3(1) \\ -t + 1 &= r_1(t^2) + r_2(t) + r_3(1) \\ t^2 + t &= s_1(t^2) + s_2(t) + s_3(1) \end{aligned}$$

Solving these equations yields the transition matrix

$$P = [[t + 1]_{\#_1} \quad [-t + 1]_{\#_1} \quad [t^2 + t]_{\#_1}] = \begin{bmatrix} 0 & 0 & 1 \\ 1 & -1 & 1 \\ 1 & 1 & 0 \end{bmatrix}$$

By Example 24, $[t^2 + 3]_{\#_2} = \begin{bmatrix} 1 \\ 2 \\ 1 \end{bmatrix}$, and it is clear that $[t^2 + 3]_{\#_1} = \begin{bmatrix} 1 \\ 0 \\ 3 \end{bmatrix}$.

Demonstrating the property of P, we have

$$P[t^2 + 3]_{\#_2} = \begin{bmatrix} 0 & 0 & 1 \\ 1 & -1 & 1 \\ 1 & 1 & 0 \end{bmatrix} \begin{bmatrix} 1 \\ 2 \\ 1 \end{bmatrix} = \begin{bmatrix} 1 \\ 0 \\ 3 \end{bmatrix} = [t^2 + 3]_{\#_1}$$

□

We saw above how the transition matrix changes coordinates of a single vector. The next example shows how the transition matrix can be used to change the coordinates of a set of vectors.

EXAMPLE 28

OPTIONAL

For $\mathbf{R}^2$, let

$$\#_1 = \left(\begin{bmatrix} 1 \\ 0 \end{bmatrix}, \begin{bmatrix} 0 \\ 1 \end{bmatrix} \right) \quad \text{and} \quad \#_2 = \left(\begin{bmatrix} \frac{1}{\sqrt{2}} \\ \frac{1}{\sqrt{2}} \end{bmatrix}, \begin{bmatrix} -\frac{1}{\sqrt{2}} \\ \frac{1}{\sqrt{2}} \end{bmatrix} \right)$$

Consider the set S of points in $\mathbf{R}^2$ whose coordinates satisfy $3x_1^2 - 2x_1x_2 + 3x_2^2 = 8$. We will find the equation satisfied by the coordinates in $(\mathbf{R}^2)_{\#_2}$ of these same points. The transition matrix

$$P = \begin{bmatrix} \frac{1}{\sqrt{2}} & -\frac{1}{\sqrt{2}} \\ \frac{1}{\sqrt{2}} & \frac{1}{\sqrt{2}} \end{bmatrix}$$

changes coordinates in $(\mathbf{R}^2)_{\#_2}$ into coordinates in $\mathbf{R}^2$. Let $\mathbf{x} = \begin{bmatrix} x_1 \\ x_2 \end{bmatrix}$ be a point satisfying the equation. For simplicity, define $\begin{bmatrix} y_1 \\ y_2 \end{bmatrix} = [\mathbf{x}]_{\#_2}$. Then, by Theorem 5.5, $P\begin{bmatrix} y_1 \\ y_2 \end{bmatrix} = \begin{bmatrix} x_1 \\ x_2 \end{bmatrix}$, and it follows that

$$\begin{aligned} \frac{1}{\sqrt{2}} y_1 - \frac{1}{\sqrt{2}} y_2 &= x_1 \\ \frac{1}{\sqrt{2}} y_1 + \frac{1}{\sqrt{2}} y_2 &= x_2 \end{aligned}$$

Substituting these expressions into the equation yields

$$\tfrac{3}{2}(y_1 - y_2)^2 - (y_1 - y_2)(y_1 + y_2) + \tfrac{3}{2}(y_1 + y_2)^2 = 8$$

or

$$y_1^2 + 2y_2^2 = 4$$

This equation describes the points in S using $\#_2$ as the coordinate system. The graphs of both equations, in their respective coordinate systems, are shown in Figure 5.17. Note that, although the coordinate descriptions are different, the graphs, being the sets of points described, are identical. □

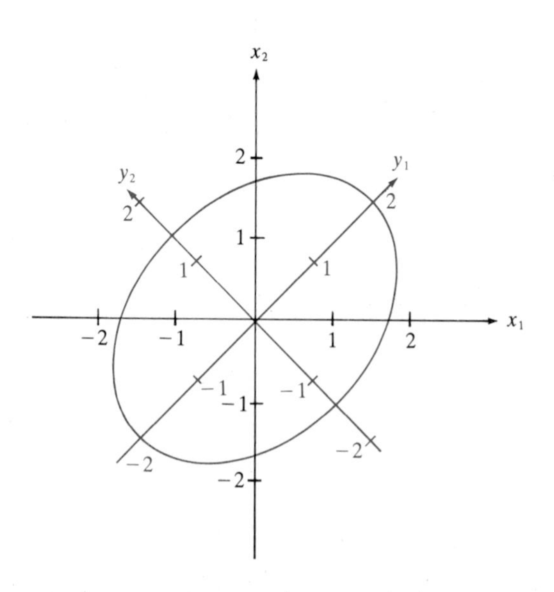

FIGURE 5.17

This completes the development of coordinate systems for finite-dimensional vector spaces. In the next section we show how to quantify linear transformations by describing them in terms of the coordinate vectors in a coordinate space.

EXERCISES FOR SECTION 5.2

COMPUTATIONAL EXERCISES

In exercises 1 through 4, graph the given coordinate system #, being sure to label axes to indicate their orientation. For the given vector $\mathbf{x}$, find the coordinate vector $[\mathbf{x}]_{\#}$. Locate and plot $\mathbf{x}$ by using the coordinates and the coordinate system #.

1. $\# = \left(\begin{bmatrix}1\\1\end{bmatrix}, \begin{bmatrix}-1\\1\end{bmatrix}\right), \mathbf{x} = \begin{bmatrix}2\\4\end{bmatrix}$

2. $\# = \left(\begin{bmatrix}1\\1\end{bmatrix}, \begin{bmatrix}1\\-1\end{bmatrix}\right), \mathbf{x} = \begin{bmatrix}5\\1\end{bmatrix}$

3. $\# = ([-1,-1],[-1,1]), \mathbf{x} = [3,-1]$

4. $\# = ([3,1],[1,2]), \mathbf{x} = [0,5]$

In exercises 5 through 10, find $[\mathbf{x}]_{\#}$.

5. $\mathbf{V} = \mathbf{R}_3$, $\# = ([1,1,0],[0,1,1],[1,0,1]), \mathbf{x} = [2,2,2]$

6. $\mathbf{V} = \mathbf{R}_3$, $\# = ([-1,1,0],[0,-1,0],[-1,0,1])$, $\mathbf{x} = [0,1,-2]$

7. $\mathbf{V} = \mathbf{P}_1$, $\# = (2t+1, 3t-1), \mathbf{x} = 3t+4$

8. $\mathbf{V} = \mathbf{P}_1$, $\# = (-t+1, t+1), \mathbf{x} = -t+3$

9. $\mathbf{V} = \mathbf{P}_2$, $\# = (t, -t+1, t^2), \mathbf{x} = 3t^2+2$

10. $\mathbf{V} = \mathbf{P}_2$, $\# = (3, 2t, 4t^2), \mathbf{x} = t^2+t+1$

In exercises 11 through 18, find $\mathbf{x}$ for the given coordinate system # and the given $[\mathbf{x}]_{\#}$. For exercises 11 through 14, plot $\mathbf{x}$ directly and plot $\mathbf{x}$ using the coordinates $[\mathbf{x}]_{\#}$. Note that $[\mathbf{x}]_{\#}$ locates $\mathbf{x}$ using #.

11. $\# = \left(\begin{bmatrix}1\\-1\end{bmatrix}, \begin{bmatrix}-2\\1\end{bmatrix}\right), [\mathbf{x}]_{\#} = \begin{bmatrix}1\\1\end{bmatrix}$

12. $\# = \left(\begin{bmatrix}1\\1\end{bmatrix}, \begin{bmatrix}1\\-1\end{bmatrix}\right), [\mathbf{x}]_{\#} = \begin{bmatrix}1\\3\end{bmatrix}$

13. $\# = ([0,1],[1,0]), [\mathbf{x}]_{\#} = \begin{bmatrix}1\\2\end{bmatrix}$

14. $\# = ([1,-1],[-2,1]), [\mathbf{x}]_{\#} = \begin{bmatrix}-2\\3\end{bmatrix}$

15. $\# = ([1,-1,2],[3,1,0],[2,1,1]), [\mathbf{x}]_{\#} = \begin{bmatrix}1\\-1\\2\end{bmatrix}$

16. $\# = ([1,2,3],[1,2,2],[1,1,2]), [\mathbf{x}]_{\#} = \begin{bmatrix}2\\1\\-3\end{bmatrix}$

17. $\# = (t+2, t-1), [\mathbf{x}]_{\#} = \begin{bmatrix}2\\-1\end{bmatrix}$

18. $\# = (2t+3, 3t-1), [\mathbf{x}]_{\#} = \begin{bmatrix}-1\\2\end{bmatrix}$

In exercises 19 through 22, given coordinate systems $\#_1$ and $\#_2$ for $\mathbf{R}_2$, find the transition matrix P from $(\mathbf{R}_2)_{\#_2}$ to $(\mathbf{R}_2)_{\#_1}$. For the given $\mathbf{x} \in \mathbf{R}_2$, find $[\mathbf{x}]_{\#_1}$ and $[\mathbf{x}]_{\#_2}$, noting that they are different coordinate vectors for the same point. Show that $P[\mathbf{x}]_{\#_2} = [\mathbf{x}]_{\#_1}$. Demonstrate your work geometrically, as in Figure 5.16.

19. $\#_1 = ([1,0],[0,1]), \#_2 = ([1,1],[-1,1]), \mathbf{x} = [2,4]$

20. $\#_1 = ([1,1],[-1,1]), \#_2 = ([1,0],[0,1]), \mathbf{x} = [1,3]$

21. $\#_1 = ([1,1],[-1,3]), \#_2 = ([-1,1],[-1,-1])$, $\mathbf{x} = [0,2]$

22. $\#_1 = ([2,1],[1,2]), \#_2 = ([1,0],[0,1]), \mathbf{x} = [3,3]$

In exercises 23 through 26, given the vector space $\mathbf{V}$ and coordinate systems $\#_1$ and $\#_2$ for $\mathbf{V}$, find the transition matrix P from $\mathbf{V}_{\#_2}$ to $\mathbf{V}_{\#_1}$. For the given $\mathbf{x} \in \mathbf{V}$, find $[\mathbf{x}]_{\#_1}$ and $[\mathbf{x}]_{\#_2}$. Show that $P[\mathbf{x}]_{\#_2} = [\mathbf{x}]_{\#_1}$.

23. $\mathbf{V} = \mathbf{R}_3$, $\#_1 = ([1,0,0],[0,1,0],[0,0,1])$, $\#_2 = ([1,1,0],[-1,1,0],[0,0,1]), \mathbf{x} = [1,1,1]$

24. $\mathbf{V} = \mathbf{R}_3$, $\#_1 = ([1,1,0],[-1,1,0],[0,0,1])$, $\#_2 = ([1,0,0],[0,1,0],[0,0,1]), \mathbf{x} = [1,2,3]$

25. $\mathbf{V} = \mathbf{P}_1$, $\#_1 = (-t+1, t+1), \#_2 = (1, t), \mathbf{x} = t+3$

26. $\mathbf{V} = \mathbf{P}_1$, $\#_1 = (1, t), \#_2 = (-t+1, t+1), \mathbf{x} = -3t+1$

In exercises 27 and 28, let S be the set of points in $\mathbf{R}_2$ whose coordinates satisfy the given equation. As in Example 28, describe this set of points using coordinates from the given coordinate system. Graph S in $\mathbf{R}_2$ and in $(\mathbf{R}_2)_{\#}$. Note that these graphs are identical.

27. $3x_1^2 - 12x_1x_2 + 12x_2^2 + 10x_1 + 5x_2 = 0$ in $\# = ([2,1],[-1,2])$

28. $4x_1x_2 - 3x_2^2 = 20$ in $\# = ([1,-2],[2,1])$

COMPLEX NUMBERS

29. Find the coordinates of the given vector $\mathbf{x}$ with respect to the given coordinate system.
(*a*) $\mathbf{x} = [1+i, 1-i, 2+3i]$, $\#_1 = ([i,0,0],[0,i,0],[0,0,i])$
(*b*) $\mathbf{x} = [4+6i, 1+i, 5+3i]$, $\#_2 = ([1+i, 1+i, 0]$, $[0, 1+i, 1+i], [1+i, 0, 1+i])$

30. Using the coordinate systems $\#_1$ and $\#_2$ of $\mathbf{C}_3$ given in exercise 29, parts (a) and (b), find the transition matrix from $(\mathbf{C}_3)_{\#_2}$ to $(\mathbf{C}_3)_{\#_1}$ and verify your calculation using the vector $\mathbf{x}$ of part (b).

THEORETICAL EXERCISES

31. Prove Lemma 5.4, parts (ii) and (iii).

32. For the coordinate system $\# = ([1,1],[1,-2])$ and the vectors $\mathbf{x} = [1,2]$ and $\mathbf{y} = [2,1]$, find $[\mathbf{x}]_\#$ and $[\mathbf{y}]_\#$. Compute the distance and angle between $\mathbf{x}$ and $\mathbf{y}$ using these coordinates. Note that the results are not those obtained by using the coordinate system $(\mathbf{e}_1, \mathbf{e}_2)$. Explain why not.

APPLICATIONS EXERCISE

33. Graph $x_3 = x_1^2 + 2x_1x_2 + x_2^2$ in $\mathbf{R}^3$. Set

$$\# = \left(\begin{bmatrix} \frac{1}{\sqrt{2}} \\ \frac{1}{\sqrt{2}} \\ 0 \end{bmatrix}, \begin{bmatrix} \frac{-1}{\sqrt{2}} \\ \frac{1}{\sqrt{2}} \\ 0 \end{bmatrix}, \begin{bmatrix} 0 \\ 0 \\ 1 \end{bmatrix} \right)$$

Then

$$P = \begin{bmatrix} \frac{1}{\sqrt{2}} & \frac{-1}{\sqrt{2}} & 0 \\ \frac{1}{\sqrt{2}} & \frac{1}{\sqrt{2}} & 0 \\ 0 & 0 & 1 \end{bmatrix}$$

is the transition matrix from $(\mathbf{R}^3)_\#$ to $\mathbf{R}^3$. Thus, setting $[\mathbf{x}]_\# = \mathbf{y}$ yields

$$x_1 = \frac{1}{\sqrt{2}}y_1 - \frac{1}{\sqrt{2}}y_2$$
$$x_2 = \frac{1}{\sqrt{2}}y_1 + \frac{1}{\sqrt{2}}y_2$$
$$x_3 = y_3$$

Substituting into the equation yields

$$y_3 = \left(\frac{1}{\sqrt{2}}y_1 - \frac{1}{\sqrt{2}}y_2\right)^2 + 2\left(\frac{1}{\sqrt{2}}y_1 - \frac{1}{\sqrt{2}}y_2\right) \cdot \left(\frac{1}{\sqrt{2}}y_1 + \frac{1}{\sqrt{2}}y_2\right) + \left(\frac{1}{\sqrt{2}}y_1 + \frac{1}{\sqrt{2}}y_2\right)^2 = 2y_1^2$$

Graph $y_3 = 2y_1^2$ in $(\mathbf{R}^3)_\#$. Note that the two graphs are identical. Which was easier to draw? (Sets of points can often be described more simply by using a change of coordinates such as above. This technique is very useful in solving algebraic, difference, differential, etc., equations.)

COMPUTER EXERCISE

34. Let $\mathbf{V}$ be a vector space and $\#$ a coordinate system for $\mathbf{V}$. Let

$$S = \{\mathbf{x}_1, \ldots, \mathbf{x}_n\} \subseteq \mathbf{V}$$

and

$$S' = \{[\mathbf{x}_1]_\#, \ldots, [\mathbf{x}_n]_\#\} \subseteq \mathbf{V}_\#$$

(*a*) Show that S is linearly independent if and only if S' is linearly independent.
(*b*) Show that S is a basis for $\mathbf{V}$ if and only if S' is a basis for $\mathbf{V}_\#$.
(These results give a completely numerical way of deciding whether a set is linearly independent and whether it is a basis.)

5.3 MATRIX REPRESENTATIONS OF LINEAR TRANSFORMATIONS

In the last section we showed that a coordinate system $\#$ of a finite-dimensional vector space $\mathbf{V}$ can be used to associate a grid to $\mathbf{V}$ and thus a coordinate vector to every vector in $\mathbf{V}$. In this section we show how a linear transformation L from $\mathbf{V}$ to $\mathbf{V}$ can be described quantitatively in terms of these coordinate vectors. This, of course, implies that a linear equation $L(\mathbf{x}) = \mathbf{b}$ can also be described in terms of coordinate vectors. As will be shown in Chapters 6 and 7, a judicious choice of coordinate system can then lead to a simple coordinate description of the linear transformation L and the linear equation $L(\mathbf{x}) = \mathbf{b}$, thus making the linear transformation easier to understand and the linear equation easier to solve.

Although this work can be done more generally (see exercise 28), we only show how to describe, in terms of coordinates, linear transformations from a finite-dimensional vector space to itself. The example below shows the basic ideas involved.

EXAMPLE 29

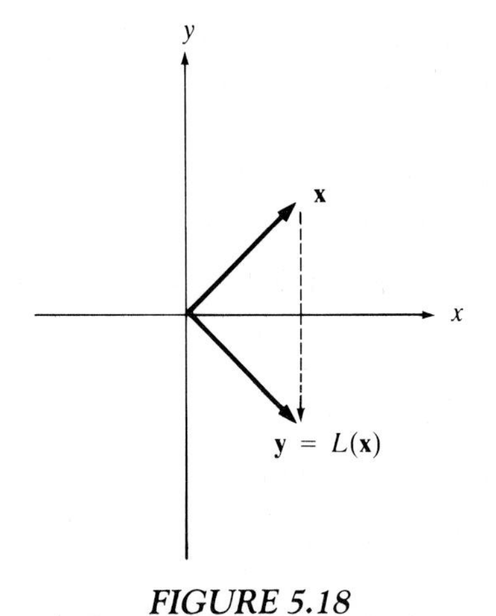

FIGURE 5.18

Let L be the linear transformation from $\mathbf{R}^2$ to $\mathbf{R}^2$ that reflects each fixed vector about the x-axis, as shown in Figure 5.18. Note that L is described verbally. To find a numerical expression for L, we choose the coordinate system $\# = (\mathbf{e}^1, \mathbf{e}^2)$. We will describe L numerically in terms of the coordinate vectors of $\mathbf{R}^2$.

Let $\mathbf{x} \in \mathbf{R}^2$ and $L(\mathbf{x}) = \mathbf{y}$. Set $[\mathbf{x}]_\# = \begin{bmatrix} x_1 \\ x_2 \end{bmatrix}$, and define the function $L_\#$ from $(\mathbf{R}^2)_\#$ to $(\mathbf{R}^2)_\#$ by $L_\#([\mathbf{x}]_\#) = [\mathbf{y}]_\#$.* We need to compute an expression for the function $L_\#$.

$$\begin{aligned} L_\#([\mathbf{x}]_\#) = [\mathbf{y}]_\# &= [L(\mathbf{x})]_\# = [L(x_1\mathbf{e}^1 + x_2\mathbf{e}^2)]_\# = [x_1 L(\mathbf{e}^1) + x_2 L(\mathbf{e}^2)]_\# \\ &= x_1[L(\mathbf{e}^1)]_\# + x_2[L(\mathbf{e}^2)]_\# = [[L(\mathbf{e}^1)]_\# \quad [L(\mathbf{e}^2)]_\#]\begin{bmatrix} x_1 \\ x_2 \end{bmatrix} \\ &= [[L(\mathbf{e}^1)]_\# \quad [L(\mathbf{e}^2)]_\#][\mathbf{x}]_\# = [\mathbf{e}^1 \quad -\mathbf{e}^2][\mathbf{x}]_\# = \begin{bmatrix} 1 & 0 \\ 0 & -1 \end{bmatrix}[\mathbf{x}]_\# \end{aligned}$$

This says that $L_\#$ is given by the matrix

$$A = [[L(\mathbf{e}^1)]_\# \quad [L(\mathbf{e}^2)]_\#] = \begin{bmatrix} 1 & 0 \\ 0 & -1 \end{bmatrix}$$

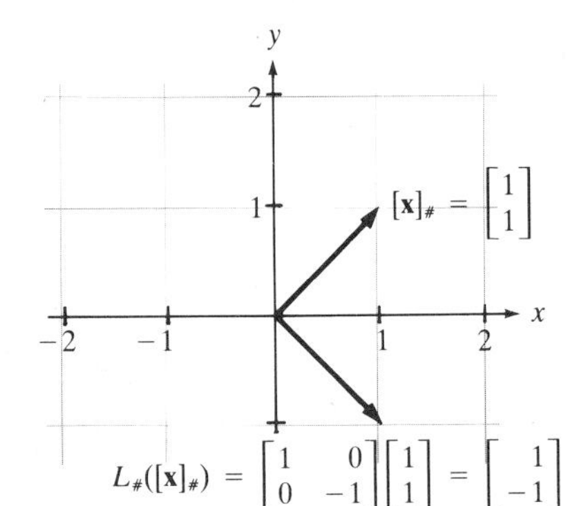

FIGURE 5.19

and that L can be described in terms of coordinates by using A. That is, $L(\mathbf{x}) = \mathbf{y}$ if and only if, in terms of coordinates,

$$L_\#([\mathbf{x}]_\#) = A[\mathbf{x}]_\# = [\mathbf{y}]_\#$$

as depicted in Figure 5.19. Note that if $\mathbf{x} = \begin{bmatrix} 1 \\ 1 \end{bmatrix}$, then verbally $L(\mathbf{x}) = \mathbf{y}$, where $\mathbf{y} = \begin{bmatrix} 1 \\ -1 \end{bmatrix}$, while numerically

$$L_\#\left[\begin{bmatrix} 1 \\ 1 \end{bmatrix}\right] = \begin{bmatrix} 1 & 0 \\ 0 & -1 \end{bmatrix}\begin{bmatrix} 1 \\ 1 \end{bmatrix} = \begin{bmatrix} 1 \\ -1 \end{bmatrix}$$

□

The general situation is as given in the example. That is, a linear transformation L can always be described in terms of coordinates by $L_\#$,

* The symbol $L_\#$ is read "L grid" or "L in grid."

where

$$L_{\#}([\mathbf{x}]_{\#}) = A[\mathbf{x}]_{\#}$$

for some matrix A, called the *matrix representation* of the transformation. This matrix can be computed, as demonstrated in the example, by the theorem below.

THEOREM 5.6

(to find the matrix A that represents L from $\mathbf{V}_{\#}$ to $\mathbf{V}_{\#}$) Let $\mathbf{V}$ be a finite-dimensional vector space and let L be a linear transformation from $\mathbf{V}$ to $\mathbf{V}$. Choose any coordinate system $\# = (\mathbf{x}_1, \ldots, \mathbf{x}_n)$ for $\mathbf{V}$. Compute

$$A = [[L(\mathbf{x}_1)]_{\#} \quad [L(\mathbf{x}_2)]_{\#} \quad \cdots \quad [L(\mathbf{x}_n)]_{\#}]$$

Define $L_{\#}([\mathbf{x}]_{\#}) = A[\mathbf{x}]_{\#}$. Then $L(\mathbf{x}) = \mathbf{y}$ if and only if $L_{\#}([\mathbf{x}]_{\#}) = [\mathbf{y}]_{\#}$ for all $\mathbf{x} \in \mathbf{V}$.

PROOF

See exercise 26. ■

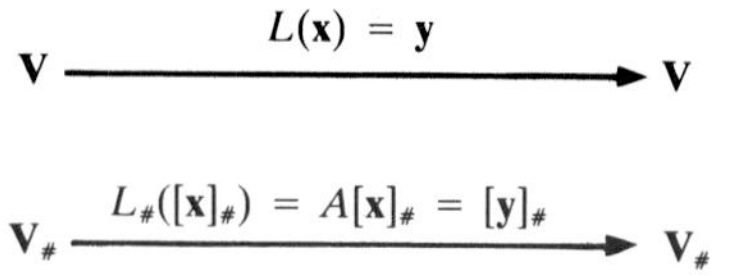

$$\mathbf{V} \xrightarrow{L(\mathbf{x}) = \mathbf{y}} \mathbf{V}$$

$$\mathbf{V}_{\#} \xrightarrow{L_{\#}([\mathbf{x}]_{\#}) = A[\mathbf{x}]_{\#} = [\mathbf{y}]_{\#}} \mathbf{V}_{\#}$$

FIGURE 5.20

From this theorem we see that $L(\mathbf{x}) = \mathbf{y}$ if and only if

$$L_{\#}([\mathbf{x}]_{\#}) = A[\mathbf{x}]_{\#} = [\mathbf{y}]_{\#}$$

Thus, as depicted in Figure 5.20, $L_{\#}$ does for coordinate vectors what L does for the vectors.

EXAMPLE 30

Let L be the θ-rotation linear mapping of Example 4, Section 5.1. Note that L is described verbally. We give a coordinate description of L.

We choose $\# = (\mathbf{e}^1, \mathbf{e}^2)$ as the coordinate system for $\mathbf{R}^2$. To calculate the matrix described in the theorem, we need to compute $L(\mathbf{e}^1)$ and $L(\mathbf{e}^2)$. Let $L\left(\begin{bmatrix} 1 \\ 0 \end{bmatrix}\right) = \begin{bmatrix} y_1 \\ y_2 \end{bmatrix}$. Then $\begin{bmatrix} y_1 \\ y_2 \end{bmatrix}$ is a rotation of $\begin{bmatrix} 1 \\ 0 \end{bmatrix}$ through an angle of θ

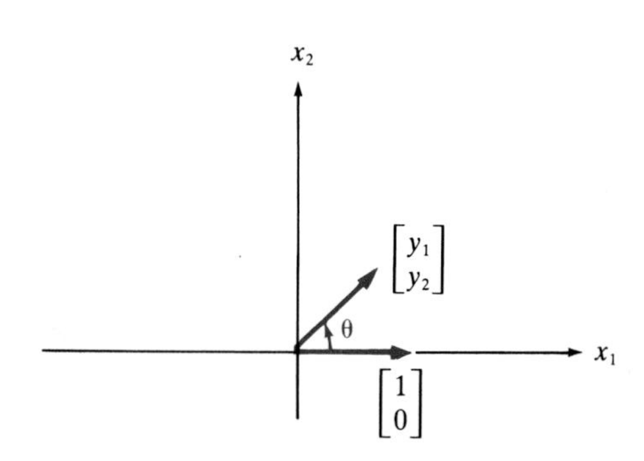

FIGURE 5.21

radians, as shown in Figure 5.21. Thus $y_1 = \cos\theta$, $y_2 = \sin\theta$, and

$$L\left(\begin{bmatrix}1\\0\end{bmatrix}\right) = \begin{bmatrix}y_1\\y_2\end{bmatrix} = \begin{bmatrix}\cos\theta\\\sin\theta\end{bmatrix}$$

Similarly,

$$L\left(\begin{bmatrix}0\\1\end{bmatrix}\right) = \begin{bmatrix}-\sin\theta\\\cos\theta\end{bmatrix}$$

Hence,

$$A = \left[\left[L\left(\begin{bmatrix}1\\0\end{bmatrix}\right)\right]_\#, \left[L\left(\begin{bmatrix}0\\1\end{bmatrix}\right)\right]_\#\right] = \begin{bmatrix}\cos\theta & -\sin\theta\\\sin\theta & \cos\theta\end{bmatrix}$$

and

$$L_\#([\mathbf{x}]_\#) = \begin{bmatrix}\cos\theta & -\sin\theta\\\sin\theta & \cos\theta\end{bmatrix}[\mathbf{x}]_\#$$

To demonstrate the theorem, let $\theta = \pi/4$. Then

$$A = \begin{bmatrix}\frac{\sqrt{2}}{2} & -\frac{\sqrt{2}}{2}\\\frac{\sqrt{2}}{2} & \frac{\sqrt{2}}{2}\end{bmatrix}$$

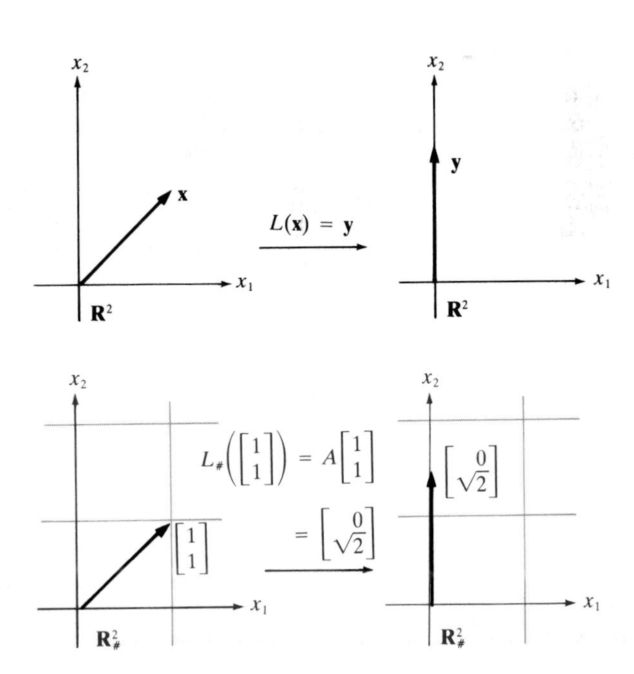

FIGURE 5.22

If we let $\mathbf{x} = \begin{bmatrix} 1 \\ 1 \end{bmatrix}$, then from the verbal description $L(\mathbf{x}) = \mathbf{y}$, where $\mathbf{y} = \begin{bmatrix} 0 \\ \sqrt{2} \end{bmatrix}$, and in terms of coordinates

$$L_{\#}\left(\begin{bmatrix} 1 \\ 1 \end{bmatrix}\right) = A\begin{bmatrix} 1 \\ 1 \end{bmatrix} = \begin{bmatrix} 0 \\ \sqrt{2} \end{bmatrix}$$

The geometry is depicted in Figure 5.22. □

EXAMPLE 31

Let $L(a_2t^2 + a_1t + a_0) = 2a_2t + a_1$ for all $a_2t^2 + a_1t + a_0 \in \mathbf{P}_2$. Then L is a linear transformation from $\mathbf{P}_2$ to $\mathbf{P}_2$. Let $\# = (t^2, t, 1)$.

To compute the matrix A of $L_{\#}$, set

$$\begin{aligned} L(t^2) &= 2t = a_{21}t + a_{31}(1) \\ L(t) &= \ 1 = a_{22}t + a_{32}(1) \\ L(1) &= \ 0 = a_{32}t + a_{33}(1) \end{aligned}$$

Solving for coefficients yields

$$[L(t^2)]_{\#} = \begin{bmatrix} 0 \\ 2 \\ 0 \end{bmatrix}, \qquad [L(t)]_{\#} = \begin{bmatrix} 0 \\ 0 \\ 1 \end{bmatrix}, \qquad [L(1)]_{\#} = \begin{bmatrix} 0 \\ 0 \\ 0 \end{bmatrix}$$

Hence $A = \begin{bmatrix} 0 & 0 & 0 \\ 2 & 0 & 0 \\ 0 & 1 & 0 \end{bmatrix}$. Thus

$$L_{\#}([\mathbf{x}]_{\#}) = \begin{bmatrix} 0 & 0 & 0 \\ 2 & 0 & 0 \\ 0 & 1 & 0 \end{bmatrix}[\mathbf{x}]_{\#}$$

Now, to demonstrate Theorem 5.6, let $p(t) = 2t^2 - t + 1$. Then $L(p(t)) = 4t - 1$. Set $q(t) = 4t - 1$, so we have $L(p(t)) = q(t)$. Then

$$[p(t)]_{\#} = \begin{bmatrix} 2 \\ -1 \\ 1 \end{bmatrix} \quad \text{and} \quad [q(t)]_{\#} = \begin{bmatrix} 0 \\ 4 \\ -1 \end{bmatrix}$$

giving us

$$L_{\#}([p(t)]_{\#}) = L_{\#}\left(\begin{bmatrix} 2 \\ -1 \\ 1 \end{bmatrix}\right) = \begin{bmatrix} 0 & 0 & 0 \\ 2 & 0 & 0 \\ 0 & 1 & 0 \end{bmatrix}\begin{bmatrix} 2 \\ -1 \\ 1 \end{bmatrix} = \begin{bmatrix} 0 \\ 4 \\ -1 \end{bmatrix} = [q(t)]_{\#}$$

Picturing vectors in $\mathbf{P}_2$ as dots, we have the diagram in Figure 5.23. □

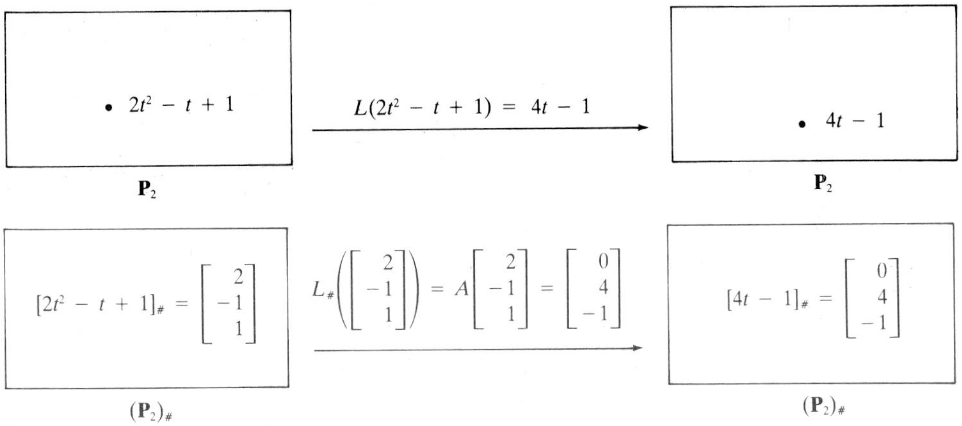

$\bullet\ 2t^2 - t + 1$ $\quad L(2t^2 - t + 1) = 4t - 1 \longrightarrow \quad \bullet\ 4t - 1$

$\mathbf{P}_2 \qquad\qquad \mathbf{P}_2$

$[2t^2 - t + 1]_\# = \begin{bmatrix} 2 \\ -1 \\ 1 \end{bmatrix} \quad L_\#\left(\begin{bmatrix} 2 \\ -1 \\ 1 \end{bmatrix}\right) = A\begin{bmatrix} 2 \\ -1 \\ 1 \end{bmatrix} = \begin{bmatrix} 0 \\ 4 \\ -1 \end{bmatrix} \longrightarrow \quad [4t - 1]_\# = \begin{bmatrix} 0 \\ 4 \\ -1 \end{bmatrix}$

$(\mathbf{P}_2)_\# \qquad\qquad (\mathbf{P}_2)_\#$

FIGURE 5.23

EXAMPLE 32

For comparison purposes, we again consider the reflection mapping of Example 29. For this example, however, we take $\# = \left(\begin{bmatrix} 1 \\ 1 \end{bmatrix}, \begin{bmatrix} 1 \\ -1 \end{bmatrix}\right)$. Then

$$A = \left[\left[L\left(\begin{bmatrix} 1 \\ 1 \end{bmatrix}\right)\right]_\#\ \left[L\left(\begin{bmatrix} 1 \\ -1 \end{bmatrix}\right)\right]_\#\right] = \left[\begin{bmatrix} 1 \\ -1 \end{bmatrix}_\#\ \begin{bmatrix} 1 \\ 1 \end{bmatrix}_\#\right] = \begin{bmatrix} 0 & 1 \\ 1 & 0 \end{bmatrix}$$

Thus $L_\#([\mathbf{x}]_\#) = \begin{bmatrix} 0 & 1 \\ 1 & 0 \end{bmatrix}[\mathbf{x}]_\#$.

To demonstrate Theorem 5.6, let $\mathbf{x} = \begin{bmatrix} 1 \\ 1 \end{bmatrix}$. Then $L(\mathbf{x}) = \begin{bmatrix} 1 \\ -1 \end{bmatrix}$. Set $\mathbf{y} = \begin{bmatrix} 1 \\ -1 \end{bmatrix}$, so that we can write $L(\mathbf{x}) = \mathbf{y}$. Now $[\mathbf{x}]_\# = \begin{bmatrix} 1 \\ 0 \end{bmatrix}$ and $[\mathbf{y}]_\# = \begin{bmatrix} 0 \\ 1 \end{bmatrix}$, giving us

$$L_\#([\mathbf{x}]_\#) = L_\#\left(\begin{bmatrix} 1 \\ 0 \end{bmatrix}\right) = \begin{bmatrix} 0 & 1 \\ 1 & 0 \end{bmatrix}\begin{bmatrix} 1 \\ 0 \end{bmatrix} = \begin{bmatrix} 0 \\ 1 \end{bmatrix} = [\mathbf{y}]_\#$$

as shown in Figure 5.24, where the axes for # are labeled y_1 and y_2. Comparing this example to Example 29, we see that the matrix A representing L in $\# = \left(\begin{bmatrix} 1 \\ 0 \end{bmatrix}, \begin{bmatrix} 0 \\ 1 \end{bmatrix}\right)$ is different from that in $\# = \left(\begin{bmatrix} 1 \\ 1 \end{bmatrix}, \begin{bmatrix} 1 \\ -1 \end{bmatrix}\right)$. □

Often a linear transformation is already described in one coordinate system. In such a case, for purposes of simplification, it is sometimes advisable to redescribe the transformation in another coordinate system, as shown in the next example.

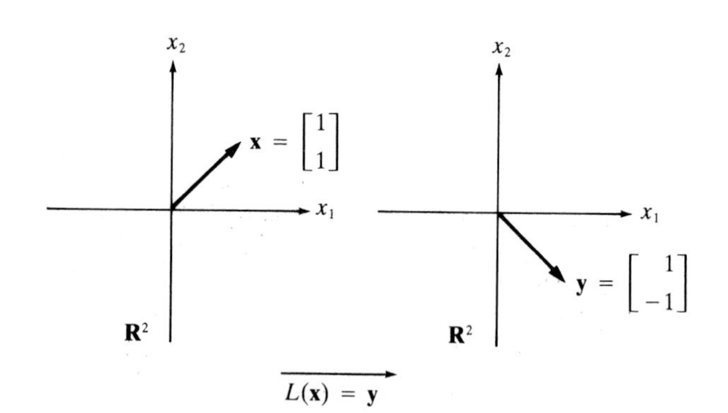

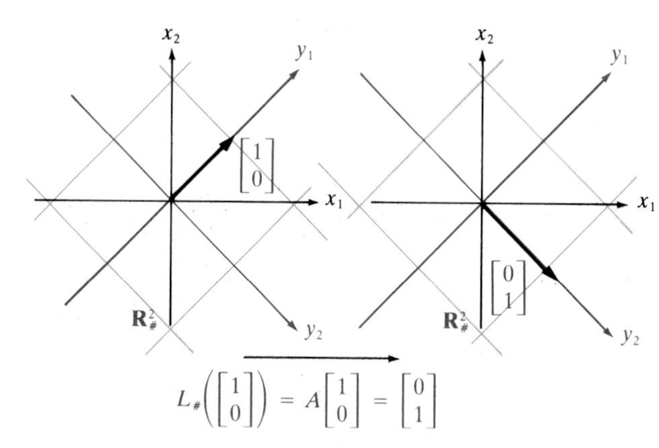

FIGURE 5.24

EXAMPLE 33

Let $L(\mathbf{x}) = \frac{1}{5}\begin{bmatrix}3 & 4\\4 & -3\end{bmatrix}\mathbf{x}$, a linear mapping from points in $\mathbf{R}^2$ to points in $\mathbf{R}^2$. Let $\# = \left(\begin{bmatrix}2\\1\end{bmatrix}, \begin{bmatrix}-1\\2\end{bmatrix}\right)$. To compute the matrix A that represents L from $(\mathbf{R}^2)_{\#}$ to $(\mathbf{R}^2)_{\#}$, set

$$L\left(\begin{bmatrix}2\\1\end{bmatrix}\right) = \begin{bmatrix}2\\1\end{bmatrix} = r_1\begin{bmatrix}2\\1\end{bmatrix} + r_2\begin{bmatrix}-1\\2\end{bmatrix}$$

and

$$L\left(\begin{bmatrix}-1\\2\end{bmatrix}\right) = \begin{bmatrix}1\\-2\end{bmatrix} = s_1\begin{bmatrix}2\\1\end{bmatrix} + s_2\begin{bmatrix}-1\\2\end{bmatrix}$$

Solving yields

$$A = \left[\left[L\left(\begin{bmatrix}2\\1\end{bmatrix}\right)\right]_{\#} \quad \left[L\left(\begin{bmatrix}-1\\2\end{bmatrix}\right)\right]_{\#}\right] = \begin{bmatrix}1 & 0\\0 & -1\end{bmatrix}$$

This result is shown geometrically in Figure 5.25, where the axes for $\#$ are labeled y_1 and y_2.

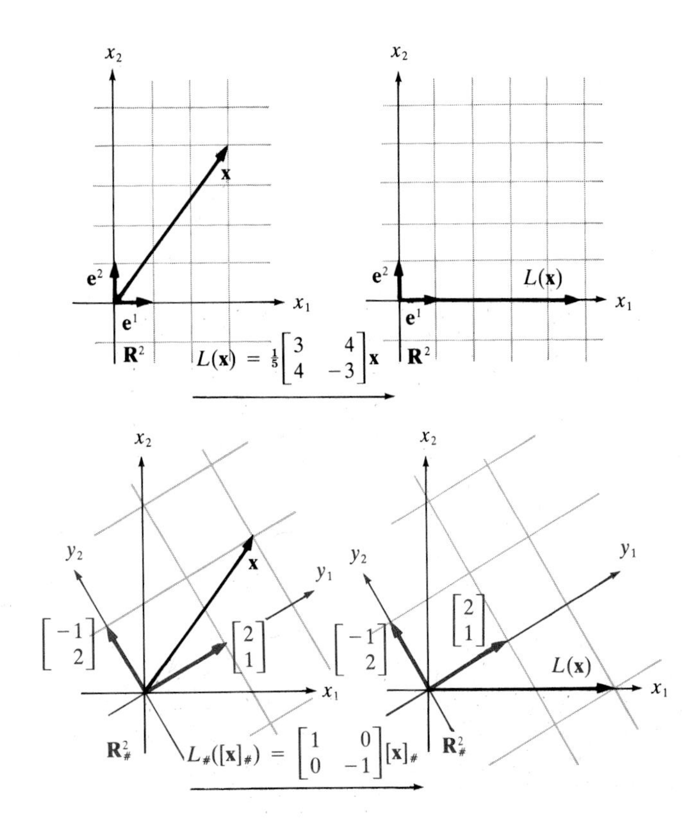

FIGURE 5.25

Note that L is more easily viewed in $(\mathbf{R}^2)_{\#}$. If we set $\mathbf{y} = [\mathbf{x}]_{\#}$, then

$$L_{\#}(\mathbf{y}) = \begin{bmatrix} 1 & 0 \\ 0 & -1 \end{bmatrix} \mathbf{y} = \begin{bmatrix} y_1 \\ -y_2 \end{bmatrix}$$

and we can view L geometrically as the mapping that reflects the plane about the y_1-axis. □

A linear transformation can be represented by a matrix for each coordinate system #. To develop a technique for finding a coordinate system in which the transformation is expressed most simply, it is important to know how the various matrix representations are related.

THEOREM 5.7

Two matrices A and B represent the same linear transformation L in different coordinate systems if and only if there is an invertible matrix P such that $B = P^{-1}AP$.

PROOF

Suppose matrices A and B represent a linear transformation L in coordinate systems $\#_1$ and $\#_2$, respectively. Let P be the transition matrix from $\mathbf{V}_{\#_2}$ to $\mathbf{V}_{\#_1}$, as shown in Figure 5.26. Let $\mathbf{x} \in \mathbf{V}$ and let $L(\mathbf{x}) = \mathbf{y}$. Now

$$L_{\#_2}([\mathbf{x}]_{\#_2}) = [\mathbf{y}]_{\#_2} = P^{-1}[\mathbf{y}]_{\#_1} = P^{-1}A[\mathbf{x}]_{\#_1} = P^{-1}AP[\mathbf{x}]_{\#_2}$$

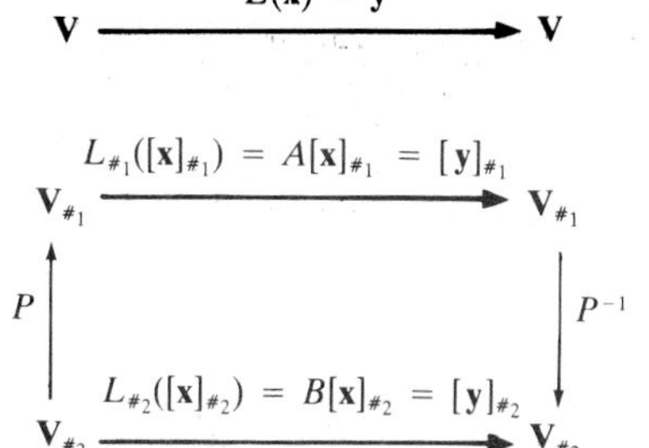

FIGURE 5.26

Since $L_{\#_2}([\mathbf{x}]_{\#_2}) = B[\mathbf{x}]_{\#_2}$, we have that

$$P^{-1}AP[\mathbf{x}]_{\#_2} = B[\mathbf{x}]_{\#_2}$$

for all $\mathbf{x} \in \mathbf{V}$. Since $[\mathbf{y}_i]_{\#_2} = \mathbf{e}^i$ for all i, we see that

$$P^{-1}AP[\mathbf{y}_i]_{\#_2} = B[\mathbf{y}_i]_{\#_2}$$

implies that the i-columns of $P^{-1}AP$ and B are the same. Since this result holds for all i, $P^{-1}AP = B$.

Also, if A, B, and P are matrices such that $B = P^{-1}AP$ and A represents a linear transformation L in a coordinate space $\mathbf{V}_{\#_1}$, we can find a coordinate space $\mathbf{V}_{\#_2}$ such that B represents L in $\#_2$ (exercise 27). ■

In Chapters 6 and 7, we will use this theorem to develop techniques for expressing linear transformations and linear equations simply, thus allowing the linear transformations to be more easily understood and the linear equations to be more easily solved. However, computationally we only use the following consequence of the theorem in these problems.

ALGORITHM FOR FINDING THE COORDINATE SYSTEM $\#_2$ WHEN $P^{-1}AP = B$ IS GIVEN

Let A, B, and P be $n \times n$ matrices such that $P^{-1}AP = B$. Let $\#_1 = (\mathbf{e}^1, \ldots, \mathbf{e}^n)$ and define $L(\mathbf{x}) = A\mathbf{x}$ for all $\mathbf{x} \in \mathbf{R}^n$, so that L is a linear transformation from $\mathbf{R}^n$ to $\mathbf{R}^n$. We find a coordinate system $\#_2$ such that B represents L in $(\mathbf{R}^n)_{\#_2}$ as follows.

Define $\#_2 = (\mathbf{p}^1, \ldots, \mathbf{p}^n)$. Then

$$L_{\#_2}([\mathbf{x}]_{\#_2}) = B[\mathbf{x}]_{\#_2}$$

for all $\mathbf{x} \in \mathbf{R}^n$. Thus B represents L from $(\mathbf{R}^n)_{\#_2}$ to $(\mathbf{R}^n)_{\#_2}$. Further, P is the transition matrix from $(\mathbf{R}^n)_{\#_2}$ to $\mathbf{R}^n$.

EXAMPLE 34

To demonstrate this algorithm, let

$$A = \begin{bmatrix} 1 & 2 \\ 2 & 1 \end{bmatrix}, \quad B = \begin{bmatrix} -1 & 0 \\ 0 & 3 \end{bmatrix}, \quad P = \begin{bmatrix} 1 & 1 \\ -1 & 1 \end{bmatrix}$$

Note that $P^{-1}AP = B$.

Let $L(\mathbf{x}) = A\mathbf{x}$. From the algorithm above, we have that

$$\#_2 = (\mathbf{p}^1, \mathbf{p}^2) = \left(\begin{bmatrix} 1 \\ -1 \end{bmatrix}, \begin{bmatrix} 1 \\ 1 \end{bmatrix} \right)$$

and B represents L with respect to $\#_2$.

For an illustration, let $\mathbf{x} = \begin{bmatrix} 1 \\ 1 \end{bmatrix}$. Then $L(\mathbf{x}) = \mathbf{y} = \begin{bmatrix} 3 \\ 3 \end{bmatrix}$. Further,

$$[\mathbf{x}]_{\#_2} = \begin{bmatrix} 0 \\ 1 \end{bmatrix} \quad \text{and} \quad [\mathbf{y}]_{\#_2} = \begin{bmatrix} 0 \\ 3 \end{bmatrix}$$

Note in the diagram of Figure 5.27 that

$$L(\mathbf{x}) = A\mathbf{x} = \begin{bmatrix} 1 & 2 \\ 2 & 1 \end{bmatrix} \begin{bmatrix} 1 \\ 1 \end{bmatrix} = \begin{bmatrix} 3 \\ 3 \end{bmatrix} = \mathbf{y}$$

and

$$L_{\#_2}([\mathbf{x}]_{\#_2}) = B[\mathbf{x}]_{\#_2} = \begin{bmatrix} -1 & 0 \\ 0 & 3 \end{bmatrix} \begin{bmatrix} 0 \\ 1 \end{bmatrix} = \begin{bmatrix} 0 \\ 3 \end{bmatrix} = [\mathbf{y}]_{\#_2}$$

□

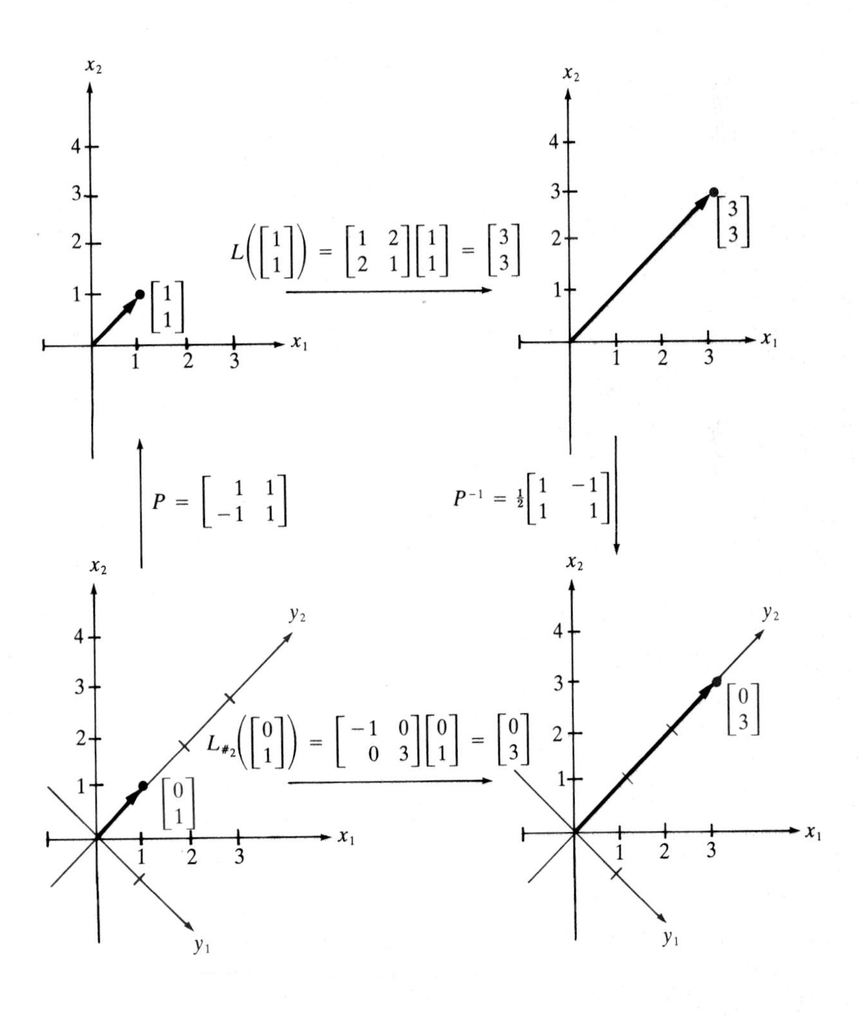

FIGURE 5.27

In Chapter 6 we will apply Theorem 5.7 to develop an approach by which changes in coordinate descriptions are used to solve problems involving linear transformations. We close this chapter with a brief view of how this will be done.

EXAMPLE 35

OPTIONAL

Using the notation and results of Example 34, note that $B = P^{-1}AP$, or $A = PBP^{-1}$. Suppose we want to solve the linear equation $\mathbf{x}(k+1) = A\mathbf{x}(k)$. Substituting, we have

$$\mathbf{x}(k+1) = PBP^{-1}\mathbf{x}(k)$$

or

$$P^{-1}\mathbf{x}(k+1) = BP^{-1}\mathbf{x}(k)$$

Since P is the transition matrix from $(\mathbf{R}^2)_{\#_2}$ to $\mathbf{R}^2$, we have

$$P^{-1}\mathbf{x}(k) = [\mathbf{x}(k)]_{\#_2}$$

for all k. For simplicity, we define $\mathbf{y}(k) = [\mathbf{x}(k)]_{\#_2}$. Thus, viewed in $(\mathbf{R}^2)_{\#_2}$, the equation becomes

$$\mathbf{y}(k+1) = B\mathbf{y}(k) = \begin{bmatrix} -1 & 0 \\ 0 & 3 \end{bmatrix}\mathbf{y}(k)$$

or

$$y_1(k+1) = (-1)y_1(k)$$
$$y_2(k+1) = 3y_2(k)$$

These equations are simple and can be solved for $y_1(k)$ and $y_2(k)$ to yield

$$y_1(k) = a_1(-1)^k \qquad \text{and} \qquad y_2(k) = a_2(3^k)$$

where a_1 and a_2 are any numbers. Thus we have

$$\begin{aligned}\mathbf{x}(k) = P\mathbf{y}(k) &= y_1(k)\mathbf{p}^1 + y_2(k)\mathbf{p}^2 \\ &= a_1(-1)^k\begin{bmatrix} 1 \\ -1 \end{bmatrix} + a_2(3^k)\begin{bmatrix} 1 \\ 1 \end{bmatrix}\end{aligned}$$

where a_1 and a_2 are any numbers. □

Note that, given a problem involving a matrix A, the approach is to find P and B such that $P^{-1}AP = B$ and such that the matrix B contains sufficiently many zeros that the equation involving it is easily solved. In Chapter 6 we will show how to obtain such matrices P and B.

EXERCISES FOR SECTION 5.3

COMPUTATIONAL EXERCISES

In exercises 1 through 6, find the matrix representation of the verbally described linear transformation L with respect to the given coordinate system $\#$. Demonstrate your work geometrically, as in Figure 5.22.

1. L rotates the plane $\mathbf{R}_2$ counterclockwise by $\pi/3$ radians, $\# = ([1,0],[0,1])$.

2. L rotates the plane $\mathbf{R}_2$ counterclockwise by $\pi/6$ radians, $\# = ([1,0],[0,1])$.

3. L reflects the plane $\mathbf{R}_2$ in the line $x_1 = x_2$, $\# = ([1,1],[1,-1])$.

4. L reflects the plane $\mathbf{R}_2$ in the line $x_1 = -x_2$, $\# = ([1,1],[1,-1])$.

5. L reflects $\mathbf{R}_3$ in the plane $x_1 + x_2 - x_3 = 0$, $\# = ([1,0,0],[0,1,0],[0,0,1])$.

6. L reflects $\mathbf{R}_3$ in the plane $x_1 - x_2 + 2x_3 = 0$, $\# = ([1,0,0],[0,1,0],[0,0,1])$.

In exercises 7 through 14, find the matrix representation A of the given linear transformation L with respect to the coordinate system $\#$. Demonstrate your work by making a diagram for the linear transformation, as in Figure 5.23 or Figure 5.25. Use the given vector $\mathbf{x}$ for the diagram.

7. $L([x_1,x_2]) = [x_2,x_1]$, $\# = ([1,1],[-1,1])$, $\mathbf{x} = [2,0]$

8. $L([x_1,x_2]) = [x_1+x_2,x_1+x_2]$, $\# = ([1,1],[-1,1])$, $\mathbf{x} = [-3,1]$

9. $L([x_1,x_2]) = [2x_1+x_2,x_1+2x_2]$, $\# = ([1,2],[1,1])$, $\mathbf{x} = [1,3]$

10. $L([x_1,x_2]) = [x_1-x_2,x_1+x_2]$, $\# = ([2,1],[1,2])$, $\mathbf{x} = [5,-2]$

11. $L([x_1,x_2,x_3]) = [x_3,x_1,x_2]$, $\# = ([1,1,0],[0,1,1],[1,0,1])$, $\mathbf{x} = [3,0,1]$

12. $L(a_2t^2 + a_1t + a_0) = 2a_2t + a_1$, $\# = (t^2,t,1)$ [note that $L(p(t)) = \dfrac{d}{dt}p(t)$], $\mathbf{x} = 3t^2 + t - 1$

13. $L(\mathbf{x}) = \begin{bmatrix}1 & 1\\ 1 & 1\end{bmatrix}\mathbf{x}$,

$$\# = \left(\begin{bmatrix}1 & 0\\ 0 & 0\end{bmatrix},\begin{bmatrix}0 & 1\\ 0 & 0\end{bmatrix},\begin{bmatrix}0 & 0\\ 1 & 0\end{bmatrix},\begin{bmatrix}0 & 0\\ 0 & 1\end{bmatrix}\right),\ \mathbf{x} = \begin{bmatrix}1 & 2\\ -1 & 1\end{bmatrix}$$

14. $L(\mathbf{x}) = \begin{bmatrix}1 & 2\\ 2 & 1\end{bmatrix}\mathbf{x}$,

$$\# = \left(\begin{bmatrix}1 & 0\\ 0 & 0\end{bmatrix},\begin{bmatrix}0 & 1\\ 0 & 0\end{bmatrix},\begin{bmatrix}0 & 0\\ 1 & 0\end{bmatrix},\begin{bmatrix}0 & 0\\ 0 & 1\end{bmatrix}\right),\ \mathbf{x} = \begin{bmatrix}2 & 3\\ 1 & -2\end{bmatrix}$$

In exercises 15 through 18, for the given matrices A, B, and P, show that $B = P^{-1}AP$. Letting $L(\mathbf{x}) = A\mathbf{x}$, find $\#_2$ such that B represents L from $(\mathbf{R}^n)_{\#_2}$ to $(\mathbf{R}^n)_{\#_2}$. In exercises 15 and 16, demonstrate your work using the vector $\mathbf{x} = \begin{bmatrix}1\\ 1\end{bmatrix}$.

15. $A = \begin{bmatrix}1 & 2\\ 2 & 2\end{bmatrix},\ B = \begin{bmatrix}2 & 2\\ 2 & 1\end{bmatrix},\ P = \begin{bmatrix}0 & 1\\ 1 & 0\end{bmatrix}$

16. $A = \begin{bmatrix}2 & 1\\ 1 & 2\end{bmatrix},\ B = \begin{bmatrix}3 & 0\\ 0 & 1\end{bmatrix},\ P = \begin{bmatrix}1 & -1\\ 1 & 1\end{bmatrix}$

17. $A = \begin{bmatrix}2 & 1 & 0\\ 1 & 2 & 0\\ 0 & 1 & 2\end{bmatrix},\ B = \begin{bmatrix}3 & 0 & 0\\ 0 & 2 & 0\\ 0 & 0 & 1\end{bmatrix},\ P = \begin{bmatrix}1 & 0 & 1\\ 1 & 0 & -1\\ 1 & 1 & 1\end{bmatrix}$

18. $A = \begin{bmatrix}1 & 0 & 0\\ 0 & 3 & 2\\ 0 & 1 & 4\end{bmatrix},\ B = \begin{bmatrix}1 & 0 & 0\\ 0 & 5 & 0\\ 0 & 0 & 2\end{bmatrix},\ P = \begin{bmatrix}1 & 0 & 0\\ 0 & 1 & 2\\ 0 & 1 & -1\end{bmatrix}$

In exercises 19 through 22, matrices A and P are given. Compute $B = P^{-1}AP$. Solve $\mathbf{x}(k+1) = A\mathbf{x}(k)$ by substitution, as shown in Example 35, obtaining $\mathbf{y}(k+1) = B\mathbf{y}(k)$. Then use P to convert the solution $\mathbf{y}(k)$ into the solution $\mathbf{x}(k)$ to $\mathbf{x}(k+1) = A\mathbf{x}(k)$.

19. $A = \begin{bmatrix}-5 & 12\\ -2 & 5\end{bmatrix},\ P = \begin{bmatrix}3 & 2\\ 1 & 1\end{bmatrix}$

20. $A = \begin{bmatrix}-5 & 8\\ -6 & 9\end{bmatrix},\ P = \begin{bmatrix}-1 & 4\\ -1 & 3\end{bmatrix}$

21. $A = \begin{bmatrix}10 & 14 & -34\\ 8 & 15 & -31\\ 6 & 10 & -22\end{bmatrix},\ P = \begin{bmatrix}6 & -1 & 2\\ 1 & 3 & 1\\ 2 & 1 & 1\end{bmatrix}$

22. $A = \begin{bmatrix}-1 & 4 & 6\\ -12 & 57 & 81\\ 8 & -38 & -54\end{bmatrix},\ P = \begin{bmatrix}1 & 0 & 2\\ 3 & -3 & -1\\ -2 & 2 & 1\end{bmatrix}$

COMPLEX NUMBERS

In exercises 23 and 24, find the matrix representations of the given linear transformations with respect to the given coordinate systems. Verify your result using the given vector $\mathbf{x}$.

23. $L([z_1,z_2]) = [z_1+z_2,z_1-z_2]$, $\# = ([1+i,1-i],[2-2i,-1-3i])$, $\mathbf{x} = [1,i]$

24. $L([z_1,z_2]) = [z_1-2z_2,2z_1-z_2]$, $\# = ([1+2i,3+i],[2+i,4-i])$, $\mathbf{x} = [4i,4+5i]$

25. Let $L([z_1, z_2]) = [z_1 + z_2, z_1 - z_2]$ and $\#_1 = ([1, i], [i, 1])$. Find the matrix A that represents L with respect to $\#_1$ and the matrix B that represents L with respect to $\#_2 = ([i, 1], [i, -1])$. Find the transition matrix P from $(\mathbf{C}_2)_{\#_2}$ to $(\mathbf{C}_2)_{\#_1}$. Show that $A = PBP^{-1}$.

THEORETICAL EXERCISES

26. Prove Theorem 5.6.

27. Prove the second part of Theorem 5.7.

28. Let L be a linear transformation from a finite-dimensional vector space $\mathbf{V}$ to a finite-dimensional vector space $\mathbf{W}$. Let

$$\#_1 = (\mathbf{x}_1, \ldots, \mathbf{x}_n) \qquad \text{and} \qquad \#_2 = (\mathbf{y}_1, \ldots, \mathbf{y}_m)$$

be coordinate systems for $\mathbf{V}$ and $\mathbf{W}$, respectively. Show that L can be described in terms of these coordinates by the matrix

$$A = [[L(\mathbf{x}_1)]_{\#_2} \quad [L(\mathbf{x}_2)]_{\#_2} \quad \cdots \quad [L(\mathbf{x}_n)]_{\#_2}]$$

29. Let $L(X) = AX + XB$, a linear transformation from $\mathbf{R}_{2,2}$ to $\mathbf{R}_{2,2}$. Let

$$\# = \left(\begin{bmatrix} 1 & 0 \\ 0 & 0 \end{bmatrix}, \begin{bmatrix} 0 & 1 \\ 0 & 0 \end{bmatrix}, \begin{bmatrix} 0 & 0 \\ 1 & 0 \end{bmatrix}, \begin{bmatrix} 0 & 0 \\ 0 & 1 \end{bmatrix}\right)$$

Find the matrix that represents L from $(\mathbf{R}_{2,2})_\#$ to $(\mathbf{R}_{2,2})_\#$. (This matrix is called the *Kronecker product* of A and B and is written $A \otimes B$.)

APPLICATIONS EXERCISE

30. In the pages of this text, you often see sketches of various objects in $\mathbf{R}_3$. They are usually drawn from a vantage point in a plane parallel to the plane π given by $3x + 2y + 2z = 0$. Assume that these sketches are obtained by a linear transformation L that projects $\mathbf{R}_3$ into π, which means that $[x, y, z]$ is mapped to the point $[x', y', z']$ in π such that the line through these two points is perpendicular to π. Find the matrix representation for L. Then map the box

$$\{[x, y, z] : 0 \leqslant x, y, z \leqslant 1\}$$

into π and sketch it. *Hint:* Note that

$$L([1,1,1]) = [-\tfrac{4}{17}, \tfrac{3}{17}, \tfrac{3}{17}], \qquad L([1,1,0]) = [\tfrac{2}{17}, \tfrac{7}{17}, -\tfrac{10}{17}],$$
$$L([1,0,1]) = [\tfrac{2}{17}, -\tfrac{10}{17}, \tfrac{7}{17}]$$

etc., and recall exercise 64 of Section 5.1.

COMPUTER EXERCISES

31. Let L be a linear transformation from a finite-dimensional vector space $\mathbf{V}$ to $\mathbf{V}$, and let $\mathbf{b} \in \mathbf{V}$ be given. Discuss how you could numerically solve $L(\mathbf{x}) = \mathbf{b}$ for $\mathbf{x}$ using a coordinate system $\#$. *Hint:* Consider $A[\mathbf{x}]_\# = [\mathbf{b}]_\#$.

32. Let L be the linear transformation that reflects $\mathbf{R}^2$ about the x-axis. Using the ideas of exercise 31, numerically compute all vectors $\mathbf{x}$ for which $L(\mathbf{x}) = -\mathbf{x}$. Check your answer geometrically.

CHAPTER SIX

SIMILARITY AND PROBLEM SOLVING

In the last chapter we studied linear transformations and linear equations defined on vector spaces. We saw that, for a finite-dimensional vector space **V** and a given coordinate system for **V**, a linear transformation can be described in terms of coordinates by a matrix. Different coordinate systems for **V** can lead to different matrix representations. This suggests that a problem involving a linear transformation and described in a coordinate space $\mathbf{V}_{\#_1}$ might be described more simply in another coordinate space $\mathbf{V}_{\#_2}$. The problem can then be solved more easily in $\mathbf{V}_{\#_2}$, and, if desired, the solution in $\mathbf{V}_{\#_2}$ converted to one in $\mathbf{V}_{\#_1}$. Diagrammatically, this idea is shown in Figure 6.1.

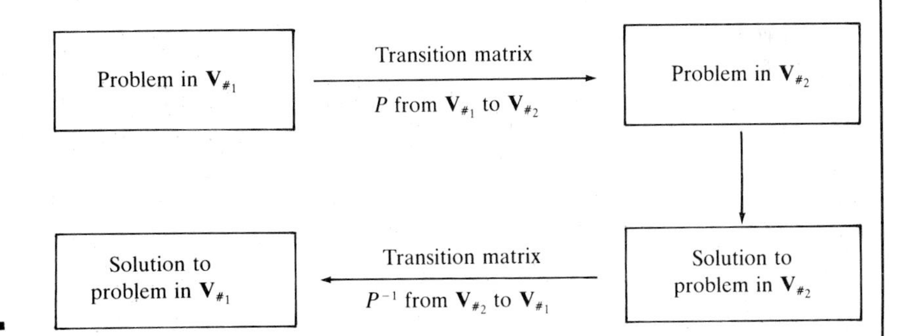

FIGURE 6.1

In this chapter we find special simple matrix representations for linear transformations. We also see how to use these matrix representations to solve problems involving linear transformations, as diagrammed in Figure 6.1, thus developing a useful problem-solving technique.

6.1 EIGENVALUES AND EIGENVECTORS

In this section we define what is meant by a special simple matrix representation for a linear transformation and start the work of showing how it can be found. Before giving this definition, we need to develop some background material.

Let A and B be $n \times n$ matrices. If there is an $n \times n$ invertible matrix P that satisfies $P^{-1}AP = B$, we say that A is *similar* to B. This way of relating matrices satisfies the following properties.

Lemma 6.1

For $A, B, C \in \mathbf{R}_{n,n}$,

(*i*) A is similar to A (reflexive property).

(*ii*) If A is similar to B, then B is similar to A (symmetric property).

(*iii*) If A is similar to B and if B is similar to C, then A is similar to C (transitive property).

Proof

For (i), $P^{-1}AP = A$ when $P = I_n$, so A is similar to A. For (ii), if $P^{-1}AP = B$, then

$$A = PBP^{-1} = Q^{-1}BQ$$

for $Q = P^{-1}$. Hence the symmetric property holds. Finally, for (iii), if $P^{-1}AP = B$ and $Q^{-1}BQ = C$, then by substitution

$$C = Q^{-1}P^{-1}APQ = (PQ)^{-1}A(PQ) = R^{-1}AR$$

where $R = PQ$. Hence the transitive property holds. ■

Property (ii) of the lemma allows us to say that two matrices are similar, rather than saying that one matrix is similar to the other.

Example 1

Some examples of similar matrices follow.

(*i*) Since $P\mathbf{0}P^{-1} = \mathbf{0}$ for all invertible matrices P, it follows that the only matrix similar to $\mathbf{0}$ is $\mathbf{0}$. In the same way, I_n is the only matrix similar to I_n.

(*ii*) The matrices $A = \begin{bmatrix} 1 & 0 \\ 0 & 0 \end{bmatrix}$ and $B = \begin{bmatrix} 1 & -1 \\ 0 & 0 \end{bmatrix}$ are similar, since $P^{-1}AP = B$ for $P = \begin{bmatrix} 1 & -1 \\ 0 & 1 \end{bmatrix}$. For $P = \begin{bmatrix} 2 & -2 \\ 0 & 3 \end{bmatrix}$, we also have that $P^{-1}AP = B$. In fact, there can be infinitely many choices for P (exercise 33). □

Example 2

Not all matrices of the same size are similar. For example,

$$A = \begin{bmatrix} 1 & 1 \\ 1 & 1 \end{bmatrix} \quad \text{and} \quad B = \begin{bmatrix} 1 & 2 \\ 3 & 4 \end{bmatrix}$$

are not similar since, if there were an invertible P such that $P^{-1}AP = B$, then we would have

$$-2 = \det B = \det(P^{-1}AP) = (\det P^{-1})(\det A)(\det P) = \det A = 0 \quad □$$

By Theorem 5.7, if L is a linear transformation and A is a matrix that represents L, then all matrices similar to A also represent L, and conversely. Thus, to find a simple matrix representation for a linear transformation L that has a matrix representation A, we need to find a simple matrix similar to A. By a simple matrix, or *canonical form*, we mean a matrix most of whose entries are 0. Canonical forms can yield simple descriptions of linear transformations.

An especially useful canonical form is a diagonal matrix

$$D = \begin{bmatrix} \lambda_1 & 0 & \cdots & 0 & 0 \\ 0 & \lambda_2 & \cdots & 0 & 0 \\ \vdots & \vdots & & \vdots & \vdots \\ 0 & 0 & \cdots & \lambda_{n-1} & 0 \\ 0 & 0 & \cdots & 0 & \lambda_n \end{bmatrix}$$

where each $\lambda_i \in \mathbf{R}$. In the remainder of this section and in the next section, we will find those matrices which are similar to diagonal matrices. Further, given such a matrix, we will show how to find the diagonal matrix.

To see how this can be done, let A be an $n \times n$ matrix. Then A is similar to a diagonal matrix D if and only if there is an invertible matrix $P = [\mathbf{p}^1 \quad \cdots \quad \mathbf{p}^n]$ that satisfies $P^{-1}AP = D$, or $AP = PD$. Computing columns yields

$$[A\mathbf{p}^1 \quad \cdots \quad A\mathbf{p}^n] = [\lambda_1\mathbf{p}^1 \quad \cdots \quad \lambda_n\mathbf{p}^n]$$

Thus

$$\begin{aligned} A\mathbf{p}^1 &= \lambda_1\mathbf{p}^1 \\ &\vdots \\ A\mathbf{p}^n &= \lambda_n\mathbf{p}^n \end{aligned}$$

Hence, to decide if A is similar to a diagonal matrix D and, if so, to find D, we need to solve an equation such as

$$A\mathbf{p} = \lambda\mathbf{p}$$

for scalars $\lambda_1, \ldots, \lambda_n$ and nonzero vectors $\mathbf{p}^1, \ldots, \mathbf{p}^n$. Further, in order for the matrix $P = [\mathbf{p}^1 \quad \cdots \quad \mathbf{p}^n]$ to be invertible, we know from Corollary 3.14C that the set $\{\mathbf{p}^1, \ldots, \mathbf{p}^n\}$ must be linearly independent. The problem of finding such $\lambda_1, \ldots, \lambda_n$ and the set $\{\mathbf{p}^1, \ldots, \mathbf{p}^n\}$ will be considered here and in the next section.

The technical terms associated with the above problem are given in the following definition.

DEFINITION

Let A be an $n \times n$ matrix. If there is a scalar λ and a nonzero vector $\mathbf{p}$ that satisfy the equation $A\mathbf{p} = \lambda\mathbf{p}$, we call λ an *eigenvalue* for A and $\mathbf{p}$ an *eigenvector* corresponding to, or belonging to, λ.*

* Except in Example 19 of Section 6.2 and a few exercises labeled as such, the matrices A in this text have only real numbers λ and vectors $\mathbf{p}$ with real number entries that satisfy this equation. In practice, however, there are numerous matrices A that have complex numbers λ and vectors with complex number entries that satisfy the equation. In such cases, the work developed in this chapter must be done using complex number arithmetic.

An eigenvalue is also called a *characteristic value* or *latent root*. An eigenvector is also called a *characteristic vector* or *latent vector*.

Example 3

Let $A = \begin{bmatrix} 1 & 1 \\ 1 & 1 \end{bmatrix}$.

(*i*) $A\begin{bmatrix} 1 \\ 1 \end{bmatrix} = 2\begin{bmatrix} 1 \\ 1 \end{bmatrix}$, so $\lambda = \lambda_1 = 2$ is an eigenvalue for A and $\mathbf{p} = \mathbf{p}^1 = \begin{bmatrix} 1 \\ 1 \end{bmatrix}$ is an eigenvector belonging to $\lambda_1 = 2$. Also,

$$A\begin{bmatrix} 2 \\ 2 \end{bmatrix} = 2\begin{bmatrix} 2 \\ 2 \end{bmatrix} \quad \text{and} \quad A\begin{bmatrix} -1 \\ -1 \end{bmatrix} = 2\begin{bmatrix} -1 \\ -1 \end{bmatrix}$$

so both $\begin{bmatrix} 2 \\ 2 \end{bmatrix}$ and $\begin{bmatrix} -1 \\ -1 \end{bmatrix}$ are also eigenvectors belonging to $\lambda_1 = 2$. In fact, since

$$A\left(c\begin{bmatrix} 1 \\ 1 \end{bmatrix}\right) = 2\left(c\begin{bmatrix} 1 \\ 1 \end{bmatrix}\right)$$

for all $c \in \mathbf{R}$, it follows that any nonzero scalar multiple of $\begin{bmatrix} 1 \\ 1 \end{bmatrix}$ is an eigenvector for $\lambda_1 = 2$.

(*ii*) $A\begin{bmatrix} -1 \\ 1 \end{bmatrix} = 0\begin{bmatrix} -1 \\ 1 \end{bmatrix}$, so $\lambda = \lambda_2 = 0$ is an eigenvalue for A and $\mathbf{p} = \mathbf{p}^2 = \begin{bmatrix} -1 \\ 1 \end{bmatrix}$ is an eigenvector belonging to $\lambda_2 = 0$. As in (i), any nonzero scalar multiple of $\begin{bmatrix} -1 \\ 1 \end{bmatrix}$ is also an eigenvector of A belonging to $\lambda_2 = 0$. □

Example 4

Let $A = \begin{bmatrix} 2 & 3 & 1 \\ 0 & -1 & 2 \\ 0 & 1 & 0 \end{bmatrix}$

(*i*) $A\begin{bmatrix} 1 \\ 0 \\ 0 \end{bmatrix} = 2\begin{bmatrix} 1 \\ 0 \\ 0 \end{bmatrix}$, so $\lambda = \lambda_1 = 2$ is an eigenvalue for A and $\mathbf{p} = \mathbf{p}^1 = \begin{bmatrix} 1 \\ 0 \\ 0 \end{bmatrix}$ is an eigenvector belonging to $\lambda_1 = 2$. Also, $c\begin{bmatrix} 1 \\ 0 \\ 0 \end{bmatrix}$ for any $c \neq 0$ is an eigenvector for A belonging to $\lambda_1 = 2$.

(*ii*) $A\begin{bmatrix} -4 \\ 1 \\ 1 \end{bmatrix} = 1\begin{bmatrix} -4 \\ 1 \\ 1 \end{bmatrix}$, so $\lambda = \lambda_2 = 1$ is an eigenvalue for A and $\mathbf{p} = \mathbf{p}^2 = \begin{bmatrix} -4 \\ 1 \\ 1 \end{bmatrix}$ is an eigenvector belonging to $\lambda_2 = 1$. Also, $c\begin{bmatrix} -4 \\ 1 \\ 1 \end{bmatrix}$ is an eigenvector for λ_2 for all $c \neq 0$.

(*iii*) $A\begin{bmatrix} 5 \\ -8 \\ 4 \end{bmatrix} = -2\begin{bmatrix} 5 \\ -8 \\ 4 \end{bmatrix}$, so $\lambda = \lambda_3 = -2$ is an eigenvalue for A and $\mathbf{p} = \mathbf{p}^3 = \begin{bmatrix} 5 \\ -8 \\ 4 \end{bmatrix}$ is an eigenvector belonging to this eigenvalue. All vectors $c\begin{bmatrix} 5 \\ -8 \\ 4 \end{bmatrix}$ with $c \neq 0$ are also eigenvectors for $\lambda_3 = -2$. □

EXAMPLE 5

Let $A = \begin{bmatrix} 1 & 2 \\ 2 & 1 \end{bmatrix}$. Then $A\begin{bmatrix} 0 \\ 0 \end{bmatrix} = 2\begin{bmatrix} 0 \\ 0 \end{bmatrix}$. This does not mean that $\lambda = 2$ is an eigenvalue for A and $\mathbf{p} = \begin{bmatrix} 0 \\ 0 \end{bmatrix}$ is an eigenvector belonging to $\lambda = 2$, since eigenvectors must be nonzero. Remember, in finding eigenvectors we are trying to find an invertible matrix P, and invertible matrices have no zero columns. □

We have reduced the problem of finding diagonal canonical forms to that of finding eigenvalues and eigenvectors. We now show how to find all eigenvalues and corresponding eigenvectors for a given matrix. To get an idea of how this can be done, we work through a 2×2 example.

EXAMPLE 6

Let $A = \begin{bmatrix} 3 & 2 \\ 2 & 3 \end{bmatrix}$. To find eigenvalues and eigenvectors, we need to solve the equation

$$A\mathbf{p} = \lambda\mathbf{p}$$

for all λ and corresponding nonzero $\mathbf{p}$. Thus, we need to find λ and $\mathbf{p}$ for which

$$\begin{bmatrix} 3 & 2 \\ 2 & 3 \end{bmatrix}\mathbf{p} = \lambda\mathbf{p}$$

This equation can be rewritten as

$$\lambda\mathbf{p} - \begin{bmatrix} 3 & 2 \\ 2 & 3 \end{bmatrix}\mathbf{p} = \mathbf{0}$$

Since $\mathbf{p} = I_2\mathbf{p}$, we can write this equation as

$$\lambda I_2\mathbf{p} - \begin{bmatrix} 3 & 2 \\ 2 & 3 \end{bmatrix}\mathbf{p} = \mathbf{0}$$

or

$$\left(\lambda I_2 - \begin{bmatrix} 3 & 2 \\ 2 & 3 \end{bmatrix}\right)\mathbf{p} = \mathbf{0}$$

By Theorem 2.13, this equation has a nonzero solution **p** if and only if

$$\det\left(\lambda I_2 - \begin{bmatrix} 3 & 2 \\ 2 & 3 \end{bmatrix}\right) = 0$$

or

$$\det\begin{bmatrix} \lambda - 3 & -2 \\ -2 & \lambda - 3 \end{bmatrix} = 0$$

This gives

$$(\lambda - 3)^2 - 4 = 0$$

or

$$(\lambda - 5)(\lambda - 1) = 0$$

Thus, the only eigenvalues are

$$\lambda = \lambda_1 = 5 \qquad \text{and} \qquad \lambda = \lambda_2 = 1$$

We now calculate all eigenvectors belonging to each of these eigenvalues.

For $\lambda = \lambda_1 = 5$, we need to solve $A\mathbf{p} = 5\mathbf{p}$ for all nonzero **p**. This equation can be written as

$$[5I_2 - A]\mathbf{p} = \mathbf{0}$$

or

$$\left(5I_2 - \begin{bmatrix} 3 & 2 \\ 2 & 3 \end{bmatrix}\right)\mathbf{p} = \begin{bmatrix} 0 \\ 0 \end{bmatrix}$$

Thus we solve

$$\begin{bmatrix} 2 & -2 \\ -2 & 2 \end{bmatrix}\mathbf{p} = \begin{bmatrix} 0 \\ 0 \end{bmatrix}$$

for **p**. Solving by the Gaussian elimination algorithm yields

$$\begin{bmatrix} 2 & -2 & 0 \\ -2 & 2 & 0 \end{bmatrix} \xrightarrow{R_2 + R_1 \to R_2} \begin{bmatrix} 2 & -2 & 0 \\ 0 & 0 & 0 \end{bmatrix}$$

or

$$2p_1 - 2p_2 = 0$$

Hence p_2 is free, $p_1 = p_2$, and $\mathbf{p} = \mathbf{p}^1 = \begin{bmatrix} p_2 \\ p_2 \end{bmatrix}$ for all $p_2 \neq 0$.

For $\lambda = \lambda_2 = 1$, we solve

$$\left(1I_2 - \begin{bmatrix} 3 & 2 \\ 2 & 3 \end{bmatrix}\right)\mathbf{p} = \begin{bmatrix} 0 \\ 0 \end{bmatrix}$$

or

$$\begin{bmatrix} -2 & -2 \\ -2 & -2 \end{bmatrix}\mathbf{p} = \begin{bmatrix} 0 \\ 0 \end{bmatrix}$$

for $\mathbf{p}$. Solving this system yields $p_1 = -p_2$ with p_2 free, and so $\mathbf{p} = \mathbf{p}^2 = \begin{bmatrix} -p_2 \\ p_2 \end{bmatrix}$ for all $p_2 \neq 0$. □

The calculation of eigenvalues and eigenvectors for an arbitrary square matrix A is essentially the same as in Example 6. To calculate eigenvalues we use the *characteristic polynomial* $\det(\lambda I_n - A)$ and the *characteristic equation* $\det(\lambda I_n - A) = 0$ as described in the next theorem.

THEOREM 6.2

(for finding eigenvalues) Let A be an $n \times n$ matrix. Then β is an eigenvalue for A if and only if β is a solution to the characteristic equation

$$\det(\lambda I_n - A) = 0$$

The characteristic polynomial $c(\lambda) = \det(\lambda I_n - A)$ is a polynomial of degree n, and thus, counting multiplicities, there are n eigenvalues for A.*

PROOF

First, let β be an eigenvalue for A. Then, for some nonzero $\mathbf{p} \in \mathbf{R}^n$,

$$A\mathbf{p} = \beta\mathbf{p}$$

Using that $\mathbf{p} = I_n\mathbf{p}$, we can write this equation as

$$(\beta I_n - A)\mathbf{p} = \mathbf{0}$$

Since $\mathbf{p} \neq \mathbf{0}$, it follows from Theorem 2.13 that

$$\det(\beta I_n - A) = 0$$

so β satisfies the equation $\det(\lambda I_n - A) = 0$

Conversely, suppose β is a solution to the equation

$$\det(\lambda I_n - A) = 0$$

Then $\det(\beta I_n - A) = 0$. By Theorem 2.13, there is a nonzero $\mathbf{p} \in \mathbf{R}^n$ such that

$$(\beta I_n - A)\mathbf{p} = \mathbf{0}$$

or

$$A\mathbf{p} = \beta\mathbf{p}$$

Hence β is an eigenvalue of A.

Proving that $c(\lambda) = \det(\lambda I_n - A)$ is a polynomial of degree n is left as exercise 34. ■

EXAMPLE 7

To find all eigenvalues of $A = \begin{bmatrix} 0 & 2 \\ 2 & 0 \end{bmatrix}$, we write out the characteristic

* Solutions to this equation could be complex numbers. Counting these, with multiplicities, gives the total of exactly n solutions.

equation

$$\det(\lambda I_2 - A) = 0$$

$$\det\left(\begin{bmatrix} \lambda & 0 \\ 0 & \lambda \end{bmatrix} - \begin{bmatrix} 0 & 2 \\ 2 & 0 \end{bmatrix}\right) = 0$$

$$\det\begin{bmatrix} \lambda & -2 \\ -2 & \lambda \end{bmatrix} = 0$$

$$\lambda^2 - 4 = 0$$

Solving yields that $\lambda = \lambda_1 = 2$ and $\lambda = \lambda_2 = -2$ are the eigenvalues for A. □

EXAMPLE 8

Let $A = \begin{bmatrix} 2 & 1 \\ 0 & 2 \end{bmatrix}$. To calculate the eigenvalues of A, write out the characteristic equation

$$\det(\lambda I_2 - A) = 0$$

$$\det\left(\lambda\begin{bmatrix} 1 & 0 \\ 0 & 1 \end{bmatrix} - \begin{bmatrix} 2 & 1 \\ 0 & 2 \end{bmatrix}\right) = 0$$

$$\det\begin{bmatrix} \lambda - 2 & -1 \\ 0 & \lambda - 2 \end{bmatrix} = 0$$

$$(\lambda - 2)(\lambda - 2) = 0$$

Thus $\lambda_1 = 2$ and $\lambda_2 = 2$ are the eigenvalues of A. Note that the eigenvalue $\lambda = 2$ has multiplicity two. □

EXAMPLE 9

Let $A = \begin{bmatrix} 1 & 2 \\ 3 & 4 \end{bmatrix}$. To find the eigenvalues, write

$$\det(\lambda I_2 - A) = 0$$

$$\det\left(\lambda\begin{bmatrix} 1 & 0 \\ 0 & 1 \end{bmatrix} - \begin{bmatrix} 1 & 2 \\ 3 & 4 \end{bmatrix}\right) = 0$$

$$\det\begin{bmatrix} \lambda - 1 & -2 \\ -3 & \lambda - 4 \end{bmatrix} = 0$$

$$(\lambda - 1)(\lambda - 4) - 6 = 0$$

$$\lambda^2 - 5\lambda - 2 = 0$$

Now, by the quadratic formula,

$$\lambda = \frac{5 \pm \sqrt{25 - 4(-2)}}{2} = \frac{5 \pm \sqrt{33}}{2}$$

or $\lambda_1 = \frac{5 + \sqrt{33}}{2}$ and $\lambda_2 = \frac{5 - \sqrt{33}}{2}$. These are the eigenvalues of A. □

EXAMPLE 10

Let $A = \begin{bmatrix} 1 & 0 & 0 \\ 1 & 2 & 4 \\ 1 & 1 & 2 \end{bmatrix}$. To compute the eigenvalues, write

$$\det \begin{bmatrix} \lambda - 1 & 0 & 0 \\ -1 & \lambda - 2 & -4 \\ -1 & -1 & \lambda - 2 \end{bmatrix} = 0$$

Expanding along the first row yields

$$(\lambda - 1) \det \begin{bmatrix} \lambda - 2 & -4 \\ -1 & \lambda - 2 \end{bmatrix} = 0$$
$$(\lambda - 1)[(\lambda - 2)^2 - 4] = 0$$
$$(\lambda - 1)(\lambda^2 - 4\lambda) = 0$$
$$\lambda(\lambda - 1)(\lambda - 4) = 0$$

Thus $\lambda_1 = 0$, $\lambda_2 = 1$, and $\lambda_3 = 4$. □

Using Theorem 6.2, we can show how eigenvalues of triangular matrices are easily calculated.

COROLLARY 6.2A

If T is an $n \times n$ triangular matrix, then its eigenvalues are $t_{11}, t_{22}, \ldots, t_{nn}$, including multiplicities.

PROOF

Using the fact that the determinant of a triangular matrix is the product of the main diagonal entries, we have that the characteristic polynomial of T is

$$\det(\lambda I_n - T) = (\lambda - t_{11})(\lambda - t_{22}) \cdots (\lambda - t_{nn})$$

Thus the eigenvalues are $t_{11}, t_{22}, \ldots, t_{nn}$. ■

EXAMPLE 11

(*i*) The eigenvalues for $T = \begin{bmatrix} 1 & 2 & 3 \\ 0 & 4 & 5 \\ 0 & 0 & 6 \end{bmatrix}$ are $\lambda_1 = 1$, $\lambda_2 = 4$, and $\lambda_3 = 6$.

(*ii*) The eigenvalues for $I_3 = \begin{bmatrix} 1 & 0 & 0 \\ 0 & 1 & 0 \\ 0 & 0 & 1 \end{bmatrix}$ are $\lambda_1 = 1$, $\lambda_2 = 1$, and $\lambda_3 = 1$.

(*iii*) The eigenvalues for $T = \begin{bmatrix} 2 & 0 & 0 & 0 \\ -1 & -5 & 0 & 0 \\ 4 & -3 & 6 & 0 \\ 2 & 1 & -3 & -9 \end{bmatrix}$ are $\lambda_1 = 2$, $\lambda_2 = -5$, $\lambda_3 = 6$, and $\lambda_4 = -9$. □

We conclude this section with a result, used in the next section, about eigenvalues of similar matrices.

COROLLARY 6.2B

If A and B are similar matrices, then they have the same eigenvalues with the same multiplicities.

PROOF

Let P be an invertible matrix such that $B = P^{-1}AP$. Then

$$\begin{aligned} \det(\lambda I_n - B) &= \det(\lambda I_n - P^{-1}AP) = \det(\lambda P^{-1}I_nP - P^{-1}AP) \\ &= \det[P^{-1}(\lambda I_n - A)P] = (\det P^{-1})[\det(\lambda I_n - A)]\det P \\ &= \det(\lambda I_n - A) \end{aligned}$$

Hence the equations $\det(\lambda I_n - B) = 0$ and $\det(\lambda I_n - A) = 0$ have the same solutions, and so A and B have the same eigenvalues with the same multiplicities. ■

In the next section we continue this work by seeing how to find eigenvectors belonging to given eigenvalues and how these can be used to find diagonal canonical forms.

EXERCISES FOR SECTION 6.1

COMPUTATIONAL EXERCISES

In exercises 1 through 4, show that, for the given matrices, $B = P^{-1}AP$ and thus A and B are similar.

1. $B = \begin{bmatrix} -1 & -3 \\ 1 & 3 \end{bmatrix}$, $A = \begin{bmatrix} 1 & 1 \\ 1 & 1 \end{bmatrix}$, $P = \begin{bmatrix} 1 & 2 \\ 0 & 1 \end{bmatrix}$

2. $B = \begin{bmatrix} \frac{3}{2} & -\frac{1}{2} \\ -\frac{1}{2} & -\frac{1}{2} \end{bmatrix}$, $A = \begin{bmatrix} 1 & 1 \\ 1 & 0 \end{bmatrix}$, $P = \begin{bmatrix} 1 & -1 \\ 1 & 1 \end{bmatrix}$

3. $B = \begin{bmatrix} 3 & 0 \\ 0 & -1 \end{bmatrix}$, $A = \begin{bmatrix} 1 & 2 \\ 2 & 1 \end{bmatrix}$, $P = \begin{bmatrix} 1 & -1 \\ 1 & 1 \end{bmatrix}$

4. $B = \begin{bmatrix} 1 & 0 \\ 0 & 2 \end{bmatrix}$, $A = \begin{bmatrix} 1 & 2 \\ 0 & 2 \end{bmatrix}$, $P = \begin{bmatrix} 1 & 2 \\ 0 & 1 \end{bmatrix}$

In exercises 5 and 6, for the given matrix A find another matrix B similar to A. *Hint:* Choose an invertible matrix P and set $B = P^{-1}AP$.

5. $A = \begin{bmatrix} 1 & -1 \\ 1 & 1 \end{bmatrix}$

6. $A = \begin{bmatrix} 1 & 2 \\ 3 & 4 \end{bmatrix}$

In exercises 7 and 8, given

$$P = [\mathbf{p}^1 \quad \cdots \quad \mathbf{p}^n] \quad \text{and} \quad D = \begin{bmatrix} \lambda_1 & & 0 \\ & \ddots & \\ 0 & & \lambda_n \end{bmatrix}$$

verify that

$$PD = [P\mathbf{d}^1 \quad P\mathbf{d}^2 \quad \cdots \quad P\mathbf{d}^n] = [\lambda_1\mathbf{p}^1 \quad \cdots \quad \lambda_n\mathbf{p}^n]$$

7. $P = \begin{bmatrix} 1 & 3 \\ 2 & 4 \end{bmatrix}$; $D = \begin{bmatrix} 3 & 0 \\ 0 & 5 \end{bmatrix}$

8. $P = \begin{bmatrix} 1 & 2 & 3 \\ 4 & 5 & 6 \\ 7 & 8 & 9 \end{bmatrix}$, $D = \begin{bmatrix} 2 & 0 & 0 \\ 0 & 3 & 0 \\ 0 & 0 & 5 \end{bmatrix}$

In exercises 9 and 10, decide if the given vector $\mathbf{x}$ is an eigenvector for the given matrix A. When it is, give the corresponding eigenvalue.

9. (a) $A = \begin{bmatrix} 4 & 1 \\ -1 & 6 \end{bmatrix}$, $\mathbf{x} = \begin{bmatrix} 1 \\ 1 \end{bmatrix}$

(b) $A = \begin{bmatrix} 1 & 7 \\ 3 & 5 \end{bmatrix}, \mathbf{x} = \begin{bmatrix} 0 \\ 0 \end{bmatrix}$

(c) $A = \begin{bmatrix} 1 & 1 & 1 \\ 1 & 1 & 1 \\ 1 & 1 & 1 \end{bmatrix}, \mathbf{x} = \begin{bmatrix} 1 \\ -1 \\ 0 \end{bmatrix}$

(d) $A = \begin{bmatrix} 1 & 0 & 0 \\ 2 & 3 & 4 \\ 5 & 6 & 7 \end{bmatrix}, \mathbf{x} = \begin{bmatrix} 1 \\ 0 \\ 0 \end{bmatrix}$

10. (a) $A = \begin{bmatrix} 1 & -1 \\ 1 & 1 \end{bmatrix}, \mathbf{x} = \begin{bmatrix} 1 \\ -1 \end{bmatrix}$

(b) $A = \begin{bmatrix} 1 & -2 \\ -1 & 2 \end{bmatrix}, \mathbf{x} = \begin{bmatrix} 2 \\ 1 \end{bmatrix}$

(c) $A = \begin{bmatrix} 1 & 2 & 3 \\ 0 & 4 & 5 \\ 0 & 6 & 7 \end{bmatrix}, \mathbf{x} = \begin{bmatrix} 1 \\ 0 \\ 0 \end{bmatrix}$

(d) $A = \begin{bmatrix} 2 & 1 & 2 \\ 1 & 5 & 1 \\ 1 & 1 & 3 \end{bmatrix}, \mathbf{x} = \begin{bmatrix} 1 \\ -1 \\ 1 \end{bmatrix}$

In exercises 11 through 14, show that λ_i and $\mathbf{p}^i$ are eigenvalues and corresponding eigenvectors, respectively, for the given matrix A. In each case, construct P and D, and show that $AP = PD$ and thus $P^{-1}AP = D$.

11. $A = \begin{bmatrix} 1 & 2 \\ 8 & 1 \end{bmatrix}$

(a) $\lambda_1 = -3, \lambda_2 = 5, \mathbf{p}^1 = \begin{bmatrix} 1 \\ -2 \end{bmatrix}, \mathbf{p}^2 = \begin{bmatrix} 1 \\ 2 \end{bmatrix}$

(b) $\lambda_1 = 5, \lambda_2 = -3, \mathbf{p}^1 = \begin{bmatrix} 3 \\ 6 \end{bmatrix}, \mathbf{p}^2 = \begin{bmatrix} -1 \\ 2 \end{bmatrix}$

12. $A = \begin{bmatrix} 2 & 1 \\ 1 & 2 \end{bmatrix}$

(a) $\lambda_1 = 3, \lambda_2 = 1, \mathbf{p}^1 = \begin{bmatrix} 1 \\ 1 \end{bmatrix}, \mathbf{p}^2 = \begin{bmatrix} 1 \\ -1 \end{bmatrix}$

(b) $\lambda_1 = 1, \lambda_2 = 3, \mathbf{p}^1 = \begin{bmatrix} -1 \\ 1 \end{bmatrix}, \mathbf{p}^2 = \begin{bmatrix} -1 \\ -1 \end{bmatrix}$

13. $A = \begin{bmatrix} 2 & 0 & 0 \\ 0 & 1 & 3 \\ 0 & 3 & 1 \end{bmatrix}, \lambda_1 = 2, \lambda_2 = 4, \lambda_3 = -2,$

$\mathbf{p}^1 = \begin{bmatrix} 1 \\ 0 \\ 0 \end{bmatrix}, \mathbf{p}^2 = \begin{bmatrix} 0 \\ 1 \\ 1 \end{bmatrix}, \mathbf{p}^3 = \begin{bmatrix} 0 \\ 1 \\ -1 \end{bmatrix}$

14. $A = \begin{bmatrix} 1 & 2 & 0 \\ 1 & 2 & 0 \\ 0 & 0 & 3 \end{bmatrix}, \lambda_1 = 3, \lambda_2 = 3, \lambda_3 = 0,$

$\mathbf{p}^1 = \begin{bmatrix} 1 \\ 1 \\ 0 \end{bmatrix}, \mathbf{p}^2 = \begin{bmatrix} 0 \\ 0 \\ 1 \end{bmatrix}, \mathbf{p}^3 = \begin{bmatrix} 2 \\ -1 \\ 0 \end{bmatrix}$

In exercises 15 through 28, use Theorem 6.2 to compute the eigenvalues and count them, including multiplicities.

15. $A = \begin{bmatrix} -4 & 3 \\ -2 & 1 \end{bmatrix}$

16. $A = \begin{bmatrix} 3 & 1 \\ 0 & -1 \end{bmatrix}$

17. $A = \begin{bmatrix} -1 & 2 \\ 2 & -1 \end{bmatrix}$

18. $A = \begin{bmatrix} 2 & -3 \\ 1 & -2 \end{bmatrix}$

19. $A = \begin{bmatrix} 1 & -4 & 1 \\ 0 & 4 & 5 \\ 0 & -1 & -2 \end{bmatrix}$

20. $A = \begin{bmatrix} -5 & 0 & 0 \\ 3 & 1 & 2 \\ 7 & 6 & -3 \end{bmatrix}$

21. $A = \begin{bmatrix} 1 & 0 & 0 \\ 0 & 2 & 0 \\ 0 & 0 & 3 \end{bmatrix}$

22. $A = \begin{bmatrix} 1 & 1 & 1 \\ 0 & 2 & 2 \\ 0 & 0 & 3 \end{bmatrix}$

23. $A = \begin{bmatrix} 1 & 0 & 0 \\ 0 & -1 & 0 \\ 0 & 0 & 7 \end{bmatrix}$

24. $A = \begin{bmatrix} 1 & 0 & 0 \\ 2 & 1 & 0 \\ 3 & 4 & 1 \end{bmatrix}$

25. $A = \begin{bmatrix} 1 & 5 & 6 & 7 \\ 0 & 2 & 3 & 4 \\ 0 & 0 & 1 & 8 \\ 0 & 0 & 0 & 2 \end{bmatrix}$

26. $A = \begin{bmatrix} 5 & 0 & 0 & 0 \\ 1 & 5 & 0 & 0 \\ 2 & 3 & 5 & 0 \\ 4 & 6 & 7 & 5 \end{bmatrix}$

27. $A = \begin{bmatrix} 1 & 5 & 6 & 7 \\ 0 & 3 & 1 & 1 \\ 0 & 3 & 1 & 1 \\ 0 & -6 & 2 & 0 \end{bmatrix}$

28. $A = \begin{bmatrix} 3 & 1 & 7 & -1 \\ 0 & 4 & 3 & 1 \\ 0 & -5 & 0 & 1 \\ 0 & 7 & -1 & -2 \end{bmatrix}$

COMPLEX NUMBERS

In exercises 29 and 30, show that $P^{-1}AP = B$ for the given matrices, and hence that A and B are similar.

29. $A = \begin{bmatrix} 1 & i \\ -i & 1 \end{bmatrix}, P = \begin{bmatrix} 0 & i \\ i & 0 \end{bmatrix}, B = \begin{bmatrix} 1 & -i \\ i & 1 \end{bmatrix}$

30. $A = \begin{bmatrix} 0 & 1+i \\ 1-i & 0 \end{bmatrix}, P = \begin{bmatrix} 0 & i \\ -i & 0 \end{bmatrix},$

$B = \begin{bmatrix} 0 & -1+i \\ -1-i & 0 \end{bmatrix}$

In exercises 31 and 32, find the eigenvalues of the given matrix A.

31. $A = \begin{bmatrix} 3 & 2 \\ -1 & 5 \end{bmatrix}$

32. $A = \begin{bmatrix} 1 & -3 \\ 1 & 2 \end{bmatrix}$

THEORETICAL EXERCISES

33. Let A, B, $P \in \mathbf{R}_{n,n}$ with P invertible and $P^{-1}AP = B$. Show that if $Q = cP$ with c any real number other than 0, then $Q^{-1}AQ = B$.

34. Prove that if A is an $n \times n$ matrix, then $c(\lambda) = \det(\lambda I_n - A)$ is a polynomial of degree n. *Hint:* Write out the terms in the definition of the determinant. Note that there is only one term of degree n, which is λ^n.

35. Suppose $A = PDP^{-1}$. Prove that $A^k = PD^kP^{-1}$ for all integers $k \geqslant 0$. Further show that if A^{-1} exists, then $A^k = PD^kP^{-1}$ for all integers $k < 0$.

36. Let A be an $n \times n$ matrix. Then the function $L(\mathbf{x}) = A\mathbf{x}$ is a linear mapping by Example 3 of Section 5.1. If λ is an eigenvalue for A and if $\mathbf{y}$ is an eigenvector belonging to λ, then $L(\mathbf{y}) = A\mathbf{y} = \lambda\mathbf{y}$. Show by sketches that what this means geometrically is that eigenvectors are vectors that are (a) stretched by L, (b) shrunk by L, or (c) reflected about the origin and then stretched or shrunk by L.

37. Let $L(\mathbf{x}) = A\mathbf{x}$ be the linear mapping that rotates each fixed arrow in $\mathbf{R}^2$ by $\pi/3$ radians. Decide if A has any real eigenvalues. *Hint:* See if there are any nonzero vectors such that $L(\mathbf{y}) = \lambda\mathbf{y}$.

38. Let $\mathbf{V}$ be a vector space and L a linear transformation from $\mathbf{V}$ to $\mathbf{V}$. If there is a nonzero $\mathbf{x} \in \mathbf{V}$ and a scalar λ such that $L(\mathbf{x}) = \lambda\mathbf{x}$, we call λ an *eigenvalue* of L and $\mathbf{x}$ an *eigenvector* of L belonging to λ. For each of the following linear transformations, find at least one eigenvalue and one corresponding eigenvector.

(*a*) $L(f(k)) = E(f(k))$ in $\mathbf{V} = \mathbf{F}(\mathbf{N})$

(*b*) (calculus) $L(f(t)) = \dfrac{d}{dt} f(t)$ in $\mathbf{V} = C_1(\mathbf{R})$

(Such problems arise in studies of difference equations, differential equations, and integral equations.)

39. In $\mathbf{R}_{n,n}$, if $D_1 = P^{-1}AP$ and $D_2 = P^{-1}BP$, show that

$$D_1D_2 = P^{-1}ABP \qquad \text{and} \qquad D_1 + D_2 = P^{-1}(A + B)P$$

40. To compute eigenvalues, we need to solve a polynomial equation

$$a_n\lambda^n + \cdots + a_1\lambda + a_0 = 0$$

If $n = 2$, this equation can be solved by the well-known quadratic formula. If $n = 3$ or 4, there are other formulas that can be used to compute the solutions. For these, see any edition of *CRC Standard Mathematical Tables* published by Chemical Rubber Company. Using these formulas, write out the solutions for

$$x^3 - 6x^2 + 11x - 6 = 0$$

(There are no such formulas if $n \geqslant 5$. This is a consequence of Galois theory, which is discussed in most abstract algebra texts. For these cases, numerical approximation methods are needed to compute solutions.)

41. Let A be an $n \times n$ matrix. Apply the Gaussian elimination algorithm to A to obtain an echelon form E. Show, by giving an example, that E does not necessarily have the same eigenvalues as A. (A common mistake is to compute eigenvalues by computing an echelon form for A.)

COMPUTER EXERCISES

42. For each matrix A, graph the eigenvalues of A for ε in the interval $[0, \infty)$.

(*a*) $A = \begin{bmatrix} 1 & \varepsilon \\ 1 & 1 \end{bmatrix}$ (*b*) $A = \begin{bmatrix} 1 & \varepsilon \\ \varepsilon & 1 \end{bmatrix}$

[Note in (a) that if ε is close to 0, small changes in ε can cause much larger changes in the eigenvalues. How much larger $\lambda(\varepsilon_1) - \lambda(\varepsilon_2)$ is than $\varepsilon_1 - \varepsilon_2$ can be seen by computing the ratio $\dfrac{\lambda(\varepsilon_1) - \lambda(\varepsilon_2)}{\varepsilon_1 - \varepsilon_2}$ for small values of ε_1 and ε_2.]

43. Let A be a matrix representation of L in $\mathbf{V}_{\#}$. Explain why $L(\mathbf{x}) = \lambda\mathbf{x}$ for some scalar λ if and only if $A[\mathbf{x}]_{\#} = \lambda[\mathbf{x}]_{\#}$. Thus λ and $\mathbf{x}$ can be numerically calculated by calculating λ and $[\mathbf{x}]_{\#}$ for A.

6.2 DIAGONAL CANONICAL FORMS

In the last section we showed that the problem of finding diagonal canonical forms reduces to the problem of finding eigenvalues and corresponding eigenvectors. Further, we showed how to find eigenvalues. We now continue this work by finding corresponding eigenvectors and, from these, diagonal canonical forms.

Having found all eigenvalues of a given matrix A, we find all eigenvectors for these computed eigenvalues by solving systems of linear algebraic equations.

THEOREM 6.3

(for finding eigenvectors) Let A be a square matrix and let λ be an eigenvalue for A. Then the eigenvectors of A belonging to λ are the nonzero solutions to

$$(\lambda I_n - A)\mathbf{p} = \mathbf{0}$$

PROOF

Let $\mathbf{p}'$ be an eigenvector of A belonging to λ. Then $A\mathbf{p}' = \lambda\mathbf{p}'$. Rearranging yields $(\lambda I_n - A)\mathbf{p}' = \mathbf{0}$. Thus $\mathbf{p}'$ is a nonzero solution to $(\lambda I_n - A)\mathbf{p} = \mathbf{0}$.

Now suppose $\mathbf{p}'$ is a nonzero solution to $(\lambda I_n - A)\mathbf{p} = \mathbf{0}$. Then, substituting, we have $(\lambda I_n - A)\mathbf{p}' = \mathbf{0}$. This equation can be rearranged to $A\mathbf{p}' = \lambda\mathbf{p}'$. Thus $\mathbf{p}'$ is an eigenvector of A belonging to λ. ■

It follows that if $\mathbf{p}$ is an eigenvector for A then so is $c\mathbf{p}$ for all nonzero $c \in \mathbf{R}$. Extending this result a bit further, if $\{\mathbf{p}^1, \ldots, \mathbf{p}^s\}$ is a linearly independent set of eigenvectors belonging to the same eigenvalue, then $r_1\mathbf{p}^1 + \cdots + r_s\mathbf{p}^s$ is also an eigenvector for that eigenvalue for all $r_1, \ldots, r_s$ at least one of which is not 0 (see exercises 25 and 26).

EXAMPLE 12

CONTINUING EXAMPLE 7 OF SECTION 6.1

The eigenvalues of $A = \begin{bmatrix} 0 & 2 \\ 2 & 0 \end{bmatrix}$ are $\lambda_1 = 2$ and $\lambda_2 = -2$. The eigenvectors for $\lambda_1 = 2$ are found by solving

$$(\lambda_1 I_2 - A)\mathbf{p} = \mathbf{0}$$

for all nonzero $\mathbf{p}$. This yields

$$\left(\begin{bmatrix} 2 & 0 \\ 0 & 2 \end{bmatrix} - \begin{bmatrix} 0 & 2 \\ 2 & 0 \end{bmatrix}\right)\mathbf{p} = \mathbf{0}$$

$$\begin{bmatrix} 2 & -2 \\ -2 & 2 \end{bmatrix}\mathbf{p} = \mathbf{0}$$

Solving this system by the Gaussian elimination algorithm gives

$$\begin{bmatrix} 2 & -2 & 0 \\ -2 & 2 & 0 \end{bmatrix} \xrightarrow{\mathbf{R}_2 + \mathbf{R}_1 \to \mathbf{R}_2} \begin{bmatrix} 2 & -2 & 0 \\ 0 & 0 & 0 \end{bmatrix}$$

so $2p_1 - 2p_2 = 0$. Thus p_2 is free and $p_1 = p_2$ for all p_2. Hence $\mathbf{p}^1 = \begin{bmatrix} p_2 \\ p_2 \end{bmatrix}$ for all $p_2 \neq 0$ are the eigenvectors for A belonging to $\lambda_1 = 2$.

For the eigenvectors belonging to $\lambda_2 = -2$, we need to solve

$$\left(\begin{bmatrix} -2 & 0 \\ 0 & -2 \end{bmatrix} - \begin{bmatrix} 0 & 2 \\ 2 & 0 \end{bmatrix}\right)\mathbf{p} = \mathbf{0}$$

$$\begin{bmatrix} -2 & -2 \\ -2 & -2 \end{bmatrix}\mathbf{p} = \mathbf{0}$$

Solving this equation yields that p_2 is free and $p_1 = -p_2$ for all $p_2 \in \mathbf{R}$. Thus, $\mathbf{p}^2 = \begin{bmatrix} -p_2 \\ p_2 \end{bmatrix}$ for all $p_2 \neq 0$ are the eigenvectors for A belonging to $\lambda_2 = -2$.

□

EXAMPLE 13

CONTINUING EXAMPLE 8 OF SECTION 6.1

Let $A = \begin{bmatrix} 2 & 1 \\ 0 & 2 \end{bmatrix}$. The eigenvalues for A are $\lambda_1 = 2$ and $\lambda_2 = 2$. To compute the eigenvectors for A belonging to $\lambda_1 = \lambda_2 = 2$, we need to solve

$$(\lambda I_2 - A)\mathbf{p} = \mathbf{0}$$

$$\left(\begin{bmatrix} 2 & 0 \\ 0 & 2 \end{bmatrix} - \begin{bmatrix} 2 & 1 \\ 0 & 2 \end{bmatrix}\right)\mathbf{p} = \mathbf{0}$$

$$\begin{bmatrix} 0 & -1 \\ 0 & 0 \end{bmatrix}\mathbf{p} = \mathbf{0}$$

Solving this equation shows that $p_2 = 0$ and p_1 is free. Thus the eigenvectors of A belonging to $\lambda_1 = \lambda_2 = 2$ are $\mathbf{p} = \begin{bmatrix} p_1 \\ 0 \end{bmatrix}$ for all $p_1 \neq 0$.

It should be noted in this example that there is no set of two linearly independent eigenvectors for A, since the only eigenvectors are $\mathbf{p} = p_1\begin{bmatrix} 1 \\ 0 \end{bmatrix}$. In order to be similar to a diagonal matrix, a matrix must have a linearly independent set of two eigenvectors. Thus A of this example cannot be similar to a diagonal matrix. □

EXAMPLE 14

CONTINUING EXAMPLE 9 OF SECTION 6.1

For $A = \begin{bmatrix} 1 & 2 \\ 3 & 4 \end{bmatrix}$, we can show, in a manner similar to that of Example 13, that the eigenvectors belonging to $\lambda_1 = \dfrac{5 + \sqrt{33}}{2}$ are $p_1\begin{bmatrix} 3 - \sqrt{33} \\ -6 \end{bmatrix}$, $p_1 \neq 0$, and those belonging to $\lambda_2 = \dfrac{5 - \sqrt{33}}{2}$ are $p_2\begin{bmatrix} 3 + \sqrt{33} \\ -6 \end{bmatrix}$, $p_2 \neq 0$. □

EXAMPLE 15

CONTINUING EXAMPLE 10 OF SECTION 6.1

Let $A = \begin{bmatrix} 1 & 0 & 0 \\ 1 & 2 & 4 \\ 1 & 1 & 2 \end{bmatrix}$. The eigenvalues for A are $\lambda_1 = 0$, $\lambda_2 = 1$, and $\lambda_3 = 4$.

For the eigenvectors belonging to $\lambda_1 = 0$, we need to solve

$$(\lambda_1 I_3 - A)\mathbf{p} = \mathbf{0}$$

$$\left(\begin{bmatrix} 0 & 0 & 0 \\ 0 & 0 & 0 \\ 0 & 0 & 0 \end{bmatrix} - \begin{bmatrix} 1 & 0 & 0 \\ 1 & 2 & 4 \\ 1 & 1 & 2 \end{bmatrix}\right)\mathbf{p} = \mathbf{0}$$

$$\begin{bmatrix} -1 & 0 & 0 \\ -1 & -2 & -4 \\ -1 & -1 & -2 \end{bmatrix}\mathbf{p} = \mathbf{0}$$

Solving this equation by the Gaussian elimination algorithm after multiplying the equation by -1 yields

$$\begin{bmatrix} 1 & 0 & 0 & 0 \\ 1 & 2 & 4 & 0 \\ 1 & 1 & 2 & 0 \end{bmatrix} \xrightarrow[R_3 - R_1 \to R_3]{R_2 - R_1 \to R_2} \begin{bmatrix} 1 & 0 & 0 & 0 \\ 0 & 2 & 4 & 0 \\ 0 & 1 & 2 & 0 \end{bmatrix}$$

$$\xrightarrow{R_3 - \frac{1}{2}R_2 \to R_3} \begin{bmatrix} 1 & 0 & 0 & 0 \\ 0 & 2 & 4 & 0 \\ 0 & 0 & 0 & 0 \end{bmatrix}$$

Thus

$$\begin{aligned} p_1 \qquad\quad &= 0 \\ 2p_2 + 4p_3 &= 0 \end{aligned}$$

Hence $p_1 = 0$, p_3 is free, and $p_2 = -2p_3$ for all p_3. Thus

$$\mathbf{p}^1 = \begin{bmatrix} 0 \\ -2p_3 \\ p_3 \end{bmatrix}$$

for all $p_3 \neq 0$. Similarly, for $\lambda_2 = 1$ we obtain

$$\mathbf{p}^2 = \begin{bmatrix} p_2 \\ -p_2 \\ 0 \end{bmatrix}$$

for all $p_2 \neq 0$, and for $\lambda_3 = 4$ we obtain

$$\mathbf{p}^3 = \begin{bmatrix} 0 \\ 2p_3 \\ p_3 \end{bmatrix}$$

for all $p_3 \neq 0$. □

EXAMPLE 16

Let $A = \begin{bmatrix} 1 & 0 & 0 \\ 0 & \frac{1}{2} & \frac{1}{2} \\ 0 & \frac{1}{2} & \frac{1}{2} \end{bmatrix}$. The eigenvalues of this matrix are $\lambda_1 = 0$, $\lambda_2 = 1$, and

$\lambda_3 = 1$. The eigenvectors for $\lambda_1 = 0$ are $p_1\begin{bmatrix} 0 \\ 1 \\ -1 \end{bmatrix}$ for all $p_1 \neq 0$. For $\lambda_2 = \lambda_3 = 1$, we have

$$(\lambda_2 I_3 - A)\mathbf{p} = \begin{bmatrix} 0 & 0 & 0 \\ 0 & \frac{1}{2} & -\frac{1}{2} \\ 0 & -\frac{1}{2} & \frac{1}{2} \end{bmatrix}\mathbf{p} = \mathbf{0}$$

Applying the Gaussian elimination algorithm yields

$$\begin{bmatrix} 0 & 0 & 0 & 0 \\ 0 & \frac{1}{2} & -\frac{1}{2} & 0 \\ 0 & -\frac{1}{2} & \frac{1}{2} & 0 \end{bmatrix} \xrightarrow{R_3 + R_2 \to R_3} \begin{bmatrix} 0 & 0 & 0 & 0 \\ 0 & \frac{1}{2} & -\frac{1}{2} & 0 \\ 0 & 0 & 0 & 0 \end{bmatrix}$$

or $\frac{1}{2}p_2 - \frac{1}{2}p_3 = 0$. Thus p_1 and p_3 are free and

$$\mathbf{p} = \begin{bmatrix} p_1 \\ p_3 \\ p_3 \end{bmatrix} = p_1\begin{bmatrix} 1 \\ 0 \\ 0 \end{bmatrix} + p_3\begin{bmatrix} 0 \\ 1 \\ 1 \end{bmatrix}$$

Choosing $p_1 = 1, p_3 = 0$ and $p_1 = 0, p_3 = 1$ yields $\begin{bmatrix} 1 \\ 0 \\ 0 \end{bmatrix}$ and $\begin{bmatrix} 0 \\ 1 \\ 1 \end{bmatrix}$ as eigenvectors, the set of which is linearly independent. Thus the eigenvectors are $\mathbf{p} = p_1\begin{bmatrix} 1 \\ 0 \\ 0 \end{bmatrix} + p_3\begin{bmatrix} 0 \\ 1 \\ 1 \end{bmatrix}$ for all p_1 and p_3 not both 0. □

Putting the previous work on eigenvalues and eigenvectors together, we now show how to find diagonal canonical forms.

THEOREM 6.4

(for finding P and D such that $P^{-1}AP = D$) Let A be an $n \times n$ matrix with eigenvalues $\lambda_1, \ldots, \lambda_n$. Then A is similar to a diagonal matrix if and only if there is a linearly independent set $\{\mathbf{p}^1, \ldots, \mathbf{p}^n\}$ of n eigenvectors for A. In this case, if the eigenvector $\mathbf{p}^i$ belongs to λ_i for $i = 1, 2, \ldots, n$, then

$$P^{-1}AP = D$$

where $P = [\mathbf{p}^1 \ \cdots \ \mathbf{p}^n]$ and $D = \begin{bmatrix} \lambda_1 & \cdots & 0 \\ \vdots & \vdots & \vdots \\ 0 & \cdots & \lambda_n \end{bmatrix}$.*

* In this case, it is often said that A is diagonalizable and that P diagonalizes A.

PROOF

First, suppose that there are eigenvectors $\mathbf{p}^1, \ldots, \mathbf{p}^n$ belonging to $\lambda_1, \ldots, \lambda_n$ such that the set $\{\mathbf{p}^1, \ldots, \mathbf{p}^n\}$ is linearly independent. Define $P = [\mathbf{p}^1 \quad \cdots \quad \mathbf{p}^n]$. By Corollary 3.14C, P is invertible. Further,

$$\begin{aligned} AP &= A[\mathbf{p}^1 \quad \cdots \quad \mathbf{p}^n] = [A\mathbf{p}^1 \quad \cdots \quad A\mathbf{p}^n] = [\lambda_1\mathbf{p}^1 \quad \cdots \quad \lambda_n\mathbf{p}^n] \\ &= [\mathbf{p}^1 \quad \cdots \quad \mathbf{p}^n]\begin{bmatrix} \lambda_1 & \cdots & 0 \\ \vdots & \vdots & \vdots \\ 0 & \cdots & \lambda_n \end{bmatrix} = PD \end{aligned}$$

Thus

$$P^{-1}AP = D$$

Now suppose that A is similar to a diagonal matrix

$$D = \begin{bmatrix} \lambda_1 & \cdots & 0 \\ \vdots & \vdots & \vdots \\ 0 & \cdots & \lambda_n \end{bmatrix}$$

where by Corollaries 6.2A and 6.2B $\lambda_1, \ldots, \lambda_n$ are the eigenvalues of A. Then there is an invertible matrix P such that

$$P^{-1}AP = D$$

or

$$AP = PD$$

Writing out the individual columns yields

$$\begin{aligned} A\mathbf{p}^1 &= \lambda_1\mathbf{p}^1 \\ &\vdots \\ A\mathbf{p}^n &= \lambda_n\mathbf{p}^n \end{aligned}$$

Thus the eigenvalues $\lambda_1, \lambda_2, \ldots, \lambda_n$ of A have corresponding eigenvectors $\mathbf{p}^1, \mathbf{p}^2, \ldots, \mathbf{p}^n$, respectively. Finally, since P is invertible, by Corollary 3.14C $\{\mathbf{p}^1, \ldots, \mathbf{p}^n\}$ is a linearly independent set. ■

EXAMPLE 17

Let $A = \begin{bmatrix} 5 & 1 \\ 1 & 5 \end{bmatrix}$. Solving the characteristic equation

$$\det\begin{bmatrix} \lambda - 5 & -1 \\ -1 & \lambda - 5 \end{bmatrix} = 0$$

for the eigenvalues, we find $\lambda_1 = 6$ and $\lambda_2 = 4$.

For an eigenvector for $\lambda_1 = 6$, we solve

$$(\lambda_1 I_2 - A)\mathbf{p} = \mathbf{0}$$

$$\begin{bmatrix} 1 & -1 \\ -1 & 1 \end{bmatrix}\begin{bmatrix} p_1 \\ p_2 \end{bmatrix} = \mathbf{0}$$

Solving yields $\mathbf{p} = \begin{bmatrix} p_2 \\ p_2 \end{bmatrix}$ for all p_2. Now to choose a particular eigenvector, we let $p_2 = 1$ and obtain $\mathbf{p}^1 = \begin{bmatrix} 1 \\ 1 \end{bmatrix}$.

For an eigenvector for $\lambda_2 = 4$, we solve

$$(\lambda_2 I_2 - A)\mathbf{p} = \mathbf{0}$$
$$\begin{bmatrix} -1 & -1 \\ -1 & -1 \end{bmatrix}\begin{bmatrix} p_1 \\ p_2 \end{bmatrix} = \mathbf{0}$$

Solving and choosing a particular eigenvector, we have $\mathbf{p}^2 = \begin{bmatrix} 1 \\ -1 \end{bmatrix}$.

Since $\{\mathbf{p}^1, \mathbf{p}^2\}$ is linearly independent, A is similar to a diagonal matrix D, where

$$P = [\mathbf{p}^1 \quad \mathbf{p}^2] = \begin{bmatrix} 1 & 1 \\ 1 & -1 \end{bmatrix} \quad \text{and} \quad D = \begin{bmatrix} \lambda_1 & 0 \\ 0 & \lambda_2 \end{bmatrix} = \begin{bmatrix} 6 & 0 \\ 0 & 4 \end{bmatrix}$$

Note that the order of the eigenvectors in P must match the order of the eigenvalues in D.

Checking, we see that

$$P^{-1}AP = \tfrac{1}{2}\begin{bmatrix} 1 & 1 \\ 1 & -1 \end{bmatrix}\begin{bmatrix} 6 & 4 \\ 6 & -4 \end{bmatrix} = \begin{bmatrix} 6 & 0 \\ 0 & 4 \end{bmatrix} = D$$

□

EXAMPLE 18

Let $A = \begin{bmatrix} 1 & 2 \\ 0 & 1 \end{bmatrix}$. To calculate the eigenvalues of A, we solve

$$\det\begin{bmatrix} \lambda - 1 & -2 \\ 0 & \lambda - 1 \end{bmatrix} = 0$$

to get $\lambda = \lambda_1 = \lambda_2 = 1$.

For eigenvectors belonging to this eigenvalue, we solve

$$(\lambda I_2 - A)\mathbf{p} = \mathbf{0}$$
$$\begin{bmatrix} 0 & -2 \\ 0 & 0 \end{bmatrix}\mathbf{p} = \mathbf{0}$$

Solving this equation yields $\mathbf{p} = p_1\begin{bmatrix} 1 \\ 0 \end{bmatrix}$ for all $p_1 \in \mathbf{R}$. Hence there is no linearly independent set of two eigenvectors for A. Thus A is not similar to a diagonal matrix.

□

EXAMPLE 19

OPTIONAL—COMPLEX NUMBERS

Let $A = \begin{bmatrix} 0 & 1 \\ -1 & 0 \end{bmatrix}$. The characteristic equation is

$$\det\begin{bmatrix} \lambda & -1 \\ 1 & \lambda \end{bmatrix} = 0$$

or

$$\lambda^2 + 1 = 0$$

and so the eigenvalues are $\lambda = \lambda_1 = i$ and $\lambda = \lambda_2 = -i$.

For corresponding eigenvectors, we have the following. For $\lambda_1 = i$ we solve

$$(iI_2 - A)\mathbf{z} = \mathbf{0}$$

$$\begin{bmatrix} i & -1 \\ 1 & i \end{bmatrix}\mathbf{z} = \mathbf{0}$$

Solving by the Gaussian elimination algorithm yields

$$\begin{bmatrix} i & -1 & 0 \\ 1 & i & 0 \end{bmatrix} \xrightarrow{R_2 + iR_1 \to R_2} \begin{bmatrix} i & -1 & 0 \\ 0 & 0 & 0 \end{bmatrix}$$

Thus, $iz_1 - z_2 = 0$. Since z_2 is free, we have $z_1 = -iz_2$ and $\mathbf{z} = z_2\begin{bmatrix} -i \\ 1 \end{bmatrix}$.

For a particular eigenvector, we take $\mathbf{p}^1 = \begin{bmatrix} -i \\ 1 \end{bmatrix}$.

For $\lambda_2 = -i$, the corresponding eigenvector is $\mathbf{z} = z_2\begin{bmatrix} i \\ 1 \end{bmatrix}$, so we take $\mathbf{p}^2 = \begin{bmatrix} i \\ 1 \end{bmatrix}$. Now

$$P = [\mathbf{p}^1 \quad \mathbf{p}^2] = \begin{bmatrix} -i & i \\ 1 & 1 \end{bmatrix} \quad \text{and} \quad D = \begin{bmatrix} \lambda_1 & 0 \\ 0 & \lambda_2 \end{bmatrix} = \begin{bmatrix} i & 0 \\ 0 & -i \end{bmatrix}$$

and $P^{-1}AP = D$. □

In computing P we need to find n eigenvectors that form a linearly independent set. Checking sets of eigenvectors for linear independence is more easily accomplished by using the following theorem.

THEOREM 6.5

(for checking linear independence of eigenvectors) Let A be an $n \times n$ matrix. Suppose that A has m distinct eigenvalues $\lambda_1, \ldots, \lambda_m$ and that S_i is any linearly independent set of eigenvectors belonging to λ_i for $i = 1, \ldots, m$. If λ_i is of multiplicity k_i, then S_i contains no more than k_i vectors. Further, $S = S_1 \cup \cdots \cup S_m$ is linearly independent.

PROOF

Proving that S_i contains no more than k_i vectors is left as exercise 27. We will show that S is linearly independent.

Let $S = \{\mathbf{p}^1, \ldots, \mathbf{p}^q\}$ and suppose that S is linearly dependent. Take any smallest subset of these eigenvectors that is also linearly dependent. Reindexing the eigenvalues and eigenvectors, we can suppose that this smallest subset is $\{\mathbf{p}^1, \ldots, \mathbf{p}^r\}$; then every proper subset of this set is linearly independent. Since no eigenvector is $\mathbf{0}$, $r \geq 2$.

Write

$$a_1\mathbf{p}^1 + \cdots + a_r\mathbf{p}^r = \mathbf{0}$$

for some $a_1, \ldots, a_r \in \mathbf{R}$, not all 0. Since every proper subset of the set $\{\mathbf{p}^1, \ldots, \mathbf{p}^r\}$ is linearly independent, in fact none of $a_1, \ldots, a_r$ is 0. We note that $\mathbf{p}^1$ belongs to λ_1, $\mathbf{p}^2$ belongs to λ_1 or $\lambda_2, \ldots,$ and $\mathbf{p}^r$ belongs to λ_s for some s. Now, multiplying the above equation by $(\lambda_1 I_n - A)$, we get

$$a_1(\lambda_1 I_n - A)\mathbf{p}^1 + \cdots + a_r(\lambda_1 I_n - A)\mathbf{p}^r = \mathbf{0}$$

or

$$a_1(\lambda_1 - \lambda_1)\mathbf{p}^1 + \cdots + a_r(\lambda_1 - \lambda_s)\mathbf{p}^r = \mathbf{0}$$

Since $\{\mathbf{p}^1, \ldots, \mathbf{p}^r\}$ is linearly dependent, some $\mathbf{p}^i \notin S_1$, so we know that $\lambda_s \neq \lambda_1$. Hence $a_r(\lambda_1 - \lambda_s) \neq 0$. Since $a_1(\lambda_1 - \lambda_1) = 0$, those vectors among $\mathbf{p}^1, \ldots, \mathbf{p}^r$ with nonzero coefficient $a_i(\lambda_1 - \lambda_j)$ form a linearly dependent set smaller than $\{\mathbf{p}^1, \ldots, \mathbf{p}^r\}$. This is a contradiction from which it follows that S is linearly independent. ■

EXAMPLE 20

CONTINUING EXAMPLE 16

Let $A = \begin{bmatrix} 1 & 0 & 0 \\ 0 & \frac{1}{2} & \frac{1}{2} \\ 0 & \frac{1}{2} & \frac{1}{2} \end{bmatrix}$. The eigenvalues of A are $\lambda_1 = 0$ with multiplicity one and $\lambda_2 = 1$ with multiplicity two. As corresponding eigenvectors, for $\lambda_1 = 0$ we found $\mathbf{p}^1 = \begin{bmatrix} 0 \\ 1 \\ -1 \end{bmatrix}$, and for $\lambda_2 = 1$ we found $\mathbf{p}^2 = \begin{bmatrix} 1 \\ 0 \\ 0 \end{bmatrix}$ and $\mathbf{p}^3 = \begin{bmatrix} 0 \\ 1 \\ 1 \end{bmatrix}$. Since $S_2 = \{\mathbf{p}^2, \mathbf{p}^3\}$ is linearly independent, $S = S_1 \cup S_2 = \{\mathbf{p}^1, \mathbf{p}^2, \mathbf{p}^3\}$ is linearly independent. □

A corollary to Theorem 6.5 gives a case in which linear independence of the set of corresponding eigenvectors is assured and so needs no computational verification.

COROLLARY 6.5A

Let A be an $n \times n$ matrix. If A has n distinct eigenvalues, then any set of n corresponding eigenvectors is linearly independent.

PROOF

Note that each S_i for $i = 1, \ldots, n$ contains exactly one vector. Thus S contains n vectors, and by Theorem 6.5 S is linearly independent. ■

ALGORITHM FOR COMPUTING P AND D FOR WHICH $P^{-1}AP = D$

1. Solve the characteristic equation $\det(\lambda I_n - A) = 0$ for all eigenvalues $\lambda_1, \ldots, \lambda_m$, where λ_i has multiplicity k_i for all i.

2. For each distinct eigenvalue λ_i, solve $(\lambda_i I_n - A)\mathbf{p} = \mathbf{0}$ by the Gaussian elimination or Gauss-Jordan algorithm. Write the solution in the form

$$\mathbf{p} = p_1\mathbf{p}^{i_1} + \cdots + p_s\mathbf{p}^{i_s}$$

where $p_1, \ldots, p_s$ are the free variables and $\mathbf{p}^{i_1}, \ldots, \mathbf{p}^{i_s} \in \mathbf{R}^n$. Then $S_i = \{\mathbf{p}^{i_1}, \ldots, \mathbf{p}^{i_s}\}$ is a linearly independent set of eigenvectors for λ_i. If the number s of eigenvectors in S_i is less than k_i, the multiplicity of λ_i, then A is not similar to a diagonal matrix.

3. Using the eigenvalues $\lambda_1, \ldots, \lambda_m$ and corresponding eigenvectors in $S_1 \cup \cdots \cup S_m = \{\mathbf{p}^1, \ldots, \mathbf{p}^n\}$, form

$$P = [\mathbf{p}^1 \quad \cdots \quad \mathbf{p}^n] \qquad \text{and} \qquad D = \begin{bmatrix} \lambda_1 & 0 & \cdots & 0 & 0 \\ 0 & \lambda_2 & \cdots & 0 & 0 \\ \vdots & \vdots & & \vdots & \vdots \\ 0 & 0 & \cdots & \lambda_{n-1} & 0 \\ 0 & 0 & \cdots & 0 & \lambda_n \end{bmatrix}$$

Then $P^{-1}AP = D$.

EXAMPLE 21

Let $A = \begin{bmatrix} 2 & -2 & -1 \\ -1 & 3 & 1 \\ 2 & -4 & -1 \end{bmatrix}$. To compute the eigenvalues, set

$$\det\begin{bmatrix} \lambda - 2 & 2 & 1 \\ 1 & \lambda - 3 & -1 \\ -2 & 4 & \lambda + 1 \end{bmatrix} = 0$$

or

$$\lambda^3 - 4\lambda^2 + 5\lambda - 2 = 0$$

Then $\lambda_1 = \lambda_2 = 1$ and $\lambda_3 = 2$.

Computing eigenvectors for $\lambda_1 = \lambda_2 = 1$, set

$$\begin{bmatrix} -1 & 2 & 1 \\ 1 & -2 & -1 \\ -2 & 4 & 2 \end{bmatrix}\mathbf{p} = \mathbf{0}$$

Applying the Gaussian elimination algorithm, we find that

$$\begin{bmatrix} -1 & 2 & 1 & 0 \\ 1 & -2 & -1 & 0 \\ -2 & 4 & 2 & 0 \end{bmatrix} \xrightarrow[R_3 - 2R_1 \to R_3]{R_2 + R_1 \to R_2} \begin{bmatrix} -1 & 2 & 1 & 0 \\ 0 & 0 & 0 & 0 \\ 0 & 0 & 0 & 0 \end{bmatrix}$$

or $$-p_1 + 2p_2 + p_3 = 0$$

Now p_2 and p_3 are free and $p_1 = 2p_2 + p_3$, so

$$\mathbf{p} = \begin{bmatrix} p_1 \\ p_3 \\ p_3 \end{bmatrix} = p_2 \begin{bmatrix} 2 \\ 1 \\ 0 \end{bmatrix} + p_3 \begin{bmatrix} 1 \\ 0 \\ 1 \end{bmatrix}$$

for all $p_2, p_3 \in \mathbf{R}$. Thus $\mathbf{p}^1 = \begin{bmatrix} 2 \\ 1 \\ 0 \end{bmatrix}$ and $\mathbf{p}^2 = \begin{bmatrix} 1 \\ 0 \\ 1 \end{bmatrix}$.

Computing an eigenvector for $\lambda_3 = 2$, set

$$(\lambda_3 I_3 - A)\mathbf{p} = \mathbf{0}$$

$$\begin{bmatrix} 0 & 2 & 1 \\ 1 & -1 & -1 \\ -2 & 4 & 3 \end{bmatrix} \mathbf{p} = \mathbf{0}$$

Solving for an eigenvector, we choose $\mathbf{p}^3 = \begin{bmatrix} 1 \\ -1 \\ 2 \end{bmatrix}$.

Thus $\{\mathbf{p}^1, \mathbf{p}^2, \mathbf{p}^3\}$ is a linearly independent set, so for

$$P = \begin{bmatrix} 2 & 1 & 1 \\ 1 & 0 & -1 \\ 0 & 1 & 2 \end{bmatrix} \quad \text{and} \quad D = \begin{bmatrix} 1 & 0 & 0 \\ 0 & 1 & 0 \\ 0 & 0 & 2 \end{bmatrix}$$

we have $P^{-1}AP = D$. □

This concludes the work on finding eigenvalues and eigenvectors and using them to find diagonal canonical forms. In the next section we apply diagonal canonical forms in problem solving.

EXERCISES FOR SECTION 6.2

COMPUTATIONAL EXERCISES

In exercises 1 through 8, find all eigenvalues for the given matrix A and, for each eigenvalue λ, find all eigenvectors belonging to λ.

1. $A = \begin{bmatrix} -3 & 2 \\ -2 & 2 \end{bmatrix}$

2. $A = \begin{bmatrix} 2 & -1 \\ 2 & 5 \end{bmatrix}$

3. $A = \begin{bmatrix} 2 & 4 \\ -1 & -2 \end{bmatrix}$

4. $A = \begin{bmatrix} 3 & 2 \\ 0 & 4 \end{bmatrix}$

5. $A = \begin{bmatrix} 1 & 0 & 0 \\ 1 & 3 & -1 \\ -2 & 2 & 0 \end{bmatrix}$

6. $A = \begin{bmatrix} -1 & -2 & 3 \\ 0 & 3 & 2 \\ 0 & -2 & -2 \end{bmatrix}$

7. $A = \begin{bmatrix} -1 & 0 & 0 \\ 1 & 2 & -2 \\ 1 & -2 & 2 \end{bmatrix}$

8. $A = \begin{bmatrix} 2 & 0 & 0 \\ 0 & 0 & 2 \\ 0 & 2 & 0 \end{bmatrix}$

In exercises 9 through 18, each matrix A has distinct eigenvalues and thus is similar to a diagonal matrix. Find P and D such that $P^{-1}AP = D$.

9. $A = \begin{bmatrix} 1 & 1 \\ 1 & 1 \end{bmatrix}$

10. $A = \begin{bmatrix} 1 & -1 \\ -1 & 1 \end{bmatrix}$

11. $A = \begin{bmatrix} 2 & -3 \\ -4 & 3 \end{bmatrix}$

12. $A = \begin{bmatrix} -4 & 3 \\ -2 & 1 \end{bmatrix}$

13. $A = \begin{bmatrix} 3 & 2 \\ 1 & 1 \end{bmatrix}$

14. $A = \begin{bmatrix} 1 & 3 \\ 1 & 5 \end{bmatrix}$

15. $A = \begin{bmatrix} 1 & 0 & 0 \\ 0 & 1 & 2 \\ 0 & 2 & 1 \end{bmatrix}$

16. $A = \begin{bmatrix} 2 & 1 & 1 \\ 0 & -1 & 1 \\ 0 & 0 & 3 \end{bmatrix}$

17. $A = \begin{bmatrix} 2 & 1 & -1 \\ 1 & 2 & 1 \\ -1 & -1 & 2 \end{bmatrix}$

18. $A = \begin{bmatrix} 2 & 1 & -1 \\ -2 & -1 & 2 \\ 2 & 1 & -1 \end{bmatrix}$

In exercises 19 through 22, each matrix A has multiple eigenvalues. Decide which matrices are similar to diagonal matrices. For the ones that are, find P and D such that $P^{-1}AP = D$.

19. $A = \begin{bmatrix} 1 & 1 \\ 0 & 1 \end{bmatrix}$

20. $A = \begin{bmatrix} 2 & 0 \\ 0 & 2 \end{bmatrix}$

21. $A = \begin{bmatrix} -2 & 1 & -1 \\ 6 & -1 & 2 \\ 9 & -3 & 4 \end{bmatrix}$

22. $A = \begin{bmatrix} 3 & 2 & 0 \\ -1 & 2 & 2 \\ 1 & 2 & 1 \end{bmatrix}$

COMPLEX NUMBERS

In exercises 23 and 24, find all eigenvalues and corresponding eigenvectors for the given matrix. Using these, give a matrix P and a diagonal matrix D such that $D = P^{-1}AP$.

23. $A = \begin{bmatrix} 1 & 1 \\ -1 & 1 \end{bmatrix}$

24. $A = \begin{bmatrix} 1 & i \\ -i & 1 \end{bmatrix}$

THEORETICAL EXERCISES

25. Let A be an $n \times n$ matrix and λ an eigenvalue of A. The set

$$\mathbf{E}_\lambda = \{\mathbf{x} : (\lambda I_n - A)\mathbf{x} = \mathbf{0}\}$$

is called an *eigenspace* of A. Prove that $\mathbf{E}_\lambda$ is $\mathbf{0}$ together with the set of eigenvectors for A belonging to λ, and that $\mathbf{E}_\lambda$ is a subspace of $\mathbf{R}^n$. Graph the eigenspaces for

(a) $A = \begin{bmatrix} 1 & 2 \\ 2 & 1 \end{bmatrix}$ (b) $A = \begin{bmatrix} 1 & 2 \\ 0 & -1 \end{bmatrix}$

26. The eigenvalues for $A = \begin{bmatrix} 1 & 2 \\ 0 & 3 \end{bmatrix}$ are $\lambda_1 = 1$ and $\lambda_2 = 3$. Find an eigenvector $\mathbf{p}^1$ for λ_1 and an eigenvector $\mathbf{p}^2$ for λ_2. Show that $\mathbf{p}^1 + \mathbf{p}^2$ is not an eigenvector for A. Compare this result to the statement following Theorem 6.3.

27. Let λ be an eigenvalue of multiplicity m of a matrix A. Prove that any linearly independent set of eigenvectors of A belonging to λ has at most m elements. *Hint:* Suppose $\{\mathbf{p}^1, \ldots, \mathbf{p}^{m+1}\}$ is a linearly independent set of eigenvectors for A. Form

$$\{\mathbf{p}^1, \ldots, \mathbf{p}^{m+1}, \mathbf{e}^1, \ldots, \mathbf{e}^n\}$$

and apply the deletion algorithm to this set, obtaining

$$\{\mathbf{p}^1, \ldots, \mathbf{p}^{m+1}, \mathbf{q}^{m+2}, \ldots, \mathbf{q}^n\}$$

Construct the matrix

$$Q = [\mathbf{p}^1 \;\; \cdots \;\; \mathbf{p}^{m+1} \;\; \mathbf{q}^{m+2} \;\; \cdots \;\; \mathbf{q}^n]$$

Show that $AQ = QT$, where

$$T = \begin{bmatrix} \lambda_1 & \cdots & 0 & & & \\ \vdots & \vdots & \vdots & & & \\ 0 & \cdots & \lambda_{m+1} & \mathbf{t}^{m+2} & \cdots & \mathbf{t}^n \\ 0 & \cdots & 0 & & & \\ \vdots & \vdots & \vdots & & & \\ 0 & \cdots & 0 & & & \end{bmatrix}$$

and $\lambda = \lambda_1 = \cdots = \lambda_{m+1}$

Now the eigenvalues of A are the same as those of T. Show that T has λ as an eigenvalue of multiplicity at least $m + 1$.

28. For A in exercises 15 and 16, choose several different invertible matrices P that give different orderings of the eigenvalues in D. (A common mistake is to find a matrix P and construct a matrix D such that the order of eigenvalues in D is different from that of the corresponding eigenvectors in P.)

29. For A in exercises 15 and 16, choose several different invertible matrices P that give the same orderings of the eigenvalues in D.

30. For A in exercises 17 and 18, compute $P^{-1}AP$ and PAP^{-1}. Which ordering of the factors in the products gives a diagonal matrix? (A common mistake is to forget which ordering is correct. Note that since P contains the eigenvectors of A as columns, P is to the right of A.)

Computer Exercise

31. Graph the eigenvectors $\mathbf{x}$, holding $x_1 = 1$, for each eigenvalue of each matrix A below. These are the same matrices used in exercise 42 of Section 6.1. [Note in (a) that if ε is close to 0, small changes in ε can cause much larger changes in the eigenvector.]

(*a*) $A = \begin{bmatrix} 1 & \varepsilon \\ 1 & 1 \end{bmatrix}$ (*b*) $A = \begin{bmatrix} 1 & \varepsilon \\ \varepsilon & 1 \end{bmatrix}$

6.3 Applications of Canonical Forms (Optional)

In this section we show how diagonal canonical forms are used in a collection of problems. Notice throughout these problems that we always trade the problem for a simpler problem involving the diagonal canonical form, as diagrammed in the introduction to this chapter.

I. Geometrically Viewing Linear Transformations

Let A be an $n \times n$ matrix that is similar to a diagonal matrix D and let P be an invertible matrix such that $A = PDP^{-1}$. Consider the linear transformation

$$L(\mathbf{x}) = A\mathbf{x}$$

from $\mathbf{R}^n$ to $\mathbf{R}^n$. To view this transformation geometrically, we redescribe it in another coordinate system. First we substitute PDP^{-1} for A to get

$$L(\mathbf{x}) = PDP^{-1}\mathbf{x}$$

or

$$P^{-1}L(\mathbf{x}) = DP^{-1}\mathbf{x}$$

Using the algorithm of Section 5.3, we change coordinates by setting $\mathbf{y} = P^{-1}\mathbf{x}$ to get the redescription

$$L_{\#}(\mathbf{y}) = D\mathbf{y}$$

where $\# = (\mathbf{p}^1, \ldots, \mathbf{p}^n)$ and $\mathbf{y}$ is the coordinate vector of $\mathbf{x}$ in $(\mathbf{R}^n)_{\#}$ (that is, $\mathbf{y} = [\mathbf{x}]_{\#}$). This change is shown in Figure 6.2 for $n = 2$. Since

$$L_{\#}\left(\begin{bmatrix} y_1 \\ \vdots \\ y_n \end{bmatrix}\right) = \begin{bmatrix} \lambda_1 y_1 \\ \vdots \\ \lambda_n y_n \end{bmatrix}$$

we see this transformation as a stretching or shrinking and possibly a reflecting about the origin, as determined by the eigenvalues, along the directions of the axes determined by the eigenvectors $\mathbf{p}^1, \ldots, \mathbf{p}^n$.

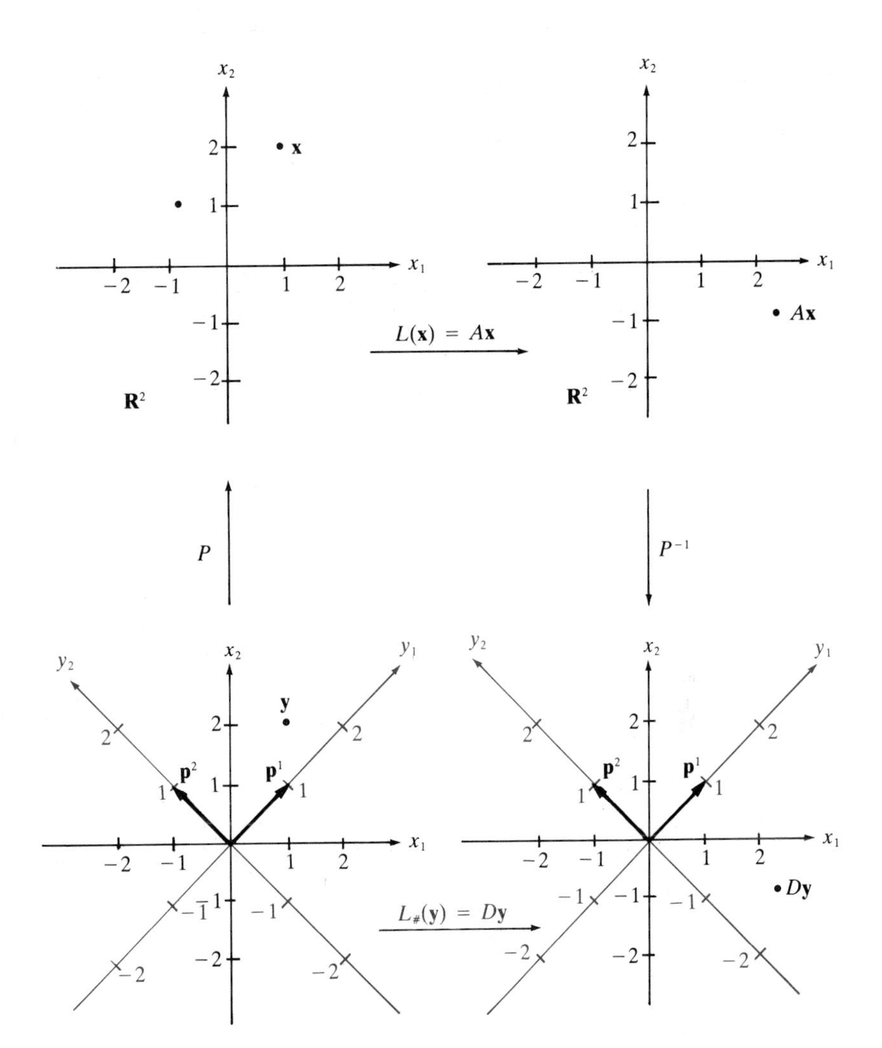

FIGURE 6.2

EXAMPLE 22

Let $A = \begin{bmatrix} 5 & 1 \\ 1 & 5 \end{bmatrix}$. Then, from Example 17 we have $D = \begin{bmatrix} 6 & 0 \\ 0 & 4 \end{bmatrix}$ and $P = \begin{bmatrix} 1 & 1 \\ 1 & -1 \end{bmatrix}$. Thus, $L(\mathbf{x}) = A\mathbf{x}$ viewed through $(\mathbf{R}^2)_\#$, where $\# = \left(\begin{bmatrix} 1 \\ 1 \end{bmatrix}, \begin{bmatrix} 1 \\ -1 \end{bmatrix}\right)$, is

$$L_\#(\mathbf{y}) = \begin{bmatrix} 6 & 0 \\ 0 & 4 \end{bmatrix}\mathbf{y} = \begin{bmatrix} 6y_1 \\ 4y_2 \end{bmatrix}$$

Hence, the mapping can be viewed as the plane stretched 6 units in the direction of the axis determined by $\mathbf{p}^1$ and stretched 4 units in the direction of the axis determined by $\mathbf{p}^2$, as shown in Figure 6.3. □

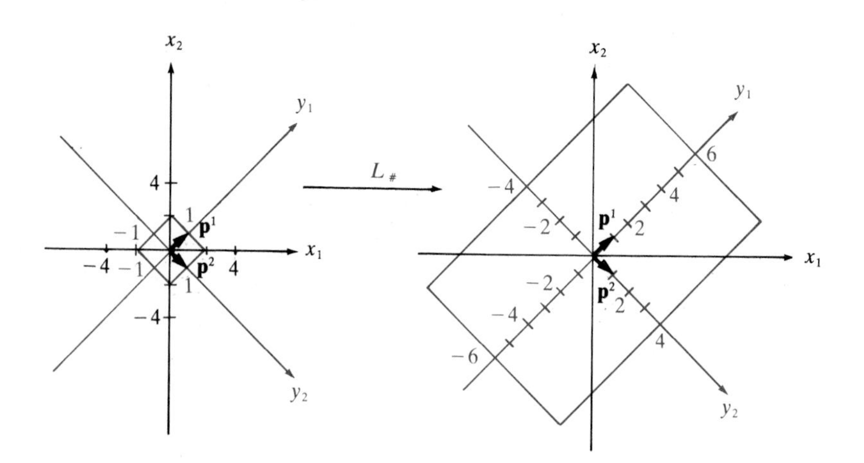

FIGURE 6.3

II. POWERS OF LINEAR MAPS

Let A be an $n \times n$ matrix that is similar to a diagonal matrix D and has the invertible matrix P. Thus $A = PDP^{-1}$. Let

$$L(\mathbf{x}) = A\mathbf{x}$$

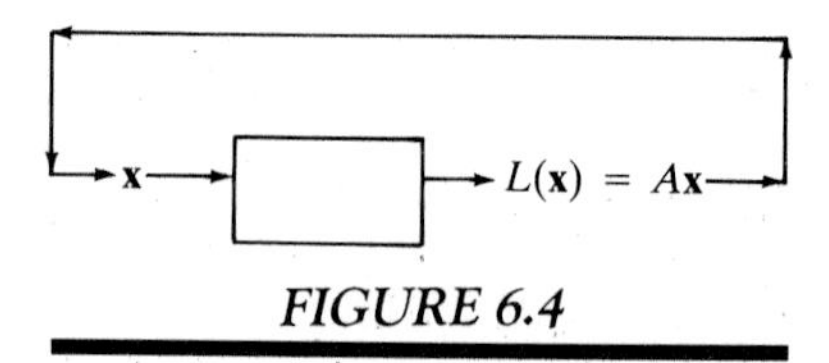

FIGURE 6.4

be a linear transformation. Interpreting this transformation as giving an output $L(\mathbf{x})$ for an input $\mathbf{x}$, as diagrammed in Figure 6.4, we continue to take the output as the new input, generating the outputs $A\mathbf{x}$, $A^2\mathbf{x}$, $A^3\mathbf{x}$, $A^4\mathbf{x}, \ldots$. In this problem, we want to know if these vectors *stabilize*—that is, get close to a fixed vector, called a *steady state.*

We can view the sequence of vectors more clearly if we describe them in another coordinate system. Substituting PDP^{-1} for A, we get

$$A\mathbf{x} = PDP^{-1}\mathbf{x}, \qquad A^2\mathbf{x} = (PDP^{-1})^2\mathbf{x} = PD^2P^{-1}\mathbf{x},$$
$$A^3\mathbf{x} = (PDP^{-1})^3\mathbf{x} = PD^3P^{-1}\mathbf{x}, \qquad \ldots$$

Multiplying through by P^{-1} yields

$$P^{-1}(A\mathbf{x}) = D(P^{-1}\mathbf{x}), \qquad P^{-1}(A^2\mathbf{x}) = D^2(P^{-1}\mathbf{x}),$$
$$P^{-1}(A^3\mathbf{x}) = D^3(P^{-1}\mathbf{x}), \qquad \ldots$$

Now, changing coordinates by setting $\mathbf{y} = P^{-1}\mathbf{x}$, we get

$$(A\mathbf{x})_\# = D\mathbf{y}, \qquad (A^2\mathbf{x})_\# = D^2\mathbf{y}, \qquad (A^3\mathbf{x})_\# = D^3\mathbf{y}, \qquad \ldots$$

where $\# = (\mathbf{p}^1, \ldots, \mathbf{p}^n)$. Thus, viewed in $(\mathbf{R}^n)_\#$, the sequence of vectors is

$$D\mathbf{y}, D^2\mathbf{y}, D^3\mathbf{y}, \ldots$$

Since

$$D^k = \begin{bmatrix} \lambda_1^k & \cdots & 0 \\ \vdots & & \vdots \\ 0 & \cdots & \lambda_n^k \end{bmatrix}$$

for all $k = 1, 2, \ldots$, it follows that we can expect stability if and only if

(i) $|\lambda_i| < 1$ or (ii) $\lambda_i = 1$

for each i. In this case, since $A^k\mathbf{x} = PD^kP^{-1}\mathbf{x}$, we can compute the steady state vector this sequence approaches by replacing D^k with the matrix D_0, which D^k tends toward as k increases.

EXAMPLE 23

Let $A = \frac{1}{3}\begin{bmatrix} 1 & 2 \\ 2 & 1 \end{bmatrix}$. The eigenvalues of A are $\lambda_1 = 1$ and $\lambda_2 = -\frac{1}{3}$. Eigenvectors belonging to these eigenvalues are $\mathbf{p}^1 = \begin{bmatrix} 1 \\ 1 \end{bmatrix}$ and $\mathbf{p}^2 = \begin{bmatrix} 1 \\ -1 \end{bmatrix}$, respectively. Thus,

$$D = \begin{bmatrix} 1 & 0 \\ 0 & -\frac{1}{3} \end{bmatrix}, \qquad P = \begin{bmatrix} 1 & 1 \\ 1 & -1 \end{bmatrix}, \qquad P^{-1} = \frac{1}{2}\begin{bmatrix} 1 & 1 \\ 1 & -1 \end{bmatrix}$$

Since $D^k = \begin{bmatrix} 1 & 0 \\ 0 & (-\frac{1}{3})^k \end{bmatrix}$, D^k approaches $D_0 = \begin{bmatrix} 1 & 0 \\ 0 & 0 \end{bmatrix}$ as k increases. Thus $D\mathbf{y}$, $D^2\mathbf{y}, D^3\mathbf{y}, \ldots$ approaches $D_0\mathbf{y}$ for any $\mathbf{y}$. Hence

$$A\mathbf{x} = PDP^{-1}\mathbf{x} = PD\mathbf{y}, \qquad A^2\mathbf{x} = PD^2P^{-1}\mathbf{x} = PD^2\mathbf{y}, \qquad \ldots,$$
$$A^k\mathbf{x} = PD^kP^{-1}\mathbf{x} = PD^k\mathbf{y}$$

approaches

$$PD_0\mathbf{y} = PD_0P^{-1}\mathbf{x} = \begin{bmatrix} 1 & 1 \\ 1 & -1 \end{bmatrix}\begin{bmatrix} 1 & 0 \\ 0 & 0 \end{bmatrix}(\tfrac{1}{2})\begin{bmatrix} 1 & 1 \\ 1 & -1 \end{bmatrix}\mathbf{x} = \frac{1}{2}\begin{bmatrix} 1 & 1 \\ 1 & 1 \end{bmatrix}\mathbf{x}$$

From this computation we see that, for any $\mathbf{x} \in \mathbf{R}^n$, $A\mathbf{x}, A^2\mathbf{x}, A^3\mathbf{x}, \ldots$ stabilizes and tends to the steady state vector

$$\frac{1}{2}\begin{bmatrix} 1 & 1 \\ 1 & 1 \end{bmatrix}\mathbf{x} = \begin{bmatrix} \dfrac{x_1 + x_2}{2} \\ \dfrac{x_1 + x_2}{2} \end{bmatrix}$$

as k increases. Thus, if $\mathbf{x} = \begin{bmatrix} 1 \\ 3 \end{bmatrix}$, the steady state vector is $\begin{bmatrix} 2 \\ 2 \end{bmatrix}$. □

EXAMPLE 24

Consider an exchange system as described in Example 36 of Section 1.4. Figure 6.5 shows the diagram for one such exchange system. The matrix determined from the diagram is $A = \frac{1}{3}\begin{bmatrix} 1 & 2 \\ 2 & 1 \end{bmatrix}$. Then, as above, for any $\mathbf{x}$ the sequence* $\mathbf{x}, \mathbf{x}A, \mathbf{x}A^2, \ldots$ stabilizes and goes to the steady state vector

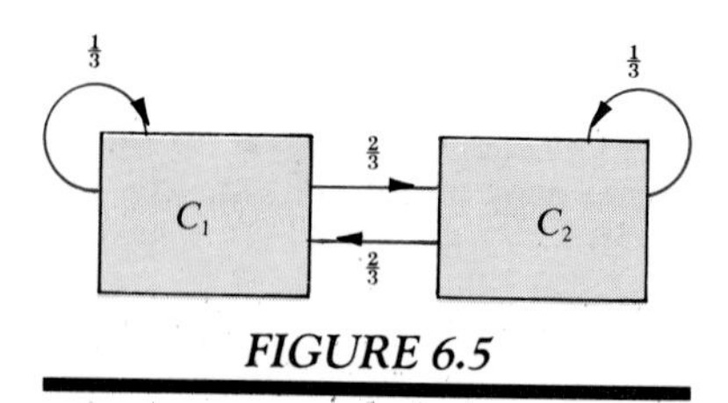

FIGURE 6.5

$$\frac{1}{2}\mathbf{x}\begin{bmatrix} 1 & 1 \\ 1 & 1 \end{bmatrix} = [\tfrac{1}{2}(x_1 + x_2), \tfrac{1}{2}(x_1 + x_2)]$$

Thus, regardless of the initial distribution, as exchanges increase, the numbers of objects in each set get closer to being equal. □

III. MATRIX EQUATIONS

Let A be an $n \times n$ matrix similar to a diagonal matrix D through an invertible matrix P. To find the matrices that commute with A, we need to find all $n \times n$ matrices X such that

$$AX = XA$$

To solve this equation, we redescribe it in another coordinate system (see exercise 25). First substitute $PDP^{-1} = A$ to get

$$PDP^{-1}X = XPDP^{-1}$$

Multiplying through by P^{-1} and P, we have

$$D(P^{-1}XP) = (P^{-1}XP)D$$

Set $Y = P^{-1}XP$ to get $DY = YD$, or

$$[\lambda_i y_{ij}] = [y_{ij}\lambda_j]$$

We can find all Y that satisfy this equation, and hence we can find all $X = PYP^{-1}$ for which $AX = XA$.

EXAMPLE 25

Solve $\begin{bmatrix} 5 & 1 \\ 1 & 5 \end{bmatrix} X = X \begin{bmatrix} 5 & 1 \\ 1 & 5 \end{bmatrix}$. By Example 17,

$$\begin{bmatrix} 5 & 1 \\ 1 & 5 \end{bmatrix} = PDP^{-1}$$

* The order of $\mathbf{x}$ and A^k in $\mathbf{x}A^k$ can be changed by using the transpose, if desired.

where $D = \begin{bmatrix} 6 & 0 \\ 0 & 4 \end{bmatrix}$ and $P = \begin{bmatrix} 1 & 1 \\ 1 & -1 \end{bmatrix}$. Then, by substitution,

$$PDP^{-1}X = XPDP^{-1}$$

or

$$D(P^{-1}XP) = (P^{-1}XP)D$$

Setting $Y = P^{-1}XP$ yields $DY = YD$, or

$$\begin{bmatrix} 6 & 0 \\ 0 & 4 \end{bmatrix} Y = Y \begin{bmatrix} 6 & 0 \\ 0 & 4 \end{bmatrix}$$

$$\begin{bmatrix} 6y_{11} & 6y_{12} \\ 4y_{21} & 4y_{22} \end{bmatrix} = \begin{bmatrix} 6y_{11} & 4y_{12} \\ 6y_{21} & 4y_{22} \end{bmatrix}$$

Equating the corresponding entries and solving yields

$$\begin{aligned} y_{11} &= \alpha, \text{ any number} \\ y_{12} &= y_{21} = 0 \\ y_{22} &= \beta, \text{ any number} \end{aligned}$$

Thus $Y = \begin{bmatrix} \alpha & 0 \\ 0 & \beta \end{bmatrix}$ for all $\alpha, \beta \in \mathbf{R}$, so

$$\begin{aligned} X = PYP^{-1} &= \begin{bmatrix} 1 & 1 \\ 1 & -1 \end{bmatrix} \begin{bmatrix} \alpha & 0 \\ 0 & \beta \end{bmatrix} \begin{bmatrix} \frac{1}{2} & \frac{1}{2} \\ \frac{1}{2} & -\frac{1}{2} \end{bmatrix} = \tfrac{1}{2} \begin{bmatrix} 1 & 1 \\ 1 & -1 \end{bmatrix} \begin{bmatrix} \alpha & \alpha \\ \beta & -\beta \end{bmatrix} \\ &= \tfrac{1}{2} \begin{bmatrix} \alpha + \beta & \alpha - \beta \\ \alpha - \beta & \alpha + \beta \end{bmatrix} = \tfrac{1}{2}\alpha \begin{bmatrix} 1 & 1 \\ 1 & 1 \end{bmatrix} + \tfrac{1}{2}\beta \begin{bmatrix} 1 & -1 \\ -1 & 1 \end{bmatrix} \end{aligned}$$

for all choices of $\alpha, \beta \in \mathbf{R}$. Note that this means that the matrices that commute with $A = \begin{bmatrix} 5 & 1 \\ 1 & 5 \end{bmatrix}$ yield a vector space of dimension 2. □

IV. SYSTEMS OF DIFFERENCE EQUATIONS

Let A be an $n \times n$ matrix that is similar to a diagonal matrix D and has invertible matrix P. Consider the difference equation

$$\mathbf{x}(k + 1) = A\mathbf{x}(k)$$

for $k = 0, 1, 2, \ldots$. To solve this equation, we redescribe it in another coordinate system. First substitute $PDP^{-1} = A$ to get

$$\mathbf{x}(k + 1) = PDP^{-1}\mathbf{x}(k)$$

or

$$P^{-1}\mathbf{x}(k + 1) = D[P^{-1}\mathbf{x}(k)]$$

Change coordinates by setting $\mathbf{y}(k) = P^{-1}\mathbf{x}(k)$ for all k. Then the equation is

$\mathbf{y}(k+1) = D\mathbf{y}(k)$, which, written entrywise, is

$$\begin{aligned} y_1(k+1) &= \lambda_1 y_1(k) \\ y_2(k+1) &= \lambda_2 y_2(k) \\ &\vdots \\ y_n(k+1) &= \lambda_n y_n(k) \end{aligned}$$

From Section 3.3 we know that the solutions of these equations are

$$\begin{aligned} y_1(k) &= a_1 \lambda_1^k \\ y_2(k) &= a_2 \lambda_2^k \\ &\vdots \\ y_n(k) &= a_n \lambda_n^k \end{aligned}$$

where $a_1, a_2, \ldots, a_n$ are numbers that can be chosen arbitrarily. For convenience in representing the solution, we let $0^0 = 1$ in these formulas. Then

$$\begin{aligned} \mathbf{x}(k) = P\mathbf{y}(k) &= y_1(k)\mathbf{p}^1 + y_2(k)\mathbf{p}^2 + \cdots + y_n(k)\mathbf{p}^n \\ &= a_1\lambda_1^k\mathbf{p}^1 + a_2\lambda_2^k\mathbf{p}^2 + \cdots + a_n\lambda_n^k\mathbf{p}^n \end{aligned}$$

for all choices of $a_1, a_2, \ldots, a_n$.

EXAMPLE 26

Let $A = \begin{bmatrix} 1 & 0 & 0 \\ 1 & 2 & 4 \\ 1 & 1 & 2 \end{bmatrix}$. The eigenvalues of A are $\lambda_1 = 0$, $\lambda_2 = 1$, and $\lambda_3 = 4$, with eigenvectors $\mathbf{p}^1 = \begin{bmatrix} 0 \\ -2 \\ 1 \end{bmatrix}$, $\mathbf{p}^2 = \begin{bmatrix} 1 \\ -1 \\ 0 \end{bmatrix}$, and $\mathbf{p}^3 = \begin{bmatrix} 0 \\ 2 \\ 1 \end{bmatrix}$, respectively, as found in Example 15.

The solution to the difference equation $\mathbf{x}(k+1) = A\mathbf{x}(k)$ can be written in terms of these eigenvalues and eigenvectors as

$$\mathbf{x}(k) = a_1 0^k \begin{bmatrix} 0 \\ -2 \\ 1 \end{bmatrix} + a_2 1^k \begin{bmatrix} 1 \\ -1 \\ 0 \end{bmatrix} + a_3 4^k \begin{bmatrix} 0 \\ 2 \\ 1 \end{bmatrix}$$

where a_1, a_2, and a_3 are arbitrarily chosen numbers. □

A particular solution can be obtained by adding some initial condition, say

$$\mathbf{x}(0) = \mathbf{a}$$

where $\mathbf{a} \in \mathbf{R}^n$. In this case we need to pick $a_1, \ldots, a_n$ so that

$$\mathbf{a} = \mathbf{x}(0) = a_1\lambda_1^0\mathbf{p}^1 + a_2\lambda_2^0\mathbf{p}^2 + \cdots + a_n\lambda_n^0\mathbf{p}^n$$

or $$\mathbf{a} = a_1\mathbf{p}^1 + a_2\mathbf{p}^2 + \cdots + a_n\mathbf{p}^n$$

Since P is invertible, its columns form a basis for $\mathbf{R}^n$, and so there is a unique solution of a_i's to this equation. This solution provides a unique solution to the system of difference equations that also satisfies the initial condition.

EXAMPLE 27

CONTINUING EXAMPLE 26

To solve

$$\mathbf{x}(k+1) = \begin{bmatrix} 1 & 0 & 0 \\ 1 & 2 & 4 \\ 1 & 1 & 2 \end{bmatrix} \mathbf{x}(k)$$

such that $\mathbf{x}(0) = \begin{bmatrix} 1 \\ -1 \\ 2 \end{bmatrix}$, we have

$$\mathbf{x}(k) = a_1 0^k \begin{bmatrix} 0 \\ -2 \\ 1 \end{bmatrix} + a_2 1^k \begin{bmatrix} 1 \\ -1 \\ 0 \end{bmatrix} + a_3 4^k \begin{bmatrix} 0 \\ 2 \\ 1 \end{bmatrix}$$

Now set

$$\begin{bmatrix} 1 \\ -1 \\ 2 \end{bmatrix} = \mathbf{x}(0) = a_1 \begin{bmatrix} 0 \\ -2 \\ 1 \end{bmatrix} + a_2 \begin{bmatrix} 1 \\ -1 \\ 0 \end{bmatrix} + a_3 \begin{bmatrix} 0 \\ 2 \\ 1 \end{bmatrix}$$

Solving this system of linear algebraic equations yields $a_1 = a_2 = a_3 = 1$. Hence

$$\mathbf{x}(k) = 0^k \begin{bmatrix} 0 \\ -2 \\ 1 \end{bmatrix} + 1^k \begin{bmatrix} 1 \\ -1 \\ 0 \end{bmatrix} + 4^k \begin{bmatrix} 0 \\ 2 \\ 1 \end{bmatrix}$$

solves the equation and initial condition. □

The system of linear difference equations is homogeneous, and thus its solution set is a vector space $\mathbf{W}$. As shown in Section 3.7, $\dim \mathbf{W} = n$. Thus we can find a basis for $\mathbf{W}$ as follows. Let

$$\mathbf{x}_1(k) = \lambda_1^k \mathbf{p}^1, \qquad \mathbf{x}_2(k) = \lambda_2^k \mathbf{p}^2, \qquad \ldots, \qquad \mathbf{x}_n(k) = \lambda_n^k \mathbf{p}^n$$

Then $\mathbf{x}(k) \in \mathbf{W}$ if and only if

$$\mathbf{x}(k) = a_1\mathbf{x}_1(k) + a_2\mathbf{x}_2(k) + \cdots + a_n\mathbf{x}_n(k)$$

for some $a_1, a_2, \ldots, a_n$. Thus

$$\mathbf{W} = \text{span}\{\mathbf{x}_1(k), \mathbf{x}_2(k), \ldots, \mathbf{x}_n(k)\}$$

Hence, $\{\mathbf{x}_1(k), \mathbf{x}_2(k), \ldots, \mathbf{x}_n(k)\}$ is a basis for $\mathbf{W}$.

As we have seen in the sampling of problems above, similarity can be used in solving numerous problems. In the next section we consider invariant functions, which constitute useful tools for working with similarity.

EXERCISES FOR SECTION 6.3

COMPUTATIONAL EXERCISES

In exercises 1 through 6, geometrically describe each linear transformation as in Example 22.

1. $L(\mathbf{x}) = \begin{bmatrix} 1 & 2 \\ 2 & 4 \end{bmatrix}\mathbf{x}$

2. $L(\mathbf{x}) = \begin{bmatrix} -3 & 1 \\ 2 & -4 \end{bmatrix}\mathbf{x}$

3. $L(\mathbf{x}) = \begin{bmatrix} -4 & 2 \\ 2 & -4 \end{bmatrix}\mathbf{x}$

4. $L(\mathbf{x}) = \begin{bmatrix} 1 & 4 \\ 3 & 2 \end{bmatrix}\mathbf{x}$

5. $L(\mathbf{x}) = \begin{bmatrix} 1 & 0 & 0 \\ 0 & 2 & 1 \\ 0 & 1 & 2 \end{bmatrix}\mathbf{x}$

6. $L(\mathbf{x}) = \begin{bmatrix} 2 & 1 & 1 \\ 0 & 1 & -1 \\ 0 & -1 & 1 \end{bmatrix}\mathbf{x}$

In exercises 7 through 12, compute A^k and the matrix that it approaches as k increases.

7. $A = \begin{bmatrix} \frac{1}{2} & \frac{1}{4} \\ \frac{1}{4} & \frac{1}{2} \end{bmatrix}$

8. $A = \begin{bmatrix} \frac{1}{2} & \frac{1}{2} \\ \frac{1}{2} & \frac{1}{2} \end{bmatrix}$

9. $A = \begin{bmatrix} \frac{1}{4} & \frac{3}{4} \\ \frac{3}{4} & \frac{1}{4} \end{bmatrix}$

10. $A = \begin{bmatrix} \frac{1}{6} & \frac{2}{3} \\ \frac{2}{3} & \frac{1}{6} \end{bmatrix}$

11. $A = \begin{bmatrix} \frac{2}{3} & 0 & 0 \\ \frac{2}{3} & \frac{2}{3} & \frac{1}{3} \\ \frac{2}{3} & \frac{1}{3} & \frac{2}{3} \end{bmatrix}$

12. $A = \begin{bmatrix} \frac{1}{2} & 0 & 0 \\ \frac{1}{2} & \frac{1}{4} & \frac{3}{4} \\ \frac{1}{2} & \frac{3}{4} & \frac{1}{4} \end{bmatrix}$

In exercises 13 through 18, solve for the matrix X.

13. $\begin{bmatrix} 1 & 0 \\ 2 & 3 \end{bmatrix}X = X\begin{bmatrix} 1 & 0 \\ 2 & 3 \end{bmatrix}$

14. $\begin{bmatrix} 1 & 2 \\ 0 & 3 \end{bmatrix}X = X\begin{bmatrix} 1 & 2 \\ 0 & 3 \end{bmatrix}$

15. $\begin{bmatrix} -1 & 3 \\ 4 & -2 \end{bmatrix}X = X\begin{bmatrix} -1 & 3 \\ 4 & -2 \end{bmatrix}$

16. $\begin{bmatrix} 4 & -1 \\ 2 & 1 \end{bmatrix}X = X\begin{bmatrix} 4 & -1 \\ 2 & 1 \end{bmatrix}$

17. $\begin{bmatrix} 2 & 4 & -8 \\ 0 & 3 & -2 \\ 0 & 2 & -2 \end{bmatrix}X = X\begin{bmatrix} 2 & 4 & -8 \\ 0 & 3 & -2 \\ 0 & 2 & -2 \end{bmatrix}$

18. $\begin{bmatrix} 1 & 0 & 0 \\ 1 & 2 & 2 \\ 1 & 3 & 1 \end{bmatrix}X = X\begin{bmatrix} 1 & 0 & 0 \\ 1 & 2 & 2 \\ 1 & 3 & 1 \end{bmatrix}$

In exercises 19 through 22, solve for $\mathbf{x}(k)$.

19. $\mathbf{x}(k+1) = \begin{bmatrix} 0 & 1 \\ 1 & 0 \end{bmatrix}\mathbf{x}(k)$, $\mathbf{x}(0) = \begin{bmatrix} 3 \\ -1 \end{bmatrix}$

20. $\mathbf{x}(k+1) = \begin{bmatrix} 1 & -1 \\ -1 & 1 \end{bmatrix}\mathbf{x}(k)$, $\mathbf{x}(0) = \begin{bmatrix} 2 \\ 0 \end{bmatrix}$

21. $\mathbf{x}(k+1) = \begin{bmatrix} 4 & 0 & 0 \\ -2 & 2 & 5 \\ 2 & 2 & -1 \end{bmatrix}\mathbf{x}(k)$, $\mathbf{x}(0) = \begin{bmatrix} 7 \\ -5 \\ 5 \end{bmatrix}$

22. $\mathbf{x}(k+1) = \begin{bmatrix} 2 & 1 & 1 \\ 0 & 2 & -1 \\ 0 & -1 & 2 \end{bmatrix}\mathbf{x}(k)$, $\mathbf{x}(0) = \begin{bmatrix} -3 \\ 1 \\ 3 \end{bmatrix}$

COMPLEX NUMBERS

23. Given that $A = \begin{bmatrix} 0 & 1 \\ -1 & 0 \end{bmatrix}$, compute A^k using the form PD^kP^{-1}.

24. Find a diagonal canonical form for $A = \begin{bmatrix} 2 & i \\ i & 2 \end{bmatrix}$. Then solve $\mathbf{x}(k+1) = A\mathbf{x}(k)$ with $\mathbf{x}(0) = \begin{bmatrix} 1 \\ 5 \end{bmatrix}$.

THEORETICAL EXERCISES

25. Let A be an $n \times n$ matrix. Consider the equation $AX = XA$. Let $\# = (\mathbf{p}^1, \ldots, \mathbf{p}^n)$ be a coordinate system for

$\mathbf{R}^n$. Prove that $L_1(\mathbf{x}) = A\mathbf{x}$ and $L_2(\mathbf{x}) = X\mathbf{x}$ are

$$L_{1\#}(\mathbf{y}) = P^{-1}AP\mathbf{y}$$

and

$$L_{2\#}(\mathbf{y}) = P^{-1}XP\mathbf{y}$$

in $(\mathbf{R}^n)_\#$, and that the equation $AX = XA$ is

$$(P^{-1}AP)(P^{-1}XP) = (P^{-1}XP)(P^{-1}AP)$$

there.

26. Let A be an $n \times n$ matrix that is similar to a diagonal matrix. Suppose the eigenvalues of A are all nonnegative. Show how to find a square root of A; that is, find a matrix B such that $B^2 = A$. Then find a square root for $\begin{bmatrix} 2 & 1 \\ 1 & 2 \end{bmatrix}$.

Hint: Trade this problem for a simpler one.

27. Let A and B be $n \times n$ matrices, each of which is similar to a diagonal matrix. Describe how you would solve

$$AX = XB$$

Show how your method applies to

$$\begin{bmatrix} 1 & 3 \\ 3 & 1 \end{bmatrix} X = X \begin{bmatrix} 4 & 1 \\ 0 & -1 \end{bmatrix}$$

Hint: Substitute $A = PD_1P^{-1}$ and $B = QD_2Q^{-1}$.

28. Let A be an $n \times n$ matrix that is similar to a diagonal matrix. Describe how to solve

$$A\mathbf{x}(k + 1) = \mathbf{x}(k)$$

(Note that A may not be invertible.) Show how your method applies to

$$\begin{bmatrix} 1 & 1 \\ 1 & 1 \end{bmatrix} \mathbf{x}(k + 1) = \mathbf{x}(k)$$

Hint: Exchange this problem for a simpler one.

29. (nonhomogeneous systems) Let A be an $n \times n$ matrix that is similar to a diagonal matrix. Consider the nonhomogeneous difference equation

$$\mathbf{x}(k + 1) = A\mathbf{x}(k) + \mathbf{u}(k)$$

(*a*) Show how, by substituting $A = PDP^{-1}$, this equation can be written as

$$\mathbf{y}(k + 1) = D\mathbf{y}(k) + \mathbf{w}(k)$$

where $\mathbf{w}(k) = P^{-1}\mathbf{u}(k)$. [This equation can be solved for $\mathbf{y}(k)$, and thus $\mathbf{x}(k)$ can be obtained.]

(*b*) Solve $\mathbf{x}(k + 1) = \begin{bmatrix} 2 & 1 \\ 1 & 2 \end{bmatrix} \mathbf{x}(k) + \begin{bmatrix} 1 \\ 1 \end{bmatrix}$ with $\mathbf{x}(0) = \begin{bmatrix} 1 \\ 0 \end{bmatrix}$, using the above ideas.

30. Let A be an $n \times n$ matrix that is similar to a diagonal matrix. Let

$$c(\lambda) = \det(\lambda I_n - A) = \lambda^n + a_{n-1}\lambda^{n-1} + \cdots + a_1\lambda + a_0$$

Prove that

$$c(A) = A^n + a_{n-1}A^{n-1} + \cdots + a_1 A + a_0 I_n = \mathbf{0}$$

Hint: Substitute $A = PDP^{-1}$ and note that $c(\lambda_i) = 0$ for all i. (This result, known as the *Cayley-Hamilton Theorem*, can be proved for any $n \times n$ matrix A.)

31. Let A be an $n \times n$ matrix that is similar to a diagonal matrix. Let the eigenvalues of A be $\lambda_1, \lambda_2, \ldots, \lambda_n$, and let k be any positive integer. Prove the following:

(*a*) The eigenvalues of A^k are $\lambda_1^k, \lambda_2^k, \ldots, \lambda_n^k$.

(*b*) The eigenvalues of cA^k are $c\lambda_1^k, c\lambda_2^k, \ldots, c\lambda_n^k$.

(*c*) The eigenvalues of $p(A)$ are $p(\lambda_1), p(\lambda_2), \ldots, p(\lambda_n)$, where $p(\lambda)$ is any polynomial.

Hint: Trade this problem for a simpler one. (This result can be proved for any $n \times n$ matrix.)

APPLICATIONS EXERCISES

In exercises 32 and 33, write the equations for the water tower system shown and determine the levels for any k.

32.

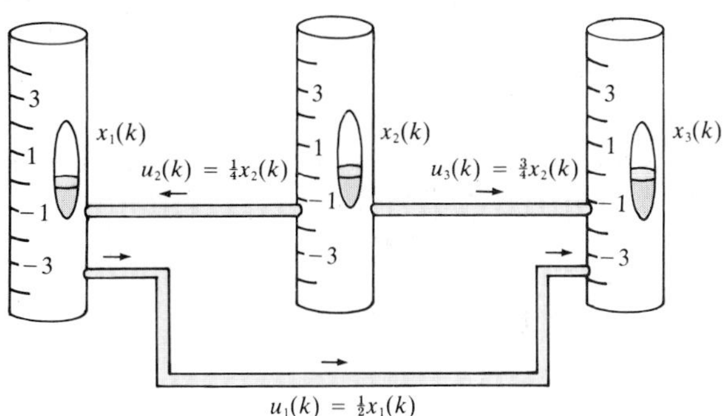

33.

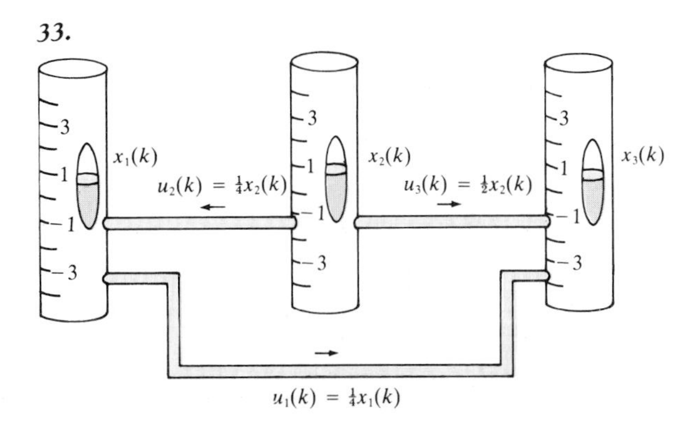

In exercises 34 and 35, write the matrix A for the given diagram (see Example 24). For initial distribution $\mathbf{x}$, compute the vector that $\mathbf{x}A^k$ approaches as k increases.

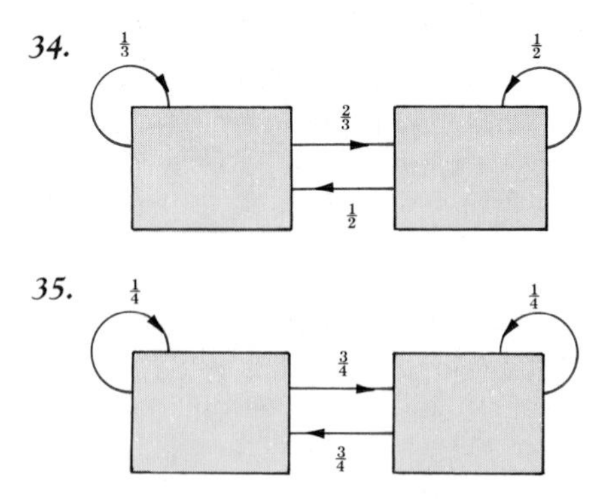

36. Consider two piles of sand, with x_1 pounds of sand in the first and x_2 pounds of sand in the second. Suppose that at times $t = 1, 2, \ldots$, one-third of the sand in pile 1 and one-half of the sand in pile 2 is exchanged between the piles. As t increases, which pile will have more sand?

37. A population of animals is grouped according to age into three categories: childhood, youth, and adulthood, with a, b, c animals, respectively, in the age groups. At the end of the kth year, for $k = 1, 2, \ldots$, a count is made of the number of animals in each category and is recorded as $x_1(k)$, $x_2(k)$, $x_3(k)$, respectively.

Suppose it is observed that each year 9% of the animals in childhood die, 1% of those in childhood change to youth, 12% of those in youth die and 1% change to adulthood, and 10% of those in adulthood die. It is also observed that adults reproduce at a rate of 2% of the total number of adults.

Write the difference equations that describe this situation. As time passes, what happens to the number of animals in each category?

6.4 SOME INVARIANTS FOR SIMILARITY (OPTIONAL)

In the previous section we saw that similarity arises in solving numerous problems. In this section, we define invariant functions, a tool often used in working with similarity. An *invariant for similarity* is a function f from $\mathbf{R}_{n,n}$ to $\mathbf{R}$ that is constant on sets of similar matrices.* Thus, if A and B are similar, then $f(A) = f(B)$.

EXAMPLE 28

Let $f(A) = 1$ for all $A \in \mathbf{R}_{n,n}$. Then f is an invariant for similarity. □

We also give an example of a function that is not an invariant for similarity.

EXAMPLE 29

For each $A \in \mathbf{R}_{2,2}$, let

$$f(A) = a_{11} + a_{12} + a_{21} + a_{22}$$

By Example 17 of Section 6.2, $A = \begin{bmatrix} 5 & 1 \\ 1 & 5 \end{bmatrix}$ is similar to $B = \begin{bmatrix} 6 & 0 \\ 0 & 4 \end{bmatrix}$. But $f(A) = 12$ while $f(B) = 10$, so this f is not an invariant for similarity. □

Two useful invariants, the determinant and the trace, are given in the next two examples.

* "Invariant" means "not varying" or "unchanging."

EXAMPLE 30

Let $f(A) = \det A$. To show that f is an invariant for similarity, let A be similar to B. Then there is an invertible matrix P such that $A = P^{-1}BP$. Thus

$$f(A) = \det A = \det(P^{-1}BP) = (\det P^{-1})(\det B)(\det P) = \det B = f(B)$$

Hence this f does not change values within sets of similar matrices, and so the determinant is an invariant for similarity. □

Define

$$\operatorname{trace} A = \sum_{k=1}^{n} a_{kk}$$

for any $A \in \mathbf{R}_{n,n}$. Then, for any $B \in \mathbf{R}_{n,n}$, trace AB = trace BA (exercise 23).

EXAMPLE 31

Let $f(A) = \operatorname{trace} A$. To show that f is an invariant for similarity, suppose A is similar to B. Then there is an invertible matrix P such that $A = P^{-1}BP$. Thus

$$f(A) = \operatorname{trace} A = \operatorname{trace} P^{-1}(BP) = \operatorname{trace}(BP)P^{-1} = \operatorname{trace} B = f(B)$$

Hence f does not change values within a set of similar matrices, and so the trace is an invariant for similarity. □

This pair of invariants cannot be used to show that two matrices are similar, as the following example illustrates.

EXAMPLE 32

Let $A = \begin{bmatrix} 2 & 1 \\ 0 & 2 \end{bmatrix}$ and $B = \begin{bmatrix} 2 & 0 \\ 0 & 2 \end{bmatrix}$. Then $\det A = \det B$ and trace A = trace B. However, Example 13 shows that A is not similar to any diagonal matrix, so A is not similar to B. □

Nevertheless, invariants can be used to show when two matrices are not similar.

EXAMPLE 33

Let $A = \begin{bmatrix} 4 & 2 \\ 2 & 4 \end{bmatrix}$. Suppose a computer is used to calculate a diagonal matrix similar to A and gives the result $D = \begin{bmatrix} 3 & 0 \\ 0 & 2 \end{bmatrix}$. Has round-off or some other error contaminated the computation?

Since trace $A = 8$ and trace $D = 5$, we conclude that the computation was not done accurately. □

Another use of these invariants is obtained by linking them to eigenvalues.

THEOREM 6.6

Let A be an $n \times n$ matrix with eigenvalues $\lambda_1, \ldots, \lambda_n$. Then
(*i*) $\det A = \lambda_1 \lambda_2 \cdots \lambda_n$
(*ii*) $\operatorname{trace} A = \lambda_1 + \lambda_2 + \cdots + \lambda_n$

PROOF

We obtain these two expressions by computing $\det(\lambda I_n - A)$ in two different ways. First, write

$$\det(\lambda I_n - A) = \lambda^n + b_{n-1}\lambda^{n-1} + \cdots + b_0$$

Then setting $\lambda = 0$, we get $b_0 = \det(-A) = (-1)^n \det A$. Expressing $\det(\lambda I_n - A)$ according to the definition and then adding all terms of degree $n - 1$, we have

$$b_{n-1} = -a_{11} - a_{22} - \cdots - a_{nn} = -\operatorname{trace} A$$

Second, using the Fundamental Theorem of Algebra, factor $\det(\lambda I_n - A)$ using the eigenvalues of A to get

$$\det(\lambda I_n - A) = (\lambda - \lambda_1)(\lambda - \lambda_2)\cdots(\lambda - \lambda_n)$$

Then

$$b_0 = (-\lambda_1)(-\lambda_2)\cdots(-\lambda_n) = (-1)^n \lambda_1 \lambda_2 \cdots \lambda_n$$

and

$$b_{n-1} = -\lambda_1 - \lambda_2 - \cdots - \lambda_n = -(\lambda_1 + \lambda_2 + \cdots + \lambda_n)$$

Equating expressions for b_0 and b_{n-1} yields

$$\det A = \lambda_1 \lambda_2 \cdots \lambda_n \quad \text{and} \quad \operatorname{trace} A = \lambda_1 + \lambda_2 + \cdots + \lambda_n$$ ■

Some uses of this theorem follow.

EXAMPLE 34

Let $A = \begin{bmatrix} 1 & 1 \\ 1 & 1 \end{bmatrix}$. We can compute the eigenvalues of A as follows. Since $\lambda_1 \lambda_2 = \det A = 0$, we know that one of the eigenvalues, say λ_1, is 0. Since $\lambda_1 + \lambda_2 = \operatorname{trace} A = 2$, we see that the other eigenvalue, λ_2, is 2. (See exercise 25.) □

EXAMPLE 35

Let $A = \begin{bmatrix} 1 & 2 & 3 \\ 4 & 5 & 6 \\ 7 & 8 & 9 \end{bmatrix}$. Suppose a computer is used to compute the eigenvalues of A and gives $\lambda_1 = 1$, $\lambda_2 = 3$, and $\lambda_3 = 6$. Has the computer done the work accurately?

Since trace $A = 1 + 5 + 9 = 15$ whereas $\lambda_1 + \lambda_2 + \lambda_3 = 1 + 3 + 6 = 10$, the answer is "no." □

This concludes our work on similarity. In summary, we should say that most matrices (see exercise 24) are similar to diagonal matrices, which, as we saw throughout the chapter, provide useful canonical forms. However, other matrices are not similar to diagonal matrices (see Example 13 of Section 6.2). For such matrices, the established canonical form is called the Jordan canonical form. This topic is covered in more advanced texts on linear algebra. However, we will show in the next chapter that every matrix is similar to another canonical form, an upper triangular matrix. Thus, this latter type of canonical form can be used when needed.

EXERCISES FOR SECTION 6.4

COMPUTATIONAL EXERCISES

In exercises 1 through 4, compute the trace of the matrix A.

1. $A = \begin{bmatrix} 1 & 7 & 3 \\ 2 & -1 & 2 \\ 0 & 0 & 5 \end{bmatrix}$

2. $A = \begin{bmatrix} 6 & -3 & 2 \\ 4 & -2 & 8 \\ 5 & 5 & 4 \end{bmatrix}$

3. $A = \begin{bmatrix} -3 & 2 & 3 & 8 \\ 2 & 12 & 7 & -8 \\ 4 & 3 & -1 & 4 \\ -8 & 6 & -7 & -5 \end{bmatrix}$

4. $A = \begin{bmatrix} 17 & 2 & 3 & 5 \\ 61 & 3 & 2 & 3 \\ 5 & -9 & 8 & \frac{2}{3} \\ -5 & -2 & 2 & -16 \end{bmatrix}$

In exercises 5 through 10, use the trace and determinant of A to find the eigenvalues of A.

5. $A = \begin{bmatrix} 2 & 2 \\ 2 & 2 \end{bmatrix}$

6. $A = \begin{bmatrix} 1 & 4 \\ -2 & -8 \end{bmatrix}$

7. $A = \begin{bmatrix} 3 & -5 \\ 2 & -4 \end{bmatrix}$

8. $A = \begin{bmatrix} 5 & 2 \\ -3 & -2 \end{bmatrix}$

9. $A = \begin{bmatrix} 7 & -1 \\ 9 & -3 \end{bmatrix}$

10. $A = \begin{bmatrix} 4 & 8 \\ 5 & 7 \end{bmatrix}$

In exercises 11 through 16, each of the matrices A has real eigenvalues. Determine the signs of these eigenvalues by using invariants.

11. $A = \begin{bmatrix} 1 & 3 \\ 3 & 1 \end{bmatrix}$

12. $A = \begin{bmatrix} 2 & 4 \\ 4 & -2 \end{bmatrix}$

13. $A = \begin{bmatrix} 4 & 2 \\ 2 & 4 \end{bmatrix}$

14. $A = \begin{bmatrix} 6 & -1 \\ -1 & 6 \end{bmatrix}$

15. $A = \begin{bmatrix} -1 & 2 \\ 2 & -1 \end{bmatrix}$

16. $A = \begin{bmatrix} 0 & 1 \\ 1 & -2 \end{bmatrix}$

In exercises 17 through 20, given the matrix A and the proposed eigenvalues, decide if the eigenvalues are correct.

17. $A = \begin{bmatrix} 3 & -1 & 2 \\ 5 & 1 & -2 \\ 4 & 1 & 2 \end{bmatrix}$; $\lambda = 4, 1, 1$

18. $A = \begin{bmatrix} 40 & -20 & 10 \\ 30 & 10 & 10 \\ 198 & -10 & 20 \end{bmatrix}$; $\lambda = -20, 45 + \sqrt{105}, 45 - \sqrt{105}$

19. $A = \begin{bmatrix} 6 & 18 & -6 \\ 12 & 6 & 24 \\ 4 & 6 & 12 \end{bmatrix}$; $\lambda = 6, 9 + \sqrt{345}, 9 - \sqrt{345}$

20. $A = \begin{bmatrix} 8 & 7 & 9 \\ 6 & 4 & 13 \\ 12 & 2 & 5 \end{bmatrix}$; $\lambda = 9, 4 + \sqrt{71}, 4 - \sqrt{71}$

Complex Numbers

In exercises 21 and 22, decide if the given numbers are the eigenvalues of the given matrix A.

21. $A = \begin{bmatrix} 1+i & i \\ 1 & 2-i \end{bmatrix}$; $\lambda_1 = 1 - i, \lambda_2 = 3 + 2i$

22. $A = \begin{bmatrix} 1 & i \\ i & 1 \end{bmatrix}$; $\lambda_1 = 1 + i, \lambda_2 = 1 - i$

Theoretical Exercises

23. Let $A, B \in \mathbf{R}_{n,n}$. Prove that trace AB = trace BA. *Hint:* Write the expression for the i,i-entry of AB and sum for $i = 1, 2, \ldots, n$. Do the same for BA. Then compare.

24. Show that if you arbitrarily select a 2×2 matrix, you should expect it to have two distinct eigenvalues. *Hint:* Let $A = \begin{bmatrix} a & b \\ c & d \end{bmatrix}$ and assume A has eigenvalues $\lambda_1 = \lambda_2 = \lambda$. Using invariants, find λ in terms of the entries of A. Note the rather strong relations imposed on these entries. (This result can be extended to $n \times n$ matrices.)

25. If A is a 3×3 matrix, can you determine the eigenvalues of this matrix by using the determinant and the trace?

26. If A is a 3×3 matrix with real eigenvalues, can you determine the signs of these numbers by using the determinant and the trace?

Computer Exercises

27. Let $A = \begin{bmatrix} 1 & 1 & 1 & 1 \\ 1 & 1 & 1 & 1 \\ 1 & 1 & 1 & 1 \\ 1 & 1 & 1 & 1 \end{bmatrix}$. A computer calculates that

$$D = \begin{bmatrix} 10^{-6} & 0 & 0 & 0 \\ 0 & 10^{-6} & 0 & 0 \\ 0 & 0 & 0 & 0 \\ 0 & 0 & 0 & 4 \end{bmatrix}$$

is similar to A. Is this calculation correct? Would you guess it is close to being correct?

28. A computer calculates the eigenvalues of

$$A = \begin{bmatrix} 1 & 1 & 1 & 1 \\ 1 & 1 & 1 & 1 \\ 1 & 1 & 1 & 1 \\ 1 & 1 & 1 & 1 \end{bmatrix}$$

and gets $\lambda_1 = 1$, $\lambda_2 = 1$, $\lambda_3 = 1$, $\lambda_4 = 1$. Is the calculation accurate? Why?

29. Suppose the eigenvalues of a 20×20 matrix are calculated. Which is the easier check for error—the determinant or the trace?

CHAPTER SEVEN

ORTHOGONAL SIMILARITY AND PROBLEM SOLVING

As shown in the last chapter, problems described in one coordinate system can often be redescribed more simply by using a different coordinate system. This redescription is an important problem-solving technique. In this chapter we consider a more special, but similar, problem-solving technique which can be of particular use in problems where geometry is important.

The vector space $\mathbf{R}^n$ is an inner product space with the standard inner product

$$(\mathbf{x}, \mathbf{y}) = x_1 y_1 + x_2 y_2 + \cdots + x_n y_n$$

where $\mathbf{x}, \mathbf{y} \in \mathbf{R}^n$. In this inner product space, we have the standard geometrical concepts such as distance, length, and angle. In some problems the geometry of the problem is important. In such cases we want to change the problem to a simpler one in some coordinate space of $\mathbf{R}^n$ so that the geometry is preserved. The diagram in Figure 7.1 depicts this situation. In this chapter we show how this change can be made.

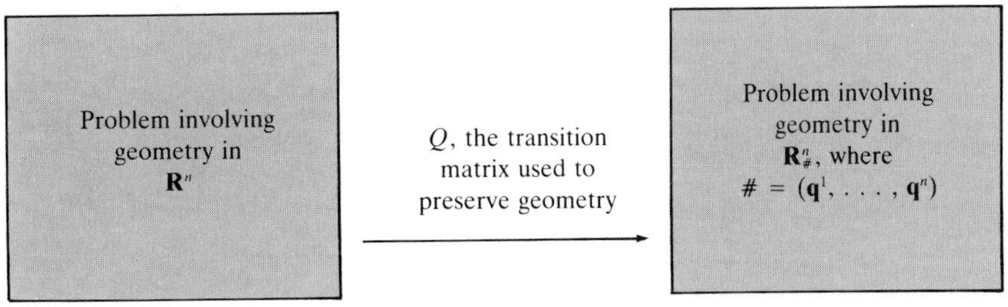

Problem involving geometry in $\mathbf{R}^n$	Q, the transition matrix used to preserve geometry →	Problem involving geometry in $\mathbf{R}^n_\#$, where $\# = (\mathbf{q}^1, \ldots, \mathbf{q}^n)$

FIGURE 7.1

7.1 ORTHOGONAL MATRICES

In this section we describe $n \times n$ matrices Q such that the linear transformations $L(\mathbf{x}) = Q\mathbf{x}$ preserve distances, lengths, and angles. We first look at an example of such a matrix.

EXAMPLE 1

Let Q be a 2×2 invertible matrix. This matrix is the transition matrix from $(\mathbf{R}^2)_\#$ to $\mathbf{R}^2$, where $\# = (\mathbf{q}^1, \mathbf{q}^2)$. Thus if $\mathbf{x} \in \mathbf{R}^2$ and $[\mathbf{x}]_\# = \mathbf{y}$, then $L(\mathbf{y}) = Q\mathbf{y} = \mathbf{x}$ is the linear transformation that changes the coordinates $\mathbf{y}$ in $(\mathbf{R}^2)_\#$ of vector $\mathbf{x}$ to the coordinates $Q\mathbf{y} = \mathbf{x}$ in $\mathbf{R}^2$ of the same vector. Now, if L preserves distances, lengths, and angles in the space, these calculations are the same whether done using the coordinates in $\mathbf{R}^2$ or those in $(\mathbf{R}^2)_\#$. Thus we can measure the lengths of $\mathbf{q}^1$ and $\mathbf{q}^2$ using the coordinates in $\mathbf{R}^2$ or the coordinates $[\mathbf{q}^1]_\# = \mathbf{e}^1$ and $[\mathbf{q}^2]_\# = \mathbf{e}^2$ in $(\mathbf{R}^2)_\#$. Doing this, we have

$$\|\mathbf{q}^1\| = \|\mathbf{e}^1\| = 1 \qquad \text{and} \qquad \|\mathbf{q}^2\| = \|\mathbf{e}^2\| = 1$$

Further, since we can calculate the angle θ between $\mathbf{q}^1$ and $\mathbf{q}^2$ using the coordinates $[\mathbf{q}^1]_\# = \mathbf{e}^1$ and $[\mathbf{q}^2]_\# = \mathbf{e}^2$ in $(\mathbf{R}^2)_\#$, we have that

$$\cos\theta = \frac{(\mathbf{q}^1, \mathbf{q}^2)}{\|\mathbf{q}^1\|\|\mathbf{q}^2\|} = \frac{(\mathbf{e}^1, \mathbf{e}^2)}{\|\mathbf{e}^1\|\|\mathbf{e}^2\|} = 0$$

so $\mathbf{q}^1$ and $\mathbf{q}^2$ are orthogonal. Thus the set $\{\mathbf{q}^1, \mathbf{q}^2\}$ of columns of Q must be an orthonormal set, as shown in Figure 7.2. (See exercise 31 for another viewpoint.) □

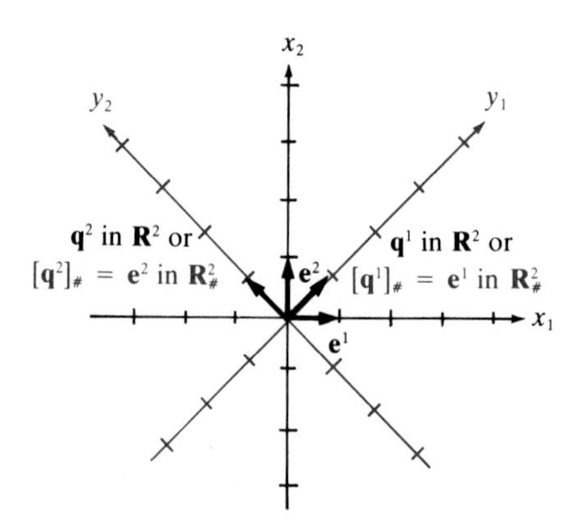

FIGURE 7.2

Based on this example, we will define an $n \times n$ *orthogonal matrix* $Q = [\mathbf{q}^1 \ \mathbf{q}^2 \ \ldots \ \mathbf{q}^n]$ as any matrix whose columns $\mathbf{q}^1, \mathbf{q}^2, \ldots, \mathbf{q}^n$ form an orthonormal basis for $\mathbf{R}^n$.

EXAMPLE 2

For $n = 2$, the following are orthogonal:

$$\begin{bmatrix} 1 & 0 \\ 0 & 1 \end{bmatrix}, \quad \begin{bmatrix} 0 & 1 \\ 1 & 0 \end{bmatrix}, \quad \frac{1}{\sqrt{2}}\begin{bmatrix} 1 & 1 \\ 1 & -1 \end{bmatrix}$$

and
$$\begin{bmatrix} \sin\theta & \pm\cos\theta \\ \cos\theta & \mp\sin\theta \end{bmatrix}$$

for any fixed θ, where both top or both bottom signs are chosen (exercise 30). □

EXAMPLE 3

For $n = 3$,

$$\begin{bmatrix} 1 & 0 & 0 \\ 0 & 1 & 0 \\ 0 & 0 & 1 \end{bmatrix}, \quad \begin{bmatrix} 0 & 1 & 0 \\ 0 & 0 & 1 \\ 1 & 0 & 0 \end{bmatrix}, \quad \begin{bmatrix} \frac{1}{\sqrt{3}} & \frac{1}{\sqrt{2}} & \frac{1}{\sqrt{6}} \\ \frac{1}{\sqrt{3}} & -\frac{1}{\sqrt{2}} & \frac{1}{\sqrt{6}} \\ \frac{1}{\sqrt{3}} & 0 & -\frac{2}{\sqrt{6}} \end{bmatrix}$$

are orthogonal. □

Note that $(\mathbf{x}, \mathbf{y}) = \mathbf{x}^t\mathbf{y}$ for all $\mathbf{x}, \mathbf{y} \in \mathbf{R}^n$ expresses the inner product as a matrix product. Using this equation, we can rewrite the definition of an orthogonal matrix in equation form, which is often a more useful form.

LEMMA 7.1

An $n \times n$ matrix Q is orthogonal if and only if

$$Q^tQ = I_n$$

PROOF

Note that $Q^tQ = [(\mathbf{q}^i)^t\mathbf{q}^j] = [(\mathbf{q}^i, \mathbf{q}^j)]$. Now, if Q is orthogonal, then $\{\mathbf{q}^1, \ldots, \mathbf{q}^n\}$ is an orthonormal set, so $(\mathbf{q}^i, \mathbf{q}^j) = 0$ if $i \neq j$ and $(\mathbf{q}^i, \mathbf{q}^j) = 1$ if $i = j$. Thus $Q^tQ = I_n$. On the other hand, if $Q^tQ = I_n$, then $[(\mathbf{q}^i, \mathbf{q}^j)] = I_n$. Hence $\{\mathbf{q}^1, \ldots, \mathbf{q}^n\}$ is an orthonormal set. Thus Q is orthogonal. ■

This lemma means that if Q is orthogonal, then Q is invertible and $Q^{-1} = Q^t$. Thus we also have

$$QQ^t = I_n$$

which means that the rows of Q form an orthonormal basis for $\mathbf{R}_n$. The previous examples demonstrate this property.

This equation form of the definition of orthogonal matrices can be used to show some basic arithmetic results about orthogonal matrices.

THEOREM 7.2

Let $\mathbf{Q}_{n,n}$ be the set of $n \times n$ orthogonal matrices.

(*i*) If $Q_1, Q_2 \in \mathbf{Q}_{n,n}$, then $Q_1Q_2 \in \mathbf{Q}_{n,n}$.

(*ii*) If $Q \in \mathbf{Q}_{n,n}$, then $Q^{-1} = Q^t \in \mathbf{Q}_{n,n}$.

PROOF

To prove (i), we show that Q_1Q_2 satisfies the equation of the lemma. Since

$$(Q_1Q_2)^t(Q_1Q_2) = Q_2^tQ_1^tQ_1Q_2 = Q_2^tI_nQ_2 = Q_2^tQ_2 = I_n$$

$Q_1Q_2 \in \mathbf{Q}_{n,n}$.

The proof of (ii) is left as exercise 32. ∎

We now show that orthogonal matrices are matrices that preserve distance, length, and angle measurements.

THEOREM 7.3

Let Q be an $n \times n$ orthogonal matrix. Then for all $\mathbf{x}, \mathbf{y} \in \mathbf{R}^n$,

(*i*) $\|Q\mathbf{x}\| = \|\mathbf{x}\|$, so length is preserved.

(*ii*) $d(Q\mathbf{x}, Q\mathbf{y}) = d(\mathbf{x}, \mathbf{y})$, so distance is preserved.

(*iii*) $(Q\mathbf{x}, Q\mathbf{y}) = (\mathbf{x}, \mathbf{y})$. Thus, as a consequence,

$$\cos\theta = \frac{(Q\mathbf{x}, Q\mathbf{y})}{\|Q\mathbf{x}\|\,\|Q\mathbf{y}\|} = \frac{(\mathbf{x}, \mathbf{y})}{\|\mathbf{x}\|\,\|\mathbf{y}\|}$$

and so angles are unchanged.

PROOF

For (i),

$$\|Q\mathbf{x}\|^2 = (Q\mathbf{x}, Q\mathbf{x}) = (Q\mathbf{x})^tQ\mathbf{x} = \mathbf{x}^tQ^tQ\mathbf{x} = \mathbf{x}^tI_n\mathbf{x} = \mathbf{x}^t\mathbf{x} = \|\mathbf{x}\|^2$$

so $\|Q\mathbf{x}\| = \|\mathbf{x}\|$.

For (ii),

$$d(Q\mathbf{x}, Q\mathbf{y}) = \|Q\mathbf{x} - Q\mathbf{y}\| = \|Q(\mathbf{x} - \mathbf{y})\|$$

By (i), $\|Q(\mathbf{x} - \mathbf{y})\| = \|\mathbf{x} - \mathbf{y}\|$. Hence

$$d(Q\mathbf{x}, Q\mathbf{y}) = \|\mathbf{x} - \mathbf{y}\| = d(\mathbf{x}, \mathbf{y})$$

Finally, for (iii),

$$(Q\mathbf{x}, Q\mathbf{y}) = (Q\mathbf{y})^tQ\mathbf{x} = \mathbf{y}^tQ^tQ\mathbf{x} = \mathbf{y}^tI_n\mathbf{x} = \mathbf{y}^t\mathbf{x} = (\mathbf{x}, \mathbf{y})$$ ∎

This theorem may be applied to linear transformations $L(\mathbf{x}) = Q\mathbf{x}$, which are interpreted as mappings from $\mathbf{R}^n$ to $\mathbf{R}^n$ or as transformations of coordinates from $(\mathbf{R}^n)_\#$ to $\mathbf{R}^n$. We will demonstrate these interpretations in the following two examples.

EXAMPLE 4

MAPPING INTERPRETATION

Let Q be an orthogonal matrix and let $L(\mathbf{x}) = Q\mathbf{x}$ be a linear mapping. Then

$$\|\mathbf{x}\| = \|Q\mathbf{x}\| \qquad \text{and} \qquad \cos\theta = \frac{(Q\mathbf{x}, Q\mathbf{y})}{\|Q\mathbf{x}\|\,\|Q\mathbf{y}\|} = \frac{(\mathbf{x}, \mathbf{y})}{\|\mathbf{x}\|\,\|\mathbf{y}\|}$$

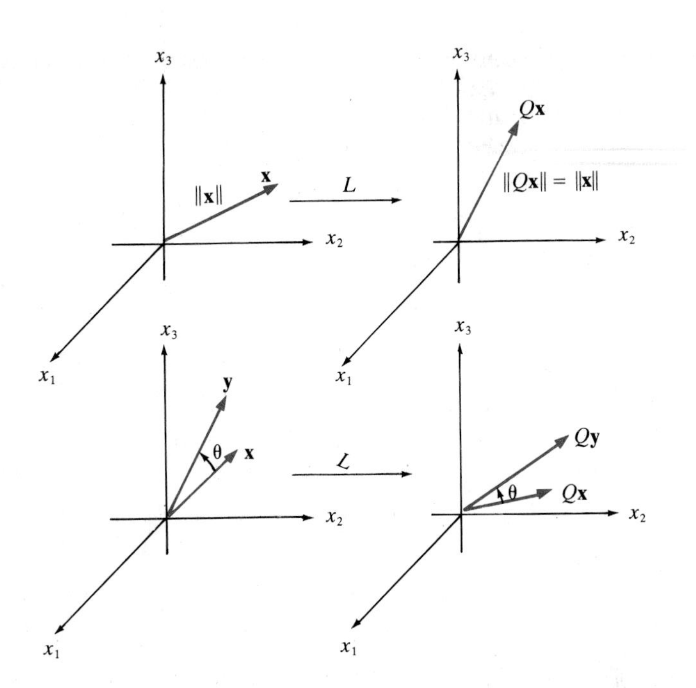

FIGURE 7.3

Thus, vectors and their images have the same lengths, and angles between vectors and between their images are the same. This situation is depicted geometrically in Figure 7.3. Further, the distance between two points is the same as the distance between the images of those points. □

EXAMPLE 5

TRANSITION MATRIX INTERPRETATION

Let Q be an orthogonal transition matrix from $(\mathbf{R}^n)_\#$ to $\mathbf{R}^n$, where $\# = (\mathbf{q}^1, \mathbf{q}^2, \ldots, \mathbf{q}^n)$. Then the length of a vector does not depend on which set of coordinates is used for the calculation. Further, the angle between vectors does not depend on which coordinate space is used for the calculation. This is demonstrated in Figure 7.4. Finally, distances between points are the same whether measured using the coordinates from $\mathbf{R}^2$ or those from $(\mathbf{R}^2)_\#$.

For a numerical demonstration, let

$$\# = (\mathbf{q}^1, \mathbf{q}^2) = \left(\frac{1}{\sqrt{2}}\begin{bmatrix}1\\1\end{bmatrix}, \frac{1}{\sqrt{2}}\begin{bmatrix}-1\\1\end{bmatrix}\right)$$

This coordinate system is composed of an orthonormal set of vectors. Further, the orthogonal matrix

$$Q = \frac{1}{\sqrt{2}}\begin{bmatrix}1 & -1\\1 & 1\end{bmatrix}$$

is the transition matrix from $(\mathbf{R}^2)_\#$ to $\mathbf{R}^2$. Let $\mathbf{x} = \begin{bmatrix}2\\1\end{bmatrix}$ and $\mathbf{x}' = \begin{bmatrix}1\\2\end{bmatrix}$. To

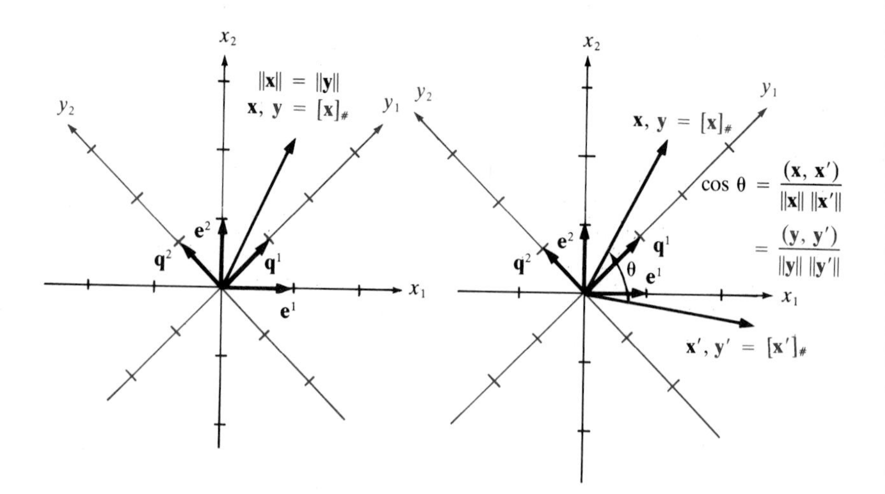

FIGURE 7.4

compute the coordinates of **x** and **x**′ in $(\mathbf{R}^2)_\#$, set

$$\mathbf{x} = c_1\begin{bmatrix}\frac{1}{\sqrt{2}}\\ \frac{1}{\sqrt{2}}\end{bmatrix} + c_2\begin{bmatrix}-\frac{1}{\sqrt{2}}\\ \frac{1}{\sqrt{2}}\end{bmatrix} \quad \text{and} \quad \mathbf{x}' = d_1\begin{bmatrix}\frac{1}{\sqrt{2}}\\ \frac{1}{\sqrt{2}}\end{bmatrix} + d_2\begin{bmatrix}-\frac{1}{\sqrt{2}}\\ \frac{1}{\sqrt{2}}\end{bmatrix}$$

Solving yields that $c_1 = \frac{3\sqrt{2}}{2}$, $c_2 = -\frac{\sqrt{2}}{2}$, $d_1 = \frac{3\sqrt{2}}{2}$, and $d_2 = \frac{\sqrt{2}}{2}$, so

$$[\mathbf{x}]_\# = \mathbf{y} = \begin{bmatrix}\frac{3\sqrt{2}}{2}\\ -\frac{\sqrt{2}}{2}\end{bmatrix} \quad \text{and} \quad [\mathbf{x}']_\# = \mathbf{y}' = \begin{bmatrix}\frac{3\sqrt{2}}{2}\\ \frac{\sqrt{2}}{2}\end{bmatrix}$$

as depicted in Figure 7.5.

Now the distance between **x** and **x**′ can be calculated in either coordinate system, as

$$d(\mathbf{x}, \mathbf{x}') = d\left(\begin{bmatrix}2\\1\end{bmatrix}, \begin{bmatrix}1\\2\end{bmatrix}\right) = \sqrt{(2-1)^2 + (1-2)^2} = \sqrt{2}$$

or

$$d(\mathbf{y}, \mathbf{y}') = d\left(\begin{bmatrix}\frac{3\sqrt{2}}{2}\\ -\frac{\sqrt{2}}{2}\end{bmatrix}, \begin{bmatrix}\frac{3\sqrt{2}}{2}\\ \frac{\sqrt{2}}{2}\end{bmatrix}\right) = \sqrt{\left(\frac{3\sqrt{2}}{2} - \frac{3\sqrt{2}}{2}\right)^2 + \left(-\frac{\sqrt{2}}{2} - \frac{\sqrt{2}}{2}\right)^2}$$
$$= \sqrt{2}$$

□

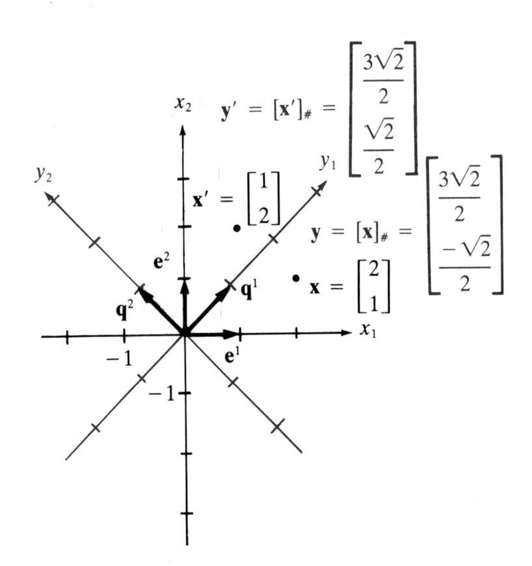

FIGURE 7.5

□

The next few examples show ways these results can be applied.

EXAMPLE 6

OPTIONAL

Let Q be a 2×2 orthogonal matrix. Let C be the set of points of a circle of radius r about point $\mathbf{x}_0$. We will show that

$$C' = \{Q\mathbf{x} : \mathbf{x} \in C\}$$

is the circle of radius r about the point $Q\mathbf{x}_0$, as shown in Figure 7.6. Thus the linear transformation $L(\mathbf{x}) = Q\mathbf{x}$ maps a circle to a congruent circle.

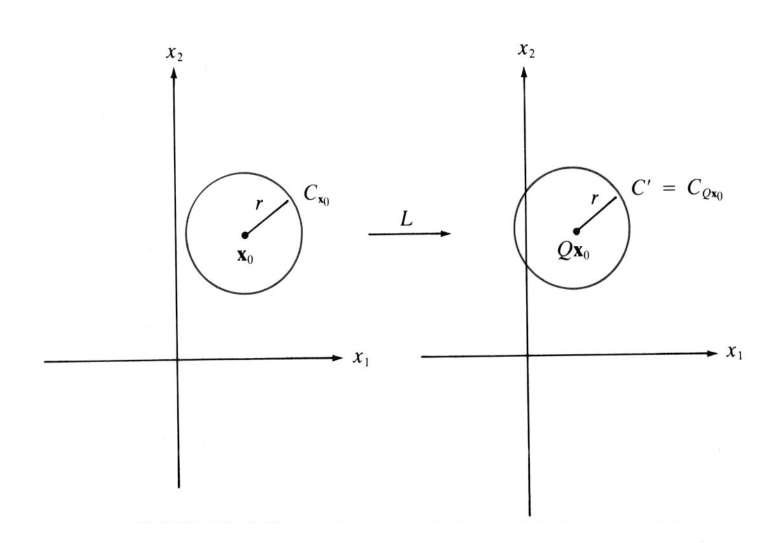

FIGURE 7.6

Let K be the circle of radius r about $Q\mathbf{x}_0$. We need to show that $C' = K$. Pick $\mathbf{x} \in C$. Then

$$r = d(\mathbf{x}, \mathbf{x}_0) = \|\mathbf{x} - \mathbf{x}_0\| = \|Q(\mathbf{x} - \mathbf{x}_0)\| = \|Q\mathbf{x} - Q\mathbf{x}_0\| = d(Q\mathbf{x}, Q\mathbf{x}_0)$$

Thus $Q\mathbf{x} \in K$, and since $\mathbf{x}$ was arbitrary, $C' \subseteq K$. Now pick $\mathbf{x}' \in K$. Then, since Q^t is orthogonal,

$$r = d(\mathbf{x}', Q\mathbf{x}_0) = \|\mathbf{x}' - Q\mathbf{x}_0\| = \|Q^t(\mathbf{x}' - Q\mathbf{x}_0)\| = \|Q^t\mathbf{x}' - \mathbf{x}_0\| = d(Q^t\mathbf{x}', \mathbf{x}_0)$$

Thus $Q^t\mathbf{x}' \in C$. Since $L(Q^t\mathbf{x}') = Q(Q^t\mathbf{x}') = \mathbf{x}'$, it follows that $\mathbf{x}' \in C'$. Hence $K \subseteq C'$ and so $K = C'$. □

EXAMPLE 7

OPTIONAL

Let A be a 2×2 matrix. We will show that

$|\det A|$ = the area of the parallelogram formed by the fixed arrows $\mathbf{a}^1$ and $\mathbf{a}^2$

We show this by considering cases. If $\mathbf{a}^1 = \mathbf{0}$ or $\mathbf{a}^2 = \mathbf{0}$, the result is immediate. Thus we may assume that neither $\mathbf{a}^1$ nor $\mathbf{a}^2$ is $\mathbf{0}$.

CASE 1 (Special Position): Suppose $\mathbf{a}^1$ lies on the positive part of the x_1-axis, as shown in Figure 7.7. Then $\mathbf{a}^2$ is either above or below this axis. In either case, it is easy to see that the area of the parallelogram is $|a_{11}a_{22}|$, the absolute value assuring that the number is positive regardless of the position of $\mathbf{a}^2$. Since $A = \begin{bmatrix} a_{11} & a_{12} \\ 0 & a_{22} \end{bmatrix}$, it follows that

$|\det A|$ = the area of the parallelogram

CASE 2 (General Position): Recall from Example 30 of Section 5.3 that the linear transformation

$$L(\mathbf{x}) = Q\mathbf{x} = \begin{bmatrix} \cos\theta & -\sin\theta \\ \sin\theta & \cos\theta \end{bmatrix}\mathbf{x}$$

seen as a mapping, rotates the plane by θ radians. Note that the matrix Q is orthogonal. Pick θ so that $Q\mathbf{a}^1$ lies on the x_1-axis as in the special case. Then, since Q preserves lengths and angles,

$|\det[Q\mathbf{a}^1 \quad Q\mathbf{a}^2]|$ = the area of the parallelogram formed from $\mathbf{a}^1$ and $\mathbf{a}^2$

Now, $[Q\mathbf{a}^1 \quad Q\mathbf{a}^2] = QA$, and by exercise 33 $|\det Q| = 1$, so it follows that

$|\det A|$ = the area of the parallelogram formed from $\mathbf{a}^1$ and $\mathbf{a}^2$

as shown in Figure 7.8. □

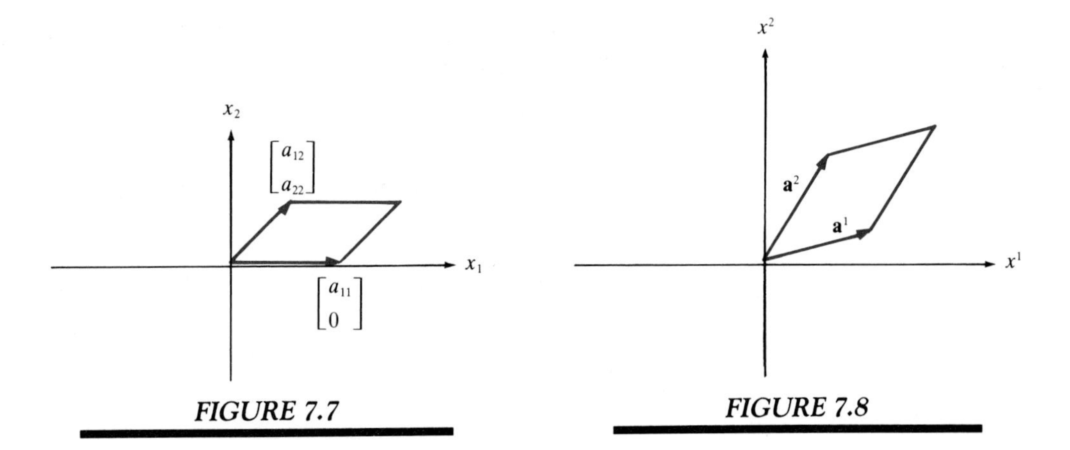

FIGURE 7.7 *FIGURE 7.8*

By Theorem 7.3, we can use orthogonal matrices when transforming a problem in which we want to preserve geometry. To develop the technique for redescribing a problem given in $\mathbf{R}^n$ as a simpler one in an orthonormal coordinate system $\# = (\mathbf{q}^1, \ldots, \mathbf{q}^n)$, we relate matrices in $\mathbf{R}_{n,n}$ as follows. Let A and B be $n \times n$ matrices such that there is an orthogonal matrix Q satisfying

$$Q^t A Q = B$$

Then we say that B is *orthogonally similar* to A. As in Section 6.1, we can show (exercise 34) the following:

(*i*) A is orthogonally similar to A (reflexive property).

(*ii*) If A is orthogonally similar to B, then B is orthogonally similar to A (symmetric property).

(*iii*) If A is orthogonally similar to B and if B is orthogonally similar to C, then A is orthogonally similar to C (transitive property).

Thus, by property (ii) we can say that A and B are orthogonally similar, rather than saying that one of them is orthogonally similar to the other.

In order to change a problem involving a linear transformation $L(\mathbf{x}) = A\mathbf{x}$ into a simpler one and yet preserve geometry, we need to find a simple matrix, or canonical form, orthogonally similar to A. Then we can change a problem as diagrammed in the introduction. In the next section we show how to find these canonical forms.

EXERCISES FOR SECTION 7.1

COMPUTATIONAL EXERCISES

In exercises 1 through 4, graph columns to decide which of the given matrices are orthogonal and then use Lemma 7.1 to show your answers computationally. Repeat the exercises using the rows of A, and compare your results.

1. $A = \begin{bmatrix} \frac{1}{\sqrt{5}} & \frac{2}{\sqrt{5}} \\ -\frac{2}{\sqrt{5}} & \frac{1}{\sqrt{5}} \end{bmatrix}$

2. $A = \begin{bmatrix} 1 & 0 \\ 0 & 2 \end{bmatrix}$

3. $A = \begin{bmatrix} \frac{1}{\sqrt{2}} & \frac{1}{\sqrt{2}} \\ \frac{1}{\sqrt{5}} & \frac{1}{\sqrt{5}} \end{bmatrix}$

4. $A = \frac{1}{\sqrt{34}} \begin{bmatrix} 3 & -5 \\ 5 & 3 \end{bmatrix}$

In exercises 5 through 8, show that the given matrices are orthogonal, and compute their inverses. Note that inverses of orthogonal matrices are easily computed.

5. (a) $Q = \begin{bmatrix} 0 & 1 \\ 1 & 0 \end{bmatrix}$

(b) $Q = \begin{bmatrix} 0 & \frac{1}{\sqrt{2}} & \frac{1}{\sqrt{2}} \\ 0 & -\frac{1}{\sqrt{2}} & \frac{1}{\sqrt{2}} \\ 1 & 0 & 0 \end{bmatrix}$

6. (a) $Q = \begin{bmatrix} -1 & 0 \\ 0 & 1 \end{bmatrix}$

(b) $Q = \begin{bmatrix} \frac{1}{\sqrt{3}} & \frac{1}{\sqrt{3}} & \frac{1}{\sqrt{3}} \\ \frac{1}{\sqrt{2}} & -\frac{1}{\sqrt{2}} & 0 \\ \frac{1}{\sqrt{6}} & \frac{1}{\sqrt{6}} & -\frac{2}{\sqrt{6}} \end{bmatrix}$

7. $Q = \begin{bmatrix} \frac{1}{\sqrt{2}} & -\frac{1}{\sqrt{2}} & 0 & 0 \\ \frac{1}{\sqrt{2}} & \frac{1}{\sqrt{2}} & 0 & 0 \\ 0 & 0 & 0 & 1 \\ 0 & 0 & 1 & 0 \end{bmatrix}$

8. $Q = \begin{bmatrix} \frac{1}{\sqrt{2}} & 0 & \frac{1}{\sqrt{2}} & 0 \\ 0 & \frac{1}{\sqrt{2}} & 0 & -\frac{1}{\sqrt{2}} \\ -\frac{1}{\sqrt{2}} & 0 & \frac{1}{\sqrt{2}} & 0 \\ 0 & \frac{1}{\sqrt{2}} & 0 & \frac{1}{\sqrt{2}} \end{bmatrix}$

In exercises 9 and 10, given the orthogonal matrix Q and the vector $\mathbf{x}$, graph $\mathbf{x}$ and $Q\mathbf{x}$. Then show that $\|Q\mathbf{x}\| = \|\mathbf{x}\|$.

9. $Q = \begin{bmatrix} \frac{1}{\sqrt{17}} & \frac{4}{\sqrt{17}} \\ -\frac{4}{\sqrt{17}} & \frac{1}{\sqrt{17}} \end{bmatrix}, \mathbf{x} = \begin{bmatrix} 1 \\ 2 \end{bmatrix}$

10. $Q = \begin{bmatrix} \frac{1}{\sqrt{2}} & -\frac{1}{\sqrt{2}} \\ \frac{1}{\sqrt{2}} & \frac{1}{\sqrt{2}} \end{bmatrix}, \mathbf{x} = \begin{bmatrix} -1 \\ 1 \end{bmatrix}$

In exercises 11 and 12, given the orthogonal matrix Q and the vectors $\mathbf{x}$ and $\mathbf{y}$, graph the pairs $\mathbf{x}$, $\mathbf{y}$ and $Q\mathbf{x}$, $Q\mathbf{y}$. Compute and compare the angle θ between $\mathbf{x}$ and $\mathbf{y}$ and the angle θ' between $Q\mathbf{x}$ and $Q\mathbf{y}$.

11. $Q = \frac{1}{\sqrt{2}} \begin{bmatrix} 1 & -1 \\ 1 & 1 \end{bmatrix}, \mathbf{x} = \begin{bmatrix} 1 \\ 1 \end{bmatrix}, \mathbf{y} = \begin{bmatrix} 0 \\ 1 \end{bmatrix}$

12. $Q = \frac{1}{\sqrt{5}} \begin{bmatrix} -2 & 1 \\ 1 & 2 \end{bmatrix}, \mathbf{x} = \begin{bmatrix} 1 \\ 1 \end{bmatrix}, \mathbf{y} = \begin{bmatrix} -1 \\ 1 \end{bmatrix}$

In exercises 13 and 14, given a basis # of orthogonal vectors, find the transition matrix Q from $(\mathbf{R}^2)_\#$ to $\mathbf{R}^2$. Graph the given vector $\mathbf{x}$ in both $(\mathbf{R}^2)_\#$ and $\mathbf{R}^2$, and show that $\|\mathbf{x}\| = \|[\mathbf{x}]_\#\|$.

13. $\# = \left(\begin{bmatrix} \frac{1}{\sqrt{2}} \\ \frac{1}{\sqrt{2}} \end{bmatrix}, \begin{bmatrix} -\frac{1}{\sqrt{2}} \\ \frac{1}{\sqrt{2}} \end{bmatrix} \right), \mathbf{x} = \begin{bmatrix} 2 \\ 1 \end{bmatrix}$

14. $\# = \left(\begin{bmatrix} \frac{1}{\sqrt{5}} \\ \frac{2}{\sqrt{5}} \end{bmatrix}, \begin{bmatrix} -\frac{2}{\sqrt{5}} \\ \frac{1}{\sqrt{5}} \end{bmatrix} \right), \mathbf{x} = \begin{bmatrix} 1 \\ 1 \end{bmatrix}$

In exercises 15 through 18, given the orthonormal coordinate system #, find the transition matrix Q from $(\mathbf{R}^2)_\#$ to $\mathbf{R}^2$. Graph $\mathbf{x}$ and $\mathbf{y}$ in both $\mathbf{R}^2$ and $(\mathbf{R}^2)_\#$. Compute and compare the angles between $\mathbf{x}$ and $\mathbf{y}$ using the coordinates from $\mathbf{R}^2$ and those from $(\mathbf{R}^2)_\#$.

15. $\# = \left(\begin{bmatrix} -\frac{1}{\sqrt{2}} \\ \frac{1}{\sqrt{2}} \end{bmatrix}, \begin{bmatrix} \frac{1}{\sqrt{2}} \\ \frac{1}{\sqrt{2}} \end{bmatrix} \right), \mathbf{x} = \begin{bmatrix} 1 \\ 0 \end{bmatrix}, \mathbf{y} = \begin{bmatrix} 1 \\ 1 \end{bmatrix}$

16. $\# = \left(\begin{bmatrix} -1 \\ 0 \end{bmatrix}, \begin{bmatrix} 0 \\ -1 \end{bmatrix} \right), \mathbf{x} = \begin{bmatrix} 1 \\ 1 \end{bmatrix}, \mathbf{y} = \begin{bmatrix} 1 \\ -1 \end{bmatrix}$

17. $\# = \left(\begin{bmatrix} 0 \\ -1 \end{bmatrix}, \begin{bmatrix} 1 \\ 0 \end{bmatrix} \right), \mathbf{x} = \begin{bmatrix} \sqrt{3} \\ 1 \end{bmatrix}, \mathbf{y} = \begin{bmatrix} 1 \\ \sqrt{3} \end{bmatrix}$

18. $\# = \left(\begin{bmatrix} \frac{3}{\sqrt{10}} \\ -\frac{1}{\sqrt{10}} \end{bmatrix}, \begin{bmatrix} \frac{1}{\sqrt{10}} \\ \frac{3}{\sqrt{10}} \end{bmatrix} \right), \mathbf{x} = \begin{bmatrix} 1 \\ 0 \end{bmatrix}, \mathbf{y} = \begin{bmatrix} 1 \\ \sqrt{3} \end{bmatrix}$

19. For the coordinate system and points given in exercise 15, compute the distance between $\mathbf{x}$ and $\mathbf{y}$, first in the

coordinates of $\mathbf{R}^2$ and then in those of $(\mathbf{R}^2)_\#$, and compare the results.

20. Carry out exercise 19 using the coordinate system and points given in exercise 16.

In exercises 21 through 24, use the determinant to find the area of the parallelogram with the given sides. Sketch the parallelogram, indicating the area.

21. $\mathbf{x} = [5,2]$, $\mathbf{y} = [3,-4]$

22. $\mathbf{x} = [-4,1]$, $\mathbf{y} = [-3,-6]$

23. From $[1,1]$ to $[2,2]$, from $[1,1]$ to $[3,5]$

24. From $[1,-2]$ to $[3,-4]$, from $[-1,2]$ to $[2,5]$

In exercises 25 and 26, use the determinant to find the area of the triangle formed by the given points. Graph this triangle. *Hint:* A triangle is half of a parallelogram.

25. Points $[1,1]$, $[2,1]$, and $[1,2]$

26. Points $[-2,3]$, $[2,1]$, and $[-1,-1]$

COMPLEX NUMBERS

Using the inner product $(\mathbf{z},\mathbf{w}) = z_1\bar{w}_1 + \cdots + z_n\bar{w}_n$ for all $\mathbf{z}, \mathbf{w} \in \mathbf{C}^n$, define an $n \times n$ *unitary matrix* A as any matrix whose columns form an orthonormal set. In exercises 27 and 28, decide which of the given matrices are unitary.

27. (*a*) $\dfrac{1}{\sqrt{2}}\begin{bmatrix} 1 & -1 \\ 1 & 1 \end{bmatrix}$ (*b*) $\begin{bmatrix} \dfrac{1-i}{2} & \dfrac{-1+i}{2} \\ \dfrac{1+i}{2} & \dfrac{1+i}{2} \end{bmatrix}$

28. (*a*) $\dfrac{1}{\sqrt{2}}\begin{bmatrix} 1 & -i \\ i & 1 \end{bmatrix}$ (*b*) $\begin{bmatrix} i & 0 & 0 \\ 0 & -i & 0 \\ 0 & 0 & 1 \end{bmatrix}$

29. Prove that A is unitary if and only if $A^*A = I_n$, where $A^* = [\bar{a}_{ij}]^t$, as defined in exercise 41 of Section 2.1.

THEORETICAL EXERCISES

30. Prove that $\begin{bmatrix} \sin\theta & \pm\cos\theta \\ \cos\theta & \mp\sin\theta \end{bmatrix}$ is orthogonal for any $\theta \in \mathbf{R}$.

31. Let $L(\mathbf{x}) = A\mathbf{x}$ be a mapping from $\mathbf{R}^2$ to $\mathbf{R}^2$. Suppose that L preserves distance, length, and angles. Show that since $L(\mathbf{e}^1) = \mathbf{a}^1$ and $L(\mathbf{e}^2) = \mathbf{a}^2$, it follows that $\{\mathbf{a}^1, \mathbf{a}^2\}$ must be an orthonormal set.

32. Prove that if $Q \in \mathbf{Q}_{n,n}$, then $Q^{-1} = Q^t$ and $Q^{-1} \in \mathbf{Q}_{n,n}$.

33. Prove that $\det\begin{bmatrix} \cos\theta & -\sin\theta \\ \sin\theta & \cos\theta \end{bmatrix} = 1$ for any $\theta \in \mathbf{R}$.

34. Prove that, for $n \times n$ matrices A, B, and C,
(*a*) A is orthogonally similar to A.
(*b*) If A is orthogonally similar to B, then B is orthogonally similar to A.
(*c*) If A is orthogonally similar to B and if B is orthogonally similar to C, then A is orthogonally similar to C.

35. Prove that $A \in \mathbf{R}_{2,2}$ is orthogonal if and only if

$$A = \begin{bmatrix} \cos\theta & \sin\theta \\ -\sin\theta & \cos\theta \end{bmatrix} \quad \text{or} \quad A = \begin{bmatrix} \cos\theta & \sin\theta \\ \sin\theta & -\cos\theta \end{bmatrix}$$

for some $\theta \in \mathbf{R}$.

36. Let Q be an orthogonal matrix. Prove that $\det Q = \pm 1$.

37. Prove that the area of the parallelogram formed by using the rows of $A \in \mathbf{R}_{2,2}$ as fixed arrows is $|\det A|$. *Hint:* Transpose the matrix.

38. Let Q be a 2×2 orthogonal matrix.
(*a*) Prove that the points on a square in $\mathbf{R}^2$, when multiplied by Q, give a congruent square.
(*b*) Do the same for a triangle in $\mathbf{R}^2$.

39. Let $L(\mathbf{x}) = A\mathbf{x}$ for $A \in \mathbf{R}_{2,2}$. Prove that the square formed by the fixed arrows $\mathbf{e}^1$ and $\mathbf{e}^2$ is mapped to the parallelogram formed by $\mathbf{a}^1$ and $\mathbf{a}^2$. Thus $|\det A|$ shows how much the area of the square is changed by L. (This result is true for any unit square, and so the determinant appears in substitution formulas for integration of functions of two variables. In this case, the determinant is called the *Jacobian* of L.)

COMPUTER EXERCISE

40. Let A be an $m \times n$ matrix. Some computing algorithms involving A can be described as a procedure for multiplying A by a sequence of matrices $P_1, \ldots, P_r$, say in the fashion $P_r \cdots P_1 A$. (The Gaussian elimination and Gauss-Jordan algorithms can be viewed in this way by using the representation in exercise 47 of Section 2.2. Some other algorithms, such as the QR-algorithm for computing eigenvalues and eigenvectors, can also be viewed in this way.)

Explain why choosing an orthogonal matrix $P = P_i$ (when possible) assures that round-off error is not magnified. *Hint:* Suppose that at some step of the algorithm we have calculated $A + E$ instead of A, where $E = [\varepsilon_{ij}]$ represents error. Then in the next multiplication we have

$$P(A + E) = PA + PE$$

where PE is the new error, before rounding again. Note that

$$PE = [P\boldsymbol{\varepsilon}^1 \quad \cdots \quad P\boldsymbol{\varepsilon}^n]$$

How can we ensure that $\|P\boldsymbol{\varepsilon}^i\|$ does not exceed $\|\boldsymbol{\varepsilon}^i\|$?

If two matrices A and B in $\mathbf{R}_{n,n}$ are orthogonally similar to some orthogonal matrix Q, then, since $Q^t = Q^{-1}$,

$$B = Q^tAQ = Q^{-1}AQ$$

It follows that A and B are similar. Since not all matrices are similar to diagonal canonical forms, it follows that not all matrices are orthogonally similar to such matrices. Hence we need to find another type of canonical form that is useful for changing a problem into a simpler problem. We will use an upper triangular matrix.

THEOREM 7.4

Let A be an $n \times n$ matrix with real eigenvalues $\lambda_1, \lambda_2, \ldots, \lambda_n$. Then A is orthogonally similar* to an upper triangular matrix T such that $t_{11} = \lambda_1, t_{22} = \lambda_2, \ldots, t_{nn} = \lambda_n$.

PROOF

Let $\mathbf{p}^1$ be an eigenvector for λ_1. This eigenvector can be taken to be real because λ_1 is real. Form the set $\{\mathbf{p}^1, \mathbf{e}^1, \ldots, \mathbf{e}^n\}$, and use the deletion algorithm to construct a linearly independent set $\{\mathbf{p}^1, \mathbf{x}^2, \ldots, \mathbf{x}^n\}$ of n vectors. Apply the Gram-Schmidt algorithm to these vectors to obtain the orthonormal set $\{\mathbf{q}^1, \mathbf{q}^2, \ldots, \mathbf{q}^n\}$. Set

$$Q_1 = [\mathbf{q}^1 \quad \mathbf{q}^2 \quad \cdots \quad \mathbf{q}^n]$$

Since $A\mathbf{p}^1 = \lambda_1 \mathbf{p}^1$ and $\mathbf{q}^1 = \dfrac{1}{\|\mathbf{p}^1\|}\mathbf{p}^1$, it follows that $A\mathbf{q}^1 = \lambda_1 \mathbf{q}^1$. Since $A\mathbf{q}^i \in \mathbf{R}^n$ and $\mathbf{R}^n = \text{span}\{\mathbf{q}^1, \ldots, \mathbf{q}^n\}$, it follows that we can write

$$A\mathbf{q}^1 = \lambda_1 \mathbf{q}^1 \qquad \text{and} \qquad A\mathbf{q}^i = b_{1i}\mathbf{q}^1 + b_{2i}\mathbf{q}^2 + \cdots + b_{ni}\mathbf{q}^n$$

for any $i > 1$ and some $b_{1i}, b_{2i}, \ldots, b_{ni}$. Writing these equations as matrix equations, we have that $AQ_1 = Q_1 B_1$, where

$$B_1 = \begin{bmatrix} \lambda_1 & & & \\ 0 & & & \\ \vdots & \mathbf{b}^2 & \cdots & \mathbf{b}^n \\ 0 & & & \end{bmatrix}$$

Write

$$B_1 = \left[\begin{array}{c|ccc} \lambda_1 & b_{12} & \cdots & b_{1n} \\ \hline 0 & & & \\ \vdots & & A_1 & \\ 0 & & & \end{array}\right]$$

* If "orthogonally similar" is replaced by "unitarily similar," this theorem, called Schur's Lemma, is true even when eigenvalues are complex numbers. The proof is essentially that given here.

Then, since Q_1 is orthogonal,

$$Q_1^t A Q_1 = B_1 = \left[\begin{array}{c|ccc} \lambda_1 & b_{12} & \cdots & b_{1n} \\ \hline 0 & & & \\ \vdots & & A_1 & \\ 0 & & & \end{array}\right]$$

Note that since A and B_1 are similar, they have the same eigenvalues. Further, evaluating along the first column, we have

$$\det(\lambda I_n - B_1) = (\lambda - \lambda_1)\det(\lambda I_{n-1} - A_1)$$

Thus A_1 has eigenvalues $\lambda_2, \ldots, \lambda_n$.

Now apply the same technique to A_1 to obtain an $(n-1) \times (n-1)$ orthogonal matrix Q_2' such that $(Q_2')^t A_1 Q_2' = B_2$, where

$$B_2 = \left[\begin{array}{c|ccc} \lambda_2 & b_{23} & \cdots & b_{2n} \\ \hline 0 & & & \\ \vdots & & A_2 & \\ 0 & & & \end{array}\right]$$

Set

$$Q_2 = \left[\begin{array}{c|ccc} 1 & 0 & \cdots & 0 \\ \hline 0 & & & \\ \vdots & & Q_2' & \\ 0 & & & \end{array}\right]$$

Since Q_2' is orthogonal, Q_2 is also. Further,

$$(Q_1 Q_2)^t A (Q_1 Q_2) = (Q_2)^t B_1 Q_2 = C = \left[\begin{array}{c|c|ccc} \lambda_1 & c_{12} & c_{13} & \cdots & c_{1n} \\ \hline 0 & \lambda_2 & c_{23} & \cdots & c_{2n} \\ \hline 0 & 0 & & & \\ \vdots & \vdots & & A_2 & \\ 0 & 0 & & & \end{array}\right]$$

Continuing this procedure, we see that we can find orthogonal matrices $Q_1, \ldots, Q_{n-1}$ such that

$$(Q_1 \cdots Q_{n-1})^t A (Q_1 \cdots Q_{n-1}) = T$$

where T is upper triangular and $t_{11} = \lambda_1, t_{22} = \lambda_2, \ldots, t_{nn} = \lambda_n$. ∎

Algorithm for Finding an Orthogonally Similar Upper Triangular Matrix

Let A be an $n \times n$ matrix.

1. Find an eigenvalue λ and a corresponding eigenvector $\mathbf{p}^1$.

2. Apply the deletion algorithm to $\{\mathbf{p}^1, \mathbf{e}^1, \ldots, \mathbf{e}^n\}$ to obtain a basis $\{\mathbf{p}^1, \mathbf{x}^2, \ldots, \mathbf{x}^n\}$.

Apply the Gram-Schmidt algorithm to this set to obtain $\{\mathbf{p}^1, \mathbf{p}^2, \ldots, \mathbf{p}^n\}$. Divide $\mathbf{p}^i$ by its length to form $\mathbf{q}^i$ for each i. Form

$$Q = [\mathbf{q}^1 \quad \mathbf{q}^2 \quad \cdots \quad \mathbf{q}^n]$$

an orthogonal matrix.

3. Set $\lambda_1 = \lambda$, $Q_1 = Q$, and calculate

$$Q_1^t A Q_1 = \left[\begin{array}{c|ccc} \lambda_1 & * & \cdots & * \\ \hline 0 & & & \\ \vdots & & A_1 & \\ 0 & & & \end{array}\right]$$

4. Set $A = A_1$, and repeat Steps 1 and 2.

5. Set $\lambda_2 = \lambda$ and

$$Q_2 = \left[\begin{array}{c|ccc} 1 & 0 & \cdots & 0 \\ \hline 0 & & & \\ \vdots & & Q & \\ 0 & & & \end{array}\right]$$

Calculate

$$(Q_1 Q_2)^t A (Q_1 Q_2) = \left[\begin{array}{c|c|ccc} \lambda_1 & * & * & \cdots & * \\ \hline 0 & \lambda_2 & * & \cdots & * \\ \hline 0 & 0 & & & \\ \vdots & \vdots & & A_2 & \\ 0 & 0 & & & \end{array}\right]$$

6. Suppose $\lambda_1, \ldots, \lambda_k$ and $Q_1, \ldots, Q_k$ have been obtained such that

$$(Q_1 \cdots Q_k)^t A (Q_1 \cdots Q_k) = \left[\begin{array}{ccc|ccc} \lambda_1 & \cdots & * & * & \cdots & * \\ \vdots & \cdots & \vdots & \vdots & \cdots & \vdots \\ 0 & \cdots & \lambda_k & * & \cdots & * \\ \hline 0 & \cdots & 0 & & & \\ \vdots & & \vdots & & A_k & \\ 0 & \cdots & 0 & & & \end{array}\right]$$

7. Set $A = A_k$, and repeat Steps 1 and 2. Set $\lambda_k = \lambda$ and

$$Q_{k+1} = \left[\begin{array}{ccc|ccc} 1 & \cdots & 0 & 0 & \cdots & 0 \\ \vdots & \cdots & \vdots & \vdots & \cdots & \vdots \\ 0 & \cdots & 1 & 0 & \cdots & 0 \\ \hline 0 & \cdots & 0 & & & \\ \vdots & & \vdots & & Q & \\ 0 & \cdots & 0 & & & \end{array}\right]$$

where Q is $(n-k) \times (n-k)$.

8. Calculate

$$(Q_1 \cdots Q_{k+1})^t A (Q_1 \cdots Q_{k+1}) = \left[\begin{array}{ccc|ccc} \lambda_1 & \cdots & * & * & \cdots & * \\ \vdots & \cdots & \vdots & \vdots & \cdots & \vdots \\ 0 & \cdots & \lambda_{k+1} & * & \cdots & * \\ \hline 0 & \cdots & 0 & & & \\ \vdots & \cdots & \vdots & & A_{k+1} & \\ 0 & \cdots & 0 & & & \end{array}\right]$$

9. Continue until $k = n - 1$. Then set $Q = Q_1 \cdots Q_{n-1}$, the orthogonal matrix, and $T = Q^t A Q$, the upper triangular matrix.

EXAMPLE 8

Find an upper triangular matrix orthogonally similar to

$$A = \begin{bmatrix} 4 & 0 & 5 \\ 1 & 0 & 2 \\ 2 & 0 & 1 \end{bmatrix}$$

Since $\det(\lambda I_3 - A) = \lambda(\lambda + 1)(\lambda - 6)$, the eigenvalues of A are $\lambda_1 = 0$, $\lambda_2 = -1$, and $\lambda_3 = 6$. For an eigenvector for $\lambda_1 = 0$, we solve $(0I_3 - A)\mathbf{p} = \mathbf{0}$ to obtain $\mathbf{p} = p_2 \begin{bmatrix} 0 \\ 1 \\ 0 \end{bmatrix}$. Choosing $p_2 = 1$, we have $\mathbf{p}^1 = \begin{bmatrix} 0 \\ 1 \\ 0 \end{bmatrix}$. Applying the deletion and Gram-Schmidt algorithms to $\left\{ \begin{bmatrix} 0 \\ 1 \\ 0 \end{bmatrix}, \mathbf{e}^1, \mathbf{e}^2, \mathbf{e}^3 \right\}$, we find $\mathbf{p}^1 = \begin{bmatrix} 0 \\ 1 \\ 0 \end{bmatrix}$, $\mathbf{p}^2 = \begin{bmatrix} 1 \\ 0 \\ 0 \end{bmatrix}$, and $\mathbf{p}^3 = \begin{bmatrix} 0 \\ 0 \\ 1 \end{bmatrix}$. Hence

$$Q_1 = \begin{bmatrix} 0 & 1 & 0 \\ 1 & 0 & 0 \\ 0 & 0 & 1 \end{bmatrix} \quad \text{and} \quad Q_1^t A Q_1 = \begin{bmatrix} 0 & 1 & 2 \\ 0 & 4 & 5 \\ 0 & 2 & 1 \end{bmatrix}$$

Now, applying Steps 1 and 2 of the algorithm to $\begin{bmatrix} 4 & 5 \\ 2 & 1 \end{bmatrix}$, we find eigenvalues -1 and 6, with an eigenvector $\begin{bmatrix} 1 \\ -1 \end{bmatrix}$ corresponding to -1. Using the deletion and Gram-Schmidt algorithms on $\left\{ \begin{bmatrix} 1 \\ -1 \end{bmatrix}, \mathbf{e}^1, \mathbf{e}^2 \right\}$, we get

$$\mathbf{p}^1 = \begin{bmatrix} 1 \\ -1 \end{bmatrix} \text{ and } \mathbf{p}^2 = \begin{bmatrix} 1 \\ 1 \end{bmatrix}, \text{ so}$$

$$Q_2 = \begin{bmatrix} 1 & 0 & 0 \\ 0 & \frac{1}{\sqrt{2}} & \frac{1}{\sqrt{2}} \\ 0 & -\frac{1}{\sqrt{2}} & \frac{1}{\sqrt{2}} \end{bmatrix}$$

Hence we have $Q_2^t Q_1^t A Q_1 Q_2 = \begin{bmatrix} 0 & -\frac{1}{\sqrt{2}} & \frac{3}{\sqrt{2}} \\ 0 & -1 & 3 \\ 0 & 0 & 6 \end{bmatrix}$ □

Orthogonal similarity is applied most often on a special subset of $\mathbf{R}_{n,n}$ called symmetric matrices. A *symmetric matrix* is any matrix that satisfies the equation

$$A = A^t$$

In terms of the entries of A, this means that

$$a_{ij} = a_{ji}$$

for all i and j. In terms of the rows and columns of A, it means that

$$\mathbf{a}_i^t = \mathbf{a}^i$$

for all i. We denote the set of all symmetric matrices in $\mathbf{R}_{n,n}$ by $\mathbf{S}_{n,n}$.

EXAMPLE 9

The following matrices are symmetric:

$$\begin{bmatrix} 1 & 2 \\ 2 & 1 \end{bmatrix}, \quad \begin{bmatrix} 1 & 2 & 3 \\ 2 & 4 & 5 \\ 3 & 5 & 6 \end{bmatrix}, \quad \begin{bmatrix} 0 & 1 & 0 & -2 \\ 1 & 3 & -4 & 2 \\ 0 & -4 & -1 & 0 \\ -2 & 2 & 0 & 5 \end{bmatrix}$$

Some nonsymmetric matrices are

$$\begin{bmatrix} 1 & 2 \\ 3 & 4 \end{bmatrix}, \quad \begin{bmatrix} 0 & 1 & 1 \\ -1 & 0 & 1 \\ -1 & -1 & 0 \end{bmatrix}, \quad \begin{bmatrix} 0 & 1 & -1 & 0 \\ 2 & 2 & 1 & -1 \\ -1 & 3 & 1 & 1 \\ 0 & 1 & -2 & 4 \end{bmatrix}$$ □

Many applications involve symmetric matrices; some of these are shown in the next section. We intend to solve such problems using the technique diagrammed in the introduction to this chapter. Thus, we now need to find canonical forms for these matrices. This requires a preliminary lemma.

LEMMA 7.5

Let A be an $n \times n$ symmetric matrix. Then A has n real eigenvalues $\lambda_1, \lambda_2, \ldots, \lambda_n$. Further, eigenvectors belonging to distinct eigenvalues are orthogonal.

PROOF

See exercise 23. ■

From this lemma we can show that every symmetric matrix is orthogonally similar to a diagonal matrix.

THEOREM 7.6

Let A be any symmetric matrix. Then there is an orthogonal matrix Q and a diagonal matrix D such that

$$Q^tAQ = D$$

PROOF

By Theorem 7.4, there is an orthogonal matrix Q such that $Q^tAQ = T$, where T is upper triangular. Since

$$(Q^tAQ)^t = Q^tAQ$$

it follows that $Q^tAQ = T$ is symmetric. Thus, the elements above the main diagonal in T are also 0's. Hence T is a diagonal matrix. ■

The orthogonal matrix Q and diagonal matrix D for which $Q^tAQ = D$ can be computed as follows.

ALGORITHM FOR COMPUTING Q AND D

1. Find all eigenvalues of A. List only the distinct eigenvalues $\lambda_1, \ldots, \lambda_s$. Suppose λ_i has multiplicity m_i for each i.

2. For λ_i, Theorem 7.6 ensures that there is a linearly independent set of m_i eigenvectors (exercise 24). Apply the Gram-Schmidt algorithm to this set to obtain an orthonormal set

$$\{\mathbf{q}^{i,1}, \mathbf{q}^{i,2}, \ldots, \mathbf{q}^{i,m_i}\}$$

of eigenvectors for λ_i.

3. Form the matrix

$$Q = [\mathbf{q}^{1,1} \quad \mathbf{q}^{1,2} \quad \cdots \quad \mathbf{q}^{1,m_1} \quad \mathbf{q}^{2,1} \quad \cdots \quad \mathbf{q}^{2,m_2} \quad \cdots \quad \mathbf{q}^{s,1} \quad \cdots \quad \mathbf{q}^{s,m_s}]$$

By the Gram-Schmidt algorithm and Lemma 7.5, every two distinct columns of Q are orthogonal. Thus Q is an orthogonal matrix.

4. Let D be the diagonal matrix such that

$$d_{11} = \cdots = d_{m_1,m_1} = \lambda_1, \ldots, d_{m_1+\cdots+m_{s-1}+1,m_1+\cdots+m_{s-1}+1} = \cdots = d_{nn} = \lambda_s$$

5. Then $Q^tAQ = D$.*

EXAMPLE 10

Let $A = \begin{bmatrix} 1 & 2 \\ 2 & 1 \end{bmatrix}$. Computing the eigenvalues of A gives

$$\begin{aligned}\det[\lambda I_2 - A] &= \det\begin{bmatrix} \lambda - 1 & -2 \\ -2 & \lambda - 1 \end{bmatrix} = (\lambda - 1)(\lambda - 1) - 4 \\ &= \lambda^2 - 2\lambda - 3 = (\lambda - 3)(\lambda + 1) = 0\end{aligned}$$

Thus $\lambda_1 = 3$ and $\lambda_2 = -1$.

For eigenvectors we first solve

$$(\lambda_1 I_2 - A)\mathbf{p} = \begin{bmatrix} 2 & -2 \\ -2 & 2 \end{bmatrix}\mathbf{p} = \begin{bmatrix} 0 \\ 0 \end{bmatrix}$$

A solution is $\mathbf{p} = \begin{bmatrix} 1 \\ 1 \end{bmatrix}$. Applying the Gram-Schmidt algorithm, we get $\mathbf{q}^1 = \dfrac{1}{\sqrt{2}}\begin{bmatrix} 1 \\ 1 \end{bmatrix}$. We next solve

$$(\lambda_2 I_2 - A)\mathbf{p} = \begin{bmatrix} -2 & -2 \\ -2 & -2 \end{bmatrix}\mathbf{p} = \begin{bmatrix} 0 \\ 0 \end{bmatrix}$$

A solution is $\mathbf{p}^2 = \begin{bmatrix} 1 \\ -1 \end{bmatrix}$. Applying the Gram-Schmidt algorithm, we get $\mathbf{q}^2 = \dfrac{1}{\sqrt{2}}\begin{bmatrix} 1 \\ -1 \end{bmatrix}$. Hence

$$Q = [\mathbf{q}^1 \quad \mathbf{q}^2] = \frac{1}{\sqrt{2}}\begin{bmatrix} 1 & 1 \\ 1 & -1 \end{bmatrix} \quad \text{and} \quad D = \begin{bmatrix} \lambda_1 & 0 \\ 0 & \lambda_2 \end{bmatrix} = \begin{bmatrix} 3 & 0 \\ 0 & -1 \end{bmatrix} \quad \square$$

* In this case it is often said that A is orthogonally diagonalizable and that Q orthogonally diagonalizes A.

EXAMPLE 11

Let $A = \begin{bmatrix} 1 & 0 & 0 \\ 0 & -1 & 2 \\ 0 & 2 & -1 \end{bmatrix}$. To find the eigenvalues, set

$$\det(\lambda I_3 - A) = \det\begin{bmatrix} \lambda - 1 & 0 & 0 \\ 0 & \lambda + 1 & -2 \\ 0 & -2 & \lambda + 1 \end{bmatrix} = (\lambda - 1)[(\lambda + 1)^2 - 4]$$
$$= (\lambda - 1)(\lambda^2 + 2\lambda - 3) = (\lambda - 1)(\lambda + 3)(\lambda - 1) = 0$$

Thus $\lambda_1 = \lambda_2 = 1$ and $\lambda_3 = -3$.

For corresponding eigenvectors, we first solve

$$(\lambda_1 I_3 - A)\mathbf{p} = \begin{bmatrix} 0 & 0 & 0 \\ 0 & 2 & -2 \\ 0 & -2 & 2 \end{bmatrix}\mathbf{p} = \mathbf{0}$$

We apply the Gaussian elimination algorithm to get

$$\begin{bmatrix} 0 & 0 & 0 & 0 \\ 0 & 2 & -2 & 0 \\ 0 & -2 & 2 & 0 \end{bmatrix} \xrightarrow[\mathbf{R_3 + R_1 \to R_3}]{\mathbf{R_1 \leftrightarrow R_2}} \begin{bmatrix} 0 & 2 & -2 & 0 \\ 0 & 0 & 0 & 0 \\ 0 & 0 & 0 & 0 \end{bmatrix}$$

so $2p_2 - 2p_3 = 0$. Thus p_2 is bound and p_1 and p_3 are free. It follows that

$$\mathbf{p} = \begin{bmatrix} p_1 \\ p_2 \\ p_3 \end{bmatrix} = p_1\begin{bmatrix} 1 \\ 0 \\ 0 \end{bmatrix} + p_3\begin{bmatrix} 0 \\ 1 \\ 1 \end{bmatrix}$$

Hence we can take $\begin{bmatrix} 1 \\ 0 \\ 0 \end{bmatrix}$ and $\begin{bmatrix} 0 \\ 1 \\ 1 \end{bmatrix}$ for eigenvectors for $\lambda_1 = \lambda_2$. Applying the Gram-Schmidt algorithm gives

$$\mathbf{q}^1 = \begin{bmatrix} 1 \\ 0 \\ 0 \end{bmatrix} \quad \text{and} \quad \mathbf{q}^2 = \frac{1}{\sqrt{2}}\begin{bmatrix} 0 \\ 1 \\ 1 \end{bmatrix}$$

We then solve

$$(\lambda_3 I_3 - A)\mathbf{p} = \begin{bmatrix} -4 & 0 & 0 \\ 0 & -2 & -2 \\ 0 & -2 & -2 \end{bmatrix}\mathbf{p} = \mathbf{0}$$

which yields an eigenvector $\begin{bmatrix} 0 \\ 1 \\ -1 \end{bmatrix}$. Applying the Gram-Schmidt algorithm gives

$$\mathbf{q}^3 = \frac{1}{\sqrt{2}}\begin{bmatrix} 0 \\ 1 \\ -1 \end{bmatrix}$$

Thus

$$Q = [\mathbf{q}^1 \quad \mathbf{q}^2 \quad \mathbf{q}^3] = \begin{bmatrix} 1 & 0 & 0 \\ 0 & \frac{1}{\sqrt{2}} & \frac{1}{\sqrt{2}} \\ 0 & \frac{1}{\sqrt{2}} & -\frac{1}{\sqrt{2}} \end{bmatrix}$$

and

$$D = \begin{bmatrix} \lambda_1 & 0 & 0 \\ 0 & \lambda_2 & 0 \\ 0 & 0 & \lambda_3 \end{bmatrix} = \begin{bmatrix} 1 & 0 & 0 \\ 0 & 1 & 0 \\ 0 & 0 & -3 \end{bmatrix}$$ □

In the next section we show some applications of this canonical form.

EXERCISES FOR SECTION 7.2

COMPUTATIONAL EXERCISES

In exercises 1 through 6, find an orthogonal matrix Q such that $Q^tAQ = T$, where T is upper triangular.

1. $A = \begin{bmatrix} 1 & 5 \\ 2 & -2 \end{bmatrix}$

2. $A = \begin{bmatrix} 1 & 2 \\ -2 & -3 \end{bmatrix}$

3. $A = \begin{bmatrix} 7 & 3 \\ -3 & 1 \end{bmatrix}$

4. $A = \begin{bmatrix} -1 & 4 \\ 2 & -3 \end{bmatrix}$

5. $A = \begin{bmatrix} -1 & -1 & 1 \\ 1 & 1 & -1 \\ 1 & -1 & -1 \end{bmatrix}$

6. $A = \begin{bmatrix} 1 & 1 & 1 \\ 1 & 1 & -1 \\ 1 & 1 & 1 \end{bmatrix}$

In exercises 7 and 8, for the given matrix A find an orthogonal matrix Q such that $T = Q^tAQ$ is upper triangular with $t_{11} = \lambda_1$.

7. $A = \begin{bmatrix} 1 & 2 \\ 1 & 2 \end{bmatrix}$, $\lambda_1 = 3$

8. $A = \begin{bmatrix} 3 & 1 \\ 2 & 2 \end{bmatrix}$, $\lambda_1 = 1$

In exercises 9 and 10, decide which of the given matrices are symmetric.

9. (*a*) $\begin{bmatrix} 0 & 1 \\ -1 & 0 \end{bmatrix}$ (*b*) $\begin{bmatrix} 1 & 2 \\ 2 & 5 \end{bmatrix}$ (*c*) $\begin{bmatrix} 6 & 1 \\ 3 & 6 \end{bmatrix}$ (*d*) $\begin{bmatrix} 4 & -8 \\ 4 & -8 \end{bmatrix}$

10. (*a*) $\begin{bmatrix} 2 & 1 & -6 \\ 1 & 4 & 5 \\ 6 & 5 & -3 \end{bmatrix}$ (*b*) $\begin{bmatrix} 4 & 3 & -2 \\ 3 & -8 & 7 \\ -2 & 7 & 0 \end{bmatrix}$ (*c*) $\begin{bmatrix} 0 & 0 & 1 \\ 0 & 1 & 0 \\ 1 & 0 & 0 \end{bmatrix}$ (*d*) $\begin{bmatrix} 0 & 1 & 0 \\ 0 & 0 & 1 \\ 1 & 0 & 0 \end{bmatrix}$

In exercises 11 through 18, each matrix A is symmetric. Find an orthogonal matrix Q and a diagonal matrix D such that $Q^tAQ = D$.

11. $A = \begin{bmatrix} 3 & 2 \\ 2 & 0 \end{bmatrix}$ *12.* $A = \begin{bmatrix} 2 & -1 \\ -1 & 2 \end{bmatrix}$ *13.* $A = \begin{bmatrix} 7 & \frac{3}{2} \\ \frac{3}{2} & 3 \end{bmatrix}$

14. $A = \begin{bmatrix} -2 & \frac{3}{2} \\ \frac{3}{2} & -6 \end{bmatrix}$

15. $A = \begin{bmatrix} 1 & 0 & 1 \\ 0 & 2 & 0 \\ 1 & 0 & 1 \end{bmatrix}$

16. $A = \begin{bmatrix} -2 & 0 & 0 \\ 0 & 7 & -3 \\ 0 & -3 & -1 \end{bmatrix}$

17. $A = \begin{bmatrix} 1 & -1 & -1 \\ -1 & 1 & -1 \\ -1 & -1 & 1 \end{bmatrix}$

18. $A = \begin{bmatrix} 1 & 2 & -1 \\ 2 & -2 & 2 \\ -1 & 2 & 1 \end{bmatrix}$

COMPLEX NUMBERS

A matrix $A \in \mathbf{C}_{n,n}$ is called *hermitian* if it satisfies the equation $A = A^* = [\bar{a}_{ij}]^t$. In exercises 19 and 20, decide which of the given matrices are hermitian.

19. (a) $\begin{bmatrix} 1 & i \\ -i & 2 \end{bmatrix}$ (b) $\begin{bmatrix} 2 & 1-i \\ 1+i & 0 \end{bmatrix}$

20. (a) $\begin{bmatrix} i & -1 \\ -1 & 1 \end{bmatrix}$ (b) $\begin{bmatrix} 0 & 1 & i \\ 1 & 1 & 1+i \\ -i & 1-i & 2 \end{bmatrix}$

In exercises 21 and 22, find a unitary matrix U such that U^*AU is diagonal.

21. $A = \begin{bmatrix} 1 & i \\ -i & 1 \end{bmatrix}$ 22. $A = \begin{bmatrix} 0 & i \\ -i & 0 \end{bmatrix}$

THEORETICAL EXERCISES

23. (complex numbers) Let A be in $\mathbf{S}_{n,n}$.
(a) Prove that A has n real eigenvalues. *Hint:* Set $A\mathbf{x} = \lambda\mathbf{x}$. Compute $\mathbf{x}^*A\mathbf{x} = \lambda\mathbf{x}^*\mathbf{x}$, where $\mathbf{x}^* = [\bar{x}_1, \ldots, \bar{x}_n]$, with $\bar{x}_i$ being the complex conjugate of x_i. Then look at complex conjugates of both sides and use the symmetry of A to get $\lambda\mathbf{x}^*\mathbf{x} = \bar{\lambda}\mathbf{x}^*\mathbf{x}$.
(b) Prove that if $\lambda_i \neq \lambda_j$ and if $A\mathbf{x} = \lambda_i\mathbf{x}$ and $A\mathbf{y} = \lambda_j\mathbf{y}$, then $(\mathbf{x}, \mathbf{y}) = \mathbf{x}^*\mathbf{y} = 0$. *Hint:* Consider $\mathbf{y}^*A\mathbf{x} = \lambda_i\mathbf{x}^*\mathbf{y}$, $\mathbf{x}^*A\mathbf{y} = \lambda_j\mathbf{x}^*\mathbf{y}$, or, taking the conjugate transpose, $\mathbf{y}^*A\mathbf{x} = \bar{\lambda}_j\mathbf{y}^*\mathbf{x}$. It follows that $\lambda_i\mathbf{y}^*\mathbf{x} = \bar{\lambda}_j\mathbf{y}^*\mathbf{x}$.

24. Let A be a symmetric matrix, and let λ be an eigenvalue of A with multiplicity m. Use Theorem 7.6 to show that there is a linearly independent set of m eigenvectors for λ. *Hint:* Consider $Q^tAQ = D$, or $AQ = QD$. Columnwise, this is equivalent to $A\mathbf{q}^i = \lambda_i\mathbf{q}^i$. Since λ appears m times on the diagonal of D, it follows that the corresponding $\mathbf{q}^i$ are eigenvectors for λ.

25. Let $p(t) = r_nt^n + \cdots + r_1t + r_0$. Let A be an $n \times n$ symmetric matrix. Prove that

$$p(A) = r_nA^n + \cdots + r_1A + r_0I_n$$

is symmetric. Thus A^2, rA, etc., are symmetric.

26. Let $A = \begin{bmatrix} 1 & 2 \\ 1 & 2 \end{bmatrix}$. Find Q such that $Q^tAQ = L$, a lower triangular matrix. *Hint:* Apply to A^t the algorithm for finding an orthogonally similar upper triangular matrix.

27. Let A be an $n \times n$ matrix, all of whose eigenvalues are real. Using the standard inner product on matrices, show that there is an $n \times n$ matrix B such that $\|A - B\|$ is arbitrarily close to 0 and such that B has distinct eigenvalues. *Hint:* Write $A = QTQ^t$. Pick a diagonal matrix D such that $T + rD$ has distinct eigenvalues for all sufficiently small r. Consider $B = Q(T + rD)Q^t$ for r sufficiently small. (Thus, every matrix has matrices arbitrarily close that are similar to diagonal matrices. This and other analyses of this type indicate that if the entries of A are estimations, we should expect A to be similar to a diagonal matrix.)

APPLICATIONS EXERCISE

28. Show that $A = \begin{bmatrix} 2 & 1 \\ -1 & 0 \end{bmatrix}$ is not similar to a diagonal matrix. Then, using Theorem 7.4, show how to solve

$$\mathbf{x}(k+1) = A\mathbf{x}(k)$$

(Thus, when diagonal canonical forms cannot be found, you can still use upper triangular canonical forms.)

COMPUTER EXERCISE

29. Let $A = \begin{bmatrix} 0 & t \\ t & 1 \end{bmatrix}$. Find the unit eigenvectors of A. Note the sensitivity of the eigenvectors when t is near 0.

7.3 APPLICATIONS OF CANONICAL FORMS (OPTIONAL)

In this section we look at three applications which show how symmetric matrices arise in problems and how diagonal canonical forms are used to solve these problems.

I. WATER TOWER SYSTEMS

The water tower system shown in Figure 7.9 has two water towers T_1 and T_2 and two sources of input S_1 and S_2.* The inputs are operated by automatic pumping devices at the sources, which behave according to the following device rule.

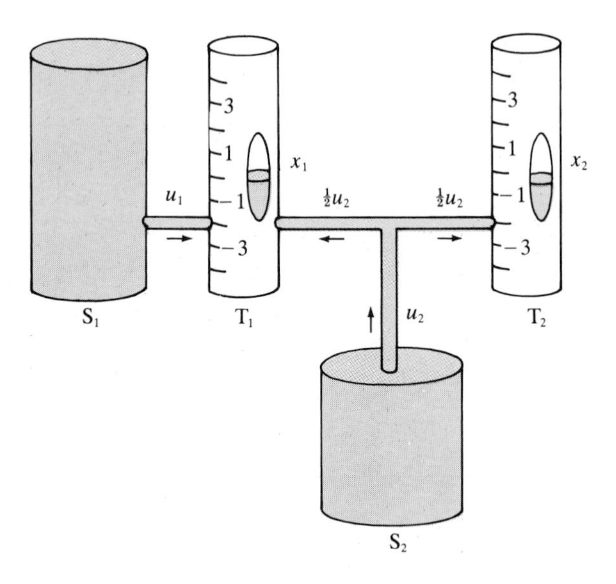

FIGURE 7.9

DEVICE RULE: Input from any source at time k is proportional to the negative of the sum of the water levels of all towers connected directly to that source at time k; that is,

$$u_j(k) = -c[\text{sum of all } x_i(k) \text{ such that } T_i \text{ is connected to } S_j]$$

Thus, if there is a cumulative excess of water in the towers monitored by a device, the device withdraws water. Similarly, the device adds water in case of a cumulative deficit of water. In Figure 7.9,

$$x_1(k+1) = x_1(k) + u_1(k) + \tfrac{1}{2}u_2(k)$$
$$x_2(k+1) = x_2(k) + \tfrac{1}{2}u_2(k)$$

and, by the device rule,

$$u_1(k) = -c_1x_1(k)$$
$$u_2(k) = -c_2[x_1(k) + x_2(k)]$$

where c_1 and c_2 are positive constants. Note that u_1 or u_2 could be negative,

* We use this problem because it requires no technical background. However, the features of the problem are seen in many applied problems, such as spring-mass systems and pendulum systems.

which would represent a withdrawal of water. Now, by substitution,

$$x_1(k+1) = x_1(k) - c_1 x_1(k) - \tfrac{1}{2}c_2[x_1(k) + x_2(k)]$$
$$x_2(k+1) = x_2(k) - \tfrac{1}{2}c_2[x_1(k) + x_2(k)]$$

which can be written as

$$x_1(k+1) = (1 - c_1 - \tfrac{1}{2}c_2)x_1(k) - \tfrac{1}{2}c_2 x_2(k)$$
$$x_2(k+1) = -\tfrac{1}{2}c_2 x_1(k) + (1 - \tfrac{1}{2}c_2)x_2(k)$$

Written as a matrix equation, this becomes

$$\mathbf{x}(k+1) = \begin{bmatrix} 1 - c_1 - \tfrac{1}{2}c_2 & -\tfrac{1}{2}c_2 \\ -\tfrac{1}{2}c_2 & 1 - \tfrac{1}{2}c_2 \end{bmatrix} \mathbf{x}(k)$$

Letting C denote the coefficient matrix in the equation yields

$$\mathbf{x}(k+1) = C\mathbf{x}(k)$$

where C is symmetric because the input u_2 applies equally to the towers.*

To solve this equation, let Q be an orthogonal matrix such that

$$C = QDQ^t$$

where D is the diagonal matrix obtained by using the eigenvalues λ_1 and λ_2 of C. Substituting gives

$$\mathbf{x}(k+1) = QDQ^t\mathbf{x}(k)$$

and multiplying by Q^t, we have

$$Q^t\mathbf{x}(k+1) = DQ^t\mathbf{x}(k)$$

Setting $\mathbf{y}(k) = Q^t\mathbf{x}(k)$ yields

$$\mathbf{y}(k+1) = D\mathbf{y}(k)$$

which describes the equation in $\# = (\mathbf{q}^1, \mathbf{q}^2)$. In terms of entries, the equations are

$$y_1(k+1) = \lambda_1 y_1(k) \qquad \text{and} \qquad y_2(k+1) = \lambda_2 y_2(k)$$

which have solutions

$$y_1(k) = a_1\lambda_1^k \qquad \text{and} \qquad y_2(k) = a_2\lambda_2^k$$

for all $a_1, a_2 \in \mathbf{R}$.

* Such symmetry occurs in various other physical systems also.

Thus

$$\mathbf{x}(k) = Q\mathbf{y}(k) = y_1(k)\mathbf{q}^1 + y_2(k)\mathbf{q}^2 = a_1\lambda_1^k\mathbf{q}^1 + a_2\lambda_2^k\mathbf{q}^2$$

Now, for any given initial $\mathbf{x}(0)$, we can determine a_1 and a_2 and thus find the exact water levels at any time k.

EXAMPLE 12

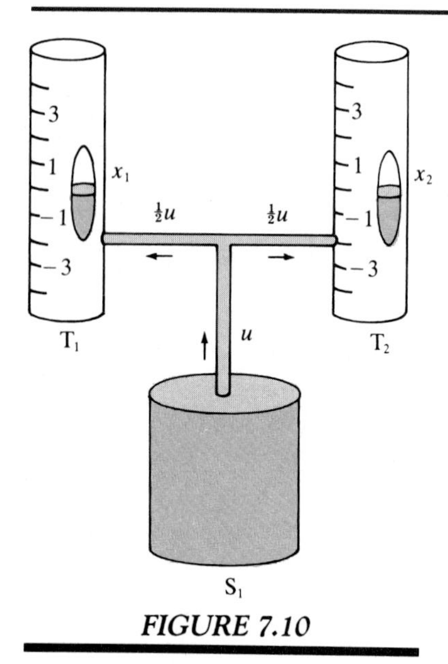

FIGURE 7.10

A symmetric water tower system is shown in Figure 7.10. Assuming that $\mathbf{x}(0) = \begin{bmatrix} 1 \\ 2 \end{bmatrix}$ and that the input behaves according to the device rule with $c = 1$, we can find the water levels at any time k. Here, the equations are

$$\begin{aligned} x_1(k+1) &= x_1(k) + \tfrac{1}{2}u(k) \\ x_2(k+1) &= x_2(k) + \tfrac{1}{2}u(k) \end{aligned}$$

Since $u(k) = -[x_1(k) + x_2(k)]$ by the device rule,

$$x_1(k+1) = \tfrac{1}{2}x_1(k) - \tfrac{1}{2}x_2(k) \qquad \text{and} \qquad x_2(k+1) = -\tfrac{1}{2}x_1(k) + \tfrac{1}{2}x_2(k)$$

or, as a matrix equation,

$$\mathbf{x}(k+1) = \tfrac{1}{2}\begin{bmatrix} 1 & -1 \\ -1 & 1 \end{bmatrix}\mathbf{x}(k)$$

The eigenvalues of the coefficient matrix are $\lambda_1 = 0$ and $\lambda_2 = 1$, with corresponding eigenvectors

$$\mathbf{q}^1 = \frac{1}{\sqrt{2}}\begin{bmatrix} 1 \\ 1 \end{bmatrix} \qquad \text{and} \qquad \mathbf{q}^2 = \frac{1}{\sqrt{2}}\begin{bmatrix} 1 \\ -1 \end{bmatrix}$$

Hence $\mathbf{x}(k) = a_1\dfrac{0^k}{\sqrt{2}}\begin{bmatrix} 1 \\ 1 \end{bmatrix} + a_2\dfrac{1^k}{\sqrt{2}}\begin{bmatrix} 1 \\ -1 \end{bmatrix}$ for any $a_1, a_2 \in \mathbf{R}$. Now, since

$$\begin{bmatrix} 1 \\ 2 \end{bmatrix} = \mathbf{x}(0) = a_1\frac{1}{\sqrt{2}}\begin{bmatrix} 1 \\ 1 \end{bmatrix} + a_2\frac{1}{\sqrt{2}}\begin{bmatrix} 1 \\ -1 \end{bmatrix}$$

we can solve for a_1 and a_2, getting $a_1 = \frac{3}{2}\sqrt{2}$ and $a_2 = -\frac{1}{2}\sqrt{2}$. Thus the levels on day k are

$$\mathbf{x}(k) = \tfrac{3}{2}0^k\begin{bmatrix} 1 \\ 1 \end{bmatrix} - \tfrac{1}{2}1^k\begin{bmatrix} 1 \\ -1 \end{bmatrix}$$

and, for $k > 0$,

$$\mathbf{x}(k) = \begin{bmatrix} -\frac{1}{2} \\ \frac{1}{2} \end{bmatrix}$$

These equations describe the behavior of the system. □

II. QUADRATIC FORMS

A *quadratic form* in two variables x_1 and x_2 is a function

$$q(x_1, x_2) = ax_1^2 + bx_1x_2 + cx_2^2$$

where $a, b, c \in \mathbf{R}$ and at least one of a, b, and c is not zero.*

EXAMPLE 13

Some examples of quadratic forms are

$$q(x_1, x_2) = x_1^2 - 2x_1x_2 + x_2^2$$
$$q(x_1, x_2) = 2x_1^2 + 4x_1x_2$$
$$q(x_1, x_2) = x_1^2 - 5x_2^2$$

□

In this application we will show how canonical forms can be used in graphing

1. the equation $q(x_1, x_2) = d$ in the coordinate plane $\mathbf{R}^2$ and
2. the function $q(x_1, x_2)$ in the coordinate space $\mathbf{R}^3$

To graph the equation $q(x_1, x_2) = d$, we plot all (x_1, x_2) that satisfy the equation.

EXAMPLE 14

The graph of $x_1^2 + 4x_2^2 = 1$ is the ellipse shown in Figure 7.11. (In general, the graph of the equation $ax_1^2 + cx_2^2 = d$ with $a > 0, c > 0$, and $d > 0$ is an ellipse.)

□

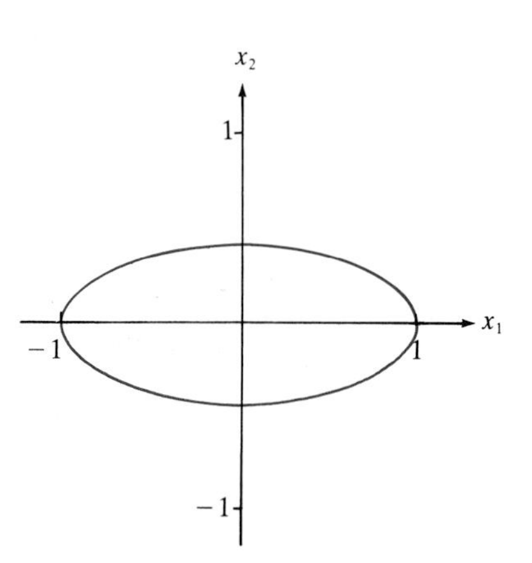

FIGURE 7.11

* Such functions arise, for example, in expressions for potential and kinetic energy. The definition of quadratic form can be extended to more variables.

EXAMPLE 15

The graph of $x_1^2 - x_2^2 = 1$ is the hyperbola shown in Figure 7.12. (In general, $ax_1^2 - cx_2^2 = \pm d$ with $a > 0$, $c > 0$, and $d > 0$ is a hyperbola.*) □

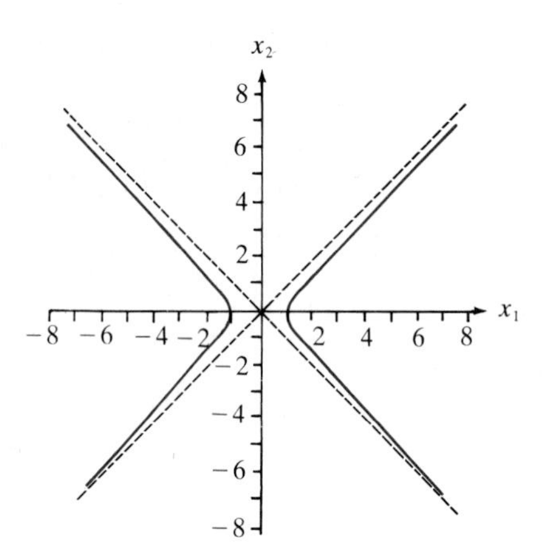

FIGURE 7.12

To graph a quadratic form q means to plot all points (x_1, x_2, x_3) in the coordinate space $\mathbf{R}^3$ that satisfy $x_3 = q(x_1, x_2)$, as shown in Figure 7.13. Graphing a function $q(x_1, x_2)$ by randomly plotting such points is usually ineffective. However, if

$$q(x_1, x_2) = ax_1^2 + cx_2^2$$

then plotting points (d, x_2, x_3), (x_1, d, x_3), and (x_1, x_2, d) that satisfy $x_3 = q(x_1, x_2)$ for various constant values of d provides curves of intersection of the graph of q in planes parallel to the coordinate planes. The resulting curves provide a sketch of the graph of q.

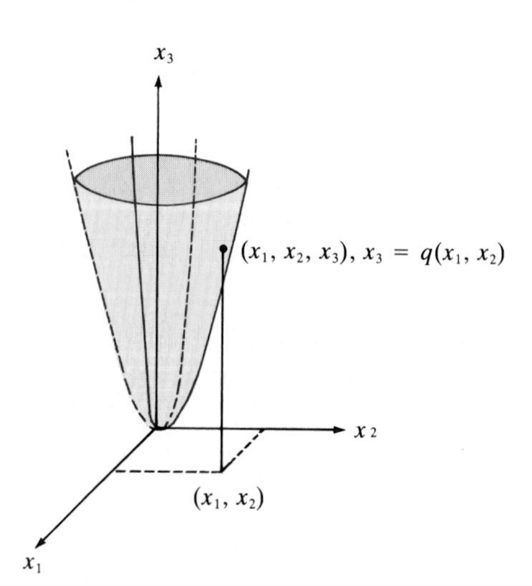

FIGURE 7.13

* This includes the form $-ax_1^2 + cx_2^2 = d$, since it can be written $ax_1^2 - cx_2^2 = -d$.

Example 16

Graph $q(x_1, x_2) = x_1^2 + 4x_2^2$. The points (x_1, x_2, x_3) that need to be plotted can be obtained by setting $q(x_1, x_2) = x_3$ and determining curves for $x_1 = d$, $x_2 = d$, and $x_3 = d$. This is done for selected values of d in Table 7.1, and the resulting curves and interpolated surface are shown in Figure 7.14. □

TABLE 7.1

$x_1^2 + 4x_2^2 = 0, x_3 = 0$	origin
$x_1^2 + 4x_2^2 = 2, x_3 = 2$	ellipse 2 units above the x_1,x_2-plane
$x_1^2 + 4x_2^2 = 4, x_3 = 4$	ellipse 4 units above the x_1,x_2-plane
$x_1^2 + 4x_2^2 = 6, x_3 = 6$	ellipse 6 units above the x_1,x_2-plane
$x_1^2 + 4x_2^2 = 8, x_3 = 8$	ellipse 8 units above the x_1,x_2-plane
$x_3 = x_1^2, x_2 = 0$	parabola in the x_1,x_3-plane
$x_3 = 4x_2^2, x_1 = 0$	parabola in the x_2,x_3-plane

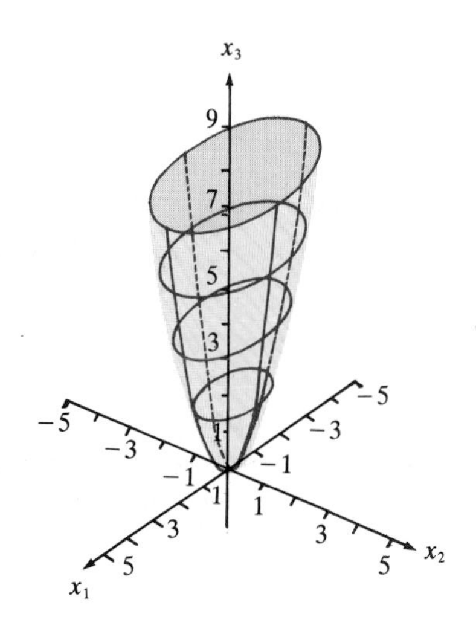

FIGURE 7.14

Example 17

Graph $q(x_1, x_2) = x_1^2 - x_2^2$. The points (x_1, x_2, x_3) satisfying $x_3 = q(x_1, x_2)$ are computed and plotted as was done in Example 16. Table 7.2 is the resulting table, and the graph is in Figure 7.15. □

TABLE 7.2

$x_3 = \frac{9}{4} - x_2^2, x_1 = \pm\frac{3}{2}$	parabolas $\frac{3}{2}$ units on either side of the x_2,x_3-plane
$x_3 = 1 - x_2^2, x_1 = \pm 1$	parabolas 1 unit on either side of the x_2,x_3-plane
$x_3 = \frac{1}{4} - x_2^2, x_1 = \pm\frac{1}{2}$	parabolas $\frac{1}{2}$ unit on either side of the x_2,x_3-plane
$x_3 = -x_2^2, x_1 = 0$	parabola in the x_2,x_3-plane
$x_3 = x_1^2 - \frac{9}{4}, x_2 = \pm\frac{3}{2}$	parabolas $\frac{3}{2}$ units on either side of the x_1,x_3-plane
$x_3 = x_1^2 - 1, x_2 = \pm 1$	parabolas 1 unit on either side of the x_1,x_3-plane
$x_3 = x_1^2 - \frac{1}{4}, x_2 = \pm\frac{1}{2}$	parabolas $\frac{1}{2}$ unit on either side of the x_1,x_3-plane
$x_3 = x_1^2, x_2 = 0$	parabola in the x_1,x_3-plane

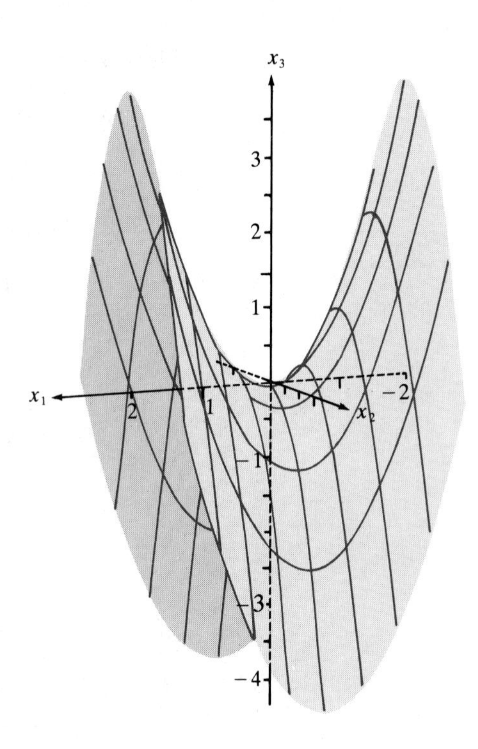

FIGURE 7.15

Graphing $q(x_1, x_2) = ax_1^2 + bx_1x_2 + cx_2^2$, or the equation $q(x_1, x_2) = d$, is more complicated if the coefficient b of the mixed term x_1x_2 is not 0. However, we intend to use canonical forms to trade these graphing problems for simpler graphing problems, such as those given above, which do not contain mixed terms. To do this, write

$$q(x_1, x_2) = [x_1 \quad x_2]\begin{bmatrix} a & \frac{b}{2} \\ \frac{b}{2} & c \end{bmatrix}\begin{bmatrix} x_1 \\ x_2 \end{bmatrix}$$

By letting A denote the coefficient matrix in this expression, we have

$$q(\mathbf{x}) = \mathbf{x}^t A\mathbf{x}$$

where $\mathbf{x} = \begin{bmatrix} x_1 \\ x_2 \end{bmatrix}$. Let the eigenvalues of A be λ_1 and λ_2. Since A is symmetric, we can find an orthogonal matrix Q such that $A = QDQ^t$, where $D = \begin{bmatrix} \lambda_1 & 0 \\ 0 & \lambda_2 \end{bmatrix}$. Substitution leads to

$$q(\mathbf{x}) = \mathbf{x}^t QDQ^t\mathbf{x}$$

We can describe the function in a second coordinate system $\# = (\mathbf{q}^1, \mathbf{q}^2)$ by changing coordinates, letting $\mathbf{y} = Q^t\mathbf{x}$. In $\#$ the function is

$$q_{\#}(\mathbf{y}) = \mathbf{y}^t D\mathbf{y} = \lambda_1 y_1^2 + \lambda_2 y_2^2$$

This means that the graphs of

$$q(x_1, x_2) = d \text{ in } \mathbf{R}^2 \qquad \text{and} \qquad q_{\#}(y_1, y_2) = d \text{ in } (\mathbf{R}^2)_{\#}$$

are identical, as are the graphs of

$$q(x_1, x_2) \text{ in } \mathbf{R}^3 \qquad \text{and} \qquad q_{\#}(y_1, y_2) \text{ in } (\mathbf{R}^3)_{\#}$$

where, in $\mathbf{R}^3$, $\# = (\mathbf{q}^1, \mathbf{q}^2, \mathbf{e}^3)$. Since graphing the expression $q_{\#}(y_1, y_2)$ is easier than graphing $q(x_1, x_2)$, we will do the former.

EXAMPLE 18

Graph $x_1^2 + 4x_1x_2 + x_2^2 = 1$. Here,

$$\begin{aligned} q(x_1, x_2) &= x_1^2 + 4x_1x_2 + x_2^2 \\ &= [x_1 \quad x_2]\begin{bmatrix} 1 & 2 \\ 2 & 1 \end{bmatrix}\begin{bmatrix} x_1 \\ x_2 \end{bmatrix} \end{aligned}$$

The eigenvalues of $A = \begin{bmatrix} 1 & 2 \\ 2 & 1 \end{bmatrix}$ are $\lambda_1 = 3$ and $\lambda_2 = -1$. Corresponding orthogonal eigenvectors of unit length are

$$\mathbf{q}^1 = \frac{1}{\sqrt{2}}\begin{bmatrix} 1 \\ 1 \end{bmatrix} \qquad \text{and} \qquad \mathbf{q}^2 = \frac{1}{\sqrt{2}}\begin{bmatrix} 1 \\ -1 \end{bmatrix}$$

Now, since $\# = (\mathbf{q}^1, \mathbf{q}^2)$ and since

$$q_{\#}(y_1, y_2) = \lambda_1 y_1^2 + \lambda_2 y_2^2 = 3y_1^2 - y_2^2$$

we can graph the hyperbola

$$3y_1^2 - y_2^2 = 1$$

in $(\mathbf{R}^2)_{\#}$. This graph is shown in Figure 7.16. □

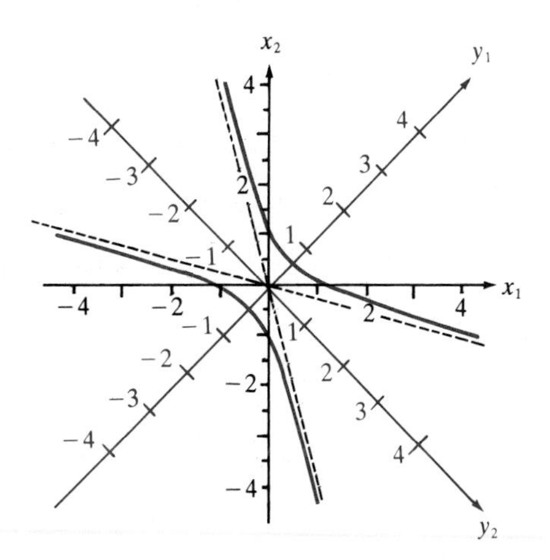

FIGURE 7.16

EXAMPLE 19

To graph

$$q(x_1, x_2) = 4x_1^2 - 2x_1x_2 + 4x_2^2 = [x_1 \quad x_2]\begin{bmatrix} 4 & -1 \\ -1 & 4 \end{bmatrix}\begin{bmatrix} x_1 \\ x_2 \end{bmatrix}$$

we first find the eigenvalues of $A = \begin{bmatrix} 4 & -1 \\ -1 & 4 \end{bmatrix}$, which are $\lambda_1 = 5$ and $\lambda_2 = 3$. Corresponding eigenvectors of unit length are

$$\mathbf{q}^1 = \frac{1}{\sqrt{2}}\begin{bmatrix} 1 \\ -1 \end{bmatrix} \quad \text{and} \quad \mathbf{q}^2 = \frac{1}{\sqrt{2}}\begin{bmatrix} 1 \\ 1 \end{bmatrix}$$

Thus, using $\# = (\mathbf{q}^1, \mathbf{q}^2, \mathbf{e}^3)$, we can graph $q(x_1, x_2)$ by graphing

$$\begin{aligned} q_{\#}(y_1, y_2) &= \lambda_1 y_1^2 + \lambda_2 y_2^2 \\ &= 5y_1^2 + 3y_2^2 \end{aligned}$$

in $(\mathbf{R}^3)_{\#}$. This graph is shown in Figure 7.17. □

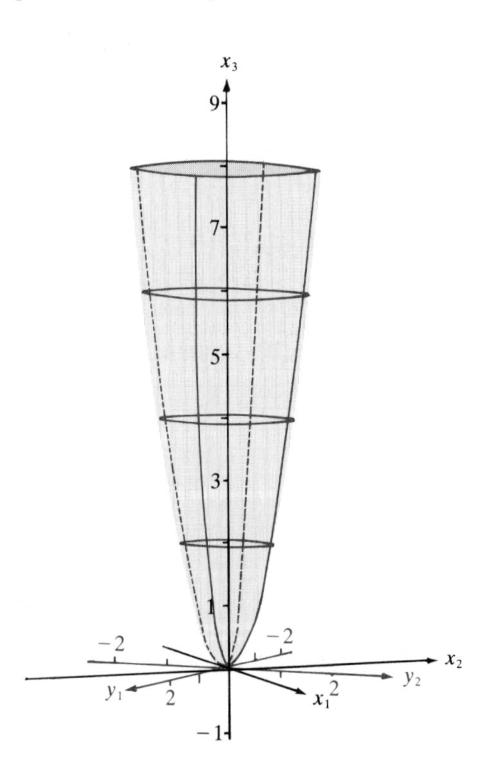

FIGURE 7.17

The basic shape of the graph of the equation $q(x_1, x_2) = d$, and hence that of $q(x_1, x_2)$, is determined by eigenvalues. Thus, the basic shape can be decided by computing invariants, which were discussed in Section 6.4.

Table 7.3 records the cases (exercise 38). Here,

$$q(x_1, x_2) = ax_1^2 + bx_1x_2 + cx_2^2 \qquad \text{and} \qquad A = \begin{bmatrix} a & \frac{b}{2} \\ \frac{b}{2} & c \end{bmatrix}$$

Although there are infinitely many algebraic equations $q(x_1, x_2) = d$, Table 7.3 shows that geometrically there are only six possibilities for the type, or shape, of the graph—namely, $\varnothing$, the origin, an ellipse, a hyperbola, one line, or two lines.

TABLE 7.3

INVARIANTS	EIGENVALUES	GRAPH $q(x_1, x_2) = d$	GRAPH $q(x_1, x_2)$
det $A > 0$ trace $A > 0$	$\lambda_1 > 0$ $\lambda_2 > 0$	origin if $d = 0$, ellipse if $d > 0$, $\varnothing$ if $d < 0$	elliptic paraboloid
det $A > 0$ trace $A < 0$	$\lambda_1 < 0$ $\lambda_2 < 0$	origin if $d = 0$, ellipse if $d < 0$, $\varnothing$ if $d > 0$	
det $A < 0$	$\lambda_1 > 0$ $\lambda_2 < 0$	two lines if $d = 0$, hyperbola if $d \neq 0$	hyperbolic paraboloid
det $A = 0$ trace $A > 0$	$\lambda_1 > 0$ $\lambda_2 = 0$	one line if $d = 0$, two lines if $d > 0$, $\varnothing$ if $d < 0$	parabolic cylinder (wedge shape)
det $A = 0$ trace $A < 0$	$\lambda_1 < 0$ $\lambda_2 = 0$	one line if $d = 0$, two lines if $d < 0$, $\varnothing$ if $d > 0$	
det $A = 0$ trace $A = 0$	$\lambda_1 = 0$ $\lambda_2 = 0$	origin if $d = 0$, $\varnothing$ if $d \neq 0$	origin

EXAMPLE 20

Identify the graph of

$$q(x_1, x_2) = 4x_1^2 - 3x_1x_2 + 2x_2^2 = 4$$

or

$$[x_1 \quad x_2]\begin{bmatrix} 4 & -\frac{3}{2} \\ -\frac{3}{2} & 2 \end{bmatrix}\begin{bmatrix} x_1 \\ x_2 \end{bmatrix} = 4$$

Since det $A = \frac{23}{4} > 0$ and trace $A = 6 > 0$, it follows that the graph is an ellipse. □

EXAMPLE 21

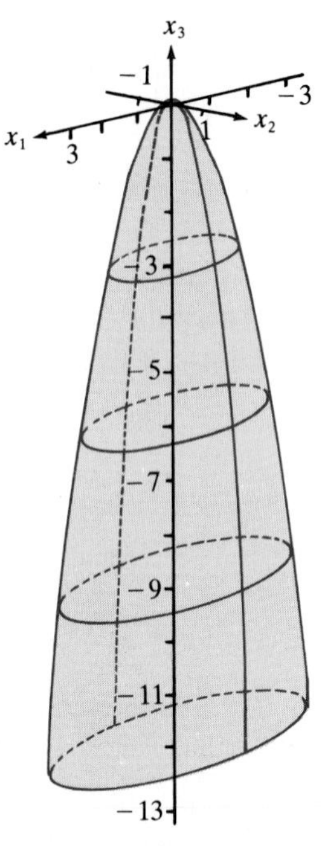

FIGURE 7.18

Identify the graph of

$$q(x_1, x_2) = -x_1^2 + 2x_1x_2 - 4x_2^2$$

Since $A = \begin{bmatrix} -1 & 1 \\ 1 & -4 \end{bmatrix}$, $\det A = 3 > 0$ and trace $A = -5 < 0$. Thus, this graph is an elliptic paraboloid, as shown in Figure 7.18. □

III. RAYLEIGH'S PRINCIPLE

Let A be an $n \times n$ symmetric matrix. Then, listing the n real eigenvalues in decreasing order, we have

$$\lambda_1 \geqslant \lambda_2 \geqslant \cdots \geqslant \lambda_n$$

In some problems it is important to obtain some simple estimate of λ_1. We will show how this can be done.

Let $\mathbf{x} = [x_1, x_2, \ldots, x_n]^t$ be a vector of n variables, and

$$q(x_1, \ldots, x_n) = \mathbf{x}^t A \mathbf{x}$$

a quadratic form. The problem

$$\max_{\|\mathbf{x}\| = 1} q(x_1, \ldots, x_n)$$

is to find a $\mathbf{z} \in \mathbf{R}^n$ such that $\|\mathbf{z}\| = 1$ and

$$q(z_1, \ldots, z_n) \geqslant q(x_1, \ldots, x_n)$$

for all $\mathbf{x}$ with $\|\mathbf{x}\| = 1$. Geometrically, this means that we want to find a $\mathbf{z}$ on the unit sphere that maximizes the value of q, as indicated in Figure 7.19. To find such a $\mathbf{z}$, we replace this problem with a simpler one, using the canonical form work of this chapter.

Let Q be an orthogonal matrix such that $A = QDQ^t$ where

$$D = \begin{bmatrix} \lambda_1 & 0 & \cdots & 0 \\ 0 & \lambda_2 & \cdots & 0 \\ \vdots & \vdots & & \vdots \\ 0 & 0 & \cdots & \lambda_n \end{bmatrix}$$

Setting $\mathbf{y} = Q^t \mathbf{x}$ and substituting gives

$$\begin{aligned} q(x_1, x_2, \ldots, x_n) &= \mathbf{x}^t A \mathbf{x} \\ &= \mathbf{x}^t Q D Q^t \mathbf{x} \\ &= \mathbf{y}^t D \mathbf{y} \\ &= q_{\#}(y_1, y_2, \ldots, y_n) \end{aligned}$$

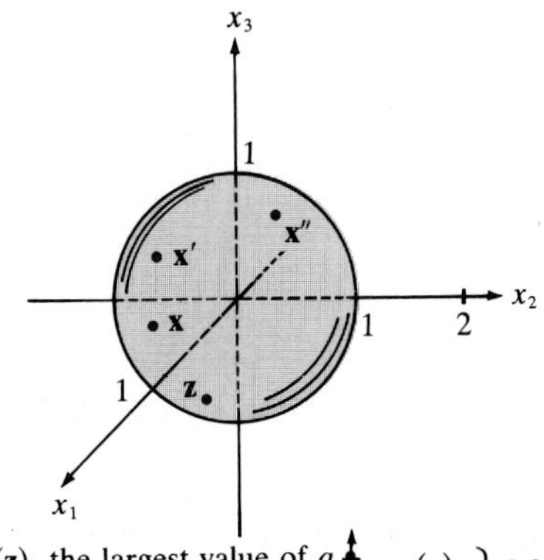

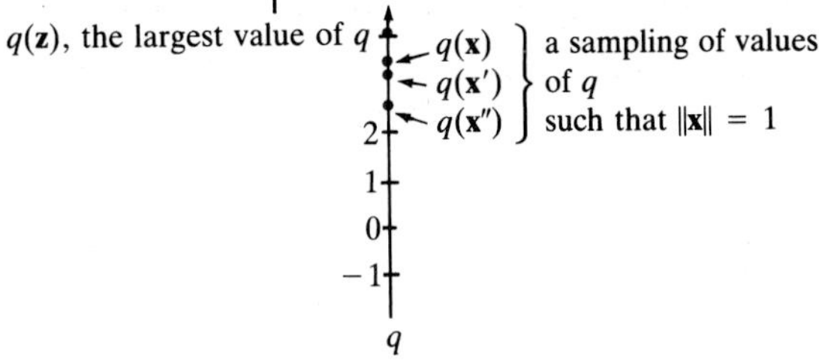

FIGURE 7.19

where $\# = (\mathbf{q}^1, \mathbf{q}^2, \ldots, \mathbf{q}^n)$. Since $\mathbf{y} = Q^t\mathbf{x}$, we have $\|\mathbf{x}\| = 1$ if and only if $\|\mathbf{y}\| = 1$; hence we can write

$$\begin{aligned}
\max_{\|\mathbf{x}\|=1} q(x_1, x_2, \ldots, x_n) &= \max_{\|\mathbf{y}\|=1} q_{\#}(y_1, y_2, \ldots, y_n) \\
&= \max_{\|\mathbf{y}\|=1} \mathbf{y}^t D \mathbf{y} \\
&= \max_{\|\mathbf{y}\|=1} (\lambda_1 y_1^2 + \lambda_2 y_2^2 + \cdots + \lambda_n y_n^2)
\end{aligned}$$

Now, for $\mathbf{y}$ such that $\|\mathbf{y}\| = 1$, we see that

$$\begin{aligned}
\lambda_1 y_1^2 + \lambda_2 y_2^2 + \cdots + \lambda_n y_n^2 &\leqslant \lambda_1 y_1^2 + \lambda_1 y_2^2 + \cdots + \lambda_1 y_n^2 \\
&= \lambda_1 (y_1^2 + y_2^2 + \cdots + y_n^2) = \lambda_1
\end{aligned}$$

Further, if $\mathbf{y} = \mathbf{e}^1$, then

$$\lambda_1 y_1^2 + \lambda_2 y_2^2 + \cdots + \lambda_n y_n^2 = \lambda_1$$

Thus

$$\max_{\|\mathbf{x}\|=1} q(x_1, x_2, \ldots, x_n) = \max_{\|\mathbf{y}\|=1} (\lambda_1 y_1^2 + \lambda_2 y_2^2 + \cdots + \lambda_n y_n^2) = \lambda_1$$

The equality

$$\max_{\|\mathbf{x}\|=1} q(x_1, x_2, \ldots, x_n) = \lambda_1$$

is called *Rayleigh's Principle.* We can use this equality to estimate λ_1. We simply calculate $q(x_1, \ldots, x_n)$ for various choices of $\mathbf{x} \in \mathbf{R}^n$ such that $\|\mathbf{x}\| = 1$, noting that $q(x_1,\ldots,x_n) \leqslant \lambda_1$ for all such choices, and then pick the largest number of this set as the estimate.*

EXAMPLE 22

Let $A = \begin{bmatrix} 1 & 2 & 0 \\ 2 & 2 & 3 \\ 0 & 3 & 3 \end{bmatrix}$. Table 7.4 contains some values for q, all smaller than λ_1. Thus $5\frac{1}{2} \leqslant \lambda_1$ and is the best estimate of λ_1 obtained. □

TABLE 7.4

x FOR WHICH $\|\mathbf{x}\| = 1$	$q(x_1, x_1, x_3)$
$[1,0,0]'$	1
$[0,1,0]'$	2
$[0,0,1]'$	3
$\frac{1}{\sqrt{2}}[1,1,0]'$	$3\frac{1}{2}$
$\frac{1}{\sqrt{2}}[0,1,1]'$	$5\frac{1}{2}$
$\frac{1}{\sqrt{3}}[1,1,1]'$	$5\frac{1}{3}$

We will use Rayleigh's Principle in the next chapter, where we will give some numerical techniques for computing eigenvalues.

EXERCISES FOR SECTION 7.3

COMPUTATIONAL EXERCISES

In exercises 1 through 4, solve for $\mathbf{x}(k)$.

1. $\mathbf{x}(k+1) = \begin{bmatrix} 1 & 3 \\ 3 & 1 \end{bmatrix}\mathbf{x}(k)$, $\mathbf{x}(0) = \begin{bmatrix} 1 \\ 3 \end{bmatrix}$

2. $\mathbf{x}(k+1) = \begin{bmatrix} 5 & 2 \\ 2 & 2 \end{bmatrix}\mathbf{x}(k)$, $\mathbf{x}(0) = \begin{bmatrix} 5 \\ -5 \end{bmatrix}$

3. $\mathbf{x}(k+1) = \begin{bmatrix} 1 & 0 & 1 \\ 0 & 2 & 0 \\ 1 & 0 & 1 \end{bmatrix}\mathbf{x}(k)$, $\mathbf{x}(0) = \begin{bmatrix} 1 \\ 1 \\ 1 \end{bmatrix}$

4. $\mathbf{x}(k+1) = \begin{bmatrix} 6 & 0 & 4 \\ 0 & -2 & 0 \\ 4 & 0 & 0 \end{bmatrix}\mathbf{x}(k)$, $\mathbf{x}(0) = \begin{bmatrix} 0 \\ 3 \\ -5 \end{bmatrix}$

In exercises 5 through 12, use canonical forms to sketch and identify the graph of the given quadratic equation.

5. $x_1^2 + 2x_1x_2 + x_2^2 = 8$

6. $7x_1^2 + 8x_1x_2 + x_2^2 = 36$

7. $2x_1x_2 = -9$

8. $8x_1^2 - 4x_1x_2 + 5x_2^2 = 36$

9. $13x_1^2 - 12x_1x_2 + 4x_2^2 = 64$

10. $8x_1^2 + 8x_1x_2 + 2x_2^2 = \frac{5}{2}$

11. $3x_1^2 - 4x_1x_2 = 4$

12. $4x_1^2 - 10x_1x_2 + 4x_2^2 = 9$

In exercises 13 through 16, write each of the quadratic forms $q(x_1, x_2)$ as $q_{\#}(y_1, y_2)$, as shown in Example 19. For each given $\mathbf{x}$, compute the corresponding coordinate vector $\mathbf{y}$. Plot

* This technique is useful in finding an initial estimate of an eigenvalue. A reasonable estimate of the eigenvalue is needed to begin some numerical algorithms for computing eigenvalues.

$\mathbf{x}$ and $\mathbf{y}$ in the respective coordinate systems. Then compute $q(x_1, x_2)$ and $q_{\#}(y_1, y_2)$, showing that these values are the same. Graph these values in the respective coordinate systems.

13. $q(x_1,x_2) = 2x_1^2 + 2x_1x_2 + 2x_2^2$;

$$\mathbf{x} = \begin{bmatrix}1\\1\end{bmatrix}, \begin{bmatrix}1\\-1\end{bmatrix}, \begin{bmatrix}-1\\1\end{bmatrix}, \begin{bmatrix}-1\\-1\end{bmatrix}$$

14. $q(x_1,x_2) = 4x_1^2 + 2x_1x_2 + 4x_2^2$;

$$\mathbf{x} = \begin{bmatrix}1\\0\end{bmatrix}, \begin{bmatrix}0\\1\end{bmatrix}, \begin{bmatrix}-1\\0\end{bmatrix}, \begin{bmatrix}0\\-1\end{bmatrix}$$

15. $q(x_1,x_2) = 5x_1^2 - 8x_1x_2 + 5x_2^2$;

$$\mathbf{x} = \begin{bmatrix}1\\1\end{bmatrix}, \begin{bmatrix}1\\-1\end{bmatrix}, \begin{bmatrix}-1\\1\end{bmatrix}, \begin{bmatrix}-1\\-1\end{bmatrix}$$

16. $q(x_1,x_2) = 2x_1^2 - 4x_1x_2 + 2x_2^2$;

$$\mathbf{x} = \begin{bmatrix}1\\0\end{bmatrix}, \begin{bmatrix}0\\1\end{bmatrix}, \begin{bmatrix}-1\\0\end{bmatrix}, \begin{bmatrix}0\\-1\end{bmatrix}$$

In exercises 17 through 22, use canonical forms to sketch and identify the graph of the given quadratic form.

17. $q(x_1,x_2) = 13x_1^2 + 12x_1x_2 + 4x_2^2$

18. $q(x_1,x_2) = 4x_1^2 - 12x_1x_2 + 9x_2^2$

19. $q(x_1,x_2) = x_1^2 + 4x_1x_2 + 4x_2^2$

20. $q(x_1,x_2) = 4x_1x_2 - 3x_2^2$

21. $q(x_1,x_2) = 7x_1^2 - 8x_1x_2 + x_2^2$

22. $q(x_1,x_2) = 10x_1^2 - 12x_1x_2 + 10x_2^2$

In exercises 23 through 30, use invariants to identify the graph of each of the given equations or quadratic forms, as in Examples 20 and 21.

23. $x_1^2 - 2x_1x_2 - x_2^2 = 2$

24. $2x_1^2 - x_1x_2 + x_2^2 = 4$

25. $x_1^2 - 2x_1x_2 + x_2^2 = 3$

26. $4x_1^2 - x_1x_2 + 2x_2^2 = -2$

27. $q(x_1,x_2) = x_1^2 + x_2^2$

28. $q(x_1,x_2) = 6x_1^2 - 2x_1x_2 + 9x_2^2$

29. $q(x_1,x_2) = x_1^2 + 4x_1x_2 - 2x_2^2$

30. $q(x_1,x_2) = 2x_1^2 - 2x_1x_2 + x_2^2$

In exercises 31 through 34, use Rayleigh's Principle to estimate the largest eigenvalue.

31. $\begin{bmatrix}2 & 3\\3 & 1\end{bmatrix}$ *33.* $\begin{bmatrix}7 & 8\\8 & 6\end{bmatrix}$

32. $\begin{bmatrix}2 & 4\\4 & 1\end{bmatrix}$ *34.* $\begin{bmatrix}6 & -1\\-1 & 5\end{bmatrix}$

35. Using unit vectors $\mathbf{e}^1$, $\mathbf{e}^2$, $\mathbf{e}^3$ and $\mathbf{e} = \dfrac{\sqrt{3}}{3}(\mathbf{e}^1 + \mathbf{e}^2 + \mathbf{e}^3)$, use Rayleigh's Principle to estimate the largest eigenvalue of

$$A = \begin{bmatrix}2 & 2 & 2\\2 & 5 & 4\\2 & 4 & 5\end{bmatrix}$$

(Note that the estimate for $\mathbf{e}^i$ is a a_{ii}, which is easily obtained. Also, the estimate for $\mathbf{e}$ is $\frac{1}{3}(\Sigma_{i,j}\, a_{ij})$, which is easily calculated.

Complex Numbers

If matrix A is hermitian, then

$$\max_{\|\mathbf{z}\|=1} \mathbf{z}^*A\mathbf{z} = \lambda$$

the largest eigenvalue of A. In exercises 36 and 37, estimate the largest eigenvalue for the given hermitian matrix.

36. $\begin{bmatrix}1 & -i\\i & 1\end{bmatrix}$ *37.* $\begin{bmatrix}0 & 1-i\\1+i & 0\end{bmatrix}$

Theoretical Exercises

38. Prove the conclusions given in the third and fourth columns of Table 7.3. (*Hint:* See the section on invariants in Chapter 6 and use canonical forms.)

39. Prove that

$$\max_{\|\mathbf{x}\|=1} \mathbf{x}^tA\mathbf{x} = \max_{\mathbf{y}\neq\mathbf{0}} \frac{\mathbf{y}^tA\mathbf{y}}{\mathbf{y}^t\mathbf{y}} = \lambda_1$$

for any $n \times n$ matrix A. *Hint:* Show that every value of $q(\mathbf{x}) = \mathbf{x}^tA\mathbf{x}$ for $\|\mathbf{x}\| = 1$ is a value of $q'(\mathbf{x}) = \dfrac{\mathbf{y}^tA\mathbf{y}}{\mathbf{y}^t\mathbf{y}}$ for $\mathbf{y} \neq \mathbf{0}$ and vice versa.

40. Prove that if A is symmetric, then

$$\lambda_n = \min_{\|\mathbf{x}\|=1} \mathbf{x}^tA\mathbf{x}$$

41. Prove that if the sign of b in

$$q(x_1,x_2) = ax_1^2 + bx_1x_2 + cx_2^2$$

is changed, the same graph is produced but the eigenvalues and eigenvectors—and hence the axes—are interchanged.

42. A symmetric $n \times n$ matrix with positive eigenvalues is called *positive definite.* Prove that if A is positive definite, then

$$q(\mathbf{x}) = \mathbf{x}^tA\mathbf{x} \geqslant 0$$

for all **x** with equality if and only if $\mathbf{x} = \mathbf{0}$. *Hint:* Substitute $A = QDQ^t$ and view the problem in the corresponding coordinate system. (This special matrix can be used to decide if a critical point for a function of n variables corresponds to a maximum, minimum, etc.)

APPLICATIONS EXERCISES

In exercises 43 through 46, for the given water tower system with the device rule, find $\mathbf{x}(k)$ for the given constants of proportionality c_i and initial state $\mathbf{x}(0)$.

43.

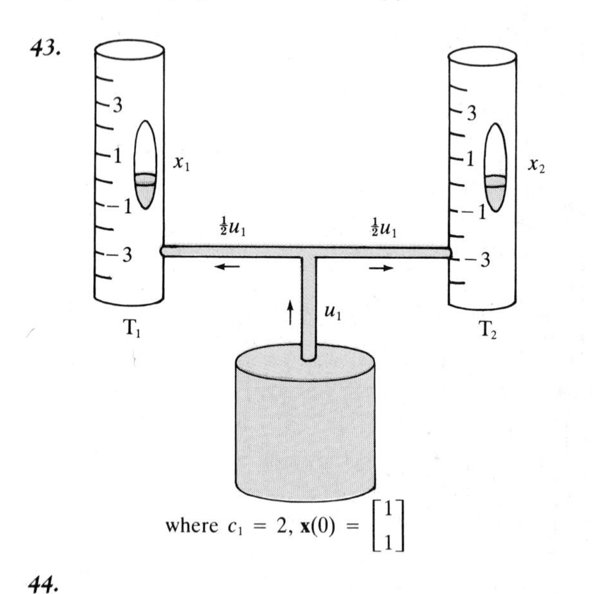

where $c_1 = 2$, $\mathbf{x}(0) = \begin{bmatrix} 1 \\ 1 \end{bmatrix}$

44.

where $c_1 = 1$, $c_2 = 2$, $c_3 = 1$, and $\mathbf{x}(0) = \begin{bmatrix} 1 \\ 1 \end{bmatrix}$

45.

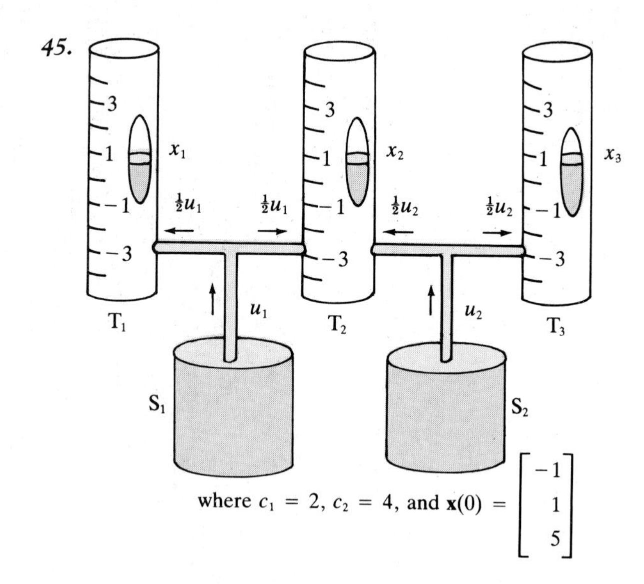

where $c_1 = 2$, $c_2 = 4$, and $\mathbf{x}(0) = \begin{bmatrix} -1 \\ 1 \\ 5 \end{bmatrix}$

46.

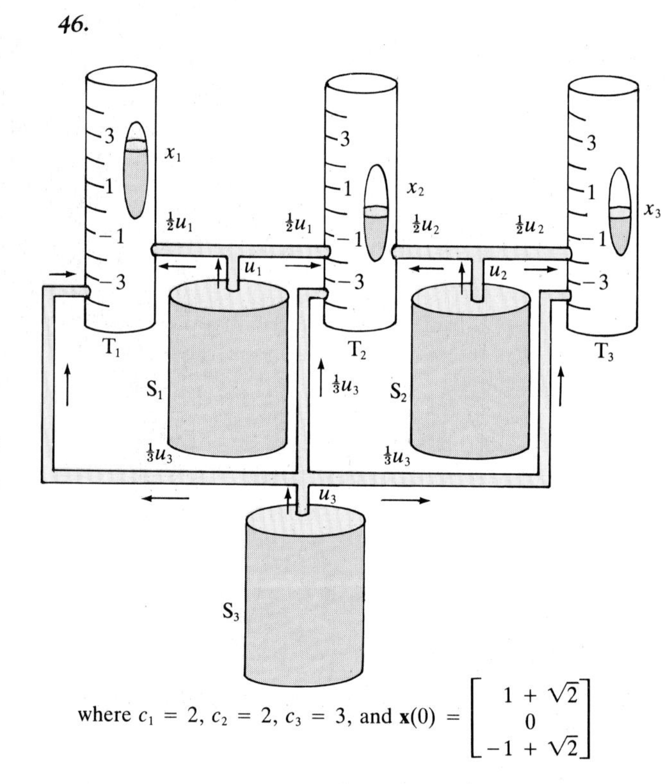

where $c_1 = 2$, $c_2 = 2$, $c_3 = 3$, and $\mathbf{x}(0) = \begin{bmatrix} 1 + \sqrt{2} \\ 0 \\ -1 + \sqrt{2} \end{bmatrix}$

47. (differential equations) Consider a spring-mass system as shown in Figure 7.20, where each of the two particles has mass l and the springs have spring constants as labeled. Associate two axes with the system so that axis x_1 has its origin at the equilibrium position of particle 1 and x_2 has its origin at the equilibrium position of particle 2.

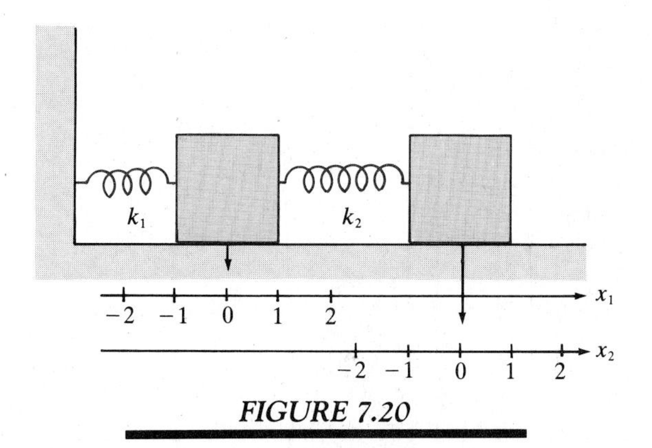

FIGURE 7.20

Let

$x_1(t) =$ the position of particle 1 on the x_1-axis
$x_2(t) =$ the position of particle 2 on the x_2-axis

Using Hooke's law that $f = cx$ (force = spring constant times displacement) and Newton's law that $f = ma$ (force = mass times acceleration), we can obtain the equations of motion

$$\frac{d^2}{dt^2}\mathbf{x}(t) + \begin{bmatrix} k_1 + k_2 & -k_2 \\ -k_2 & k_2 \end{bmatrix}\mathbf{x}(t) = \mathbf{0}$$

Note that the matrix involved is symmetric because forces due to the second spring are symmetric. This kind of symmetry occurs in numerous applied problems.

Solve the following equations:

(*a*) $\frac{d^2}{dt^2}\mathbf{x}(t) + \begin{bmatrix} 10 & -4 \\ -4 & 4 \end{bmatrix}\mathbf{x}(t) = \mathbf{0}, \mathbf{x}(0) = \begin{bmatrix} 1 \\ 1 \end{bmatrix},$
$\mathbf{x}'(0) = \begin{bmatrix} -1 \\ 1 \end{bmatrix}$

(*b*) $\frac{d^2}{dt^2}\mathbf{x}(t) + \begin{bmatrix} 5 & -2 \\ -2 & 2 \end{bmatrix}\mathbf{x}(t) = \mathbf{0}, \mathbf{x}(0) = \begin{bmatrix} 1 \\ 0 \end{bmatrix},$
$\mathbf{x}'(0) = \begin{bmatrix} 0 \\ 0 \end{bmatrix}$

Diagram the corresponding spring-mass systems and find the positions of the particles at time $t = \pi$. Locate these on the diagram.

COMPUTER EXERCISE

48. (plane rotation) Let $Q = \begin{bmatrix} \cos\theta & -\sin\theta \\ \sin\theta & \cos\theta \end{bmatrix}$. In Section 5.1 we showed that the linear transformation $L(\mathbf{x}) = A\mathbf{x}$, seen as a mapping, rotates the plane by θ radians. Explain the effect of the following linear transformations.

(*a*) $L(\mathbf{x}) = \begin{bmatrix} 1 & \mathbf{0} \\ \mathbf{0} & Q \end{bmatrix}\mathbf{x}$

(*b*) $L(\mathbf{x}) = \begin{bmatrix} Q & \mathbf{0} \\ \mathbf{0} & 1 \end{bmatrix}\mathbf{x}$

(*c*) $L(\mathbf{x}) = \begin{bmatrix} \cos\theta & 0 & -\sin\theta \\ 0 & 1 & 0 \\ \sin\theta & 0 & \cos\theta \end{bmatrix}\mathbf{x}$

Plane rotations, and more generally orthogonal matrices, are often used in numerical work to obtain special forms from a given matrix. A basic result follows.

(*d*) Let $\mathbf{x} \neq \mathbf{0}$ be in $\mathbf{R}^2$. Calculate θ such that $Q\mathbf{x} = r\mathbf{e}^1$, where $r = \sqrt{x_1^2 + x_2^2}$.
(*e*) Let $\mathbf{x} \neq \mathbf{0}$ be in $\mathbf{R}^3$. Find plane rotations P_1 and P_2 such that $P_2P_1\mathbf{x} = r\mathbf{e}^1$, where r is a scalar.
(*f*) Let $A \in \mathbf{R}_{3,3}$. Using parts (d) and (e), show how to find plane rotations P_1, P_2, and P_3 such that $P_3P_2P_1A = T$, where T is upper triangular. *Hint:* Find P_1 and P_2 such that $P_2P_1\mathbf{a}^1 = r\mathbf{e}^1$. Then choose P_3 such that the second column in P_2P_1A is changed to the form $\begin{bmatrix} x_1 \\ x_2 \\ 0 \end{bmatrix}$. (Note that multiplying A by orthogonal matrices does not magnify error.)

CHAPTER EIGHT

NUMERICAL METHODS

In previous chapters we discussed numerous computational problems. Two of the more important of these are

1. solving a system of linear algebraic equations and

2. computing eigenvalues and corresponding eigenvectors of a matrix.

In this chapter we will briefly introduce the topic of solving these two types of problems using a computer. Further work can be found in most numerical analysis texts.

8.1 ROUND-OFF AND RELATIVE ERROR

In this section we will discuss round-off in computer calculations and show how it can be measured.

Let k be a positive integer. Most computers calculate with k-digit numbers* $0.d_1 d_2 \ldots d_k \times 10^r$, called k-digit floating point numbers, where $d_1, \ldots, d_k$ are digits, $d_1 \neq 0$ unless the number itself is 0, and r is an integer. The value of k and the bounds on r are determined by the particular computer.

EXAMPLE 1

Table 8.1 lists some k-digit numbers. □

TABLE 8.1

k	k-DIGIT NUMBERS
2	$2100 = 0.21 \times 10^4$, $0.0041 = 0.41 \times 10^{-2}$, $200 = 0.20 \times 10^3$, $-0.003 = -0.30 \times 10^{-2}$
3	$413 = 0.413 \times 10^3$, $0.163 = 0.163 \times 10^0$, $0.02 = 0.200 \times 10^{-1}$, $410 = 0.410 \times 10^3$

To fix our thoughts, we will define a *k-digit computer* as a computer that can hold k-digit numbers and do arithmetic operations on these numbers, rounding to k digits after each operation.

* In most computers, these numbers are actually binary numbers, but for simplicity in this text we use decimal numbers.

EXAMPLE 2

In a three-digit computer, we have the following.

$$351 + 842 = 1190$$

since 1193 rounded to three digits is 1190.

$$1 \div 3 = 0.333$$

which is $\frac{1}{3}$ rounded to 3 digits.

$$237 + 4.21 \times 3 = 237 + 12.6 = 250$$

since $4.21 \times 3 = 12.6$ and $237 + 12.6 = 250$ rounded to three digits. Computations can depend on grouping. For example,

$$(1000 + 3) + 4 = 1000 + 4 = 1000$$

while

$$1000 + (3 + 4) = 1000 + 7 = 1010$$ □

Our k-digit computer can also compute with $2k$ digits when required to do so. This doubling of digits is called *double precision*, a useful feature in handling round-off.

EXAMPLE 3

In a two-digit computer, whereas normal computation yields $1 \div 3 = 0.33$, double precision gives

$$1 \div 3 = 0.3333$$

Similarly, whereas normal computation yields $(1 \div 6) \times 2 = 0.34$, double precision gives

$$(1 \div 6) \times 2 = 0.3334$$ □

From the above work we can see that errors in answers computed using a k-digit computer occur as a result of

1. round-off of numbers exceeding k digits when input into the computer and

2. round-off of answers after calculations.

To solve a problem on a k-digit computer, we submit to the computer a problem having solution y and an algorithm for solving it, as depicted in Figure 8.1. Since computations performed by a computer, as instructed by the algorithm, may be in error, the computed solution x to the problem is not necessarily the exact solution y. We do not necessarily know how close a computed solution is to the exact solution, although solving the problem a

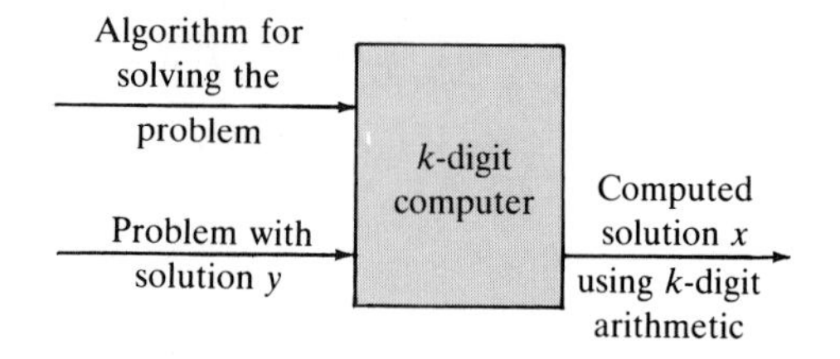

FIGURE 8.1

second time using double precision and then comparing the digits of the two solutions can give some indication of accuracy.

In this chapter we will present methods that iteratively generate a sequence of improved computed solutions $x_1, x_2, \ldots$ such that $x_1, x_2, \ldots$ get closer to y as we proceed with the computing. The process is diagrammed in Figure 8.2. Thus, we can continue to compute these solutions, comparing digits in consecutive solutions, until we have the desired accuracy.

To measure the accuracy of computed solutions, we first consider solutions that are numbers and later solutions that are vectors. To see how well a number x approximates another number $y \neq 0$, we use the *relative error* or *percentage error of* x *with respect to* y defined by

$$\text{relative error} = \frac{|x - y|}{|y|}$$

and

$$\text{percentage error} = \frac{|x - y|}{|y|} \times 100$$

Note that if $\dfrac{|x - y|}{|y|} \leqslant 10^{-r}$, where r is a positive integer, then $|x - y| \leqslant 10^{-r}|y|$. Since 10^{-r} moves the decimal point r places to the left in y, it follows that x approximates y to at least r digits.

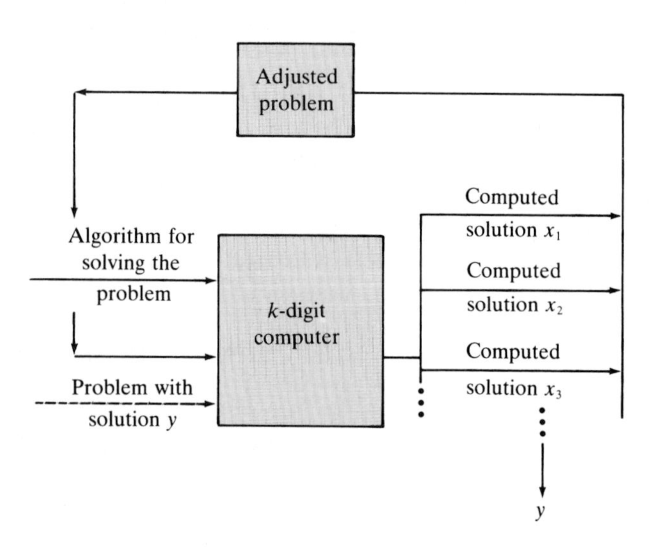

FIGURE 8.2

EXAMPLE 4

If $r = 2$ and $x = x_1x_2x_3x_4$, $y = y_1y_2y_3y_4$, then we have

$$|x - y| \leqslant |y_1y_2y_3y_4|10^{-2} = y_1y_2.y_3y_4$$

So x and y begin to differ in the third digit of y. Thus x approximates y to at least two digits. Also, x approximates y to within $10^{-2} \times 100 = 1\%$. □

Some numerical examples follow.

EXAMPLE 5

If $x = 124$ and $y = 123$, then

$$\text{relative error} = \frac{|x - y|}{|y|} = \frac{1}{123} \approx 0.008 < 10^{-2}$$

and percentage error $< 10^{-2} \times 100 = 1\%$. Here $r = 2$ and x approximates y to 2 digits. Also, x approximates y to within 1%. □

EXAMPLE 6

If $x = 0.124$ and $y = 0.123$, then

$$\text{relative error} = \frac{|x - y|}{|y|} = \frac{1}{123} \approx 0.008 < 10^{-2}$$

as in the previous example. □

Comparing Examples 5 and 6, we see that relative error and percentage error do not depend on the size of the number, but on the digits of accuracy. Since a k-digit computer holds only k-digit numbers, we are only interested in how well the computer computes the first k digits of a solution. Thus, relative error and percentage error are appropriate measures of the accuracy of a calculation. In such a computer, you should not expect more than k digits of accuracy. Hence you should not expect a relative error less than 10^{-k} or a percentage error less than 10^{-k+2} (see exercises 19, 20, and 21).

EXAMPLE 7

If $x = 0.999$ and $y = 1$, then

$$\text{relative error} = \frac{|x - y|}{|y|} = \frac{0.001}{1} = 0.001 = 10^{-3}$$

We see no agreement at all in the digits. However, since

$$|x - y| = 1.000 - 0.999 = 0.001$$

we see that x approximates y to three digits of y, the difference being in the fourth digit of y. And x approximates y to within $0.001 \times 100 = 0.1\%$. (Some computers, after some calculations, will represent 1 as 0.999....) □

If $\mathbf{x}$ and $\mathbf{y} \neq 0$ are vectors in $\mathbf{R}^n$, and if $\mathbf{x}$ approximates $\mathbf{y}$, we measure accuracy by using

$$\text{relative error} = \frac{\|\mathbf{x} - \mathbf{y}\|}{\|\mathbf{y}\|}$$

or

$$\text{percentage error} = \frac{\|\mathbf{x} - \mathbf{y}\|}{\|\mathbf{y}\|} \times 100$$

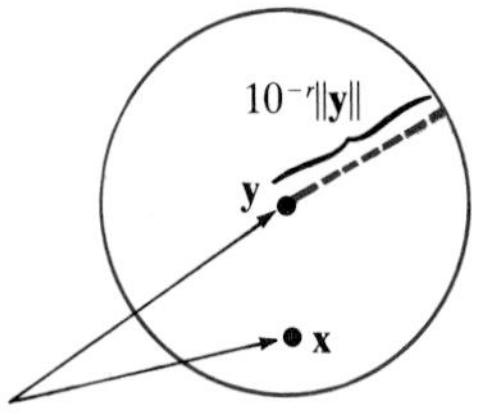

Thus, if the relative error is $\leqslant 10^{-r}$ for some positive integer r, then $\|\mathbf{x}-\mathbf{y}\| \leqslant 10^{-r}\|\mathbf{y}\|$ and $\mathbf{x}$ is within $10^{-r} \times 100 = 10^{-r+2}\%$ of $\mathbf{y}$. Hence, $\mathbf{x}$ and $\mathbf{y}$ differ by a vector whose size is less than or equal to $10^{-r}\|\mathbf{y}\|$. This situation is shown geometrically in Figure 8.3.

FIGURE 8.3

EXAMPLE 8

Let $\mathbf{x} = [124, 0.999]$ and $\mathbf{y} = [123, 1]$. Then

$$\begin{aligned}\text{relative error} &= \frac{\|[124, 0.999] - [123, 1]\|}{\|[123, 1]\|} \\ &= \frac{\|[1, -0.001]\|}{\|[123, 1]\|} \\ &= \frac{\sqrt{1.000001}}{\sqrt{15130}} \approx 0.00813 < 10^{-2}\end{aligned}$$

and percentage error $< 10^{-2} \times 100 = 1\%$. Thus x approximates y to within 1% of y. □

In the next section we develop a method, as outlined above, for solving a system of linear algebraic equations on a k-digit computer.

EXERCISES FOR SECTION 8.1

COMPUTATIONAL EXERCISES

In exercises 1 through 6, compute the given expression using a k-digit computer for the given k. Then compute the number exactly, and compare your result x to the exact result y using relative error.

1. $(24 \times 72) + 340$, $k = 2$

2. $(16 \times 12) - 98$, $k = 2$

3. $(24 \times 71) - 330$, $k = 2$

4. $(16 \times 12) + 54$, $k = 2$

5. 2^8, $k = 2$, $k = 4$

6. 3^5, $k = 2$, $k = 4$

In exercises 7 through 12, compute the relative error and the percentage error of x with respect to y. State the number of digits of y to which x approximates y.

7. $x = 976$, $y = 978$

8. $x = 1060$, $y = 1061$

9. $x = 0.001$, $y = 0.0010002$

10. $x = 0.28$, $y = 0.3003$

11. $x = 1.1$, $y = 1.0999$

12. $x = 0.9999$, $y = 1$

In exercises 13 through 16, compute the relative error and the percentage error of $\mathbf{x}$ with respect to $\mathbf{y}$.

13. $\mathbf{x} = [1372, 2163]$, $\mathbf{y} = [1371, 2164]$

14. $\mathbf{x} = [1, 5.021]$, $\mathbf{y} = [0.9999, 5.0212]$

15. $\mathbf{x} = [-1.0112, 3.6251]$, $\mathbf{y} = [-1.01119, 3.6251]$

16. $\mathbf{x} = [78361, 52914]$, $\mathbf{y} = [78362, 52913]$

THEORETICAL EXERCISES

17. For $a > 0$, the graph of $f(x) = a/x$ is shown in Figure 8.4. When $0 < x < \sqrt{a}$, a small error in x can cause a much greater error in $f(x)$, especially when x is close to 0. Hence, if possible, calculations in a computer should be done so that division by relatively small numbers is avoided. For the given functions, calculate $f(x_1)$ and $f(x_2)$. Calculate the change $|\Delta f| = |f(x_1) - f(x_2)|$ in f and the change $|\Delta x| = |x_1 - x_2|$ in x. Determine which is larger by computing $|\Delta f/\Delta x|$.

(a) $\frac{3}{x}$ for $x_1 = 1$ and $x_2 = 1.1$

(b) $\frac{5}{x}$ for $x_1 = 1$ and $x_2 = 0.9$

(c) $\frac{1}{x}$ for $x_1 = 0.001$ and $x_2 = 0.0011$

(d) $\frac{2}{x}$ for $x_1 = 0.0001$ and $x_2 = 0.00009$

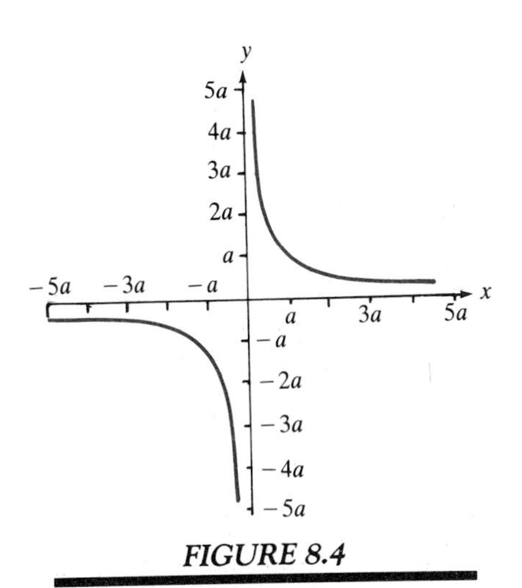

FIGURE 8.4

18. Let $\mathbf{x}$ and $\mathbf{y}$ be vectors in $\mathbf{R}^n$. Show that the relative error of $10^r\mathbf{x}$ with respect to $10^r\mathbf{y}$ does not depend on r. (In this sense, relative error does not depend on the size but on the digits of $\|\mathbf{y}\|$.)

APPLICATIONS EXERCISES

19. A ruler measures to $\frac{1}{16}$ inch. Is it reasonable to expect to be able to measure to within $\frac{1}{1000}$ foot using it? What kind of accuracy can you expect?

20. Using a two-digit computer, can you expect to calculate solutions to ten-digit accuracy? What is a reasonable expectation for the accuracy of your calculations?

21. Suppose x is a calculated solution to a problem having solution y. Suppose $y - x = 1{,}000{,}000$. Is the calculated solution a good one? *Hint:* For a three-digit computer, suppose $y = 2{,}111{,}000{,}000$ and $x = 2{,}110{,}000{,}000$.

8.2 THE EQUATION $A\mathbf{x} = \mathbf{b}$ WHERE A IS AN INVERTIBLE MATRIX

In this section we discuss solving, on a k-digit computer, the equation

$$A\mathbf{x} = \mathbf{b}$$

where A is an $n \times n$ invertible matrix. To obtain a computed solution, we will use the Gaussian elimination algorithm with an adjustment intended to reduce round-off error. The adjustment is that we replace Step 1 of the algorithm with the following.

1′. Let x be the leftmost variable of the system with a nonzero coefficient. From all equations, select one, say equation i, in which the coefficient of x is largest in absolute value. If $i \neq 1$, interchange equations i and 1.

This adjusted algorithm is called the Gaussian elimination algorithm *with partial pivoting.*

Example 9

To solve $\begin{bmatrix} 2 & 1 \\ 4 & -3 \end{bmatrix}\begin{bmatrix} x_1 \\ x_2 \end{bmatrix} = \begin{bmatrix} -2 \\ 2 \end{bmatrix}$, we apply the Gaussian elimination algorithm with partial pivoting to obtain

$$\begin{bmatrix} 2 & 1 & -2 \\ 4 & -3 & 2 \end{bmatrix} \xrightarrow{R_1 \leftrightarrow R_2} \begin{bmatrix} 4 & -3 & 2 \\ 2 & 1 & -2 \end{bmatrix} \xrightarrow{R_2 - \frac{1}{2}R_1 \to R_2} \begin{bmatrix} 4 & -3 & 2 \\ 0 & 2.5 & -3 \end{bmatrix}$$

from which we have

$$x_2 = -\frac{3}{2.5} = -1.2 \qquad \text{and} \qquad x_1 = \frac{2 + 3(-1.2)}{4} = -0.4$$

□

The purpose of the adjustment is to assure that the coefficient of x in the first row is as large in absolute value as possible. Small coefficients can magnify round-off error in the back substitution part of the algorithm, as is shown in the following example.

Example 10

If $R_2 - 1000R_1 \to R_2$ is applied to the system

$$\begin{bmatrix} 0.01 & 1.2 & 1.3 \\ 10 & 1 & 11 \end{bmatrix}$$

using a two-digit computer, we get

$$\begin{bmatrix} 0.01 & 1.2 & 1.3 \\ 0 & -1200 & -1300 \end{bmatrix}$$

Thus

$$x_2 = \frac{1300}{1200} = 1.1 \qquad \text{and} \qquad x_1 = \frac{1.3 - 1.2x_2}{0.01} = \frac{1.3 - 1.2(1.1)}{0.01} = 0$$

The exact solution is $x_2 = \frac{1289}{1199} \approx 1.08$, $x_1 = \frac{1190}{1199} \approx 0.99$; thus there is an error of about 0.02 in x_2 and about 0.99 in x_1. Since $x_1 = (1.3 - 1.2x_2)/0.01$, the error in x_2 was increased by the coefficient $1.2/0.01 = 1200$ when x_1 was multiplied by the coefficient. Thus the back substitution part of the algorithm magnified the error.

On the other hand, if we first apply $R_1 \leftrightarrow R_2$ to get

$$\begin{bmatrix} 10 & 1 & 11 \\ 0.01 & 1.2 & 1.3 \end{bmatrix}$$

and then apply $R_2 - \frac{1}{1000}R_1 \rightarrow R_2$, we get

$$\begin{bmatrix} 10 & 1 & 11 \\ 0 & 1.2 & 1.3 \end{bmatrix}$$

Thus

$$x_2 = \frac{1.3}{1.2} = 1.1 \qquad \text{and} \qquad x_1 = \frac{11 - x_2}{10} = \frac{11 - 1.1}{10} = 0.99$$

Note here that the error in x_2 was multiplied by the coefficient $-1/10$, which in fact shrank this error. □

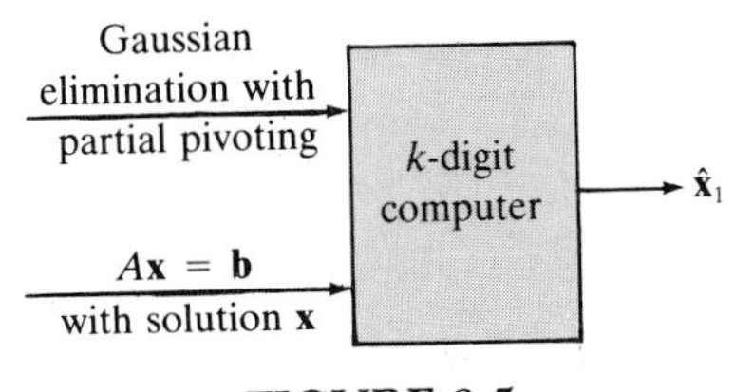

FIGURE 8.5

As a consequence of the above example, when solving systems of linear algebraic equations, we will use the Gaussian elimination algorithm with partial pivoting.

We now show how to use this algorithm iteratively to improve computed solutions, as depicted in Figure 8.5. Let $\hat{\mathbf{x}}_1$ be the first computed solution. To measure how well $\hat{\mathbf{x}}_1$ satisfies the equation, calculate

$$\mathbf{b} - A\hat{\mathbf{x}}_1 = \mathbf{r}_1$$

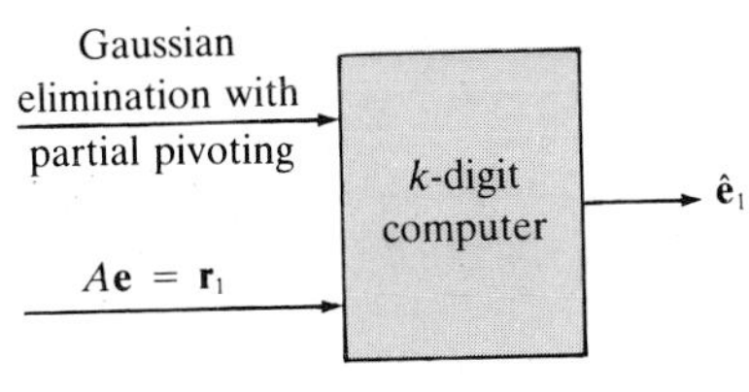

FIGURE 8.6

If $\hat{\mathbf{x}}_1$ is the exact solution, then $\mathbf{r}_1 = \mathbf{0}$. If $\mathbf{r}_1 \neq \mathbf{0}$, we improve $\hat{\mathbf{x}}_1$ as follows. As depicted in Figure 8.6, we compute a solution $\hat{\mathbf{e}}_1$ to the adjusted problem $A\mathbf{e} = \mathbf{r}_1$, called the error equation. Then we set $\hat{\mathbf{x}}_2 = \hat{\mathbf{x}}_1 + \hat{\mathbf{e}}_1$ as the next computed solution. To measure how well $\hat{\mathbf{x}}_2$ satisfies the equation, we compute

$$\mathbf{b} - A\hat{\mathbf{x}}_2 = \mathbf{b} - A(\mathbf{x}_1 + \hat{\mathbf{e}}_1) = \mathbf{b} - A\hat{\mathbf{x}}_1 - A\hat{\mathbf{e}}_1 = \mathbf{r}_1 - A\hat{\mathbf{e}}_1 = \mathbf{r}_2$$

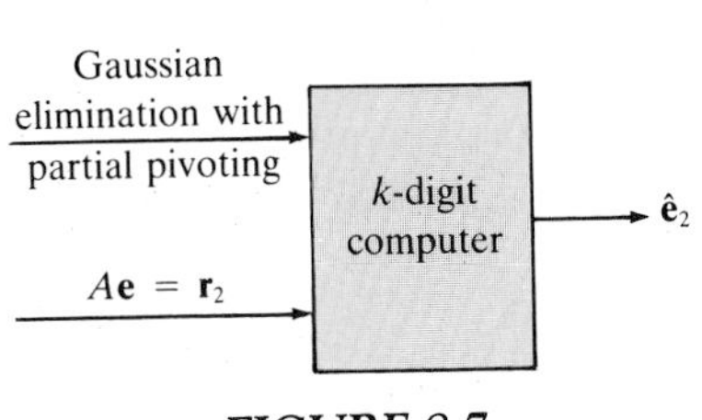

FIGURE 8.7

If $\mathbf{r}_2 = \mathbf{0}$, $\hat{\mathbf{e}}_1$ is the exact solution to the error equation and $\hat{\mathbf{x}}_2$ is the exact solution to $A\mathbf{x} = \mathbf{b}$. If $\mathbf{r}_2 \neq \mathbf{0}$, we solve the adjusted problem, or error equation, $A\mathbf{e} = \mathbf{r}_2$ for $\hat{\mathbf{e}}_2$, as shown in Figure 8.7. We then set $\hat{\mathbf{x}}_3 = \hat{\mathbf{x}}_2 + \hat{\mathbf{e}}_2$ as the next solution.

Continuing this procedure of solving adjusted problems, we obtain a sequence $\hat{\mathbf{x}}_1, \hat{\mathbf{x}}_2, \ldots$ of computed solutions. It can be shown (Stewart, 200–206) that if A is not "ill-conditioned" (very small changes in the entries of A yield very large changes in the entries in the corresponding solutions), then the computed solutions $\hat{\mathbf{x}}_1, \hat{\mathbf{x}}_2, \ldots$ approach the exact solution to the system. Thus we can compute solutions iteratively until the desired accuracy is attained.

The formal description of this method follows.

IMPROVEMENT ALGORITHM

To solve $A\mathbf{x} = \mathbf{b}$, where A is an $n \times n$ invertible matrix, choose a number $\varepsilon > 0$ to measure the desired accuracy of the solution.*

* For a computer using k-digit arithmetic, use, say, 10^{-k} or 10^{-k+1}.

1. Compute a solution $\hat{\mathbf{x}}$ to $A\mathbf{x} = \mathbf{b}$ by the Gaussian elimination algorithm with partial pivoting.

2. Calculate $\mathbf{r} = \mathbf{b} - A\hat{\mathbf{x}}$, using double precision for greater accuracy, and round to single precision. (Proofs of convergence of this algorithm and experience show that calculation of $\mathbf{r}$ needs to be done with more precision than the remaining calculations.)

3. Compute a solution $\hat{\mathbf{e}}$ to the adjusted problem, or error equation, $A\mathbf{e} = \mathbf{r}$.

4. Test to see if the relative error between the calculated solution and the previous calculated solution, $\|\hat{\mathbf{e}}\|/\|\hat{\mathbf{x}}\|$, $\leqslant \varepsilon$. If it is, set $\mathbf{x} = \hat{\mathbf{x}} + \hat{\mathbf{e}}$ as the solution. Otherwise, set $\hat{\mathbf{x}} = \hat{\mathbf{x}} + \hat{\mathbf{e}}$ and return to Step 2.*

EXAMPLE 11

On a three-digit computer, we will compute solutions to the system having the augmented matrix

$$\begin{bmatrix} 17 & 23 & 1 \\ 21 & 65 & 3 \end{bmatrix}$$

by the improvement algorithm with $\varepsilon = 0.001$. The computations follow.

First, applying $R_1 \leftrightarrow R_2$, we have

$$\begin{bmatrix} 21 & 65 & 3 \\ 17 & 23 & 1 \end{bmatrix} \xrightarrow{R_2 - \frac{17}{21}R_1 \to R_2} \begin{bmatrix} 21 & 65 & 3 \\ 0 & -29.7 & -1.43 \end{bmatrix}$$

so

$$\hat{x}_2 = \frac{1.43}{29.7} = 0.0481$$

and

$$\hat{x}_1 = \frac{3 - 65(0.0481)}{21} = \frac{3 - 3.13}{21} = -0.00619$$

For improvement, using double precision, we set

$$\begin{aligned} \mathbf{r}_1 = \mathbf{b} - A\hat{\mathbf{x}}_1 &= \begin{bmatrix} 1 \\ 3 \end{bmatrix} - \begin{bmatrix} 17 & 23 \\ 21 & 65 \end{bmatrix} \begin{bmatrix} -0.00619 \\ 0.0481 \end{bmatrix} \\ &= \begin{bmatrix} 1 \\ 3 \end{bmatrix} - \begin{bmatrix} 1.00107 \\ 2.99651 \end{bmatrix} \\ &= \begin{bmatrix} -0.00107 \\ 0.00349 \end{bmatrix} \end{aligned}$$

Then solving $A\mathbf{e} = \mathbf{r}_1$ yields

$$\begin{bmatrix} 17 & 23 & -0.00107 \\ 21 & 65 & 0.00349 \end{bmatrix} \xrightarrow{R_1 \leftrightarrow R_2} \begin{bmatrix} 21 & 65 & 0.00349 \\ 17 & 23 & -0.00107 \end{bmatrix}$$

$$\xrightarrow{R_2 - \frac{17}{21}R_1 \to R_2} \begin{bmatrix} 21 & 65 & 0.00349 \\ 0 & -29.7 & -0.00390 \end{bmatrix}$$

* Experience has shown that only a few passes through Step 2 are required.

Thus

$$\hat{e}_2 = \frac{0.00390}{29.7} = 0.000131$$

and $$\hat{e}_1 = \frac{0.00349 - 65(0.000131)}{21} = \frac{0.00349 - 0.00852}{21} = -0.00024$$

Now

$$\frac{\|\hat{\mathbf{e}}\|}{\|\hat{\mathbf{x}}_1\|} = \frac{\sqrt{(-0.00024)^2 + (0.000131)^2}}{\sqrt{(-0.00619)^2 + (0.0481)^2}} = \frac{0.000273}{0.0485} = 0.00563 > 0.001 = \varepsilon$$

Thus we set

$$\hat{\mathbf{x}}_2 = \hat{\mathbf{x}}_1 + \hat{\mathbf{e}} = \begin{bmatrix} -0.00619 \\ 0.0481 \end{bmatrix} + \begin{bmatrix} -0.00024 \\ 0.000131 \end{bmatrix} = \begin{bmatrix} -0.00643 \\ 0.0482 \end{bmatrix}$$

and return to Step 2. Again using double precision, we have

$$\begin{aligned} \mathbf{r}_2 = \mathbf{b} - A\hat{\mathbf{x}}_2 &= \begin{bmatrix} 1 \\ 3 \end{bmatrix} - \begin{bmatrix} 17 & 23 \\ 21 & 65 \end{bmatrix} \begin{bmatrix} -0.00643 \\ 0.0482 \end{bmatrix} \\ &= \begin{bmatrix} 1 \\ 3 \end{bmatrix} - \begin{bmatrix} 0.99929 \\ 2.99797 \end{bmatrix} \\ &= \begin{bmatrix} 0.00071 \\ 0.00203 \end{bmatrix} \end{aligned}$$

Then solving $A\mathbf{e} = \mathbf{r}_2$ yields

$$\begin{bmatrix} 17 & 23 & 0.00071 \\ 21 & 65 & 0.00203 \end{bmatrix} \xrightarrow{\mathbf{R}_1 \leftrightarrow \mathbf{R}_2} \begin{bmatrix} 21 & 65 & 0.00203 \\ 17 & 23 & 0.00071 \end{bmatrix}$$

$$\xrightarrow{\mathbf{R}_2 - \frac{17}{21}\mathbf{R}_1 \rightarrow \mathbf{R}_2} \begin{bmatrix} 21 & 65 & 0.00203 \\ 0 & -29.7 & -0.00093 \end{bmatrix}$$

Thus

$$\hat{e}_2 = \frac{0.00093}{29.7} = 0.0000313$$

and $$\hat{e}_1 = \frac{0.00203 - 65(0.0000313)}{21} = \frac{0.00203 - 0.00203}{21} = 0$$

Now

$$\frac{\|\hat{\mathbf{e}}\|}{\|\hat{\mathbf{x}}_2\|} = \frac{\sqrt{(0)^2 + (0.0000313)^2}}{\sqrt{(-0.00643)^2 + (0.0482)^2}} = \frac{0.0000313}{0.0486} = 0.000644 < 0.001 = \varepsilon$$

Thus we set

$$\mathbf{x} = \hat{\mathbf{x}}_2 + \hat{\mathbf{e}} = \begin{bmatrix} -0.00643 \\ 0.0482 \end{bmatrix} + \begin{bmatrix} 0 \\ 0.0000313 \end{bmatrix} = \begin{bmatrix} -0.00643 \\ 0.0482 \end{bmatrix}$$

and stop.* □

In the next section we consider the problem of computing eigenvalues and eigenvectors on a computer.

EXERCISES FOR SECTION 8.2

COMPUTATIONAL EXERCISES

In exercises 1 through 6, use the Gaussian elimination algorithm with partial pivoting to solve $A\mathbf{x} = \mathbf{b}$ for $\mathbf{x}$.

1. $A = \begin{bmatrix} 1 & 3 \\ 2 & -1 \end{bmatrix}$, $\mathbf{b} = \begin{bmatrix} -1 \\ 2 \end{bmatrix}$

2. $A = \begin{bmatrix} 3 & 7 \\ 8 & 2 \end{bmatrix}$, $\mathbf{b} = \begin{bmatrix} 3 \\ 1 \end{bmatrix}$

3. $A = \begin{bmatrix} -2 & 3 \\ 8 & 4 \end{bmatrix}$, $\mathbf{b} = \begin{bmatrix} 2 \\ 2 \end{bmatrix}$

4. $A = \begin{bmatrix} -3 & 6 \\ 12 & -1 \end{bmatrix}$, $\mathbf{b} = \begin{bmatrix} -1 \\ 1 \end{bmatrix}$

5. $A = \begin{bmatrix} 2 & 1 & 3 \\ 0 & 1 & 4 \\ -5 & -2 & 5 \end{bmatrix}$, $\mathbf{b} = \begin{bmatrix} 1 \\ -1 \\ 2 \end{bmatrix}$

6. $A = \begin{bmatrix} 0 & 1 & 2 \\ -8 & 5 & 1 \\ 2 & 3 & 0 \end{bmatrix}$, $\mathbf{b} = \begin{bmatrix} 1 \\ 8 \\ 3 \end{bmatrix}$

In exercises 7 through 10, use a two-digit computer to find an approximate solution to the system whose augmented matrix is given, first by the Gaussian elimination algorithm and then by the Gaussian elimination algorithm with partial pivoting. Compare the results to the given solution, exact to six digits, by using relative error.

7. $\begin{bmatrix} 2 & 60 & 12 \\ 4000 & 400 & 100 \end{bmatrix}$, exact solution: $x_1 \approx 0.00501672$, $x_2 \approx 0.199833$

8. $\begin{bmatrix} 0.5 & 27 & 3 \\ 50 & 16 & 8 \end{bmatrix}$, exact solution: $x_1 \approx 0.125186$, $x_2 \approx 0.108793$

9. $\begin{bmatrix} 400 & 0.1 & 300 & 20 \\ 500{,}000 & 20 & 200 & 40 \\ 3{,}000{,}000 & 0.001 & 0.05 & 0.0002 \end{bmatrix}$, exact solution: $x_1 \approx -1.48296(10^{-9})$, $x_2 \approx 1.33783$, $x_3 \approx 0.0662207$

10. $\begin{bmatrix} 1 & 0.05 & 1{,}700 & 310 \\ 0.0002 & 0.08 & 100{,}000 & 200 \\ 17{,}000 & 22 & 75.5 & 4 \end{bmatrix}$, exact solution: $x_1 \approx -8.38012$, $x_2 \approx 6475.74$, $x_3 \approx -0.00318058$

In exercises 11 through 16, solve the system as in Example 11, using a three-digit computer and the improvement algorithm, with $\varepsilon = 0.001$.

11. $\begin{bmatrix} 2.5 & 7.9 \\ 3.6 & -8.2 \end{bmatrix}\mathbf{x} = \begin{bmatrix} 79.9 \\ -82.7 \end{bmatrix}$

12. $\begin{bmatrix} 1.3 & 1.4 \\ -2.1 & 1.3 \end{bmatrix}\mathbf{x} = \begin{bmatrix} 3.1 \\ 2.4 \end{bmatrix}$

13. $\begin{bmatrix} 2 & -30 \\ 100 & 4 \end{bmatrix}\mathbf{x} = \begin{bmatrix} 1 \\ -1 \end{bmatrix}$

14. $\begin{bmatrix} 3 & 45 \\ 210 & -1 \end{bmatrix}\mathbf{x} = \begin{bmatrix} 3 \\ 2 \end{bmatrix}$

15. $\begin{bmatrix} 2.1 & 0 & 1.3 \\ 12 & -1.1 & 0 \\ 0 & 23 & 8 \end{bmatrix}\mathbf{x} = \begin{bmatrix} -1 \\ 1 \\ 1 \end{bmatrix}$

16. $\begin{bmatrix} 1.1 & 12 & -3.1 \\ 121 & -2.3 & 10 \\ 17 & 0 & 7.5 \end{bmatrix}\mathbf{x} = \begin{bmatrix} 19.1 \\ 823 \\ 93.7 \end{bmatrix}$

* The exact solution is $x_1 = -\frac{2}{311} \approx 0.006431$ and $x_2 = \frac{15}{311} \approx 0.04823$.

THEORETICAL EXERCISES

17. Suppose you solve a system of linear algebraic equations $A\mathbf{x} = \mathbf{b}$ using the improvement algorithm and you get

$$\hat{\mathbf{x}}_1 = \begin{bmatrix} 1.37 \\ -2.43 \end{bmatrix}, \quad \hat{\mathbf{x}}_2 = \begin{bmatrix} 0.936 \\ -1.98 \end{bmatrix}, \quad \hat{\mathbf{x}}_3 = \begin{bmatrix} 0.999 \\ -2.02 \end{bmatrix}, \quad \hat{\mathbf{x}}_4 = \begin{bmatrix} 0.999 \\ -2.01 \end{bmatrix}$$

Of course, you do not know the solution $\mathbf{x}$ to this system. However, using $\varepsilon = 10^{-2}$, what would you use for the numerical solution to $A\mathbf{x} = \mathbf{b}$?

18. Let the first column of A be $\mathbf{a}^1 = \begin{bmatrix} 100 \\ 1 \end{bmatrix}$. To solve $A\mathbf{x} = \mathbf{b}$, you can apply $R_1 \leftrightarrow R_2$, making the 1 the 1,1-entry of the system. Is this a good numerical strategy?

8.3 EIGENVALUES AND CORRESPONDING EIGENVECTORS

This section is intended only as an introduction to the topic of computing eigenvalues and eigenvectors on a computer. More advanced work, including methods, can be found in most numerical analysis texts.

Let A be an $n \times n$ matrix. We show an elementary method for finding all eigenvalues $\lambda_1, \lambda_2, \ldots, \lambda_n$ and simultaneously the corresponding eigenvectors for A in the special case where $|\lambda_1| > |\lambda_2| > \cdots > |\lambda_n| > 0$. However, round-off error accumulates quickly with this method, and so it is only effective for small n, such as $n = 3$ or possibly 4 or 5.

Since A has distinct eigenvalues, we know by Corollary 6.5A that A has n corresponding eigenvectors $\mathbf{x}^1, \mathbf{x}^2, \ldots, \mathbf{x}^n$, the set of which is linearly independent and hence forms a basis for $\mathbf{R}^n$. Since

$$\det(\lambda I_n - A) = \det(\lambda I_n - A)^t = \det(\lambda I_n - A^t)$$

A and A^t have the same characteristic polynomial and thus the same list of eigenvalues. Thus it follows that A^t has a set of n corresponding eigenvectors which, for later notational convenience, we write as $\mathbf{y}_1^t, \mathbf{y}_2^t, \ldots, \mathbf{y}_n^t$. The principle of biorthogonality (exercise 7) states that these vectors satisfy $\mathbf{y}_i\mathbf{x}^i \neq 0$ and $\mathbf{y}_i\mathbf{x}^j = 0$ for all $i \neq j$.

To compute simultaneously the first eigenvalue λ_1 and corresponding eigenvector, we proceed as follows. Pick $\mathbf{x}^0$ so that in expressing

$$\mathbf{x}^0 = r_1\mathbf{x}^1 + r_2\mathbf{x}^2 + \cdots + r_n\mathbf{x}^n$$

we have $r_1 \neq 0$.* Noting that $A\mathbf{x}^i = \lambda_i\mathbf{x}^i$ for all i, we can repeatedly multiply the vector $\mathbf{x}^0$ by A, which yields

$$\begin{aligned} A\mathbf{x}^0 &= r_1\lambda_1\mathbf{x}^1 + r_2\lambda_2\mathbf{x}^2 + \cdots + r_n\lambda_n\mathbf{x}^n \\ A^2\mathbf{x}^0 &= r_1\lambda_1^2\mathbf{x}^1 + r_2\lambda_2^2\mathbf{x}^2 + \cdots + r_n\lambda_n^2\mathbf{x}^n \\ &\vdots \\ A^s\mathbf{x}^0 &= r_1\lambda_1^s\mathbf{x}^1 + r_2\lambda_2^s\mathbf{x}^2 + \cdots + r_n\lambda_n^s\mathbf{x}^n \end{aligned}$$

* This condition merely requires that $\mathbf{x}^0 \notin \operatorname{span}\{\mathbf{x}^2, \ldots, \mathbf{x}^n\}$, which is almost certain if $\mathbf{x}^0$ is randomly chosen. Of course, if there is doubt, several choices can be made for $\mathbf{x}^0$ and the procedure followed for each choice.

Dividing this last equation by λ_1^s yields

$$\frac{1}{\lambda_1^s} A^s \mathbf{x}^0 = r_1 \mathbf{x}^1 + r_2 \left(\frac{\lambda_2}{\lambda_1}\right)^s \mathbf{x}^2 + \cdots + r_n \left(\frac{\lambda_n}{\lambda_1}\right)^s \mathbf{x}^n$$

Since $\left|\frac{\lambda_j}{\lambda_1}\right| < 1$ for all $j > 1$, it follows that $\left(\frac{\lambda_j}{\lambda_1}\right)^s$ approaches 0 as s increases. Thus, by taking s sufficiently large, we can make $\frac{1}{\lambda_1^s} A^s \mathbf{x}^0$ arbitrarily close to $r_1 \mathbf{x}^1$. Hence $\frac{1}{\lambda_1^s} A^s \mathbf{x}^0$ approaches an eigenvector $r_1 \mathbf{x}^1$ for λ_1. This situation is depicted in Figure 8.8.

In actual practice, however, we do not know λ_1 before or during calculations, and thus we cannot compute the vector $\frac{1}{\lambda_1^s} A^s \mathbf{x}^0$. However, $\frac{1}{\lambda_1^s}$ is a scalar, and so we know that its effect is to reflect, shrink, or stretch the vector $A^s \mathbf{x}^0$. Thus the basic effect of $\frac{1}{\lambda_1^s}$ on $A^s \mathbf{x}^0$ is to assure that $\frac{1}{\lambda_1^s} A^s \mathbf{x}^0$ approaches a nonzero vector. This effect, however, can also be produced by other scalings. For example, after each multiplication by A, we can divide the resulting vector by its length as follows.

Compute $A\mathbf{x}^0$. Set $\mathbf{w}^1 = \frac{1}{\|A\mathbf{x}^0\|} A\mathbf{x}^0$.

Compute $A\mathbf{w}^1$. Set $\mathbf{w}^2 = \frac{1}{\|A\mathbf{w}^1\|} A\mathbf{w}^1 = \frac{1}{\|A^2\mathbf{x}^0\|} A^2\mathbf{x}^0$.

$\vdots$

Compute $A\mathbf{w}^{s-1}$. Set $\mathbf{w}^s = \frac{1}{\|A\mathbf{w}^{s-1}\|} A\mathbf{w}^{s-1} = \frac{1}{\|A^s\mathbf{x}^0\|} A^s\mathbf{x}^0$.

Then $\mathbf{w}^s = \frac{A^s\mathbf{x}^0}{\|A^s\mathbf{x}^0\|} = \frac{\lambda_1^s}{\|A^s\mathbf{x}^0\|} \frac{A^s\mathbf{x}^0}{\lambda_1^s}$. Since $\frac{A^s\mathbf{x}^0}{\lambda_1^s}$ approaches $r_1\mathbf{x}^1$ as s

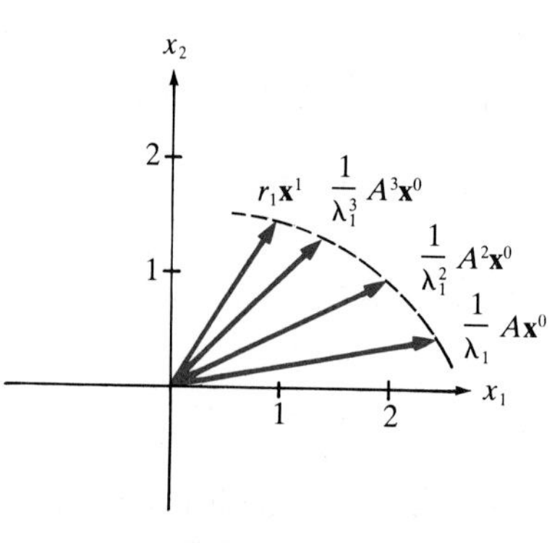

FIGURE 8.8

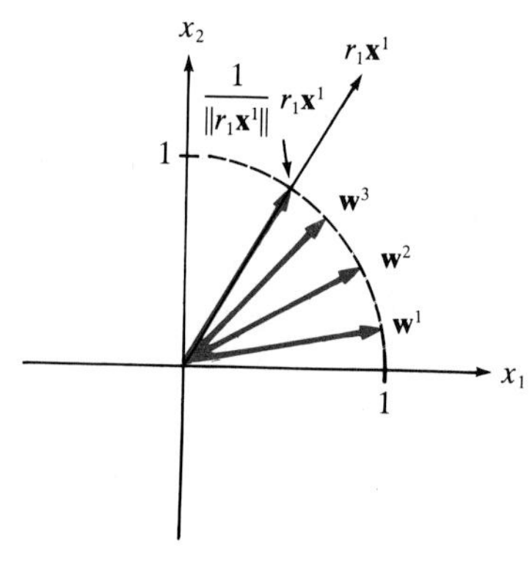

FIGURE 8.9

increases, $\dfrac{|\lambda_1^s|}{\|A^s\mathbf{x}^0\|}$ approaches $\dfrac{1}{\|r_1\mathbf{x}^1\|}$ as s increases. Hence it follows that

(*i*) If $\lambda_1 > 0$, then $\mathbf{w}^s$ approaches $\dfrac{1}{\|r_1\mathbf{x}^1\|}r_1\mathbf{x}^1$.

(*ii*) If $\lambda_1 < 0$, then $\mathbf{w}^{2s}$ approaches $\dfrac{1}{\|r_1\mathbf{x}^1\|}r_1\mathbf{x}^1$.

Thus an eigenvector $\dfrac{1}{\|r_1\mathbf{x}^1\|}r_1\mathbf{x}^1$ for λ_1 can be computed iteratively. A diagram is shown in Figure 8.9.

After computing a vector $\mathbf{w}^s$ sufficiently close to $\dfrac{1}{\|r_1\mathbf{x}^1\|}r_1\mathbf{x}^1$, we can compute an approximation to λ_1 as follows. We choose λ so that the right side of $A\mathbf{w}^s = \lambda\mathbf{w}^s$ is as close to the left side as possible. Thus, we take λ so that

$$\|A\mathbf{w}^s - \lambda\mathbf{w}^s\|^2 = (A\mathbf{w}^s - \lambda\mathbf{w}^s, A\mathbf{w}^s - \lambda\mathbf{w}^s) = \lambda^2 - 2\lambda(\mathbf{w}^s, A\mathbf{w}^s) + \|A\mathbf{w}^s\|^2$$

is as small as possible. Since this expansion is a quadratic in λ, it follows that the smallest value occurs when

$$\lambda = (\mathbf{w}^s, A\mathbf{w}^s)$$

Thus we set $\lambda_1 = (\mathbf{w}^s, A\mathbf{w}^s)$.

A description of this method—an algorithm for finding the largest eigenvalue and a corresponding eigenvector—follows. It should be noted that our derivation only requires that $|\lambda_1| > |\lambda_2| > \cdots > |\lambda_n|$ and that A have a basis of eigenvectors. Thus, the algorithm can be used to calculate λ_1 and a corresponding $\mathbf{x}^1$ for any matrix with these properties.

POWER ALGORITHM

1. Choose some ε to measure the desired accuracy.*
2. Pick some $\mathbf{x}^0 \neq \mathbf{0} \in \mathbf{R}^n$. Set $\mathbf{w} = \dfrac{1}{\|\mathbf{x}^0\|}\mathbf{x}^0$.
3. Compute $A\mathbf{w}$.
4. Compute $\lambda = (A\mathbf{w}, \mathbf{w})$.
5. Test to see if $\dfrac{\|A\mathbf{w} - \lambda\mathbf{w}\|}{\|\lambda\mathbf{w}\|} < \varepsilon$. If not, set $\mathbf{w} = \dfrac{1}{\|A\mathbf{w}\|}A\mathbf{w}$ and return to Step 3.
6. Set $\mathbf{x}^1 = \mathbf{w}$ and $\lambda_1 = \lambda$.

EXAMPLE 12

Let $A = \begin{bmatrix} 3 & 1 & 1 \\ 0 & 1 & 1 \\ 0 & 0 & 2 \end{bmatrix}$, $\mathbf{x}^0 = \begin{bmatrix} 5 \\ 1 \\ 7 \end{bmatrix}$, $k = 4$, and $\varepsilon = 0.001$. Some of the iterations for the power algorithm are listed in Table 8.2. □

* If a k-digit computer is being used, choose, say, $\varepsilon = 10^{-k}$ or $\varepsilon = 10^{-k+1}$.

TABLE 8.2

ITERATION	w	z = Aw	λ	
1	$\begin{bmatrix} 0.5774 \\ 0.1155 \\ 0.8083 \end{bmatrix}$	$\begin{bmatrix} 2.656 \\ 0.9238 \\ 1.617 \end{bmatrix}$	2.948	
3	$\begin{bmatrix} 0.9316 \\ 0.2252 \\ 0.2866 \end{bmatrix}$	$\begin{bmatrix} 3.307 \\ 0.5118 \\ 0.5732 \end{bmatrix}$	3.36	
5	$\begin{bmatrix} 0.9902 \\ 0.09762 \\ 0.1031 \end{bmatrix}$	$\begin{bmatrix} 3.172 \\ 0.2007 \\ 0.2062 \end{bmatrix}$	3.182	
7	$\begin{bmatrix} 0.9984 \\ 0.04094 \\ 0.04148 \end{bmatrix}$	$\begin{bmatrix} 3.077 \\ 0.08242 \\ 0.08296 \end{bmatrix}$	3.078	
9	$\begin{bmatrix} 0.9997 \\ 0.01759 \\ 0.01765 \end{bmatrix}$	$\begin{bmatrix} 3.035 \\ 0.03524 \\ 0.0353 \end{bmatrix}$	3.036	
11	$\begin{bmatrix} 1.0 \\ 0.007685 \\ 0.007692 \end{bmatrix}$	$\begin{bmatrix} 3.016 \\ 0.01538 \\ 0.01538 \end{bmatrix}$	3.016	
13	$\begin{bmatrix} 1.0 \\ 0.003389 \\ 0.003389 \end{bmatrix}$	$\begin{bmatrix} 3.006 \\ 0.006778 \\ 0.006778 \end{bmatrix}$	3.006	
15	$\begin{bmatrix} 1.0 \\ 0.001501 \\ 0.001501 \end{bmatrix}$	$\begin{bmatrix} 3.004 \\ 0.003002 \\ 0.003002 \end{bmatrix}$	3.004	(last iteration*)

Exact results: $\mathbf{w} = \begin{bmatrix} 1 \\ 0 \\ 0 \end{bmatrix}$, $A\mathbf{w} = \begin{bmatrix} 3 \\ 0 \\ 0 \end{bmatrix}$, $\lambda = 3$

Of course, if we replace A with A^t, the power algorithm can also be used to calculate an eigenvector $\mathbf{y}_1^t$, belonging to λ_1, for A^t. We will suppose this eigenvector has been computed and multiplied by a scalar so that $\mathbf{y}_1\mathbf{x}^1 = 1$. This scaling is necessary to the definition of the matrix B given below.

To compute λ_2 and a corresponding eigenvector, given $\mathbf{x}^1$ and $\mathbf{y}_1$ such that $\mathbf{y}_1\mathbf{x}^1 = 1$, define

$$B = A - \lambda_1\mathbf{x}^1\mathbf{y}_1$$

LEMMA 8.1

The eigenvalues of B are $0, \lambda_2, \ldots, \lambda_n$ with corresponding eigenvectors $\mathbf{x}^1, \mathbf{x}^2, \ldots, \mathbf{x}^n$. Further, B^t has corresponding eigenvectors $\mathbf{y}_1^t, \mathbf{y}_2^t, \ldots, \mathbf{y}_n^t$.

* Depending on the accuracy desired, the power algorithm can require hundreds of iterations.

PROOF

Note that

$$B\mathbf{x}^i = (A - \lambda_1 \mathbf{x}^1 \mathbf{y}_1)\mathbf{x}^i = A\mathbf{x}^i - \lambda_1 \mathbf{x}^1 \mathbf{y}_1 \mathbf{x}^i$$

Thus, for $i = 1$,

$$B\mathbf{x}^1 = \lambda_1 \mathbf{x}^1 - \lambda_1 \mathbf{x}^1 = 0\mathbf{x}^1$$

For $i > 1$, using the principle of biorthogonality,

$$B\mathbf{x}^i = \lambda_i \mathbf{x}^i - \lambda_1 \mathbf{x}^1(\mathbf{y}_1 \mathbf{x}^i) = \lambda_i \mathbf{x}^i$$

The corresponding result for B^t is obtained in the same way. ■

Hence we can apply the power algorithm, with some adjustments, to B to compute λ_2, $\mathbf{x}^2$, and $\mathbf{y}_2$ such that $\mathbf{y}_2\mathbf{x}^2 = 1$. The revised algorithm follows. It is stated in terms of $\mathbf{x}^j$ and λ_j rather than $\mathbf{x}^2$ and λ_2 because we will use it later for finding further eigenvalues and eigenvectors.

POWER ALGORITHM FOR B

This algorithm computes the remaining eigenvalues and corresponding eigenvectors of A. If the power algorithm for B is being used for B^t, then A^t should be used wherever A appears and $\mathbf{y}_j$ should be used wherever $\mathbf{x}^j$ appears.

1. Choose some ε to measure the desired accuracy.
2. Pick some $\mathbf{x}^0 \neq \mathbf{0} \in \mathbf{R}^n$. Set $\mathbf{w} = \dfrac{1}{\|\mathbf{x}^0\|}\mathbf{x}^0$.
3. Compute $B\mathbf{w}$.
4. Compute $\lambda = (A\mathbf{w}, \mathbf{w})$.
5. Test to see if $\dfrac{\|A\mathbf{w} - \lambda\mathbf{w}\|}{\|\lambda\mathbf{w}\|} < \varepsilon$. If not, set $\mathbf{w} = \dfrac{1}{\|B\mathbf{w}\|}B\mathbf{w}$ and return to Step 3.
6. Set $\mathbf{x}^j = \mathbf{w}$ and $\lambda_j = \lambda$.

Should this algorithm not converge, then we must conclude that A does not satisfy the hypotheses given in the introduction.

In general, suppose $\lambda_1, \lambda_2, \ldots, \lambda_{j-1}$; $\mathbf{x}^1, \mathbf{x}^2, \ldots, \mathbf{x}^{j-1}$; and $\mathbf{y}_1, \mathbf{y}_2, \ldots, \mathbf{y}_{j-1}$ have been computed such that $\mathbf{y}_i\mathbf{x}^i = 1$. Set

$$B = A - \lambda_1\mathbf{x}^1\mathbf{y}_1 - \lambda_2\mathbf{x}^2\mathbf{y}_2 - \cdots - \lambda_{j-1}\mathbf{x}^{j-1}\mathbf{y}_{j-1}$$

LEMMA 8.2

The eigenvalues of B are $0, \ldots, 0$, $\lambda_j, \ldots, \lambda_n$ with corresponding eigenvectors $\mathbf{x}^1, \ldots, \mathbf{x}^n$. Further, B^t has corresponding eigenvectors $\mathbf{y}_1^t, \mathbf{y}_2^t, \ldots, \mathbf{y}_n^t$.

PROOF

As that of Lemma 8.1. ■

TABLE 8.3

COMPUTED	EXACT
$B = \begin{bmatrix} 3 & 1 & 1 \\ 0 & 1 & 1 \\ 0 & 0 & 2 \end{bmatrix}$	$B = \begin{bmatrix} 3 & 1 & 1 \\ 0 & 1 & 1 \\ 0 & 0 & 2 \end{bmatrix}$
For B, $\mathbf{w} = \begin{bmatrix} 1.0 \\ 0.001501 \\ 0.001501 \end{bmatrix}$, $\lambda = 3.004$	For B, $\mathbf{w} = \begin{bmatrix} 1 \\ 0 \\ 0 \end{bmatrix}$, $\lambda = 3$
For B^t, $\mathbf{w} = \begin{bmatrix} 0.5363 \\ 0.2681 \\ 0.8003 \end{bmatrix}$, $\lambda = 3.004$	For B^t, $\mathbf{w} = \frac{1}{\sqrt{14}} \begin{bmatrix} 2 \\ 1 \\ 3 \end{bmatrix} \approx \begin{bmatrix} 0.5345 \\ 0.2673 \\ 0.8018 \end{bmatrix}$
$B = \begin{bmatrix} 0.005 & -0.497 & -3.47 \\ -0.004494 & 0.9978 & 0.9933 \\ -0.004494 & -0.002247 & 1.993 \end{bmatrix}$	$B = \begin{bmatrix} 0 & -0.5 & -3.5 \\ 0 & 1 & 1 \\ 0 & 0 & 2 \end{bmatrix}$
For B, $\mathbf{w} = \begin{bmatrix} -0.8152 \\ 0.4089 \\ 0.4101 \end{bmatrix}$, $\lambda = 2.000$	For B, $\mathbf{w} = \frac{1}{\sqrt{6}} \begin{bmatrix} -2 \\ 1 \\ 1 \end{bmatrix} \approx \begin{bmatrix} -0.8165 \\ 0.4082 \\ 0.4082 \end{bmatrix}$, $\lambda = 2$
For B^t, $\mathbf{w} = \begin{bmatrix} -0.002254 \\ 0.0001873 \\ 1.000 \end{bmatrix}$, $\lambda = 2.001$	For B^t, $\mathbf{w} = \begin{bmatrix} 0 \\ 0 \\ 1 \end{bmatrix}$
$B = \begin{bmatrix} -0.003924 & -0.4963 & 0.488 \\ -0.000018 & 0.9974 & -0.9927 \\ -0.000004 & -0.00262 & 0.0009999 \end{bmatrix}$	$B = \begin{bmatrix} 0 & -0.5 & 0.5 \\ 0 & 1 & -1 \\ 0 & 0 & 0 \end{bmatrix}$
For B, $\mathbf{w} = \begin{bmatrix} 0.4443 \\ -0.8964 \\ 0.002352 \end{bmatrix}$, $\lambda = 1.001$	For B, $\mathbf{w} = \frac{1}{\sqrt{5}} \begin{bmatrix} 1 \\ -2 \\ 0 \end{bmatrix} \approx \begin{bmatrix} 0.4472 \\ -0.8944 \\ 0 \end{bmatrix}$, $\lambda = 1$
For B^t, $\mathbf{w} = \begin{bmatrix} 0.00001426 \\ -0.7093 \\ 0.7048 \end{bmatrix}$, $\lambda = 0.9998$	For B^t, $\mathbf{w} = \frac{1}{\sqrt{2}} \begin{bmatrix} 0 \\ -1 \\ 1 \end{bmatrix} \approx \begin{bmatrix} 0 \\ -0.7071 \\ 0.7071 \end{bmatrix}$

Hence λ_j and $\mathbf{x}^j$ and $\mathbf{y}_j$ such that $\mathbf{y}_j\mathbf{x}^j = 1$ can be computed by the power algorithm for B.

EXAMPLE 13

EXAMPLE 12 CONTINUED
Let A, $\mathbf{x}^0$, k, and ε be as in Example 12. The matrices B and the eigenvalues and corresponding eigenvectors found by the algorithm, together with the exact eigenvalues and corresponding eigenvectors, are shown in Table 8.3. □

In conclusion, it should be noted that error in $\lambda_1, \mathbf{x}^1, \mathbf{y}_1, \ldots, \lambda_p, \mathbf{x}^p, \mathbf{y}_p$ causes error in $B = A - \lambda_1\mathbf{x}^1\mathbf{y}_1 - \lambda_2\mathbf{x}^2\mathbf{y}_2 - \cdots - \lambda_p\mathbf{x}^p\mathbf{y}_p$. Because of this accumulated error, the power algorithm for B becomes less effective as p increases, as mentioned on page 353.

Two of the more popular methods for computing eigenvalues and eigenvectors are the QR-algorithm and, for matrices that are symmetric, the Jacobi method. Descriptions and studies of these algorithms can be found in most numerical analysis texts.

EXERCISES FOR SECTION 8.3

COMPUTATIONAL EXERCISES

In exercises 1 and 2, use the power algorithm with $k = 3$ and $\varepsilon = 0.01$ to find an approximation to the largest eigenvalue and a corresponding eigenvector.

1. $A = \begin{bmatrix} 3 & 9 \\ 1 & -8 \end{bmatrix}$, $\mathbf{x}^0 = \begin{bmatrix} -4 \\ 5 \end{bmatrix}$

2. $A = \begin{bmatrix} 7 & 2 \\ 1 & -6 \end{bmatrix}$, $\mathbf{x}^0 = \begin{bmatrix} 12 \\ 1 \end{bmatrix}$

In exercises 3 through 6, a matrix A and its eigenvalues are given. Find eigenvectors for matrices A and A^t belonging to λ_1, and use these to compute matrix B_1. Verify that B_1 has eigenvalues $0, \lambda_2$ or $0, \lambda_2, \lambda_3$. In exercises 5 and 6, then find eigenvectors for B_1 and B_1^t belonging to λ_2, use these to compute matrix B_2, and verify that B_2 has eigenvalues $0, 0, \lambda_3$.

3. $A = \begin{bmatrix} 2 & 7 \\ 1 & -4 \end{bmatrix}$, $\lambda_1 = -5, \lambda_2 = 3$

4. $A = \begin{bmatrix} -10 & -7 \\ 15 & 12 \end{bmatrix}$, $\lambda_1 = 5, \lambda_2 = -3$

5. $A = \begin{bmatrix} 2 & -1 & 3 \\ 1 & 2 & 1 \\ 1 & 1 & 0 \end{bmatrix}$, $\lambda_1 = 3, \lambda_2 = 2, \lambda_3 = -1$

6. $A = \begin{bmatrix} 2 & 0 & 1 \\ 1 & 1 & 1 \\ -1 & 1 & 3 \end{bmatrix}$, $\lambda_1 = 3, \lambda_2 = 2, \lambda_3 = 1$

THEORETICAL EXERCISES

7. (principle of biorthogonality) Let A be an $n \times n$ matrix such that A has n distinct eigenvalues, let A have eigenvectors $\mathbf{x}^1, \ldots, \mathbf{x}^n$, and let A^t have eigenvectors $\mathbf{y}_1^t, \ldots, \mathbf{y}_n^t$ corresponding to these eigenvalues. Show that $\mathbf{y}_i\mathbf{x}^j \neq 0$ if $i = j$ and $\mathbf{y}_i\mathbf{x}^j = 0$ if $i \neq j$. *Hint:* Consider $A\mathbf{x}^j = \lambda_j\mathbf{x}^j$ and $\mathbf{y}_iA = \lambda_i\mathbf{y}_i$. This leads to $\lambda_i\mathbf{y}_i\mathbf{x}^j = \mathbf{y}_iA\mathbf{x}^j = \lambda_j\mathbf{y}_i\mathbf{x}^j$. To show that $\mathbf{y}_i\mathbf{x}^j \neq 0$, write $\mathbf{y}_i = r_1\mathbf{x}^1 + \cdots + r_n\mathbf{x}^n$ and compute the inner product of each side with $\mathbf{y}_i$.

8. Explain why the power algorithm may converge very slowly if $\dfrac{|\lambda_2|}{|\lambda_1|}$ is near 1. *Hint:* Look at $\left(\dfrac{|\lambda_2|}{|\lambda_1|}\right)^k$ as k increases, and think about how slowly it tends to 0.

9. Let $A = \begin{bmatrix} 1 & 1 \\ 0 & 1 \end{bmatrix}$. Note that the eigenvalues of A are both 1. Apply the power algorithm to A and observe what occurs. Recall that the algorithm requires that $|\lambda_1| > |\lambda_2|$.

10. Let $A = \begin{bmatrix} 1 & 1 \\ 1 & 1 \end{bmatrix}$. Show that the eigenvalues of A are 2 and 0. Apply the power algorithm to A and observe what happens. Recall that the algorithm requires that $|\lambda_n| > 0$.

11. Let $A = \begin{bmatrix} 0 & 1 \\ -1 & 0 \end{bmatrix}$. Show that A has no real eigenvalues. Apply the power algorithm to A and observe what happens. Recall that the algorithm requires that $|\lambda_1| > \cdots > |\lambda_n| > 0$.

APPLICATIONS EXERCISE

12. In practice the power algorithm is rarely used to compute all eigenvalues and eigenvectors for a matrix. However, it is often used to find the eigenvalue with the largest absolute value and a corresponding eigenvector. In some problems, such as the one shown below, this is basically all the information desired.

If A is similar to a diagonal matrix, then the difference equation $\mathbf{x}(j+1) = A\mathbf{x}(j)$ has the solution

$$\mathbf{x}(j) = r_1\lambda_1^j\mathbf{p}^1 + \cdots + r_n\lambda_n^j\mathbf{p}^n$$

where $\lambda_1, \ldots, \lambda_n$ are the eigenvalues of A and $\mathbf{p}^1, \ldots, \mathbf{p}^n$ are the corresponding eigenvectors. If $|\lambda_1| > |\lambda_j|$ for $j > 1$, then

$$\frac{1}{\lambda_1^j}\mathbf{x}(j) = r_1\mathbf{p}^1 + r_2\left(\frac{\lambda_2}{\lambda_1}\right)^j\mathbf{p}^2 + \cdots + r_n\left(\frac{\lambda_n}{\lambda_1}\right)^j\mathbf{p}^n$$

and, as j increases, $\frac{1}{\lambda_1^j}\mathbf{x}(j)$ approaches $r_1\mathbf{p}^1$. Thus, for large j, $\mathbf{x}(j)$ looks almost like $r_1\lambda_1^j\mathbf{p}^1$, which is determined by λ_1 and $\mathbf{p}^1$.

Solve $\mathbf{x}(j+1) = \frac{1}{3}\begin{bmatrix} 1 & 2 \\ 2 & 1 \end{bmatrix}\mathbf{x}(j)$ and demonstrate the above remarks.

CHAPTER NINE

LINEAR PROGRAMMING

A programming problem is essentially a problem that involves finding a best, or optimal, solution from among a set of possible solutions.*

EXAMPLE 1

An oil company has refineries in Houston and Tulsa. Orders for gasoline produced in these plants come from Dallas, Los Angeles, and Atlanta. The numbers of gallons ordered in a given week and the shipping costs are given in Table 9.1 below. Equipment limits production at the Houston refinery to 300,000 gallons per week, and at the Tulsa refinery to 400,000 gallons per week. Of course, there are many possible solutions to the problem of how to ship the required amounts of gasoline to their destinations, one of which appears in Figure 9.1. (Also see exercise 23 of Section 9.1.) However, what is important here is to find a shipping solution for which the cost is least. For this problem, such a solution would be a best, or optimal, solution. □

TABLE 9.1

CITY	NUMBER OF GALLONS ORDERED	COST PER 1000 GALLONS FROM HOUSTON	COST PER 1000 GALLONS FROM TULSA
Atlanta	200,000	$500	$450
Los Angeles	150,000	$550	$600
Dallas	250,000	$50	$75

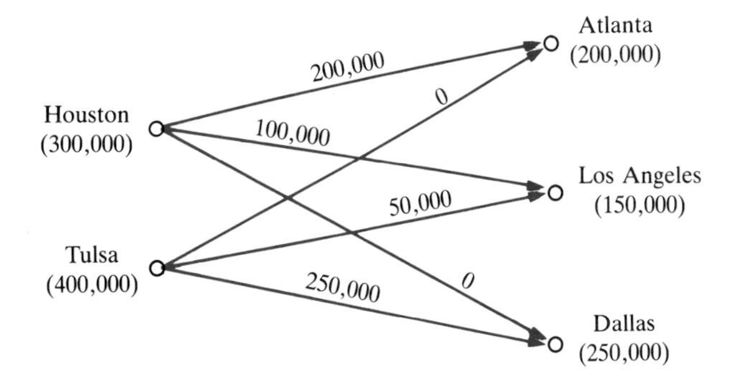

FIGURE 9.1

* "Programming" is not used here in the same sense as in computer programming. Originally, "programming" referred to allocation, or scheduling, and it is this meaning we use in this chapter.

In this chapter we will show how to solve a linear programming problem such as that of Example 1. The form in which such a problem is stated and some of the associated technical language follow.

Let $f(x_1,\dots,x_n) = c_1x_1 + \cdots + c_nx_n$ be a linear function. For any number b, the inequality

$$c_1x_1 + \cdots + c_nx_n \leqslant b \tag{1}$$

is called a *linear inequality*. A set of such inequalities is called a *system of linear inequalities*.

A *linear program* is written:

$$\begin{aligned}
\text{optimize}\quad & z = a_1x_1 + a_2x_2 + \cdots + a_nx_n \\
\text{subject to}\quad & a_{11}x_1 + a_{12}x_2 + \cdots + a_{1n}x_n \leqslant b_1 \\
& \qquad\vdots \\
& a_{i1}x_1 + a_{i2}x_2 + \cdots + a_{in}x_n \leqslant b_i \\
& \qquad\vdots \\
& a_{j1}x_1 + a_{j2}x_2 + \cdots + a_{jn}x_n \geqslant b_j \\
& \qquad\vdots \\
& a_{k1}x_1 + a_{k2}x_2 + \cdots + a_{kn}x_n = b_k \\
& \qquad\vdots \\
& x_1 \geqslant 0, x_2 \geqslant 0, \dots, x_n \geqslant 0
\end{aligned}$$

The word "optimize" may be replaced by either "maximize" or minimize," depending on the application.*

The linear programming problem is to find, from among all solutions to the system of linear inequalities, a solution $x_1, x_2, \dots, x_n$ at which z is *optimal*—that is, at which z has a maximum value or a minimum value, whichever is desired. The function $z = a_1x_1 + \cdots + a_nx_n$ in the linear program is called the *objective function*, and each equation and inequality following "subject to" is called a *constraint*. A point $[x_1, x_2, \dots, x_n] \in \mathbf{R}_n$ that satisfies all constraints is called a *feasible solution*, and the set of all feasible solutions is called the *feasible set*.†

If there is an $\hat{\mathbf{x}} = [\hat{x}_1, \hat{x}_2, \dots, \hat{x}_n]$ in the feasible set such that z at $\hat{\mathbf{x}}$ is optimal, then $\hat{\mathbf{x}}$ is called an *optimal solution* to the linear program, and the value of z at $\hat{\mathbf{x}}$ is called the *optimal value*.

EXAMPLE 2

The linear program

$$\begin{aligned}
\text{max}\quad & z = 0.5x_1 \\
\text{s.t.}\quad & 3x_1 \leqslant 4 \\
& 2x_1 \geqslant 1 \\
& x_1 \geqslant 0
\end{aligned}$$

* In problem statements we will abbreviate "maximize," "minimize," and "subject to" as "max," "min," and "s.t.," respectively.

† "Feasible" means "possible."

can be viewed geometrically as shown in Figure 9.2. The feasible set is the interval $[\frac{1}{2}, \frac{4}{3}]$. The optimal solution is $x_1 = \frac{4}{3}$, and the optimal value is $z = \frac{2}{3}$. □

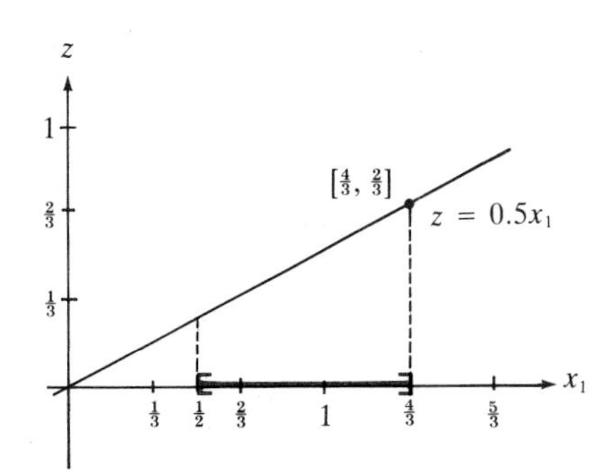

FIGURE 9.2

EXAMPLE 3

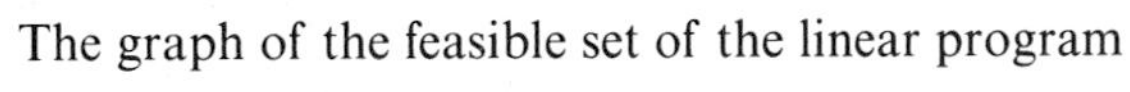

The graph of the feasible set of the linear program

$$\begin{aligned} \min \quad & z = x_1 + x_2 \\ \text{s.t.} \quad & x_1 + 2x_2 \leqslant 2 \\ & 2x_1 + x_2 \leqslant 2 \\ & x_1 \geqslant 0, x_2 \geqslant 0 \end{aligned}$$

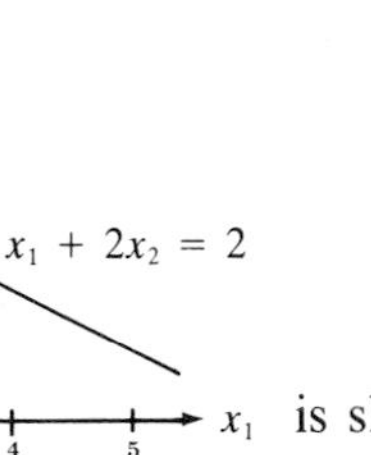

FIGURE 9.3

is shown as the shaded region in Figure 9.3. Since $z = x_1 + x_2 \geqslant 0$ for all (x_1, x_2) in the feasible set, it follows that the optimal solution is the point $[0, 0]$ and the optimal value is $z = 0$. □

That the problem described in Example 1 is a linear programming problem is shown next.

EXAMPLE 4

CONTINUING EXAMPLE 1

In Example 1, let x_{ij} be the number of thousands of gallons of gasoline shipped from one city to another, as shown in tabular form in Table 9.2 and in diagram form in Figure 9.4.

Each row and each column of the table translates into a constraint, as does each point of departure and each point of destination in the diagram. For

TABLE 9.2

FROM	TO Atlanta	Los Angeles	Dallas	TOTAL AVAILABLE
Houston	x_{11}	x_{12}	x_{13}	300
Tulsa	x_{21}	x_{22}	x_{23}	400
Total Required	200	150	250	

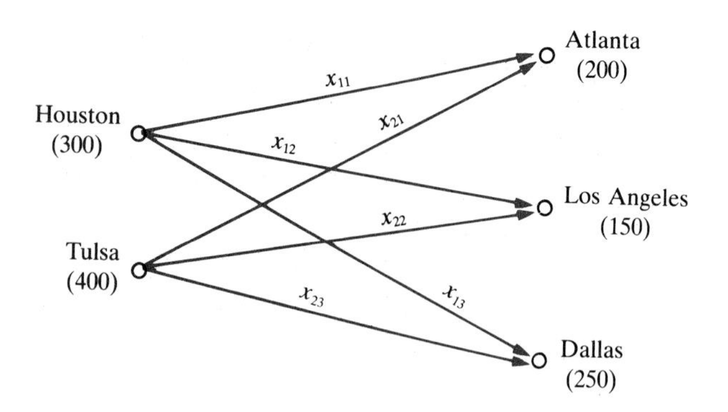

FIGURE 9.4

the objective function, we let z denote the shipping cost. Then, using the cost figures from Table 9.1, we need to solve

$$\begin{aligned}
\text{min} \quad & z = 500x_{11} + 550x_{12} + 50x_{13} + 450x_{21} + 600x_{22} + 75x_{23} \\
\text{s.t.} \quad & x_{11} + x_{21} \geqslant 200 \\
& x_{12} + x_{22} \geqslant 150 \\
& x_{13} + x_{23} \geqslant 250 \\
& x_{11} + x_{12} + x_{13} \leqslant 300 \\
& x_{21} + x_{22} + x_{23} \leqslant 400 \\
& x_{11} \geqslant 0, x_{12} \geqslant 0, x_{13} \geqslant 0, x_{21} \geqslant 0, x_{22} \geqslant 0, x_{23} \geqslant 0
\end{aligned}$$

Exercise 27 of Section 9.4 asks you to solve this linear program. □

In this chapter we show how to solve linear programming problems. In the first section we present a graphical method which can be applied when there are exactly two variables. In the remaining sections we develop a more general solution method.

9.1 A GRAPHICAL METHOD

In this section we describe a geometric method which can be used to solve a linear program involving two variables x_1 and x_2. Before describing this technique, we list two geometrical preliminaries.

PRELIMINARY 1

The equation $ax_1 + bx_2 = c$, where at least one of a or b is not 0, is a line in the coordinate plane. This line separates the plane into two sets:

(*i*) those points $[x_1, x_2]$ that satisfy $ax_1 + bx_2 < c$ and

(*ii*) those points $[x_1, x_2]$ that satisfy $ax_1 + bx_2 > c$ (exercise 15).

Each of these sets, together with the bounding line $ax_1 + bx_2 = c$, is called a *half-plane.*

EXAMPLE 5

The line $x_1 + x_2 = 1$ separates the coordinate plane as shown in Figure 9.5. Since the point $[0, 0]$ satisfies $x_1 + x_2 < 1$, all points $[x_1, x_2]$ below the line satisfy this inequality, and those points $[x_1, x_2]$ above the line satisfy $x_1 + x_2 > 1$. □

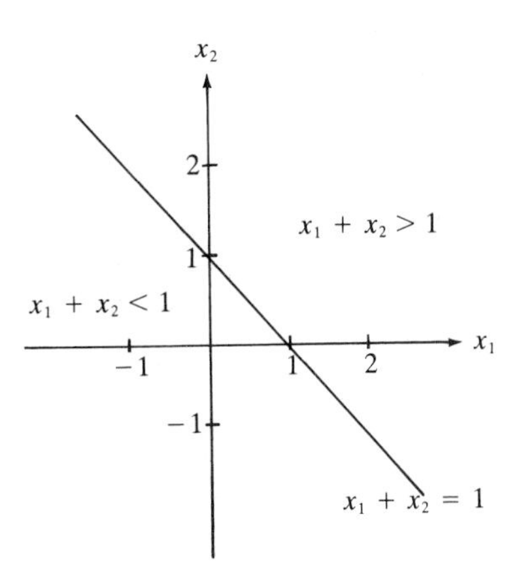

FIGURE 9.5

PRELIMINARY 2

The equation $ax_1 + bx_2 + cx_3 = d$, where at least one of a, b, or c is not 0, is the equation of a plane in $\mathbf{R}_3$. This plane separates the space into two sets of points:

(*i*) those points $[x_1, x_2, x_3]$ that satisfy $ax_1 + bx_2 + cx_3 < d$ and

(*ii*) those points $[x_1, x_2, x_3]$ that satisfy $ax_1 + bx_2 + cx_3 > d$ (exercise 16).

Each of these sets, together with the bounding plane $ax_1 + bx_2 + cx_3 = d$, is called a *half-space.*

EXAMPLE 6

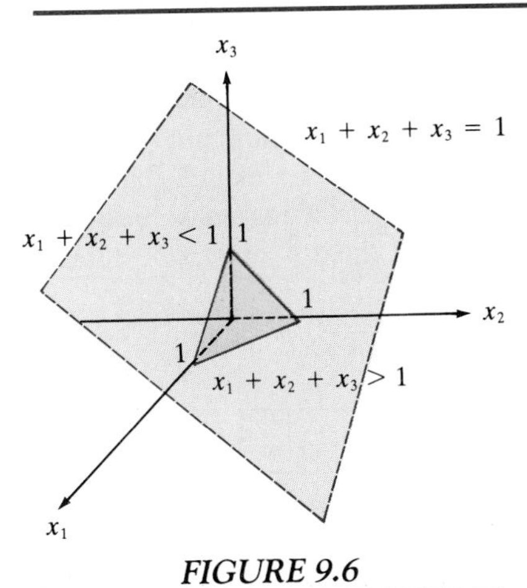

FIGURE 9.6

The plane $x_1 + x_2 + x_3 = 1$ separates the coordinate space as shown in Figure 9.6. Since $[0, 0, 0]$ satisfies $x_1 + x_2 + x_3 < 1$, all points $[x_1, x_2, x_3]$ below this plane satisfy $x_1 + x_2 + x_3 < 1$, and those points $[x_1, x_2, x_3]$ above the plane satisfy $x_1 + x_2 + x_3 > 1$. □

Using these preliminaries, we can show how to view a linear program in two variables geometrically. Consider the linear program

$$\begin{aligned} \max \quad & z = a_1x_1 + a_2x_2 \\ \text{s.t.} \quad & a_{11}x_1 + a_{12}x_2 \leqslant b_1 \\ & a_{21}x_1 + a_{22}x_2 \leqslant b_2 \\ & x_1 \geqslant 0, x_2 \geqslant 0 \end{aligned}$$

Graphing the four constraints in $\mathbf{R}_2$ gives four half-planes (indicated by arrows pointing into the half-planes from their boundary lines) whose intersection is shaded in Figure 9.7. The *vertices* of a feasible set S in $\mathbf{R}_2$ are those points in S (labeled A, B, C, and D in the figure) at the intersection of lines bounding the

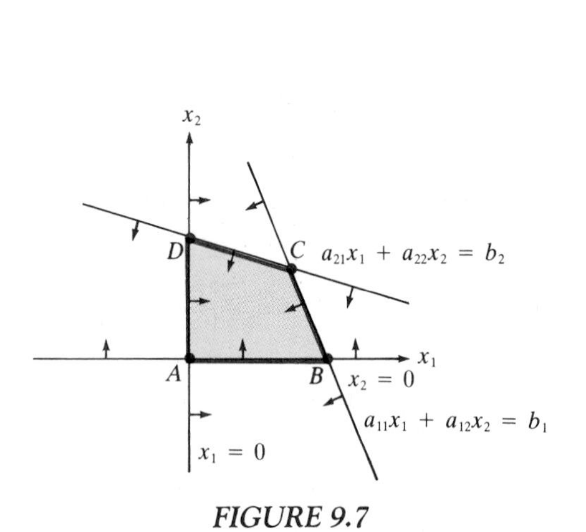

FIGURE 9.7

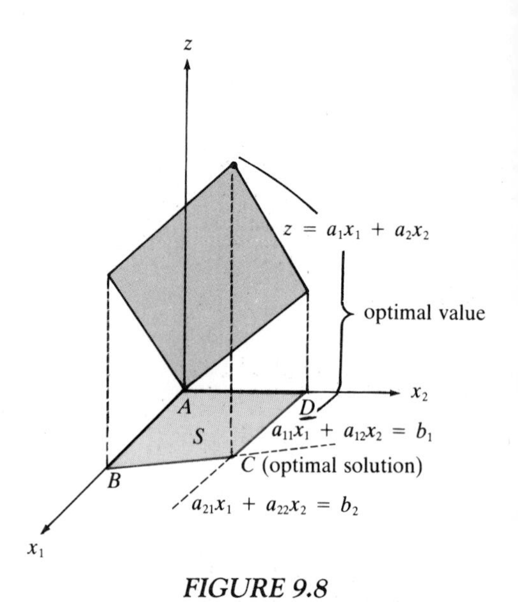

FIGURE 9.8

feasible set. Now graphing the objective function in $\mathbf{R}_3$ yields a plane, as shown in Figure 9.8.

We want to find the point $[x_1, x_2]$ in the feasible set S (shaded in color in Figure 9.8) at which $z = a_1x_1 + a_2x_2$ is largest. Geometrically it is clear that the optimal solution $[x_1, x_2]$ occurs at the vertex C in the figure of the feasible set, corresponding to the highest point of the plane $z = a_1x_1 + a_2x_2$ above the feasible set.

Using this picture, we can also see that there are two ways in which a linear program can fail to have an optimal solution. First, the feasible set can be empty; without any feasible solutions, there can be no optimal solution. Second, since a half-plane obtained from a constraint is of infinite extent, it can happen that taking the intersection of half-planes yields a feasible set that is also of infinite extent (unbounded) and the objective function could increase or decrease without limit in the direction of the infinite extent. Then the corresponding maximization or minimization problem has no solution. It should be noted that in an unbounded feasible set, the boundaries of infinite extent are rays (exercise 17).

EXAMPLE 7

Consider the linear program

$$
\begin{array}{ll}
\max & z = 3x_1 + 2x_2 \\
\text{s.t.} & x_1 + x_2 \leqslant 2 \\
 & -2x_1 + x_2 \leqslant -5 \\
 & x_1 \geqslant 0, x_2 \geqslant 0
\end{array}
$$

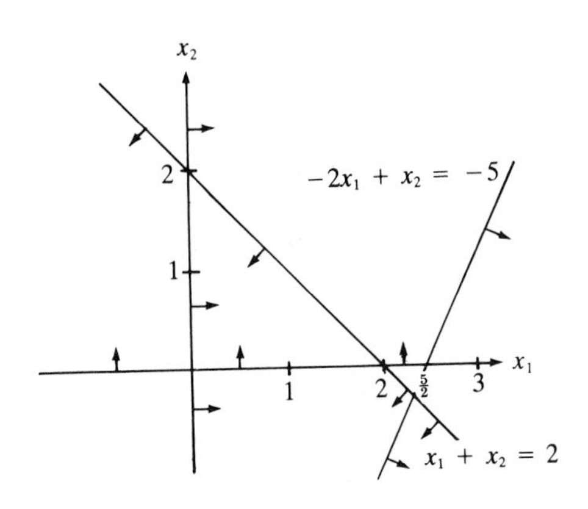

FIGURE 9.9

Graphing the feasible set, as shown in Figure 9.9, we see that it is empty. □

EXAMPLE 8

Consider the linear program

$$\begin{array}{ll} \max & z = 3x_1 + 2x_2 \\ \text{s.t.} & -2x_1 + x_2 \leqslant 3 \\ & x_1 - 2x_2 \leqslant 4 \\ & x_1 \geqslant 0, x_2 \geqslant 0 \end{array}$$

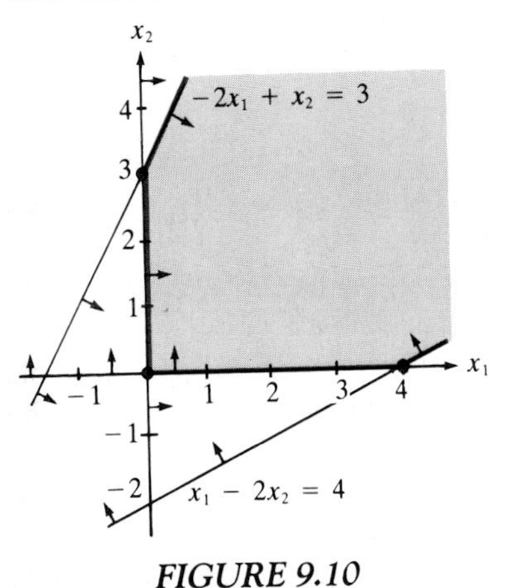

FIGURE 9.10

As Figure 9.10 shows, the feasible set in this case is unbounded. Further, the boundaries of infinite extent are the rays $x_2 = 2x_1 + 3$, $x_1 \geqslant 0$, and $x_2 = \frac{1}{2}x_1 - 2$, $x_1 \geqslant 4$. Since

$$z = 3x_1 + 2x_2 = 3x_1 + 2(2x_1 + 3) = 7x_1 + 6$$

along the first of these rays, we can see that z increases without bound in the feasible set. Thus there is no solution to the linear program. □

Based on the material just presented, we have a way of solving linear programs involving two variables or of determining that no solution exists when that is the case.

GEOMETRIC ALGORITHM FOR SOLVING LINEAR PROGRAMS IN TWO VARIABLES

1. Graph the feasible set in $\mathbf{R}_2$. If the set is empty, stop; there is no solution.

2. If the feasible set is unbounded, for every boundary of infinite extent, solve the equation of that ray for one of the variables and substitute the result into the objective function to obtain a function of one variable. If any such resulting expression for z has no optimum value, no optimal solution exists for the linear program.

3. Otherwise, using the graph to determine points that are vertices, compute all the vertices of the feasible set. Calculate the value of the objective function at each vertex.

4. For the maximization problem, the largest value is the optimal value and the vertex at which it occurs is an optimal solution. For the minimization problem, the smallest value is the optimal value and the vertex at which it occurs is an optimal solution.

EXAMPLE 9

Solve the linear program

$$\begin{aligned} \min \quad & z = x_1 - x_2 \\ \text{s.t.} \quad & 2x_1 + x_2 \leqslant 3 \\ & x_1 + 2x_2 \leqslant 3 \\ & x_1 \geqslant 0, x_2 \geqslant 0 \end{aligned}$$

First we graph the feasible set, as shown in Figure 9.11, labeling the vertices as A, B, C, and D. Note that the feasible set is bounded. We compute the coordinates of the vertices A, B, C, and D as listed in Table 9.3.

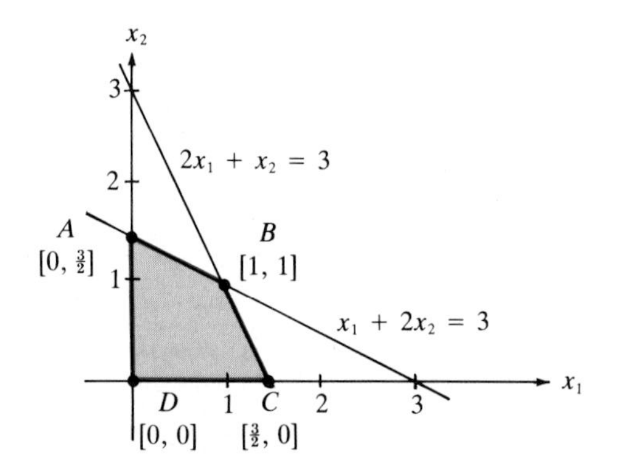

FIGURE 9.11

TABLE 9.3

VERTEX	INTERSECTING LINES	POINT
A	$x_1 = 0, x_1 + 2x_2 = 3$	$[0, \frac{3}{2}]$
B	$2x_1 + x_2 = 3, x_1 + 2x_2 = 3$	$[1, 1]$
C	$x_2 = 0, 2x_1 + x_2 = 3$	$[\frac{3}{2}, 0]$
D	$x_1 = 0, x_2 = 0$	$[0, 0]$

Finally, computing the value of the objective function at each vertex gives the results shown in Table 9.4.

TABLE 9.4

VERTEX	VALUE OF z
A	$-\frac{3}{2}$
B	0
C	$\frac{3}{2}$
D	0

Hence the optimal solution is $[0, \frac{3}{2}]$ and the optimal value is $z = -\frac{3}{2}$. □

EXAMPLE 10

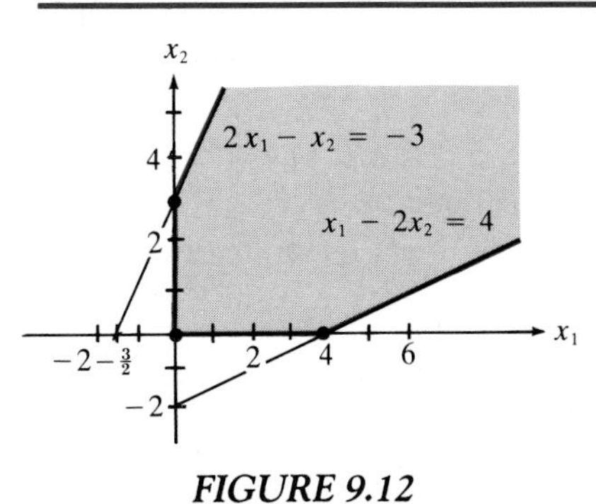

FIGURE 9.12

Solve the linear program

$$\begin{aligned} \min \quad & z = 3x_1 - 2x_2 \\ \text{s.t.} \quad & 2x_1 - x_2 \geqslant -3 \\ & x_1 - 2x_2 \leqslant 4 \\ & x_1 \geqslant 0, x_2 \geqslant 0 \end{aligned}$$

To solve this problem, we graph the feasible set, as shown in Figure 9.12. We see that the feasible set is unbounded, so we check the objective function along the boundary rays $x_1 - 2x_2 = 4$, $x_1 \geqslant 4$, and $2x_1 - x_2 = -3$, $x_1 \geqslant 0$. On the first of these,

$$z = 3x_1 - 2x_2 = 3x_1 - x_1 + 4 = 2x_1 + 4$$

which does not decrease without bound as x_1 increases. On the second ray,

$$z = 3x_1 - 2x_2 = 3x_1 - 4x_1 - 6 = -x_1 - 6$$

which does decrease without bound as x_1 increases. Hence no minimum value for z exists in the feasible set, and the linear program has no optimal solution. □

EXAMPLE 11

An oil company has two different processes for manufacturing gasoline from crude oil. Using the data from Table 9.5, set up the linear program for maximizing profit and solve it.

TABLE 9.5

PROCESS	CRUDE OIL NEEDED PER RUN	GASOLINE OUTPUT PER RUN	PROFIT PER RUN
1	10 tons	5 tons	\$100
2	8 tons	3 tons	\$70
	140 tons (available)	60 tons (needed)	

These data can be used to make sketches of the processes as shown in Figure 9.13.

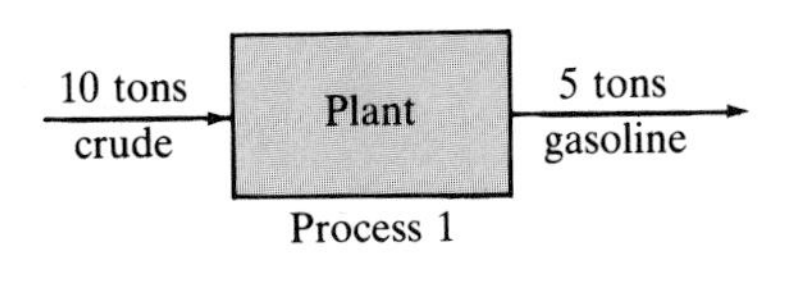

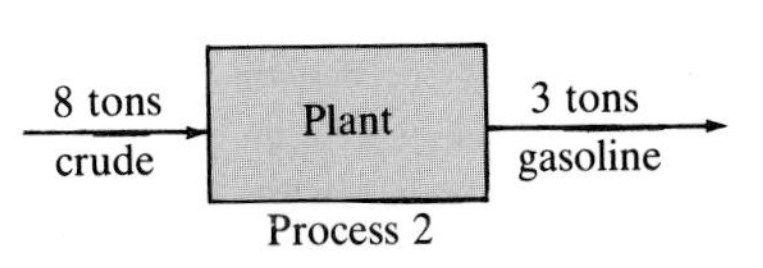

FIGURE 9.13

For $i = 1, 2$, we let x_i be the number of runs of process i needed. Then each output yields a constraint. The coefficients for the objective function are listed in the last column of Table 9.5. Hence, the problem we need to solve is

$$\begin{aligned} \max \quad & z = 100x_1 + 70x_2 \\ \text{s.t.} \quad & 10x_1 + 8x_2 \leqslant 140 \\ & 5x_1 + 3x_2 \geqslant 60 \\ & x_1 \geqslant 0, x_2 \geqslant 0 \end{aligned}$$

To solve this problem, we graph the feasible set, as shown in Figure 9.14, and label the vertices A, B, C, and D. Note that the feasible set is bounded. We

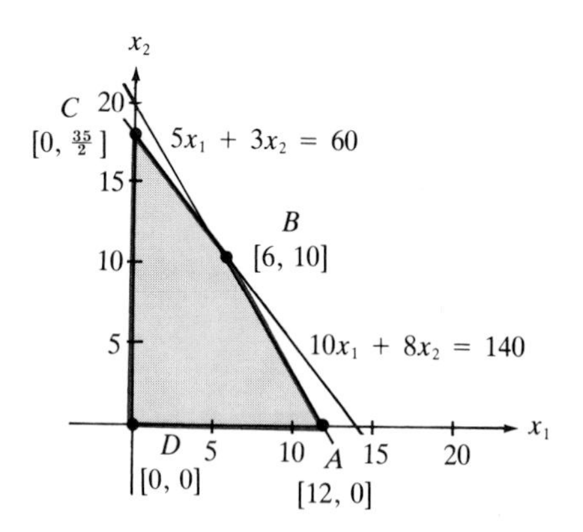

FIGURE 9.14

TABLE 9.6

VERTEX	INTERSECTING LINES	POINT
A	$x_2 = 0, 5x_1 + 3x_2 = 60$	$[12, 0]$
B	$10x_1 + 8x_2 = 140, 5x_1 + 3x_2 = 60$	$[6, 10]$
C	$x_1 = 0, 10x_1 + 8x_2 = 140$	$[0, \frac{35}{2}]$
D	$x_1 = 0, x_2 = 0$	$[0, 0]$

compute the coordinates of the vertices A, B, C, and D, as listed in Table 9.6.

Finally, computing the value of the objective function at each vertex gives the results shown in Table 9.7.

TABLE 9.7

VERTEX	VALUE OF z
A	\$1200
B	\$1300
C	\$1225
D	\$0

Hence the optimal solution is to make 6 runs using process 1 and 10 runs using process 2. This procedure will yield an optimal profit of $z = \$1300$. □

In the next sections we will show a method for solving linear programs involving more than two variables.

EXERCISES FOR SECTION 9.1

COMPUTATIONAL EXERCISES

In exercises 1 through 4, identify each of the following for the given linear program: objective function, constraints, feasible set, optimal solution, and optimal value of z.

1. max $\quad z = 3x \quad$ s.t. $\quad 2x \leqslant 3, x \geqslant 0$

2. min $\quad z = 5x \quad$ s.t. $\quad 3x \leqslant 8, x \geqslant 0$

3. min $\quad z = 8x \quad$ s.t. $\quad 4x \leqslant 7, x \geqslant 0$

4. max $\quad z = 2x \quad$ s.t. $\quad 2x \leqslant 9, x \geqslant 0$

In exercises 5 through 14, use the geometric algorithm to solve the given linear program for both the maximum and the minimum value of z, or explain why one or both of these do not exist. Using graph paper will help you keep your work neat.

5. optimize $z = 2x_1 - x_2$
s.t.
$$\begin{aligned} 2x_1 - x_2 &\geq -21 \\ 2x_1 - x_2 &\leq 15 \\ x_1 \geq 0, x_2 &\geq 0 \end{aligned}$$

6. optimize $z = x_1 - x_2$
s.t.
$$\begin{aligned} 5x_1 + 7x_2 &\leq 35 \\ x_1 \geq 0, x_2 &\geq 0 \end{aligned}$$

7. optimize $z = 8x_1 + x_2$
s.t.
$$\begin{aligned} 8x_1 + x_2 &\leq 8 \\ x_1 \geq 0, x_2 &\geq 0 \end{aligned}$$

8. optimize $z = 3x_1 - x_2$
s.t.
$$\begin{aligned} 3x_1 - x_2 &\geq 0 \\ 3x_1 + x_2 &\leq 24 \\ x_1 \geq 0, x_2 &\geq 0 \end{aligned}$$

9. optimize $z = 3x_1 + x_2$
s.t.
$$\begin{aligned} x_1 - x_2 &\leq 2 \\ 3x_1 + 4x_2 &\leq 48 \\ x_1 \geq 0, x_2 &\geq 0 \end{aligned}$$

10. optimize $z = 5x_1 - x_2$
s.t.
$$\begin{aligned} 4x_1 - x_2 &\geq -8 \\ x_1 - 2x_2 &\leq 6 \\ x_1 \geq 0, x_2 &\geq 0 \end{aligned}$$

11. optimize $z = x_1 - 2x_2$
s.t.
$$\begin{aligned} x_1 + 3x_2 &\leq 54 \\ x_1 + 2x_2 &\leq 39 \\ x_1 + x_2 &\leq 29 \\ 2x_1 + 3x_2 &\geq 15 \\ x_1 \geq 0, x_2 &\geq 0 \end{aligned}$$

12. optimize $z = x_1 + x_2$
s.t.
$$\begin{aligned} x_1 + 4x_2 &\leq 40 \\ 2x_1 + x_2 &\leq 38 \\ 3x_1 + 5x_2 &\leq 56 \\ 3x_1 + x_2 &\geq 5 \\ x_1 \geq 0, x_2 &\geq 0 \end{aligned}$$

13. optimize $z = x_3$
s.t.
$$\begin{aligned} 2x_1 + x_2 + 2x_3 &\leq 10 \\ 2x_1 + x_2 + x_3 &\leq 7 \\ x_1 \geq 0, x_2 \geq 0, x_3 &\geq 0 \end{aligned}$$

14. optimize $z = x_2$
s.t.
$$\begin{aligned} 4x_1 + x_2 + 2x_3 &\leq 8 \\ 4x_1 + 3x_2 + 2x_3 &\leq 12 \\ x_1 \geq 0, x_2 \geq 0, x_3 &\geq 0 \end{aligned}$$

THEORETICAL EXERCISES

15. Prove Preliminary 1 as stated in this section. *Hint:* Note that $[a, b] \neq \mathbf{0}$. Show that all $[x, y]$ orthogonal to $[a, b]$ through a point $[x_0, y_0]$ satisfy

$$([a, b], [x - x_0, y - y_0]) = 0$$

that is,

$$a(x - x_0) + b(y - y_0) = 0$$

Define $c = ax_0 + by_0$, so this line can be written as $ax + by = c$. If $[x, y]$ is not on the line, then $[x - x_0, y - y_0]$ is a vector whose angle θ with $[a, b]$ is between 0 and $\pi/2$ if $[x, y]$ is on one side of the line and between $\pi/2$ and π if $[x, y]$ is on the other side of the line. Then consider

$$([a, b], [x - x_0, y - y_0]) = \|[a, b]\| \, \|[x - x_0, y - y_0]\| \cos\theta$$

16. Prove Preliminary 2 as stated in this section. *Hint:* The plane through $[d_1, d_2, d_3] \in \mathbf{R}_3$ and orthogonal to $[a, b, c] \in \mathbf{R}_3$, where $[a, b, c] \neq \mathbf{0}$, is described by

$$a(x_1 - d_1) + b(x_2 - d_2) + c(x_3 - d_3) = 0$$

Using this result, follow the pattern of the hint for exercise 15.

17. Show by sketches that an unbounded region formed by an intersection of half-planes has either one or two rays on its boundary.

18. Write a linear program with an empty feasible set.

19. Write a linear program with an unbounded feasible set. Can such a linear program have an optimal value?

20. Write a linear program that has infinitely many optimal solutions. Is it also possible to have a linear program with infinitely many optimal values?

21. Write a linear program whose feasible set has vertices $[1, 0]$, $[0, 1]$, and $[1, 1]$ and whose maximum optimal value is at $[0, 1]$.

22. Must every linear program in two variables with a nonempty feasible set have either a maximum value or a minimum value, or can there be a linear program that has neither?

APPLICATIONS EXERCISES

23. Find three different shipping solutions for the problem given in Example 1.

24. A plant makes two plastic products which are manufactured in two stages, mixing and curing, as diagrammed in Figure 9.15. Use the data from Table 9.8 to write a linear program to maximize profit. Solve the linear program.

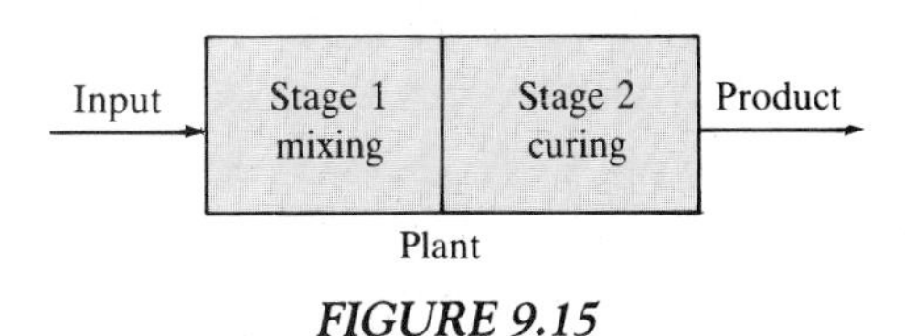

FIGURE 9.15

TABLE 9.8

STAGE	TIME PER STAGE IN MINUTES FOR ONE UNIT OF PRODUCT 1	2	MINUTES STAGE CAN RUN IN ONE DAY
Mixing	2	3	330
Curing	12	5	1200
Profit per unit	30¢	20¢	

25. A chemical processing plant makes two products, a rodent poison and an insecticide, in three processing steps. Use the data from Table 9.9 to write a linear program to maximize the plant's profit, and solve the linear program using the geometric algorithm.

26. A small oil company has one refinery from which it can obtain gasoline and diesel fuel by two different processes. The amounts of gasoline and diesel fuel obtainable and the costs of the processes are given in Table 9.10, together with the demand for each of the fuels. Use the data in the table to write a linear program to minimize the cost of meeting the specified demands, and solve the linear program.

27. A glassworks makes ordinary window glass and plate glass. To satisfy customer demand, the factory must make at least 8 tons of ordinary window glass and at least 6 tons of plate glass per day. The factory has a maximum capacity of 19 tons of glass per day. Suppose the profit on ordinary window glass is $1200 per ton and on plate glass is $1500 per ton. Write a linear program to determine how many tons of each type should be produced in a day to maximize profit. Solve the program.

28. It is expected that mining on the moon will involve loading surface material into capsules to be shot through a linear accelerator into space, for processing there. (Physicist Gerard O'Neill has designed and built working models of the accelerator, called a mass driver. His book is listed in the References.) Suppose a moon miner is required to ship up minerals containing a total of at least 420 tons of oxygen and 198 tons of aluminum. He has available one site whose minerals contain 500 pounds of oxygen and 100 pounds of aluminum per ton, and another site whose minerals contain 300 pounds of oxygen and 300 pounds of aluminum per ton. For this operation, silicon is considered a waste material. Suppose the minerals from the first site contain 800 pounds of silicon per ton and the minerals from the second site contain 1000 pounds of silicon per ton. Write a linear program to determine how many tons of each site's minerals should be sent to minimize waste, and use the geometric algorithm to solve the program.

TABLE 9.9

STAGE	TIME PER STAGE IN MINUTES Poison	Insecticide	MINUTES STAGE RUNS IN ONE DAY
1	2	9	450
2	1	2	125
3	1	1	105
Profit per Gallon	$3	$4	
Number of Gallons Made	x_1	x_2	

TABLE 9.10

PROCESS	AMOUNT OBTAINABLE (IN BARRELS) Gasoline	Diesel Fuel	COST PER RUN	NUMBER OF RUNS
1	5000	5000	$8000	x_1
2	3000	2000	$4000	x_2
Demand in Barrels	45000	40000		

9.2 THE SIMPLEX ALGORITHM

In the remaining three sections of this chapter we will develop a method for solving any linear programming problem, regardless of the number of variables. In this section we give a method for solving the special linear program

$$\begin{array}{ll} \max & z = a_1x_1 + a_2x_2 + \cdots + a_nx_n \\ \text{s.t.} & a_{11}x_1 + a_{12}x_2 + \cdots + a_{1n}x_n \leqslant b_1 \\ & a_{21}x_1 + a_{22}x_2 + \cdots + a_{2n}x_n \leqslant b_2 \\ & \vdots \\ & a_{m1}x_1 + a_{m2}x_2 + \cdots + a_{mn}x_n \leqslant b_m \\ & x_1 \geqslant 0, x_2 \geqslant 0, \ldots, x_n \geqslant 0 \end{array}$$

where $b_1 \geqslant 0, b_2 \geqslant 0, \ldots, b_m \geqslant 0$.* The following section provides proofs for the results on which this method rests. Then in the last section we show how the method developed here can be extended to solve any linear programming problem.

The method presented in this section is different from the geometric method given in the previous section in that it is an algebraic technique. The first step in the algebraic technique is to convert the linear program into a special system of linear algebraic equations, called the *linear program system*:

$$\begin{array}{llllll} a_{11}x_1 + a_{12}x_2 + \cdots + a_{1n}x_n + s_1 & & & & = b_1 \\ a_{21}x_1 + a_{22}x_2 + \cdots + a_{2n}x_n & + s_2 & & & = b_2 \\ & & & & \vdots \\ a_{m1}x_1 + a_{m2}x_2 + \cdots + a_{mn}x_n & & + s_m & & = b_m \\ -a_1x_1 - a_2x_2 - \cdots - a_nx_n & & & + z & = 0 \end{array}$$

Note that all variables are on the left sides of these equations, and all constants are on the right sides. The variables $s_1, s_2, \ldots, s_m$—called the *slack variables* because they make up the difference, or slack, between the right and left sides of constraints in the linear program—are added to each constraint in forming the corresponding equation. Finally, the objective function is listed as the last equation of the system.

A solution to the linear program system such that

$$x_1 \geqslant 0, x_2 \geqslant 0, \ldots, x_n \geqslant 0, s_1 \geqslant 0, \ldots, s_m \geqslant 0$$

is called a *feasible solution.* A feasible solution for which z is largest is called an *optimal solution.*

* The program is special in that (1) it maximizes, (2) all constraints except those of the type $x_i \geqslant 0$ are of the type $\leqslant$, and (3) each $b_i \geqslant 0$.

EXAMPLE 12

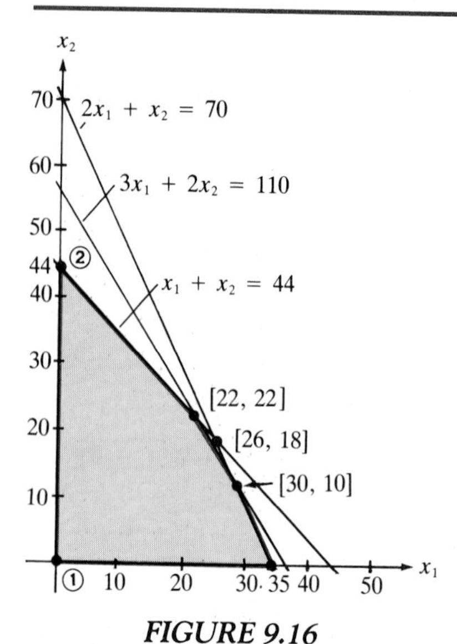

FIGURE 9.16

The linear program

$$\begin{aligned} \max \quad & z = 5x_1 + 3x_2 \\ \text{s.t.} \quad & x_1 + x_2 \leqslant 44 \\ & 3x_1 + 2x_2 \leqslant 110 \\ & 2x_1 + x_2 \leqslant 70 \\ & x_1 \geqslant 0, x_2 \geqslant 0 \end{aligned}$$

whose feasible set is shaded in Figure 9.16, has the optimal solution [30, 10] and the optimal value $z = 180$.

The corresponding linear program system is

$$\begin{aligned} x_1 + x_2 + s_1 \qquad\qquad\qquad &= 44 \\ 3x_1 + 2x_2 \qquad + s_2 \qquad\qquad &= 110 \\ 2x_1 + x_2 \qquad\qquad + s_3 \qquad &= 70 \\ -5x_1 - 3x_2 \qquad\qquad\qquad + z &= 0 \end{aligned}$$

In finding an optimal solution, we look for a feasible solution such that z is largest. To do this, we need to find $x_1 \geqslant 0, x_2 \geqslant 0, s_1 \geqslant 0, s_2 \geqslant 0$, and $s_3 \geqslant 0$ such that

$$\begin{aligned} x_1 + x_2 + s_1 \qquad\qquad &= 44 \\ 3x_1 + 2x_2 \qquad + s_2 \qquad &= 110 \\ 2x_1 + x_2 \qquad\qquad + s_3 &= 70 \end{aligned}$$

and $z = 5x_1 + 3x_2$, obtained from $-5x_1 - 3x_2 + z = 0$, is largest. Since $s_1 \geqslant 0$, $s_2 \geqslant 0$, and $s_3 \geqslant 0$, this involves finding $x_1 \geqslant 0$ and $x_2 \geqslant 0$ such that

$$\begin{aligned} x_1 + x_2 &\leqslant 44 \\ 3x_1 + 2x_2 &\leqslant 110 \\ 2x_1 + x_2 &\leqslant 70 \end{aligned}$$

and $z = 5x_1 + 3x_2$ is optimal. The solution of the linear program, given above, shows these values to be $x_1 = 30$ and $x_2 = 10$, in which case $z = 180$. Substituting into the equations in the linear program system, we have $s_1 = 4$ and $s_2 = s_3 = 0$, and thus [30, 10, 4, 0, 0, 180] is the optimal solution to the linear program system. □

As can be seen in Example 12 (see exercise 25), a linear program is linked to the linear program system as follows.

THEOREM 9.1

A point $[x_1, \ldots, x_n] \in \mathbf{R}_n$ is a feasible solution to the linear program and z is the corresponding value if and only if there are slack variables $s_1 \geqslant 0, \ldots, s_m \geqslant 0$ such that $[x_1, \ldots, x_n, s_1, \ldots, s_m, z]$ is a feasible solution to the linear program system. Further, $[x_1, \ldots, x_n, s_1, \ldots, s_m, z]$ is an optimal solution to the linear program system if and only if $[x_1, \ldots, x_n]$ is an optimal solution and z is the optimal value to the linear program.

PROOF See Exercise 26. ■

Thus, instead of solving the linear program, we will find an optimal solution $[x_1, \ldots, x_n, s_1, \ldots, s_m, z]$ to the linear program system. Then, ignoring the entries $s_1, \ldots, s_m$, we have that $[x_1, \ldots, x_n]$ is an optimal solution and z is the optimal value of the linear program.

As a computational convenience in doing this work, we write the linear program system in augmented matrix form:

$$\begin{array}{ccccccccc|c} x_1 & x_2 & & x_n & s_1 & s_2 & & s_m & z & \\ a_{11} & a_{12} & \cdots & a_{1n} & 1 & 0 & \cdots & 0 & 0 & b_1 \\ a_{21} & a_{22} & \cdots & a_{2n} & 0 & 1 & \cdots & 0 & 0 & b_2 \\ \vdots & \vdots & & \vdots & \vdots & \vdots & & \vdots & \vdots & \vdots \\ a_{m1} & a_{m2} & \cdots & a_{mn} & 0 & 0 & \cdots & 1 & 0 & b_m \\ \hline -a_1 & -a_2 & \cdots & -a_n & 0 & 0 & \cdots & 0 & 1 & 0 \end{array}$$

Note that the size of this matrix is $(m + 1) \times (n + m + 2)$.

We now describe the method we will use to find the optimal solution for a linear program system. Considering the above system, pick z and any other m variables from among $x_1, \ldots, x_n$, $s_1, \ldots, s_m$; for example, $x_{i_1}, \ldots, x_{i_r}$, $s_{i_{r+1}}, \ldots, s_{i_m}$, z. If possible, use Gaussian operations to solve the system of equations for these variables, obtaining a matrix with exactly one nonzero entry, that entry being a 1, in the columns for $x_{i_1}, \ldots, x_{i_r}$, $s_{i_{r+1}}, \ldots, s_{i_m}$, and z, such that one of these 1's is in each row of the matrix. As will be seen, this can be done by using a variation of the Gauss-Jordan algorithm. If the first m constants in the last column of the matrix are nonnegative, we call this matrix a *vertex matrix* for the system.* The bound variables $x_{i_1}, \ldots, x_{i_r}$, $s_{i_{r+1}}, \ldots, s_{i_m}$ are called *basic*, and the remaining, free, variables are called *nonbasic*. By setting the nonbasic variables to 0 and solving for the basic variables, we get a feasible solution called a *basic feasible solution.*

EXAMPLE 13

Using the linear program system for the linear program of Example 12,

$$\begin{array}{c} \\ s_1 \\ s_2 \\ s_3 \\ z \end{array} \begin{array}{cccccc|c} x_1 & x_2 & s_1 & s_2 & s_3 & z & \\ 1 & 1 & 1 & 0 & 0 & 0 & 44 \\ 3 & 2 & 0 & 1 & 0 & 0 & 110 \\ 2 & 1 & 0 & 0 & 1 & 0 & 70 \\ \hline -5 & -3 & 0 & 0 & 0 & 1 & 0 \end{array}$$

we have a vertex matrix with basic variables s_1, s_2, s_3, and z listed to its left. Setting the nonbasic variables x_1 and x_2 equal to 0 yields $s_1 = 44$, $s_2 = 110$, $s_3 = 70$, $z = 0$ and the basic feasible solution $[0, 0, 44, 110, 70, 0]$. Hence the feasible solution corresponding to this vertex matrix is $[0, 0]$, the vertex labeled 1 in Figure 9.16, where $z = 0$.

* The vertex matrix is also called a *simplex tableau* or even a *tableau*. We prefer the term *vertex matrix* since it relates to the corresponding geometrical solution.

Solving for x_2, s_2, s_3, and z by applying $R_2 - 2R_1 \to R_2, R_3 - R_1 \to R_3, R_4 + 3R_1 \to R_4$ yields the vertex matrix

$$\begin{array}{c} \\ x_2 \\ s_2 \\ s_3 \\ z \end{array} \begin{array}{c} \begin{array}{cccccc} x_1 & x_2 & s_1 & s_2 & s_3 & z \end{array} \\ \left[\begin{array}{rrrrrr|r} 1 & 1 & 1 & 0 & 0 & 0 & 44 \\ 1 & 0 & -2 & 1 & 0 & 0 & 22 \\ 1 & 0 & -1 & 0 & 1 & 0 & 26 \\ \hline -2 & 0 & 3 & 0 & 0 & 1 & 132 \end{array}\right] \end{array}$$

with basic variables x_2, s_2, s_3, and z. Setting nonbasic variables x_1 and s_1 to 0 yields $x_2 = 44$, $s_2 = 22$, $s_3 = 26$, $z = 132$ and the basic feasible solution $[0, 44, 0, 22, 26, 132]$. The corresponding feasible solution to the linear program is $[0, 44]$, the vertex labeled 2 in Figure 9.16, where $z = 132$.

Solving for x_1, s_2, s_3, and z yields

$$\begin{array}{c} \\ x_1 \\ s_2 \\ s_3 \\ z \end{array} \begin{array}{c} \begin{array}{cccccc} x_1 & x_2 & s_1 & s_2 & s_3 & z \end{array} \\ \left[\begin{array}{rrrrrr|r} 1 & 1 & 1 & 0 & 0 & 0 & 44 \\ 0 & -1 & -3 & 1 & 0 & 0 & -22 \\ 0 & -1 & -2 & 0 & 1 & 0 & -18 \\ \hline 0 & 2 & 5 & 0 & 0 & 1 & 220 \end{array}\right] \end{array}$$

Since the 2,7-entry is -22, this matrix is not a vertex matrix. □

It can be shown that for linear programs with two variables, a basic feasible solution always yields a vertex in the feasible set, as it did in the previous example.* We should also note here that, after a linear program system is solved for a specified collection of variables, a necessary and sufficient condition for this solution to be a basic feasible solution is that the first m constants in the last column of the matrix are nonnegative.

In the next section we will show that if a linear program system has an optimal solution, then at least one such solution is a basic feasible solution. We will further show that there are only finitely many basic feasible solutions. Thus, when an optimal solution exists, if we compute all of the basic feasible solutions we can select the optimal solution from this set, just as we did in Section 9.1.

EXAMPLE 14

Using the linear program from Example 12, we have the linear program system

$$\begin{array}{c} \\ s_1 \\ s_2 \\ s_3 \\ z \end{array} \begin{array}{c} \begin{array}{cccccc} x_1 & x_2 & s_1 & s_2 & s_3 & z \end{array} \\ \left[\begin{array}{rrrrrr|r} 1 & 1 & 1 & 0 & 0 & 0 & 44 \\ 3 & 2 & 0 & 1 & 0 & 0 & 110 \\ 2 & 1 & 0 & 0 & 1 & 0 & 70 \\ \hline -5 & -3 & 0 & 0 & 0 & 1 & 0 \end{array}\right] \end{array}$$

* Thus a reasonable alternative name for a basic feasible solution, though not used in the literature, is vertex solution. Such a choice would have much greater geometric appeal.

We solve for x_1, x_2, s_1, z; x_1, x_2, s_2, z; x_1, x_2, s_3, z; x_1, s_1, s_2, z; x_1, s_1, s_3, z; x_1, s_2, s_3, z; x_2, s_1, s_2, z; x_2, s_1, s_3, z; x_2, s_2, s_3, z; and s_1, s_1, s_3, z. The solutions are $[30, 10, 4, 0, 0, 180]$, $[26, 18, 0, -4, 0, 184]$, $[22, 22, 0, 0, 4, 176]$, $[35, 0, 9, 5, 0, 175]$, $[\frac{110}{3}, 0, \frac{22}{3}, 0, -\frac{10}{3}, \frac{550}{3}]$, $[44, 0, 0, -22, -18, 220]$, $[0, 70, -26, -30, 0, 210]$, $[0, 55, -11, 0, 15, 165]$, $[0, 44, 0, 22, 26, 132]$, and $[0, 0, 44, 110, 70, 0]$, respectively. We eliminate the second, fifth, sixth, seventh, and eighth of these, since the negative numbers among the first five entries show that they are not feasible solutions. Among the remaining solutions, $[30, 10, 4, 0, 0, 180]$ has the largest z value. It follows, as we saw in Example 12, that this solution is an optimal solution of the linear program system. □

This method can require computation of hundreds of solutions to the system, making the approach impractical. To eliminate this necessity, we will give a strategy for selecting a sequence of vertex matrices having *adjacent* basic feasible solutions—that is, solutions differing in exactly one basic and one nonbasic variable—with increasing z values. Thus, many basic feasible solutions need not be computed. The adjacent feasible solutions will be computed systematically until an optimal solution is found, as indicated in Figure 9.17.

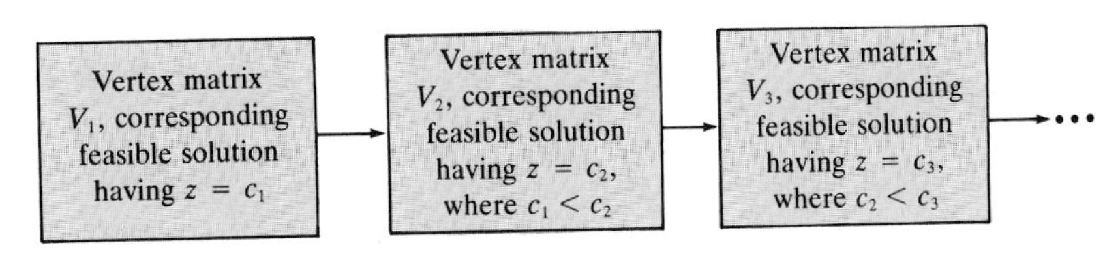

FIGURE 9.17

As we will prove in the next section, an optimal solution is obtained when the above sequence reaches a vertex matrix having exclusively nonnegative entries in the first $m + n$ positions of the last row.

EXAMPLE 15

The vertex matrix

$$\begin{array}{c} \\ s_1 \\ x_2 \\ x_1 \\ z \end{array} \begin{array}{c} \begin{array}{cccccccc} x_1 & x_2 & s_1 & s_2 & s_3 & z & & \end{array} \\ \left[\begin{array}{cccccc|c} 0 & 0 & 1 & -1 & 1 & 0 & 4 \\ 0 & 1 & 0 & 2 & -3 & 0 & 10 \\ 1 & 0 & 0 & -1 & 2 & 0 & 30 \\ \hline 0 & 0 & 0 & 1 & 1 & 1 & 180 \end{array}\right] \end{array}$$

can be obtained by a sequence of Gaussian operations from the linear program system for Example 12. Since the coefficients of the variables in the last row are all nonnegative, this vertex matrix yields an optimal solution which can be obtained by setting the nonbasic variables s_2 and s_3 to 0 and solving for the basic variables to get $x_1 = 30$, $x_2 = 10$, $s_1 = 4$, and $z = 180$. Note that increasing either of the nonbasic variables could only have the effect of reducing z, since $z = 180 - s_2 - s_3$. Hence the optimal solution to the system is $[30, 10, 4, 0, 0, 180]$ and to the linear program is $[30, 10]$ with optimal value $z = 180$. □

For a vertex matrix that has a negative coefficient in the last row, we will now show how to find an adjacent feasible solution with a z value at least as large as that already seen. To do this, we find a nonbasic variable, called the *entering variable*, which we make basic in the next vertex matrix, and a basic variable, called the *departing variable*, which we make nonbasic.

To describe the strategy, let

$$C = \left[\begin{array}{c|c} A & \mathbf{b} \\ \hline \mathbf{a} & c \end{array}\right]$$

be the given vertex matrix, where A is $m \times (n + m + 1)$.

CHOOSING THE ENTERING NONBASIC VARIABLE

If $a_i \geqslant 0$ for all i, we have a vertex matrix from which an optimal solution can be obtained. If $a_k < 0$ for some k, then choose the nonbasic variable in a column j for which $a_j < 0$ and $|a_j|$ is largest.* (This choice is made so as to find the adjacent basic feasible solution with the largest z value. See exercise 27 for the reasoning behind this choice.)

EXAMPLE 16

For the vertex matrix

$$\begin{array}{c} \\ s_1 \\ s_2 \\ s_3 \\ z \end{array}
\begin{array}{c} \begin{array}{cccccc} x_1 & x_2 & s_1 & s_2 & s_3 & z \end{array} \\
\left[\begin{array}{cccccc|c} 1 & 1 & 1 & 0 & 0 & 0 & 44 \\ 3 & 2 & 0 & 1 & 0 & 0 & 110 \\ 2 & 1 & 0 & 0 & 1 & 0 & 70 \\ \hline -5 & -3 & 0 & 0 & 0 & 1 & 0 \end{array}\right] \end{array}$$

of Examples 12 and 13, -5 is the most negative number in $\mathbf{a} = [-5, -3, 0, 0, 0, 1]$, so x_1 is chosen as the entering nonbasic variable. □

CHOOSING THE DEPARTING BASIC VARIABLE

Suppose the jth variable has been chosen as the entering nonbasic variable. For all $a_{kj} > 0$, form the ratios $b_k/a_{kj} \geqslant 0$. If b_i/a_{ij} is the smallest such ratio, the basic variable determined by the ith row is chosen as the departing variable.* (This choice ensures that the b_k's remain nonnegative, so that the matrix obtained by solving the equations for the new basic variable is a vertex matrix. See exercise 28 for the reasoning behind this.) The entry a_{ij} is called the *pivot*. If $a_{kj} \leqslant 0$ for all k, then there is no maximum value of z (exercise 29).

EXAMPLE 17

In Example 16, the chosen entering nonbasic variable was x_1. Forming the ratios b_k/a_{k1} gives $b_1/a_{11} = 44$, $b_2/a_{21} = 110/3 \approx 36.7$, and $b_3/a_{31} = 35$.

* If there are several, choose any one of them.

Since b_3/a_{31} is smallest, s_3, the basic variable determined by row 3, is chosen as the departing basic variable. □

In computing the next vertex matrix in a sequence, we want the chosen entering nonbasic variable to be basic and the chosen departing basic variable to be nonbasic. We obtain this matrix as follows. Since the chosen smallest ratio b_i/a_{ij} has the property that the pivot $a_{ij} > 0$, we use this entry and Gaussian operations to obtain 0's in all other entries of this column. Then we divide row i by a_{ij}, obtaining a 1 as the i,j-entry.*

EXAMPLE 18

Continuing Example 16 and circling the pivot a_{ij}, we have

$$
\begin{array}{c} \\ s_1 \\ s_2 \\ s_3 \\ z \end{array}
\begin{array}{c}
\begin{array}{cccccc|c} x_1 & x_2 & s_1 & s_2 & s_3 & z & \end{array} \\
\left[\begin{array}{cccccc|c}
1 & 1 & 1 & 0 & 0 & 0 & 44 \\
3 & 2 & 0 & 1 & 0 & 0 & 110 \\
\textcircled{2} & 1 & 0 & 0 & 1 & 0 & 70 \\
\hline
-5 & -3 & 0 & 0 & 0 & 1 & 0
\end{array}\right]
\end{array}
$$

$$
\xrightarrow[R_4+\frac{5}{2}R_3 \to R_4]{\substack{R_1-\frac{1}{2}R_3 \to R_1 \\ R_2-\frac{3}{2}R_3 \to R_2}}
\begin{array}{c}
\begin{array}{cccccc|c} x_1 & x_2 & s_1 & s_2 & s_3 & z & \end{array} \\
\left[\begin{array}{cccccc|c}
0 & \frac{1}{2} & 1 & 0 & -\frac{1}{2} & 0 & 9 \\
0 & \frac{1}{2} & 0 & 1 & -\frac{3}{2} & 0 & 5 \\
\textcircled{2} & 1 & 0 & 0 & 1 & 0 & 70 \\
\hline
0 & -\frac{1}{2} & 0 & 0 & \frac{5}{2} & 1 & 175
\end{array}\right]
\end{array}
$$

$$
\xrightarrow{\frac{1}{2}R_3 \to R_3}
\begin{array}{c} \\ s_1 \\ s_2 \\ x_1 \\ z \end{array}
\begin{array}{c}
\begin{array}{cccccc|c} x_1 & x_2 & s_1 & s_2 & s_3 & z & \end{array} \\
\left[\begin{array}{cccccc|c}
0 & \frac{1}{2} & 1 & 0 & -\frac{1}{2} & 0 & 9 \\
0 & \frac{1}{2} & 0 & 1 & -\frac{3}{2} & 0 & 5 \\
1 & \frac{1}{2} & 0 & 0 & \frac{1}{2} & 0 & 35 \\
\hline
0 & -\frac{1}{2} & 0 & 0 & \frac{5}{2} & 1 & 175
\end{array}\right]
\end{array}
$$

as the next vertex matrix. □

Summarizing the above remarks yields an algorithm for solving a linear program system.

THE SIMPLEX ALGORITHM

For a given linear program, write out the linear program system. Let $C = \left[\begin{array}{c|c} A & \mathbf{b} \\ \hline \mathbf{a} & c \end{array}\right]$ be the vertex matrix for this system.

1. If $a_i \geqslant 0$ for all i, then the corresponding basic feasible solution is an optimal solution. If $a_i < 0$ for some i, then choose the most negative a_j. The nonbasic variable for this column is chosen to be made basic in the next vertex matrix.

* Of course, you can divide row i by a_{ij} first and then apply the remaining Gaussian operations.

2. Compute all ratios b_k/a_{kj} for which $a_{kj} > 0$. If there are no such ratios, there is no largest z and the linear program has no optimal solution. Otherwise, choose any smallest such ratio, say b_i/a_{ij}. The basic variable for this row is chosen to be made nonbasic in the next vertex matrix.

3. Use the entry a_{ij} and Gaussian operations to obtain the next vertex matrix $C' = \left[\begin{array}{c|c} A' & \mathbf{b}' \\ \hline \mathbf{a}' & c' \end{array}\right]$. (Make a partial check on your work by verifying that $b_i \geqslant 0$ for all i and that $c' \geqslant c$.)

4. Go to Step 1 with $C = C'$.

Applying this algorithm yields a sequence of vertex matrices with nondecreasing z. If no $b_i = 0$ arises in the row of a departing basic variable, then z increases in each vertex matrix of the sequence. Since there are only finitely many such matrices, we must eventually obtain a largest z and thus a solution to the linear program system.

If some $b_k = 0$ at some iteration of the algorithm, we say the problem has become *degenerate*. If the pivot is in row k in the next iteration, z will not change. In this event, the algorithm continues, and at a later iteration an increase in z may again occur. In some cases, however, a cycling of basic feasible solutions may occur, so the largest value of z is not obtained (exercise 3 of Section 9.3). These cases, however, are not known to have arisen in practice. Means of solving such problems exist (see References), but they are beyond the scope of this book.

EXAMPLE 19

Solve the linear program given in Example 12:

$$\begin{aligned} \max \quad & z = 5x_1 + 3x_2 \\ \text{s.t.} \quad & x_1 + x_2 \leqslant 44 \\ & 3x_1 + 2x_2 \leqslant 110 \\ & 2x_1 + x_2 \leqslant 70 \\ & x_1 \geqslant 0, x_2 \geqslant 0 \end{aligned}$$

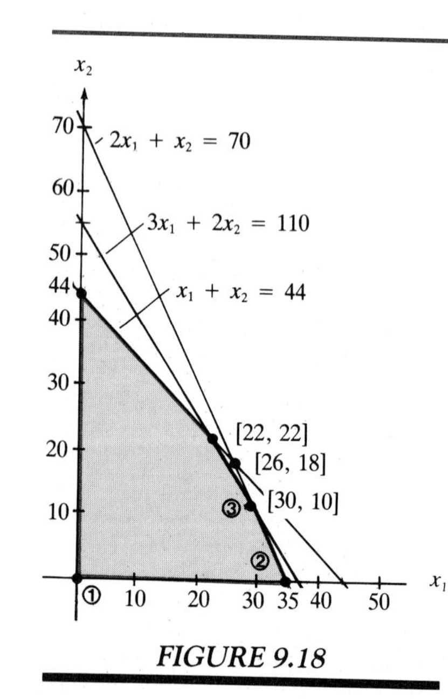

FIGURE 9.18

The feasible set for this linear program is shown in Figure 9.18.

In Example 13 we found the first vertex matrix for this program, and in Examples 16 through 18 we applied Steps 1, 2, and 3 of the simplex algorithm to this vertex matrix to obtain

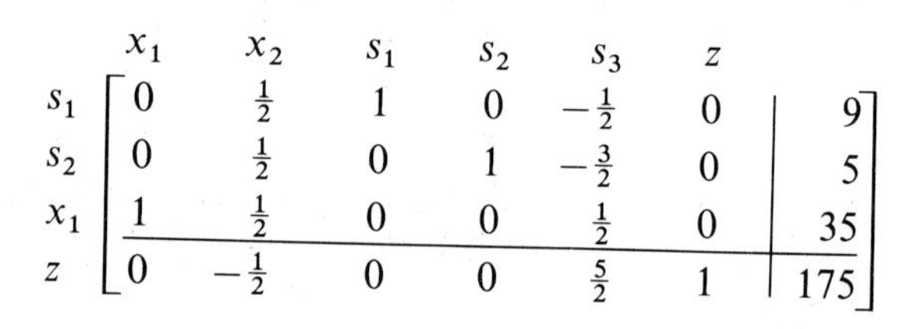

$$\begin{array}{c} \\ s_1 \\ s_2 \\ x_1 \\ z \end{array} \begin{array}{c} \begin{array}{cccccc|c} x_1 & x_2 & s_1 & s_2 & s_3 & z & \end{array} \\ \left[\begin{array}{cccccc|c} 0 & \frac{1}{2} & 1 & 0 & -\frac{1}{2} & 0 & 9 \\ 0 & \frac{1}{2} & 0 & 1 & -\frac{3}{2} & 0 & 5 \\ 1 & \frac{1}{2} & 0 & 0 & \frac{1}{2} & 0 & 35 \\ \hline 0 & -\frac{1}{2} & 0 & 0 & \frac{5}{2} & 1 & 175 \end{array}\right] \end{array}$$

The corresponding basic feasible solution is [35, 0, 9, 5, 0, 175]. Thus, [35, 0] is the feasible solution at the vertex labeled 2 in Figure 9.18, where $z = 175$.

Now, continuing with the algorithm, we see that $-\frac{1}{2}$ is the only negative term in the last row. Thus x_2 is the entering variable. Calculating ratios, we have $9/\frac{1}{2} = 18$, $5/\frac{1}{2} = 10$, and $35/\frac{1}{2} = 70$. Since 10 is the smallest ratio, s_2 is the departing variable and a_{22} is the pivot. Applying $R_1 - R_2 \to R_1$, $R_3 - R_2 \to R_3$, $R_4 + R_2 \to R_4$, and $2R_2 \to R_2$, we obtain

$$\begin{array}{c} \\ s_1 \\ x_2 \\ x_1 \\ z \end{array} \begin{array}{c} \begin{array}{cccccc|c} x_1 & x_2 & s_1 & s_2 & s_3 & z & \end{array} \\ \left[\begin{array}{cccccc|c} 0 & 0 & 1 & -1 & 1 & 0 & 4 \\ 0 & 1 & 0 & 2 & -3 & 0 & 10 \\ 1 & 0 & 0 & -1 & 2 & 0 & 30 \\ \hline 0 & 0 & 0 & 1 & 1 & 1 & 180 \end{array}\right] \end{array}$$

Since no $a_i < 0$, this vertex matrix yields the optimal solution $[30, 10, 4, 0, 0, 180]$. Hence $[30, 10]$, labeled 3 in Figure 9.18, is an optimal solution and $z = 180$ is the optimal value for the linear program. □

EXAMPLE 20

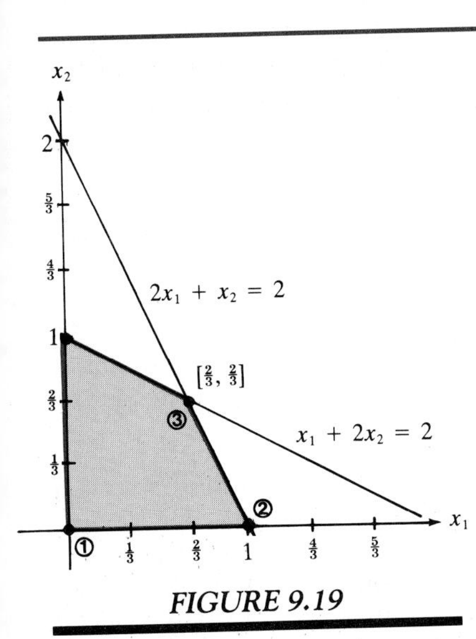

FIGURE 9.19

Solve

$$\begin{aligned} \max \quad & z = x_1 + x_2 \\ \text{s.t.} \quad & 2x_1 + x_2 \leqslant 2 \\ & x_1 + 2x_2 \leqslant 2 \\ & x_1 \geqslant 0, x_2 \geqslant 0 \end{aligned}$$

whose feasible set is shown in Figure 9.19.

The linear program system is

$$\begin{aligned} 2x_1 + x_2 + s_1 \qquad\qquad &= 2 \\ x_1 + 2x_2 \qquad + s_2 \qquad &= 2 \\ -x_1 - x_2 \qquad\qquad + z &= 0 \end{aligned}$$

or, as an augmented matrix,

$$\begin{array}{c} \\ s_1 \\ s_2 \\ z \end{array} \begin{array}{c} \begin{array}{ccccc|c} x_1 & x_2 & s_1 & s_2 & z & \end{array} \\ \left[\begin{array}{ccccc|c} 2 & 1 & 1 & 0 & 0 & 2 \\ 1 & 2 & 0 & 1 & 0 & 2 \\ \hline -1 & -1 & 0 & 0 & 1 & 0 \end{array}\right] \end{array}$$

The basic variables are s_1, s_2, and z. The corresponding basic feasible solution is $[0, 0, 2, 2, 0]$. The feasible solution $[0, 0]$ is the vertex labeled 1 in Figure 9.19, where $z = 0$.

The -1's in the last row are the most negative a_i. Arbitrarily, we choose x_1 as the entering nonbasic variable. Calculating ratios yields $b_1/a_{11} = 1$ and $b_2/a_{21} = 2$. Hence s_1 is the departing basic variable. Now, applying

$R_2 - \frac{1}{2}R_1 \to R_2$ and $R_3 + \frac{1}{2}R_1 \to R_3$ gives

$$\begin{array}{ccccc|c} x_1 & x_2 & s_1 & s_2 & z & \\ 2 & 1 & 1 & 0 & 0 & 2 \\ 0 & \frac{3}{2} & -\frac{1}{2} & 1 & 0 & 1 \\ \hline 0 & -\frac{1}{2} & \frac{1}{2} & 0 & 1 & 1 \end{array}$$

and applying $\frac{1}{2}R_1 \to R_1$ gives

$$\begin{array}{c|ccccc|c} & x_1 & x_2 & s_1 & s_2 & z & \\ x_1 & 1 & \frac{1}{2} & \frac{1}{2} & 0 & 0 & 1 \\ s_2 & 0 & \frac{3}{2} & -\frac{1}{2} & 1 & 0 & 1 \\ \hline z & 0 & -\frac{1}{2} & \frac{1}{2} & 0 & 1 & 1 \end{array}$$

which is the next vertex matrix. The corresponding basic feasible solution is $[1, 0, 0, 1, 1]$, which yields the vertex $[1, 0]$ labeled 2 in Figure 9.19, where $z = 1$. Since there is a negative a_j, the z value need not be optimal. Now, since $a_2 = -\frac{1}{2}$ is the only negative a_j, the entering variable is x_2. Computing ratios produces $b_1/a_{12} = 2$ and $b_2/a_{22} = \frac{2}{3}$. Thus the departing basic variable is s_2. Applying $R_1 - \frac{1}{3}R_2 \to R_1$, $R_3 + \frac{1}{3}R_2 \to R_3$, and $\frac{2}{3}R_2 \to R_2$ yields

$$\begin{array}{c|ccccc|c} & x_1 & x_2 & s_1 & s_2 & z & \\ x_1 & 1 & 0 & \frac{2}{3} & -\frac{1}{3} & 0 & \frac{2}{3} \\ x_2 & 0 & 1 & -\frac{1}{3} & \frac{2}{3} & 0 & \frac{2}{3} \\ \hline z & 0 & 0 & \frac{1}{3} & \frac{1}{3} & 1 & \frac{4}{3} \end{array}$$

Since each $a_i \geq 0$, we stop. Setting nonbasic variables s_1 and s_2 to 0 and solving for the basic variables, we find $x_1 = \frac{2}{3}$, $x_2 = \frac{2}{3}$, and $z = \frac{4}{3}$. Thus $[\frac{2}{3}, \frac{2}{3}, 0, 0, \frac{4}{3}]$ is an optimal solution to the linear program system, $[\frac{2}{3}, \frac{2}{3}]$ is an optimal solution to the linear program, and $z = \frac{4}{3}$ is the optimal value. The vertex $[\frac{2}{3}, \frac{2}{3}]$ is labeled 3 in Figure 9.19. □

In the next section we state and prove the results used in this section, verifying the simplex algorithm.

EXERCISES FOR SECTION 9.2

COMPUTATIONAL EXERCISES

In exercises 1 through 4, convert the given linear program to a linear program system, and then write the system in augmented matrix form.

1. max $z = 3x_1 + x_2 - 4x_3$
s.t. $x_1 + x_2 - x_3 \leq 5$
$2x_1 - x_2 + 3x_3 \leq 7$
$4x_1 + 2x_2 + x_3 \leq 8$
$x_1 \geq 0, x_2 \geq 0, x_3 \geq 0$

2. max $z = 2x_2 - x_3 + x_4$
s.t. $x_1 - x_3 \leq 12$
$x_2 + 4x_4 \leq 18$
$x_1 + x_2 + x_3 \leq 12$
$x_1 \geq 0, x_2 \geq 0, x_3 \geq 0, x_4 \geq 0$

3. max $z = x_1 + x_2 + x_3 + x_4$
s.t. $2x_1 - 3x_2 + x_4 \leq 17$
$3x_1 + 5x_2 - x_3 + 2x_4 \leq 12$
$-x_1 + 2x_2 + x_3 \leq 8$
$x_1 \geq 0, x_2 \geq 0, x_3 \geq 0, x_4 \geq 0$

4. max $z = x_1 + x_2 + x_3$

s.t.
$$\begin{aligned} 2x_1 + 3x_2 - x_3 &\leq 8 \\ -x_1 + 4x_2 + x_3 &\leq 10 \\ 3x_1 - x_2 + 3x_3 &\leq 11 \\ x_1 \geq 0, x_2 \geq 0, x_3 &\geq 0 \end{aligned}$$

In exercises 5 and 6, convert the given vertex matrix into a linear program system, and then convert the linear program system into a linear program.

5.
$$\begin{array}{ccccccc|c} x_1 & x_2 & x_3 & s_1 & s_2 & s_3 & z & \\ 1 & -2 & 0 & 1 & 0 & 0 & 0 & 12 \\ 2 & 1 & 3 & 0 & 1 & 0 & 0 & 14 \\ 1 & 0 & -1 & 0 & 0 & 1 & 0 & 4 \\ \hline -1 & 3 & -4 & 0 & 0 & 0 & 1 & 0 \end{array}$$

6.
$$\begin{array}{cccccccc|c} x_1 & x_2 & x_3 & x_4 & s_1 & s_2 & s_3 & z & \\ 1 & 1 & -1 & 1 & 1 & 0 & 0 & 0 & 5 \\ 1 & -1 & 1 & -1 & 0 & 1 & 0 & 0 & 8 \\ 2 & 0 & 3 & 0 & 0 & 0 & 1 & 0 & 12 \\ \hline -1 & -2 & -1 & -2 & 0 & 0 & 0 & 1 & 0 \end{array}$$

In exercises 7 and 8, write the basic feasible solution to the linear program system, the corresponding feasible solution to the linear program, and the value of the objective function at that point.

7.
$$\begin{array}{cccccccc|c} x_1 & x_2 & x_3 & x_4 & s_1 & s_2 & s_3 & z & \\ 1 & 0 & -1 & 0 & 1 & 0 & 0 & 0 & 12 \\ -\frac{1}{4} & 0 & -\frac{1}{4} & 1 & 0 & \frac{1}{4} & -\frac{1}{4} & 0 & \frac{3}{2} \\ 1 & 1 & 1 & 0 & 0 & 0 & 1 & 0 & 12 \\ \hline \frac{7}{4} & 0 & \frac{11}{4} & 0 & 0 & \frac{1}{4} & \frac{7}{4} & 1 & \frac{51}{2} \end{array}$$

8.
$$\begin{array}{cccccccc|c} x_1 & x_2 & x_3 & x_4 & s_1 & s_2 & s_3 & z & \\ 0 & -\frac{19}{3} & \frac{2}{3} & -\frac{1}{3} & 1 & -\frac{2}{3} & 0 & 0 & 9 \\ 1 & \frac{5}{3} & -\frac{1}{3} & \frac{2}{3} & 0 & \frac{1}{3} & 0 & 0 & 4 \\ 0 & \frac{11}{3} & \frac{2}{3} & \frac{2}{3} & 0 & \frac{1}{3} & 1 & 0 & 12 \\ \hline 0 & \frac{2}{3} & -\frac{4}{3} & -\frac{1}{3} & 0 & \frac{1}{3} & 0 & 1 & 4 \end{array}$$

In exercises 9 and 10, write the corresponding linear program system. Decide whether each given point is a feasible solution to the linear program. If it is, find the corresponding value of the objective function z. Then, using Theorem 9.1, write out the corresponding feasible solution to the linear program system.

9. max $z = -x_1 - 2x_2 - x_3$

s.t.
$$\begin{aligned} x_1 + 2x_2 + x_3 &\leq 7 \\ x_1 - 2x_2 + x_3 &\leq 6 \\ 2x_1 + x_2 - x_3 &\leq 4 \\ x_1 \geq 0, x_2 \geq 0, x_3 &\geq 0 \end{aligned}$$

Points $[1, 1, 0]$, $[1, -1, 1]$, $[1, 2, 1]$

10. max $z = 2x_1 - x_2 + x_3$

s.t.
$$\begin{aligned} x_1 + x_2 + x_3 &\leq 5 \\ x_2 + x_3 &\leq 4 \\ x_1 + x_2 \quad &\leq 3 \\ x_1 \geq 0, x_2 \geq 0, x_3 &\geq 0 \end{aligned}$$

Points $[1, 1, 1]$, $[1, 2, 0]$, $[0, 2, 3]$

In exercises 11 through 14, write the augmented matrix for the linear program system and solve the system for the variables indicated, as in Example 13. State whether the final matrix in the calculation is a vertex matrix.

11.
$$\begin{aligned} 2x_1 + 3x_2 + s_1 \qquad\qquad &= 2 \\ x_1 + 2x_2 \qquad + s_2 \qquad &= 2 \\ x_1 - x_2 \qquad\qquad + z &= 0 \end{aligned}$$
for x_1, x_2, z

12.
$$\begin{aligned} 3x_1 - x_2 + s_1 \qquad\qquad &= 3 \\ x_1 + 2x_2 \qquad + s_2 \qquad &= 2 \\ x_1 + x_2 \qquad\qquad + z &= -2 \end{aligned}$$
for x_1, x_2, z

13.
$$\begin{aligned} x_1 - x_2 + s_1 \qquad\qquad &= 1 \\ x_1 + x_2 \qquad + s_2 \qquad &= 2 \\ 2x_1 - 3x_2 \qquad\qquad + z &= 1 \end{aligned}$$
for x_1, s_2, z

14.
$$\begin{aligned} -x_1 + 2x_2 + s_1 \qquad\qquad &= -1 \\ x_1 + x_2 \qquad + s_2 \qquad &= 0 \\ 2x_1 + 4x_2 \qquad\qquad + z &= 2 \end{aligned}$$
for x_2, s_1, z

In exercises 15 through 20, graph the feasible region for the linear program. Then solve the program by the simplex algorithm. Indicate the computed vertex for each vertex matrix by numbering the vertices consecutively on the graph. Using graph paper will help you keep rows and columns in order during calculations.

15. max $z = x_1 + x_2$

s.t.
$$\begin{aligned} 4x_1 + 7x_2 &\leq 28 \\ x_1 \geq 0, x_2 &\geq 0 \end{aligned}$$

16. max $z = x_1 + 3x_2$

s.t.
$$\begin{aligned} 2x_1 + x_2 &\leq 14 \\ x_1 \geq 0, x_2 &\geq 0 \end{aligned}$$

17. max $z = x_1 + 3x_2$

s.t.
$$\begin{aligned} -2x_1 + x_2 &\leq 0 \\ x_1 + x_2 &\leq 15 \\ x_1 \geq 0, x_2 &\geq 0 \end{aligned}$$

18. max $z = 3x_1 - x_2$

s.t.
$$\begin{aligned} x_1 + x_2 &\leq 24 \\ x_1 - 2x_2 &\leq 0 \\ x_1 \geq 0, x_2 &\geq 0 \end{aligned}$$

19. max $z = 3x_1 + 2x_2$

s.t.
$$\begin{aligned} -3x_1 + x_2 &\leq 0 \\ x_1 + x_2 &\leq 24 \\ 2x_1 + x_2 &\leq 35 \\ x_1 - 2x_2 &\leq 0 \\ x_1 \geq 0, x_2 &\geq 0 \end{aligned}$$

20. max $z = 5x_1 + x_2$

s.t.
$$\begin{aligned} x_1 - 3x_2 &\leq 0 \\ 3x_1 - x_2 &\leq 72 \\ x_1 + x_2 &\leq 48 \\ -2x_1 + x_2 &\leq 0 \\ x_1 \geq 0, x_2 &\geq 0 \end{aligned}$$

In exercises 21 through 24, solve by using the simplex algorithm.

21. max $z = 8x_1 + 4x_2 - 6x_3$

s.t.
$$\begin{aligned} x_1 + 2x_2 + x_3 &\leq 6 \\ 2x_1 \qquad + x_3 &\leq 4 \\ x_1 \geq 0, x_2 \geq 0, x_3 &\geq 0 \end{aligned}$$

22. max $z = 12x_1 - 18x_2 + 4x_3$
s.t.
$$x_1 + x_2 \leqslant 5$$
$$x_1 + x_2 + x_3 \leqslant 8$$
$$x_1 \geqslant 0, x_2 \geqslant 0, x_3 \geqslant 0$$

23. max $z = x_1 + x_2 + 2x_3 + x_4$
s.t.
$$x_1 + x_2 + x_3 \leqslant 3$$
$$x_1 + x_3 + x_4 \leqslant 7$$
$$x_1 + x_2 + x_4 \leqslant 3$$
$$x_1 + x_2 + x_3 + x_4 \leqslant 5$$
$$x_1 \geqslant 0, x_2 \geqslant 0, x_3 \geqslant 0, x_4 \geqslant 0$$

24. max $z = x_1 + 2x_2 + x_3 + 3x_5$
s.t.
$$x_1 + x_2 + x_5 \leqslant 5$$
$$x_1 + x_3 \leqslant 1$$
$$x_2 + x_4 \leqslant 3$$
$$x_2 + x_5 \leqslant 3$$
$$x_1 + x_4 \leqslant 4$$
$$x_1 \geqslant 0, x_2 \geqslant 0, x_3 \geqslant 0, x_4 \geqslant 0, x_5 \geqslant 0$$

THEORETICAL EXERCISES

25. Explain why Theorem 9.1 is true, using the linear program of Example 12.

26. Prove Theorem 9.1.

27. Explain the rationale behind the choice of max $|a_k|$ in the strategy for choosing the entering variable. *Hint:* The choice of the largest $|a_k|$ is not essential for the working of the algorithm. However, it improves the chances of finding the next basic feasible solution for which the corresponding value of z is largest. Write $z = -a_1x_1 - \cdots - a_{m+n}s_m$ and explain why this choice of the entering basic variable seems intuitively to be a good one for obtaining this result.

28. Explain why our choice of the departing basic variable leads to a next vertex matrix. To do this, write the ith equation of the linear program system in the form

$$y = b_i - a_{ik_1}x_{k_1} - \cdots - a_{ik_r}x_{k_r} - a_{il_1}s_{l_1} - \cdots - a_{il_t}s_{l_t}$$

where y is the basic variable determined by that equation and $x_{k_1}, x_{k_2}, \ldots, x_{k_r}, s_{l_1}, \ldots, s_{l_t}$ are the nonbasic variables and so have been set to 0. Recall that the entering, or jth, variable is one of these variables. *Hint:* In the ith constraint of the linear program, if $a_{ij} > 0$, what is the largest amount by which we could increase the value of the jth variable without leaving the feasible set? What is the effect of increasing the value of the jth variable in a constraint in which $a_{ij} \leqslant 0$?

29. Show that if $a_j < 0$ and $a_{kj} \leqslant 0$ for all $k = 1, 2, \ldots, m$, then z can be increased without bound.

APPLICATIONS EXERCISES

30. A small construction company makes three types of houses, A, B, and C. The houses are constructed in three phases, framing (including sheathing), roofing, and exterior painting. The time requirements for each phase as well as the profit for each type of house are given in Table 9.11.

TABLE 9.11

PHASES	WORKER-HOURS REQUIRED PER HOUSE A	B	C	WORKER-HOURS AVAILABLE PER WEEK
Framing	200	100	100	500
Roofing	50	100	50	250
Painting	50	50	0	100
Profit	\$4000	\$3000	\$2000	

Find the average number of each type of house the company should construct in a week in order to receive the largest profit. *Hint:* It is acceptable to scale the inequalities in the linear program with a positive constant before starting the simplex algorithm.

31. A manufacturer makes videorecorders, copiers, and computers in two phases, subassembly and finishing. The time requirements and profit per item are given in Table 9.12.

TABLE 9.12

PHASES	HOURS REQUIRED PER UNIT Videorecorder	Copier	Computer	TIME AVAILABLE PER WEEK (IN HOURS)
Subassembly	5	2	4	700
Finishing	2	$\frac{1}{2}$	$\frac{3}{2}$	240
Profit per Unit	\$50	\$65	\$75	

Determine the average number of videorecorders, copiers, and computers that should be made in a week in order to maximize the manufacturer's profit.

32. A 700-acre farm is to be planted in cotton, corn, peas, and clover; some part may be left fallow. The farm can get a loan of up to $450,000 to cover planting, cultivating, and harvesting expenses. The expenses and predicted income from each crop are listed in Table 9.13.

TABLE 9.13

	COTTON	CORN	PEAS	CLOVER
Expense per Acre	$1290	$870	$1100	$710
Income per Acre	$1700	$1200	$1520	$980

How many acres of each crop should be planted to obtain the largest profit? (Recall that profit = income − expenses.)

9.3 PROOF OF THE SIMPLEX ALGORITHM (OPTIONAL)

In this section we prove the results used in obtaining the simplex algorithm of the previous section. These results were as follows:

1. If a linear program system has an optimal solution, it has a basic feasible solution that is an optimal solution.

2. There are only finitely many basic feasible solutions.

3. The matrices produced by the simplex algorithm are vertex matrices.

4. The simplex algorithm never reduces the value of z.

5. The stopping criterion in Step 2 of the simplex algorithm is valid.

Recall that the simplex algorithm provides a method for solving the special linear program

$$\begin{array}{ll} \max & z = a_1x_1 + a_2x_2 + \cdots + a_nx_n \\ \text{s.t.} & a_{11}x_1 + a_{12}x_2 + \cdots + a_{1n}x_n \leqslant b_1 \\ & a_{21}x_1 + a_{22}x_2 + \cdots + a_{2n}x_n \leqslant b_2 \\ & \vdots \\ & a_{m1}x_1 + a_{m2}x_2 + \cdots + a_{mn}x_n \leqslant b_m \\ & x_1 \geqslant 0, x_2 \geqslant 0, \ldots, x_n \geqslant 0 \end{array}$$

where $b_1 \geqslant 0, b_2 \geqslant 0, \ldots, b_m \geqslant 0$. This is done by writing the linear program as the linear program system

$$\begin{array}{c} \begin{array}{cccccccccc} x_1 & x_2 & \cdots & x_n & s_1 & s_2 & \cdots & s_m & z & \end{array} \\ \left[\begin{array}{ccccccccc|c} a_{11} & a_{12} & \cdots & a_{1n} & 1 & 0 & \cdots & 0 & 0 & b_1 \\ a_{21} & a_{22} & \cdots & a_{2n} & 0 & 1 & \cdots & 0 & 0 & b_2 \\ \vdots & \vdots & & \vdots & \vdots & \vdots & & \vdots & \vdots & \vdots \\ a_{m1} & a_{m2} & \cdots & a_{mn} & 0 & 0 & \cdots & 1 & 0 & b_m \\ \hline -a_1 & -a_2 & \cdots & -a_n & 0 & 0 & \cdots & 0 & 1 & 0 \end{array}\right] \end{array}$$

which is a vertex matrix yielding the basic feasible solution

$$[0, \ldots, 0, b_1, \ldots, b_m, 0]$$

Theorem 9.2 is a statement of result 1.

THEOREM 9.2

If a linear program system has an optimal solution, it has a basic feasible solution that is an optimal solution.

PROOF

Choose an optimal solution **s**, with fewest nonzero entries, to the linear program system. By rearranging the variables if necessary, we can ensure that the variables among $x_1, \ldots, x_n, s_1, \ldots, s_m$ belonging to the nonzero entries correspond to the first r columns of the augmented matrix of the linear program system. Now we delete the last row of this matrix and apply the Gauss-Jordan algorithm to the resulting matrix to obtain an echelon form having **s** as its solution. Since the augmented matrix corresponding to the linear program system has rank $m + 1$, it follows that either the echelon form has I_m in the first m columns or one of the first m variables is free. In the first case, we see that the optimal solution **s** must be a basic feasible solution. We now show that the second case cannot occur.

Suppose variable j is the left-most free variable in the echelon form. Then the echelon form looks like

$$\begin{bmatrix} 1 & 0 & 0 & \cdots & 0 & c_{1j} \\ 0 & 1 & 0 & \cdots & 0 & c_{2j} \\ \vdots & \vdots & \vdots & & \vdots & \vdots \\ 0 & 0 & 0 & \cdots & 1 & c_{j-1,j} \\ 0 & 0 & 0 & \cdots & 0 & 0 \\ \vdots & \vdots & \vdots & & \vdots & \vdots \\ 0 & 0 & 0 & \cdots & 0 & 0 \end{bmatrix}$$

in the first j columns. In this case, we can find another optimal solution by decreasing the value of the variable in the jth column and adjusting the values of variables in columns 1 through $j - 1$ so that equality holds in equations 1 through $j-1$. It is clear that the values in equations j through m are not affected by this process. We end the adjustment process when one of the variables in columns 1 through j is sent to 0, thus producing an optimal solution with fewer nonzero entries. This is a contradiction of our choice of **s**; thus the first case is the only possible case, which proves the theorem. ■

By this theorem, a linear program system that has an optimal solution has one that is a basic feasible solution. We now give the descriptive features of a vertex matrix from which this basic feasible solution can be obtained, verifying result 5.

THEOREM 9.3

Let $\left[\begin{array}{c|c} A & \mathbf{b} \\ \hline \mathbf{a} & c \end{array}\right]$ be a vertex matrix for a linear program system, where A is an $m \times (m + n + 1)$ matrix, $\mathbf{a} \in \mathbf{R}_{m+n+1}$, $\mathbf{b} \in \mathbf{R}^m$, and $c \in \mathbf{R}$. If $a_1 \geqslant 0, \ldots, a_{m+n+1} \geqslant 0$, then the corresponding basic feasible solution is an optimal solution to the linear program system.

PROOF

The only equation involving z is the last equation. In this equation, all variables have nonnegative coefficients, with the coefficients for the basic variables other than z being 0. Thus the solution with largest z is obtained by setting the nonbasic variables to 0 and solving for the basic variables. ■

We found an optimal solution for a linear program system by searching through the vertex matrices for one that fits the description in the theorem. As the proof of the next theorem shows, verifying result 2, this search is done on a finite set.

THEOREM 9.4

For a given linear program system, there are only finitely many different vertex matrices, and consequently only finitely many basic feasible solutions.

PROOF

Suppose first that we have two vertex matrices; for example,

$$[I_{m+1} \quad R] \qquad \text{and} \qquad [I_{m+1} \quad S]$$

where R and S are $(m + 1) \times (n + 1)$ matrices for a given linear program system. Consider the equations

$$I_{m+1}\mathbf{x} = \mathbf{r}^j \qquad \text{and} \qquad I_{m+1}\mathbf{x} = \mathbf{s}^j$$

Since $[I_{m+1} \quad S]$ can be obtained from $[I_{m+1} \quad R]$ by a sequence of Gaussian operations, the equations above have the same solution set:

$$\mathbf{x} = \mathbf{r}^j = \mathbf{s}^j$$

Since this is true for any j, $R = S$ and $[I_{m+1} \quad R] = [I_{m+1} \quad S]$

Now, using this same technique, we can show that if two vertex matrices have equal columns corresponding to basic variables, then the two vertex matrices are equal. Since there can be only finitely many sets of basic variables and each column for a basic variable has only m positions in which the 1 can be placed, it follows that there are only finitely many vertex matrices for each linear program system. ■

Let

$$C' = \left[\begin{array}{c|c} A' & \mathbf{b}' \\ \hline \mathbf{a}' & c' \end{array}\right]$$

be the matrix obtained from the vertex matrix

$$C = \left[\begin{array}{c|c} A & \mathbf{b} \\ \hline \mathbf{a} & c \end{array}\right]$$

by exchanging a basic and a nonbasic variable as we search through this finite set by the simplex algorithm. Some observations can be made about this new matrix. The proofs of these assumptions verify results 3 and 4.

THEOREM 9.5

With C and C' defined as above,

(*i*) $b'_k \geqslant 0$ for all k, and so C' is a vertex matrix.

(*ii*) $c' \geqslant c$ with equality only when $b_i = 0$; thus, if $b_i \neq 0$, C' gives a larger z value than does C.

PROOF

For (i), note that

$$b'_k = b_k - \frac{a_{kj}}{a_{ij}} b_i = b_k - a_{kj} \frac{b_i}{a_{ij}}$$

Thus, if $a_{kj} \leqslant 0$, the result follows. If $a_{kj} > 0$, then

$$b'_k = a_{kj}\left(\frac{b_k}{a_{kj}} - \frac{b_i}{a_{ij}}\right) \geqslant 0$$

by the choice of the ratio b_i/a_{ij}.

For (ii), note that

$$c' = c - \frac{a_j}{a_{ij}} b_i = c - a_j \frac{b_i}{a_{ij}}$$

Since $a_j < 0$ and $b_i/a_{ij} \geqslant 0$, $c' - c \geqslant 0$ with equality only if $b_i = 0$. ■

In the next section we will show how to solve general linear programming problems.

EXERCISES FOR SECTION 9.3

THEORETICAL EXERCISES

1. If the objective function of a linear program is

$$z = a_1 x_1 + a_2 x_2 + \cdots + a_n x_n$$

and if b is a nonzero constant, explain the effect on the solution to the linear program of replacing the objective function with

$$z = a_1 x_1 + a_2 x_2 + \cdots + a_n x_n + b$$

2. In exercise 30 of Section 9.2, it is claimed that scaling the constraints by a positive constant will have no effect on the solution to the linear program. Explain why.

3. Try to solve the following linear program:

$$\begin{array}{ll} \max & z = \frac{3}{4}x_1 - 150x_2 + \frac{1}{50}x_3 - 6x_4 \\ \text{s.t.} & \frac{1}{4}x_1 - 60x_2 - \frac{1}{25}x_3 + 9x_4 \leqslant 0 \\ & \frac{1}{2}x_1 - 90x_2 - \frac{1}{50}x_3 + 3x_4 \leqslant 0 \\ & x_3 \leqslant 1 \\ & x_1 \geqslant 0, x_2 \geqslant 0, x_3 \geqslant 0, x_4 \geqslant 0 \end{array}$$

When choices of entering variable and departing variable arise, use the left-most and the upper-most, respectively. (The simplex algorithm cycles from this problem. The problem, given by E. M. L. Beale in "Cycling in the Dual Simplex Algorithm," *Naval Research Logistics Quarterly* 2 (1955), 269–276, is arranged to have this property. No such cycling is known to have occurred in a problem that arose in practice, and so cycling appears to be extremely rare.)

9.4 Solving the General Linear Programming Problem

In this section we describe how to solve any linear program. The method used is the simplex algorithm given in Section 9.2, with certain adjustments. We will explain these various adjustments in subsections.

Adjusting a Minimization Problem to a Maximization Problem

The simplex algorithm as given in Section 9.2 and extended in this section is designed to solve a maximization problem. If the linear program is a minimization problem, we solve the problem by first converting it to a maximization problem as follows. Consider the linear program

$$
\begin{aligned}
\mathrm{P_m}: \min \quad & z = a_1x_1 + a_2x_2 + \cdots + a_nx_n \\
\text{s.t.} \quad & a_{11}x_1 + a_{12}x_2 + \cdots + a_{1n}x_n \leqslant b_1 \\
& \vdots \\
& a_{i1}x_1 + a_{i2}x_2 + \cdots + a_{in}x_n \leqslant b_i \\
& \vdots \\
& a_{j1}x_1 + a_{j2}x_2 + \cdots + a_{jn}x_n \geqslant b_j \\
& \vdots \\
& a_{k1}x_1 + a_{k2}x_2 + \cdots + a_{kn}x_n = b_k \\
& \vdots \\
& x_1 \geqslant 0, x_2 \geqslant 0, \ldots, x_n \geqslant 0
\end{aligned}
$$

Write the linear program

$$
\begin{aligned}
\mathrm{P_M}: \max \quad & z' = -a_1x_1 - a_2x_2 - \cdots - a_nx_n \\
\text{s.t.} \quad & a_{11}x_1 + a_{12}x_2 + \cdots + a_{1n}x_n \leqslant b_1 \\
& \vdots \\
& a_{i1}x_1 + a_{i2}x_2 + \cdots + a_{in}x_n \leqslant b_i \\
& \vdots \\
& a_{j1}x_1 + a_{j2}x_2 + \cdots + a_{jn}x_n \geqslant b_j \\
& \vdots \\
& a_{k1}x_1 + a_{k2}x_2 + \cdots + a_{kn}x_n = b_k \\
& \vdots \\
& x_1 \geqslant 0, x_2 \geqslant 0, \ldots, x_n \geqslant 0
\end{aligned}
$$

Note that $z' = -z$, the only change in expressions, is easily calculated.

It can be shown that the optimal solutions to $\mathrm{P_m}$ and $\mathrm{P_M}$ are precisely the same and that the optimal values z and z' for $\mathrm{P_m}$ and $\mathrm{P_M}$, respectively, are related by $-z = z'$ (exercise 23).

EXAMPLE 21

Solve

$$\begin{aligned} P_m\colon \min \quad & z = 2x_1 - x_2 \\ \text{s.t.} \quad & x_1 + x_2 \leqslant 1 \\ & x_1 - x_2 \leqslant 0 \\ & x_1 \geqslant 0, x_2 \geqslant 0 \end{aligned}$$

Here,

$$\begin{aligned} P_M\colon \max \quad & z' = -2x_1 + x_2 \\ \text{s.t.} \quad & x_1 + x_2 \leqslant 1 \\ & x_1 - x_2 \leqslant 0 \\ & x_1 \geqslant 0, x_2 \geqslant 0 \end{aligned}$$

We solve P_M by the simplex algorithm, as follows. The first vertex matrix is

$$\begin{array}{c} \\ s_1 \\ s_2 \\ z' \end{array} \begin{array}{c} \begin{array}{cccccc} x_1 & x_2 & s_1 & s_2 & z' & \end{array} \\ \left[\begin{array}{ccccc|c} 1 & 1 & 1 & 0 & 0 & 1 \\ 1 & -1 & 0 & 1 & 0 & 0 \\ \hline 2 & -1 & 0 & 0 & 1 & 0 \end{array}\right] \end{array}$$

Since $a_2 = -1$ is the only negative a_j, the entering nonbasic variable is x_2. Calculating ratios yields only $b_1/a_{12} = 1$. Thus s_1 is the departing basic variable. Applying $R_2 + R_1 \to R_2$ and $R_3 + R_1 \to R_3$, we obtain the vertex matrix

$$\begin{array}{c} \\ x_2 \\ s_2 \\ z' \end{array} \begin{array}{c} \begin{array}{cccccc} x_1 & x_2 & s_1 & s_2 & z' & \end{array} \\ \left[\begin{array}{ccccc|c} 1 & 1 & 1 & 0 & 0 & 1 \\ 2 & 0 & 1 & 1 & 0 & 1 \\ \hline 3 & 0 & 1 & 0 & 1 & 1 \end{array}\right] \end{array}$$

Since each $a_j \geqslant 0$, we stop. Setting the nonbasic variables x_1 and s_1 to 0 and solving for the basic variables, we get $x_2 = 1$, $s_2 = 1$, and $z' = 1$. Thus $[0, 1, 0, 1, 1]$ is an optimal solution, $[0, 1]$ is an optimal solution to P_M, and $z' = 1$ is the optimal value for P_M. It follows that $[0, 1]$ is an optimal solution to P_m, with $z = -z' = -1$ the optimal value.

ADJUSTING ARBITRARY CONSTRAINTS TO EQUATIONS IN THE LINEAR PROGRAM SYSTEM

In converting a constraint in a linear program into an equation in a linear program system, we first write the constraint so that the constant on the right side is nonnegative. This may require multiplying the constraint by -1. In the following remarks we will suppose this has already been done.

As shown in the previous sections, the constraint

$$a_{i1}x_1 + \cdots + a_{in}x_n \leqslant b_i$$

is adjusted by adding a slack variable $s_i \geqslant 0$ to get the equation

$$a_{i1}x_1 + \cdots + a_{in}x_n + s_i = b_i$$

In the constraint

$$a_{j1}x_1 + \cdots + a_{jn}x_n \geqslant b_j$$

the left side is larger than the right, so we subtract the *surplus variable* $S_j \geqslant 0$ to get

$$a_{j1}x_1 + \cdots + a_{jn}x_n - S_j = b_j$$

By converting these inequality constraints into equations as just described, we can convert a linear program

$$\begin{array}{ll} \text{P: max} & z = a_1x_1 + a_2x_2 + \cdots + a_nx_n \\ \text{s.t.} & a_{11}x_1 + a_{12}x_2 + \cdots + a_{1n}x_n \leqslant b_1 \\ & \qquad\vdots \\ & a_{i1}x_1 + a_{i2}x_2 + \cdots + a_{in}x_n \leqslant b_i \\ & \qquad\vdots \\ & a_{j1}x_1 + a_{j2}x_2 + \cdots + a_{jn}x_n \geqslant b_j \\ & \qquad\vdots \\ & a_{k1}x_1 + a_{k2}x_2 + \cdots + a_{kn}x_n = b_k \\ & \qquad\vdots \\ & x_1 \geqslant 0, x_2 \geqslant 0, \ldots, x_n \geqslant 0 \end{array}$$

into a linear program system

$$\begin{array}{ll} \text{PS:} & a_{11}x_1 + a_{12}x_2 + \cdots + a_{1n}x_n + s_1 = b_1 \\ & \qquad\vdots \\ & a_{i1}x_1 + a_{i2}x_2 + \cdots + a_{in}x_n + s_i = b_i \\ & \qquad\vdots \\ & a_{j1}x_1 + a_{j2}x_2 + \cdots + a_{jn}x_n - S_j = b_j \\ & \qquad\vdots \\ & a_{k1}x_1 + a_{k2}x_2 + \cdots + a_{kn}x_n = b_k \\ & \qquad\vdots \\ & -a_1x_1 - a_2x_2 - \cdots - a_nx_n + z = 0 \end{array}$$

It can be shown that $[x_1, \ldots, x_n, \ldots, s_i, \ldots, S_j, \ldots, z]$ is an optimal solution to PS if and only if $[x_1, \ldots, x_n]$ is an optimal solution to P with z the optimal value (exercise 24). Thus, to solve the linear program P we can solve the linear program system PS for an optimal solution.

To solve this system, however, we need a vertex matrix. As the following example demonstrates, the augmented matrix for PS may not be a vertex matrix.

EXAMPLE 22

Consider

$$\begin{aligned} \max \quad & z = 2x_1 - x_2 \\ \text{s.t.} \quad & x_1 + x_2 \leqslant 4 \\ & x_1 - x_2 \leqslant -1 \\ & x_1 \geqslant 0, x_2 \geqslant 0 \end{aligned}$$

Multiplying through the second constraint to obtain a positive b_2, we get $-x_1 + x_2 \geqslant 1$. Then writing the problem as a linear program system yields

$$\begin{aligned} x_1 + x_2 + s_1 \qquad\qquad &= 4 \\ -x_1 + x_2 \qquad - S_1 \qquad &= 1 \\ -2x_1 + x_2 \qquad\qquad + z &= 0 \end{aligned}$$

or

$$\begin{array}{c} \begin{matrix} x_1 & x_2 & s_1 & S_1 & z \end{matrix} \\ \left[\begin{array}{rrrrr|r} 1 & 1 & 1 & 0 & 0 & 4 \\ -1 & 1 & 0 & -1 & 0 & 1 \\ \hline -2 & 1 & 0 & 0 & 1 & 0 \end{array}\right] \end{array}$$

Note that this matrix is not a vertex matrix, since $a_{24} = -1$. If we set x_1 and x_2 to 0, we get $s_1 = 4$, $S_1 = -1$, $z = 0$. Since S_1 is not nonnegative, a feasible solution cannot be obtained in this manner. □

To obtain a starting vertex matrix, we need to solve an artificial problem called *Phase I*. Once this starting vertex matrix for PS has been found, we enter *Phase II*, in which we apply the simplex algorithm to this vertex matrix to find an optimal solution to PS.

PHASE I DESCRIPTION

For the linear program P, convert the constraints to equations as follows: Change

$$a_{i1}x_1 + \cdots + a_{in}x_n \leqslant b_i \qquad \text{to} \qquad a_{i1}x_1 + \cdots + a_{in}x_n + s_i = b_i$$

that is, add a slack variable $s_i \geqslant 0$. Change

$$a_{j1}x_1 + \cdots + a_{jn}x_n \geqslant b_j \qquad \text{to} \qquad a_{j1}x_1 + \cdots + a_{jn}x_n - S_j + A_j = b_j$$

that is, subtract a surplus variable $S_j \geqslant 0$ and add an *artificial variable* $A_j \geqslant 0$. Change

$$a_{k1}x_1 + \cdots + a_{kn}x_n = b_k \qquad \text{to} \qquad a_{k1}x_1 + \cdots + a_{kn}x_n + A_k = b_k$$

that is, add an artificial variable $A_k \geqslant 0$.

This intermediate linear program system now appears as

$$
\text{IPS:}\qquad
\begin{aligned}
a_{11}x_1 + \cdots + a_{1n}x_n + s_1 &= b_1\\
&\ \vdots\\
a_{i1}x_1 + \cdots + a_{in}x_n + s_i &= b_i\\
&\ \vdots\\
a_{j1}x_1 + \cdots + a_{jn}x_n - S_j + A_j &= b_j\\
&\ \vdots\\
a_{k1}x_1 + \cdots + a_{kn}x_n + A_k &= b_k\\
&\ \vdots\\
-a_1x_1 - a_2x_2 - \cdots - a_nx_n + z &= 0
\end{aligned}
$$

or, in augmented matrix form,

$$
\begin{array}{ccccccccccccc}
x_1 & \cdots & x_n & \cdots & s_i & \cdots & S_j & \cdots & A_j & \cdots & A_k & \cdots & z
\end{array}
$$
$$
\left[\begin{array}{ccccccccccccc|c}
a_{11} & \cdots & a_{1n} & \cdots & 0 & \cdots & 0 & \cdots & 0 & \cdots & 0 & \cdots & 0 & b_1\\
\vdots & & \vdots & & \vdots & & \vdots & & \vdots & & \vdots & & \vdots & \vdots\\
a_{i1} & \cdots & a_{in} & \cdots & 1 & \cdots & 0 & \cdots & 0 & \cdots & 0 & \cdots & 0 & b_i\\
\vdots & & \vdots & & \vdots & & \vdots & & \vdots & & \vdots & & \vdots & \vdots\\
a_{j1} & \cdots & a_{jn} & \cdots & 0 & \cdots & -1 & \cdots & 1 & \cdots & 0 & \cdots & 0 & b_j\\
\vdots & & \vdots & & \vdots & & \vdots & & \vdots & & \vdots & & \vdots & \vdots\\
a_{k1} & \cdots & a_{kn} & \cdots & 0 & \cdots & 0 & \cdots & 0 & \cdots & 1 & \cdots & 0 & b_k\\
\vdots & & \vdots & & \vdots & & \vdots & & \vdots & & \vdots & & \vdots & \vdots\\
\hline
-a_1 & \cdots & -a_n & \cdots & 0 & \cdots & 0 & \cdots & 0 & \cdots & 0 & \cdots & 1 & 0
\end{array}\right]
$$

which is a vertex matrix. To obtain a starting vertex matrix for PS, however, we want to obtain from this matrix a vertex matrix in which all artificial variables are nonbasic. Then we can delete the columns corresponding to the artificial variables and obtain the desired vertex matrix. To obtain a vertex matrix in which all artificial variables are nonbasic, add to the system an additional objective function, $w = -A_1 - \cdots - A_r$, where $A_1, \ldots, A_r$ are all the artificial variables. Thus, we have an artificial linear program system

$$
\text{APS:}\quad
\begin{array}{cccccccccccccc}
x_1 & \cdots & x_n & \cdots & s_i & \cdots & S_j & \cdots & A_j & \cdots & A_k & \cdots & z & w
\end{array}
$$
$$
\left[\begin{array}{cccccccccccccc|c}
a_{11} & \cdots & a_{1n} & \cdots & 0 & \cdots & 0 & \cdots & 0 & \cdots & 0 & \cdots & 0 & 0 & b_1\\
\vdots & & \vdots & & \vdots & & \vdots & & \vdots & & \vdots & & \vdots & \vdots & \vdots\\
a_{i1} & \cdots & a_{in} & \cdots & 1 & \cdots & 0 & \cdots & 0 & \cdots & 0 & \cdots & 0 & 0 & b_i\\
\vdots & & \vdots & & \vdots & & \vdots & & \vdots & & \vdots & & \vdots & \vdots & \vdots\\
a_{j1} & \cdots & a_{jn} & \cdots & 0 & \cdots & -1 & \cdots & 1 & \cdots & 0 & \cdots & 0 & 0 & b_j\\
\vdots & & \vdots & & \vdots & & \vdots & & \vdots & & \vdots & & \vdots & \vdots & \vdots\\
a_{k1} & \cdots & a_{kn} & \cdots & 0 & \cdots & 0 & \cdots & 0 & \cdots & 1 & \cdots & 0 & 0 & b_k\\
\vdots & & \vdots & & \vdots & & \vdots & & \vdots & & \vdots & & \vdots & \vdots & \vdots\\
\hline
-a_1 & \cdots & -a_n & \cdots & 0 & \cdots & 0 & \cdots & 0 & \cdots & 0 & \cdots & 1 & 0 & 0\\
0 & \cdots & 0 & \cdots & 0 & \cdots & 0 & \cdots & 1 & \cdots & 1 & \cdots & 0 & 1 & 0
\end{array}\right]
$$

Since the artificial variables are all at least 0, $w \leqslant 0$. Further, $w = 0$ exactly when all of the artificial variables are 0. Hence, we will maximize w by the simplex algorithm. If we cannot reach 0, PS and consequently P have no feasible solution (exercise 25). If $w = 0$ is possible and we have succeeded in making all artificial variables nonbasic* (see exercise 26), then we delete all columns corresponding to artificial variables and w, and the row corresponding to the additional objective function. The result is a vertex matrix we can use to begin Phase II, the phase in which the simplex algorithm is applied to this matrix to solve PS. Methods for handling the case in which the artificial variables are not all nonbasic can be found in the References.

Note that the matrix for APS is not itself a vertex matrix. However, it can be converted into a vertex matrix by subtracting from the last row each of the rows containing a coefficient 1 for an artificial variable. Thus we get

$$\begin{array}{cccccccccccccccc|c}
x_1 & \cdots & x_n & \cdots & s_i & \cdots & S_j & \cdots & A_j & \cdots & A_k & \cdots & z & w & & \\
a_{11} & \cdots & a_{1n} & \cdots & 0 & \cdots & 0 & \cdots & 0 & \cdots & 0 & \cdots & 0 & 0 & & b_1 \\
\vdots & & \vdots & & \vdots & & \vdots & & \vdots & & \vdots & & \vdots & \vdots & & \vdots \\
a_{i1} & \cdots & a_{in} & \cdots & 1 & \cdots & 0 & \cdots & 0 & \cdots & 0 & \cdots & 0 & 0 & & b_i \\
\vdots & & \vdots & & \vdots & & \vdots & & \vdots & & \vdots & & \vdots & \vdots & & \vdots \\
a_{j1} & \cdots & a_{jn} & \cdots & 0 & \cdots & -1 & \cdots & 1 & \cdots & 0 & \cdots & 0 & 0 & & b_j \\
\vdots & & \vdots & & \vdots & & \vdots & & \vdots & & \vdots & & \vdots & \vdots & & \vdots \\
a_{k1} & \cdots & a_{kn} & \cdots & 0 & \cdots & 0 & \cdots & 0 & \cdots & 1 & \cdots & 0 & 0 & & b_k \\
\vdots & & \vdots & & \vdots & & \vdots & & \vdots & & \vdots & & \vdots & \vdots & & \vdots \\
\hline
-a_1 & \cdots & -a_n & \cdots & 0 & \cdots & 0 & \cdots & 0 & \cdots & 0 & \cdots & 1 & 0 & & 0 \\
-c_1 & \cdots & -c_n & \cdots & 0 & \cdots & 1 & \cdots & 0 & \cdots & 0 & \cdots & 0 & 1 & & -d
\end{array}$$

where $c_s = \sum_{l \geqslant j} a_{ls}$ and $d = \sum_{l \geqslant j} b_l$. This is a starting vertex matrix for Phase I. Now we can apply the simplex algorithm to obtain a solution to APS. In applying the algorithm, the last two rows are not used as constraint equations. Now if $w = 0$ and all artificial variables are nonbasic, we drop the columns corresponding to all artificial variables A, the column corresponding to w, and the last row, obtaining a vertex matrix for PS. We then go to Phase II.

PHASE II DESCRIPTION

Apply the simplex algorithm to the vertex matrix for PS obtained in Phase I to obtain an optimal solution and an optimal value for the linear program P. Describing the procedure in algorithmic form, we have the following.

* This occurs if the system is not degenerate.

PHASE I–PHASE II ALGORITHM

By adjusting constraints, convert the given linear program P into the adjusted linear program system IPS. If no artificial variables have been used, solve by the simplex algorithm. Otherwise, proceed as follows.

PHASE I

1. Add the additional objective function equation $w = -\Sigma A_i$ to IPS and add a column for the variable w, thus forming APS. (Adjustments to P can be made and APS written without going through the intermediate system IPS.)

2. Subtract all rows containing an artificial variable from the last row to obtain a starting vertex matrix for APS.

3. Apply the simplex algorithm to this vertex matrix, making sure not to use entries in the last two rows in calculating ratios. If APS has no solution, neither does PS or P. Otherwise calculate the optimal value of w.

4. If $w < 0$, there is no solution to PS or P. If $w = 0$, check to see whether all A_i's are nonbasic. If some are basic, see the References for a book by Dantzig which explains how to handle the problem. If all A_i's are nonbasic, delete the corresponding columns, the column for w, and the last row, obtaining a vertex matrix for PS.

PHASE II

5. Apply the simplex algorithm to the vertex matrix obtained in Step 4. This determines the solution to PS and consequently to P.

EXAMPLE 23

Continuing Example 22, solve

$$\begin{aligned} \max \quad & z = 2x_1 - x_2 \\ \text{s.t.} \quad & x_1 + x_2 \leqslant 4 \\ & -x_1 + x_2 \geqslant 1 \\ & x_1 \geqslant 0, x_2 \geqslant 0 \end{aligned}$$

This program requires the Phase I–Phase II technique. For Phase I, write the linear program system

$$\text{APS:} \quad \begin{array}{rcl} x_1 + x_2 + s_1 & & = 4 \\ -\ x_1 + x_2 \quad - S_1 + A_1 & & = 1 \\ -2x_1 + x_2 \qquad\qquad + z & & = 0 \\ A_1 \quad + w & & = 0 \end{array}$$

The augmented matrix for this system is

$$\begin{array}{ccccccc|c} x_1 & x_2 & s_1 & S_1 & A_1 & z & w & \\ 1 & 1 & 1 & 0 & 0 & 0 & 0 & 4 \\ -1 & 1 & 0 & -1 & 1 & 0 & 0 & 1 \\ \hline -2 & 1 & 0 & 0 & 0 & 1 & 0 & 0 \\ 0 & 0 & 0 & 0 & 1 & 0 & 1 & 0 \end{array}$$

To obtain the first vertex matrix for APS, apply $R_4 - R_2 \rightarrow R_4$ to get

$$\begin{array}{c} \\ s_1 \\ A_1 \\ z \\ w \end{array} \begin{array}{c} \begin{array}{ccccccc|c} x_1 & x_2 & s_1 & S_1 & A_1 & z & w & \end{array} \\ \left[\begin{array}{ccccccc|c} 1 & 1 & 1 & 0 & 0 & 0 & 0 & 4 \\ -1 & 1 & 0 & -1 & 1 & 0 & 0 & 1 \\ \hline -2 & 1 & 0 & 0 & 0 & 1 & 0 & 0 \\ 1 & -1 & 0 & 1 & 0 & 0 & 1 & -1 \end{array}\right] \end{array}$$

Now, solving for w, we have x_2 as the entering nonbasic variable; checking ratios gives A_1 for the departing basic variable. Applying $R_1 - R_2 \rightarrow R_1$, $R_3 - R_2 \rightarrow R_3$, and $R_4 + R_2 \rightarrow R_4$ yields the next vertex matrix,

$$\begin{array}{c} \\ s_1 \\ x_2 \\ z \\ w \end{array} \begin{array}{c} \begin{array}{ccccccc|c} x_1 & x_2 & s_1 & S_1 & A_1 & z & w & \end{array} \\ \left[\begin{array}{ccccccc|c} 2 & 0 & 1 & 1 & -1 & 0 & 0 & 3 \\ -1 & 1 & 0 & -1 & 1 & 0 & 0 & 1 \\ \hline -1 & 0 & 0 & 1 & -1 & 1 & 0 & -1 \\ 0 & 0 & 0 & 0 & 1 & 0 & 1 & 0 \end{array}\right] \end{array}$$

Here the simplex algorithm stops, and we see that $w = 0$ and the artificial variable A_1 is nonbasic. Hence we delete the columns for A_1 and w and the last row to get

$$\begin{array}{c} \\ s_1 \\ x_2 \\ z \end{array} \begin{array}{c} \begin{array}{ccccc|c} x_1 & x_2 & s_1 & S_1 & z & \end{array} \\ \left[\begin{array}{ccccc|c} 2 & 0 & 1 & 1 & 0 & 3 \\ -1 & 1 & 0 & -1 & 0 & 1 \\ \hline -1 & 0 & 0 & 1 & 1 & -1 \end{array}\right] \end{array}$$

Now for Phase II, we apply the simplex algorithm to this matrix. Since $a_1 = -1$, x_1 is the entering variable. Ratios show that s_1 is the departing basic variable. Thus we apply $R_2 + \frac{1}{2}R_1 \rightarrow R_2$, $R_3 + \frac{1}{2}R_1 \rightarrow R_3$, and $\frac{1}{2}R_1 \rightarrow R_1$ to get

$$\begin{array}{c} \\ x_1 \\ x_2 \\ z \end{array} \begin{array}{c} \begin{array}{ccccc|c} x_1 & x_2 & s_1 & S_1 & z & \end{array} \\ \left[\begin{array}{ccccc|c} 1 & 0 & \frac{1}{2} & \frac{1}{2} & 0 & \frac{3}{2} \\ 0 & 1 & \frac{1}{2} & -\frac{1}{2} & 0 & \frac{5}{2} \\ \hline 0 & 0 & \frac{1}{2} & \frac{3}{2} & 1 & \frac{1}{2} \end{array}\right] \end{array}$$

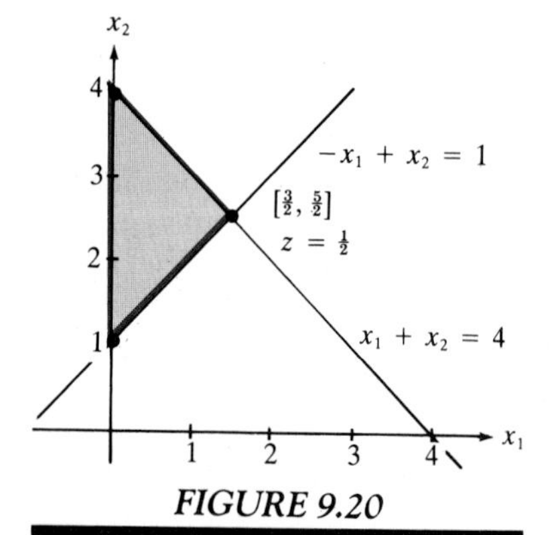

FIGURE 9.20

Since $a_i \geqslant 0$ for all i, we stop. Setting the nonbasic variables to 0 and solving for the basic variables, we find that $[\frac{3}{2}, \frac{5}{2}, 0, 0, \frac{1}{2}]$ is the solution of the linear program system, so $[\frac{3}{2}, \frac{5}{2}]$ is an optimal solution to the linear program, with $z = \frac{1}{2}$ the optimal value. Figure 9.20 shows the feasible region of the linear program and the vertex corresponding to the optimal solution. □

EXAMPLE 24

The sheet metal division of a supply company stocks strips of sheet metal that are 18 inches wide and 10,000 feet long in large rolls. Suppose a gutter manufacturer needs at least 1600 feet of 7-inch-wide strips and at least 3200 feet of 8-inch-wide strips. The supply company will measure off 18-inch-wide strips of any desired length, roll each strip into an 18-inch-long cylinder, and use a guillotine blade to cut the cylinder into pieces of lengths 7 inches and 8 inches, and perhaps some left-over shorter pieces. When the 7- and 8-inch pieces of the cylinder are unrolled, they form the strips desired by the gutter manufacturer. Pieces of less than 7 inches are waste. Given that the customer must pay for the entire 18-inch-wide strips cut for the order, minimize the total waste.

In order to solve this problem, we must first identify the possible lengthwise cuts for the strips of metal. There are three patterns available:

1. Cut two 7-inch strips, leaving 4 inches of waste.

2. Cut one 7-inch and one 8-inch strip, leaving 3 inches of waste.

3. Cut two 8-inch strips, leaving 2 inches of waste.

Let x_1 be the length in feet cut according to pattern 1, x_2 the length cut according to pattern 2, and x_3 the length cut according to pattern 3. Then we wish to minimize $z = 4x_1 + 3x_2 + 2x_3$, where z is measured in inch-feet, which are equal to 12 times the equivalent number of square feet. In order to satisfy the customer's order, it is necessary that $2x_1 + x_2 \geqslant 1600$ and $x_2 + 2x_3 \geqslant 3200$. These constraints give us the linear program

$$\begin{aligned} \text{min} \quad & z = 4x_1 + 3x_2 + 2x_3 \\ \text{s.t.} \quad & 2x_1 + x_2 \geqslant 1600 \\ & x_2 + 2x_3 \geqslant 3200 \\ & x_1 \geqslant 0, x_2 \geqslant 0, x_3 \geqslant 0 \end{aligned}$$

We can solve this linear program using the methods shown in this chapter. Doing so, we find that z is minimized for $x_1 = 0$, $x_2 = 1600$, and $x_3 = 800$. Thus a 1600-foot strip should be guillotined according to pattern 2, and an 800-foot strip should be guillotined according to pattern 3. The result will be 1600 feet of 7-inch-wide strips and 3200 feet of 8-inch-wide strips, along with 6400 inch-feet, or 1600/3 square feet, of waste in 2- and 3-inch strips. □

EXERCISES FOR SECTION 9.4

COMPUTATIONAL EXERCISES

In exercises 1 through 4, graph the feasible set. Then solve the linear program, in part (a) maximizing the given objective function and in part (b) minimizing the given objective function. In each part, sequentially number the vertices on the graph corresponding to each vertex matrix in Phase II.

1. (*a*) max $z = 2x_1 + x_2$ (*b*) min $z = x_1 + x_2$

$$\begin{aligned} \text{s.t.} \quad & x_1 + x_2 \leqslant 14 \\ & 3x_1 + x_2 \leqslant 24 \\ & 2x_1 + x_2 \geqslant 8 \\ & x_1 \geqslant 0, x_2 \geqslant 0 \end{aligned}$$

2. (*a*) max $z = -x_1 - 3x_2$ (*b*) min $z = -3x_1 - 2x_2$

$$\begin{aligned} \text{s.t.} \quad & x_1 + x_2 \leqslant 13 \\ & 2x_1 + x_2 \leqslant 18 \\ & x_1 + 2x_2 \geqslant 6 \\ & x_1 \geqslant 0, x_2 \geqslant 0 \end{aligned}$$

3. (*a*) max $z = -x_1 + 2x_2$ (*b*) min $z = -5x_1 + x_2$

$$\begin{aligned} \text{s.t.} \quad & x_1 + x_2 \leqslant 13 \\ & -x_1 + x_2 \geqslant 1 \\ & 2x_1 + x_2 \geqslant 16 \\ & x_1 \geqslant 0, x_2 \geqslant 0 \end{aligned}$$

4. (a) max $z = 6x_1 + 2x_2$ (b) min $z = 3x_1 - 2x_2$

$$\text{s.t.} \quad \begin{aligned} x_1 + 3x_2 &\leqslant 32 \\ 2x_1 + x_2 &\geqslant 14 \\ -x_1 + 2x_2 &\geqslant 8 \\ x_1 \geqslant 0, x_2 &\geqslant 0 \end{aligned}$$

In exercises 5 through 12, graph the feasible region for the linear program. Then solve the program using the methods described in this section. Sequentially number the vertices on the graph corresponding to each vertex matrix in Phase II.

5. min $z = 2x_1 - 3x_2$
$$\text{s.t.} \quad \begin{aligned} x_1 - 2x_2 &\leqslant 0 \\ 4x_1 + x_2 &\leqslant 36 \\ -2x_1 + x_2 &\leqslant 0 \\ x_1 \geqslant 0, x_2 &\geqslant 0 \end{aligned}$$

6. min $z = -8x_1 + x_2$
$$\text{s.t.} \quad \begin{aligned} x_1 - 2x_2 &\leqslant 0 \\ 4x_1 + x_2 &\leqslant 36 \\ -2x_1 + x_2 &\leqslant 0 \\ x_1 \geqslant 0, x_2 &\geqslant 0 \end{aligned}$$

7. min $z = -3x_1 - x_2$
$$\text{s.t.} \quad \begin{aligned} x_1 - 3x_2 &\leqslant 0 \\ 3x_1 - x_2 &\leqslant 72 \\ x_1 + x_2 &\leqslant 48 \\ -2x_1 + x_2 &\leqslant 0 \\ x_1 \geqslant 0, x_2 &\geqslant 0 \end{aligned}$$

8. min $z = -2x_1 + x_2$
$$\text{s.t.} \quad \begin{aligned} -3x_1 + x_2 &\leqslant 0 \\ x_1 + x_2 &\leqslant 24 \\ 2x_1 + x_2 &\leqslant 35 \\ x_1 - 2x_2 &\leqslant 0 \\ x_1 \geqslant 0, x_2 &\geqslant 0 \end{aligned}$$

9. min $z = x_1 + x_2$
$$\text{s.t.} \quad \begin{aligned} 2x_1 + x_2 &\geqslant 2 \\ x_1 + 2x_2 &\geqslant 2 \\ x_1 \geqslant 0, x_2 &\geqslant 0 \end{aligned}$$

10. min $z = -x_1 - x_2$
$$\text{s.t.} \quad \begin{aligned} -x_1 + x_2 &\leqslant 2 \\ -x_1 + x_2 &\geqslant -2 \\ x_1 \geqslant 0, x_2 &\geqslant 0 \end{aligned}$$

11. max $z = x_1 + 2x_2$
$$\text{s.t.} \quad \begin{aligned} x_1 &\leqslant 4 \\ x_2 &\leqslant 4 \\ x_1 + x_2 &\geqslant 1 \\ x_1 \geqslant 0, x_2 &\geqslant 0 \end{aligned}$$

12. max $z = 2x_1 + 3x_2$
$$\text{s.t.} \quad \begin{aligned} -x_1 + 4x_2 &\leqslant 40 \\ 3x_1 - x_2 &\leqslant 12 \\ 2x_1 + x_2 &\geqslant 2 \\ x_1 \geqslant 0, x_2 &\geqslant 0 \end{aligned}$$

In exercises 13 through 22, solve the linear program problem using the methods of this section.

13. min $z = x_1 - 2x_2 - x_3$
$$\text{s.t.} \quad \begin{aligned} 4x_1 - x_3 &\leqslant 8 \\ 4x_2 - x_3 &\leqslant 16 \\ -x_1 + x_3 &\leqslant 4 \\ x_1 \geqslant 0, x_2 \geqslant 0, x_3 &\geqslant 0 \end{aligned}$$

14. min $z = -3x_1 - 8x_2 + 4x_3$
$$\text{s.t.} \quad \begin{aligned} 4x_1 + 2x_2 + x_3 &\leqslant 4 \\ 3x_1 + x_2 + x_3 &\leqslant 3 \\ x_1 \geqslant 0, x_2 \geqslant 0, x_3 &\geqslant 0 \end{aligned}$$

15. max $z = x_1 + 2x_2 - x_3$
$$\text{s.t.} \quad \begin{aligned} x_1 + x_2 + 4x_3 &\leqslant 4 \\ x_1 + 3x_2 + 6x_3 &\leqslant 6 \\ x_1 + x_2 + x_3 &\geqslant 1 \\ x_1 \geqslant 0, x_2 \geqslant 0, x_3 &\geqslant 0 \end{aligned}$$

16. max $z = -x_1 + x_2 - x_3$
$$\text{s.t.} \quad \begin{aligned} x_1 &\geqslant 2 \\ x_2 &\geqslant 2 \\ x_1 + x_2 + x_3 &\leqslant 6 \\ x_1 \geqslant 0, x_2 \geqslant 0, x_3 &\geqslant 0 \end{aligned}$$

17. min $z = x_1 + x_2 + x_3$
$$\text{s.t.} \quad \begin{aligned} x_1 + x_2 &\geqslant 2 \\ x_2 + x_3 &\geqslant 2 \\ x_1 + x_3 &\geqslant 2 \\ x_1 \geqslant 0, x_2 \geqslant 0, x_3 &\geqslant 0 \end{aligned}$$

18. min $z = -2x_1 + x_2 + 3x_3$
$$\text{s.t.} \quad \begin{aligned} x_1 + x_2 + 4x_3 &\leqslant 4 \\ x_1 + 3x_2 + 6x_3 &\leqslant 6 \\ x_1 + x_2 + x_3 &\geqslant 1 \\ x_1 \geqslant 0, x_2 \geqslant 0, x_3 &\geqslant 0 \end{aligned}$$

19. min $z = x_1 - x_2 + x_3$
$$\text{s.t.} \quad \begin{aligned} x_2 &\geqslant 2 \\ x_1 + x_3 &\leqslant 4 \\ x_1 \geqslant 0, x_2 \geqslant 0, x_3 &\geqslant 0 \end{aligned}$$

20. min $z = 3x_1 - x_2 - 2x_3$
$$\text{s.t.} \quad \begin{aligned} x_1 &\leqslant 5 \\ x_2 + x_3 &\geqslant 2 \\ -x_1 + x_2 + 2x_3 &\geqslant 3 \\ x_1 \geqslant 0, x_2 \geqslant 0, x_3 &\geqslant 0 \end{aligned}$$

21. max $z = 40x_1 + 120x_3$
$$\text{s.t.} \quad \begin{aligned} x_1 + x_2 + 2x_3 &\leqslant 80 \\ x_1 + 6x_2 + 9x_3 &\leqslant 360 \\ x_1 + 6x_2 &= 90 \\ x_1 \geqslant 0, x_2 \geqslant 0, x_3 &\geqslant 0 \end{aligned}$$

22. max $z = 200x_2 + 50x_3$
$$\text{s.t.} \quad \begin{aligned} 3x_1 - 2x_2 + 2x_3 &\leqslant 120 \\ -x_1 + 2x_2 + 2x_3 &\leqslant 120 \\ x_1 + 2x_2 &= 60 \\ x_1 \geqslant 0, x_2 \geqslant 0, x_3 &\geqslant 0 \end{aligned}$$

THEORETICAL EXERCISES

23. Show that the optimal solutions to P_m and P_M are precisely the same and that the optimal values z and z' for P_m and P_M, respectively, are related by $-z = z'$.

24. Show that $[x_1, \ldots, x_n, \ldots, s_i, \ldots, S_j, \ldots, z]$ is an optimal solution to PS if and only if $[x_1, \ldots, x_n]$ is an optimal solution to P with z the optimal value.

25. Prove that if the optimal value of w in the solution of APS is less than 0, then there is no $\mathbf{x}$ that satisfies the constraints, and so there is no solution.

26. Solve

$$\begin{aligned} \max \quad & z = x_1 + x_2 \\ \text{s.t.} \quad & -x_1 - x_2 \geqslant 0 \\ & x_1 \geqslant 0, x_2 \geqslant 0 \end{aligned}$$

using the simplex algorithm. Note that at the end of Phase I, the artificial variable A is basic. Thus, if we delete the columns for the artificial variable and w and the last row in going to Phase II, the resulting matrix will not be a vertex matrix. For ways to handle such problems see the References.

APPLICATIONS EXERCISES

27. Solve the problem presented in Examples 1 and 4 of this chapter.

28. A beef company has two plants at which beef is processed. Beef from these plants is shipped to two supermarket chains. The costs of shipping are given in Table 9.14. Plant A and plant B can produce at most 80 tons and 70 tons of beef, respectively, per month. Chain A and chain B require at least 60 tons and 90 tons of beef, respectively, per month. What quantities of beef should be shipped to meet the requirements while minimizing cost?

TABLE 9.14

PLANT	COST OF SHIPPING TO CHAIN A	B
A	\$200/ton	\$250/ton
B	\$250/ton	\$200/ton

29. An oil company brings oil into two ports, 1 and 2, and ships it to three refineries, 3, 4, and 5. Given the costs of shipping, the demands, and the supplies shown in Table 9.15, write a linear program to find the amounts x_{ij} that must be shipped from each port i to each refinery j in order to minimize the cost of meeting the demands, but do not solve the linear program.

TABLE 9.15

PORT	COSTS OF SHIPPING TO REFINERY 3	4	5	SUPPLY (in thousands of barrels)
1	\$200	\$200	\$210	500
2	\$220	\$300	\$250	600
Demand (in thousands of barrels)	300	150	250	

30. A catering service prepares meals for several care centers in Los Angeles. The dinner meal consists of beef, potatoes, green beans, and milk. The dinner must contain at least 463 calories, 3530 milligrams of protein, 284 milligrams of calcium, and 5 milligrams of iron. If one serving of beef costs \$1.00, one potato costs \$0.25, one cup of green beans costs \$0.30, and one glass of milk costs \$0.30, write a linear program to find the portions that should be used for the most affordable dinner. Use the table in Example 2 of Chapter 1 for data.

31. Three grades of home heating fuel oil, A, B, and C, are produced in a two-step operation. Step I is a setup in which pots are scrubbed, etc., before use. In Step II the oil is mixed. The numbers of hours needed in each step and the profit for one unit of each grade of oil are given in Table 9.16.

TABLE 9.16

STEP	HOURS NEEDED FOR ONE UNIT OF GRADE A	B	C	NUMBER OF HOURS AVAILABLE
I	2	2	2	60
II	4	1	2	60
Profit per Unit	\$1000	\$900	\$4000	

Grade C is a poor but cheap fuel oil, sold in large units at a low but very profitable price. To meet a winter's demand, the number of units of grades A and B plus $\frac{3}{8}$ of the number of units of grade C must be at least 15. Write the linear program to determine the number of units of each grade of oil that should be produced to maximize total profit. Solve the linear program.

32. A rancher has been feeding his cattle a mix that is 20% protein and 8% fat; however, he wants to change to a mix that is at least 25% protein and 10% fat. The local feed store has on hand three types of mix, A, B, and C, that could be blended to obtain this desired mixture. The percentages of protein and fat and the cost per pound for each type of mix are given in Table 9.17.

TABLE 9.17

	PERCENTAGES IN MIXES A	B	C
Protein	30	20	25
Fat	10	12	5
Cost per lb	50¢	45¢	40¢

Write a linear program to determine how many pounds of each type of mix the feed store should use in order to obtain exactly 50 pounds of the special mix at the lowest cost.

33. A candy factory has two processing sections, each of which produces two kinds of candy from chocolate, sugar, and a prepared mix of milk, eggs, corn syrup, and other ingredients. Use the simplex algorithm to determine the numbers x_1 and x_2 of runs per week of the two processes that will maximize profit for the factory while meeting the demand for the two types of candy, given the information in Table 9.18 on weekly supply and demand.

TABLE 9.18

PROCESS	INGREDIENTS (TONS) Chocolate	Sugar	Mix	OUTPUT (TONS) Candy 1	Candy 2	PROFIT
1	1	12	5	20	16	\$300
2	1	4	1	5	8	\$200
	29	156 (supply)	55	110 (demand)	128	

34. A building company needs to carpet the halls of a large new office building. It has available only 16-foot-wide rolls of carpet. The requirements are 3000 yards of 5-foot-wide carpeting and 2000 yards of 8-foot-wide carpeting. Determine how the carpet rolls should be cut in order to minimize waste. *Hint:* From a given length of carpet, the company could cut 3 5-foot strips, 1 5-foot and 1 8-foot strip, or 2 8-foot strips. Assign a variable to the length of carpet devoted to each of these possibilities, and compute the waste associated with each possibility. Then write a linear program to minimize waste and solve it.

35. A chemical plant has a special order for a chemical whose manufacture requires a new and expensive catalyst based on palladium. The plant has no palladium catalyst in inventory. On investigation, the managers find they can buy it for \$50 per kilogram; they can have it purified in one week for \$20 per kilogram or in two weeks for \$10 per kilogram. The run will take 4 weeks, but a load of palladium catalyst needs to be replaced with pure material after only one week. For week i of the 4-week run, let

x_i = the amount bought at the beginning of the week for use that week

y_i = the amount sent out at the end of the week for one-week purification

w_i = the amount sent out at the end of the week for two-week purification

v_i = the amount used in week i but not sent out that week for purification

The amounts of palladium catalyst needed are 30 kilograms during week 1, 40 kilograms during week 2, 40 kilograms during week 3, and 30 kilograms during week 4. Set up the linear program to minimize the cost of meeting the requirements, but do not solve the linear program. *Hint:* In addition to ensuring that the required amounts of palladium catalyst are available, you must ensure that no more palladium catalyst is sent out for purification than has been used and not purified up to that time. This produces the two constraints $-x_1 + y_1 + w_1 + v_1 = 0$ and $-x_2 + y_2 - v_1 \leqslant 0$. Why?

APPENDIX A

COMPLEX NUMBERS

In this appendix we give a brief computational view of complex numbers.

A *complex number* is a number that can be written as

$$a + bi$$

where $a, b \in \mathbf{R}$ and i is the symbol used for $\sqrt{-1}$. The number a is called the *real part* of the complex number and the number b is called the *imaginary part*. For simplicity, we write

$$a = a + 0i, \qquad bi = 0 + bi, \qquad i = 0 + 1i, \qquad a - bi = a + (-b)i$$

EXAMPLE 1

Some examples of complex numbers are $2 + 3i$, 3, $-4 - 2i$, $6i$, and i. □

We say complex numbers $a + bi$ and $c + di$ are *equal* if $a = c$ and $b = d$. To add two complex numbers, we add the corresponding real and imaginary parts:

$$(a + bi) + (c + di) = (a + c) + (b + d)i$$

Subtraction is handled similarly:

$$(a + bi) - (c + di) = (a - c) + (b - d)i$$

Calculating rules are the same as those for the real numbers. That is, the commutative and associative laws hold.

EXAMPLE 2

(*i*) $(3 + 2i) + (6 - 4i) = 9 - 2i$

(*ii*) $i + 1 + 3i - 2 - 6i = -1 - 2i$ □

Multiplication is defined as

$$(a + bi)(c + di) = (ac - bd) + (ad + bc)i$$

Since the commutative, associative, and distributive laws hold, this computation can be done as it would be with real numbers, noting that $i \cdot i = \sqrt{-1}\sqrt{-1} = -1$.

EXAMPLE 3

(*i*) $(2 + 3i)(4 - 2i) = (2 + 3i)4 - (2 + 3i)(2i)$
$= 8 + 12i - 4i - 6i \cdot i = 8 + 8i + 6 = 14 + 8i$

(*ii*) $(1 - i)(3 + 2i) = 3 - 3i + 2i + 2 = 5 - i$ □

Finally, for division, we write

$$(a + bi) \div (c + di) = \frac{a + bi}{c + di}$$

when $c + di \neq 0$. Then, multiplying numerator and denominator by $c - di$, called the *conjugate* of $c + di$, we get

$$\frac{a + bi}{c + di} \cdot \frac{c - di}{c - di} = \frac{(ac + bd) + (bc - ad)i}{c^2 + d^2} = \frac{ac + bd}{c^2 + d^2} + \frac{bc - ad}{c^2 + d^2} i$$

EXAMPLE 4

(*i*) $\dfrac{2 + 3i}{1 - i} = \dfrac{2 + 3i}{1 - i} \cdot \dfrac{1 + i}{1 + i} = \dfrac{2 - 3 + 2i + 3i}{1 + 1} = \dfrac{-1 + 5i}{2} = -\dfrac{1}{2} + \dfrac{5}{2} i$

(*ii*) $\dfrac{1}{-2 + 4i} = \dfrac{1}{-2 + 4i} \cdot \dfrac{-2 - 4i}{-2 - 4i} = \dfrac{-2 - 4i}{4 + 16} = \dfrac{-2 - 4i}{20} = -\dfrac{1}{10} - \dfrac{1}{5} i$ □

EXERCISES FOR APPENDIX A

COMPUTATIONAL EXERCISES

In exercises 1 through 14, make the indicated calculations and write the answer in the form $a + bi$ with $a, b \in \mathbf{R}$.

1. $(3 + 2i) + (5 - 3i)$

2. $(-4 + i) + (-2 + i)$

3. $(7 - 5i) - (-4 - 2i)$

4. $(2 + 3i) - (2 - 2i)$

5. $(2 + 3i)(1 - i)$

6. $(3 - 4i)(2 - i)$

7. $(3 - 4i)(3 + 4i)$

8. $(5 + 6i)(5 - 6i)$

9. $(1 - 2i)^3$

10. $(2 - i)^3$

11. $\dfrac{1 - i}{1 + i}$

12. $\dfrac{2 + i}{2 - i}$

13. $\dfrac{3 - 2i}{1 - 2i}$

14. $\dfrac{4 + 3i}{3 + i}$

In exercises 15 and 16, find the square root of the given complex number z. *Hint:* Find $a + bi$ such that $(a + bi)^2 = z$.

15. $z = 3 - 4i$

16. $z = -8 + 6i$

APPENDIX B

REFERENCES

THEORETICAL

1. Beale, E. M. L., "Cycling in the Dual Simplex Algorithm," *Naval Research Logistics Quarterly*, 2 (1955), 269–276.
2. Gass, S. I., *Linear Programming Methods and Applications*, McGraw-Hill Book Co., New York, 1958.
3. Halmos, P., *Finite Dimensional Vector Spaces*, D. Van Nostrand Co., Princeton, N.J., 1958.
4. Jacobson, N., *Lectures in Abstract Algebra*, D. Van Nostrand Co., Princeton, N.J., 1964 (three volumes).
5. Liu, C. L., *Introduction to Combinatorial Mathematics*, McGraw-Hill Book Co., New York, 1968.

See also references 1, 5, 7, 8, and 11 under Applications.

APPLICATIONS

1. Bellman, Richard, *Introduction to Matrix Analysis*, McGraw-Hill Book Co., New York, 1960.
2. Berge, C., *The Theory of Graphs*, John Wiley and Sons, New York, 1964.
3. Bondy, J. A., and U. S. R. Murty, *Graph Theory with Applications*, American Elsevier Publ. Co., New York, 1976.
4. Busacker, R. G., and T. L. Saaty, *Finite Graphs and Networks*, McGraw-Hill Book Co., New York, 1965.
5. Dantzig, G. B., *Linear Programming and Extensions*, Princeton University Press, Princeton, N.J., 1963.
6. Deo, N., *Graph Theory with Applications to Engineering and Computer Science*, Prentice-Hall, Englewood Cliffs, N.J., 1974.
7. Ford, L. R., and D. R. Fulkerson, *Flows in Networks*, Princeton University Press, Princeton, N.J., 1962.
8. Franklin, J. N., *Matrix Theory*, Prentice-Hall, Englewood Cliffs, N.J., 1968.
9. Leontief, W., "The Choice of Technology," *Scientific American*, 252 (1985), 37–45, 136.
10. McEliece, R. J., "The Reliability of Computer Memories," *Scientific American*, 252 (1985), 88–95, 120.
11. Noble, B., and J. W. Daniel, *Applied Linear Algebra*, D. Van Nostrand Co., Princeton, N.J., 1958.

12. G. O'Neill, *The High Frontier*, Anchor Press/Doubleday, Garden City, N.Y., 1982.
13. *The Radio Amateur's Handbook*, 36th edition, American Radio Relay League, West Hartford, Conn., 1959.

COMPUTATIONAL

1. Conte, S. D., and C. de Boor, *Elementary Numerical Analysis—An Algorithmic Approach*, McGraw-Hill Book Co., New York, 1972.
2. Golub, G. H., and C. F. Van Loan, *Matrix Computations*, The Johns Hopkins University Press, Baltimore, Md., 1983.
3. Stewart, G. W., *Introduction to Matrix Computations*, Academic Press, New York, 1973.
4. Wilkerson, J. H., *The Algebraic Eigenvalues Problem*, Oxford University Press, New York, 1965.

HISTORY

1. Bezout, E., *Histoire de l'Académie des Sciences*, Paris, 1764, 288–388.
2. Cauchy, A., "Memoire sur les fonctions qui ne peuvent obtenir que deux valeurs egales et de signes contraires par suite des transpositions operées entre les variables qu'elles renferment," *Journal de l'Ecole Polytechnique*, 10 (1815), 29–112, republished in *Oeuvres* (Gauthier Villars, Paris), Series 2, Vol. 1 (1905), 91–169.
3. Cayley, A., "Remarques sur la notation des fonctions algebraiques," *Journal fur reine und angewante Mathematik*, 50 (1855), 282–285.
4. Cayley, A., "A Memoir on the Theory of Matrices," *Royal Society of London Philosophical Transactions*, 148 (1858), 17–37.
5. Cramer, G., *Introduction à l'analyse des lignes courbes algebraiques*, Frères Cramer, Geneva, 1750.
6. Feldman, R. W., Jr., "Historically Speaking," *The Mathematics Teacher*, 55–56 (October 1962–March 1963).
7. Kline, M., *Mathematical Thought from Ancient to Modern Times*, Oxford University Press, New York, 1972.
8. Leibniz, G. W., "Mathematische Schriften," Vol. 2, 229, letter of 1693.
9. Maclaurin, C., *Treatise of Algebra*, printed for A. Millar and J. Nourse, London, 1748.

MISCELLANEOUS

1. James, G., and R. C. James, *Mathematics Dictionary*, 2nd edition, D. Van Nostrand Co., Princeton, N.J., 1959.
2. Lipschutz, S., *Linear Algebra*, Schaum's Outline Series, McGraw-Hill Book Co., New York, 1968. Contains many additional exercises.
3. Any recent edition of *Standard Mathematical Tables, Student Edition*, Chemical Rubber Publishing Co., Cleveland.

ANSWERS TO SELECTED EXERCISES

SECTION 1.1

1. (*a*) No (*b*) Yes *3.* Yes
7. (*a*) No (*b*) Yes (*c*) Yes
9. $\{(2-3y, y): y \text{ is any number}\}$

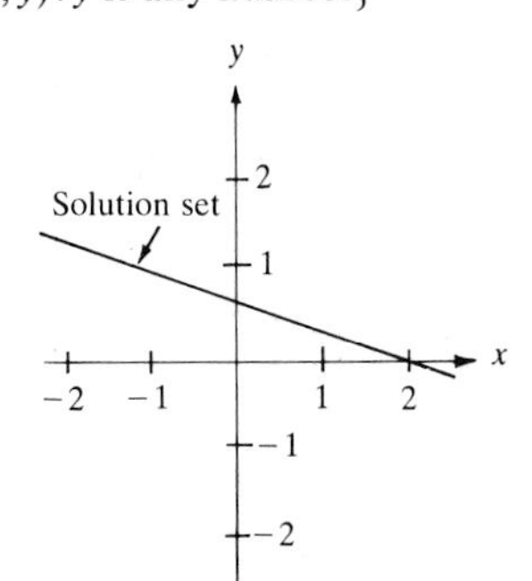

11. $\{(1,0)\}$

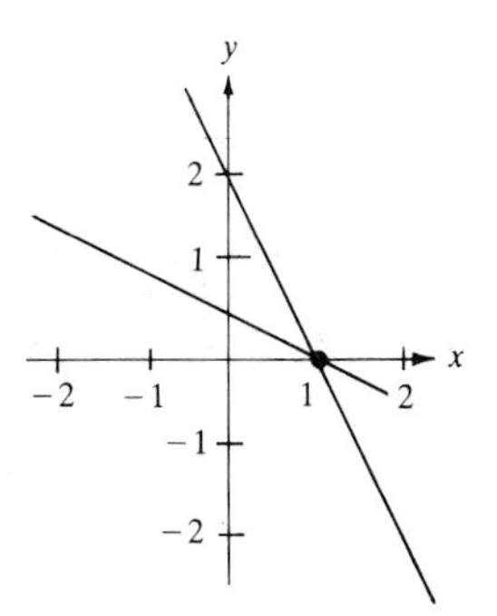

13. $\varnothing$

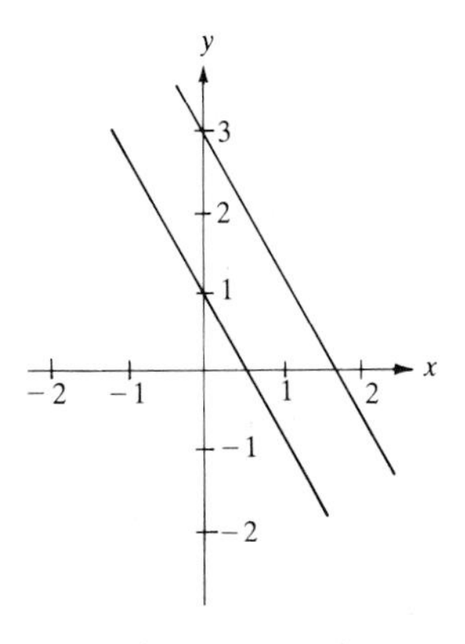

15. $\{(1+y, y): y \text{ is any number}\}$

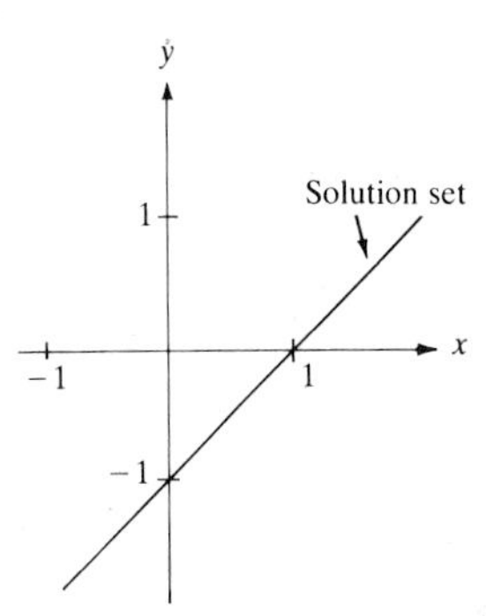

17.
$$\begin{aligned}
4c + 2e + f &= 0\\
4c - 2e + f &= 0\\
4a + 2d + f &= 0\\
4a - 2d + f &= 0\\
\tfrac{4}{3}a + \tfrac{4}{3}b + \tfrac{4}{3}c + \tfrac{2}{3}\sqrt{3}d + \tfrac{2}{3}\sqrt{3}e + f &= 0\\
\tfrac{4}{3}a - \tfrac{4}{3}b + \tfrac{4}{3}c - \tfrac{2}{3}\sqrt{3}d + \tfrac{2}{3}\sqrt{3}e + f &= 0
\end{aligned}$$

19.
$$\begin{aligned}
a_4 + a_3 + a_2 + a_1 + a_0 &= 0\\
a_0 &= -3\\
16a_4 + 8a_3 + 4a_2 + 2a_1 + a_0 &= 33\\
a_1 &= 0\\
4a_4 + 3a_3 + 2a_2 + a_1 &= 10\\
32a_4 + 12a_3 + 4a_2 + a_1 &= 68
\end{aligned}$$

21. $x_1 = 2y_1 - 3y_2$
$x_2 = y_1 - 2y_2$

22. (*a*) $z = 3 - i, w = 4 + 2i$ (*b*) $z = 1 - 2i, w = 2i$

24. 0, 1, or ∞. We get 0 if the lines are distinct and parallel, 1 if the lines are distinct and intersect, and ∞ if the lines are identical.

25. We get zero solutions if $a = 0$ and $b \neq 0$, one solution if $a \neq 0$, and infinitely many solutions if $a = b = 0$.

26. (*a*) $p = -\frac{1}{5}x + 4$; $p = 4$ gives $x = 0$
(*b*) $p = \frac{1}{5}x + 1$; $p = 4$ gives $x = 15$
(*c*) The lines intersect at $x = \frac{15}{2}$, $p = \frac{5}{2}$. If the price of the commodity is set at p, then x represents both the amount of the commodity that suppliers will provide and the amount that consumers will demand.

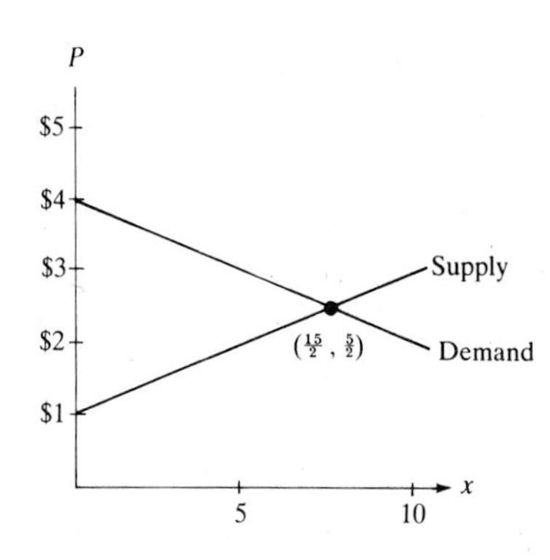

27. (*a*) $p = 1.7x - 3158.3$, $p(1927) \approx 117.6$ million
(*b*) $p = 1.38x - 2543.9$, $p(1912) \approx 94.7$ million
28. $x_1 = \$8600$, $x_2 = \$5700$
29. $x_1 = \$7000$, $x_2 = \$7500$
30. (*a*) $y = 0.99$, $x = 0$ (*b*) $x = y = 1$
(*c*) With (a): percentage error in y is 1%, percentage error in x is 100%. With (b): percentage error in both is 0%.

SECTION 1.2

1. $\begin{aligned} x_1 - 3x_2 + x_3 &= 1 \\ 2x_1 + x_2 - x_3 &= 3 \end{aligned}$ *3.* $\begin{aligned} x + 2y - 4z &= 6 \\ 2x - 3y + z &= 5 \end{aligned}$
5. $(2y - 2, y, -1)$ *7.* $(-2, 1, 3)$ *9.* $\varnothing$
11. $(-4z/19, 7z/19, z)$ *13.* $(8, 10, 6)$
15. $(-3/2 - 7z/4, -1 - z/2, z)$
17. $(x_3 - 51, -32, x_3, -8)$ *19.* $p = -x^2 + 4x - 2$
21. x_1 "free": $x_2 = 2 - x_1$, $x_3 = 0$
x_2 free: $x_1 = 2 - x_2$, $x_3 = 0$
23. $z = 2 + 5i$, $w = 6 - i$
27. $D = x + 2y + 3$, $D(8, 5) = 21$
28. Since we may take $a = 1$ or $b = 1$ by scaling, 5 points will do.
29. $y(x) = -\frac{1}{5}x^4 + \frac{4}{5}x^3 - \frac{8}{5}x$
30. Two-digit: $x_1 = 5$, $x_2 = -3$. Four-digit: $x_1 = 6.667$, $x_2 = -4.667$. Exact answer: $x_1 = \frac{20}{3} \approx 6.667$, $x_2 = -\frac{14}{3} \approx -4.667$. Percentage error on two-digit computer: 25% for x_1, 35.7% for x_2. Percentage error on four-digit computer: 0.005% for x_1, 0.007% for x_2.
31. (*a*) $\begin{aligned} x - 2y &= -3 \\ 5y &= 11 \end{aligned}$ (*b*) $\begin{aligned} x - 4y &= 9 \\ 11y &= -19 \\ 0 &= -6 \end{aligned}$

(*c*) $\begin{aligned} x - y + z &= -6 \\ y - 9z &= 13 \\ 47z &= -50 \end{aligned}$
32. (*a*) System to be solved:

$$\begin{aligned} u_{11} &= \tfrac{1}{4}(u_{10} + u_{12} + u_{21} + u_{01}) \\ u_{21} &= \tfrac{1}{4}(u_{11} + u_{20} + u_{31} + u_{22}) \\ u_{12} &= \tfrac{1}{4}(u_{02} + u_{11} + u_{22} + u_{13}) \\ u_{22} &= \tfrac{1}{4}(u_{21} + u_{12} + u_{32} + u_{23}) \end{aligned}$$

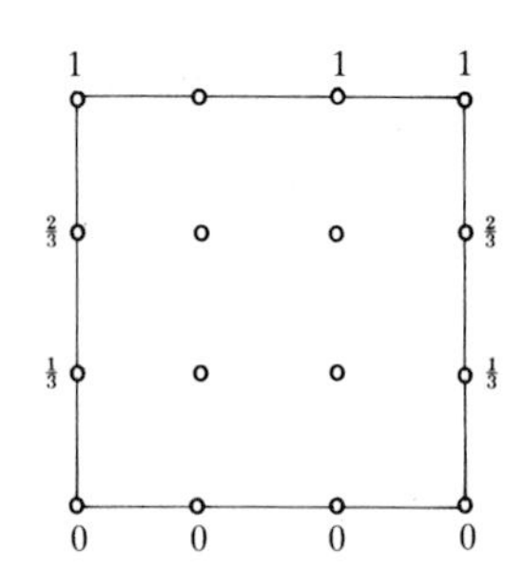

Solution:

$$\begin{aligned} u_{11} &= \tfrac{1}{24}(7u_{10} + 7u_{01} + 2u_{20} + 2u_{31} \\ &\quad + 2u_{02} + 2u_{13} + u_{32} + u_{23}) \\ u_{21} &= \tfrac{1}{24}(2u_{10} + 2u_{01} + 7u_{20} + 7u_{31} \\ &\quad + u_{02} + u_{13} + 2u_{32} + 2u_{23}) \\ u_{12} &= \tfrac{1}{24}(2u_{10} + 2u_{01} + u_{20} + u_{31} \\ &\quad + 7u_{02} + 7u_{13} + 2u_{32} + 2u_{23}) \\ u_{22} &= \tfrac{1}{24}(u_{10} + u_{01} + 2u_{20} + 2u_{31} \\ &\quad + 2u_{02} + 2u_{13} + 7u_{32} + 7u_{23}) \end{aligned}$$

(*b*) $(n - 1)^2$

SECTION 1.3

1. $((2 - 5y)/3, y)$

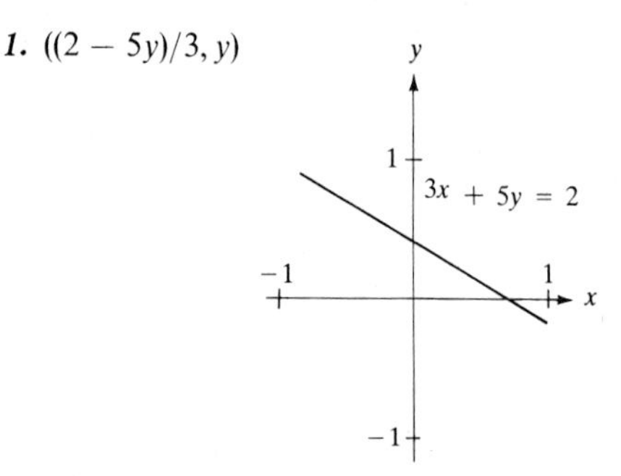

3. $(\frac{30}{23}, \frac{1}{23})$

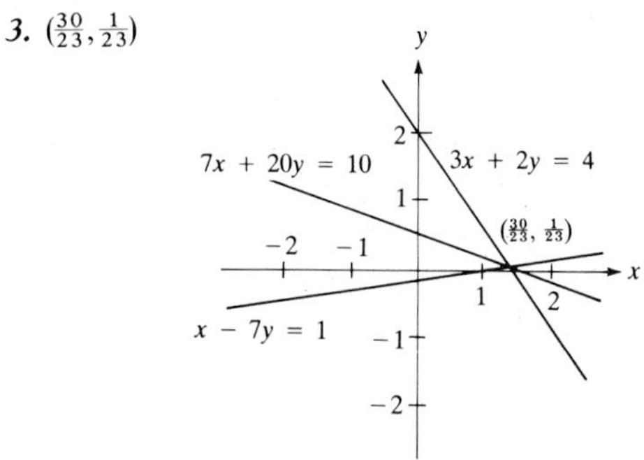

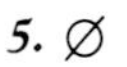

5. $\varnothing$

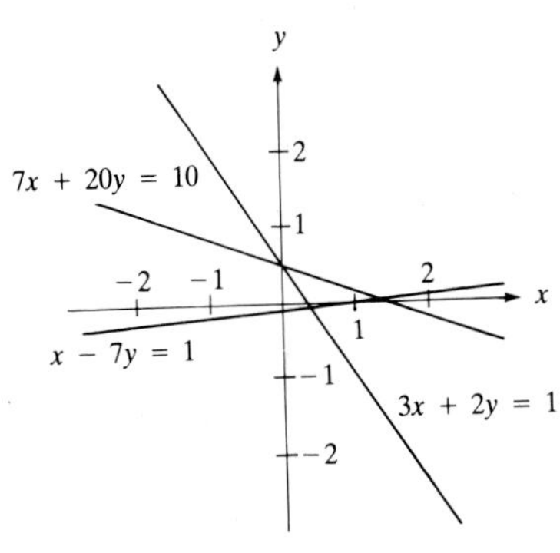

7. $x = y = 0$ 9. $(-\frac{33}{2}, -9, 4)$ 11. $\varnothing$
13. $(7 - 2z, 4 - z, z)$ 15. $\varnothing$ 17. $(0, x_3, x_3)$
19. $(0, 0, 0)$ 21. $z = 3i, w = 1 - 2i$
24. No, Theorem 1.2 states that there are either 0, 1, or infinitely many solutions.
25. $n + 1$ 26. $I_1 = 3, I_2 = 1, I_3 = 2$
27. $I_1 = 3, I_2 = 1, I_3 = 2, I_4 = 3$
28.
$$\begin{aligned} 2I_1 \qquad\qquad\qquad &= 56 \\ 10I_2 \qquad + 2I_4 &= 56 \\ 3I_3 + 2I_4 &= 56 \\ I_2 + I_3 - I_4 &= 0 \end{aligned}$$
$I_1 = 28, I_2 = 3, I_3 = 10, I_4 = 13$ amp

29.
$$\begin{aligned} 2I_1 + 2I_2 \qquad\qquad\qquad &= 16 \\ 2I_1 \qquad + 2I_3 + 2I_4 \qquad &= 16 \\ 2I_1 \qquad + 2I_3 \qquad + 2I_5 &= 16 \\ I_1 - I_2 - I_3 \qquad\qquad &= 0 \\ I_3 - I_4 - I_5 &= 0 \end{aligned}$$
$I_1 = 5, I_2 = 3, I_3 = 2, I_4 = I_5 = 1$ amp

30. Two-digit: $x = 1.5$, $y = 0.51$. Four-digit: $x = 1.5$, $y = 0.5001$. The slopes differ by quite a lot, so small changes in slopes and intercepts result in small changes in the intersection point of the lines.
31. (a) 17 (b) 21. Since (b) takes more multiplications and divisions than (a), it is slower.

SECTION 1.4

1. 2, 3, 6, $[3 \quad 4]$, $\begin{bmatrix} 1 \\ 3 \\ 5 \end{bmatrix}$

3. 2, 7, 13, 11, $[1 \quad 2 \quad 3 \quad 4]$, $[9 \quad 10 \quad 11 \quad 12]$, $\begin{bmatrix} 2 \\ 6 \\ 10 \\ 14 \end{bmatrix}$, $\begin{bmatrix} 4 \\ 8 \\ 12 \\ 16 \end{bmatrix}$

5. (a) 1×3 (b) 1×1 (c) 2×2 (d) 3×1 (e) 2×3

7. (a) $\begin{bmatrix} 2 & -3 & 6 \\ 5 & 6 & -4 \end{bmatrix}$

(b) $\begin{bmatrix} -1 & 1 & -3 & 0 & 5 \\ 4 & 0 & -3 & 0 & 6 \\ 0 & 3 & 0 & 2 & 1 \end{bmatrix}$

8. (a)
$$\begin{aligned} x_1 - x_2 \qquad\quad + 2x_4 &= 1 \\ 4x_1 + 2x_2 - 3x_3 \qquad &= 6 \\ -x_1 + 5x_2 + 2x_3 - x_4 &= 1 \end{aligned}$$

9. $(-2, 1, 3)$ 11. $\varnothing$ 13. $(8, 10, 6)$
15. $(x_3 - 51, -32, x_3, -8)$ 17. $(-\frac{33}{2}, -9, 4)$ 19. $\varnothing$
21. $(7 - 2z, 4 - z, z)$
23. $(0, x_3, x_3)$ 25. $(0, 0, 0)$

27. (a) $\begin{bmatrix} 1 & 2 & -1 \\ 0 & 3 & 0 \\ 0 & 0 & -1 \\ 0 & 0 & 0 \end{bmatrix}$ leading to $\begin{bmatrix} 1 & 2 & -1 \\ 0 & 1 & 0 \\ 0 & 0 & 1 \\ 0 & 0 & 0 \end{bmatrix}$

28. (a) $\begin{bmatrix} 1 & 0 & 0 & -\frac{7}{15} & \frac{2}{5} \\ 0 & 1 & 0 & \frac{2}{15} & \frac{3}{5} \\ 0 & 0 & 1 & -\frac{8}{15} & -\frac{2}{5} \end{bmatrix}$

29. (a) Inconsistent (b) Consistent (c) Consistent
31. $z = 1 + i, w = 2 - i$

33. $\begin{bmatrix} 0.9 & 0 & 0.1 \\ 0.2 & 0.7 & 0.1 \\ 0.1 & 0.1 & 0.8 \end{bmatrix}$ 34. n rows

35. (a) No, we would not expect the numbers in the augmented matrix to satisfy all $m - n$ of the constant equations $0 = b$ in the echelon form.
(b) Yes, we would expect that the coefficients in A would be unlikely to cause a variable to be free before we had exhausted the available equations.
36. (a) 3 (b) 3 (c) 2
37. (a) 2 (b) 1 (c) 2

38. $A =$

Carburetor	Generator	Alternator	
2	0	3	Z-10
1	1	0	Z-12
3	2	0	Z-16

39. $A = \begin{bmatrix} 0 & 1 & 1 & 0 \\ 0 & 0 & 0 & 1 \\ 0 & 1 & 0 & 0 \\ 0 & 0 & 1 & 0 \end{bmatrix}$

40. (a) $\begin{bmatrix} \frac{1}{2} & \frac{1}{2} \\ \frac{3}{4} & \frac{1}{4} \end{bmatrix}$ (b) $\begin{bmatrix} 0.7 & 0.3 \\ 0.8 & 0.2 \end{bmatrix}$ 41. (a) $\begin{bmatrix} \frac{9}{10} & \frac{1}{10} \\ \frac{1}{20} & \frac{19}{20} \end{bmatrix}$

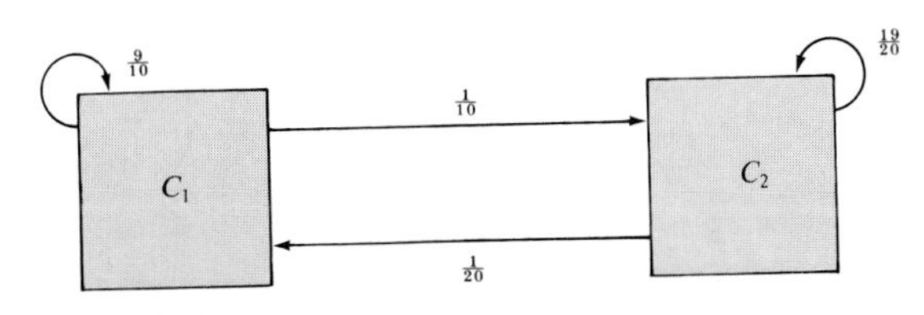

(*b*) $\begin{bmatrix} \frac{99}{100} & \frac{1}{100} \\ \frac{1}{150} & \frac{149}{150} \end{bmatrix}$ (*c*) $\begin{bmatrix} \frac{7}{8} & \frac{1}{8} \\ \frac{1}{6} & \frac{5}{6} \end{bmatrix}$

42. $I_2 = \dfrac{10(x-y)}{2xy+7x+7y}$, so $I_2 = 0$ exactly when $x = y$.

43. (*a*) No, the system has a free variable.
(*b*) One person is needed to count the single free variable. Since the last equation reduces to $0 = 0$, one counter can be eliminated.

44. (*a*) Store $a_{11} = 2$, Store $a_{12} = 1$, Store $a_{13} = 3$, Store $a_{21} = 0$, Store $a_{22} = 1$, Store $a_{23} = 2$, Store $a_{31} = 3$, Store $a_{32} = 0$, Store $a_{33} = 0$
(*b*) Store $a_{11} = 2$, Store $a_{12} = 0$, Store $a_{13} = 3$, Store $a_{21} = 1$, Store $a_{22} = 1$, Store $a_{23} = 0$, Store $a_{31} = 3$, Store $a_{32} = 2$, Store $a_{33} = 0$
(*c*) Store $a_{11} = 1$, Store $a_{23} = 1$

45. Store $a_{11} = 1$, Store $a_{12} = -0.33$, Store $a_{13} = 0.14$, Store $a_{21} = 0.25$, Store $a_{22} = -0.17$, Store $a_{23} = 0.38$

46. (*a*) Two-digit: $x = -18$, $y = 77$. Four-digit: $x = -14.29$, $y = 64.02$. Exact answer: $x = -\frac{100}{7} \approx -14.286$, $y = \frac{12100}{189} \approx 64.021$. Percentage error on two-digit computer: 26% for x, 20% for y. Percentage error on four-digit computer: 0% for x, 0% for y. The slopes of the lines here are quite close.
(*b*) Two-digit answer: $x = 1.4$, $y = 1.3$. Four-digit answer: $x = 1.395$, $y = 1.269$. Exact answer: $x = \frac{462}{331} \approx 1.3958$, $y = \frac{420}{331} \approx 1.2689$. Percentage error on two-digit computer: 0% for x, 2% for y. Percentage error on four-digit computer: 0% for x, 0% for y. The slopes here are quite far apart.

SECTION 2.1

1. Upper triangular, triangular

3. Upper triangular, lower triangular, triangular, diagonal

5. $\begin{bmatrix} 1 & 2 \\ 3 & 1 \end{bmatrix}$ *7.* $\begin{bmatrix} 1 & 2 & -1 \\ 1 & 1 & 2 \\ -2 & -3 & 0 \end{bmatrix}$ *9.* (*a*) 1 (*b*) 2 (*c*) 3

11. $\begin{bmatrix} 1 & 2 \\ 3 & 4 \end{bmatrix}$ *13.* $\begin{bmatrix} 0 & -3 \\ -4 & -5 \end{bmatrix}$

15. Cannot; sizes are not the same. *17.* $\begin{bmatrix} -2 & -5 \\ -8 & -11 \\ 2 & 2 \end{bmatrix}$

19. $\begin{bmatrix} 7 & 5 & -4 & 4 \\ 6 & 2 & -5 & 8 \\ 13 & 0 & 3 & 0 \\ 4 & -10 & 11 & 9 \end{bmatrix}$ *21.* $[-10]$

23. $\begin{bmatrix} -6 & -18 \\ 9 & -6 \end{bmatrix}$ *25.* $\begin{bmatrix} -2 & 6 & -10 & -4 \\ -4 & -2 & 2 & 0 \\ -12 & 0 & 8 & -16 \end{bmatrix}$

27. $\begin{bmatrix} 3 & -1 & 2 \\ -5 & 6 & 1 \end{bmatrix}$ *29.* $5A + 5B$

31. $x = -1, y = 7, z = -4$ *33.* $x = -1, y = 2, z = -\frac{21}{8}$

35. $2\begin{bmatrix} 3 & -1 \\ 0 & 6 \end{bmatrix}$ *37.* $\begin{bmatrix} 3 & 17-2i \\ -26-24i & -11+7i \end{bmatrix}$

39. $\begin{bmatrix} 13 & 32+9i \\ 13+26i & -7+9i \end{bmatrix}$ *41.* $\begin{bmatrix} 1+i & 2-i \\ 2+3i & 3-2i \end{bmatrix}$

48. Two-digit: $\begin{bmatrix} 0.46 & 0.5 \\ 0.75 & 5 \end{bmatrix}$. Four-digit: $\begin{bmatrix} 0.4583 & 0.5 \\ 0.75 & 5 \end{bmatrix}$.

49. If A is $m \times n$, $(rs)A$ requires $mn + 1$ multiplications and $r(sA)$ requires $2mn$ multiplications.

SECTION 2.2

1. $[0]$ *3.* $\begin{bmatrix} 0 & 0 & 0 \\ -1 & -3 & 2 \\ 2 & 6 & -4 \end{bmatrix}$ *5.* $\begin{bmatrix} 5 & 15 & -6 \\ -9 & -31 & 14 \end{bmatrix}$

7. $\begin{bmatrix} -36 & 4 & -8 \\ 4 & 0 & 5 \end{bmatrix}$ *9.* Cannot; sizes do not match.

11. $\begin{bmatrix} 10 & 4 & -9 \\ -11 & -10 & 7 \\ 31 & 2 & 5 \end{bmatrix}$ *13.* $\begin{bmatrix} 3 & -3 & -2 & 6 \\ 24 & -1 & -4 & 1 \\ 2 & 4 & 1 & 10 \\ 25 & 15 & -9 & 6 \end{bmatrix}$

15. $[1 \quad 1]B = [0 \quad 0 \quad 0]$, $[-1 \quad 1]B = [2 \quad -4 \quad -6]$, $A\begin{bmatrix} -1 \\ 1 \end{bmatrix} = \begin{bmatrix} 0 \\ 2 \end{bmatrix}$, $A\begin{bmatrix} 3 \\ -3 \end{bmatrix} = \begin{bmatrix} 0 \\ -6 \end{bmatrix}$

17. $[2 \quad 1]B = [4 \quad 6]$, $[1 \quad 2]B = [-1 \quad 9]$, $A\begin{bmatrix} 3 \\ -2 \end{bmatrix} = \begin{bmatrix} 4 \\ -1 \\ 5 \end{bmatrix}$, $A\begin{bmatrix} 1 \\ 4 \end{bmatrix} = \begin{bmatrix} 6 \\ 9 \\ -3 \end{bmatrix}$

19. (*a*) $\begin{bmatrix} 38 & -50 & 40 \\ -26 & 37 & -27 \\ -35 & 60 & -35 \end{bmatrix}$ (*b*) $\begin{bmatrix} -16 & 9 & 21 \\ 33 & -18 & -57 \\ -54 & 30 & 80 \end{bmatrix}$

21. $3AB$ *23.* $15BC - 9BA$ *25.* $A^2 + AB + BA + B^2$

27. $A^3 + A^2B + ABA + AB^2 + BA^2 + BAB + B^2A + B^3$

29. (*a*) $z_1 = -3x_1 - 8x_2$
$z_2 = 2x_1 + 3x_2$

(*b*) $z = \begin{bmatrix} -2 & 3 \\ 1 & -1 \end{bmatrix}\begin{bmatrix} 3 & 1 \\ 1 & -2 \end{bmatrix}\mathbf{x} = \begin{bmatrix} -3 & -8 \\ 2 & 3 \end{bmatrix}\mathbf{x}$

The results are essentially the same.

31. $\begin{bmatrix} -2 & 2 \\ -2 & -2 \end{bmatrix}$ **33.** $\begin{bmatrix} 2 & 1 & 1 \\ 0 & 1 & 0 \\ 1 & 1 & 1 \end{bmatrix}$

35. $\begin{bmatrix} 6i & 8+4i \\ 14+2i & 13-12i \end{bmatrix}$

42. $(A_1A_2\cdots A_k)^t = A_k^t \cdots A_2^t A_1^t$ **43.** No

44. $\begin{bmatrix} 0 & b \\ 0 & 0 \end{bmatrix}, \begin{bmatrix} 0 & 0 \\ c & 0 \end{bmatrix}$, or $\begin{bmatrix} a & b \\ c & d \end{bmatrix}$ with $a + d = 0, bc = -a^2$, and no zero entries

47. (a) $E_r \cdots E_1 = \begin{bmatrix} 1 & 0 & -1 \\ 0 & 1 & 0 \\ 0 & 0 & 1 \end{bmatrix}\begin{bmatrix} 1 & 0 & 0 \\ 0 & 1 & 0 \\ 0 & 0 & -1 \end{bmatrix} \times$

$\begin{bmatrix} 1 & 0 & 0 \\ 0 & 1 & 0 \\ 0 & -1 & 1 \end{bmatrix}\begin{bmatrix} 1 & 0 & 0 \\ 0 & 0 & 1 \\ 0 & 1 & 0 \end{bmatrix}\begin{bmatrix} 1 & 0 & 0 \\ 0 & 1 & 0 \\ -1 & 0 & 1 \end{bmatrix}\begin{bmatrix} 1 & 0 & 0 \\ -1 & 1 & 0 \\ 0 & 0 & 1 \end{bmatrix}$

(b) $E_r \cdots E_1 = \begin{bmatrix} 1 & 0 & 0 \\ 0 & \frac{1}{4} & 0 \\ 0 & 0 & 1 \end{bmatrix}\begin{bmatrix} -1 & 0 & 0 \\ 0 & 1 & 0 \\ 0 & 0 & 1 \end{bmatrix} \times$

$\begin{bmatrix} 1 & -\frac{1}{2} & 0 \\ 0 & 1 & 0 \\ 0 & 0 & 1 \end{bmatrix}\begin{bmatrix} 1 & 0 & 0 \\ 0 & 1 & 1 \\ 0 & 0 & 1 \end{bmatrix}\begin{bmatrix} 1 & 0 & -1 \\ 0 & 1 & 0 \\ 0 & 0 & 1 \end{bmatrix}\begin{bmatrix} 1 & 0 & 0 \\ 0 & 1 & 0 \\ 0 & 0 & \frac{1}{3} \end{bmatrix} \times$

$\begin{bmatrix} 1 & 0 & 0 \\ 0 & 1 & 0 \\ 0 & -2 & 1 \end{bmatrix}\begin{bmatrix} 1 & 0 & 0 \\ 0 & 1 & 0 \\ 3 & 0 & 1 \end{bmatrix}\begin{bmatrix} 1 & 0 & 0 \\ 2 & 1 & 0 \\ 0 & 0 & 1 \end{bmatrix}$

48. $\begin{bmatrix} y_1 \\ y_2 \end{bmatrix} = \begin{bmatrix} 34 & 23 \\ 46 & 31 \end{bmatrix}\begin{bmatrix} u_1 \\ u_2 \end{bmatrix}$

49. $A = \begin{bmatrix} 0 & 1 & 1 & 0 \\ 0 & 0 & 0 & 1 \\ 0 & 0 & 0 & 0 \\ 1 & 0 & 0 & 0 \end{bmatrix}, A^2 = \begin{bmatrix} 0 & 0 & 0 & 1 \\ 1 & 0 & 0 & 0 \\ 0 & 0 & 0 & 0 \\ 0 & 1 & 1 & 0 \end{bmatrix},$

$A^3 = \begin{bmatrix} 1 & 0 & 0 & 0 \\ 0 & 1 & 1 & 0 \\ 0 & 0 & 0 & 0 \\ 0 & 0 & 0 & 1 \end{bmatrix}$

50. (a) $\mathbf{x} = [2 \text{ million}, 5 \text{ million}], S = \begin{bmatrix} \frac{7}{8} & \frac{1}{8} \\ \frac{1}{10} & \frac{9}{10} \end{bmatrix},$

$\mathbf{x}S^2 = [\frac{391}{160} \text{ million}, \frac{729}{160} \text{ million}]$

(b) $\mathbf{x} = [15 \text{ million}, 20 \text{ million}], S = \begin{bmatrix} \frac{4}{5} & \frac{1}{5} \\ \frac{1}{4} & \frac{3}{4} \end{bmatrix},$

$\mathbf{x}S^2 = [18.1 \text{ million}, 16.9 \text{ million}]$

(c) $\mathbf{x} = [600, 400], S = \begin{bmatrix} \frac{4}{5} & \frac{1}{5} \\ \frac{1}{4} & \frac{3}{4} \end{bmatrix}, \mathbf{x}S^2 = [569, 431]$

51. $A(B + C)$

53. Two-digit: $\begin{bmatrix} 0.086 & 0.62 \\ 0.57 & 6.2 \end{bmatrix}$

Four-digit: $\begin{bmatrix} 0.08334 & 0.6112 \\ 0.5625 & 6.167 \end{bmatrix}$

SECTION 2.3

3. $A^{-1} = \begin{bmatrix} -1 & 1 \\ -2 & 1 \end{bmatrix}$ **5.** A^{-1} does not exist

7. A^{-1} does not exist **9.** $A^{-1} = \frac{1}{20}\begin{bmatrix} 16 & -9 & -31 \\ 8 & -7 & -13 \\ 4 & -1 & 1 \end{bmatrix}$

11. $A^{-1} = \begin{bmatrix} 1 & -1 & 0 & -1 \\ -1 & 1 & 1 & 0 \\ 1 & 0 & -1 & 0 \\ 0 & 0 & 0 & 1 \end{bmatrix}$ **13.** $\begin{bmatrix} 3 & -4 \\ -2 & 3 \end{bmatrix}$

15. $\mathbf{x} = \mathbf{0}$ **17.** $\mathbf{x} = \mathbf{b}A^{-1}$ **19.** $X = A^{-1}BA$

21. $X = \frac{1}{3}(C - B)A^{-1}$

23. $A^{-1} = \begin{bmatrix} 0 & 1 \\ 1 & -1 \end{bmatrix}, \mathbf{x} = \begin{bmatrix} 1 \\ 0 \end{bmatrix}, \begin{bmatrix} 1 \\ -2 \end{bmatrix}, \begin{bmatrix} 0 \\ 0 \end{bmatrix}$

25. $\begin{aligned} 5x_1 + 6x_2 &= -1 \\ 7x_1 \quad &= 3 \end{aligned}$ **27.** $\begin{aligned} x_1 + 4x_2 + 7x_3 &= -1 \\ 3x_1 + 5x_2 + 8x_3 &= 2 \\ -6x_1 - 4x_2 - 2x_3 &= 3 \end{aligned}$

29. $[1 \quad 3]\begin{bmatrix} x \\ y \end{bmatrix} = [2]$ **31.** $\begin{bmatrix} 1 & -1 \\ -2 & 2 \end{bmatrix}\begin{bmatrix} x \\ y \end{bmatrix} = \begin{bmatrix} 1 \\ -2 \end{bmatrix}$

33. $\mathbf{x} = \begin{bmatrix} 1 & 1 \\ 1 & -1 \end{bmatrix}\mathbf{u}, \mathbf{y} = \begin{bmatrix} 1 & 2 \\ -1 & 1 \end{bmatrix}\mathbf{x};$

$\mathbf{y} = \begin{bmatrix} 1 & 2 \\ -1 & 1 \end{bmatrix}\begin{bmatrix} 1 & 1 \\ 1 & -1 \end{bmatrix}\mathbf{u}$, so $\mathbf{u} = -\frac{1}{6}\begin{bmatrix} -2 & 1 \\ 0 & 3 \end{bmatrix}\mathbf{y}$.

35. $\begin{bmatrix} 2-a & 3-b \\ a & b \end{bmatrix}$, where a and b are arbitrary numbers

37. $\begin{bmatrix} -1 & 0 \\ 1 & \frac{1}{2} \end{bmatrix}$ **39.** $A^{-1} = \begin{bmatrix} 1+2i & 1-i \\ 2+i & -i \end{bmatrix}$

43. $(A_1A_2\cdots A_k)^{-1} = A_k^{-1}\cdots A_2^{-1}A_1^{-1}$ **44.** A^{-1}

45. A^t

46. D^{-1} is a diagonal matrix with the i,i-term $1/d_{ii}$

47. $T^{-1} = \begin{bmatrix} \frac{1}{t_{11}} & \frac{-t_{12}}{t_{11}t_{22}} & \frac{t_{12}t_{23} - t_{13}t_{22}}{t_{11}t_{22}t_{33}} \\ 0 & \frac{1}{t_{22}} & \frac{-t_{23}}{t_{22}t_{33}} \\ 0 & 0 & \frac{1}{t_{33}} \end{bmatrix}$

48. $\mathbf{x} = ABC\mathbf{z}$, $\mathbf{z} = C^{-1}B^{-1}A^{-1}\mathbf{x}$
49. $B = P^{-1}AP$, $\mathbf{w} = P^{-1}\mathbf{u}$
50. $R = P^{-1}AP$, $S = QBQ^{-1}$
51. $\mathbf{x}(k+1) = A\mathbf{x}(k)$ 52. $\frac{d}{dt}\mathbf{x}(t) = A\mathbf{x}(t)$
53. (a) 44, 31, 76, 57, 56, 47, 56, 37, 64, 48, 18, 9
(b) "Correct"
55. Two-digit: $A^{-1} \approx \begin{bmatrix} 1.1 & -0.20 \\ -0.40 & 1.2 \end{bmatrix}$.
Four-digit: $A^{-1} \approx \begin{bmatrix} 1.068 & -0.2034 \\ -0.4068 & 1.220 \end{bmatrix}$. (Unlike this example, in Section 2.6 we will see that occasionally two-digit or four-digit calculations of inverses are very poor.)

SECTION 2.4

1. (a) +1 (b) +1 (c) −1 (d) −1 (e) +1
3. −1 5. −15 7. −107
9. (a) 6 (b) −72
11. $\lambda = 1$ 13. $\lambda = -1, 3$ 15. $18 + 6i$
18. For the 3×3 case, it is sufficient to see what is produced by the pattern—det A. For $n \times n$ with $n \geqslant 4$, the pattern produces $2n$ terms, all in the definition of the determinant, and omits the other $n! - 2n > 0$ terms.
20. 37—yes, 42—no; the number of interchanges can be any odd number larger than 11
22. Two-digit: −1.1. Four-digit: −1.056
23. The definition uses a product involving numerous factors. The corollary uses interchanges and thus is simpler to use.

SECTION 2.5

1. det $A = 8$
(a) $B = \begin{bmatrix} 2 & 2 \\ 3 & -1 \end{bmatrix}$, det $B = -\det A = -8$
(b) $B = \begin{bmatrix} 9 & -1 \\ 6 & 2 \end{bmatrix}$, det $B = 3 \det A = 24$
(c) $B = \begin{bmatrix} -1 & -5 \\ 0 & -8 \end{bmatrix}$, det $B = \det A = 8$
3. det $A = 14$
(a) $B = \begin{bmatrix} 2 & -2 & 6 \\ 0 & 3 & 4 \\ 0 & -1 & 1 \end{bmatrix}$, det $B = \det A = 14$
(b) $B = \begin{bmatrix} 1 & 0 & 6 \\ 0 & 2 & 6 \\ 0 & 0 & 7 \end{bmatrix}$, det $B = \det A = 14$
(c) $B = \begin{bmatrix} 0 & 0 & 7 \\ 0 & 1 & 4 \\ -1 & 0 & 1 \end{bmatrix}$, det $B = \frac{1}{2} \det A = 7$
5. det $C = -2$ 7. det $C = -12$ 9. 60
11. −14 13. $\frac{45}{7}$ 15. $\frac{3}{20}$ 17. −296
19. $-\frac{1}{225}$ 21. $\begin{bmatrix} 1 & 2 \\ 2 & 4 \end{bmatrix}$ and infinitely many others
23. $\begin{bmatrix} 2 & 0 \\ 0 & 4 \end{bmatrix}$ and infinitely many others
25. This cannot be done, since the determinant will be 0.
27. det $A = 1$, det $B = 3$, det $rA = 4$, det $AB = 3$
29. det $A^2 = 25$ 31. $-22 - 21i$
42. No; det $A = 1$, so det $A^{150} = 1$, but
$\det \begin{bmatrix} 56 & -2 \\ -2 & 51 \end{bmatrix} = 2852.$
43. The algorithm requires 23 multiplications and divisions, vs. 72 for the definition. The algorithm is faster.
44. Two digit: −0.0037, for 9.6% error. Four-digit: −0.004097, for 0.1% error

SECTION 2.6

1. (a) −41 (b) 21 3. (a) 159 (b) 583
5. −102 7. −365 9. 32 11. −4 13. $\frac{18}{385}$
15. 12 17. $\begin{bmatrix} 3 & -2 \\ 2 & -1 \end{bmatrix}$ 19. $\begin{bmatrix} -2 & -1 \\ 0 & 2 \end{bmatrix}$
21. $\begin{bmatrix} 12 & -8 & 12 \\ 12 & 12 & -8 \\ -8 & 12 & 12 \end{bmatrix}$
23. det $A = 1$, so $A^{-1} = \begin{bmatrix} 3 & -2 \\ 2 & -1 \end{bmatrix}$.
25. det $A = -4$, so $A^{-1} = -\frac{1}{4}\begin{bmatrix} -2 & -1 \\ 0 & 2 \end{bmatrix}$.
27. det $A = 80$, so $A^{-1} = \frac{1}{80}\begin{bmatrix} 12 & -8 & 12 \\ 12 & 12 & -8 \\ -8 & 12 & 12 \end{bmatrix}$.
29. $x = 1$ 31. $x_2 = -\frac{19}{23}$ 33. $y = \frac{705}{338}$, $z = \frac{107}{338}$
35. $A^{-1} = \begin{bmatrix} \frac{1}{1-2\lambda} & \frac{-2}{1-2\lambda} \\ \frac{-\lambda}{1-2\lambda} & \frac{1}{1-2\lambda} \end{bmatrix}$, so A^{-1} is sensitive to λ near $\lambda = \frac{1}{2}$.

37. $\mathbf{x} = \begin{bmatrix} \dfrac{1-2\lambda}{1-\lambda} \\ \dfrac{1}{1-\lambda} \end{bmatrix}$, so $\mathbf{x}$ is sensitive to λ near $\lambda = 1$ and both variables are affected by λ.

39. $-27 - 36i$

41. $A^{-1} = -\dfrac{1}{13}\begin{bmatrix} 6+9i & -5-i \\ 1-5i & -3+2i \end{bmatrix}$

43. $A^{-1} = -\dfrac{1}{13}\begin{bmatrix} 16-2i \\ -6-9i \end{bmatrix}$ 47. 60

53. For R_1, all three currents decrease. For R_2, I_1 and I_2 decrease while I_3 increases. For R_3, I_1 and I_3 decrease while I_2 increases.

54. $p = 3\varepsilon/5 + 2/5$, which increases with increasing ε, and $x = -\varepsilon/5 + 1/5$, which decreases with increasing ε.

55. $x_1 = l + \dfrac{b-3l}{1+\dfrac{k_1}{k_2}+\dfrac{k_1}{k_3}}$ and $x_2 = b - l - \dfrac{b-3l}{1+\dfrac{k_3}{k_1}+\dfrac{k_3}{k_2}}$.

Thus, if $b < 3l$, x_1 shrinks and x_2 increases as k_2 increases. If $b = 3l$, nothing happens. If $b > 3l$, x_1 increases and x_2 shrinks as k_2 increases.

56. Using determinants, 280 operations are required for 4×4 and 39 for 3×3. Using the algorithm, 88 operations are required for 4×4 and 36 for 3×3. For $n \geq 3$, the algorithm is faster.

57. $\det A = 0.11$, so $A^{-1} = \dfrac{1}{0.11}\begin{bmatrix} 4.11 & -2 \\ -2 & 1 \end{bmatrix} \approx \begin{bmatrix} 37.36 & -18.18 \\ -18.18 & 9.09 \end{bmatrix}$. Using a two-digit computer, we have $\det A = 0.1$, so $A^{-1} = \dfrac{1}{0.1}\begin{bmatrix} 4.1 & -2 \\ -2 & 1 \end{bmatrix} = \begin{bmatrix} 41 & -20 \\ -20 & 10 \end{bmatrix}$. Thus the entries of A^{-1} are sensitive to roundoff.

SECTION 3.1

1. $\sqrt{5}$ 3. $2\sqrt{3}$ 5. $\sqrt{3}$ 7. $[4, -4]$

9. $[-7, -4]$ 11. $[-1, 1, -2]$ 13. Isosceles, right

15. Equilateral

17. No; $[0, 2]$, $[2, 4]$, and $[4, 6]$ are collinear.

19. $[x, y, z] = [-1, 1, -1] + t[2, -3, 1]$, $\dfrac{x+1}{2} = \dfrac{y-1}{-3} = \dfrac{z+1}{1}$

21. $(x-1) + (y-1) + (z-1) = 0$

23. The sum is $[3, 3]$.

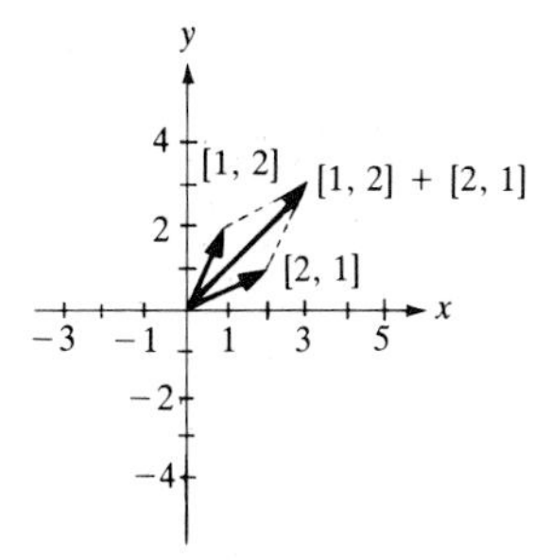

25. $[1, -2]$, $[-4, -6]$, $[-3, -8]$

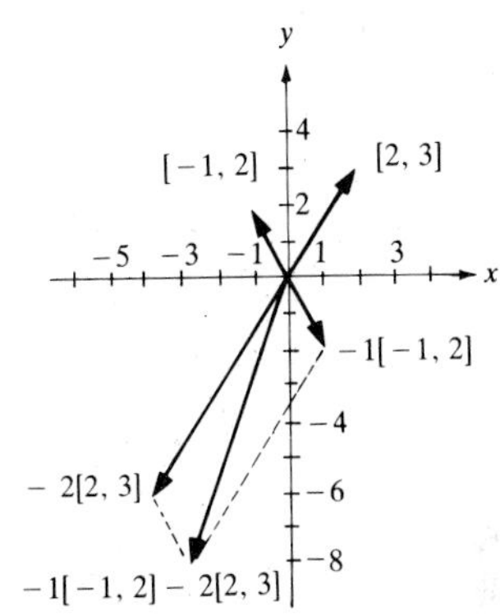

27. $[2, 2, 0]$, $[0, 3, 3]$, $[2, 5, 3]$

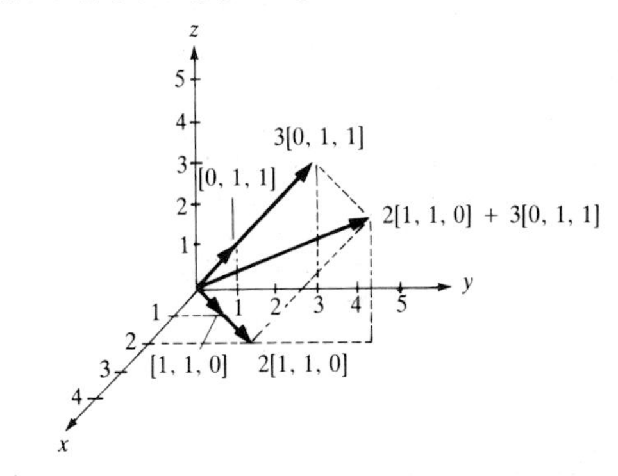

29. (a) (b)

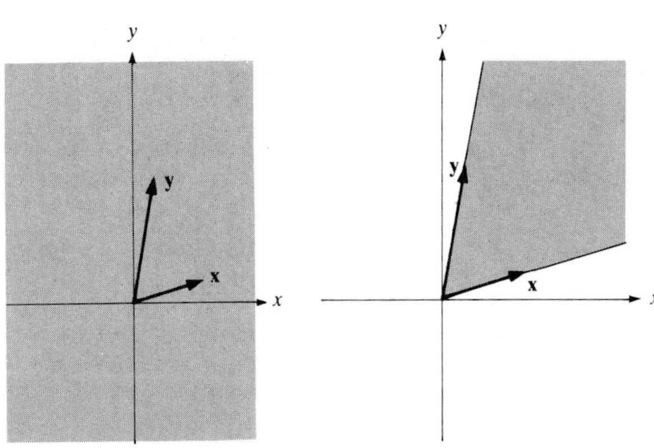

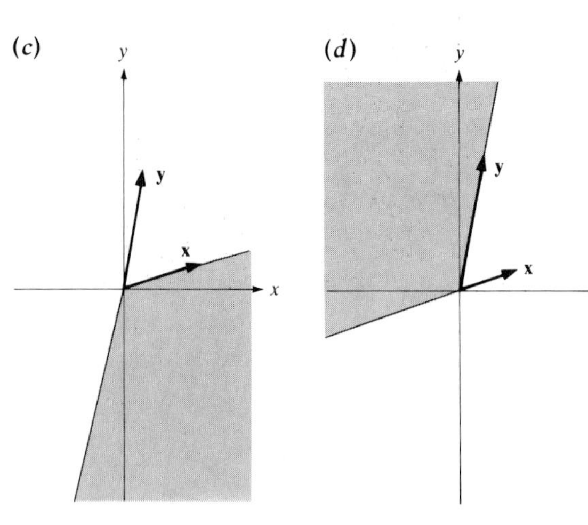

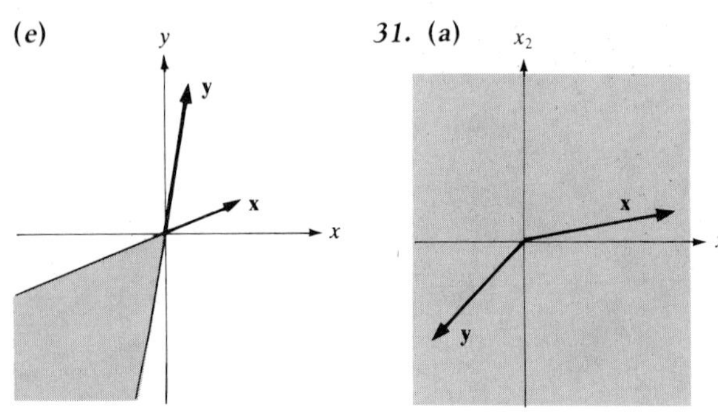

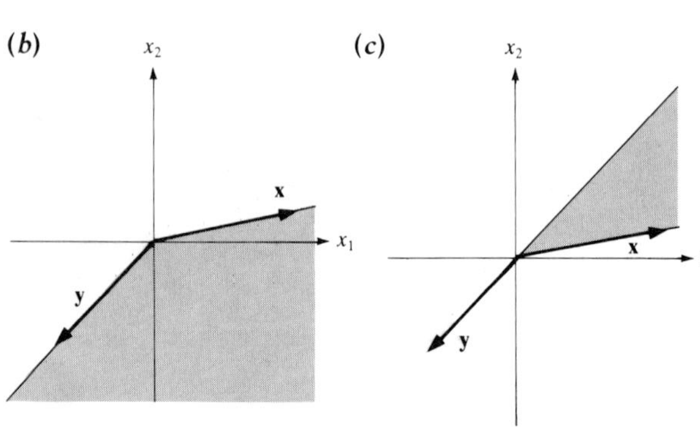

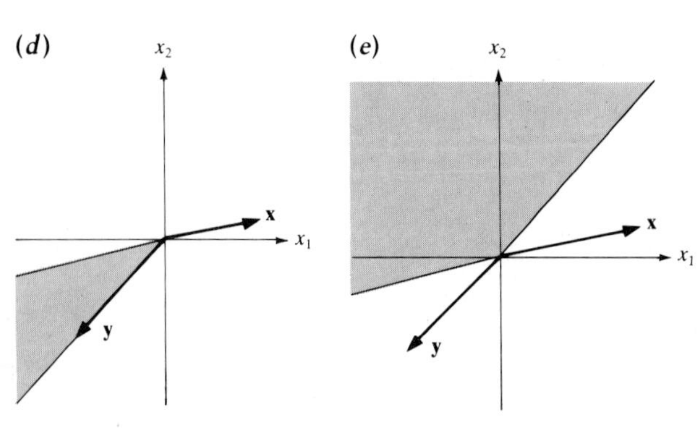

33. (*a*) -1 (*b*) 0 (*c*) 0
35. (*a*) $\sqrt{37}$ (*b*) $\sqrt{2}$
37. (*a*) $\sqrt{6}$ (*b*) $\sqrt{29}$
39. $\cos\theta = -1/\sqrt{2}$, $\theta = 3\pi/4$
41. $\cos\theta = -8/\sqrt{65}$, $\theta \approx 3.02$
43. $\cos\theta \approx 0.99272$, $\theta \approx 0.12$
45. $\cos\theta = 0$, $\theta = \pi/2$
47. $\cos\theta = -1/2$, $\theta = 2\pi/3$
51. $\|\mathbf{x}\| \cos\theta = \|\mathbf{x}\| \dfrac{(\mathbf{x}, \mathbf{y})}{\|\mathbf{x}\| \|\mathbf{y}\|} = \dfrac{(\mathbf{x}, \mathbf{y})}{\|\mathbf{y}\|}$ (*a*) 1 (*b*) -1
(*c*) $3/\sqrt{2}$ (*d*) $9/\sqrt{29}$
54. Direction [10, 150], speed $10\sqrt{226} \approx 150.3$ miles per hour
55. Direction [94.5, 90.1], speed ≈ 130.6 miles per hour
56. Direction [-7.50, 249.89], time 2.0009 hours
57. Direction [13, 2], force 186 pounds
58. Direction [45, -28, 5], force 21.7 pounds

SECTION 3.2

1. $-7\mathbf{x} + 3\mathbf{y}$ *3.* $-3\mathbf{x} - \mathbf{y} + \mathbf{z}$
5. $\mathbf{x} = -2\mathbf{a} + \frac{2}{3}\mathbf{b}$ *7.* $\mathbf{x} = \frac{5}{7}\mathbf{a} + \frac{2}{7}\mathbf{b}$
9. Exactly (i), (ii), 3, and 4 fail.
11. Only (ii) fails. Systems of this kind are useful for showing that the properties (i) and (ii) and 1 through 8 are independent in the sense that no one of them follows from the others.
13. Only (i) fails. *15.* Only 5 fails. *17.* Yes.
19. No; only (ii) and 4 fail. *21.* Yes.
23. No; (i), (ii), and 3 fail, but 4 does not fail.
27. Only three properties need to be checked to verify that an object is a subspace, whereas 10 need to be checked for a vector space. However, subspace verification can be used only when the set is a subset of a vector space, whereas vector space verification can always be used.
39. (1) Linear algebra students often talk about general vector spaces rather than particular ones. In this sense, they do not know what they are talking about. (2) Sometimes they have not learned the subject.
40. In each case we do not want to name a particular (number or vector space).
41. As a mathematical entity, an exchange system can be interpreted in terms of many different physical and abstract systems. For example, as seen in Chapter 1, we can view maze running by rats, economic systems, and population movements as exchange systems. This is useful in the same sense that all of mathematics is useful: results obtained in general can be applied over and over in many different particular situations.

Section 3.3

1. Column space of $A = \left\{x_1\begin{bmatrix}1\\1\end{bmatrix} + x_2\begin{bmatrix}1\\-1\end{bmatrix} : x_1, x_2 \in \mathbf{R}\right\} = \left\{\begin{bmatrix}1 & 1\\1 & -1\end{bmatrix}\mathbf{x} : \mathbf{x} \in \mathbf{R}^2\right\}$

3. Column space of $A = \left\{x_1\begin{bmatrix}1\\2\\-1\end{bmatrix} + x_2\begin{bmatrix}0\\1\\4\end{bmatrix} + x_3\begin{bmatrix}-1\\3\\2\end{bmatrix} : x_1, x_2, x_3 \in \mathbf{R}\right\} = \left\{\begin{bmatrix}1 & 0 & -1\\2 & 1 & 3\\-1 & 4 & 2\end{bmatrix}\mathbf{x} : \mathbf{x} \in \mathbf{R}^3\right\}$

5. $\begin{bmatrix}2 & 1\\1 & -1\\-3 & 2\end{bmatrix}\mathbf{x} = \mathbf{0}$. Yes.

7. $\begin{bmatrix}1 & -2 & 1\\2 & 0 & -1\end{bmatrix}\mathbf{x} = \begin{bmatrix}1\\2\end{bmatrix}$. No.

9. $[1 \quad 1 \quad 1]\mathbf{x} = \mathbf{0}$. Yes. *11.* No solution.

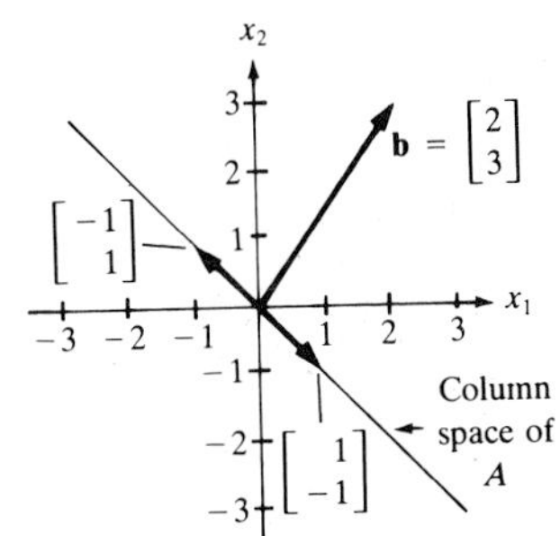

13. (*a*) $x(0) = 1$, $x(1) = 2$, $x(k+1) = 2^{k+1}$, $x(k+1) - x(k) = 2^k$
(*b*) $x(0) = 7$, $x(1) = 10$, $x(k+1) = 3k + 10$, $x(k+1) - x(k) = 3$
(*c*) $x(0) = 1$, $x(1) = 1/2$, $x(k+1) = 1/(k+2)$, $x(k+1) - x(k) = -1/(k^2 + 3k + 2)$

15. (*a*)

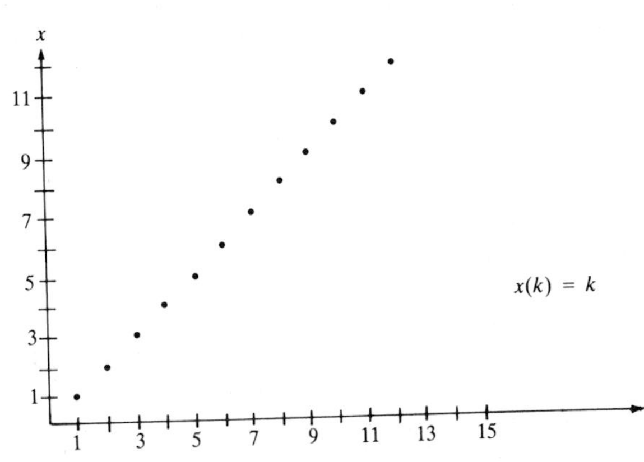

(*b*)

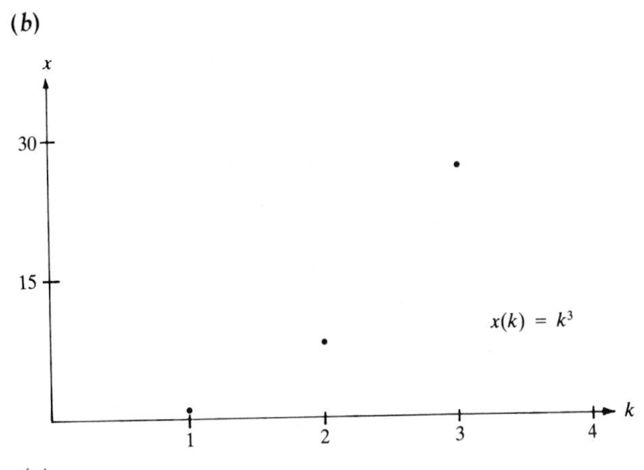

(*c*)

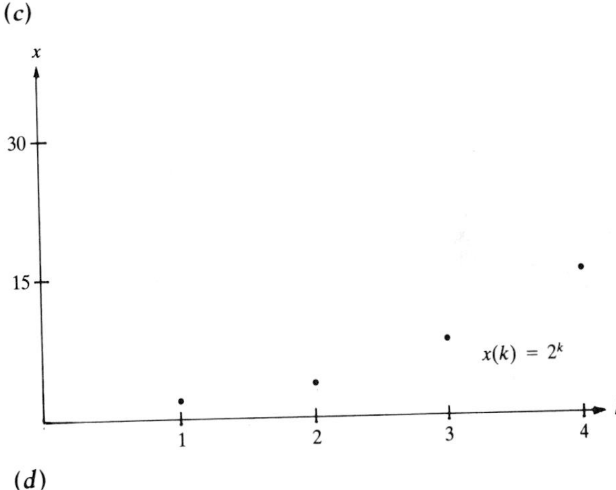

(*d*)

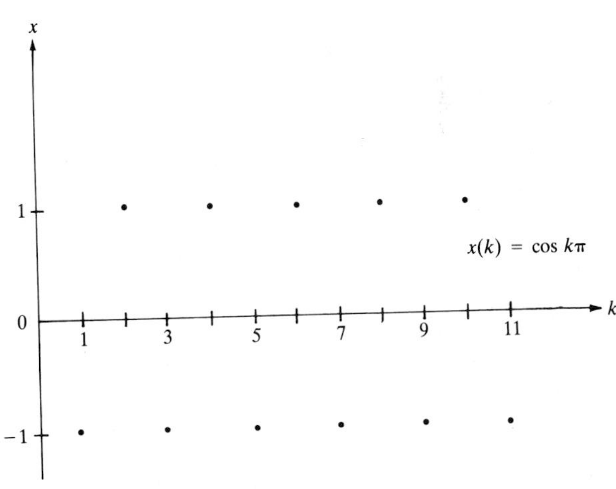

17. Yes *19.* No *21.* $x(k) = x(0)$
23. $x(k) = (-\frac{1}{2})^k x(0)$ *27.* Exactly (i), (ii), 3, and 4 fail.
29. Both (ii) and (iii) of the test for subspaces fail.
31. Both (ii) and (iii) of the test for subspaces fail.
35. $x(k) = (0.99)^k(10^7)$ *36.* $x(k) = (1.2)^k(10^3)$

37. $x_1(k+1) = \frac{1}{10}x_1(k) + 1000$,
$x_2(k+1) = \frac{9}{10}x_1(k) + \frac{1}{10}x_2(k)$. In matrix form,

$$\mathbf{x}(k+1) = \mathbf{x}(k)\begin{bmatrix} \frac{1}{10} & \frac{9}{10} \\ 0 & \frac{1}{10} \end{bmatrix} + [1000, 0].$$

$\mathbf{x}(1) = [1100, 990]$, $\mathbf{x}(2) = [1110, 1089]$,
$\mathbf{x}(3) \approx [1111, 1108]$.
The sizes are increasing and seem to be converging to a steady state.

38. $x_1(k+1) = \frac{1}{10}x_2(k) + 100$, $x_2(k+1) = \frac{9}{10}x_1(k)$. In matrix form, $\mathbf{x}(k+1) = \mathbf{x}(k)\begin{bmatrix} 0 & \frac{9}{10} \\ \frac{1}{10} & 0 \end{bmatrix} + [100, 0]$.

SECTION 3.4

1. (a) Yes. $\mathbf{z} = \frac{1}{2}\mathbf{x} + \frac{5}{4}\mathbf{y}$ (b) Yes. $\mathbf{z} = 2\mathbf{x} - \mathbf{y}$

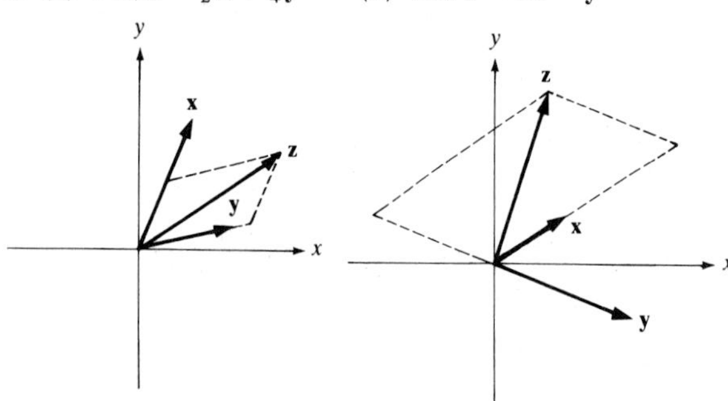

(c) No parallelogram (d) Yes. $\mathbf{z} = (-\frac{3}{2})\mathbf{x} - \mathbf{y}$. The other solutions use the pairs $\{\mathbf{x}, \mathbf{w}\}$ and $\{\mathbf{y}, \mathbf{w}\}$

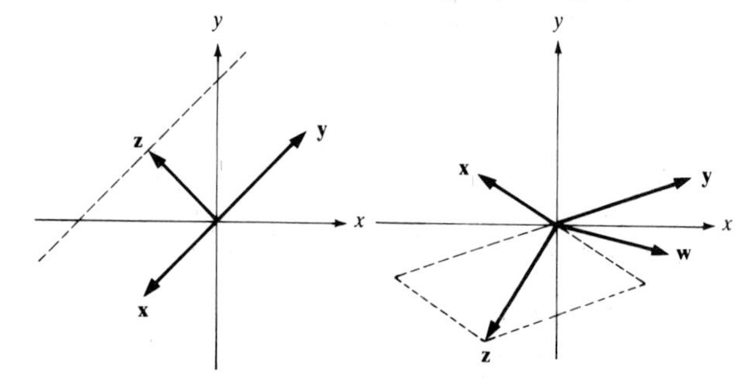

3. (a) 2, 1 (b) 2, −3 (c) 7, −2

5. (a) $[4, -2, 0] = 2[1, 1, 1] - [1, 1, -1] + 3[1, -1, -1]$
(b) $[-1, -5, -1] = -1[1, 1, 1] - 2[1, 1, -1] + 2[1, -1, -1]$

7. (a) $[2, 2, 3, 2] = [1, 1, 1, 0] + [0, 1, 1, 1] + [1, 0, 1, 1]$ (b) No

9. (a) $t^2 + 6t + 3 = (1 + 2r)(3t^2 - 2t - 1) + r(-6t^2 + 4t + 2) - 2(t^2 - 4t - 2)$ for any real number r
(b) No

11. $\mathbf{R}_2$; the subset is $\{\mathbf{x}_1, \mathbf{x}_2\}$

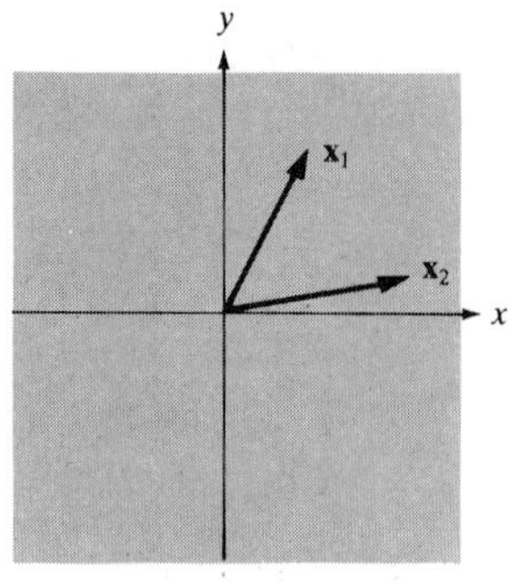

13. The plane of the vectors; the subsets are $\{\mathbf{x}_1, \mathbf{x}_2\}$, $\{\mathbf{x}_1, \mathbf{x}_3\}$, and $\{\mathbf{x}_2, \mathbf{x}_3\}$

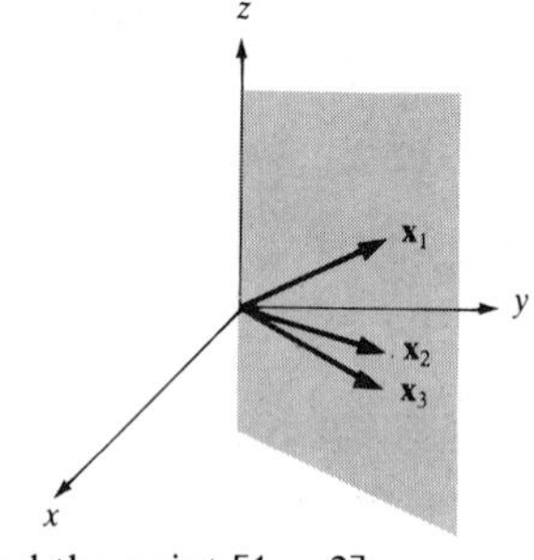

15. Line through the origin and the point $[1, -2]$

17. The plane of the vectors

19. $\mathbf{x} = -2[2, 1, 3] + [4, 7, -1]$

21. $\mathbf{x} = [1, -1, 0, 1] + [0, 1, -1, 1]$ 23. No

25. Yes 27. Yes 29. No 31. Yes

33. $t^3 + t^2 + t + 1 = \frac{4}{3}p_0(t) + \frac{8}{5}p_1(t) + \frac{2}{3}p_2(t) + \frac{2}{5}p_3(t)$

36. (a) The line is formed by taking all scalar multiples of vector **a**. Thus the line is finite dimensional. (b) The plane is formed by taking all linear combinations of vectors **a** and **b**. Thus the plane is finite dimensional.

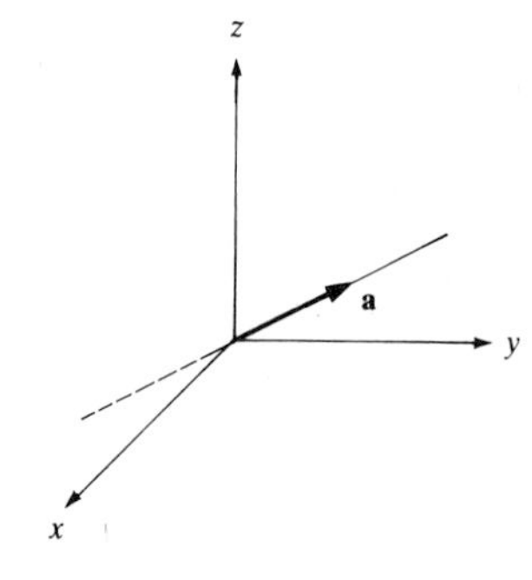

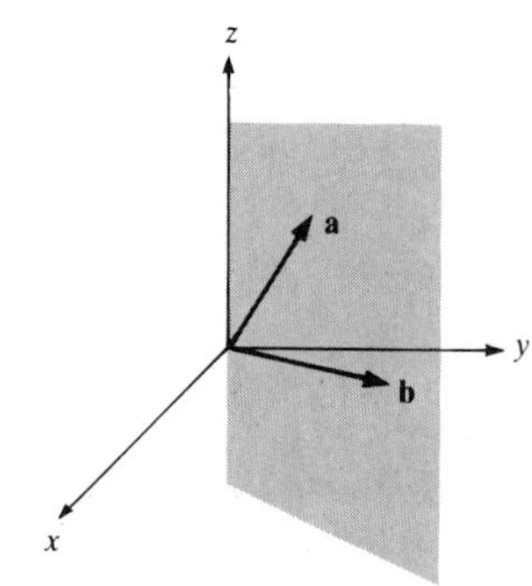

39. (a) $\mathbf{x} = [1, 1] + t[1, -1]$
(b) $\mathbf{x} = [1, 2, 0] + t[0, 0, 1]$
(c) $\mathbf{x} = [1, -1, 1, 1] + t[2, 1, -2, 3]$
(d) $\mathbf{x} = [1, 0, 1, -1, 2] + t[0, 1, -1, 3, 2]$

40. (a) Yes; $r = \frac{1}{2}$ (b) No; $\mathbf{x} = -\mathbf{y} + 2\mathbf{z}$
(c) Yes; $r = \frac{1}{2}$

41. $r_1 = 4$, $r_2 = -3$

42. $r_1 = -1$, $r_2 = 1$

43. $r_1 = -\frac{1}{2}$, $r_2 = \frac{3}{2}$

44. In $\mathbf{P}_n$, use the coefficients of each polynomial for each column of A. In $\mathbf{R}_{m,n}$, each column of A consists of all of the entries of one of the matrices.

45. (a) $u_1(0) = u_2(0) = 1$ (b) $u_1(0) = -1$, $u_2(0) = -\frac{1}{2}$

46. $\mathbf{W}(1) = \text{span}\left\{\begin{bmatrix}1\\0\\0\end{bmatrix}, \begin{bmatrix}-1\\ \frac{1}{2} \\ \frac{1}{2}\end{bmatrix}\right\}$. There is a choice of inputs such that the water levels x_1, x_2, and x_3 can be achieved.

The equations are
$x_1(k+1) = x_1(k) + u_1(k) - u_2(k)$
$x_2(k+1) = x_2(k) + \frac{1}{2}u_2(k)$
$x_3(k+1) = x_3(k) + \frac{1}{2}u_2(k)$

47. Any $\mathbf{z}'$ with tip in the sphere and not in the x,y-plane

SECTION 3.5

1. (*a*) Yes; single nonzero vector—$\mathbf{x}_1$ points out of origin.

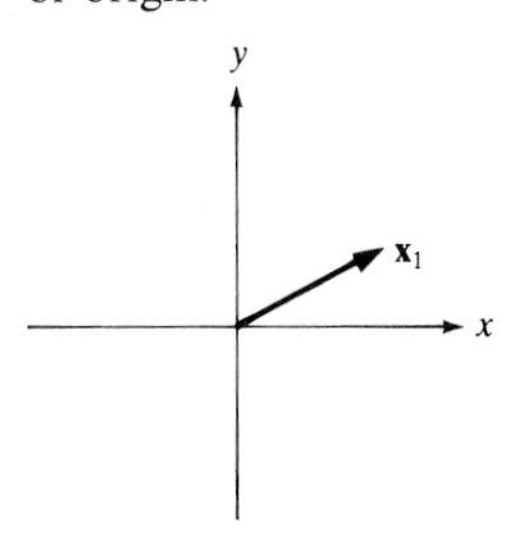

(*b*) Yes; two noncollinear vectors—$\mathbf{x}_1$ points out of l.

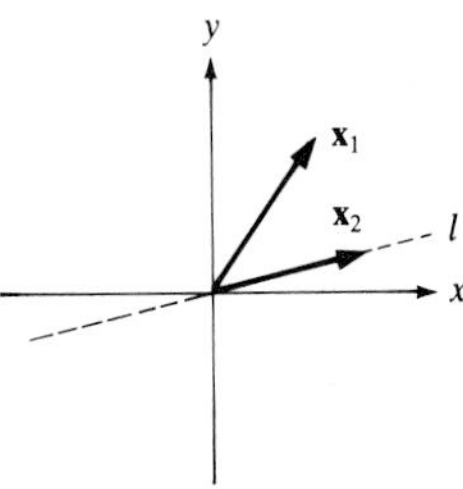

(*c*) No; two collinear vectors—$\mathbf{x}_2$ does not point out of l.

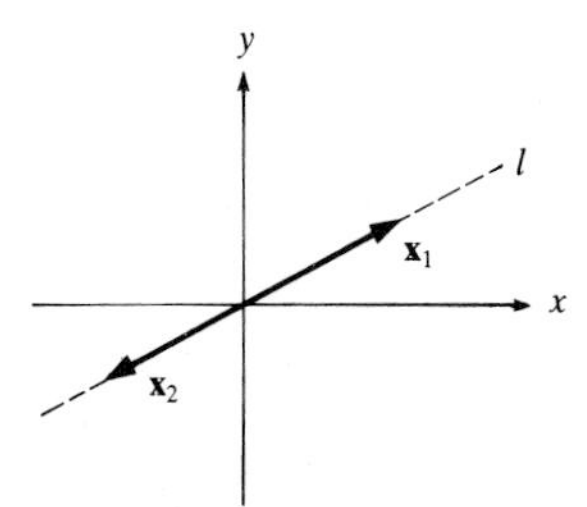

3. (*a*) span $\{\mathbf{x}_1\}$ = span $\{\mathbf{x}_2\}$ = span $\{\mathbf{x}_1, \mathbf{x}_2\} = l$

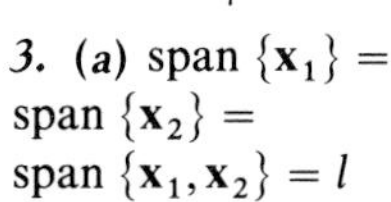

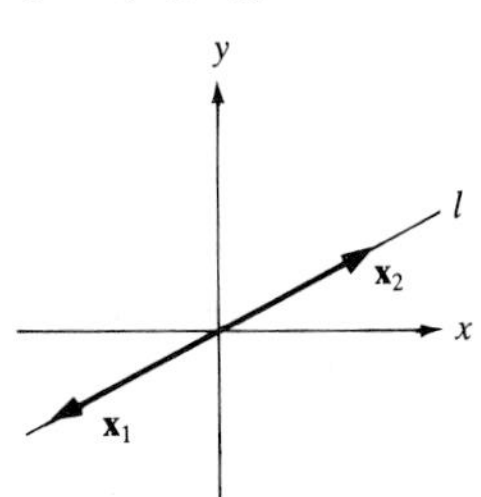

(*b*) span $\{\mathbf{x}_1, \mathbf{x}_2, \mathbf{x}_3\}$ = span $\{\mathbf{x}_1, \mathbf{x}_2\}$ = span $\{\mathbf{x}_1, \mathbf{x}_3\}$ = span $\{\mathbf{x}_2, \mathbf{x}_3\} = \mathbf{R}^2$

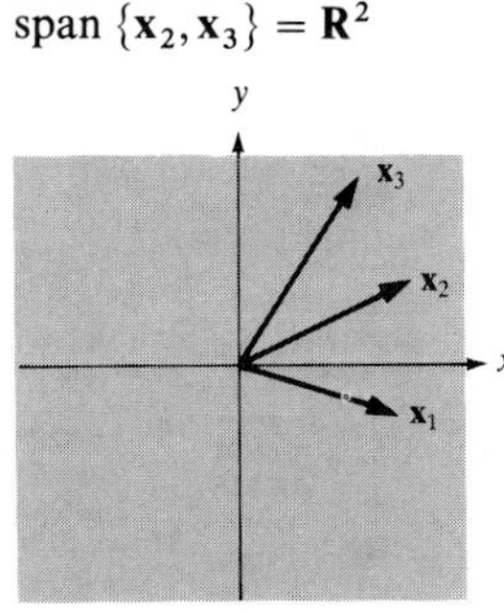

(*c*) span $\{\mathbf{x}_1\} = l$

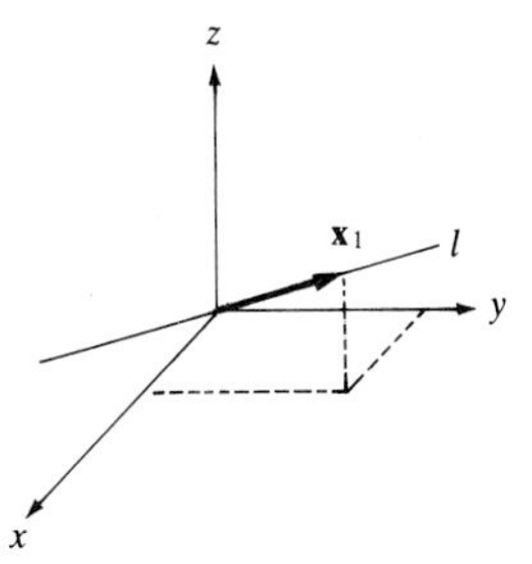

5. Linearly dependent; $[-3, 0] = -3[1, 0]$
7. Linearly independent
9. Linearly dependent; $[-5, 8] = (-\frac{31}{5})[1, -1] + (\frac{3}{5})[2, 3]$
11. Linearly independent *13.* Linearly dependent
15. Linearly independent *17.* Linearly dependent
19. Linearly independent *21.* Linearly independent
23. Linearly dependent
25. Yes; vectors are not on line, so they span $\mathbf{R}_2$. *27.* Yes; vectors are not on same plane, so they span $\mathbf{R}^3$.

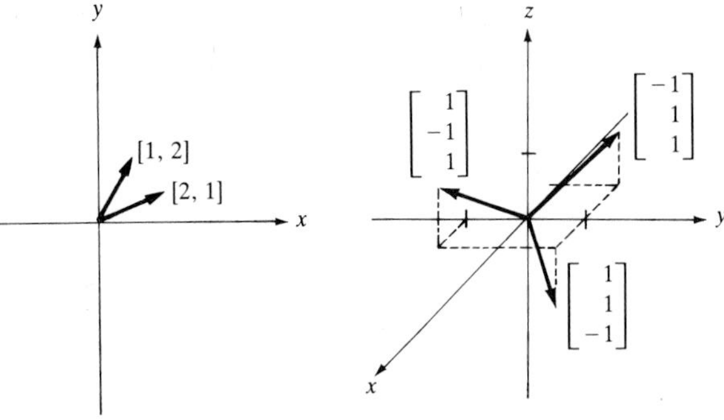

29. Yes *31.* No *33.* Yes *35.* No *37.* No
39. Yes *41.* Yes
43. $(y - z = 0) = (x - z = 0) - (x - y = 0)$
44. Linearly independent
54. (*a*) Yes (*b*) No (*c*) Yes

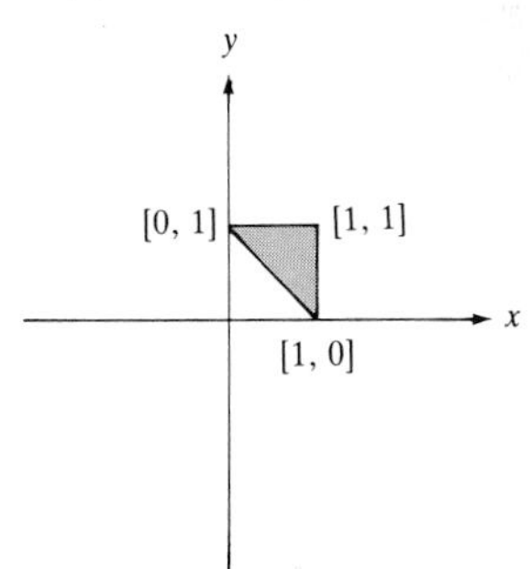

55. $\mathbf{x}(k+1) = \mathbf{x}(k) + u_1\begin{bmatrix}-1\\ \frac{1}{2}\\ \frac{1}{2}\end{bmatrix} + u_2\begin{bmatrix}1\\ -1\\ 0\end{bmatrix} + u_3\begin{bmatrix}0\\ 1\\ -1\end{bmatrix}$

Any one of the three lines could be deleted.

56.

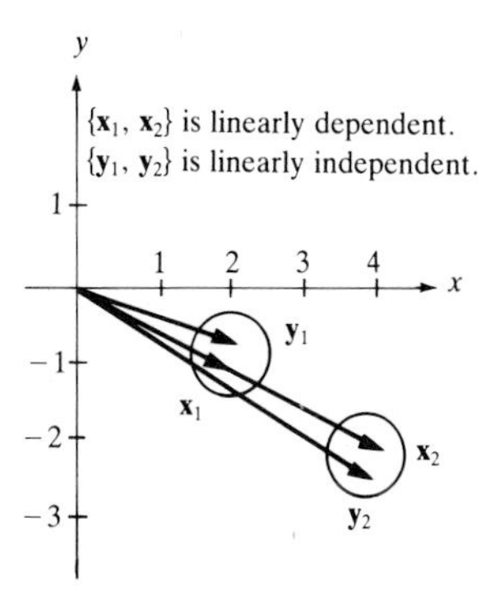

SECTION 3.6

1. $\{\mathbf{x}_1, \mathbf{x}_2\}$, $\{\mathbf{x}_1, \mathbf{x}_4\}$, $\{\mathbf{x}_2, \mathbf{x}_3\}$, $\{\mathbf{x}_3, \mathbf{x}_4\}$. All bases have the same number of vectors.

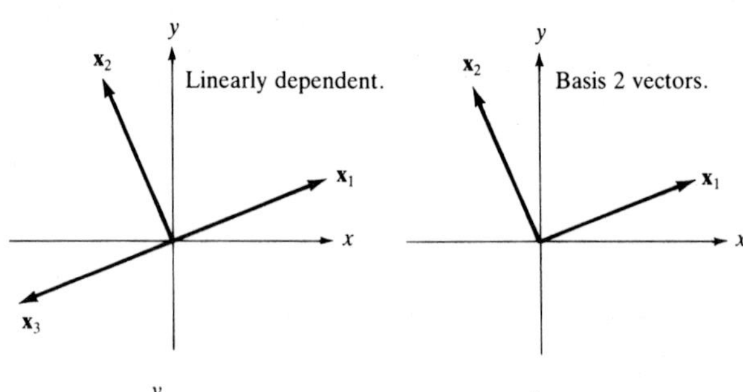

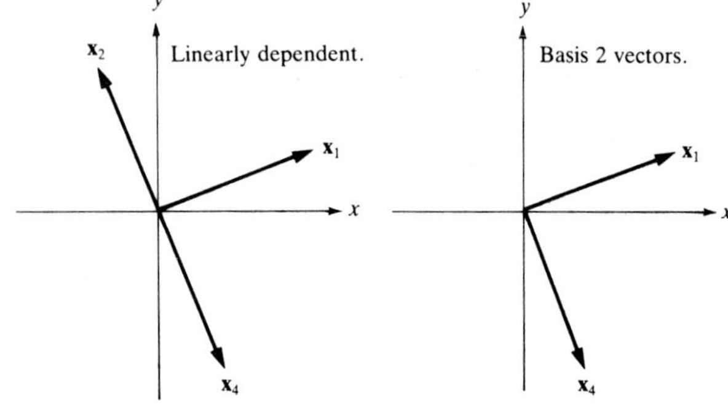

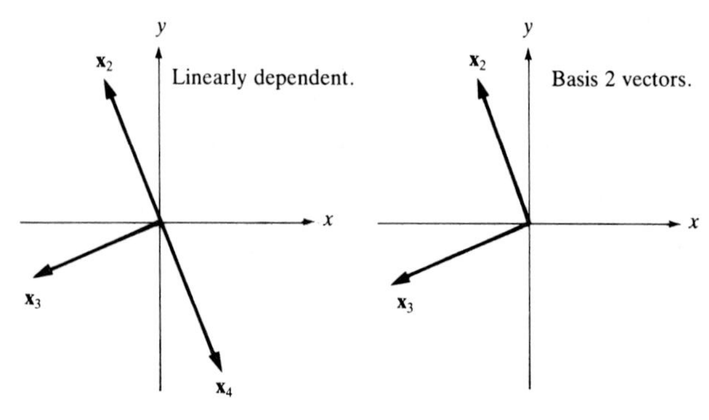

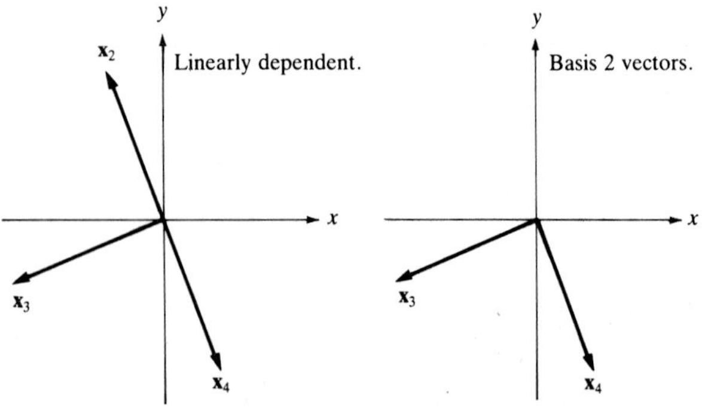

3. $\{\mathbf{x}_1, \mathbf{x}_2\}$, $\{\mathbf{x}_1, \mathbf{x}_3\}$, $\{\mathbf{x}_2, \mathbf{x}_3\}$. All bases have the same number of vectors.

5. All bases have two vectors.

(*a*)

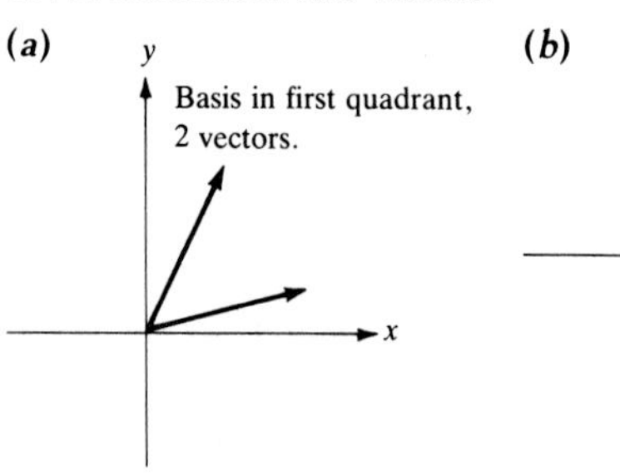

(*b*)

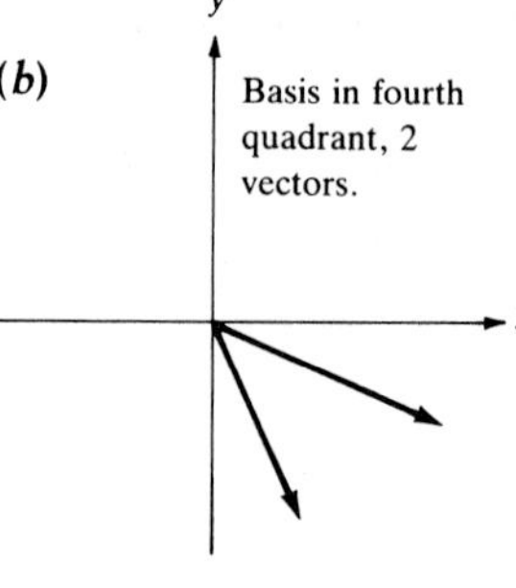

(*c*)

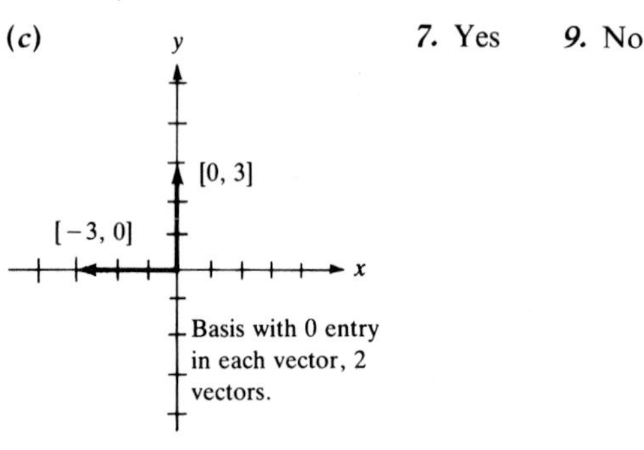

7. Yes *9.* No

11. No *13.* Yes *15.* $\{[1,2],[1,3]\}$
17. $\{[-1,1,0],[1,0,-1],[0,1,1]\}$
19. $\{[1,1,0,0],[0,1,1,0],[0,0,1,1],[1,0,1,0]\}$
21. $\{t^2+t-1, 2t^2-3\}$
23. Linearly dependent; use either of $\{1, 2^k\}$ or $\{2, 2^k\}$.
25. Linearly independent
27. The third equation is twice the first plus three times the second.
29. The third equation is equal to the first equation minus twice the second.
31. $\mathbf{R}_3$; linearly independent
33. $\mathbf{R}_3$; $\{[-2,1,-1],[1,0,-1],[0,-1,1]\}$ *40.* Yes
44. No; the vectors are coplanar. *45.* (*b*) No

SECTION 3.7

1. Row space basis: $\{[1,2]\}$; column space basis: $\left\{\begin{bmatrix}1\\2\end{bmatrix}\right\}$; dimension 1

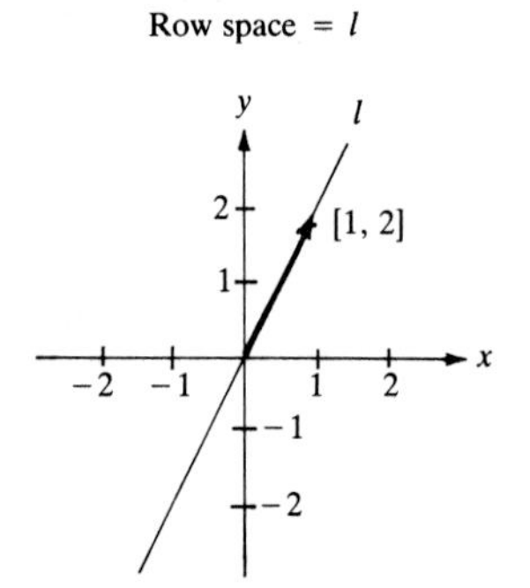

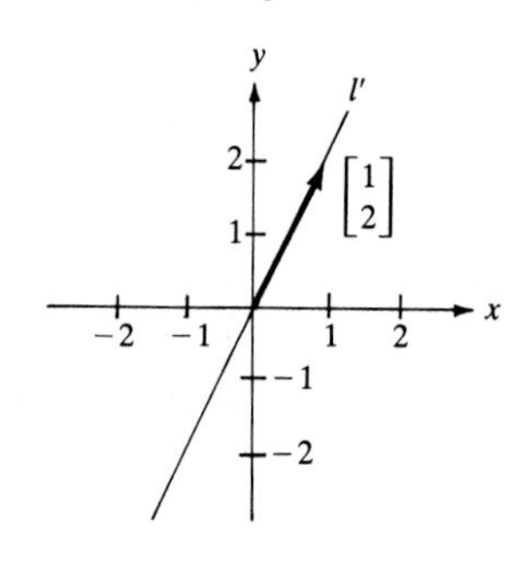

3. Row space basis: $\{[-1,3],[2,5]\}$; column space basis: $\left\{\begin{bmatrix}-1\\2\\4\end{bmatrix},\begin{bmatrix}3\\5\\-1\end{bmatrix}\right\}$; dimension 2

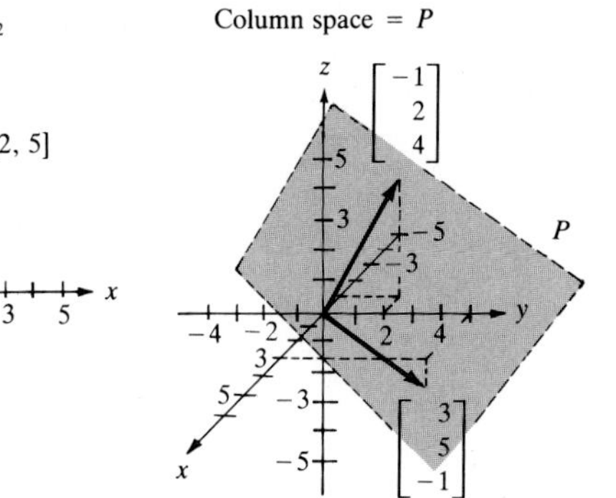

5. Rank = 1, so dimension = 1.
7. Rank = 2, so dimension = 2.
9. Rank = 3, so dimension = 3.
11. Rank A = 1.
(*a*) Rank $[A \quad \mathbf{b}] = 1$. There is a solution.
(*b*) Rank $[A \quad \mathbf{b}] = 2$. There is no solution.
13. Rank A = 2.
(*a*) Rank $[A \quad \mathbf{b}] = 2$. There is a solution.
(*b*) Rank $[A \quad \mathbf{b}] = 3$. There is no solution.
15.

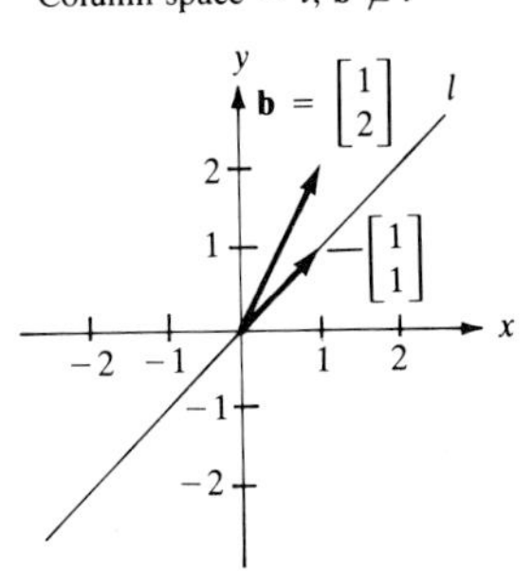

17.

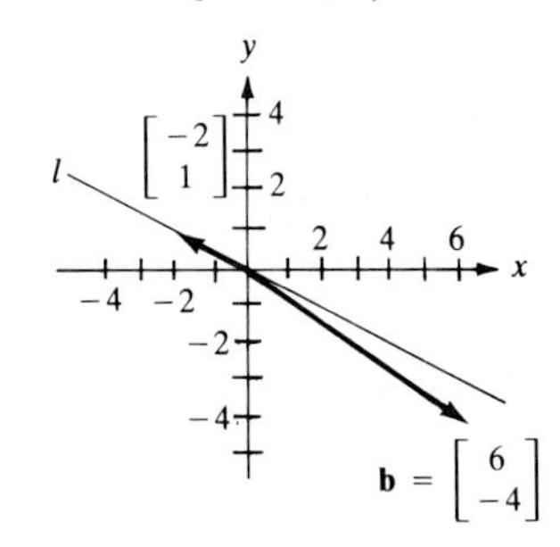

19. Rank 2 *21.* Rank 1
23. Rank 2. The solution set has dimension 1. An echelon form is $\begin{bmatrix}1&2&1\\0&1&1\\0&0&0\\0&0&0\end{bmatrix}$.
25. Rank 3. The solution set has dimension 2. An echelon form is $\begin{bmatrix}1&0&1&1&2\\0&1&-5&0&-5\\0&0&27&1&27\\0&0&0&0&0\end{bmatrix}$.

27. $\left\{\begin{bmatrix}\frac{3}{2}\\1\end{bmatrix}\right\}$, dimension 1
29. $\{[3,1,0]^t,[-7,0,1]^t\}$, dimension 2
31. $\{[-1,6,8]^t\}$, dimension 1
33. $\begin{bmatrix}1&1\\-1&1\end{bmatrix},\begin{bmatrix}1&2\\-1&1\end{bmatrix},\begin{bmatrix}1&-1\\1&1\end{bmatrix},\begin{bmatrix}1&2\\1&1\end{bmatrix},\begin{bmatrix}-1&2\\1&1\end{bmatrix},$
$\begin{bmatrix}1&1\\1&-1\end{bmatrix},\begin{bmatrix}2&1\\1&-1\end{bmatrix},\begin{bmatrix}-1&1\\1&1\end{bmatrix},\begin{bmatrix}2&1\\1&1\end{bmatrix},\begin{bmatrix}2&-1\\1&1\end{bmatrix}$
35. $x(k) = a_1(-4)^k + a_2(3^k)$
37. $x(k) = a_1(2^k) + a_2k(2^k)$
39. Dimension = 2; yes
41. Dimension = 2; no. The dimension of the described solution is 1.
43. Dimension = 3; yes
45. Rank 2
47. Rank $A = 1$, rank $[A \quad \mathbf{b}] = 2$; no solution
52. (*a*) $\left\{\begin{bmatrix}1\\0\\1\end{bmatrix},\begin{bmatrix}-1\\1\\0\end{bmatrix},\begin{bmatrix}1\\1\\1\end{bmatrix}\right\}$ (*b*) $\left\{\begin{bmatrix}2\\4\\-2\end{bmatrix},\begin{bmatrix}1\\3\\1\end{bmatrix},\begin{bmatrix}3\\1\\-2\end{bmatrix}\right\}$
57. Dimension = 1; basis is $\{1\}$.
58. Dimension = 2; basis is $\{1, t\}$.
59. Dimension = n; basis is $\{(n+1)t^n - 1, nt^{n-1} - 1, \ldots, 2t - 1\}$.
60. $x_1(k+1) = x_1(k) - u_1 + \frac{1}{2}u_2$
$x_2(k+1) = x_2(k) + \frac{1}{2}u_1 - u_2$
$x_3(k+1) = x_3(k) + \frac{1}{2}u_1 + \frac{1}{2}u_2$
$\mathbf{W}(1) = \text{span}\left\{\begin{bmatrix}-1\\\frac{1}{2}\\\frac{1}{2}\end{bmatrix},\begin{bmatrix}\frac{1}{2}\\-1\\\frac{1}{2}\end{bmatrix}\right\}$. The dimension is 2.
61. (*a*) $r(t) = \begin{cases}2 & \text{if } t \neq 1\\1 & \text{if } t = 1\end{cases}$ (*b*) $r(t) = \begin{cases}2 & \text{if } t \neq \pm 2\\1 & \text{if } t = \pm 2\end{cases}$

rank $A = \begin{cases}1 \text{ if } t = 1\\2 \text{ if } t \neq 1\end{cases}$ rank $A = \begin{cases}1 \text{ if } t = \pm 2\\2 \text{ if } t \neq \pm 2\end{cases}$

SECTION 4.1

1. Yes *3.* No. Fails: 1 only. *5.* Yes
7. No. Fails: 1 and 2 only. *9.* $(\mathbf{x},\mathbf{x}) + 2(\mathbf{x},\mathbf{y}) + (\mathbf{y},\mathbf{y})$
11. $(\mathbf{x},\mathbf{x}) + (\mathbf{x},\mathbf{z}) - (\mathbf{y},\mathbf{y}) - (\mathbf{y},\mathbf{z})$ *13.* $\sqrt{3}$ *15.* $\sqrt{39}$

17. $\sqrt{15}$ *19.* $1/\sqrt{3}$ *21.* $\sqrt{2}$ *23.* $\dfrac{1}{\sqrt{3}}[1,1,1]$

25. $\dfrac{1}{\sqrt{21}}\begin{bmatrix} 1 & -2 \\ 4 & 0 \end{bmatrix}$ *27.* $\sqrt{2}$ *29.* 6 *31.* $\sqrt{18}$

35. $d(\mathbf{x},\mathbf{z}) = 2$, $d(\mathbf{y},\mathbf{z}) = 2\sqrt{2}$, so $\mathbf{x}$ is closer to $\mathbf{z}$.
37. $d(\mathbf{x},\mathbf{z}) = 2$, $d(\mathbf{y},\mathbf{z}) = 2\sqrt{2}$, so $\mathbf{x}$ is closer to $\mathbf{z}$.
40. (*a*) $\sqrt{20}$ (*b*) $\sqrt{31}$ *48.* $d(p,q) = 1/5$
50. (*a*) BAT (*b*) CAT
52. (*a*) Error $\sqrt{0.02}$, relative error $1/\sqrt{151}$
(*b*) Error $\sqrt{0.0006}$, relative error $1/\sqrt{9901}$
53. (*a*) Error $\sqrt{0.03}$, relative error $\sqrt{3/323}$
(*b*) Error 0.2, relative error $1/\sqrt{161}$
54. A2 is better.

SECTION 4.2

1. 1.11 rad *3.* $\pi/3$ *5.* $5\pi/6$ *7.* $\pi/2$ *9.* $\pi/2$
11. $\{[1,-1],[1,1]\}$ only
13. $\{[1,1,1],[2,-3,1]\}$, $\{[3,2,0],[2,-3,1]\}$
15. $\{[1,-1,1,-1],[1,1,1,1]\}$,
$\{[1,-1,1,-1],[3,1,-2,0]\}$

17. All pairs *19.* $\left\{t, 1-\dfrac{3t}{2}\right\}, \left\{1-\dfrac{3t}{2}, t-\dfrac{2t}{3}\right\}$

21. $\{[1,1],[-1,1]\}$
23. $\{[1,1,0],[1,-1,2],[-1,1,1]\}$

25. $\left\{\dfrac{1}{\sqrt{14}}[3,-1,1], \dfrac{1}{\sqrt{532}}[-1,12,9]\right\}$

27. $\left\{\dfrac{1}{\sqrt{3}}[1,0,-1,1], \dfrac{1}{\sqrt{33}}[-2,3,2,4], \dfrac{1}{\sqrt{110}}[4,-6,7,3]\right\}$

29. $\left\{\dfrac{1}{\sqrt{3}}\begin{bmatrix} 1 & 1 \\ 1 & 0 \end{bmatrix}, \dfrac{1}{\sqrt{15}}\begin{bmatrix} 1 & 1 \\ -2 & 3 \end{bmatrix}\right\}$

31. $\{1, (2\sqrt{3})t - \sqrt{3}, (6\sqrt{5})t^2 - (6\sqrt{5})t + \sqrt{5}\}$
36. The excess vectors are reduced to **0**. Of course, **0**'s should be deleted.
38. $\mathbf{z} = t[1,-1,1], t \neq 0$ *40.* $\{1, t-1, \frac{1}{2}t^2 - 2t + 1\}$

41. $\left\{\dfrac{1}{\pi^{1/2}}, \dfrac{\sqrt{2}}{\pi^{1/2}}t, \dfrac{\sqrt{2}}{2\pi^{1/2}}(2t^2-1)\right\}$

42. $\left\{\dfrac{1}{\sqrt{\pi}}, \dfrac{\sqrt{2}}{\sqrt{\pi}}t, \dfrac{2\sqrt{2}}{\sqrt{\pi}}\left(t^2 - \dfrac{1}{2}\right)\right\}$

45. $\mathbf{x}_1, \mathbf{y}_1$ angle is 2.21 rad; $\mathbf{x}_2, \mathbf{y}_2$ angle is $\pi/2 \approx 1.57$ rad.
47. (*a*) $x_1 + x_2 + x_3 = 3$ (*b*) $x_2 = 0$
(*c*) $2x_1 - x_2 + 3x_3 + x_4 = 10$
(*d*) $3x_1 - x_2 + 2x_3 - 4x_4 + 5x_5 = -4$

SECTION 4.3

1. (*a*) $\|\varepsilon\| = 0$ (*b*) $\|\varepsilon\| = 1$ (*c*) $\|\varepsilon\| = \sqrt{13}$
3. $x = 0, y = \frac{3}{2}$; $A\mathbf{x} = [\frac{3}{2}, 3, \frac{3}{2}]^t$

5. $x = -\frac{5}{7}, y = \frac{40}{7}$, $A\mathbf{x} = \begin{bmatrix} \frac{30}{7} \\ \frac{25}{7} \\ \frac{15}{7} \end{bmatrix}$

7. $x = 2 - 2y - 3z$; y, z free; $A\mathbf{x} = [2]$
9. $y = \frac{37}{70}x - \frac{4}{7}$

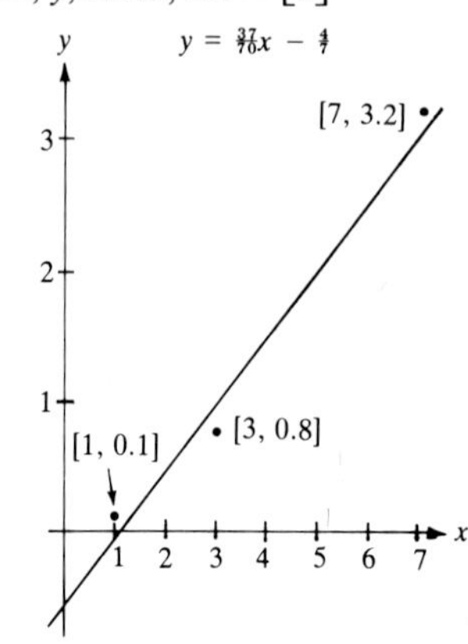

11. $y = \frac{3}{4}x^2 + \frac{69}{20}x + \frac{77}{20}$ *13.* $z = -\frac{5}{3}x + \frac{5}{2}y + \frac{3}{2}$
15. $\mathbf{u} = [\frac{1}{2}, \frac{1}{2}]$ *17.* $\mathbf{u} = [\frac{1}{3}, \frac{1}{3}, \frac{1}{3}]$

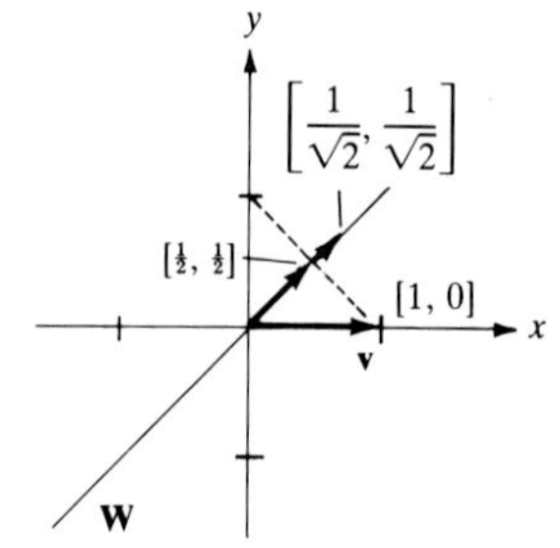

19. $\frac{3}{5}t$ *24.* $t \approx -\dfrac{2}{\pi}(\sin t) - \dfrac{1}{\pi}\sin 2t$

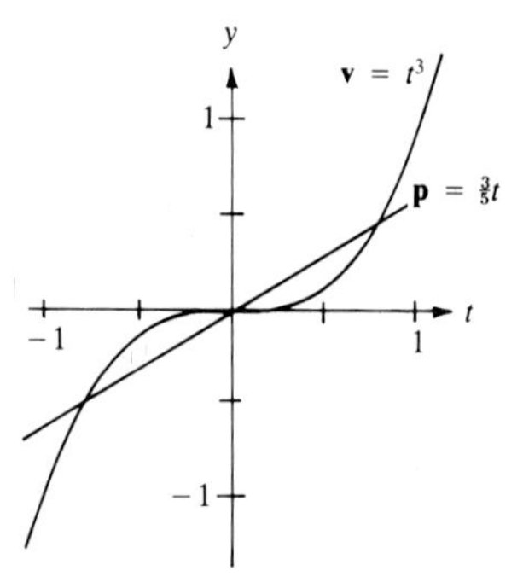

25. $y = ax + \frac{3}{2} - a$ for any real number a. *Note:* This gives all lines through the point $[1, \frac{3}{2}]$ having slope.
28. $p = -\frac{675}{122}x + \frac{44925}{61}$ cents. $p = 500$ implies $x \approx 43$.

29. $p = k^2 - 8k + 17$, $p(6) = 5$. Reason: The vertex of the parabola is at $[4, 1]$, so after $k = 4$ the price increases.
30. $h = -0.025d^2 + 0.42d + 1.99$
31. $p = 1.494y - 2761.72$, $p(1940) \approx 136.6$, $p(1950) \approx 151.6$

SECTION 5.1

1. $\begin{bmatrix} 0 \\ 0 \end{bmatrix}, \begin{bmatrix} 3 \\ 3 \end{bmatrix}, \begin{bmatrix} 6 \\ 6 \end{bmatrix}, \begin{bmatrix} -6 \\ -6 \end{bmatrix}, \begin{bmatrix} 3 \\ 3 \end{bmatrix}, \begin{bmatrix} 6 \\ 6 \end{bmatrix}, \begin{bmatrix} -6 \\ -6 \end{bmatrix}$

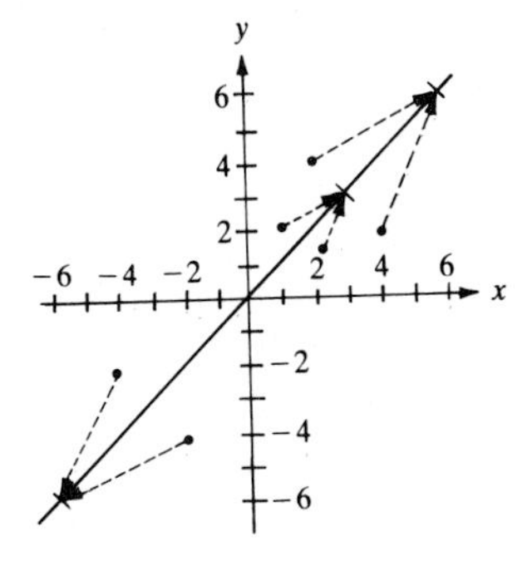

3. No; $L(\mathbf{0}) \neq \mathbf{0}$ 5. Yes 7. Yes
9. Yes 11. Yes
13. No; $L(x(k) + y(k)) = L(x(k)) + L(y(k)) + 2x(k)y(k)$
15. Yes 17. No; $L(ry) = r^2 L(y)$
19. (a) No; neither computational property is valid.
(b) No; neither computational property is valid.

21. (a) $L(\mathbf{x}) = \begin{bmatrix} 1 & 1 \\ 1 & -1 \end{bmatrix} \mathbf{x}$

(b) $L(\mathbf{x}) = \begin{bmatrix} 3 & -1 & 1 \\ 2 & 1 & 3 \\ 0 & 5 & 7 \end{bmatrix} \mathbf{x}$

23. $L(\mathbf{x}) = [-3, 5]$, $L(\mathbf{y}) = [5, -7]$
25. $L(\mathbf{x}) = 6(2^k) + 6(-3)^k$, $L(\mathbf{y}) = -2(2^k) - 12(-3)^k$
27. (a) $3\text{Å} - 2\Lambda + 2\psi$ (b) $2\Lambda - \psi + 3\Pi$
28. (a) $2\Xi - 3\Gamma + 2£$
29.

31. Linear, not homogeneous, form $\mathbf{x}_0 + \mathbf{K}$
33. Linear, homogeneous, form $\mathbf{K}$
35. Linear, not homogeneous, form $\mathbf{x}_0 + \mathbf{K}$
37. Not linear 39. Linear, homogeneous, form $\mathbf{K}$

41. $\begin{bmatrix} 2 \\ 0 \end{bmatrix} + t \begin{bmatrix} -1 \\ 1 \end{bmatrix}$

43. 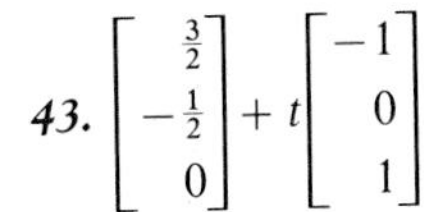
$\begin{bmatrix} \frac{3}{2} \\ -\frac{1}{2} \\ 0 \end{bmatrix} + t \begin{bmatrix} -1 \\ 0 \\ 1 \end{bmatrix}$

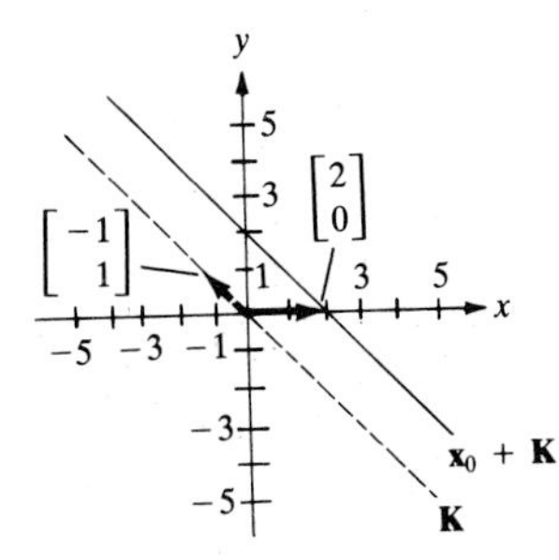

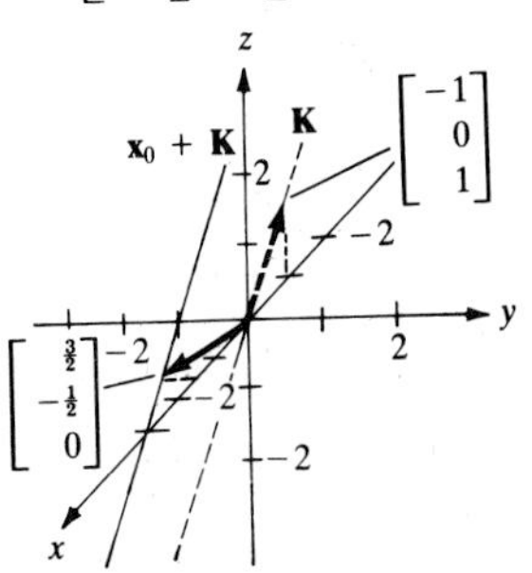

45. $-1 + r_1(2^k) + r_2(-2)^k$
47. $y = -\frac{3}{4} + r_1 e^{2t} + r_2 e^{-2t}$
49. (a) Yes; both properties are valid.
(b) No; neither property is valid. 50. (a) Yes

51. $\begin{bmatrix} -i \\ 1 + i \\ 0 \end{bmatrix} + z \begin{bmatrix} -3 \\ 1 - 2i \\ 1 \end{bmatrix}$

66. (a) kernel $L = \text{span}\left\{\begin{bmatrix} 1 \\ -1 \end{bmatrix}\right\}$, range $L = \text{span}\left\{\begin{bmatrix} 1 \\ 1 \end{bmatrix}\right\}$

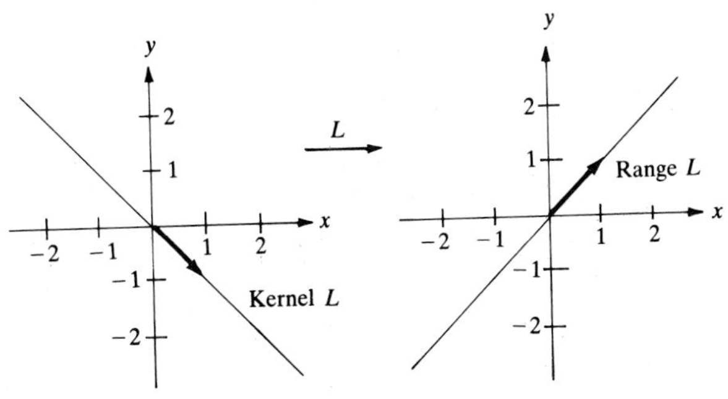

(b) kernel $L = \{0\}$, range $L = \text{span}\left\{\begin{bmatrix} 1 \\ 1 \\ 0 \end{bmatrix}, \begin{bmatrix} 0 \\ 1 \\ 1 \end{bmatrix}\right\}$

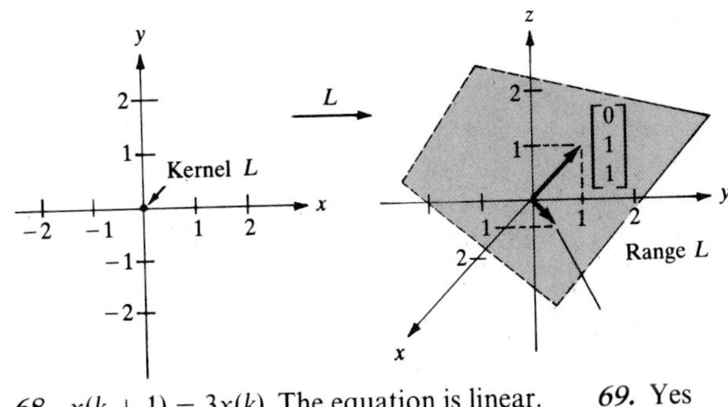

68. $x(k + 1) = 3x(k)$. The equation is linear. 69. Yes
71. No; $(A + B)^2 \neq A^2 + B^2$

SECTION 5.2

1. $[\mathbf{x}]_\# = \begin{bmatrix} 3 \\ 1 \end{bmatrix}$

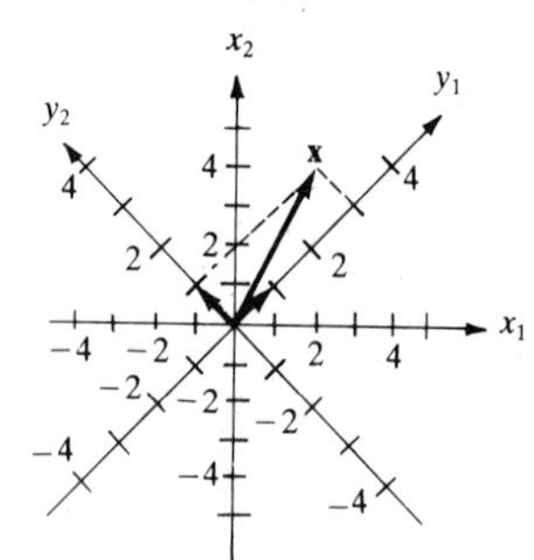

3. $[\mathbf{x}]_\# = \begin{bmatrix} -1 \\ -2 \end{bmatrix}$

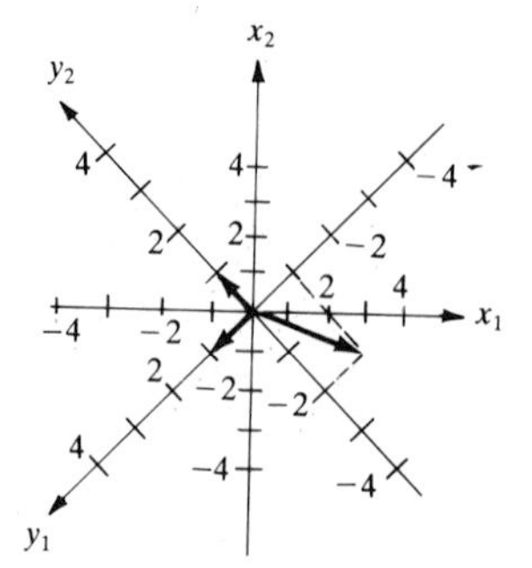

5. $[\mathbf{x}]_\# = \begin{bmatrix} 1 \\ 1 \\ 1 \end{bmatrix}$ 7. $[\mathbf{x}]_\# = \begin{bmatrix} 3 \\ -1 \end{bmatrix}$ 9. $[\mathbf{x}]_\# = \begin{bmatrix} 2 \\ 2 \\ 3 \end{bmatrix}$

11. $\mathbf{x} = \begin{bmatrix} -1 \\ 0 \end{bmatrix}$

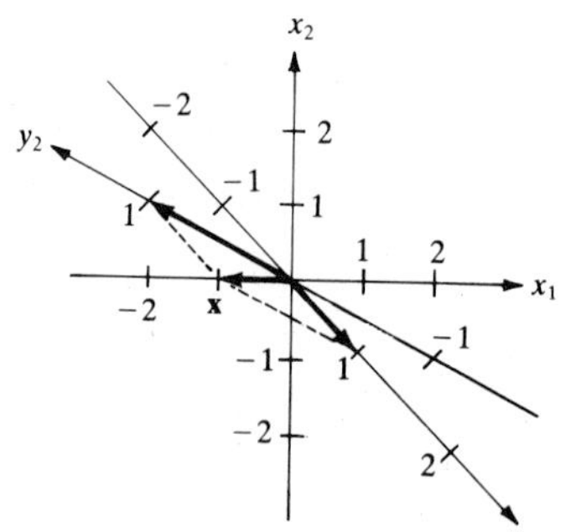

13. $\mathbf{x} = [2, 1]$ 15. $\mathbf{x} = [2, 0, 4]$ 17. $\mathbf{x} = t + 5$

19. $P = \begin{bmatrix} 1 & -1 \\ 1 & 1 \end{bmatrix}$, $[\mathbf{x}]_{\#_2} = \begin{bmatrix} 3 \\ 1 \end{bmatrix}$, $[\mathbf{x}]_{\#_1} = \begin{bmatrix} 2 \\ 4 \end{bmatrix}$,

$\begin{bmatrix} 1 & -1 \\ 1 & 1 \end{bmatrix}\begin{bmatrix} 3 \\ 1 \end{bmatrix} = \begin{bmatrix} 2 \\ 4 \end{bmatrix}$ 21. $P = \frac{1}{2}\begin{bmatrix} -1 & -2 \\ 1 & 0 \end{bmatrix}$,

$[\mathbf{x}]_{\#_2} = \begin{bmatrix} 1 \\ -1 \end{bmatrix}$, $[\mathbf{x}]_{\#_1} = \frac{1}{2}\begin{bmatrix} 1 \\ 1 \end{bmatrix}$, $\frac{1}{2}\begin{bmatrix} -1 & -2 \\ 1 & 0 \end{bmatrix}\begin{bmatrix} 1 \\ -1 \end{bmatrix} = \begin{bmatrix} \frac{1}{2} \\ \frac{1}{2} \end{bmatrix}$

23. $P = \begin{bmatrix} 1 & -1 & 0 \\ 1 & 1 & 0 \\ 0 & 0 & 1 \end{bmatrix}$, $[\mathbf{x}]_{\#2} = \begin{bmatrix} 1 \\ 0 \\ 1 \end{bmatrix}$, $[\mathbf{x}]_{\#_1} = \begin{bmatrix} 1 \\ 1 \\ 1 \end{bmatrix}$,

$\begin{bmatrix} 1 & -1 & 0 \\ 1 & 1 & 0 \\ 0 & 0 & 1 \end{bmatrix}\begin{bmatrix} 1 \\ 0 \\ 1 \end{bmatrix} = \begin{bmatrix} 1 \\ 1 \\ 1 \end{bmatrix}$

25. $P = \frac{1}{2}\begin{bmatrix} 1 & -1 \\ 1 & 1 \end{bmatrix}$, $[\mathbf{x}]_{\#_2} = \begin{bmatrix} 3 \\ 1 \end{bmatrix}$, $[\mathbf{x}]_{\#_1} = \begin{bmatrix} 1 \\ 2 \end{bmatrix}$,

$\frac{1}{2}\begin{bmatrix} 1 & -1 \\ 1 & 1 \end{bmatrix}\begin{bmatrix} 3 \\ 1 \end{bmatrix} = \begin{bmatrix} 1 \\ 2 \end{bmatrix}$

27. $y_1 = -3y_2^2$ 29. (a) $\begin{bmatrix} 1 - i \\ -1 - i \\ 3 - 2i \end{bmatrix}$ (b) $\begin{bmatrix} 1 + i \\ -i \\ 4 \end{bmatrix}$

32. $[\mathbf{x}]_\# = \begin{bmatrix} \frac{4}{3} \\ -\frac{1}{3} \end{bmatrix}$, $[\mathbf{y}]_\# = \begin{bmatrix} \frac{5}{3} \\ \frac{1}{3} \end{bmatrix}$. $d(\mathbf{x}, \mathbf{y}) = \sqrt{2}$, and the angle between $\mathbf{x}$ and $\mathbf{y}$ is approximately 0.64 rad. $d([\mathbf{x}]_\#, [\mathbf{y}]_\#) = (\sqrt{5})/3$, and the angle between $[\mathbf{x}]_\#$ and $[\mathbf{y}]_\#$ is approximately 0.44 rad. The # system is not a rotation of the natural system, so measurements taken in the grid determined by # will be different from those taken in the grid determined by the natural system.

33.

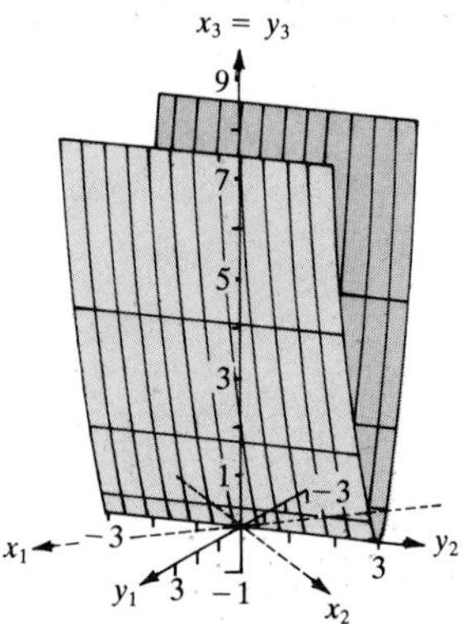

SECTION 5.3

1. $\frac{1}{2}\begin{bmatrix} 1 & -\sqrt{3} \\ \sqrt{3} & 1 \end{bmatrix}$

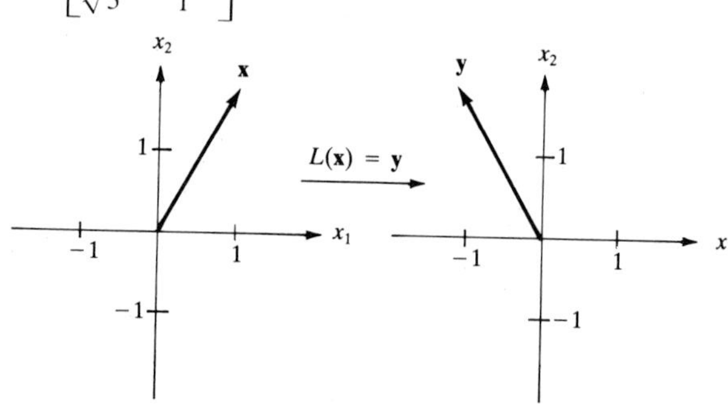

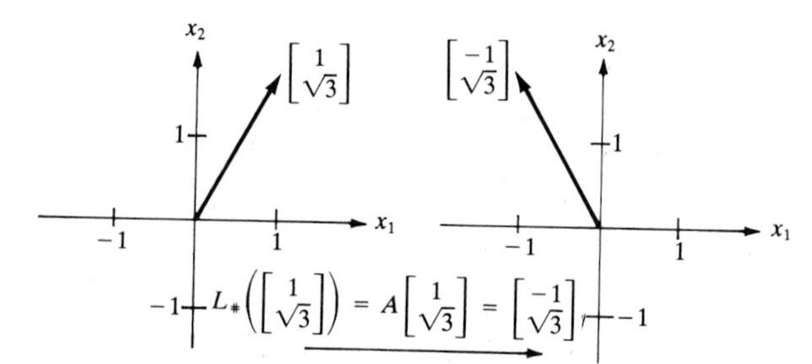

3. $\begin{bmatrix} 1 & 0 \\ 0 & -1 \end{bmatrix}$ 5. $\frac{1}{3}\begin{bmatrix} 1 & -2 & 2 \\ -2 & 1 & 2 \\ 2 & 2 & 1 \end{bmatrix}$

7. $L(\mathbf{x}) = [0, 2]$, $A = \begin{bmatrix} 1 & 0 \\ 0 & -1 \end{bmatrix}$, $[\mathbf{x}]_{\#} = \begin{bmatrix} 1 \\ -1 \end{bmatrix}$,

$[L(\mathbf{x})]_{\#} = \begin{bmatrix} 1 \\ 1 \end{bmatrix}$

9. $L(\mathbf{x}) = [5, 7]$, $A = \begin{bmatrix} 1 & 0 \\ 3 & 3 \end{bmatrix}$, $[\mathbf{x}]_{\#} = \begin{bmatrix} 2 \\ -1 \end{bmatrix}$,

$[L(\mathbf{x})]_{\#} = \begin{bmatrix} 2 \\ 3 \end{bmatrix}$

11. $L(\mathbf{x}) = [1, 3, 0]$, $A = \begin{bmatrix} 0 & 0 & 1 \\ 1 & 0 & 0 \\ 0 & 1 & 0 \end{bmatrix}$, $[\mathbf{x}]_{\#} = \begin{bmatrix} 1 \\ -1 \\ 2 \end{bmatrix}$,

$[L(\mathbf{x})]_{\#} = \begin{bmatrix} 2 \\ 1 \\ -1 \end{bmatrix}$

13. $L(\mathbf{x}) = \begin{bmatrix} 0 & 3 \\ 0 & 3 \end{bmatrix}$, $A = \begin{bmatrix} 1 & 0 & 1 & 0 \\ 0 & 1 & 0 & 1 \\ 1 & 0 & 1 & 0 \\ 0 & 1 & 0 & 1 \end{bmatrix}$, $[\mathbf{x}]_{\#} = \begin{bmatrix} 1 \\ 2 \\ -1 \\ 1 \end{bmatrix}$,

$[L(\mathbf{x})]_{\#} = \begin{bmatrix} 0 \\ 3 \\ 0 \\ 3 \end{bmatrix}$

15. $\#_2 = \left(\begin{bmatrix} 0 \\ 1 \end{bmatrix}, \begin{bmatrix} 1 \\ 0 \end{bmatrix}\right)$. $A\mathbf{x} = \begin{bmatrix} 3 \\ 4 \end{bmatrix}$ and $B[\mathbf{x}]_{\#_2} = B\begin{bmatrix} 1 \\ 1 \end{bmatrix} =$ $\begin{bmatrix} 4 \\ 3 \end{bmatrix}$, so $L(\mathbf{x}) = 4\begin{bmatrix} 0 \\ 1 \end{bmatrix} + 3\begin{bmatrix} 1 \\ 0 \end{bmatrix} = \begin{bmatrix} 3 \\ 4 \end{bmatrix}$.

17. $\#_2 = \left(\begin{bmatrix} 1 \\ 1 \\ 1 \end{bmatrix}, \begin{bmatrix} 0 \\ 0 \\ 1 \end{bmatrix}, \begin{bmatrix} 1 \\ -1 \\ 1 \end{bmatrix}\right)$

19. $B = \begin{bmatrix} -1 & 0 \\ 0 & 1 \end{bmatrix}$, $P^{-1} = \begin{bmatrix} 1 & -2 \\ -1 & 3 \end{bmatrix}$, $\mathbf{y}(k) = \begin{bmatrix} a_1(-1)^k \\ a_2 \end{bmatrix}$,

$\mathbf{x}(k) = \begin{bmatrix} 3a_1(-1)^k + 2a_2 \\ a_1(-1)^k + a_2 \end{bmatrix}$

21. $B = \begin{bmatrix} 1 & 0 & 0 \\ 0 & 2 & 0 \\ 0 & 0 & 0 \end{bmatrix}$, $P^{-1} = \begin{bmatrix} 2 & 3 & -7 \\ 1 & 2 & -4 \\ -5 & -8 & 19 \end{bmatrix}$,

$\mathbf{y}(k) = \begin{bmatrix} a_1 \\ a_2(2^k) \\ 0 \end{bmatrix}$, $\mathbf{x}(k) = \begin{bmatrix} 6a_1 - a_2(2^k) \\ a_1 + 3a_2(2^k) \\ 2a_1 + a_2(2^k) \end{bmatrix}$

23. $A = \begin{bmatrix} -3-5i & -12+3i \\ -2+2i & 3+5i \end{bmatrix}$, $[\mathbf{x}]_{\#} = \begin{bmatrix} -\frac{3}{2} - \frac{5}{2}i \\ -1+i \end{bmatrix}$,

$A[\mathbf{x}]_{\#} = [L(\mathbf{x})]_{\#} = \begin{bmatrix} 1 \\ 0 \end{bmatrix}$

25. $A = B = \begin{bmatrix} 0 & 1+i \\ 1-i & 0 \end{bmatrix}$, $P = \begin{bmatrix} 0 & i \\ 1 & 0 \end{bmatrix}$,

$P^{-1} = \begin{bmatrix} 0 & 1 \\ -i & 0 \end{bmatrix}$

29. $\begin{bmatrix} a_{11}+b_{11} & b_{21} & a_{12} & 0 \\ b_{12} & a_{11}+b_{22} & 0 & a_{12} \\ a_{21} & 0 & a_{22}+b_{11} & b_{21} \\ 0 & a_{21} & b_{12} & a_{22}+b_{22} \end{bmatrix}$

30. $A = \frac{1}{17}\begin{bmatrix} 8 & -6 & -6 \\ -6 & 13 & -4 \\ -6 & -4 & 13 \end{bmatrix}$ 32. $\left\{\begin{bmatrix} 0 \\ x_2 \end{bmatrix} : x_2 \in \mathbf{R}\right\}$

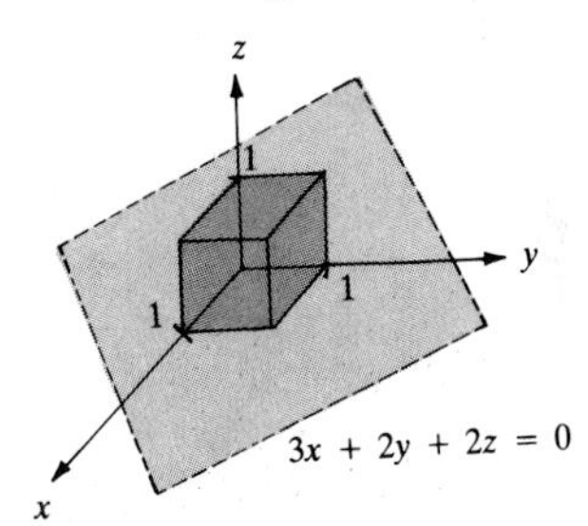

SECTION 6.1

5. $B = \begin{bmatrix} -2 & -5 \\ 2 & 4 \end{bmatrix}$, and infinitely many others

9. (*a*) Yes; $\lambda = 5$ (*b*) No (*c*) Yes; $\lambda = 0$ (*d*) No

11. (*a*) $P = \begin{bmatrix} 1 & 1 \\ -2 & 2 \end{bmatrix}$, $D = \begin{bmatrix} -3 & 0 \\ 0 & 5 \end{bmatrix}$

(*b*) $P = \begin{bmatrix} 3 & -1 \\ 6 & 2 \end{bmatrix}$, $D = \begin{bmatrix} 5 & 0 \\ 0 & -3 \end{bmatrix}$

13. $P = \begin{bmatrix} 1 & 0 & 0 \\ 0 & 1 & 1 \\ 0 & 1 & -1 \end{bmatrix}$, $D = \begin{bmatrix} 2 & 0 & 0 \\ 0 & 4 & 0 \\ 0 & 0 & -2 \end{bmatrix}$

15. $\lambda = -2, -1$; two eigenvalues
17. $\lambda = -3, 1$; two eigenvalues
19. $\lambda = -1, 1, 3$; three eigenvalues
21. $\lambda = 1, 2, 3$; three eigenvalues
23. $\lambda = 1, -1, 7$; three eigenvalues
25. $\lambda = 1$ (multiplicity two), 2 (multiplicity two); four eigenvalues

27. $\lambda = 2$ (multiplicity two) 1, 0; four eigenvalues
31. $\lambda_1 = 4 + i, \lambda_2 = 4 - i$
38. **(a)** For any real number r, $\lambda = r$ and $\mathbf{p} = r^k$.
(b) For any real number r, $\lambda = r$ and $\mathbf{p} = e^{rt}$.
40. $x = 1, 2, 3$

41. Example: $A = \begin{bmatrix} 1 & 1 \\ 1 & 2 \end{bmatrix} \xrightarrow{R_2 - R_1 \to R_2} \begin{bmatrix} 1 & 1 \\ 0 & 1 \end{bmatrix} = B.$

The eigenvalues of A are $\frac{3}{2} \pm \frac{1}{2}\sqrt{5}$, whereas the eigenvalues of B are 1 with multiplicity 2. Infinitely many other examples exist.

42. **(a)** $\lambda = 1 \pm \sqrt{\varepsilon}$ **(b)** $\lambda = 1 \pm \varepsilon$

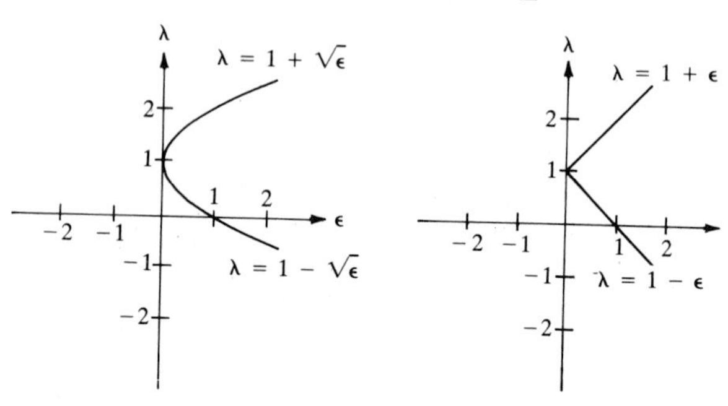

In (a), $\dfrac{\lambda(\varepsilon_1) - \lambda(\varepsilon_2)}{\varepsilon_1 - \varepsilon_2} = \pm \dfrac{\sqrt{\varepsilon_1} - \sqrt{\varepsilon_2}}{\varepsilon_1 - \varepsilon_2} = \dfrac{\pm 1}{\sqrt{\varepsilon_1} + \sqrt{\varepsilon_2}}.$

If ε is near 0, a small difference between ε_1 and ε_2 leaves a sum $\sqrt{\varepsilon_1} + \sqrt{\varepsilon_2}$ close to 0, so the quotient can be quite large.

SECTION 6.2

1. $\lambda = 1$, eigenvectors $y\begin{bmatrix} \frac{1}{2} \\ 1 \end{bmatrix}, y \neq 0;$

$\lambda = -2$, eigenvectors $y\begin{bmatrix} 2 \\ 1 \end{bmatrix}, y \neq 0$

3. $\lambda = 0$, multiplicity 2, eigenvectors $y\begin{bmatrix} -2 \\ 1 \end{bmatrix}, y \neq 0$

5. $\lambda = 1$, multiplicity 2, eigenvectors $z\begin{bmatrix} 0 \\ 1 \\ 2 \end{bmatrix}, z \neq 0;$

$\lambda = 2$, eigenvectors $z\begin{bmatrix} 0 \\ 1 \\ 1 \end{bmatrix}, z \neq 0$

7. $\lambda = -1$, eigenvectors $z\begin{bmatrix} -1 \\ 1 \\ 1 \end{bmatrix}, z \neq 0;$

$\lambda = 0$, eigenvectors $z\begin{bmatrix} 0 \\ 1 \\ 1 \end{bmatrix}, z \neq 0;$

$\lambda = 4$, eigenvectors $z\begin{bmatrix} 0 \\ -1 \\ 1 \end{bmatrix}, z \neq 0$

9. $\lambda = 0, 2;\ P = \begin{bmatrix} 1 & 1 \\ -1 & 1 \end{bmatrix};\ D = \begin{bmatrix} 0 & 0 \\ 0 & 2 \end{bmatrix}$

11. $\lambda = -1, 6;\ P = \begin{bmatrix} 1 & 3 \\ 1 & -4 \end{bmatrix};\ D = \begin{bmatrix} -1 & 0 \\ & 0 & 6 \end{bmatrix}$

13. $\lambda = 2 \pm \sqrt{3};\ P = \begin{bmatrix} 1 + \sqrt{3} & 1 - \sqrt{3} \\ 1 & 1 \end{bmatrix};$

$D = \begin{bmatrix} 2 + \sqrt{3} & 0 \\ 0 & 2 - \sqrt{3} \end{bmatrix}$

15. $\lambda = 1, 3, -1;\ P = \begin{bmatrix} 1 & 0 & 0 \\ 0 & 1 & 1 \\ 0 & 1 & -1 \end{bmatrix};\ D = \begin{bmatrix} 1 & 0 & 0 \\ 0 & 3 & 0 \\ 0 & 0 & -1 \end{bmatrix}$

17. $\lambda = 1, 2, 3;\ P = \begin{bmatrix} 1 & -1 & 1 \\ -1 & 1 & 0 \\ 0 & 1 & -1 \end{bmatrix};\ D = \begin{bmatrix} 1 & 0 & 0 \\ 0 & 2 & 0 \\ 0 & 0 & 3 \end{bmatrix}$

19. Not similar to a diagonal matrix

21. $P = \begin{bmatrix} 1 & 0 & -1 \\ 3 & 1 & 2 \\ 0 & 1 & 3 \end{bmatrix},\ D = \begin{bmatrix} 1 & 0 & 0 \\ 0 & 1 & 0 \\ 0 & 0 & -1 \end{bmatrix}$

23. $\lambda = 1 + i$, eigenvectors $c\begin{bmatrix} 1 \\ i \end{bmatrix}, c \neq 0;$

$\lambda = 1 - i$, eigenvectors $c\begin{bmatrix} 1 \\ -i \end{bmatrix}, c \neq 0;$

$P = \begin{bmatrix} 1 & 1 \\ i & -i \end{bmatrix};\ D = \begin{bmatrix} 1 + i & 0 \\ 0 & 1 - i \end{bmatrix}$

25. **(a)** $\lambda_1 = -1, \mathbf{p}^1 = \begin{bmatrix} 1 \\ -1 \end{bmatrix};\ \lambda_2 = 3, \mathbf{p}^2 = \begin{bmatrix} 1 \\ 1 \end{bmatrix}$

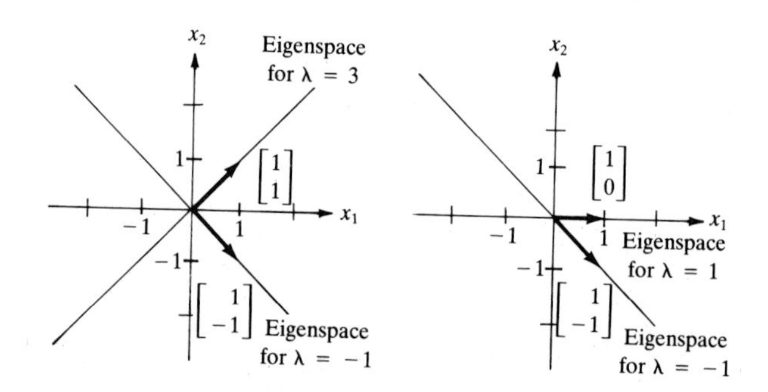

26. $\mathbf{p}^1 = \begin{bmatrix} 1 \\ 0 \end{bmatrix}, \mathbf{p}^2 = \begin{bmatrix} 1 \\ 1 \end{bmatrix}, \mathbf{p}^1 + \mathbf{p}^2 = \begin{bmatrix} 2 \\ 1 \end{bmatrix},$

but $\begin{bmatrix} 1 & 2 \\ 0 & 3 \end{bmatrix}\begin{bmatrix} 2 \\ 1 \end{bmatrix} = \begin{bmatrix} 4 \\ 3 \end{bmatrix} \neq c\begin{bmatrix} 2 \\ 1 \end{bmatrix}$ for any number c.

30. For exercise 17, $PAP^{-1} = \begin{bmatrix} -2 & -2 & -2 \\ 2 & 3 & 0 \\ 4 & 2 & 5 \end{bmatrix},$

$P^{-1}AP = \begin{bmatrix} 1 & 0 & 0 \\ 0 & 2 & 0 \\ 0 & 0 & 3 \end{bmatrix}$. $P^{-1}AP$ is the correct ordering.

31. (a) $\lambda = 1 + \sqrt{\varepsilon}, \mathbf{p} = \begin{bmatrix} 1 \\ \frac{1}{\sqrt{\varepsilon}} \end{bmatrix}; \lambda = 1 - \sqrt{\varepsilon}, \mathbf{p} = \begin{bmatrix} 1 \\ \frac{-1}{\sqrt{\varepsilon}} \end{bmatrix}$

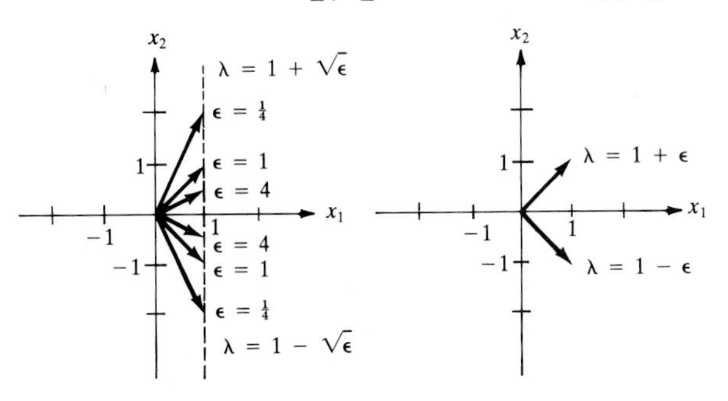

SECTION 6.3

1. Collapsed parallel to $\begin{bmatrix} 2 \\ -1 \end{bmatrix}$ into the axis determined by $\begin{bmatrix} 1 \\ 2 \end{bmatrix}$ and stretched 5 units in the direction of that axis

3. Reflected in the direction determined by $\begin{bmatrix} 1 \\ -1 \end{bmatrix}$ and stretched 6 units in that direction, and reflected in the direction determined by $\begin{bmatrix} 1 \\ 1 \end{bmatrix}$ and stretched 2 units in that direction

5. Left unchanged in the directions of $\begin{bmatrix} 1 \\ 0 \\ 0 \end{bmatrix}$ and $\begin{bmatrix} 0 \\ -1 \\ 1 \end{bmatrix}$ and stretched 3 units in the direction of $\begin{bmatrix} 0 \\ 1 \\ 1 \end{bmatrix}$

7. $A^k = (\frac{1}{4})^k(\frac{1}{2})\begin{bmatrix} 3^k + 1 & 3^k - 1 \\ 3^k - 1 & 3^k + 1 \end{bmatrix} \to \begin{bmatrix} 0 & 0 \\ 0 & 0 \end{bmatrix}$ as $k \to \infty$

9. $A^k = \frac{1}{2}\begin{bmatrix} 1 + (-\frac{1}{2})^k & 1 - (-\frac{1}{2})^k \\ 1 - (-\frac{1}{2})^k & 1 + (-\frac{1}{2})^k \end{bmatrix} \to \frac{1}{2}\begin{bmatrix} 1 & 1 \\ 1 & 1 \end{bmatrix}$ as $k \to \infty$

11. $A^k = \frac{1}{2}\begin{bmatrix} 2(\frac{2}{3})^k & 0 & 0 \\ 4 - 4(\frac{2}{3})^k & 1 + (\frac{1}{3})^k & 1 - (\frac{1}{3})^k \\ 4 - 4(\frac{2}{3})^k & 1 - (\frac{1}{3})^k & 1 + (\frac{1}{3})^k \end{bmatrix} \to \frac{1}{2}\begin{bmatrix} 0 & 0 & 0 \\ 4 & 1 & 1 \\ 4 & 1 & 1 \end{bmatrix}$

13. $X = \begin{bmatrix} y_{11} & 0 \\ y_{22} - y_{11} & y_{22} \end{bmatrix}$

15. $X = \frac{1}{7}\begin{bmatrix} 4y_{11} + 3y_{22} & 3y_{11} - 3y_{22} \\ 4y_{11} - 4y_{22} & 3y_{11} + 4y_{22} \end{bmatrix}$

17. $\frac{1}{3}\begin{bmatrix} 3y_{11} & 4y_{11} + 2y_{12} - 4y_{33} & -8y_{11} - y_{12} + 8y_{33} \\ 6y_{21} & 8y_{21} + 4y_{22} - y_{33} & -16y_{21} - 2y_{22} + 2y_{33} \\ 3y_{21} & 4y_{21} + 2y_{22} - 2y_{33} & -8y_{21} - y_{22} + 4y_{33} \end{bmatrix} = X$

19. $\mathbf{x}(k) = 2(-1)^k\begin{bmatrix} 1 \\ -1 \end{bmatrix} + 1(1)^k\begin{bmatrix} 1 \\ 1 \end{bmatrix}$

21. $\mathbf{x}(k) = -2(4^k)\begin{bmatrix} -1 \\ 1 \\ 0 \end{bmatrix} + (4^k)\begin{bmatrix} 5 \\ 0 \\ 2 \end{bmatrix} - 3(-3)^k\begin{bmatrix} 0 \\ 1 \\ -1 \end{bmatrix}$

23. $A^k = \frac{1}{2}\begin{bmatrix} i^k[1 + (-1)^k] & i^{k+1}[(-1)^k - 1] \\ i^{k+1}[1 - (-1)^k] & i^{k+2}[(-1)^{k+1} - 1] \end{bmatrix}$. *Note:* All entries in this answer are real for k either even or odd.

26. A square root of $\begin{bmatrix} 2 & 1 \\ 1 & 2 \end{bmatrix}$ is $\begin{bmatrix} \frac{\sqrt{3} + 1}{2} & \frac{\sqrt{3} - 1}{2} \\ \frac{\sqrt{3} - 1}{2} & \frac{\sqrt{3} + 1}{2} \end{bmatrix}$.

27. Find Y such that $D_1 Y = YD_2$. Then $X = PYQ^{-1}$.
$X = y_{22}\begin{bmatrix} 5 & 1 \\ 5 & 1 \end{bmatrix}$.

28. Let $A = PDP^{-1}$ and $\mathbf{y}(k) = P^{-1}\mathbf{x}(k)$. Solve $D\mathbf{y}(k + 1) = \mathbf{y}(k)$. Then $\mathbf{x}(k) = P\mathbf{y}(k)$. $\mathbf{x}(k) = \begin{bmatrix} a(\frac{1}{2})^k \\ a(\frac{1}{2})^k \end{bmatrix}$ for any real number a.

29. (a) $\mathbf{x}(k + 1) = PDP^{-1}\mathbf{x}(k) + \mathbf{u}(k)$ implies $P^{-1}\mathbf{x}(k + 1) = DP^{-1}\mathbf{x}(k) + P^{-1}\mathbf{u}(k)$. Let $\mathbf{y}(k) = P^{-1}\mathbf{x}(k)$. Then $\mathbf{y}(k + 1) = D\mathbf{y}(k) + \mathbf{w}(k)$ and $\mathbf{x}(k) = P\mathbf{y}(k)$.

(b) $\mathbf{x}(k) = \begin{bmatrix} 3^k \\ 3^k - 1 \end{bmatrix}$

33. $\mathbf{x}(k + 1) = \begin{bmatrix} \frac{3}{4} & \frac{1}{4} & 0 \\ 0 & \frac{1}{4} & 0 \\ \frac{1}{4} & \frac{1}{2} & 1 \end{bmatrix}\mathbf{x}(k)$, so

$$\mathbf{x}(k) = \begin{bmatrix} (\frac{3}{4})^k x_1(0) + \frac{1}{2}[(\frac{3}{4})^k - (\frac{1}{4})^k]x_2(0) \\ (\frac{1}{4})^k x_2(0) \\ [1 - (\frac{3}{4})^k]x_1(0) + [1 - \frac{1}{2}(\frac{1}{4})^k - \frac{1}{2}(\frac{3}{4})^k]x_2(0) + x_3(0) \end{bmatrix}$$

35. $[x_1 \quad x_2]A^k$ approaches $\frac{1}{2}(x_1 + x_2)[1 \quad 1]$.

36. The first pile is larger. $\mathbf{x}A^k$ approaches $\frac{1}{5}(x_1 + x_2)[3 \quad 2]$.

37. $x_1(k+1) = 0.9x_1(k) + 0.02x_3(k)$
$x_2(k+1) = 0.01x_1(k) + 0.87x_2(k)$
$x_3(k+1) = 0.01x_2(k) + 0.9x_3(k)$

$$\mathbf{x}(k+1) = \begin{bmatrix} 0.9 & 0 & 0.02 \\ 0.01 & 0.87 & 0 \\ 0 & 0.01 & 0.9 \end{bmatrix}\mathbf{x}(k),$$

$$P = \begin{bmatrix} 2 & 1+\sqrt{3} & 1-\sqrt{3} \\ 1 & -1+\sqrt{3} & -1-\sqrt{3} \\ -1 & 1 & 1 \end{bmatrix},$$

$$P^{-1} = \tfrac{1}{6}\begin{bmatrix} 2 & -2 & -4 \\ 1 & -1+\sqrt{3} & 1+\sqrt{3} \\ 1 & -1-\sqrt{3} & 1-\sqrt{3} \end{bmatrix},$$

$$D = \begin{bmatrix} 0.89 & 0 & 0 \\ 0 & 0.89+0.01\sqrt{3} & 0 \\ 0 & 0 & 0.89-0.01\sqrt{3} \end{bmatrix},$$

$$\begin{aligned} 6x_1(k) = {} & [4(0.89)^k + (1+\sqrt{3})(0.89+0.01\sqrt{3})^k \\ & + (1-\sqrt{3})(0.89-0.01\sqrt{3})^k]a + [-4(0.89)^k \\ & + 2(0.89+0.01\sqrt{3})^k + 2(0.89-0.01\sqrt{3})^k]b \\ & + [-8(0.89)^k + (4+2\sqrt{3})(0.89+0.01\sqrt{3})^k \\ & + (4-2\sqrt{3})(0.89-0.01\sqrt{3})^k]c \end{aligned}$$

$$\begin{aligned} 6x_2(k) = {} & [2(0.89)^k + (-1+\sqrt{3})(0.89+0.01\sqrt{3})^k \\ & - (1+\sqrt{3})(0.89-0.01\sqrt{3})^k]a + [-2(0.89)^k \\ & + (4-2\sqrt{3})(0.89+0.01\sqrt{3})^k \\ & + (4+2\sqrt{3})(0.89-0.01\sqrt{3})^k]b + [-4(0.89)^k \\ & + 2(0.89+0.01\sqrt{3})^k + 2(0.89-0.01\sqrt{3})^k]c \end{aligned}$$

$$\begin{aligned} 6x_3(k) = {} & [-2(0.89)^k + (0.89+0.01\sqrt{3})^k \\ & + (0.89-0.01\sqrt{3})^k]a + [2(0.89)^k \\ & + (-1+\sqrt{3})(0.89+0.01\sqrt{3})^k \\ & - (1+\sqrt{3})(0.89-0.01\sqrt{3})^k]b + [4(0.89)^k \\ & + (1+\sqrt{3})(0.89+0.01\sqrt{3})^k \\ & + (1-\sqrt{3})(0.89-0.01\sqrt{3})^k]c \end{aligned}$$

The population of all age groups is tending toward 0.

SECTION 6.4

1. 5 3. 3 5. $\lambda = 0, 4$ 7. $\lambda = -2, 1$
9. $\lambda = -2, 6$ 11. +, − 13. +, + 15. +, −
17. No; $\det A = 32$, not 4. 19. Yes
21. No; trace $A \neq \lambda_1 + \lambda_2$.
25. No 26. No
27. Trace $A = 4$. Since trace $D = 4 + 2(10^{-6})$, the calculation is incorrect. However, it could be close, and in fact it is close.
28. No; $\det A = 0$ whereas $\lambda_1\lambda_2\lambda_3\lambda_4 = 1$.
29. The trace

SECTION 7.1

1. Orthogonal

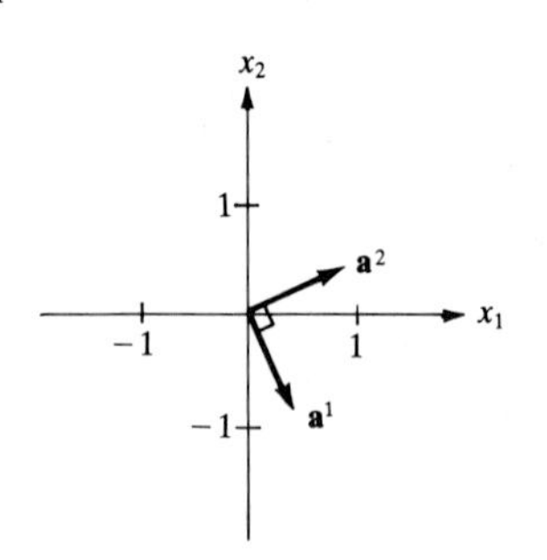

3. Not orthogonal; $\left\| \begin{bmatrix} \frac{1}{\sqrt{2}} \\ \frac{1}{\sqrt{5}} \end{bmatrix} \right\| \neq 1$

5. (a) $Q^{-1} = \begin{bmatrix} 0 & 1 \\ 1 & 0 \end{bmatrix}$ (b) $Q^{-1} = \begin{bmatrix} 0 & 0 & 1 \\ \frac{1}{\sqrt{2}} & -\frac{1}{\sqrt{2}} & 0 \\ \frac{1}{\sqrt{2}} & \frac{1}{\sqrt{2}} & 0 \end{bmatrix}$

7. $Q^{-1} = \begin{bmatrix} \frac{1}{\sqrt{2}} & \frac{1}{\sqrt{2}} & 0 & 0 \\ -\frac{1}{\sqrt{2}} & \frac{1}{\sqrt{2}} & 0 & 0 \\ 0 & 0 & 0 & 1 \\ 0 & 0 & 1 & 0 \end{bmatrix}$

9. $Q\mathbf{x} = \begin{bmatrix} \frac{9}{\sqrt{17}} \\ -\frac{2}{\sqrt{17}} \end{bmatrix}$, $\|\mathbf{x}\| = \|Q\mathbf{x}\| = \sqrt{5}$

11. $Q\mathbf{x} = \frac{1}{\sqrt{2}}\begin{bmatrix} 0 \\ 2 \end{bmatrix}$, $Q\mathbf{y} = \frac{1}{\sqrt{2}}\begin{bmatrix} -1 \\ 1 \end{bmatrix}$, $\theta = \pi/4$, $\theta' = \pi/4$

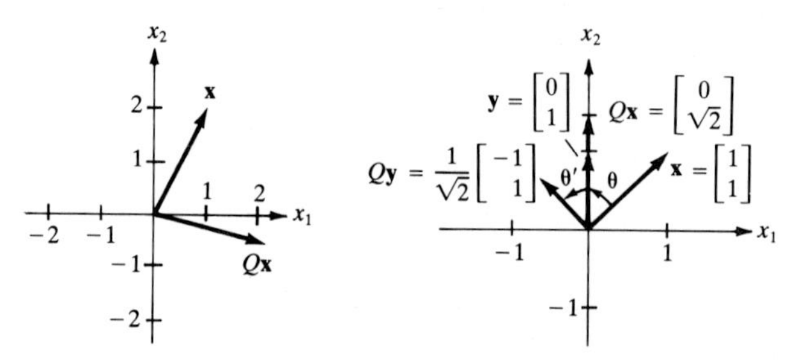

13. $Q = \frac{1}{\sqrt{2}}\begin{bmatrix} 1 & -1 \\ 1 & 1 \end{bmatrix}$, $Q\mathbf{x} = \frac{1}{\sqrt{2}}\begin{bmatrix} 1 \\ 3 \end{bmatrix}$, $\|\mathbf{x}\| = \|[\mathbf{x}]_{\#}\| = \sqrt{5}$

15. $Q = \frac{1}{\sqrt{2}}\begin{bmatrix} -1 & 1 \\ 1 & 1 \end{bmatrix}$, $[\mathbf{x}]_\# = \frac{1}{\sqrt{2}}\begin{bmatrix} -1 \\ 1 \end{bmatrix}$, $[\mathbf{y}]_\# = \frac{1}{\sqrt{2}}\begin{bmatrix} 0 \\ 2 \end{bmatrix}$, both angles $= \frac{\pi}{4}$

17. $Q = \begin{bmatrix} 0 & 1 \\ -1 & 0 \end{bmatrix}$, $[\mathbf{x}]_\# = \begin{bmatrix} -1 \\ \sqrt{3} \end{bmatrix}$, $[\mathbf{y}]_\# = \begin{bmatrix} -\sqrt{3} \\ 1 \end{bmatrix}$,

both angles $= \frac{\pi}{6}$

19. $\|\mathbf{x} - \mathbf{y}\| = \left\|\begin{bmatrix} 0 \\ -1 \end{bmatrix}\right\| = 1$,

$$\|Q^{-1}\mathbf{x} - Q^{-1}\mathbf{y}\| = \left\|\begin{bmatrix} -\frac{1}{\sqrt{2}} \\ -\frac{1}{\sqrt{2}} \end{bmatrix}\right\| = 1$$

21. Area = 26 23. Sides are [1, 1] and [2, 4]; area = 2

25. Sides are [1, 0], [0, 1], and [−1, 1]; area = $\frac{1}{2}$

27. (*a*) Unitary (*b*) Unitary

SECTION 7.2

1. $Q = \frac{1}{\sqrt{29}}\begin{bmatrix} 5 & 2 \\ 2 & -5 \end{bmatrix}$, $T = \begin{bmatrix} 3 & -3 \\ 0 & -4 \end{bmatrix}$

3. $Q = \frac{1}{\sqrt{2}}\begin{bmatrix} 1 & 1 \\ -1 & 1 \end{bmatrix}$, $T = \begin{bmatrix} 4 & 6 \\ 0 & 4 \end{bmatrix}$

5. $Q = \begin{bmatrix} \frac{1}{\sqrt{2}} & 0 & \frac{1}{\sqrt{2}} \\ 0 & 1 & 0 \\ \frac{1}{\sqrt{2}} & 0 & -\frac{1}{\sqrt{2}} \end{bmatrix}$, $T = \begin{bmatrix} 0 & -\sqrt{2} & 0 \\ 0 & 1 & \sqrt{2} \\ 0 & 0 & -2 \end{bmatrix}$

7. $Q = \frac{1}{\sqrt{2}}\begin{bmatrix} 1 & 1 \\ 1 & -1 \end{bmatrix}$, $T = \begin{bmatrix} 3 & -1 \\ 0 & 0 \end{bmatrix}$

9. Only (b) is symmetric.

11. $Q = \frac{1}{\sqrt{5}}\begin{bmatrix} 1 & 2 \\ -2 & 1 \end{bmatrix}$, $D = \begin{bmatrix} -1 & 0 \\ 0 & 4 \end{bmatrix}$

13. $Q = \frac{1}{\sqrt{10}}\begin{bmatrix} 3 & 1 \\ 1 & -3 \end{bmatrix}$, $D = \begin{bmatrix} \frac{15}{2} & 0 \\ 0 & \frac{5}{2} \end{bmatrix}$

15. $Q = \begin{bmatrix} \frac{1}{\sqrt{2}} & 0 & \frac{1}{\sqrt{2}} \\ 0 & 1 & 0 \\ -\frac{1}{\sqrt{2}} & 0 & \frac{1}{\sqrt{2}} \end{bmatrix}$, $D = \begin{bmatrix} 0 & 0 & 0 \\ 0 & 2 & 0 \\ 0 & 0 & 2 \end{bmatrix}$

17. $Q = \begin{bmatrix} -\frac{1}{\sqrt{2}} & -\frac{1}{\sqrt{6}} & \frac{1}{\sqrt{3}} \\ \frac{1}{\sqrt{2}} & -\frac{1}{\sqrt{6}} & \frac{1}{\sqrt{3}} \\ 0 & \frac{2}{\sqrt{6}} & \frac{1}{\sqrt{3}} \end{bmatrix}$, $D = \begin{bmatrix} 2 & 0 & 0 \\ 0 & 2 & 0 \\ 0 & 0 & -1 \end{bmatrix}$

19. (a) and (b) are Hermitian.

21. $D = \begin{bmatrix} 0 & 0 \\ 0 & 2 \end{bmatrix}$, $U = \frac{1}{\sqrt{2}}\begin{bmatrix} 1 & i \\ i & 1 \end{bmatrix}$

26. $Q = \frac{1}{\sqrt{2}}\begin{bmatrix} 1 & 1 \\ -1 & 1 \end{bmatrix}$, $T = \begin{bmatrix} 0 & 0 \\ -1 & 3 \end{bmatrix}$

28. $T = \begin{bmatrix} 1 & 2 \\ 0 & 1 \end{bmatrix}$, $Q = \frac{1}{\sqrt{2}}\begin{bmatrix} 1 & 1 \\ -1 & 1 \end{bmatrix}$. If $\mathbf{x}(0) = \begin{bmatrix} a \\ b \end{bmatrix}$, then

$$\mathbf{x}(k) = \frac{1}{\sqrt{2}}\begin{bmatrix} a + b + 2kb \\ -a + b - 2kb \end{bmatrix}.$$

29. Eigenvectors are

$$\begin{bmatrix} \frac{-1 + \sqrt{1 + 4t}}{2t} \\ 1 \end{bmatrix} \text{ and } \begin{bmatrix} \frac{-1 - \sqrt{1 + 4t}}{2t} \\ 1 \end{bmatrix}.$$

These are sensitive near 0 because of the term $-1/2t$ in the x-coordinate of each.

SECTION 7.3

1. $\mathbf{x}(k) = 2(4^k)\begin{bmatrix} 1 \\ 1 \end{bmatrix} - (-2)^k\begin{bmatrix} 1 \\ -1 \end{bmatrix}$ for all k

3. $\mathbf{x}(k) = 2^k\begin{bmatrix} 1 \\ 1 \\ 1 \end{bmatrix}$ for all k

5. $Q = \frac{1}{\sqrt{2}}\begin{bmatrix} 1 & 1 \\ -1 & 1 \end{bmatrix}$, $y_2 = \pm 2$, 2 straight lines

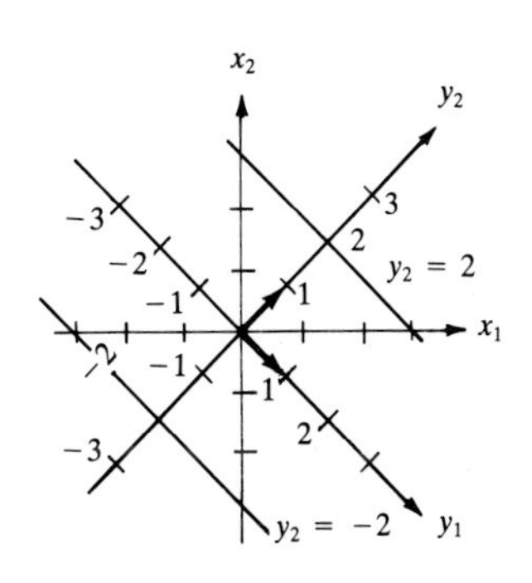

7. $Q = \dfrac{1}{\sqrt{2}}\begin{bmatrix} 1 & 1 \\ -1 & 1 \end{bmatrix}$, $y_1^2 - y_2^2 = 9$, hyperbola

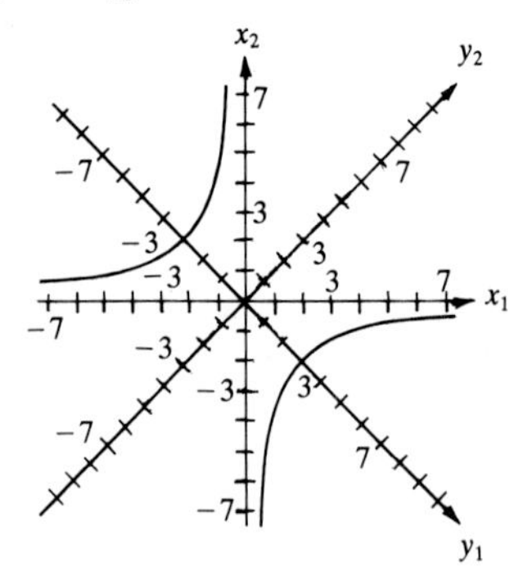

9. $Q = \dfrac{1}{\sqrt{5}}\begin{bmatrix} 1 & -2 \\ 2 & 1 \end{bmatrix}$,

$y_1^2 + 16y_2^2 = 64$, ellipse

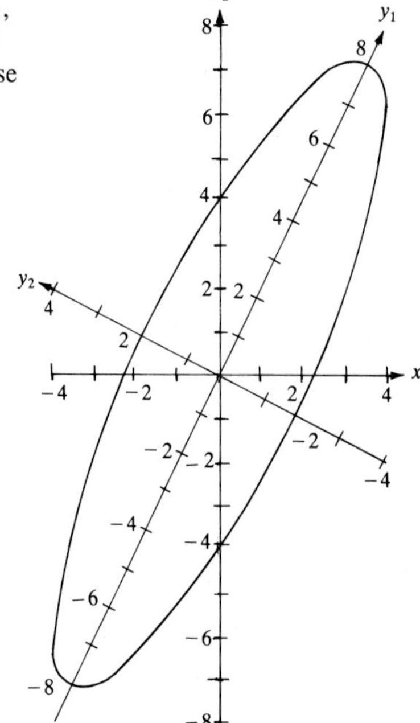

11. $Q = \dfrac{1}{\sqrt{5}}\begin{bmatrix} 1 & -2 \\ 2 & 1 \end{bmatrix}$, $-y_1^2 + 4y_2^2 = 4$, hyperbola

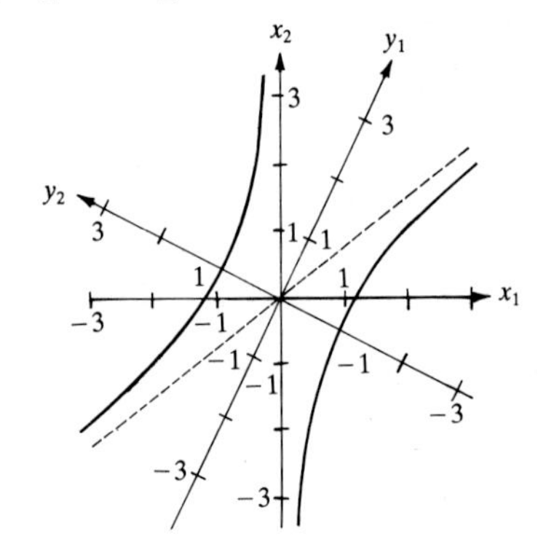

13. $Q = \dfrac{1}{\sqrt{2}}\begin{bmatrix} 1 & 1 \\ -1 & 1 \end{bmatrix}$, $q_{\#} = y_1^2 + 3y_2^2$, $\mathbf{y} = \begin{bmatrix} 0 \\ \sqrt{2} \end{bmatrix}, \begin{bmatrix} \sqrt{2} \\ 0 \end{bmatrix}, \begin{bmatrix} -\sqrt{2} \\ 0 \end{bmatrix}, \begin{bmatrix} 0 \\ -\sqrt{2} \end{bmatrix}$. In order, the values $q(x_1, x_2) = q_{\#}(y_1, y_2)$ are 6, 2, 2, 6.

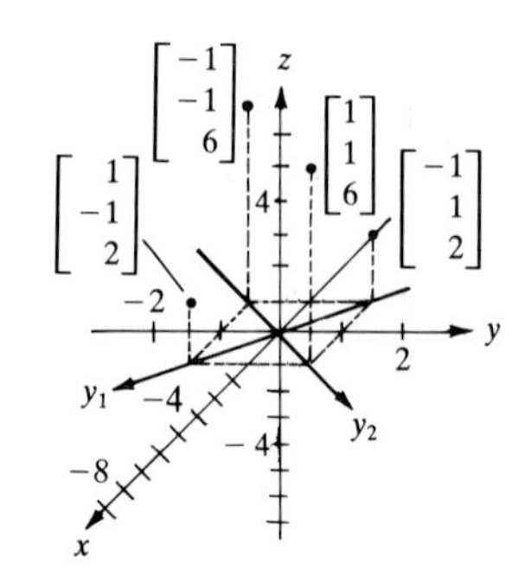

15. $Q = \dfrac{1}{\sqrt{2}}\begin{bmatrix} 1 & 1 \\ -1 & 1 \end{bmatrix}$, $q_{\#} = 9y_1^2 + y_2^2$, $\mathbf{y} = \begin{bmatrix} 0 \\ \sqrt{2} \end{bmatrix}, \begin{bmatrix} \sqrt{2} \\ 0 \end{bmatrix}, \begin{bmatrix} -\sqrt{2} \\ 0 \end{bmatrix}, \begin{bmatrix} 0 \\ -\sqrt{2} \end{bmatrix}$. In order, the values $q(x_1, x_2) = q_{\#}(y_1, y_2)$ are 2, 18, 18, 2.

17. $Q = \dfrac{1}{\sqrt{5}}\begin{bmatrix} 2 & 1 \\ 1 & -2 \end{bmatrix}$,

$q_{\#} = 16y_1^2 + y_2^2$, elliptic paraboloid

19. $Q = \dfrac{1}{\sqrt{5}}\begin{bmatrix} 1 & -2 \\ 2 & 1 \end{bmatrix}$,

$q_{\#} = 5y_1^2$, parabolic cylinder

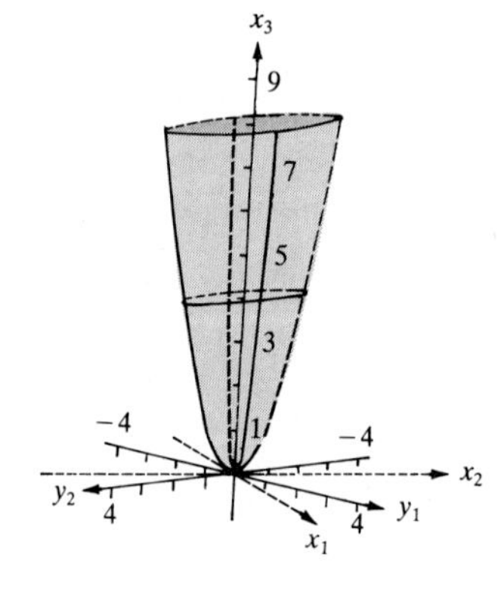

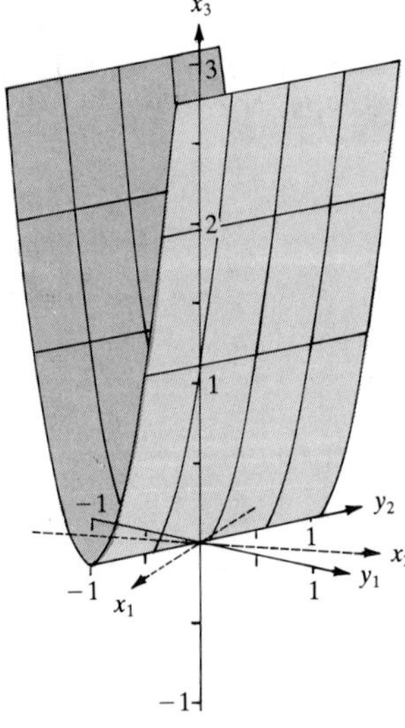

21. $Q = \frac{1}{\sqrt{5}}\begin{bmatrix} -2 & 1 \\ 1 & 2 \end{bmatrix}$, $q_\# = 9y_1^2 - y_2^2$, hyperbolic paraboloid

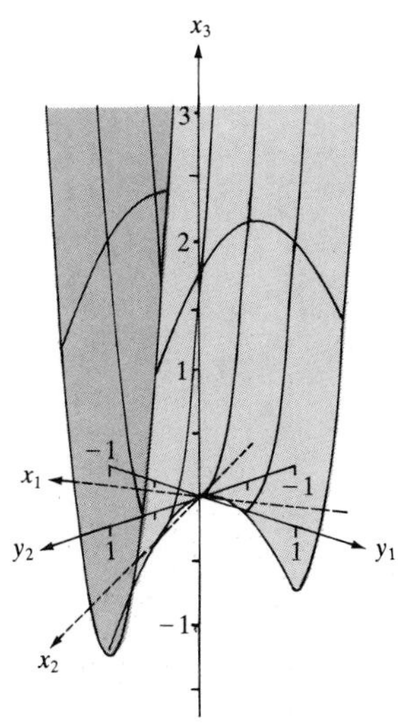

23. det $A = -2$, trace $A = 0$, hyperbola
25. det $A = 0$, trace $A = 2$, two lines
27. det $A = 1$, trace $A = 2$, elliptic paraboloid opening upward, having x_3-axis as axis and having vertex at origin
29. det $A = -6$, trace $A = -1$, hyperbolic paraboloid

In exercises 31 and 33, using unit vectors $\begin{bmatrix} 1 \\ 0 \end{bmatrix}, \begin{bmatrix} 0 \\ 1 \end{bmatrix}$, $\frac{1}{\sqrt{2}}\begin{bmatrix} 1 \\ 1 \end{bmatrix}, \frac{1}{\sqrt{2}}\begin{bmatrix} 1 \\ -1 \end{bmatrix}, \frac{1}{\sqrt{5}}\begin{bmatrix} 1 \\ 2 \end{bmatrix}, \frac{1}{\sqrt{5}}\begin{bmatrix} -1 \\ 2 \end{bmatrix}$, we have the following:

31. Estimate 4.5, exact eigenvalue $\frac{3 + \sqrt{37}}{2} \approx 4.5414$

33. Estimate 14.5, exact eigenvalue $\frac{13 + \sqrt{257}}{2} \approx 14.5156$

35. Estimates 2, 5, 5, $\frac{28}{3} \approx 9.3333$, exact eigenvalue 10

37. Using unit vectors $\begin{bmatrix} 1 \\ 0 \end{bmatrix}, \begin{bmatrix} 0 \\ 1 \end{bmatrix}, \frac{1}{\sqrt{2}}\begin{bmatrix} 1 \\ 1 \end{bmatrix}, \frac{1}{\sqrt{2}}\begin{bmatrix} 1 \\ -1 \end{bmatrix}$, $\frac{1}{\sqrt{5}}\begin{bmatrix} 1 \\ 2 \end{bmatrix}, \frac{1}{\sqrt{5}}\begin{bmatrix} -1 \\ 2 \end{bmatrix}, \begin{bmatrix} i \\ 0 \end{bmatrix}, \begin{bmatrix} 0 \\ i \end{bmatrix}, \frac{1}{\sqrt{2}}\begin{bmatrix} -i \\ i \end{bmatrix}, \frac{1}{2}\begin{bmatrix} 1-i \\ 1+i \end{bmatrix}$, we have estimate 1, exact eigenvalue $\sqrt{2} \approx 1.4142$.

43. $\mathbf{x}(k) = (-1)^k\begin{bmatrix} 1 \\ 1 \end{bmatrix}$ for all k

45. $Q = \frac{1}{\sqrt{3}}\begin{bmatrix} 1 & -(1+\sqrt{3})/2 & -(1-\sqrt{3})/2 \\ -1 & (1-\sqrt{3})/2 & (1+\sqrt{3})/2 \\ 1 & 1 & 1 \end{bmatrix}$,

$\mathbf{x}(k) =$

$$\begin{bmatrix} 1 - (1+\sqrt{3})(-2+\sqrt{3})^k - (1-\sqrt{3})(-2-\sqrt{3})^k \\ -1 + (1-\sqrt{3})(-2+\sqrt{3})^k + (1+\sqrt{3})(-2-\sqrt{3})^k \\ 1 + 2(-2+\sqrt{3})^k + 2(-2-\sqrt{3})^k \end{bmatrix}$$

47. (*a*) $\mathbf{x}(t) =$

$$\begin{bmatrix} \frac{1}{5\sqrt{2}}\sin\sqrt{2}t + \frac{3}{5}\cos\sqrt{2}t - \frac{\sqrt{3}}{5}\sin 2\sqrt{3}t + \frac{2}{5}\cos 2\sqrt{3}t \\ \frac{\sqrt{2}}{5}\sin\sqrt{2}t + \frac{6}{5}\cos\sqrt{2}t + \frac{\sqrt{3}}{10}\sin 2\sqrt{3}t - \frac{1}{5}\cos 2\sqrt{3}t \end{bmatrix}$$

(*b*) $\mathbf{x}(t) = \frac{1}{5}\begin{bmatrix} \cos t + 4\cos\sqrt{6}t \\ 2\cos t - 2\cos\sqrt{6}t \end{bmatrix}$

48. (*a*) Rotates the y,z-plane θ radians about the x-axis. Q affects the second and third rows of $\mathbf{x}$.
(*b*) Rotates the x,y-plane θ radians about the z-axis. Q affects the first and second rows of $\mathbf{x}$.
(*c*) Rotates the x,z-plane θ radians about the y-axis. Q affects the first and third rows of $\mathbf{x}$.
(*d*) Let θ_1 be the angle $\mathbf{x}$ makes with the positive x-axis. Then $\theta = -\theta_1$.
(*e*) Let θ_2 be the angle $\mathbf{x}$ makes with the x_1,x_2-plane.

Then $P_1 = \begin{bmatrix} 1 & \mathbf{0} \\ \mathbf{0} & Q_1 \end{bmatrix}$, where Q_1 rotates by $-\theta_2$ radians.

Let θ_3 be the angle $P_1\mathbf{x}$ makes with the positive x-axis.

Then $P_2 = \begin{bmatrix} Q_2 & \mathbf{0} \\ \mathbf{0} & 1 \end{bmatrix}$, where Q_2 rotates by $-\theta_3$ radians.

(*f*) By the definition of matrix multiplication, if $\mathbf{x}$ is the first column of A and if $P_2P_1\mathbf{x} = r\mathbf{e}^1$, then the first column of P_2P_1A is $r\mathbf{e}^1$. Use the solution of part (e) to find P_1 and P_2. Letting θ_2 be the angle the second column of P_2P_1A makes with the y-axis, form P_3 by using θ_2 in the matrix of part (a). Then, by part (d), $P_3P_2P_1A = T$, as required.

SECTION 8.1

1. $x = 1700 + 340 = 2000$, exact $y = 2068$; relative error $= 0.033$
3. $x = 1700 - 330 = 1400$, exact $y = 1374$; relative error $= 0.019$
5. Two-digit $x = 260$, four-digit $x = 256$, exact $y = 256$; two-digit relative error $= 0.016$, four-digit relative error $= 0.0$
7. Relative error $\approx 0.0020 < 10^{-2}$, percentage error 0.2%, so two digits
9. Relative error $\approx 0.00020 < 10^{-3}$, percentage error 0.02%, so three digits

11. Relative error $\approx 0.000091 < 10^{-4}$, percentage error 0.0091%, so four digits
13. Relative error $\approx 0.00055 < 10^{-3}$, percentage error 0.055%
15. Relative error $\approx 0.0000027 < 10^{-5}$, percentage error 0.00027%
17. (*a*) $f(x_1) = 3$, $f(x_2) = 2.727\ldots$, $|\Delta f| = 0.2727\ldots$, $|\Delta x| = 0.1$, $|\Delta f/\Delta x| \approx 2.73$
(*b*) $f(x_1) = 5$, $f(x_2) = 5.555\ldots$, $|\Delta f| = 0.555\ldots$, $|\Delta x| = 0.1$, $|\Delta f/\Delta x| \approx 5.56$
(*c*) $f(x_1) = 1000$, $f(x_2) = 909.0909\ldots$, $|\Delta f| = 90.9090\ldots$, $|\Delta x| = 0.0001$, $|\Delta f/\Delta x| \approx 909{,}090.90$
(*d*) $f(x_1) = 20{,}000$, $f(x_2) = 22{,}222.222\ldots$, $|\Delta f| = 2{,}222.222\ldots$, $|\Delta x| = 0.00001$, $|\Delta f/\Delta x| \approx 222{,}222{,}222.22$
18. $\dfrac{\|10^r\mathbf{x} - 10^r\mathbf{y}\|}{\|10^r\mathbf{y}\|} = \dfrac{10^r\|\mathbf{x} - \mathbf{y}\|}{10^r\|\mathbf{y}\|} = \dfrac{\|\mathbf{x} - \mathbf{y}\|}{\|\mathbf{y}\|}$
19. No. $1/[(16)(12)] \approx 0.0052 < 10^{-2}$. Thus we expect two-digit accuracy.
21. Although the error $|x - y|$ is large, it can be that x and y rounded to the first k digits are close and so accuracy for a k-digit machine is good. For example, for the given x and y, $|x - y| = 1{,}000{,}000$, yet the numbers rounded to the first three digits show total agreement.

SECTION 8.2

1. $\begin{bmatrix} \frac{5}{7} \\ -\frac{4}{7} \end{bmatrix}$ *3.* $\begin{bmatrix} -\frac{1}{16} \\ \frac{5}{8} \end{bmatrix}$ *5.* $\begin{bmatrix} \frac{26}{21} \\ -\frac{61}{21} \\ \frac{10}{21} \end{bmatrix}$

7. Without pivoting, $x_1 = 0$, $x_2 = 0.20$, with relative errors of 1 and 0.0008, respectively. With pivoting, $x_1 = 0.0050$, $x_2 = 0.20$, with relative errors of 0.003 and 0.0008, respectively.
9. Without pivoting, $x_1 = 0.0$, $x_2 = 0.0$, $x_3 = 0.067$, with relative errors of 1, 1, and 0.01, respectively. With pivoting, $x_1 = -1.5(10^{-9})$, $x_2 = 1.4$, $x_3 = 0.067$, with relative errors of 0.01, 0.05, and 0.01, respectively.
11. $\mathbf{x} = \begin{bmatrix} 0.0378 \\ 10.1 \end{bmatrix}$ in two passes through the algorithm
13. $\mathbf{x} = \begin{bmatrix} -0.00864 \\ -0.0339 \end{bmatrix}$ in one pass through the algorithm
15. $\mathbf{x} = \begin{bmatrix} 0.118 \\ 0.377 \\ -0.96 \end{bmatrix}$ in two passes through the algorithm
17. $\begin{bmatrix} 0.999 \\ -2.0 \end{bmatrix}$, the digits of agreement of $\hat{\mathbf{x}}_3$ and $\hat{\mathbf{x}}_4$
18. No; errors can be magnified by this approach.

SECTION 8.3

1. In three iterations, $\lambda = -8.74$, $\mathbf{x}^1 = \begin{bmatrix} -0.611 \\ 0.792 \end{bmatrix}$.

Exact answer: $\lambda_1 = \dfrac{-5 - \sqrt{157}}{2} \approx -8.76$,

$$\mathbf{x}^1 = \frac{\sqrt{2}}{(141 - 11\sqrt{157})^{1/2}} \begin{bmatrix} \frac{11 - \sqrt{157}}{2} \\ 1 \end{bmatrix} \approx \begin{bmatrix} -0.608 \\ 0.794 \end{bmatrix}$$

3. $B_1 = \begin{bmatrix} \frac{21}{8} & \frac{21}{8} \\ \frac{3}{8} & \frac{3}{8} \end{bmatrix}$

5. $B_1 = \begin{bmatrix} \frac{1}{2} & -1 & \frac{3}{2} \\ -2 & 2 & -2 \\ -\frac{1}{2} & 1 & -\frac{3}{2} \end{bmatrix}$, $B_2 = \begin{bmatrix} -\frac{1}{6} & -\frac{1}{3} & \frac{5}{6} \\ 0 & 0 & 0 \\ \frac{1}{6} & \frac{1}{3} & -\frac{5}{6} \end{bmatrix}$

12. $\mathbf{x}(j) = r_1(1^j)\begin{bmatrix} 1 \\ 1 \end{bmatrix} + r_2(-\frac{1}{3})^j\begin{bmatrix} 1 \\ -1 \end{bmatrix}$. For j large, $(-\frac{1}{3})^j$ is small and so $r_1(1^j)(1) \pm r_2(-\frac{1}{3})^j \approx r_1$, or $\mathbf{x}(j) \approx r_1(1^j)\begin{bmatrix} 1 \\ 1 \end{bmatrix}$.

SECTION 9.1

1. Objective function $z = 3x$, constraints $2x \leqslant 3$ and $x \geqslant 0$, feasible set the interval $[0, \frac{3}{2}]$, optimal solution $x = \frac{3}{2}$, optimal value $z = \frac{9}{2}$
3. Objective function $z = 8x$, constraints $4x \leqslant 7$ and $x \geqslant 0$, feasible set the interval $[0, \frac{7}{4}]$, optimal solution $x = 0$, optimal value $z = 0$
5. $\min z = -21$ on line $2x_1 - x_2 = -21$; $\max z = 15$ on line $2x_1 - x_2 = 15$

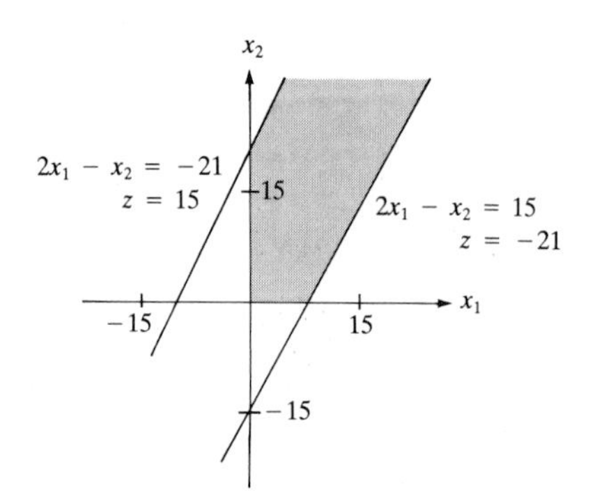

7. $\min z = 0$ at $[0, 0]$; $\max z = 8$ on line $8x_1 + x_2 = 8$
9. $\min z = 0$ at $[0, 0]$; $\max z = 30$ at $[8, 6]$
11. $\min z = -36$ at $[0, 18]$; $\max z = 29$ at $[29, 0]$

13. min $z = 0$ on x_1, x_2-plane in the triangle bounded by $x_1 = 0, x_2 = 0$, and $2x_1 + x_2 = 7$; max $z = 5$ at $[0, 0, 5]$

18. One such program is max $z = x_1 + x_2$ s.t. $-x_1 + x_2 \leqslant 1, -x_1 + x_2 \geqslant 2, x_1 \geqslant 0, x_2 \geqslant 0$.
19. One such program is given in exercise 5. Yes.
20. One such program is given in exercise 7 with max z. No.
21. One such program is max $z = -x_1 + x_2$ s.t. $x_1 + x_2 \geqslant 1, x_1 \leqslant 1, x_2 \leqslant 1, x_1 \geqslant 0, x_2 \geqslant 0$.
22. The linear program optimize $z = -x_1 + x_2$ s.t. $x_1 \geqslant 0, x_2 \geqslant 0$ has neither a maximum nor a minimum optimal value.
23. $[x_{11}, x_{12}, x_{13}, x_{21}, x_{22}, x_{23}] = [50, 0, 150, 150, 150, 100]$, $[50, 50, 150, 150, 100, 100]$, $[50, 150, 0, 150, 0, 250]$, and infinitely many others
25. max $z = 3x_1 + 4x_2$ s.t. $2x_1 + 9x_2 \leqslant 450, x_1 + 2x_2 \leqslant 125, x_1 + x_2 \leqslant 105, x_1 \geqslant 0, x_2 \geqslant 0$. Then the maximum is $z = \$335$ at $[85, 20]$.
27. max $z = 1200x_1 + 1500x_2$ s.t. $x_1 \geqslant 8, x_2 \geqslant 6, x_1 + x_2 \leqslant 19, x_1 \geqslant 0, x_2 \geqslant 0$. Then the factory earns a maximum of \$26,100 making 8 tons of ordinary window glass and 11 tons of plate glass.

SECTION 9.2

1. Linear program system:

$$\begin{array}{rcl} x_1 + x_2 - x_3 + s_1 & = & 5 \\ 2x_1 - x_2 + 3x_3 + s_2 & = & 7 \\ 4x_1 + 2x_2 + x_3 + s_3 & = & 8 \\ -3x_1 - x_2 + 4x_3 + z & = & 0 \end{array}$$

Augmented matrix form:

$$\begin{array}{c} \begin{array}{cccccccc} x_1 & x_2 & x_3 & s_1 & s_2 & s_3 & z & \end{array} \\ \left[\begin{array}{rrrrrrr|r} 1 & 1 & -1 & 1 & 0 & 0 & 0 & 5 \\ 2 & -1 & 3 & 0 & 1 & 0 & 0 & 7 \\ 4 & 2 & 1 & 0 & 0 & 1 & 0 & 8 \\ \hline -3 & -1 & 4 & 0 & 0 & 0 & 1 & 0 \end{array}\right] \end{array}$$

3. Linear program system:

$$\begin{array}{rcl} 2x_1 - 3x_2 + x_4 + s_1 & = & 17 \\ 3x_1 + 5x_2 - x_3 + 2x_4 + s_2 & = & 12 \\ -x_1 + 2x_2 + x_3 + s_3 & = & 8 \\ -x_1 - x_2 - x_3 - x_4 + z & = & 0 \end{array}$$

Augmented matrix form:

$$\begin{array}{c} \begin{array}{ccccccccc} x_1 & x_2 & x_3 & x_4 & s_1 & s_2 & s_3 & z & \end{array} \\ \left[\begin{array}{rrrrrrrr|r} 2 & -3 & 0 & 1 & 1 & 0 & 0 & 0 & 17 \\ 3 & 5 & -1 & 2 & 0 & 1 & 0 & 0 & 12 \\ -1 & 2 & 1 & 0 & 0 & 0 & 1 & 0 & 8 \\ \hline -1 & -1 & -1 & -1 & 0 & 0 & 0 & 1 & 0 \end{array}\right] \end{array}$$

5. Linear program system:

$$\begin{array}{rcl} x_1 - 2x_2 + s_1 & = & 12 \\ 2x_1 + x_2 + 3x_3 + s_2 & = & 14 \\ x_1 - x_3 + s_3 & = & 4 \\ -x_1 + 3x_2 - 4x_3 + z & = & 0 \end{array}$$

Linear program: max $z = x_1 - 3x_2 + 4x_3$ s.t. $x_1 - 2x_2 \leqslant 12$, $2x_1 + x_2 + 3x_3 \leqslant 14$, $x_1 - x_3 \leqslant 4$, $x_1 \geqslant 0, x_2 \geqslant 0, x_3 \geqslant 0$

7. $[0, 12, 0, \frac{3}{2}, 12, 0, 0, \frac{51}{2}]$; $[0, 12, 0, \frac{3}{2}]$; $z = \frac{51}{2}$
9. $[1, 1, 0, 4, 7, 1, -3]$; no; $[1, 2, 1, 1, 8, 1, -6]$
11.

$$\begin{array}{c} \begin{array}{cccccc} x_1 & x_2 & s_1 & s_2 & z & \end{array} \\ \left[\begin{array}{rrrrr|r} 1 & 0 & 2 & -3 & 0 & -2 \\ 0 & 1 & -1 & 2 & 0 & 2 \\ \hline 0 & 0 & -3 & 5 & 1 & 4 \end{array}\right] \end{array}$$

; final matrix is not a vertex matrix.

13.

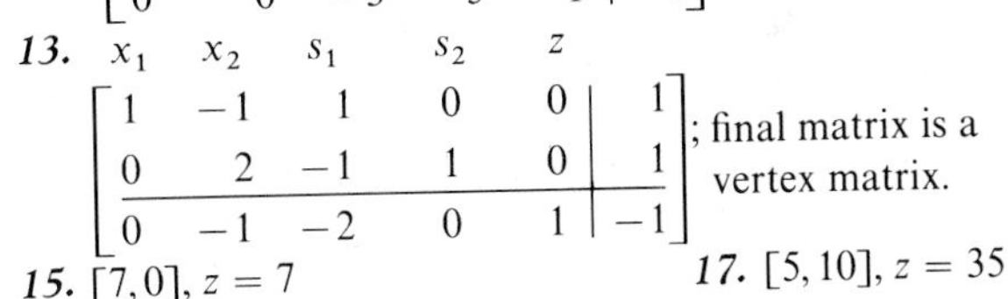

$$\begin{array}{c} \begin{array}{cccccc} x_1 & x_2 & s_1 & s_2 & z & \end{array} \\ \left[\begin{array}{rrrrr|r} 1 & -1 & 1 & 0 & 0 & 1 \\ 0 & 2 & -1 & 1 & 0 & 1 \\ \hline 0 & -1 & -2 & 0 & 1 & -1 \end{array}\right] \end{array}$$

; final matrix is a vertex matrix.

15. $[7, 0]$, $z = 7$

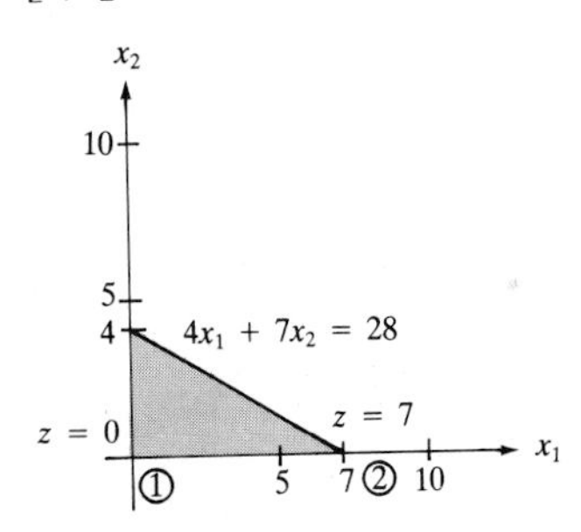

17. $[5, 10]$, $z = 35$

19. $[11, 13]$, $z = 59$

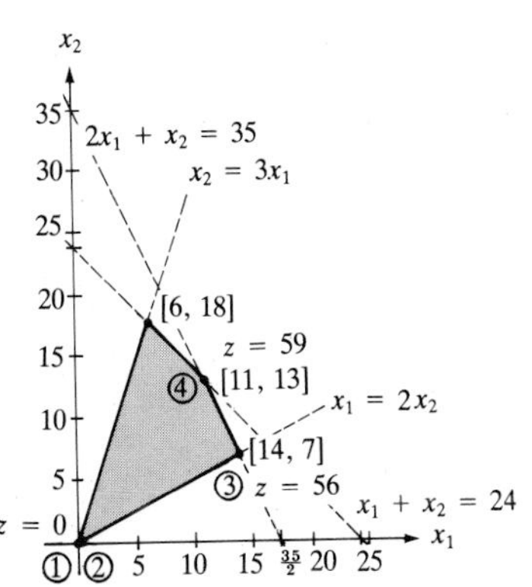

21. [2, 2, 0], $z = 24$ *23.* [0, 0, 3, 2], $z = 8$

28. Let the jth variable be called y. If we set all of x_{k_1}, $x_{k_2}, \ldots, x_{k_r}, s_{l_1}, \ldots, s_{l_t}$ except y to 0, we get $y = b_i - a_{ij}y$. But y must be at least 0, so $b_i \geqslant a_{ij}y$. If $a_{ij} > 0$, then $y \leqslant b_i/a_{ij}$ is required to ensure that $y \geqslant 0$. If $a_{ij} = 0$, changing y will have no effect on y. If $a_{ij} < 0$, increasing y will increase y. Thus our choice causes the entries in the last column to remain positive.

31. Making 350 copiers and no videorecorders or computers will provide a maximum of \$22,750 per week.

SECTION 9.3

3. The first and seventh vertex matrices are identical.

SECTION 9.4

1. (*a*) [5, 9], $z = 19$ (*b*) [4, 0], $z = 4$

3. (*a*) [3, 10], $z = 17$

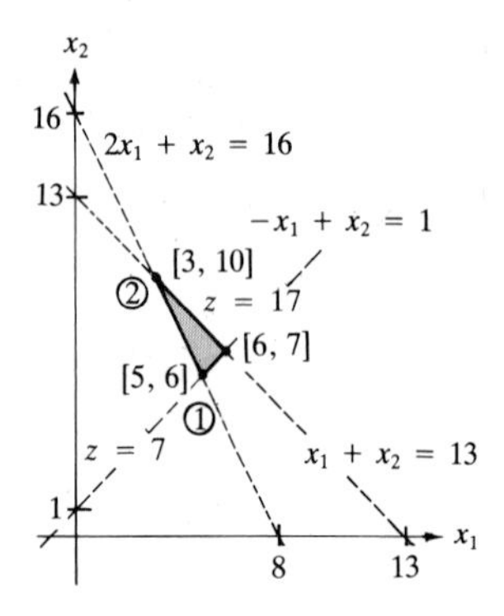

(*b*) [6, 7], $z = -23$

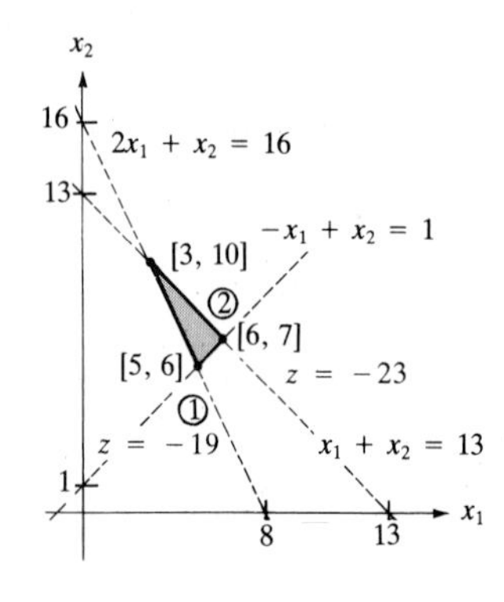

5. [6, 12], $z = -24$ *7.* [30, 18], $z = -108$

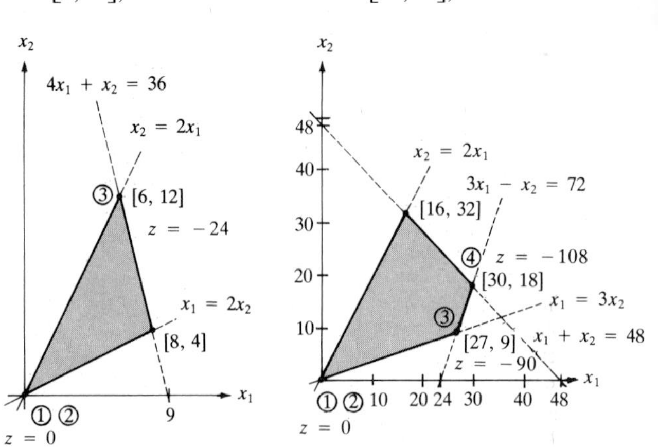

9. $[\frac{2}{3}, \frac{2}{3}]$, $z = \frac{4}{3}$ *11.* [4, 4], $z = 12$

13. [4, 6, 8], $z = -16$ *15.* [3, 1, 0], $z = 5$

17. [1, 1, 1], $z = 3$ *19.* No minimum

21. [6, 14, 30], $z = 3840$

26. The resulting matrix is $\left[\begin{array}{cccc|c} -1 & -1 & -1 & 0 & 0 \\ -1 & -1 & 0 & 1 & 0 \end{array}\right]$.

27. The minimum cost is \$187,500 at [0, 150, 150, 200, 0, 100], measured in thousands of barrels.

29. $\min z = 200x_{11} + 200x_{12} + 210x_{13} + 220x_{21} + 300x_{22} + 250x_{23}$ s.t. $x_{11} + x_{12} + x_{13} \leqslant 500$, $x_{21} + x_{22} + x_{23} \leqslant 600$, $x_{11} + x_{21} \geqslant 300$, $x_{12} + x_{22} \geqslant 150$, $x_{13} + x_{23} \geqslant 250$, $x_{11} \geqslant 0$, $x_{12} \geqslant 0$, $x_{13} \geqslant 0$, $x_{21} \geqslant 0$, $x_{22} \geqslant 0$, $x_{23} \geqslant 0$

31. $\max z = 1000x_1 + 900x_2 + 4000x_3$ s.t. $x_1 + x_2 + x_3 \leqslant 30$, $4x_1 + x_2 + 2x_3 \leqslant 60$, $8x_1 + 8x_2 + 3x_3 \geqslant 120$, $x_1 \geqslant 0$, $x_2 \geqslant 0$, $x_3 \geqslant 0$. The maximum is \$101,600 for 2 units of grade A, 4 units of grade B, and 24 units of grade C.

33. [5, 24] for $z = \$6300$

35. $\min z = 50(x_1 + x_2 + x_3 + x_4) + 20(y_1 + y_2) + 10w_1$ s.t. $x_1 \geqslant 30$, $x_2 \geqslant 40$, $x_3 + y_1 \geqslant 40$, $x_4 + y_2 + w_1 \geqslant 30$, $-x_1 + y_1 + w_1 + v_1 = 0$, $-x_2 + y_2 - v_1 \leqslant 0$, $x_1 \geqslant 0$, $x_2 \geqslant 0$, $x_3 \geqslant 0$, $x_4 \geqslant 0$, $y_1 \geqslant 0$, $y_2 \geqslant 0$, $w_1 \geqslant 0$
The source of the last two constraints is that (1) the amount sent for cleaning plus the amount held at the end of the first week must equal the amount bought and (2) the amount sent for cleaning at the end of the second week cannot exceed the amount bought that week plus the amount held over from the previous week.

APPENDIX A

1. $8 - i$ *3.* $11 - 3i$ *5.* $5 + i$ *7.* 25

9. $-11 + 2i$ *11.* $-i$ *13.* $\frac{7}{5} + \frac{4}{5}i$ *15.* $2 - i$

INDEX

Note: Boldface type is used to indicate the most complete definition or description of the word or idea.